Acta Numerica 2012

Acta Numerica

Volume 21 2012

CAMBRIDGE
UNIVERSITY PRESS

Published by the Press Syndicate of the University of Cambridge
The Pitt Building, Trumpington Street, Cambridge CB2 1RP
32 Avenue of the Americas, New York, NY 10013-2473, USA
10 Stamford Road, Oakleigh, Melbourne 3166, Australia

First published 2012

Printed in Great Britain at the University Press, Cambridge

Library of Congress cataloguing in publication data available

A catalogue record for this book is available from the British Library

ISBN 978-1-107-026155

ISSN 0962-4929

Contents

The heterogeneous multiscale method 1
Assyr Abdulle, Weinan E, Björn Engquist and Eric Vanden-Eijnden

Numerical-asymptotic boundary integral methods in high-frequency acoustic scattering 89
Simon N. Chandler-Wilde, Ivan G. Graham, Stephen Langdon and Euan A. Spence

Mathematical cancer therapy planning in deep regional hyperthermia 307
Peter Deuflhard, Anton Schiela and Martin Weiser

High-performance computing systems: Status and outlook .. 379
J. J. Dongarra and A. J. van der Steen

Secrets of image denoising cuisine 475
M. Lebrun, M. Colom, A. Buades and J. M. Morel

Pricing futures by deterministic methods 577
Olivier Pironneau

Acta Numerica (2012), pp. 1–87
doi:10.1017/S0962492912000025

The heterogeneous multiscale method*

Assyr Abdulle
ANMC, Mathematics Section,
École Polytechnique Fédérale de Lausanne,
Lausanne, Switzerland
E-mail: assyr.abdulle@epfl.ch

Weinan E
Beijing International Center for Mathematical Research,
Peking University,
Beijing, China
and
Department of Mathematics and PACM,
Princeton University,
Princeton, USA
E-mail: weinan@math.princeton.edu

Björn Engquist
Department of Mathematics,
University of Texas,
Austin, USA
E-mail: engquist@ices.utexas.edu

Eric Vanden-Eijnden
Courant Institute of Mathematical Sciences,
New York University,
New York, USA
E-mail: eve2@cims.nyu.edu

The heterogeneous multiscale method (HMM), a general framework for designing multiscale algorithms, is reviewed. Emphasis is given to the error analysis that comes naturally with the framework. Examples of finite element and finite difference HMM are presented. Applications to dynamical systems and stochastic simulation algorithms with multiple time scales, spall fracture and heat conduction in microprocessors are discussed.

* Colour online for monochrome figures available at journals.cambridge.org/anu.

CONTENTS

1	Introduction	2
2	The HMM framework	4
3	ODEs and dynamical systems	16
4	Finite element HMM	37
5	Finite volume methods	67
6	Conclusion	78
	References	80

1. Introduction

The heterogeneous multiscale method (HMM) is a general framework for designing multiscale algorithms for a wide variety of applications (E and Engquist 2002, E and Engquist 2003, E *et al.* 2007*a*). The word 'heterogeneous' was used in order to emphasize the fact that the algorithm may involve macro and micro models of very different natures: for example, the micro model may come from molecular dynamics and the macro model may come from continuum theory. In fact, at a very rough level, HMM can be thought of as a way of blending together models of very heterogeneous types.

Most problems that we encounter in nature have a multiscale character. The multiscale character can occur in a variety of ways. Take, for example, problems from materials science, where many properties, such as conductivity, have a multiscale nature. This is the case for composites. It could also be that the material can be viewed at different levels of detail: as a continuous medium, in which case one naturally applies the principles of continuum mechanics, or at the atomic scale, in which case one naturally applies various atomistic models of molecular dynamics or quantum mechanics. Each viewpoint has its merits and drawbacks. Continuum models are quite efficient but sometimes their accuracy is inadequate, particularly when defects are involved. Atomic models are typically more accurate, but much less efficient. This situation is not limited to materials science but is quite common in most areas of science and engineering. One of the main motivations for multiscale modelling is to develop models that have accuracy close to that of microscopic models and efficiency close to that of macroscopic models.

From the viewpoint of numerical algorithms, we are interested in extracting useful information from the microscopic model, which in principle has the required accuracy. If we use the traditional viewpoint, then we would have to solve the microscopic model in full detail, which is practically impossible for engineering applications. In terms of computational complexity, the best one can do with such an approach is to have linear scaling algorithms: the complexity scales as the number of microscopic degrees of freedom. However, in many cases, we are not interested in the full microscopic solution or we cannot afford the cost of computing it. Instead, we are only interested in the behaviour of some macroscopic variables, or the microscopic behaviour in very small parts of the system, for example near defects. The question is: Can we develop much more efficient algorithms, such as *sublinear* scaling algorithms, that would give us such information? To develop these new types of algorithms, we have to compress not only the operators, as has been done in multigrid methods, but also the variables. We have to be content with getting information about only a subset of the system variables. These types of algorithms cannot be completely general: one has to explore special features of the problem in order to construct such algorithms.

From the viewpoint of analysis, many analytical techniques have been developed in order to derive simplified models. Examples include averaging methods, homogenization methods, matched asymptotics, WKB methods, geometric optics approximations, and renormalization group methods (E 2011). The principles of such techniques are quite general, but in practice they only give us explicit analytical models in very limited situations. In other situations, it is natural to ask whether one can devise efficient computational techniques based on these principles. This is the case that we are interested in, and it was one of the main motivations for developing HMM.

This was the background against which HMM was proposed. Of course, prior to HMM, there were already many techniques of a similar spirit in many different fields. Well-known examples include:

- Car–Parrinello molecular dynamics, in which electronic structure models are used together with molecular dynamics to predict the dynamics of nuclei (Car and Parrinello 1985),
- the quasicontinuum method, in which atomistic models are used to analyse the mechanical deformation of crystalline solids (Tadmor, Ortiz and Phillips 1996),
- superparametrization models, in which cloud-resolving models are used to capture large-scale tropical dynamics of the atmosphere (Grabowski 2001, Xing, Grabowski and Majda 2009).

HMM was proposed as a general framework that can be used for a variety of problems. It was not the only attempt. Other notable examples include the extended multigrid method and the equation-free approach (Brandt 2002, Kevrekidis *et al.* 2003). A common theme of these approaches is that the microscopic models are used throughout the computational process. These should be contrasted with techniques such as model reduction methods, wavelet-based homogenization and variational multiscale methods, in which the microscale model is only used at the beginning of the computation to obtain compressed effective operators.

In spite of competing efforts, HMM was the only general attempt based on a top-down philosophy, which at the time was not the most popular viewpoint. In fact, in the early days of multiscale modelling, most efforts were devoted to a bottom-up approach, seeking strategies that would give us the information needed by working only with the microscale model, without any prior information about the system at the macroscale. This certainly sounds very attractive, and may at first sight seem the only correct approach. In one way, a key insight of HMM was the recognition that the bottom-up approach is bound to have technical difficulties, and will for some time be limited to rather simple applications. One can appreciate such difficulties by noticing the fact that, even if the effective macroscale model is explicitly available, designing stable and accurate numerical algorithms for such macroscale models is still a non-trivial task. Important constraints, such as conservation properties or upwinding, have to be implemented in order to guarantee that the algorithms give rise to the correct numerical solutions. Implementing such constraints at the level of microscale models, in the absence of any explicit knowledge about the macroscale model, seems to be next to impossible. Therefore compromises are necessary: for many problems we have to guess the form of the macroscale model to start from. Fortunately, in many cases we do have some prior knowledge of the macroscale behaviour of the system under consideration, and this knowledge is often sufficient for us to make an adequate guess.

Since multiscale modelling is a vast subject, touching almost all aspects of modelling, we will not be able to do justice to all the work that has been done on this subject. Instead we will focus on HMM. For a general introduction to multiscale modelling, we refer to the monograph by E (2011).

2. The HMM framework

2.1. The main components of HMM

We will use U to describe the set of macroscopic variables, and u the set of microscopic variables. They are related by

$$U = Q(u), \tag{2.1}$$

where Q is called the compression operator. Any operator that reconstructs u from U is called a reconstruction operator:

$$u = R(U). \tag{2.2}$$

For consistency, Q and R should satisfy the relation

$$Q(R(U)) = U. \tag{2.3}$$

In HMM, we assume that we have an incomplete macroscale model to begin with:

$$F(U, D) = 0. \tag{2.4}$$

Figure 2.1. Schematics of the HMM framework.

Here D represents the missing part of the model. For example, if this is a model in continuum mechanics, then D might be the constitutive relation for the stress. If it is a model for molecular dynamics, then D might be the inter-atomic forces. If it is a model for heat conduction in composite materials, then D might be the macroscale effective conductivity tensor.

HMM proceeds by estimating the missing data on the fly using the microscale model, at each location where some missing data is needed. To do this, the microscale model has to be constrained so that its macrostate is the same as the macrostate we are interested in, that is,

$$f(u, d(U)) = 0. \tag{2.5}$$

Here $d(U)$ represents the constraint for the microscale model. For example, if the microscale model is the canonical ensemble of molecular dynamics, d might be the average density, momentum and energy.

If we use H and h to denote the macro and micro numerical parameters, such as mesh size, one can write HMM abstractly in the following form:

$$\begin{aligned} F_H(U_H, D_H(u_h)) &= 0, \\ f_h(u_h, d_h(U_H)) &= 0. \end{aligned} \tag{2.6}$$

In practical terms, the basic components of HMM are as follows.

1 *A macroscopic solver.* Based on knowledge of the macroscale behaviour of the system, we make an assumption about the form of the macroscale model, for which we select a suitable macroscale solver. For example, if we are dealing with a variational problem, we may use a finite element method as the macroscale solver.

2 *A procedure for estimating the missing macroscale data D using the microscale model.* This is typically done in two steps.

   (a) *Constrained microscale simulation.* At each point where macroscale data are needed, perform a series of microscopic simulations which are constrained so that they are consistent with the local value of the macro variable.

(b) *Data processing.* Use the results from the microscopic simulations to extract the macroscale data needed in the macroscale solver.

For dynamical problems, we can state the HMM procedure formally as follows. At each macro time step:

1 Given the current state of the macro variables U^n, re-initialize the micro variables:

$$u^{n,0} = RU^n. \tag{2.7}$$

2 Evolve the micro variables for some micro time steps:

$$u^{n,m+1} = \mathcal{S}_{\delta t}(u^{n,m}; U^n), \quad m = 0, \ldots, M-1. \tag{2.8}$$

3 Estimate D:

$$D^n = \mathcal{D}_M(u^{n,0}, u^{n,1}, \ldots, u^{n,M}). \tag{2.9}$$

4 Evolve the macro variables for one macro time step using the macro solver:

$$U^{n+1} = S_{\Delta t}(U^n; D^n). \tag{2.10}$$

Here R is some reconstruction operator which plays the same role as the interpolation or prolongation operators in the multigrid method; $\mathcal{S}_{\delta t}$ is the micro solver, which also depends on U^n through the constraints, as indicated; and $\mathcal{D}_M$ is some data processing operator, which in general involves spatial/temporal/ensemble averaging. This is sometimes referred to as the data estimator. Finally, $S_{\Delta t}$ is the macro solver.

For static problems, the procedure is very similar, particularly in the context of iterative algorithms: we simply replace macro time step by macro iteration step.

For dynamic problems, there are two important time scales that we need to consider. The first, denoted by t_M, is the time scale for the dynamics of the macro variables. The second, denoted by τ_ε, is the relaxation time for the microscopic model. We will need to distinguish between two different cases. The first is when the two time scales are comparable, that is, $\tau_\varepsilon \sim t_M$. In this case, from the viewpoint of numerical efficiency, there is not much room to play with as far as time scales are concerned. We just have to evolve the microscale model along with the macroscale model. The second case is when $\tau_\varepsilon \ll t_M$. This is the case we will focus on. The general guideline in this case is as follows.

1 Choose Δt to accurately resolve the t_M time scale.

2 Choose M so that $M\delta t$ covers the τ_ε time scale sufficiently to allow equilibration to take place in the micro model.

2.2. Simple examples

Examples of this set-up include the following.

- ODEs with disparate time scales, where U is a complete set of slow variables and u is the full set of variables. In this case, the macroscale model could just be an ODE system for U, and the missing data could just be the force. If we know that the macroscale system has additional structure, then we can take that into account when selecting the macro solver. For example, if the macroscale system is a Hamiltonian system, then we can use a symplectic integrator as the macro solver.
- Elliptic equations with multiscale coefficients, such as those arising in the modelling of the behaviour of composite materials, where U is the averaged displacement field and u is the microscale displacement field. In this case, the macroscale model is still an elliptic equation, and the missing data could be the coefficients in the macroscale model. If we use the finite element method as the macro solver, then the missing data could just be the stiffness matrix, which has to be evaluated with the help of the full microscale model.
- Molecular dynamics models of complex fluids such as polymer fluids. Here U is the set of hydrodynamic variables, which in the simplest case would be the field of mass, momentum and energy densities, and u is the set of microscopic variables, that is, the position and momentum of all of the participating particles in a molecular dynamics model. The macroscale model might be the set of local conservation laws for U. The missing data might be the fluxes in these conservation laws.
- Microscopic pore-scale models of the pressure distribution in a porous medium. Here U is the macroscale pressure distribution and u is the set of variables in the pore-scale model, which could be a network-based model. The macroscale model is of Darcy law type. The missing data are the coefficients in this model.
- Microscale models of the moving contact line. Here U is the set of macroscopic variables (velocity and pressure fields, position of the interface between the fluid phases), and u is the set of variables in the microscopic description, say using molecular dynamics, of the contact line region. The macroscale model might just be the standard model in two-phase flows. The missing data might be the boundary conditions at the contact line.

Example 1: Stiff ODEs. Consider

$$\begin{aligned} \frac{\mathrm{d}x}{\mathrm{d}t} &= f(x,y), \\ \frac{\mathrm{d}y}{\mathrm{d}t} &= -\frac{1}{\varepsilon}(y-\varphi(x)). \end{aligned} \tag{2.11}$$

Here $U = x, u = (x, y)$. The macroscale should of course be an ODE, which can be written as

$$\frac{\mathrm{d}x}{\mathrm{d}t} = F(x) = f(x, \varphi(x)). \tag{2.12}$$

The missing data are implicitly given by F.

Let us choose the simplest solver for (2.12), the forward Euler method. HMM would then proceed as follows.

1 Initialize the micro solver, *e.g.*, $y^{n,0} = y^{n-1,M}$.

2 Apply the micro solver for M micro steps:

$$y^{n,m+1} = y^{n,m} - \frac{\delta t}{\varepsilon}(y^{n,m} - \varphi(x^n)), \tag{2.13}$$

for $m = 0, 1, \ldots, M-1$.

3 Estimate $F(x)$:

$$F^n = f(x^n, y^{n,m}). \tag{2.14}$$

4 Apply the macro solver:

$$x^{n+1} = x^n + \Delta t\, F^n. \tag{2.15}$$

Example 2: Stiff stochastic ODEs. Consider the stochastic ODE

$$\begin{aligned} \frac{\mathrm{d}x}{\mathrm{d}t} &= f(x, y), \\ \frac{\mathrm{d}y}{\mathrm{d}t} &= -\frac{1}{\varepsilon}(y - \varphi(x)) + \sqrt{\frac{2}{\varepsilon}}\dot{w}, \end{aligned} \tag{2.16}$$

where $\dot{w}(t)$ is standard white noise. Averaging theorems suggest that the effective macroscale equation should again be an ODE:

$$\frac{\mathrm{d}x}{\mathrm{d}t} = F(x). \tag{2.17}$$

HMM with forward Euler as the macro solver proceeds as follows.

1 Initialize the micro solver, *e.g.*, $y^{n,0} = y^{n-1,M}$.

2 Apply the micro solver for M micro steps:

$$y^{n,m+1} = y^{n,m} - \frac{\delta t}{\varepsilon}(y^{n,m} - \varphi(x^n)) + \sqrt{\frac{2\delta t}{\varepsilon}}\xi^{n,m}, \tag{2.18}$$

for $m = 0, 1, \ldots, M-1$. Here the $\{\xi^{n,m}\}$ are independent normal random variables with mean 0 and variance 1.

3 Estimate $F(x)$:

$$F^n = \frac{1}{M}\sum_{m=1}^{M} f(x^n, y^{n,m}). \tag{2.19}$$

4 Apply the macro solver:

$$x^{n+1} = x^n + \Delta t\, F^n. \tag{2.20}$$

Example 3: Elliptic PDEs with multiscale coefficients. Consider

$$-\nabla \cdot (a^\varepsilon(x)\nabla)u^\varepsilon(x) = f(x). \tag{2.21}$$

Abstract homogenization theory tells us that the macroscale model should be of the form

$$-\nabla \cdot (a^0(x)\nabla)U(x) = f(x). \tag{2.22}$$

The missing data are the coefficients $a^0(x)$. Naturally, for the macroscale solver we choose standard finite element methods, for example the piecewise linear finite element method over a coarse mesh. The data that need to be estimated form the stiffness matrix for the finite element method. If $a^0 = a^0(x)$ were known, we would simply follow standard practice and use numerical quadrature to compute the elements in the stiffness matrix. Since a^0 is not known, we set up a microscale simulation around each quadrature point in order to estimate the function value needed at that quadrature point. The details of this procedure will be discussed later.

Example 4: The parabolic homogenization problem. Consider

$$\partial_t u^\varepsilon = \partial_x \cdot \left(a\left(x, \frac{x}{\varepsilon}, t \right) \partial_x u^\varepsilon \right), \tag{2.23}$$

where $a(x, y, t)$ is a smooth function and is periodic in y, say with period 1. The macroscale model is of the form

$$\partial_t U = \partial_x \cdot D, \tag{2.24}$$

$$D = \left\langle a\left(x, \frac{x}{\varepsilon}, t \right) \partial_x u^\varepsilon \right\rangle, \tag{2.25}$$

where $\langle \cdot \rangle$ means taking spatial averages.

We will choose a finite volume method as the macro solver. Then D needs to be evaluated at the cell boundaries (Abdulle and E 2003). We will make the assumption that the flux D depends on the local values of U and $\partial_x U$ only. Consequently, for the micro model, we will impose the boundary condition that $u^\varepsilon(x,t) - Ax$ is periodic where $A = \partial_x U$ is evaluated at the location of interest.

Denote the micro solver by

$$u^{n+1} = \mathcal{S}_{\delta\tau,\delta x}(u^n; A). \tag{2.26}$$

Assuming that we have the numerical approximation $\{U_j^n\}$ (where $t^n = n\Delta t, U_j^n \sim U(n\Delta t, j\Delta x)$) at the nth macro time step, we obtain the numerical approximation at the next macro time step via the following procedure.

for some constant C. Therefore, we have

$$\|Qu^\varepsilon - U_{\text{HMM}}\| \leq C(\varepsilon^\alpha + (\Delta t)^k + e(\text{HMM})). \tag{2.38}$$

The key to getting concrete error estimates, and thereby giving guidelines to designing multiscale methods, lies in the estimation of $e(\text{HMM})$. This is specific to each problem. But typically, $e(\text{HMM})$ contains several contributions:

- the relaxation error, due to the influence of the transients,
- the sampling error, due to empirical averages being used to estimate the ensemble average,
- the size effect, due to a computational domain of much smaller size being used to estimate quantities on domains of much larger size.

Examples of these error components will be discussed in subsequent sections.

2.4. *Difficulties with HMM*

The most significant shortcoming of HMM is that it is based on a preconceived macroscale model. If the form of the macroscale model is chosen incorrectly, one cannot expect the resulting HMM procedure to produce accurate results. For example, if the effective macroscale model should be a stochastic ODE but one makes the assumption that it is a deterministic ODE, then the stochastic component of the macroscale solution will not be captured correctly by an HMM based on such an assumption.

There is a important reason for starting with the macro solver: even for problems for which we do have a sufficiently accurate macroscale model, finding an effective numerical algorithm for that macroscale model may still be a significant task. Indeed, this has been the focus of the computational mathematics community for more than 50 years. One example is Euler's equation in gas dynamics, whose solutions typically contain shocks, *i.e.*, discontinuities (LeVeque 1990). In this case, the numerical algorithms have to satisfy certain constraints in order to be able to solve Euler's equation accurately. Obviously this should also be a concern for multiscale methods.

For practical problems of interest, we have often accumulated some knowledge of the macroscale model. Such information can be used when making assumptions about the macroscale model used in HMM. In cases when one makes a wrong assumption, one can still argue that HMM produces an 'optimal approximation' for the macroscale behaviour of the solution in the class of the models considered. In this sense, HMM is a way of addressing the following question: What is the best one can do given the knowledge we have about the problem at all scales?

Another difficulty is that HMM, as presented, is not very 'seamless', for two reasons.

- It requires knowledge of a complete set of macroscale variables to begin with.
- It requires conversion back and forth between the macro- and microstates of the system. This can become rather difficult in actual implementations, particularly when constructing discrete microstates (*e.g.* in molecular dynamics) from continuous macroscale variables.

The seamless strategy proposed in E, Ren and Vanden-Eijnden (2009) is intended to overcome the second difficulty. This will be explained next. The first difficulty is more significant. It can be dealt with in various situations with different levels of satisfaction, as we explain in Section 3.1.2.

2.5. *Seamless HMM*

To motivate the seamless algorithm, let us consider the stiff ODE example (2.11). If we want an efficient algorithm for capturing the behaviour of x without resolving the detailed behaviour of y, we can simply change the small parameter ε to a larger value ε', the size of which is determined by the accuracy requirement

$$\begin{aligned} \frac{\mathrm{d}x}{\mathrm{d}t} &= f(x,y), \\ \frac{\mathrm{d}y}{\mathrm{d}t} &= -\frac{1}{\varepsilon'}(y-\varphi(x)). \end{aligned} \tag{2.39}$$

This is then solved using standard ODE solvers. This idea of *boosting* the parameters is very simple, but it can also be quite useful, as in Car–Parrinello molecular dynamics (Car and Parrinello 1985).

We can look at this differently. Instead of changing the value of ε, we may change the clock for the microscale model, that is, if we use $\tau = t\varepsilon/\varepsilon'$ in the second equation in (2.39), then (2.39) can be written as

$$\begin{aligned} \frac{\mathrm{d}x}{\mathrm{d}t} &= f(x,y), \\ \frac{\mathrm{d}y}{\mathrm{d}\tau} &= -\frac{1}{\varepsilon}(y-\varphi(x)). \end{aligned} \tag{2.40}$$

If we discretize this equation using standard ODE solvers but with different time step sizes for the first and second equations in (2.40), we obtain the following algorithm:

$$y^{n+1} = y^n - \frac{\delta\tau}{\varepsilon}(y^n - \varphi(x^n)), \tag{2.41}$$

$$D^{n+1} = y^{n+1}, \tag{2.42}$$

$$x^{n+1} = x^n + \Delta' t f(x^n, D^{n+1}). \tag{2.43}$$

Here $y^n \sim y(n\delta\tau)$ and $x^n \sim x(n\Delta' t)$. The value of $\delta\tau$ is the time step size we

Example: SDEs with multiple time scales. Going back to the SDE (2.16), the seamless algorithm with forward Euler scheme is simply

$$y^{n+1} = y^n - \frac{\delta\tau}{\varepsilon}(y^n - \phi(x^n)) + \sqrt{\frac{2\delta\tau}{\varepsilon}}\xi^n, \tag{2.49}$$

$$x^{n+1} = x^n + \Delta' t\, f(x^n, y^{n+1}), \tag{2.50}$$

where the $\{\xi^n\}$ are independent normal random variables with mean 0 and variance 1. Note that for HMM we have $x^n \sim x(n\Delta t)$, but for the seamless algorithm we have $x^n \sim x(n\Delta' t) = x(n\Delta t/M)$.

2.6. HMM for type A problems

There are two major types of multiscale problems: type A problems, for which multiscale modelling is needed to resolve local defects or singularities, and type B problems, for which multiscale modelling is needed to supply constitutive relations at the macroscale (E and Engquist 2003). HMM is most easily formulated for type B problems, and so far the formulations we have discussed are mostly for these problems. But HMM can also be used for type A problems, particularly in situations for which the relaxation time scale for the local structure near defects is much faster than the macroscopic dynamic time scale. In that case, the effect of the microscale structure near defects can be viewed as providing effective boundary conditions for the macroscale model at the defects.

One example is the moving contact line problem. There the macroscale model is simply the continuum two-phase flow model. The missing data are the boundary conditions at the contact line, which is then extracted from a molecular dynamics simulation of the local structure around the contact line.

3. ODEs and dynamical systems

3.1. General considerations

3.1.1. Standard HMM scheme

It is useful to revisit the ideas discussed in the previous section in the context of more general stiff ODEs. Consider a generic multiscale system of the type

$$\begin{aligned} \frac{\mathrm{d}x^\epsilon}{\mathrm{d}t} &= f(x^\epsilon, y^\epsilon), \\ \frac{\mathrm{d}y^\epsilon}{\mathrm{d}t} &= \frac{1}{\epsilon} g(x^\epsilon, y^\epsilon). \end{aligned} \tag{3.1}$$

Let us assume that, in the limit as $\epsilon \to 0$, the dynamics of this system can be described by the parametric distribution

$$\mu_{X(t)}(\mathrm{d}y), \tag{3.2}$$

where $\mu_x(\mathrm{d}y)$ is the invariant distribution of

$$\frac{\mathrm{d}Y}{\mathrm{d}t} = g(x, Y), \tag{3.3}$$

with x viewed as a fixed parameter, and $X(t)$ in (3.2) solves

$$\frac{\mathrm{d}X}{\mathrm{d}t} = F(X(t)), \tag{3.4}$$

where

$$F(x) = \int f(x, y)\mu_x(\mathrm{d}y). \tag{3.5}$$

In other words, we assume that, as $\epsilon \to 0$, x^ϵ converges pathwise to X, while y^ϵ converges in law to Y, in a fashion that is captured by the distribution (3.2). This way of phrasing the limit theorem encompasses all the situations of interest to HMM. For example, (3.1) might be a set of stiff ODEs with a slow manifold structure, in which case $\mu_x(y)$ is atomic on this manifold (*i.e.*, for any x, $\mu_x(\mathrm{d}y) = \delta_{y(x)}(\mathrm{d}x)$ for some $y(x)$). Alternatively, (3.1) might be a set of SDEs, in which case $\mu_x(y)$ is typically broad (E 2011).

The standard HMM scheme starts with an integrator for (3.4). We will denote the numerical map associated with this integrator by η_t^F. Here the superscript F makes explicit the dependence of this map on the function F defined in (3.5), which we do not know explicitly: the aim of HMM is to approximate it on the fly. Note that η_t^F could depend on the derivatives of F itself, in which case the HMM scheme will have to estimate these as well as F: the F-information needed in the map η_t^F will be denoted collectively by $(F(x), DF(x), \ldots)$. In order to obtain this input, let ζ_t^x denote the numerical map associated with (3.3), where x is now a parameter. Finally, given the output $(\zeta_{\delta t}^x(y_0), (\zeta_{\delta t}^x)^{o2}(y_0), \ldots, (\zeta_{\delta t}^x)^{oN}(y_0))$ of this map for N micro steps of size δt starting from the initial condition $Y(0) = y_0$, let

$$(\tilde{F}(x), D\tilde{F}(x), \ldots) = \mathcal{F}\big(\zeta_{\delta t}^x(y_0), (\zeta_{\delta t}^x)^{o2}(y_0), \ldots, (\zeta_{\delta t}^x)^{oN}(y_0)\big) \tag{3.6}$$

denote the estimator for $(F(x), DF(x), \ldots)$. The HMM scheme then iterates using the following procedure.

1 *Microscale solver.* Given $X_{m\Delta t}$, generate

$$(\zeta_{\delta t}^{X_{m\Delta t}}(y_0), (\zeta_{\delta t}^{X_{m\Delta t}})^{o2}(y_0), \ldots, (\zeta_{\delta t}^{X_{m\Delta t}})^{oN}(y_0)).$$

2 *Estimator.* Use these data to estimate

$$(\tilde{F}(X_{m\Delta t}), D\tilde{F}(X_{m\Delta t}), \ldots) = \\ (\zeta_{\delta t}^{X_{m\Delta t}}(y_0), (\zeta_{\delta t}^{X_{m\Delta t}})^{o2}(y_0), \ldots, (\zeta_{\delta t}^{X_{m\Delta t}})^{oN}(y_0)).$$

3 *Macroscale solver.* Set

$$X_{(m+1)\Delta t} = \eta_{\Delta t}^{\tilde{F}}(X_{m\Delta t}).$$

for some $C>0$, then that of the scheme (3.16) is

$$C(\Delta' t/\epsilon')^a + \bar{C}\epsilon'^b = C(\delta t/\epsilon)^a + \bar{C}(\epsilon\Delta' t/\delta t)^b, \tag{3.18}$$

where $\bar{C} > 0$ and the exponent b depends on the rate of convergence of the solution of (3.1) towards its limit as $\epsilon \to 0$. Note that in order to be useful, the constant C in the error estimate (3.18) needs to be independent of ϵ. This property is not standard for time-dependent problems, but is expected to hold if the map ψ_t for (3.8) is chosen appropriately, and takes advantage of the fact that the fast variables reach a stable (statistical) steady state conditional on the slow variables being fixed. The scheme (3.16) was first introduced by Fatkullin and Vanden-Eijnden (2004) and further developed by E *et al.* (2009). It is a seamless version of HMM in which the slow and fast variables are evolved concurrently, and no re-initialization of the fast variables is required. This last feature is especially handy in situations where the values taken by the fast variables are constrained by the slow variables, which arises in molecular dynamics simulations, for example (for applications in this context see, *e.g.*, Maragliano and Vanden-Eijnden (2006, 2007), Abrams and Vanden-Eijnden (2010)).

The scheme (3.16) is so simple, both in its implementation and interpretation, that it may seem like a trivialization of HMM: just boost the value of ϵ in the original system (3.1), adjust the time step accordingly, and we are done! Yet it opens the door to more sophisticated strategies. Indeed, since $\bar{C}(\epsilon\Delta' t/\delta t)^b$ is the only extra term appearing in the error estimate (3.18), and this term is controlled by the rate of convergence of (3.1) towards its limit as $\epsilon \to 0$, this suggests modifying (3.1) in such a way that the limit is unchanged but convergence is accelerated. How to actually achieve this should be examined on a case-by-case basis.

The seamless version of HMM also opens the door to schemes in which the slow variables do not need to be identified beforehand, as was first noted by Vanden-Eijnden (2007). To see why, suppose that (3.1) is replaced by

$$\frac{\mathrm{d}z^\epsilon}{\mathrm{d}t} = F(z^\epsilon) + \frac{1}{\epsilon}G(z^\epsilon), \tag{3.19}$$

and let us assume that there exists some diffeomorphism that maps z^ϵ onto the variables (x^ϵ, y^ϵ) satisfying (3.1). More specifically, we require that $x^\epsilon = \theta(z^\epsilon)$ for some θ such that

$$D\theta(z)G(z) = 0, \tag{3.20}$$

so that

$$\frac{\mathrm{d}x^\epsilon}{\mathrm{d}t} = D\theta(z^\epsilon)F(z^\epsilon), \tag{3.21}$$

that is, the variables $x^\epsilon = \theta(z^\epsilon)$ are indeed slow. We also require that no other (hidden) slow variable exists, that is, $x^\epsilon = \theta(z^\epsilon)$ are those entering the limit theorem on which the HMM scheme is built. Even if we do not know the explicit form of θ we can make use of its existence. To see how, let Φ_t denote the map that approximates the solution to

$$\frac{\mathrm{d}Z}{\mathrm{d}t} = F(Z), \tag{3.22}$$

and let Ψ_t denote the map that approximates the solution to

$$\frac{\mathrm{d}Z}{\mathrm{d}t} = G(Z). \tag{3.23}$$

In terms of these maps, a vanilla time-splitting integrator for (3.19) would, for example, read (*cf.* (3.9))

$$\Phi_{\delta t} \circ \Psi_{\delta t/\epsilon} \tag{3.24}$$

or (*cf.* (3.10))

$$\Phi_{M\delta t} \circ \Psi_{\delta t/\epsilon}^{\circ M}. \tag{3.25}$$

This last scheme updates the variables by $M\delta t$ at every iterate, and the discussion above suggests replacing it instead by (*cf.* (3.11))

$$\Phi_{\Delta t} \circ \Psi_{\delta t/\epsilon}^{\circ M} \tag{3.26}$$

or

$$(\Phi_{\Delta' t} \circ \Psi_{\delta t/\epsilon})^{\circ M}, \tag{3.27}$$

where again $\Delta' t = \Delta t/M$. In the limit (3.14), both (3.26) and (3.27) are consistent with a version of (3.19) where ϵ has been boosted to a higher value $\epsilon' = \epsilon\alpha = \epsilon\Delta t/M\delta t$, that is,

$$\frac{\mathrm{d}z^{\epsilon'}}{\mathrm{d}t} = F(z^{\epsilon'}) + \frac{1}{\epsilon'}G(z^{\epsilon'}), \tag{3.28}$$

which can again be used to explain why the scheme is accurate: essentially (3.15) needs to hold, and if the numerical error of the vanilla time-splitting scheme (3.24) is $C(\delta t/\epsilon)^a$, for some $C > 0$, then that of the scheme (3.27) is again given by (3.18).

The scheme (3.27) was first proposed by Vanden-Eijnden (2007). Later a very close variant of this scheme was proposed by Tao, Owhadi and Marsden (2010) under the name 'FLAVORS'. Specifically, if we let $\chi_t^{1/\epsilon}$ denote the map that approximates numerically the solution of (3.19), and assume that we can control the size of ϵ in this map, then the scheme proposed by Tao *et al.* (2010) is (in our notation)

$$\left(\chi^0_{\Delta' t - \delta t} \circ \chi^{1/\epsilon}_{\delta t}\right)^{\circ M}. \tag{3.29}$$

As can be seen, compared to (3.27), this scheme amounts to a slight modification of the way the time-splitting is implemented: (3.27) evolves the

Such systems are widely studied in the construction of multiscale methods; see, for instance, Vanden-Eijnden (2003) or E (2003). In this case, it is reasonable to use y as the macroscopic variable, that is, $U = \bar{y} \simeq y$, and $\mathcal{R}(U, D_R) = (x^*, y)$, where D_R gives the value $x^* \in \mathcal{M}$. For example, x^* may be taken from the x values in the previous microscale simulation. The compression $\mathcal{Q}$ may simply be $\mathcal{Q}(x, y) = y$.

However, if there are resonances among the oscillations, $x(t)$ is unlikely to remain on a fixed invariant manifold (Kreiss 1992), and more sophistication in the algorithm is needed. One can see the essence of this problem from the simple example

$$\begin{cases} \dfrac{\mathrm{d}x}{\mathrm{d}t} = \frac{\mathrm{i}}{\epsilon}x + g(x), \ x(0) = 1, \\ \dfrac{\mathrm{d}y}{\mathrm{d}t} = h(x), \ y(0) = y_0. \end{cases} \Longrightarrow \begin{cases} \dfrac{\mathrm{d}w}{\mathrm{d}t} = \mathrm{e}^{-\frac{\mathrm{i}}{\epsilon}t} g(\mathrm{e}^{\frac{\mathrm{i}}{\epsilon}t}w), \ x(t) = \mathrm{e}^{\frac{\mathrm{i}}{\epsilon}t}w(t), \\ \dfrac{\mathrm{d}y}{\mathrm{d}t} = h(\mathrm{e}^{\frac{\mathrm{i}}{\epsilon}t}w), \ y(0) = y_0. \end{cases}$$

Let us formally decompose $\mathrm{e}^{-\frac{\mathrm{i}}{\epsilon}t} g(\mathrm{e}^{\frac{\mathrm{i}}{\epsilon}t}w) = \bar{g}(w) + \alpha(\mathrm{e}^{\frac{\mathrm{i}}{\epsilon}t}, w)$, where $\bar{g}$ does not depend on any fast oscillations. If $\bar{g} \equiv 0$, then $w(t)$ stays close to 1, due to strong self-averaging in α. Thus (3.32) corresponds to averaging h over the unit circle, and $\mathrm{d}\mu$ is the arc-length element. Resonance in this system corresponds to the case where $\bar{g}$ is non-zero. Consequently, the averaging has to be performed with the correct measure:

$$\frac{\mathrm{d}\bar{y}}{\mathrm{d}t} = \bar{h}(y, t) := \int_{\mathcal{M}(t)} h(x, y, t)\, \mathrm{d}\mu(x; t).$$

For example, if $g(x) = x$, then $\bar{g}(w) = w$, and $\alpha \equiv 0$. Consequently, $\mathcal{M}(t)$ is a circle with radius equal to $w(t) = \exp(t)$. Without knowledge of $w(t)$, it is impossible to define a consistent reconstruction operator $\mathcal{R}$, and consequently it is impossible to build a convergent multiscale algorithm. In some of the literature, the issue caused by resonance is referred to as the system having hidden slow variables (Fatkullin and Vanden-Eijnden 2004, Engquist and Tsai 2005, Ariel, Engquist, Kreiss and Tsai 2009*c*, Tao *et al.* 2010). It is essential that a multiscale method should compute the effect of the hidden slow variables accurately.

It is possible to design multiscale algorithms that compute the effective behaviour of highly oscillatory dynamical systems by using slow variables (Ariel, Engquist and Tsai 2009*a*). We continue our discussion using the previous example, but instead we rewrite the equation for x as a system in $\mathbb{R}^2$:

$$\frac{\mathrm{d}x_1}{\mathrm{d}t} = \frac{1}{\epsilon}x_2 + x_1,$$

$$\frac{\mathrm{d}x_2}{\mathrm{d}t} = -\frac{1}{\epsilon}x_1 + x_2,$$

with initial conditions $(x_1(0), x_2(0)) = (0, 1)$. Thus

$$(x_1(t), x_2(t)) = (\mathrm{e}^t \sin \epsilon^{-1} t,\ \mathrm{e}^t \cos \epsilon^{-1} t).$$

Taking $I = x_1^2 + x_2^2$, we notice that I has a bounded derivative along the trajectory of the solution, that is, $I' := (\mathrm{d}/\mathrm{d}t) I(x_1(t), x_2(t)) = 2I$ is independent of ϵ. For this particular example one can easily solve for I, $I(t) = I(0)\,\mathrm{e}^{2t}$. In fact, the uniform bound on I' indicates the *slow* nature of $I(x_1(t), x_2(t))$ when compared with the fast oscillations in $(x_1(t), x_2(t))$. This type of characterization of the effective dynamics of highly oscillation systems is commonly used in the literature. In this example it was easy to find the slow quantity I. In general it is difficult, and the proposed research aims to avoid this difficulty. In classical mechanics, I is referred to as the action variable, and there is a corresponding angle variable whose gradient is τ.

We say that $\xi : \mathcal{D}_0 \subset \mathbb{R}^d \mapsto \mathbb{R}$ has a bounded derivative along $x^\epsilon(t)$ if, for $0 < \epsilon < \epsilon_0$,

$$\sup_{x^\epsilon \in \mathcal{D}_0, \epsilon \in (0, \epsilon_0)} |\nabla \xi(x^\epsilon) \cdot \dot{x}^\epsilon| \leq C. \tag{3.33}$$

Such functions are commonly referred to as slow variables of the system; see, for example, Kreiss (1979, 1992), Kreiss and Lorenz (1993), Fatkullin and Vanden-Eijnden (2004), Gear and Kevrekidis (2005) and Ariel *et al.* (2009*a*, 2009*b*). Other approaches to finding slow variables include, for example, those of Artstein, Kevrekidis, Slemrod and Titi (2007*a*) and Artstein, Linshiz and Titi (2007*b*).

Typically, one may expect the trace of a slow variable along the dynamical system's solutions, $\xi \circ x^\epsilon$, to converge to a smooth curve as $\epsilon \to 0$. We shall denote this limit by $\bar{\xi}(t; x_0)$. In designing multiscale algorithms for this type of problem, it is often convenient to aim at constructing accurate approximation of $\bar{\xi}$ by suitable filtering of the oscillations in $x^\epsilon(t)$.

3.2.3. Algorithms with hidden slow variables

Consider an ODE system of the general form

$$\frac{\mathrm{d}x}{\mathrm{d}t} = \frac{1}{\epsilon} f(x) + g(x), \quad x(0) = x_0, \tag{3.34}$$

where $0 < \epsilon \leq \epsilon_0$ is a small parameter that characterizes the separation of time scales in the problem. Let $\zeta(t)$ denote the solution of

$$\frac{\mathrm{d}\zeta}{\mathrm{d}t} = \bar{F}(\zeta) = \int_{S^1} F(\zeta, \sigma, \gamma = 0)\,\mathrm{d}\sigma, \quad \zeta(0) = \xi(0). \tag{3.35}$$

The first step in our algorithm is to identify the slow variables $\xi(x)$. The slow variables can be determined analytically or numerically by identifying coefficients in a predetermined form. For details, see Ariel *et al.* (2009*a*).

Then the ODE (3.34) is integrated using a two-level algorithm: each level corresponds to the integration of (3.34) in a different time scale. The first is a macro solver, which integrates the averaged equation (3.35). The second level is a micro solver that is invoked whenever the macro solver calls for it. Each time the micro solver is invoked, it computes a short time solution of (3.34) using suitable initial data.

1 *Construction of slow variables.* Find functions $\xi_1(x), \ldots, \xi_r(x)$ such that $|\nabla_x \xi \dot{x}| \leq C_0$ and $\operatorname{rank}(\partial\xi/\partial x) = r$. (See Ariel *et al.* (2009*a*) for details.)

2 *Multiscale evolution.*

(a) Initial conditions: $x(0) = x_0$ and $n = 0$.

(b) *Force estimation.*

(i) Microscale solver: solve (3.34) in $t \in [t_n, t_n + 2\eta]$ with initial conditions x_n.

(ii) Averaging: approximate $\dot{\xi}(t_n + \eta)$ by

$$\langle \dot{\xi} \rangle_\eta (t_n + \eta) = (K_\eta * \dot{\xi})(t_n + \eta) = (-\dot{K}_\eta * \xi)(t_n + \eta).$$

(c) Macroscale solver (forward Euler example): $x_{n+1} = x_n^{M/2} + H\delta x$, where δx is the least-squares solution to the linear system

$$\delta x \cdot \nabla \xi_i = \bar{F}_i(\xi) = \langle \dot{\xi}_i \rangle_\eta,$$

for all $i = 1, \ldots, r$.

(d) $n = n + 1$. Repeat steps (b) and (c) until time T.

Note that there is no need to actually change the original ODE (3.34) to a form with explicit use of the slow variables.

3.2.4. A Fermi–Pasta–Ulam model

The Fermi–Pasta–Ulam model (Fermi, Pasta and Ulam 1955) is a simple system of unit mass particles connected by springs. The springs alternate between stiff linear and soft non-linear ones. Recently, this model was considered by Hairer *et al.* (2002) as a benchmark problem for studying the long-time properties of numerical solutions to stiff ODEs using geometric integrators. The model is derived from the Hamiltonian

$$H(p,q) = \frac{1}{2}\sum_{i=1}^{2k} p_i^2 + \frac{1}{4}\epsilon^{-2}\sum_{i=1}^{k}(q_{2i} - q_{2i-1})^2 + \sum_{i=0}^{k}(q_{2i+1} - q_{2i})^4. \qquad (3.36)$$

The following linear change of variables is convenient, since it separates the elongations of the stiff springs and associated momentum,

$$x_i = \epsilon^{-1}(q_{2i-1} - q_{2i})/\sqrt{2}, \quad v_i = (p_{2i-1} - p_{2i})/\sqrt{2}, \qquad (3.37)$$

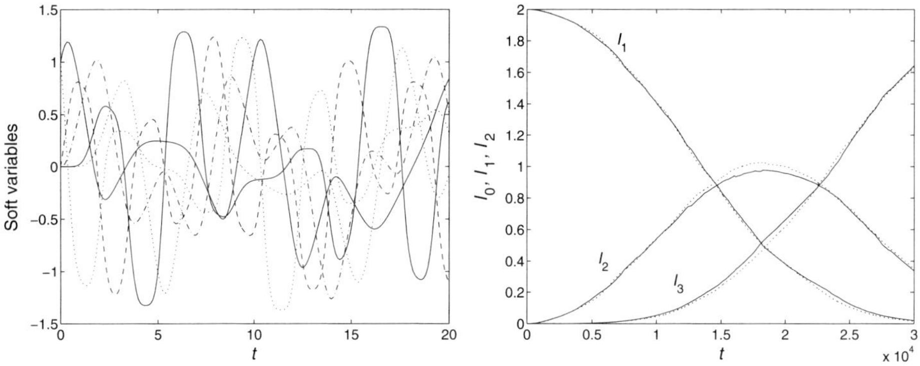

(a) (b)

Figure 3.2. Comparison of the HMM approximation for the solution of the Fermi–Pasta–Ulam equations of motion (3.39) with three stiff springs, $k = 3$, with that obtained using the Verlet method with step size of the order of ϵ. (a) Soft variables on an $O(1)$ time scale, and (b) I_1, I_2 and I_3 on an $O(\epsilon^{-1})$ scale. With the above parameters the HMM algorithm runs an order of magnitude faster than the Verlet one. The ratio between running times increases with smaller ϵ.

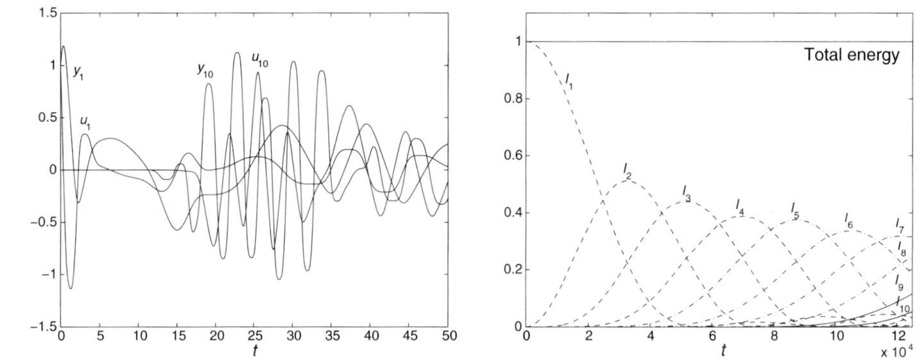

(a) (b)

Figure 3.3. The HMM approximation for the solution of the Fermi–Pasta–Ulam equations of motion (3.39) with ten stiff springs, $k = 10$. (a) y_1, u_1, y_{10} and u_{10} on an $O(1)$ time scale, and (b) $I_1, \ldots, I_{10}$ on an $O(\epsilon^{-1})$ scale. The Verlet method takes too long to integrate.

In this algorithm, r_1 is used to update the clock and r_2 is used to select the particular reaction to be executed.

Now assume the rate functions have the form

$$a(x) = \left(a^s(x), \frac{1}{\varepsilon} a^f(x)\right), \tag{3.46}$$

where $\varepsilon \ll 1$ represents the ratio of the fast and slow time scales of the system. The corresponding reactions and the associated stoichiometric vectors can be grouped accordingly:

$$R^s = \{(a^s, \nu^s)\}, \quad R^f = \left\{\left(\frac{1}{\varepsilon} a^f, \nu^f\right)\right\}. \tag{3.47}$$

We call R^s the slow reactions and R^f the fast reactions. We have made the simplifying assumption that the rates are divided into two groups. Clearly the algorithm can be easily extended to situations when there are multiple groups. The more serious assumption here is that the groups of slow and fast reactions do not change in time. Some adaptive strategies have to be introduced in order to remove this restriction.

Our interest is in the dynamics of the slow processes, not the detailed dynamics of the fast processes. To this end, an effective system can be derived using standard averaging methods (Kurtz 1973, Papanicolaou 1977). For this purpose it is helpful to introduce an auxiliary process, called the *virtual fast process* (Cao, Gillespie and Petzold 2005). This auxiliary process retains the fast reactions only: all slow reactions are turned off. Intuitively, each realization of the SSA consists of a sequence of realizations of the virtual fast process, punctuated by occasional firing of the slow reactions. Due to the time scale separation, with high probability, the virtual fast process has enough time to relax to equilibrium before another slow reaction takes place. Therefore the effective slow rates for the slow dynamics should simply be the original slow rates averaged over the equilibrium distributions associated with the virtual fast process.

Let μ_x denote the equilibrium distribution of the virtual fast process when it is initialized at x, and let

$$\tilde{a}_i^s(x) = \sum_{z \in \mathcal{X}} a_i^s(z) \mu_x(z). \tag{3.48}$$

It can be shown that the effective dynamics is governed by the set of reactions:

$$\bar{R} = (\bar{a}^s(x), \bar{\nu}^s). \tag{3.49}$$

The effective master equation is given by

$$\frac{\partial u}{\partial t}(x,t) = \sum_{i=1}^{M_s} \tilde{a}_i^s(x) \big(u(x + \nu_i^s, t) - u(x,t)\big). \tag{3.50}$$

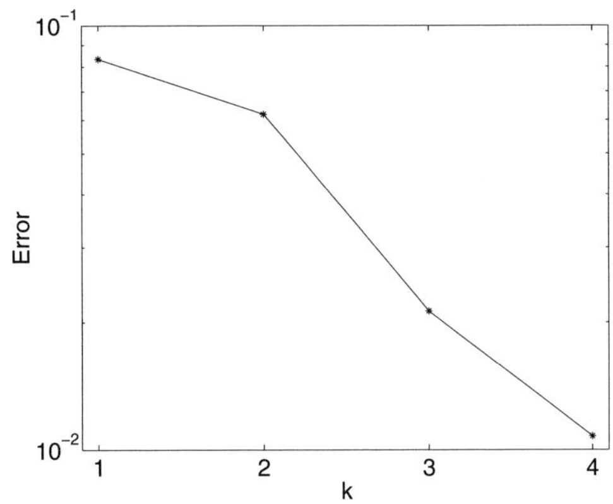

Figure 3.4. Relative errors of $\overline{template}$ using the nested SSA for the virus infection model (courtesy of Di Liu).

which is consistent with the results in Figure 3.4. Table 3.3 gives the total CPU time and the obtained values of $\overline{template}$ and var($template$), with the parameters of inner SSA chosen according to (3.63) and using $N_0 = 10^6$ realizations of the outer SSA (the same as in the direct SSA).

4. Finite element HMM

Finite element heterogeneous multiscale methods (FE-HMM) for homogenization problems, first proposed by E and Engquist (2003), have been developed for elliptic problems (Abdulle 2005*b*, E, Ming and Zhang 2005*b*, Abdulle 2006, Abdulle and Schwab 2005), elastic problems (Abdulle 2006), parabolic problems (Ming and Zhang 2007, Abdulle and Vilmart 2012*a*), wave problems (Abdulle and Grote 2011), and advection–diffusion problems (Abdulle 2005*a*, Henning and Ohlberger 2010). Further developments for elliptic problems include coupling of finite element macroscopic methods with spectral methods (Abdulle and Engquist 2007), with reduced basis methods (Abdulle and Bai 2012), FE-HMM with special quadrature formulas (Du and Ming 2010), and discontinuous Galerkin FE-HMM (Abdulle 2008, Abdulle 2012).

Various numerical methods have been derived in the past few years for multiscale partial differential equations, mostly for elliptic problems. We do not attempt to review the literature on the subject, and just give a few references. Numerical approaches for homogenization problems were pioneered by Babuška (1976) for static problems and Engquist (1987) for dynamic problems. For multiscale elliptic problems, an important early development is the generalized finite element method proposed by Babuška and Osborn (1983), who developed the idea of adapting the finite element space to the particular fine-scale features of the problem. Dorobantu and Engquist

(1998) and Engquist and Runborg (2002) proposed a method based on multiresolution analysis, Neuss, Jäger and Wittum (2001) combined the multigrid method with homogenization in the coarsening process, Hou, Wu and Cai (1999) proposed the multiscale finite element method (MsFEM) based on modified basis functions obtained from the fine-scale equations (see Efendiev and Hou (2009) for a review), Matache, Babuška and Schwab (2000) and Matache and Schwab (2002) developed the two-scale finite element method, and Viet Ha Hoang and Schwab (2005) proposed the high-dimensional finite element method. We also mention the huge literature in the structural mechanics and engineering communities concerned with micro–macro methods based on representative volume elements (RVEs). Such methods have been proposed for various types of problems, but often without convergence analysis. We mention the methods of Terada and Kikuchi (2001), Miehe, Schröder and Bayreuther (2002) and Geers, Kouznetsova and Brekelmans (2010).

4.1. General methodology

We start by explaining the methodology of the FE-HMM. We consider a general multiscale problem of the form

$$L^{\varepsilon}(u^{\varepsilon}, a^{\varepsilon}) = f \quad \text{in } \Omega,$$

where L^{ε} is a differential operator, a^{ε} denotes the data of the problem, Ω is an open bounded subset of $\mathbb{R}^d$, $f : \Omega \to \mathbb{R}$ is given and $u^{\varepsilon} : \bar{\Omega} \to \mathbb{R}$ is the solution of the above problem for which appropriate boundary conditions are specified. To emphasize the multiscale nature of the data of the above problem, we put a superscript ε (representing the typical size of a small scale in the considered problem) on a, L and u. Here, for simplicity, in view of numerical discretization, we assume that $\Omega \subset \mathbb{R}^d$ ($d = 1, 2, 3$) is a polygonal

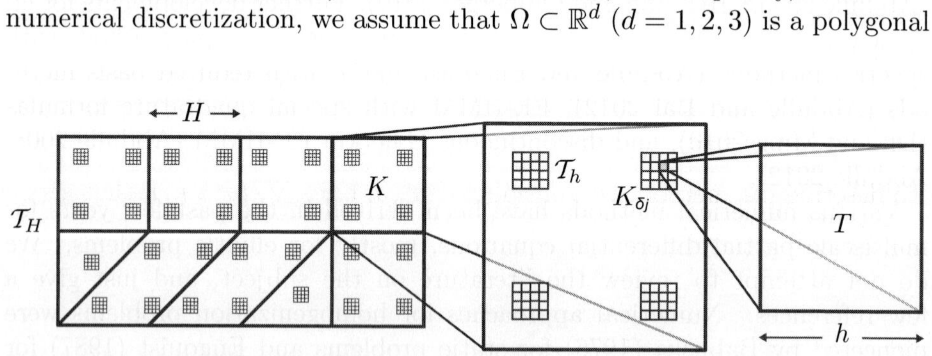

Figure 4.1. Sketch of the FE-HMM algorithm: macro triangulation $\mathcal{T}_H$ (quadrilateral and simplicial elements) with sampling domains within the macro elements. A close-up of the sampling domains shows the micro triangulation $\mathcal{T}_h$, chosen here to be quadrilaterals.

domain. The effective problem, assumed to exist, is of the form

$$L^0(u^0, a^0) = f \quad \text{in } \Omega,$$

where $u^0 : \bar{\Omega} \to \mathbb{R}$ is the effective solution. One can think of u^0 as being the limit (in an appropriate sense) of the solution u^ε as $\varepsilon \to 0$. The weak solutions of any of the above problems are assumed to belong to an appropriate Sobolev space, denoted by $H(\Omega)$. The FE-HMM can be summarized by the following steps.

- *Macroscale solver.*
  - *Macro triangulation.* Define a macroscopic partition $\Omega = \bigcup_{K\in\mathcal{T}_H} K$ and a macroscopic finite-dimensional subspace $S(\Omega, \mathcal{T}_H)$ of $H(\Omega)$.
  - *Macro method.* Let $u_H \in S(\Omega, \mathcal{T}_H)$ be the solution of $L_H(u_H) = f$, where L_H is an unknown approximation of the effective differential operator L^0.
- *Microscale solver.*
  - *Constrained micro simulations.* For every $K \in \mathcal{T}_H$, consider a suitable *quadrature formula* $\{x_{K_j}, \omega_{K_j}\}_{j=1}^J$, and *sampling domains* $K_{\delta_j} = x_{K_j} + \delta I$, $I = (-1/2, 1/2)^d$ $(\delta \geq \varepsilon)$. Define a micro triangulation $\bigcup_{T\in\mathcal{T}_h} T = K_{\delta_j}$ and a micro finite-dimensional space $S(K_{\delta_j}, \mathcal{T}_h)$. Compute micro solutions $u_h \in S(K_{\delta_j}, \mathcal{T}_h)$ constrained by the macro variable using the original fine-scale operator.
  - *Data processing.* Recover $L_H(u_H)|_K$ locally by a suitable average of u_h in K_{δ_j}.

We notice that the method described here could be used to solve problems with a right-hand side depending on ε (*e.g.*, f^ε) with minor changes (see Abdulle and Schwab (2005)).

4.2. Model problem

To describe the method, we consider $L^\varepsilon(u^\varepsilon, a^\varepsilon) = -\nabla\cdot(a^\varepsilon \nabla u^\varepsilon)$, that is, the linear elliptic homogenization problem seen in Section 2,

$$-\nabla\cdot(a^\varepsilon \nabla u^\varepsilon) = f \quad \text{in } \Omega, \qquad u^\varepsilon = 0 \quad \text{on } \partial\Omega, \tag{4.1}$$

where the family of tensors $a^\varepsilon(x) \in (L^\infty(\Omega))^{d\times d}$ (indexed by ε) is assumed to be uniformly elliptic and bounded, that is,

$$\exists \lambda, \Lambda > 0 \quad \text{such that} \quad \lambda|\xi|^2 \leq a^\varepsilon(x)\xi\cdot\xi \quad \text{and} \quad |a^\varepsilon(x)\xi| \leq \Lambda|\xi|, \tag{4.2}$$
$$\forall \xi \in \mathbb{R}^d, \quad \forall\varepsilon, \quad \text{a.e. } x \in \Omega,$$

where ε is a microscopic scale that characterizes the multiscale nature of the tensor $a^\varepsilon(x)$. Here we take zero Dirichlet boundary conditions for simplicity.[1] We also assume that $f \in L^2(\Omega)$ ($f \in H^{-1}(\Omega)$ would also be possible). By the Lax–Milgram theorem, owing to the uniform ellipticity and boundedness of the tensor a^ε, the weak form of (4.1) possesses for each ε a unique solution, and we can thus consider a family of solutions $\{u^\varepsilon\}$ (indexed by ε) which are bounded in $H_0^1(\Omega)$ by the same constant.

Homogenization results. The goal of homogenization theory is to find an 'averaged equation' corresponding to (4.1). Using H-convergence theory (Murat and Tartar 1997) – or G-convergence theory in the symmetric case (De Giorgi and Spagnolo 1973) – and without further assumptions on the heterogeneities of the tensor $a^\varepsilon(x)$, it is possible to show that there exists a tensor $a^0(x)$ (again uniformly elliptic and bounded) and a subsequence of $\{u^\varepsilon\}$ which weakly converges in $H_0^1(\Omega)$ to a function $u^0 \in H_0^1(\Omega)$, such that

$$-\nabla \cdot (a^0 \nabla u^0) = f \quad \text{in } \Omega, \quad u^0 = 0 \quad \text{on } \partial\Omega. \tag{4.3}$$

Returning to the general description of Section 4.1, we see that

$$L^0(u^0, a^0) = -\nabla \cdot (a^0 \nabla u^0).$$

If $a^\varepsilon(x)$ has a more specific structure, for example if $a^\varepsilon(x) = a(x, x/\varepsilon)$ and is periodic in its second argument, then classical results show that the whole sequence $\{u^\varepsilon\}$ weakly converges to $u^0 \in H_0^1(\Omega)$, and the homogenized tensor $a^0(x)$ at $x \in \Omega$ can be characterized as the average of the solutions of 'cell problems' (*i.e.*, d boundary value problems on a domain of size ε^d involving the fine-scale tensor $a^\varepsilon(x) = a(x, x/\varepsilon)$). For details we refer, for example, to Bensoussan, Lions and Papanicolaou (1978), Jikov, Kozlov and Oleinik (1994) and Cioranescu and Donato (1999). Notice that even in this fortunate case, we generally still have an infinite number of cell problems to solve, due to the slow variation in the tensor. Classical numerical approaches consist of pre-computing an approximation of the effective tensor $a^0(x_i)$ at predefined sampling points $x_i \in \Omega$, $i = 1, \dots, p$, and using a standard finite element method based on these pre-computed data. Although straightforward, this sequential strategy does not allow for a natural way to derive *a priori* or *a posteriori* error control of the overall discretization (as it depends on the accuracy of the numerically pre-computed homogenized tensors). Further, this procedure does not give a straightforward discretization for non-periodic or non-linear problems, and makes it difficult to switch locally to a fine-scale approximation (desirable in some applications).

[1] The algorithm and results stated in this section are valid for other boundary conditions (non-zero Dirichlet, Neumann, mixed, *etc.*) with obvious changes.

4.3. FE-HMM: algorithm

In this section we give a detailed description of the FE-HMM. We restrict ourselves to the model problem (4.1). In Section 4.3.3, generalizations to more involved problems (*e.g.*, non-linear problems) will be described.

4.3.1. Macroscale solver

Let $\mathcal{T}_H$ denote a family of (macro) partitions[2] of Ω in simplicial or rectangular elements K. The diameter of an element $K \in \mathcal{T}_H$ is denoted by H_K and we set $H = \max_{K\in\mathcal{T}_H} H_K$. We then consider a macro finite element space

$$S_0^\ell(\Omega, \mathcal{T}_H) = \{v_H \in H_0^1(\Omega) : v_H|_K \in \mathcal{R}^\ell(K),\ \forall K \in \mathcal{T}_H\}, \tag{4.4}$$

where $\mathcal{R}^\ell(K)$ is the space $\mathcal{P}^\ell(K)$ of polynomials on K of total degree at most ℓ, if K is a simplicial finite element, or the space $\mathcal{Q}^\ell(K)$ of polynomials on K of degree at most ℓ in each variable, if K is a rectangular finite element. For every K we consider the quadrature formula $\{x_{K_j}, \omega_{K_j}\}_{j=1}^J$ (described further in Section 4.3.2), and sampling domains K_{δ_j}, defined as $K_{\delta_j} = x_{K_j} + \delta I,\ I = (-1/2, 1/2)^d\ \ (\delta \geq \varepsilon)$.

Multiscale method. Find $u_H \in S_0^\ell(\Omega, \mathcal{T}_H)$ such that

$$B_H(u_H, v_H) = \int_\Omega f v_H \,\mathrm{d}x, \quad \forall v_H \in S_0^\ell(\Omega, \mathcal{T}_H), \tag{4.5}$$

where for arbitrary $v_H, w_H \in S_0^\ell(\Omega, \mathcal{T}_H)$ the bilinear form $B_H(\cdot,\cdot)$ is defined by

$$B_H(v_H, w_H) := \sum_{K\in\mathcal{T}_H} \sum_{j=1}^J \frac{\omega_{K_j}}{|K_{\delta_j}|} \int_{K_{\delta_j}} a^\varepsilon(x) \nabla v_{h,K_j} \cdot \nabla w_{h,K_j} \,\mathrm{d}x, \tag{4.6}$$

where $v_{h,K_j} = R_{K_j}(v_H)$, $w_{h,K_j} = R_{K_j}(w_H)$ and $R_{K_j} = R|_{K_j}$ is the reconstruction operator (in the terminology of Section 2) restricted to the sampling domain K_{δ_j}. In the notation of Section 4.1, we have

$$L_H :\ S_0^\ell(\Omega, \mathcal{T}_H) \longrightarrow L^2(\Omega),$$

where

$$(L_H(v_H), w_H) = B_H(v_H, w_H),$$

and $(\cdot,\cdot)$ is the L^2 inner product. In order to assemble the bilinear form $B_H(v_H, w_H)$ we need to compute the functions $R_{K_j}(v_H), R_{K_j}(w_H)$, as described in the next subsection.

[2] By macro partition we mean that $H \gg \varepsilon$ is allowed.

4.3.2. Microscale solver

The location of the sampling domains K_{δ_j} and the choice of the weights ω_{K_j} $(j = 1, \ldots, J)$ in the definition of $B_H(\cdot, \cdot)$ rely on the definition of appropriate quadrature formulas. We consider $\hat{K}$ to be the reference element, and for every element K of the triangulation we let F_K denote the $\mathcal{C}^1$-diffeomorphism such that $K = F_K(\hat{K})$. For every K we consider the quadrature formula $x_{K_j} = F_K(\hat{x}_j)$, $\omega_{K_j} = \hat{\omega}_j |\det(\partial F_K)|$, $j = 1, \ldots, J$, where $\{\hat{x}_j, \hat{\omega}_j\}_{j=1}^J$ is a quadrature formula on $\hat{K}$. We then consider a (micro) partition $\mathcal{T}_h$ of each sampling domain K_{δ_j} in simplicial or rectangular elements T and define a micro finite element space

$$S^q(K_{\delta_j}, \mathcal{T}_h) = \{z_h \in W(K_{\delta_j}) : z_h|_T \in \mathcal{R}^q(T),\ \forall T \in \mathcal{T}_h\}, \tag{4.7}$$

where $h = \max_{T \in \mathcal{T}_h} h_T$ (h_T is the diameter of the element T) and $W(K_{\delta_j})$ is a given Sobolev space.

Coupling. Various spaces $W(K_{\delta_j})$ can be chosen for the micro numerical method, for example

$$W(K_{\delta_j}) = W^1_{\text{per}}(K_{\delta_j}) = \left\{ z \in H^1_{\text{per}}(K_{\delta_j}) : \int_{K_{\delta_j}} z \,\mathrm{d}x = 0 \right\}, \tag{4.8}$$

for a periodic coupling, or

$$W(K_{\delta_j}) = H^1_0(K_{\delta_j}) \tag{4.9}$$

for a coupling via Dirichlet boundary conditions. Other coupling conditions, constraining the averaged gradient of a microscale solution, can be used. We refer to Yue and E (2007) for further discussion.

Micro method. For each $v_H \in S^\ell_0(\Omega, \mathcal{T}_H)$ we define a micro function v_{h,K_j}, satisfying $(v_{h,K_j} - v_{H,\text{lin}}) \in S^q(K_{\delta_j}, \mathcal{T}_h)$, which is a solution of

$$\int_{K_{\delta_j}} a^\varepsilon(x) \nabla v_{h,K_j} \cdot \nabla z_h \,\mathrm{d}x = 0, \quad \forall z_h \in S^q(K_{\delta_j}, \mathcal{T}_h), \tag{4.10}$$

where $v_{H,\text{lin}}|_{K_{\delta_j}} = v_H(x_{K_j}) + (x - x_{K_j}) \cdot \nabla v_H(x_{K_j})$ is the linearization of v_H at the quadrature point x_{K_j}. As explained in Section 4.4.1, the bilinear form (4.6) can be rewritten as

$$B_H(v_H, w_H) = \sum_{K \in \mathcal{T}_H} \sum_{j=1}^{J} \omega_{K_j} a^0_K(x_{K_j}) \nabla v_H(x_{K_j}) \cdot \nabla w_H(x_{K_j}),$$

where $a^0_K(x_{K_j})$ is the macroscale data (effective tensor) at the quadrature point x_{K_j} recovered by the microscale simulations. This is the data processing step mentioned in Section 4.1, which is implicitly computed by defining the bilinear form via the quantities

$$\frac{1}{|K_{\delta_j}|}\int_{K_{\delta_j}} a^{\varepsilon}(x)\nabla v_{h,K_j}\cdot\nabla w_{h,K_j}\,\mathrm{d}x.$$

Remark 4.1. Conditions on the quadrature formula are needed to ensure that the optimal convergence rates for elliptic finite element methods are obtained when using numerical quadrature. We make the following classical assumptions on the quadrature formula $\{\hat{x}_j,\hat{\omega}_j\}_{j=1}^J$ on the reference element $\hat{K}$ (see Ciarlet and Raviart (1972)):

(Q1) $\hat{\omega}_j>0,\ j=1,\dots,J,\ \sum_{j=1}^J\hat{\omega}_j|\nabla\hat{p}(\hat{x}_j)|^2\geq\hat{\lambda}\|\nabla\hat{p}\|^2_{L^2(\hat{K})}$, for all $\hat{p}(\hat{x})\in\mathcal{R}^{\ell}(\hat{K})$, $\hat{\lambda}>0$,

(Q2) $\int_{\hat{K}}\hat{p}(\hat{x})\,\mathrm{d}\hat{x}=\sum_{j=1}^J\hat{\omega}_j\hat{p}(\hat{x}_j)$, for all $\hat{p}(\hat{x})\in\mathcal{R}^{\sigma}(\hat{K})$, for which $\sigma=\max(2\ell-2,\ell)$ if $\hat{K}$ is a simplicial finite element, or $\sigma=\max(2\ell-1,\ell+1)$ if $\hat{K}$ is a rectangular finite element.

4.3.3. Implementation and numerical illustration

The macro–micro methodology described in Section 4.3 allows easy implementation and design of a code whose structure follows the classical finite element implementation at the macro level. In particular, elementwise assembly for each macro element K can be computed to find the additive contribution from (4.6) to the macroscopic stiffness matrix. Abdulle and Nonnenmacher (2009*b*) propose a short (less than 200 lines) and flexible MATLAB implementation, capable of handling two- and three-dimensional elliptic and parabolic problems. All the numerical experiments presented in these subsections have been made with the code presented by Abdulle and Nonnenmacher (2009*b*) or a variation of it (*e.g.*, Abdulle and Vilmart (2011*a*)). These public-domain codes are available at http://anmc.epfl.ch. Additional numerical experiments can be found in Abdulle (2009) and Ming and Yue (2006).

Macro–micro mesh refinement. Some care is needed in the choice of the macroscopic mesh $\mathcal{T}_H$ (a triangulation of the physical domain Ω) and the microscopic mesh $\mathcal{T}_h$ (a triangulation of the sampling domain K_{δ_j}). Taking N_{mic} elements in each space dimension for the discretization of the sampling domain K_{δ_j}, we have $h=\delta/N_{\mathrm{mic}}$ and thus $\hat{h}:=h/\varepsilon=(\delta/\varepsilon)\cdot(1/N_{\mathrm{mic}})$. Since δ scales with ε, typically $\delta=C\varepsilon$ (where C is a constant of moderate size), we have $\hat{h}=C/N_{\mathrm{mic}}$. We let $M_{\mathrm{mic}}=O(\hat{h}^{-d})$ denote the number of degrees of freedom for the micro finite element method and let M_{mac} denote the number of degrees of freedom for the macro finite element method. For quasi-uniform macro meshes, the macro mesh size H (sometimes denoted

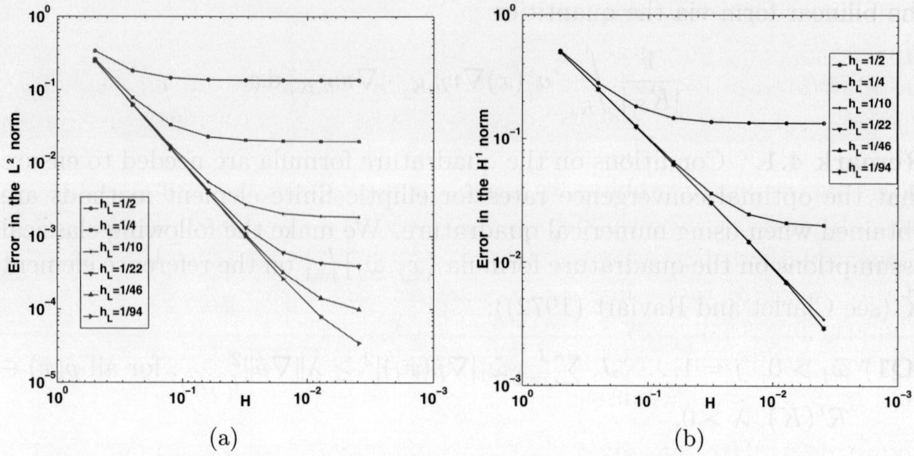

Figure 4.2. Error with respect to the homogenized solution u^0 for the FE-HMM applied to a two-dimensional periodic problem: (a) L^2 error, (b) H^1 error. The horizontal axis represents the macro mesh size (decreasing from left to right). Each curve represents the error obtained by macro mesh refinement for fixed micro meshes $h/\varepsilon = 1/2, 1/4, 1/10, 1/22, 1/46, 1/94$. Horizontal curves indicate a dominant micro error.

$H = 1/N_{\text{mac}}$) and the micro mesh size $\hat{h}$ are related to M_{mac} and M_{mic} by

$$H = O(M_{\text{mac}}^{-1/d}), \quad \hat{h} = O(M_{\text{mic}}^{-1/d}).$$

A priori error analysis for the fully discrete method (4.5), first given in Abdulle (2005*b*), provides an insight into this meshing problem, and reveals that the optimal refinement of the micro mesh is given by

$$\hat{h} := \frac{h}{\varepsilon} \simeq H^{\frac{\ell}{2q}} \quad (H^1 \text{ norm}), \quad \hat{h} := \frac{h}{\varepsilon} \simeq H^{\frac{\ell+1}{2q}} \quad (L^2 \text{ norm}).$$

The corresponding complexity, in terms of macro degrees of freedom, satisfies

$$\underbrace{H^{-d}}_{M_{\text{mac}}} \cdot \underbrace{H^{\frac{-d\ell}{2q}}}_{M_{\text{mic}}} \cdot n_s = (M_{\text{mac}})^{1+\frac{\ell}{2q}} \cdot n_s \qquad \text{for the } H^1 \text{ norm},$$

$$\underbrace{H^{-d}}_{M_{\text{mac}}} \underbrace{H^{\frac{-d(\ell+1)}{2q}}}_{M_{\text{mic}}} \cdot n_s = (M_{\text{mac}})^{1+\frac{\ell+1}{2q}} \cdot n_s \qquad \text{for the } L^2 \text{ norm},$$

where n_s denotes the number of sampling domains per macro element $K \in \mathcal{T}_H$.[3] For example, using piecewise linear polynomials on simplicial finite elements, assuming quasi-uniform macro and micro meshes, and that the

[3] Notice that as the micro problems are solved independently, the method is well suited to parallel implementation, which can reduce significantly the complexity of the FE-HMM.

complexity is proportional to the total degrees of freedom, we obtain a cost of $O(M_{\mathrm{mac}}^{3/2})$ (H^1 norm) and $O(M_{\mathrm{mac}}^2)$ (L^2 norm). These convergence rates are illustrated in Figure 4.2 on a very simple multiscale problem, for which reference solutions can easily be computed to high precision (see Abdulle (2005*b*)). In contrast, the memory demand is proportional to only $M_{\mathrm{mac}} + M_{\mathrm{mic}}$, as the micro problems, being independent, can be solved one at a time. Finally, we note that by using spectral methods or high-order finite element methods for the micro solvers, it is possible to reduce the total cost of the FE-HMM to log-linear complexity in the macro degrees of freedom. This was investigated by Abdulle and Engquist (2007). Such an approach, however, requires high regularity of the oscillating tensor a^ε, which may not hold for some applications, for example in materials science.

Example 1: Homogenization problem with non-periodic tensor. We consider problem (4.1), with a log-normal conductivity tensor a^ε (taken from Abdulle and Nonnenmacher (2009*b*)). The domain Ω consists of a semicircle and a rectangle, meshed with 1137 nodes using 576 triangles and 784 quadrilaterals, respectively: see Figure 4.3(a). We use the moving ellipse average method (Wallstrom *et al.* 1999, Section 4.1) to generate the realization of the log-normal stochastic field with mean zero and variance $\sigma = 1$. The correlation lengths of the stochastic field are set to be $\varepsilon_1 = 0.01$ and $\varepsilon_2 = 0.02$ in the x, y direction, respectively. A snapshot of this tensor is shown in Figure 4.3(a).

As this problem does not have an explicit analytical solution, we compute a fine-scale solution using a standard finite element method with a fine mesh of about 10^6 degrees of freedom in order to resolve the microscale for a given realization: see Figure 4.3(b). We compare this reference solution with the FE-HMM on the coarse macro grid with about 1100 degrees of freedom for the same realization. For the FE-HMM we present results for various sizes of the sampling domains ($\delta = 0.02$ and $\delta = 0.06$). We can see in Figures 4.3(c) and 4.3(d) that the FE-HMM solution is closer to the reference solution, as the sampling domain contains more correlation lengths of the random field.

Table 4.1. Energy norms of the FE-HMM and fine-scale solutions of the problem described in Example 1 with a random tensor. FE-HMM results are given for various micro sampling-domain sizes $\delta \times \delta$ and degrees of freedom N_{mic}^2 of the micro problems ($N_{\mathrm{mic}} = 1/\hat{h}$).

N_{mic}	4	8	16	32	64	Fine-scale
$\delta = 0.02$	0.2352	0.2415	0.2439	0.2449	0.2454	0.2583
$\delta = 0.06$	0.2313	0.2454	0.2520	0.2551	0.2567	0.2583

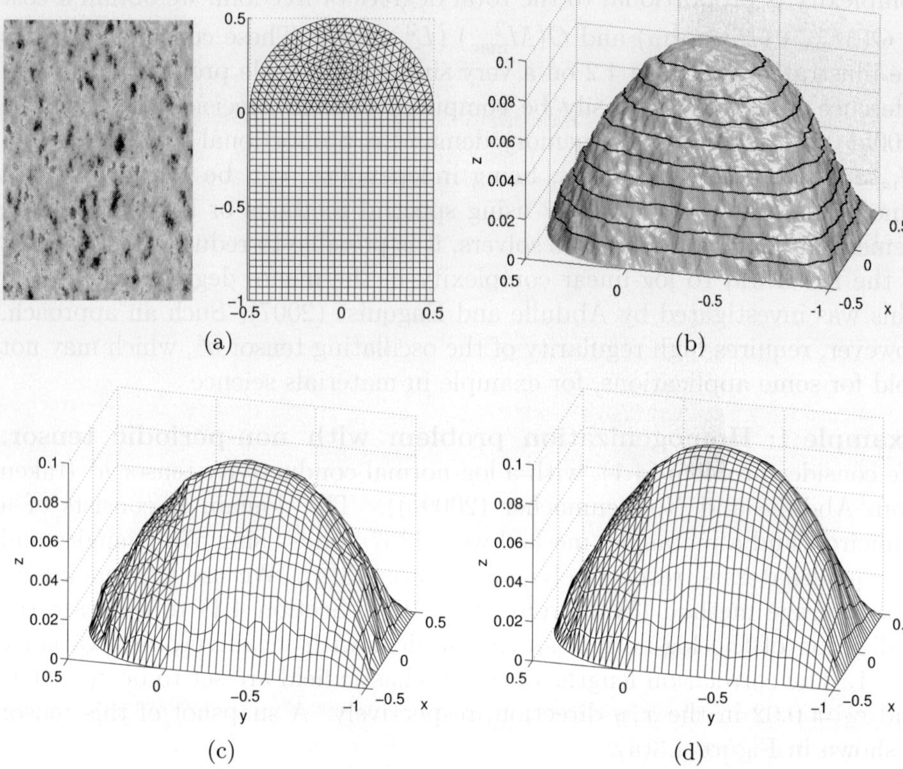

Figure 4.3. FE-HMM and fine-scale solution of the random problem described in Example 1. (a) Snapshot of the random tensor and computational domain. (b) Fine-scale solution computed with a standard finite element method on grid with 10^6 degrees of freedom. (c) FE-HMM with $N_{\text{mic}} = 64$ and micro domain size $\delta = 0.02$ on a macro grid with 1100 degrees of freedom. (d) FE-HMM with $N_{\text{mic}} = 64$ and micro domain size $\delta = 0.06$ on a macro grid with 1100 degrees of freedom.

This observation can also be seen in Table 4.1 when comparing the energy norm of the various solutions obtained with the FE-HMM to the energy norm of the reference solution.

Example 2: A non-linear problem, the Richards equation. We consider the steady-state Richards equation, which describes the fluid pressure $u(x,t)$ in an unsaturated porous medium,

$$-\nabla \cdot \big(a^\varepsilon(u^\varepsilon(x))\nabla(u^\varepsilon(x) - x_2)\big) = f(x) \quad \text{in } \Omega = (0,1)^2, \tag{4.11}$$

where x_2 is the vertical coordinate, and f corresponds to possible sources or sinks. Here a^ε is a multiscale permeability tensor that depends on the pressure u^ε, hence equation (4.11) is a non-linear non-monotone problem.

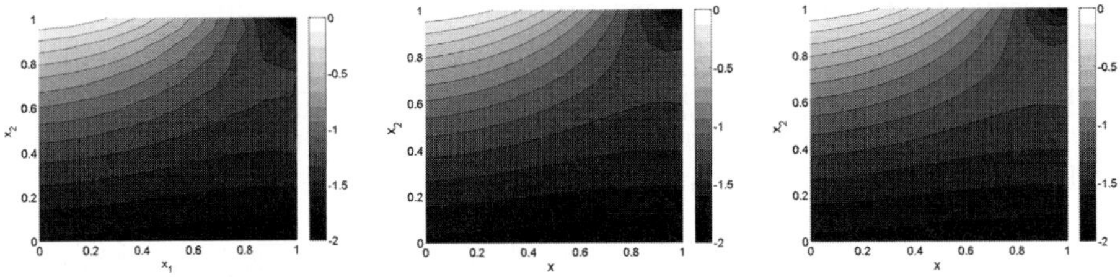

(a) FE-HMM 64 macro DOF (b) FE-HMM macro 256 DOF (c) FE-HMM macro 1024 DOF

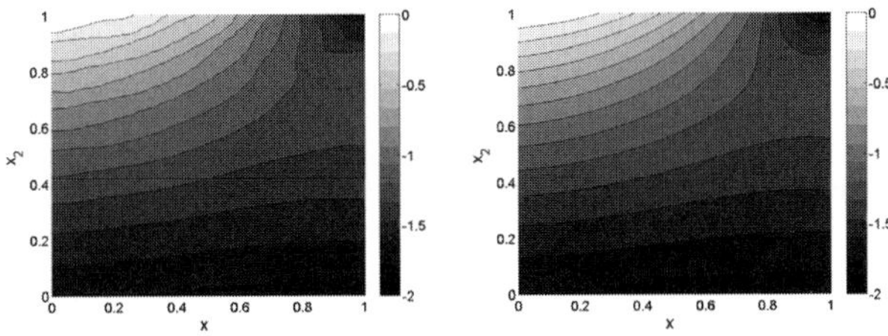

(d) FEM 1024 DOF (e) resolved FEM 1.5×10^6 DOF

Figure 4.4. The Richards problem (4.11) with mixed boundary conditions (Dirichlet and Neumann). (a–c) Level curves of the FE-HMM solutions with $N_{\text{mic}} = N_{\text{mac}}$. The macro degrees of freedom are indicated. (d, e) Level curves of a standard finite element method with respectively unresolved (1024 degrees of freedom) and resolved (1.5×10^6 degrees of freedom) meshes for the given microscale.

The application of the FE-HMM to such problems was first considered by E *et al.* (2005*b*). A complete analysis has recently been given by Abdulle and Vilmart (2011*b*, 2011*a*), and is summarized in Section 4.4.2. The FE-HMM for this problem reads as follows. Find $u_H \in S_0^\ell(\Omega, \mathcal{T}_H)$ such that

$$B_H(u_H; u_H, w_H) = F_H(w_H), \ \forall w_H \in S_0^\ell(\Omega, \mathcal{T}_H), \tag{4.12}$$

where

$$\begin{aligned} &B_H(u_H; v_H, w_H) \\ &:= \sum_{K \in \mathcal{T}_H} \sum_{j=1}^{J} \frac{\omega_{K_j}}{|K_{\delta_j}|} \int_{K_{\delta_j}} a^\varepsilon(x, u_H(x_{K_j})) \nabla v_{h,K_j}^{u_H(x_{K_j})}(x) \cdot \nabla w_{h,K_j}^{u_H(x_{K_j})}(x)\, \mathrm{d}x, \end{aligned} \tag{4.13}$$

and $v_{h,K_j}^{u_H(x_{K_j})}$ (and similarly $w_{h,K_j}^{u_H(x_{K_j})}$) is a solution of a linear micro problem similar to (4.10). For the parameter $s = u_H(x_{K_j})$ it reads: find v_{h,K_j}^s such that $(v_{h,K_j}^s - v_{H,\text{lin}}) \in S^q(K_{\delta_j}, \mathcal{T}_h)$ and

$$\int_{K_{\delta_j}} a^\varepsilon(x, s) \nabla v_{h,K_j}^s(x) \cdot \nabla z_h(x)\, \mathrm{d}x = 0, \quad \forall z_h \in S^q(K_{\delta_j}, \mathcal{T}_h). \tag{4.14}$$

We observe that while the macro problem (4.12) is non-linear, the micro problems (4.14) are linear. Indeed, in the implementation $s = u_H(x_{K_j})$ is given by the current (available) state of the macro solution (*e.g.*, using a Newton method to solve the macro problem). For the numerical simulation of (4.11) we set $f(x) \equiv 0$ for simplicity, and consider the boundary conditions

$$u^\varepsilon(x) = -1.9x_1^2 \quad \text{on } \partial\Omega_D = [0,1] \times \{1\},$$
$$n \cdot (a^\varepsilon(u^\varepsilon(x))\nabla(u^\varepsilon(x) - x_2)) = 0 \quad \text{on } \partial\Omega_N = \{0,1\} \times [0,1] \cup [0,1] \times \{0\},$$

and an exponential model for the permeability tensor a^ε, similar to that in Chen, Deng and Ye (2005*b*, Section 5.1),

$$a^\varepsilon(x,s) = \alpha^\varepsilon(x)\,\mathrm{e}^{\alpha^\varepsilon(x)s} \quad \text{where } \alpha^\varepsilon(x) = \frac{1/117.4}{2 + 1.8\sin(2\pi(2x_2/\varepsilon - x_1/\varepsilon))}. \tag{4.15}$$

Numerical solutions are shown in Figure 4.4. Figures 4.4(a)–4.4(c) are solutions of the FE-HMM with $\mathcal{P}^1$-triangular finite elements and a decreasing macro mesh size (uniform macro meshes with $H = 1/8, 1/16, 1/32$). The meshes of the micro solver are refined according to the 'optimal L^2 refinement strategy' $h/\varepsilon \simeq H$. The numerical results obtained with the FE-HMM are compared to a reference solution of the problem (4.11) (obtained by a resolved standard finite element method) plotted in Figure 4.4(e) (for $\varepsilon = 10^{-2}$, $\sim 10^6$ degrees of freedom are used). For comparison, we also plot in Figure 4.4(d) the result obtained by a standard finite element method on a coarse 32×32 mesh that does not resolve the fine oscillations. We observe that this unresolved finite element method does not give a correct qualitative result. In contrast, the FE-HMM captures the correct behaviour of the problem at a much lower computational cost.

Example 3: A crack problem and adaptive FE-HMM. In our next experiment we consider a crack problem in a heterogeneous medium, characterized by a highly oscillating conductivity tensor,

$$-\nabla \cdot (a^\varepsilon(x)\nabla u^\varepsilon) = 1 \quad \text{in } \Omega, \tag{4.16}$$
$$u^\varepsilon = g_D \quad \text{on } \Gamma_D = \partial\Omega, \tag{4.17}$$

on a domain $\Omega = \{|x| + |y| < 1\} \backslash \{0 \le x \le 1,\ y = 0\}$ with a crack along the positive x-axis (see Figure 4.5). As the solution of the homogenized problem is not in $H^2(\Omega)$, it is well known that a standard piecewise linear finite element method for the effective problem will give sub-optimal convergence rates. In this situation, adaptive mesh refinement is needed. By using local error indicators, one identifies, marks and refines those elements that contribute the most to the global error in order to better equidistribute the error in the finite element mesh. Overall, the procedure consists of the

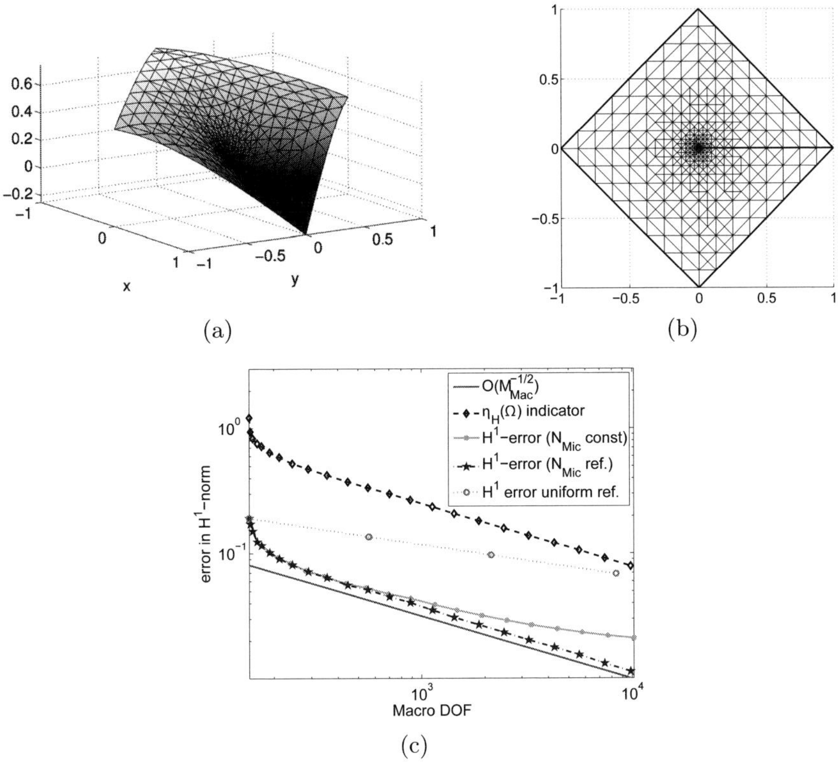

Figure 4.5. (a, b) FE-HMM solution and mesh after 10 iterations for the crack problem described in Example 3, using highly oscillating locally periodic coefficients. (c) Rate of decay for various errors and for the error indicator in the H^1 norm.

following cycle:

$$\text{SOLVE} \longrightarrow \text{ESTIMATE} \longrightarrow \text{MARK} \longrightarrow \text{REFINE}.$$

An energy-based *a posteriori* error analysis for a standard finite element method usually relies on local error indicators η_h comprising two components: the jump residual, which measures the discontinuity of the normal flux across elements interfaces, and the element residual, which measures the accuracy of the solution inside each element. Both quantities depend on the (conductivity) tensor of the elliptic problem. Residual-based adaptive FE-HMMs for the energy norm were proposed and analysed by Abdulle and Nonnenmacher (2009*a*, 2011*a*) and also by Ohlberger (2005), who derived an *a posteriori* error estimate in a 'two-scale norm'. For the FE-HMM, data involved in the usual error indicator are not available beforehand but computed during the integration process. An important ingredient for adaptive FE-HMM is an effective jump based on 'multiscale fluxes', first defined by

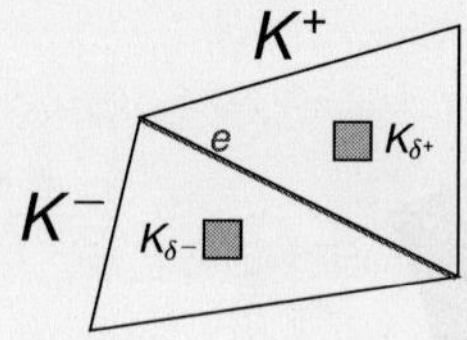

Figure 4.6. Sketch of the ingredients for the multiscale flux jump defined in (4.18).

Abdulle (2008, 2012). Let $\mathcal{T}_H$ denote a conformal mesh and let $\mathcal{E}_H$ be the set of interfaces. Two elements sharing an interface $e \in \mathcal{E}_H$ are labelled K^+ and K^-. Consider the micro functions u_{h,K^+} and u_{h,K^-} satisfying (4.10) in the two sampling domains K_δ^+ and K_δ^- of the elements K^+ and K^-, respectively, constrained by the macro solution u_H of (4.5). The multiscale flux jump is then defined by

$$\overline{[\![a^\varepsilon(x)\nabla u_h]\!]_e} := \qquad (4.18)$$
$$\left(\frac{1}{|K_\delta^+|}\int_{K_\delta^+} a^\varepsilon(x)\nabla u_{h,K^+}\,\mathrm{d}x - \frac{1}{|K_\delta^-|}\int_{K_\delta^-} a^\varepsilon(x)\nabla u_{h,K^-}\,\mathrm{d}x\right)\cdot n_e,$$

where the unit outward normal n_e is chosen to be $n_e = n^+$ (see Figure 4.6). We omit the index K_δ for the micro solutions u_h in $\overline{[\![a^\varepsilon(x)\nabla u_h]\!]_e}$, as the jump over e involves two sampling domains in adjacent elements. We also assume for simplicity that (4.5) is solved using a piecewise linear macro finite element space (see Abdulle and Nonnenmacher (2011*b*) and Nonnenmacher (2011) for a generalization of the 'multiscale flux jump' and the *a posteriori* error analysis for higher-order macro finite elements). Then, the local error indicator $\eta_H(K)$ on an element K is defined by (Abdulle and Nonnenmacher 2009*a*, 2011*a*)

$$\eta_H(K)^2 := H_K^2\|f_H\|_{L^2(K)}^2 + \frac{1}{2}\sum_{e\subset\partial K} H_e\|\,\overline{[\![a^\varepsilon\nabla u_h]\!]_e}\,\|_{L^2(e)}^2, \qquad (4.19)$$

where f_H is a piecewise constant approximation of f. The quantity $\eta_H(K)^2$ is a measure of the local error $\|u^0 - u_H\|_{H^1(K)}^2$, and is used to identify the elements that contribute most to the error. These elements are then marked for refinement. In order to prove the reliability and the efficiency of an *a posteriori* error estimate, Abdulle and Nonnenmacher (2009*a*, 2011*a*) have derived rigorous upper and lower bounds of the error in terms of the residual. For such bounds, one needs to introduce the data approximation error $\xi_H(K)$ on an element K by

$$\xi_H(K)^2 := H_K^2\|f_H - f\|_{L^2(K)}^2 + \|(a_K^0 - a^0(x))\nabla u_H\|_{L^2(K)}^2, \qquad (4.20)$$

where $a^0(x)$ is the unknown homogenized tensor of problem (4.3) and a_K^0

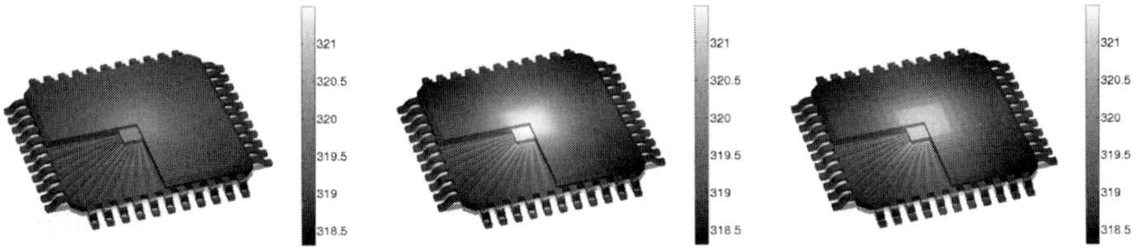

(a) (b) (c)

Figure 4.10. Steady-state heat distribution in the three-dimensional microprocessor problem of Example 4, with $\varepsilon = 5 \times 10^{-4}$ m, 81 000 degrees of freedom in the macro mesh. Comparison of various solutions. (a) Averaged tensor (arithmetic mean), 81 000 degrees of freedom, (b) FE-HMM solution, 81 000 macro degrees of freedom, (c) fine-scale solution, 3.9×10^6 degrees of freedom. The bar indicates the temperature in Kelvin (K).

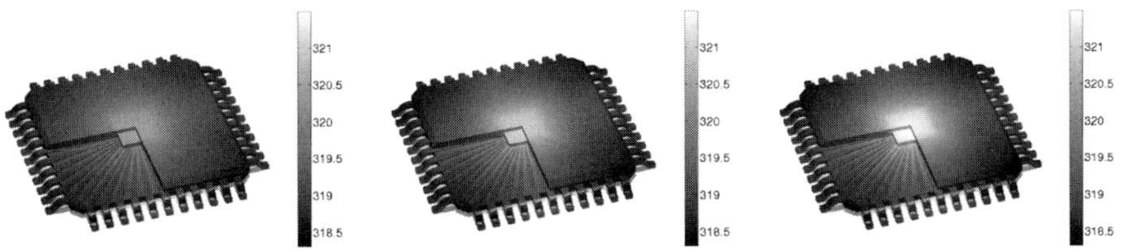

(a) (b) (c)

Figure 4.11. Steady-state heat distribution in the three-dimensional microprocessor problem of Example 4, with $\varepsilon = 10^{-6}$ m, 81 000 degrees of freedom in the macro mesh. Comparison of FE-HMM solutions with different resolutions of the micro problems: (a) $N_{\text{mic}} = 4$, (b) $N_{\text{mic}} = 8$, (c) $N_{\text{mic}} = 16$. The bar indicates the temperature in Kelvin (K).

using a naive averaging procedure for the micro structure (here the arithmetic average). We observe that the conductivity is overestimated in this situation, leading to incorrect qualitative behaviour for the heat distribution in the microprocessor. To get a rough estimate of the quality of the various experiments performed in this section, we provide the value of the energy norm of the FE-HMM calculation in Table 4.3, for various resolutions of the micro structure, and the energy norm of the resolved solution.

Finally, in Figure 4.11 we present computations with the realistic value $\varepsilon = 10^{-6}$ m. We notice that a computation with this ε is no more expensive than previous computations with a coarser value of ε (recall that the FE-HMM captures the effective solution). The computational cost, however, depends on the resolution of the micro structure in the sampling domains,

Table 4.3. Energy norm of various solutions of the heat transfer problem in a microprocessor (FE-HMM solution, resolved solution, solution obtained using arithmetic average for the various tensors). Here micro degrees of freedom on the sampling domains are given by N_{mic}^d.

		$N_{\text{mic}} = 4$	$N_{\text{mic}} = 8$	$N_{\text{mic}} = 16$	Fine-scale	Average
$\varepsilon = 5 \times 10^{-4}$	$\|u\|_A$	0.3906	0.4961	0.5514	0.5189	0.3122
$\varepsilon = 10^{-6}$	$\|u\|_A$	0.3894	0.4963	0.5512	–	0.3122

that is, the number of points per wavelength. Indeed, as can be seen in Figure 4.11, the effect of the micro error at the macroscale is not negligible (see again Abdulle (2005*b*) and Section 4.4 for an analysis of this behaviour).

4.3.4. More sophisticated coupling

In all of the above examples, we used a standard finite element method at the micro and macro levels. The structure of the FE-HMM algorithm, which couples the macro and micro solvers only via constraints, allows us to use other types of method.

For many problems, local conservation properties in the numerical approximation and flexibility in meshing (*e.g.*, hanging nodes, local refinements) are desirable. For single-scale problems, these requirements have triggered the development of discontinuous Galerkin (DG) finite element methods. Such methods have been extensively studied for hyperbolic problems, advection–diffusion and diffusion problems (see Arnold, Brezzi, Cockburn and Marini (2001/2002) for a review). While a large body of literature is available for DG methods applied to single-scale problems, the construction and analysis of multiscale DG methods have rarely been addressed. In the context of the HMM, a DG method for hyperbolic and parabolic one-dimensional scalar problems was proposed in Chen, E and Shu (2005*a*). For multiscale elliptic problems, a DG method based on HMM was proposed in Abdulle (2008, 2012).

Another issue, as seen in Section 4.3.3, is the cost of the repeated micro solutions in sampling domains in the FE-HMM. If the micro structure is regular enough, one can take advantage of fast micro solvers (based on spectral methods, for example) to speed up the computation time considerably. The simultaneous refinement of macro and micro meshes can be avoided in this situation (see Abdulle and Engquist (2007)).

If, in contrast, there is only low regularity of the micro solutions, one can try to avoid repeated micro problems by suitable interpolation of the micro solutions in a few representative sampling domains. Provided that there is some smoothness in the micro solution with respect to the macro variables, this strategy can be successful. The use of reduced basis techniques for numerical homogenization was first explored by Boyaval (2008) and recently integrated into the HMM methodology of Abdulle and Bai (2012), who have proposed and analysed a reduced basis FE-HMM.

4.4. Error estimates

In this subsection we discuss *a priori* and *a posteriori* error analysis for the FE-HMM. We present the main steps of the *a priori* error analysis for linear elliptic problems in Section 4.4.1. We only discuss the analysis for the FE-HMM based on macro and micro finite element methods, and we refer to papers by Abdulle and Engquist (2007) for an analysis using micro pseudo-spectral methods, Abdulle (2012) for an analysis using discontinuous Galerkin macro methods, and Abdulle and Bai (2012) for an analysis incorporating reduced basis methods at the micro level. In Section 4.4.2 we briefly mention how the results obtained in the linear case can be generalized to a class of non-linear problems. For details on the analysis of such problems, we refer to Abdulle and Vilmart (2011*b*, 2011*a*). Finally we also discuss *a posteriori* error analysis for the FE-HMM in Section 4.4.3, following Abdulle and Nonnenmacher (2009*a*, 2011*a*). Unless otherwise specified, we shall assume that the macroscopic triangulation $\mathcal{T}_H$ is conformal and shape-regular.

4.4.1. A priori error analysis

We collect here a few preliminary results that are useful for the analysis of the FE-HMM.

Energy equivalence and coercivity (Abdulle 2005*b*, E *et al.* 2005*b*, Abdulle 2009). As an application of the following lemma, we deduce the coercivity of the bilinear form (4.6), hence the existence and uniqueness of problem (4.5).

Lemma 4.2. Let $v_H \in S_0^\ell(\Omega, \mathcal{T}_H)$, and let v_{h,K_j} be the corresponding solution of (4.10) in $S^q(K_{\delta_j}, \mathcal{T}_h)$, with boundary conditions given by (4.8) or (4.9). Assume that (4.2) holds. Then we have

$$\|\nabla v_{H,\text{lin}}\|_{L^2(K_{\delta_j})} \le \|\nabla v_{h,K_j}\|_{L^2(K_{\delta_j})} \le C\|\nabla v_{H,\text{lin}}\|_{L^2(K_{\delta_j})}, \tag{4.24}$$

where $v_{H,lin}$ is defined in Section 4.3.2.

Reformulation of the FE-HMM (Abdulle 2009, Abdulle 2012, E *et al.* 2005*b*). The bilinear form (4.6) has the reformulation

$$B_H(v_H, w_H) = \sum_{K\in\mathcal{T}_H} \sum_{j=1}^{J} \omega_{K_j} a_K^0(x_{K_j}) \nabla v_H(x_{K_j}) \cdot \nabla w_H(x_{K_j}). \tag{4.25}$$

In order for the expression (4.6) to be equal to the expression (4.25), we see that the following identity must hold:

$$\frac{1}{|K_{\delta_j}|}\int_{K_{\delta_j}} a^{\varepsilon}(x)\nabla v_{h,K_j}\cdot\nabla w_{h,K_j}\,\mathrm{d}x = \frac{1}{|K|}\int_K a^0_K(x_{K_j})\nabla v_{H,\mathrm{lin}}\cdot\nabla w_{H,\mathrm{lin}}\,\mathrm{d}x, \tag{4.26}$$

for all $v_H, w_H \in S^{\ell}_0(\Omega,\mathcal{T}_H)$ and for all v_{h,K_j} (resp. w_{h,K_j}) solutions of problem (4.10). The following lemma gives an appropriate definition for $a^0_K(x_{K_j})$, so that the above equation is valid (see Abdulle (2012)).

Lemma 4.3. For e_i, $i = 1,\ldots,d$ (the canonical basis of $\mathbb{R}^d$) consider the following problem. Find $\psi^i_{h,K_j} \in S^q(K_{\delta_j},\mathcal{T}_h)$ such that

$$\int_{K_{\delta_j}} a^{\varepsilon}(x)\nabla\psi^i_{h,K_j}\cdot\nabla z_h\,\mathrm{d}x = -\int_{K_{\delta_j}} a^{\varepsilon}(x)e_i\cdot\nabla z_h\,\mathrm{d}x,\ \forall z_h\in S^q(K_{\delta_j},\mathcal{T}_h), \tag{4.27}$$

where $S^q(K_{\delta_j},\mathcal{T}_h)$ is defined in (4.7) with either periodic or Dirichlet boundary conditions. If $a^0_K(x_{K_j})$ is defined by

$$a^0_K(x_{K_j}) = \frac{1}{|K_{\delta_j}|}\int_{K_{\delta_j}} a^{\varepsilon}(x)\big(I + J^T_{\psi_{h,K_j}}(x)\big)\,\mathrm{d}x, \tag{4.28}$$

where $J_{\psi_{h,K_j}}(x)$ is the $d\times d$ matrix with entries

$$\big(J_{\psi_{h,K_j}}(x)\big)_{i\ell} = \frac{\partial\psi^i_{h,K_j}}{\partial x_\ell}(x),$$

then (4.26) holds.

Remark 4.4. We observe that a standard finite element method with numerical quadrature for the effective problem (4.3) reads as follows (see Ciarlet and Raviart (1972)). Find $u_{0,H}\in S^{\ell}_0(\Omega,\mathcal{T}_H)$ such that

$$B_{0,H}(u_{0,H},v_H) = \int_{\Omega} f v_H\,\mathrm{d}x,\quad \forall v_H\in S^{\ell}_0(\Omega,\mathcal{T}_H), \tag{4.29}$$

where

$$B_{0,H}(v_H,w_H) = \sum_{K\in\mathcal{T}_H}\sum_{j=1}^{J}\omega_{j,K}a^0(x_{K_j})\nabla v_H(x_{K_j})\cdot\nabla w_H(x_{K_j}). \tag{4.30}$$

Thus, (4.6) can be seen as a finite element method with numerical quadrature for a modified effective problem. We expect (4.25) to be close to (4.30) and hence u_H to be close to $u_{0,H}$, if the procedure to recover the macro data from the microscale computation is accurate enough. This will be seen in the error analysis presented below.

The first theorem gives the macroscopic error of the FE-HMM (Abdulle 2005*a*, Abdulle 2011, E *et al.* 2005*b*). It does not rely on structure assumption (*e.g.*, periodicity) of the multiscale problem (4.1), but requires sufficient smoothness of the data and the solution of the effective problem (4.3). For a given natural number ℓ and $\mu = 0$ or 1, we assume that the solution u^0 and the tensor a^0 of (4.3) satisfy:

(H1) $u^0 \in H^{\ell+1}(\Omega)$, $a^0_{ij} \in W^{\ell+\mu,\infty}(\Omega)$ for $i, j = 1, \ldots, d$.

We notice that hypothesis (**H1**) is the usual assumption for a finite element method with numerical quadrature to converge with optimal rate H^ℓ and $H^{\ell+1}$ in the H^1 and L^2 norms, respectively.

Theorem 4.5. (Macro error) Let u^0, u_H be the solutions of problems (4.3) and (4.5), respectively. Suppose that (4.2), (**Q1**), (**Q2**) and (**H1**) hold.[4] Then we have the following estimates:

$$\|u^0 - u_H\|_{H^1(\Omega)} \leq C\big(H^l + e(\text{HMM})\big), \tag{4.31}$$

$$\|u^0 - u_H\|_{L^2(\Omega)} \leq C\big(H^{l+1} + e(\text{HMM})\big), \tag{4.32}$$

where C is independent of H and h and

$$e(\text{HMM}) = \sup_{K\in\mathcal{T}_H, x_{K_j}\in K} \|a^0(x_{K_j}) - a^0_K(x_{K_j})\|_F, \tag{4.33}$$

where $a^0_K(x_{K_j})$ is defined in (4.28) and $\|\cdot\|_F$ is the Frobenius norm.

Proof. Let $u_{0,H}$ be the solution of (4.29). It is proved in Ciarlet and Raviart (1972) that provided (4.2), (**Q1**), (**Q2**) and (**H1**) hold, we have

$$\|u^0 - u_{0,H}\|_{H^1(\Omega)} \leq CH^l, \tag{4.34}$$

$$\|u^0 - u_{0,H}\|_{L^2(\Omega)} \leq CH^{l+1}, \tag{4.35}$$

where C is independent of H. Using the coercivity of the bilinear form (4.30) (which follows from (**Q1**)), we obtain

$$\|u_{0,H} - u_H\|_{H^1(\Omega)} \leq C \sup_{w_H\in S^\ell_0(\Omega,\mathcal{T}_H)} \frac{|B_{0,H}(u_H, w_H) - B_H(u_H, w_H)|}{\|w_H\|_{H^1(\Omega)}}.$$

In view of (4.30), (4.25) and (**Q2**), using the Cauchy–Schwarz inequality we see that

$$|B_{0,H}(u_H, w_H) - B_H(u_H, w_H)| \\ \leq C \sup_{K\in\mathcal{T}_H, x_{K_j}\in K} \|a^0(x_{K_j}) - a^0_K(x_{K_j})\|_F \|u_H\|_{H^1(\Omega)}\|w_H\|_{H^1(\Omega)}.$$

The above estimates together with (4.34) and (4.35) give the claimed result. □

[4] In hypothesis (**H1**), $\mu = 0$ for the estimate (4.31) and $\mu = 1$ for the estimate (4.32).

The next task is to estimate the quantity $e(\text{HMM})$. It is convenient to decompose this quantity further. We therefore consider, for each macro element $K \in \mathcal{T}_H$ and each sampling domain $K_{\delta_j} \subset K$,

$$\bar{a}_K^0(x_{j,K}) = \frac{1}{|K_{\delta_j}|}\int_{K_{\delta_j}} a^\varepsilon(x)\big(I + J^T_{\psi_{K_j}}(x)\big)\,\mathrm{d}x, \tag{4.36}$$

a tensor defined as in (4.28), but here

$$\big(J_{\psi_{K_j}}(x)\big)_{i\ell} = \frac{\partial \psi^i_{K_j}}{\partial x_\ell}(x),$$

and $\psi^i_{K_j}$ is the (exact) solution of (4.27) in (4.8) or (4.9). Then, obviously,

$$e(\text{HMM}) \le e_{\text{MIC}} + e_{\text{MOD}},$$

where

$$e_{\text{MIC}} := \sup_{K\in\mathcal{T}_H, x_{K_j}\in K} \|\bar{a}_K^0(x_{K_j}) - a_K^0(x_{K_j})\|_F, \tag{4.37}$$

$$e_{\text{MOD}} := \sup_{K\in\mathcal{T}_H, x_{K_j}\in K} \|a^0(x_{K_j}) - \bar{a}_K^0(x_{K_j})\|_F. \tag{4.38}$$

For symmetric tensors, the quantity e_{MIC} was first analysed in Abdulle (2005*a*) and the quantity e_{MOD} in E *et al.* (2005*b*). To analyse e_{MIC}, appropriate regularity of the oscillating tensor is required. As we use standard *a priori* results of finite element methods in the sampling domain, we also need appropriate regularity of $\psi^i_{K_j}$. However, the coefficient a^ε is allowed to be discontinuous (at the macroscopic level) through smooth interfaces. We therefore make the following assumption:

(H2) for $q \in \mathbb{N}$ we assume that $|\psi^i_{K_j}|_{H^{q+1}(K_{\delta_j})} \le C\,\varepsilon^{-q}\sqrt{|K_{\delta_j}|}$, with C independent of ε, of the quadrature points x_{K_j} and the domain K_{δ_j}.

Remark 4.6. If one assumes $a^\varepsilon \in W^{1,\infty}(\Omega)$ and that $|a^\varepsilon_{ij}|_{W^{1,\infty}(\Omega)} \le C\varepsilon^{-1}$ for $i,j = 1,\dots,d$, then **(H2)** can be proved for $q = 1$ and $W(K_{\delta_j}) = H^1_0(K_{\delta_j})$, using classical H^2 regularity results (Ladyzhenskaya 1985, Chapter 2.6) (in fact we only need local regularity for a^ε_{ij}). If $a^\varepsilon = a(x, x/\varepsilon) = a(x,y)$, which is Y-periodic in y, if $\delta/\varepsilon \in \mathbb{N}$, and $W(K_{\delta_j}) = W^1_{\text{per}}(K_{\delta_j})$, then **(H2)** can be established for a given q, provided that the tensor a^ε is sufficiently smooth (this follows classical regularity results for solutions of periodic boundary value problems: see, *e.g.*, Bers, John and Schechter (1957, Chapter 3)).

The following theorem was first obtained in Abdulle (2005*a*) for elliptic problems and in Abdulle (2006) for problems in elasticity (see also Abdulle (2009, 2012) for generalizations). Du and Ming (2010) extended the estimate to non-symmetric tensors (see also Abdulle and Vilmart (2011*a*, Lemma 4.6) for a short alternative proof).

Theorem 4.7. (Micro error) Assume that (4.2) and (**H2**) hold. Then

$$e_{\mathrm{MIC}} \leq C\left(\frac{h}{\varepsilon}\right)^{2q}, \tag{4.39}$$

where C is independent of H, h and ε.

As first observed in Abdulle (2005*a*, 2009), we see that a fixed number of grid points per wavelength of the smallest oscillation of the problem does not ensure robust convergence of the FE-HMM. Indeed, the above theorem shows that for fixed micro meshes, the errors due to the micro solver are dominant for sufficiently fine macro meshes. Comparing the rate of convergence of the macro solver (Theorem 4.5) with the rate of convergence of the micro solver (see Theorem 4.7) gives a criterion to obtain the optimal (macro) convergence rate with minimal computational cost. Indeed, if the macro problem (4.5) is solved in $S_0^\ell(\Omega, \mathcal{T}_H)$, with micro problems (4.10) solved in $S^q(K_{\delta_j}, \mathcal{T}_h)$, then, assuming uniform macro and micro meshes, we have

$$\hat{h} \simeq H^{\frac{\ell}{2q}} \quad \text{(optimal refinement in } H^1 \text{ norm)}, \tag{4.40}$$

$$\hat{h} \simeq H^{\frac{\ell+1}{2q}} \quad \text{(optimal refinement in } L^2 \text{ norm)}, \tag{4.41}$$

and we recall that $\hat{h} = h/\varepsilon$ is independent of ε (see Section 4.3.3).

The last task is to estimate the modelling error. Here we require structure assumptions on the tensor such as local periodicity or random stationarity. For deterministic tensors we make the following assumption:

(**H3**) $a^\varepsilon(x) = a(x, x/\varepsilon) = a(x, y)$ is Y-periodic in y, and

$$a_{ij}(x, y) \in \mathcal{C}\big(\bar{\Omega}; W^{1,\infty}_{\mathrm{per}}(Y)\big),$$

for all $i, j = 1, \ldots, d$,

where we set $Y = (0,1)^d$ for simplicity. For such a tensor the variables x and y are usually referred to as slow and fast variables, respectively. First we note the following result, obtained by Abdulle and Schwab (2005).

Lemma 4.8. Assume (**H3**) and that the micro problems (4.10) are solved in $S^q(K_{\delta_j}, \mathcal{T}_h) \subset W^1_{\mathrm{per}}(K_{\delta_j})$, with $\delta/\varepsilon \in \mathbb{N}$. Assume further that the slow variable tensor $a(x, x/\varepsilon)$ is collocated at the quadrature points x_{K_j}, that is, $a(x_{K_j}, x/\varepsilon)$ is used in the problem (4.10) and in the bilinear form (4.6). Then $e_{\mathrm{MOD}} = 0$.

A consequence of this result is that the FE-HMM converges with a robust rate (*i.e.*, independent of ε) to the homogenized solution. In a more general situation, for example when the size of the period is unknown, a modelling error arises due to a mismatch of the sampling domain size and ε (*e.g.*, $\delta/\varepsilon \notin \mathbb{N}$) and to artificial boundary conditions. This error is often called cell resonance in the literature (see, *e.g.*, Hou *et al.* (1999)). The following result was obtained by E *et al.* (2005*b*) (see also Yue and E (2007)).

Theorem 4.9. (Modelling error) Assume (**H3**) and that the micro problems (4.10) are solved in $S^q(K_{\delta_j}, \mathcal{T}_h) \subset H_0^1(K_{\delta_j})$, with $\delta > \varepsilon$. Then

$$e_{\text{MOD}} \leq C\left(\frac{\varepsilon}{\delta} + \delta\right),$$

where C is independent of H, h and ε.

We conclude this section by mentioning that a new approach to cell resonance has recently been proposed by Gloria (2011). By modifying the cell problem (4.10) and adding a zeroth-order term, we obtain a micro problem with better error decay due to artificial boundary conditions, because of the faster decay of the associated Green's function. The bias introduced by modifying the cell problem can be controlled by tuning the constant associated with the zeroth-order term. Gloria (2011) reports a rate of $(\varepsilon/\delta)^p$, $p < 4$. Numerical experiments, however, show that this asymptotic rate is only obtained for $\delta \gg \varepsilon$. Nevertheless, this constitutes a promising approach for the issue of cell resonance error.

Elliptic problems with random coefficients. The FE-HMM for elliptic problems with random conductivity tensors was investigated by E *et al.* (2005*b*), who analysed the modelling error e_{MOD} under a stationarity assumption. Convergence rates have been derived for dimensions $d = 1$ and $d = 3$ under the additional assumption that the random tensors satisfy a uniform mixing condition (Yurinskiĭ 1986).

4.4.2. Example of a priori analysis for non-linear problems

We consider a class of non-linear non-monotone multiscale problems described in Example 2:

$$-\nabla \cdot (a^\varepsilon(x, u^\varepsilon(x)))\nabla u^\varepsilon(x)) = f(x) \text{ in } \Omega, \quad u^\varepsilon(x) = 0 \text{ on } \partial\Omega, \tag{4.42}$$

where $a^\varepsilon(x, s)$ is a $d \times d$ tensor satisfying (4.2), with entries $a^\varepsilon_{ij}(x, s)$ which are continuous functions on $\overline{\Omega} \times \mathbb{R}$ and uniformly Lipschitz-continuous with respect to s and ε, that is,

$$\exists \Lambda_1 > 0, \quad \text{such that} \quad |a^\varepsilon_{ij}(x, s_1) - a^\varepsilon_{ij}(x, s_2)| \leq \Lambda_1 |s_1 - s_2|, \tag{4.43}$$
$$\forall \varepsilon,\ \forall x \in \overline{\Omega},\ \forall s_1, s_2 \in \mathbb{R},\ \forall 1 \leq i, j \leq d.$$

For such problems, it is known (see, *e.g.*, Boccardo and Murat (1981)) that there exists a subsequence of $\{a^\varepsilon(\cdot, s)\}$ (again indexed by ε) such that the corresponding sequence of solutions $\{u^\varepsilon\}$ converges weakly to u^0 in $H^1(\Omega)$, where u^0 is the solution of the so-called homogenized problem

$$-\nabla \cdot (a^0(x, u^0(x))\nabla u^0(x)) = f(x) \text{ in } \Omega, \quad u^0(x) = 0 \text{ on } \partial\Omega, \tag{4.44}$$

with a homogenized tensor $a^0(x, s)$, which can be shown to have properties similar to those assumed for $a^\varepsilon(x, s)$. The numerical method for computing an effective solution of (4.42) was described in (4.12). The coerciveness of the bilinear form $B_H(z^H; \cdot, \cdot)$ can be established similarly as in the linear case. Then the existence of a solution of problem (4.12) relies on the Brouwer fixed-point theorem applied to the non-linear map $S_H : S_0^\ell(\Omega, \mathcal{T}_H) \to S_0^\ell(\Omega, \mathcal{T}_H)$, where, for $z_H \in S_0^\ell(\Omega, \mathcal{T}_H)$, $S_H z_H$ is defined to be the solution of the linear problem

$$B_H(z_H; S_H z_H, w_H) = \int_\Omega f w_H \,\mathrm{d}x, \ \forall w_H \in S_0^\ell(\Omega, \mathcal{T}_H). \tag{4.45}$$

For details we refer, for example, to Douglas and Dupont (1975). In contrast, the uniqueness is much more involved (see below). For deriving *a priori* error estimates and rates of convergence for the error $\|u^0 - u_H\|_{H^1(\Omega)}$, we follow the methodology described in the linear case. There is, however, one main difference: *a priori* error analysis for finite element methods with numerical quadrature such as the results in Ciarlet and Raviart (1972), used in the linear case, need to be derived. Such results, which are quite technical in the non-linear non-monotone case, were derived by Abdulle and Vilmart (2011*b*, 2012*b*). Using their results, it is then possible to show the following (see Abdulle and Vilmart (2012*b*) for a proof).

Theorem 4.10. Let u^0, u_H be the solutions of problems (4.44) and (4.12), respectively. Let $\ell \geq 1$. Let $\mu = 0$ or 1. Assume (**Q1**), (**Q2**) and (**H1**) hold.[5] In addition, assume that $u^0 \in W^{1,\infty}(\Omega)$, $\partial_u a^0_{mn} \in W^{1,\infty}(\Omega \times \mathbb{R})$, and that the coefficients $a^0_{mn}(x, s)$ are twice differentiable with respect to s, with the first- and second-order derivatives continuous and bounded on $\overline{\Omega} \times \mathbb{R}$, for all $m, n = 1, \ldots, d$.

Then there exist $r_0 > 0$ and $H_0 > 0$ such that, provided

$$H \leq H_0 \quad \text{and} \quad e(\text{HMM}) \leq r_0, \tag{4.46}$$

any solution u_H of (4.12) satisfies

$$\|u^0 - u_H\|_{H^1(\Omega)} \leq C(H^\ell + e(\text{HMM})) \quad \text{if } \mu = 0, 1, \tag{4.47}$$

$$\|u^0 - u_H\|_{L^2(\Omega)} \leq C(H^{\ell+1} + e(\text{HMM})) \text{ if } \mu = 1, \tag{4.48}$$

where $e(\text{HMM})$ is given in (4.49). Here, the constants C are independent of H and $e(\text{HMM})$.

[5] Here $a^0_{ij} \in W^{\ell+\mu,\infty}(\Omega \times \mathbb{R})$ for $i, j = 1, \ldots, d$.

Remark 4.11. Note that the macroscopic family of triangulations $\{\mathcal{T}_H\}$ satisfies the inverse assumption $H/H_K \leq C$ for all $K \in \mathcal{T}_H$ and all $\mathcal{T}_H$ in Theorem 4.10; such a condition is often used for the finite element analysis of non-linear problems (Xu 1996). We also assume, for simplicity, the convexity of the domain Ω for the L^2 bound. Such a condition on the domain allows us to infer suitable regularity to use an Aubin–Nitsche-type duality argument. A weaker condition on the adjoint of the linearized problem associated to (4.44) (see Abdulle and Vilmart (2011*b*, equation (5))) is sufficient. This condition is usually assumed for L^2 estimates of linear finite element methods with numerical quadrature (see Raviart (1973, equation (2.11))).

The proof of Theorem 4.10 is more involved than the similar theorem in the linear case. We note that the additional assumptions compared to the linear case (Theorem 4.5), that is, the $W^{1,\infty}$ assumption on u^0 and the smoothness of $s \mapsto a(x, s)$, are already used for the analysis of finite element methods for single-scale non-linear problems of the type (4.44); see for example Douglas and Dupont (1975). The inverse assumption is also used in Douglas and Dupont (1975) to prove the uniqueness of a numerical solution. The quantity $e(\text{HMM})$ is defined by

$$e(\text{HMM}) := \sup_{K \in \mathcal{T}_H, x_{K_j} \in K, s \in \mathbb{R}} \|a^0(x_{K_j}, s) - a^0_{K_j}(s)\|_F, \qquad (4.49)$$

where a^0 is the homogenized tensor of (4.44) and $a^0_{K_j}$ is the tensor defined similarly to (4.28) by replacing $a^\varepsilon(x)$ with the non-linear tensor $a^\varepsilon(x, s)$ (where s plays the role of a parameter), ψ^i_{h,K_j} with $\psi^{i,s}_{h,K_j}$ and $J_{\psi_{h,K_j}}(x)$ with $J_{\psi^s_{h,K_j}}(x)$. A nice feature of the analysis in Abdulle and Vilmart (2011*a*) is that micro and macro modelling errors can be analysed, thanks to Theorem 4.10, following the analysis obtained for the linear case (estimates with a similar rate of convergence can be derived: see Abdulle and Vilmart (2011*a*, Theorems 3.6, 3.7)). The proof of the uniqueness of a numerical solution of problem (4.12) is quite involved, but can be derived without any structure assumption on the oscillation provided that

$$\begin{aligned} &s \in \mathbb{R} \mapsto a^\varepsilon(\cdot, s) \in (W^{1,\infty}(\Omega))^{d \times d} \text{ is of class } C^2 \\ &\text{and} \quad |\partial^k_u a^\varepsilon(\cdot, s)|_{W^{1,\infty}(\Omega)} \leq C\varepsilon^{-1}, \ k \leq 2, \end{aligned} \qquad (4.50)$$

where C is independent of s and ε (see Abdulle and Vilmart (2011*a*, Theorem 3.3)). Here we present a simpler uniqueness result, which can be stated solely in terms of the size of the macro and micro meshes (see Abdulle and Vilmart (2011*a*, Corollary 3.4)).

Theorem 4.12. In addition to the hypothesis of Theorem 4.10, assume (**H2**) and (**H3′**), and (4.50). Assume $W(K_{\delta_j}) = W^1_{\text{per}}(K_{\delta_j})$ (periodic coupling conditions), $\delta/\varepsilon \in \mathbb{N}$ and that the slow variable of the tensor $a(x, x/\varepsilon, s)$

is collocated at the quadrature points x_{K_j} in the problem (4.14) and in the form (4.13). Then there exists a positive constant H_0 such that, for all

$$(h/\varepsilon)^{2q} \leq H \leq H_0,$$

the solution u_H of (4.12) is unique.

Here (**H3′**) is defined similarly to (**H3**) for the non-linear tensor $a^\varepsilon_{ij}(x,s) = a_{ij}(x, x/\varepsilon, s)$, where $a_{ij}(x,y,s)$ is Y-periodic in y, and the map $(x,s) \mapsto a_{ij}(x,\cdot,s)$ is Lipschitz-continuous and bounded from $\overline{\Omega} \times \mathbb{R}$ into $W^{1,\infty}_{\text{per}}(Y)$.

4.4.3. A posteriori error analysis

The goal of the residual-based *a posteriori* analysis is to give upper and lower bounds for the error $e_H := u^0 - u_H$ in the H^1 norm for a given solution u_H of the problem (4.5) in terms of residual and data approximations, $\eta_H(\Omega), \xi_H(\Omega)$ defined in (4.19) and (4.20), respectively. As mentioned in Section 4.3.3 (see Example 3), we will concentrate here on piecewise linear simplicial macro finite elements. Compared to classical residual-based *a posteriori* error analysis, we have two additional difficulties:

- the data of the effective problem (4.3) (used in the residual η_H) are not available beforehand,
- 'variational crimes' (see Braess (2007)) are committed by replacing the true bilinear form with the FE-HMM bilinear form.

The first issue can be dealt with by introducing the multiscale flux jump for the modelling of (4.19). For the second issue, the following representation formula derived in Abdulle and Nonnenmacher (2009*a*, 2011*a*) is crucial. Let $B_0(\cdot,\cdot)$ be the bilinear form corresponding to the variational formulation of (4.3). Then we can relate the bilinear form $B_0(\cdot,\cdot)$ evaluated at e_H to the actual numerical solution u_H, the multiscale flux and data approximation as follows. Here for simplicity we consider piecewise linear macroscopic finite element functions.

Lemma 4.13. (Error representation formula) For all $v \in H^1_0(\Omega)$, we have

$$\begin{aligned} B_0(e_H, v) = &\int_\Omega f v \,\mathrm{d}x - \sum_{e \in \mathcal{E}_H} \int_e \overline{[\![a^\varepsilon(x)\nabla u_h]\!]_e} v \,\mathrm{d}s \\ &+ \sum_{K \in \mathcal{T}_H} \int_K (a^0_K(x_K) - a^0(x))\nabla u_H \cdot \nabla v \,\mathrm{d}x, \end{aligned} \tag{4.51}$$

where $\mathcal{E}_H$ denotes the set of interfaces of the conformal mesh $\mathcal{T}_H$, u_H is the solution of (4.5) and where the multiscale flux jump $\overline{[\![a^\varepsilon(x)\nabla u_h]\!]_e}$ is defined in (4.18).

Using the above formula, we obtain for $v = u^0 - u_H$

$$\begin{aligned} B_0(e_H, e_H) &= \int_\Omega f_H(e_H - I_H e_H)\,\mathrm{d}x + \int_\Omega (f - f_H)(e_H - I_H e_H)\,\mathrm{d}x \\ &\quad - \sum_{e\in\mathcal{E}_H} \int_e \overline{[\![a^\varepsilon(x)\nabla u_h]\!]_e}(e_H - I_H e_H)\,\mathrm{d}s \\ &\quad + \sum_{K\in\mathcal{T}_H} \int_K (a_K^0(x_K) - a^0(x))\nabla u_H \cdot \nabla e_H\,\mathrm{d}x, \end{aligned}$$

where $a_K^0(x_K)$ is defined in (4.28), and I_H denotes the Clément interpolation operator (see Clément (1975)). It is a linear operator $I_H : H^1(\Omega) \to S(\Omega, \mathcal{T}_H)$ having the property that, for all $v \in H^1(\Omega)$ and $K \in \mathcal{T}_H$,

$$\|v - I_H v\|_{L^2(K)} \leq C H_K \|\nabla v\|_{L^2(N(K))} \tag{4.52}$$

and

$$\|\nabla(v - I_H v)\|_{L^2(K)} \leq C\|\nabla v\|_{L^2(N(K))}, \tag{4.53}$$

where $N(K)$ is the neighbourhood of K that consists of all elements of $\mathcal{T}_H$ which have a non-empty intersection with K. Using the properties of the Clément interpolation operator, Cauchy–Schwarz and Poincaré inequalities, the coercivity of $B_0(\cdot,\cdot)$ and the finite overlapping property of the neighbourhoods $N(K)$, we can derive the following result (see Abdulle and Nonnenmacher (2011*a*) for a complete proof).

Theorem 4.14. (Upper bound) Let u^0, u_H be the solutions of problems (4.3) and (4.5), respectively. Then

$$\|u^0 - u_H\|^2_{H^1(\Omega)} \leq C\big(\eta_H(\Omega)^2 + \xi_H(\Omega)^2\big),$$

where C depends only on the shape-regularity constant, the coercivity and continuity bounds (4.2), the dimension d and the constant of the Poincaré inequality.

To derive a lower bound, bubble functions can be used, as in the classical theory (Verfürth 1996). We recall the construction of such functions. Let $\widetilde{\mathcal{T}}_H$ be a refinement of $\mathcal{T}_H$ such that every $K \in \mathcal{T}_H$ has an interior node $\tilde{x}_K \in K\backslash\partial K$ in $\widetilde{\mathcal{T}}_H$, and every edge e of $\mathcal{T}_H$ not on the boundary $\partial\Omega$ has an interior node $\tilde{x}_e \in e\backslash\partial e$ in $\widetilde{\mathcal{T}}_H$. Based on the triangulation $\widetilde{\mathcal{T}}_H$, we define a piecewise linear finite element space $S_0^1(\Omega, \widetilde{\mathcal{T}}_H) \supset S_0^1(\Omega, \mathcal{T}_H)$.

We start by estimating the so-called interior residual. For any $K \in \mathcal{T}_H$, consider an interior bubble function, that is, a function $\psi_K \in S_0^1(\Omega, \widetilde{\mathcal{T}}_H)$ such that $0 \leq \psi_K \leq 1$, $\psi_K(\tilde{x}_K) = 1$ and $\psi_K \equiv 0$ on $\Omega\backslash K$. Using the representation formula (4.51) with a test function given by $v := \psi_K f_H \in$

$H_0^1(\Omega)$, one can estimate the so-called interior residual

$$H_K^2\|f_H\|_{L^2(K)}^2 \leq C\Big(\|\nabla e_H\|_{L^2(K)}^2 + H_K^2\|f - f_H\|_{L^2(K)}^2 + \|(a_K^0(x_K) - a^0(x))\nabla u_H\|_{L^2(K)}^2\Big). \tag{4.54}$$

Next, we estimate the so-called jump residual. Let $e \in \mathcal{E}_H$ be an interior interface and let $K_1 \in \mathcal{T}_H$ and $K_2 \in \mathcal{T}_H$ be such that $K_1 \cap K_2 = e$. Furthermore, let $\tilde{x}_e \in e$ be an interior node and let $\psi_e \in S_0^1(\Omega, \widetilde{\mathcal{T}}_H)$ be a bubble function such that $\psi_e(\tilde{x}_e) = 1$ and $\psi_e \equiv 0$ on $\Omega\backslash(K_1 \cup K_2)$. Using the representation formula (4.51) with $v := \psi_e$, we find

$$H_e\|[\![\overline{a^\varepsilon(x)\nabla u_h}]\!]_e\|_{L^2(e)}^2 \leq C\Big(\|\nabla e_H\|_{L^2(\omega_e)}^2 + H_{\omega_e}^2\|f - f_H\|_{L^2(\omega_e)}^2 + \|(a_K^0(x_K) - a^0(x))\nabla u_H\|_{L^2(\omega_e)}^2\Big), \tag{4.55}$$

where $H_{\omega_e} = \max_{i=1,2} H_i$, $\omega_e = K_1 \cup K_2$. Using the estimates (4.54) and (4.55) on the interior and jump residuals, respectively, we derive the following lower bound (see Abdulle and Nonnenmacher (2011*a*) for a complete proof).

Theorem 4.15. (Lower bound) Let u^0, u_H be the solutions of problems (4.3) and (4.5), respectively. Denote by ω_K the domain which consists of all elements sharing at least one side with K. Then

$$\eta_H(K)^2 \leq C\big(\|u^0 - u_H\|_{H^1(\omega_K)}^2 + \xi_H(\omega_K)^2\big),$$

where C depends only on the shape-regularity constant, the coercivity and continuity bounds (4.2), the dimension d and the constant of the Poincaré inequality.

5. Finite volume methods

5.1. The algorithm

In this section we discuss the application of HMM in the setting of finite volume methods. Finite volume methods are generally preferred when the macroscopic models are in the form of conservation laws. In fluid and solid mechanics, the macroscale models are often derived from the conservation laws of mass, momentum and energy:

$$\begin{aligned} \partial_t\rho + \nabla\cdot q &= 0, \\ \partial_t q + \nabla\cdot\tau &= 0, \\ \partial_t E + \nabla\cdot J &= 0. \end{aligned} \tag{5.1}$$

Here ρ is the mass density of the system, $q = \rho v$ is the momentum density, with v being the velocity, and E is the total energy density; τ and J are the momentum and energy fluxes, respectively. In conventional continuum models, empirical relations are sought to express these fluxes in terms of the conserved densities. In HMM, the fluxes are computed directly from some underlying microscopic models.

An example of the microscopic model that one can consider is that of molecular dynamics, that is, Newton's equations of motion for the constituting atoms:

$$\begin{aligned} \dot{x}_i &= v_i, \\ m_i \dot{v}_i &= \sum_{j \neq i} f(x_i(t) - x_j(t)). \end{aligned} \tag{5.2}$$

Here m_i is the mass of the ith particle and $f(r) = -\nabla\phi(r)$ is the force exerted on the particles with interatomic potential $\phi(r)$. Here we have assumed that the particles interact via a pairwise potential, even though the algorithms discussed here can be easily adapted to the situation when general interatomic potentials are used.

To see how the microscopic model can be linked with the macroscopic model, observe that for (5.2) we can also write down the analogues of the equations (5.1), by defining

$$\begin{aligned} \widetilde{\rho}(x,t) &= \sum_i m_i \delta(x - x_i(t)), \\ \widetilde{q}(x,t) &= \sum_i m_i v_i(t) \delta(x - x_i(t)), \\ \widetilde{E}(x,t) &= \frac{1}{2} \sum_i \Big[m_i |v_i|^2 + \sum_{j \neq i} \phi(x_i(t) - x_j(t)) \Big] \delta(x - x_i(t)), \end{aligned} \tag{5.3}$$

and

$$\begin{aligned} \widetilde{\tau}(x,t) = {} & \sum_i m_i v_i \otimes v_i \delta(x - x_i(t)) \\ & + \frac{1}{2} \sum_i \sum_{j \neq i} f(x_i - x_j) \otimes (x_i - x_j) \\ & \times \int_0^1 \delta(x - (x_j + \lambda(x_i - x_j)))\, \mathrm{d}\lambda, \end{aligned} \tag{5.4a}$$

$$\begin{aligned} \widetilde{J}(x,t) = {} & \sum_i v_i \Big[\frac{1}{2} m_i |v_i|^2 + \frac{1}{2} \sum_{j \neq i} \phi(x_i(t) - x_j(t)) \Big] \delta(x - x_i(t)) \\ & + \frac{1}{4} \sum_{j \neq i} (v_j(t) + v_i(t)) \cdot f(x_j(t) - x_i(t))(x_i - x_j) \\ & \times \int_0^1 \delta(x - (x_j + \lambda(x_i - x_j)))\, \mathrm{d}\lambda. \end{aligned} \tag{5.4b}$$

One can easily verify that from (5.2) we have

$$\begin{aligned} \partial_t \widetilde{\rho} + \nabla_x \cdot \widetilde{q} &= 0, \\ \partial_t \widetilde{q} + \nabla_x \cdot \widetilde{\tau} &= 0, \\ \partial_t \widetilde{E} + \nabla_x \cdot \widetilde{J} &= 0. \end{aligned} \tag{5.5}$$

An important factor to be considered is the scale separation: microscopic processes usually take place at a length scale of nanometres (10^{-9} m) and a time scale of femto- or picoseconds, so one can neglect the variation of the macroscale variables. Locally, one can then think of the atomistic system as being constrained by the local macroscopic quantities, namely the local mass, momentum and energy densities. In addition, the atomistic system will stay close to local equilibrium since there is sufficient time for relaxation to take place. Consequently, one may view the models (5.1) as being the ensemble averages of (5.2). This is the ideal situation for HMM.

An HMM strategy for this setting has been developed by Li and E (2005) and Ren and E (2005). Our presentation follows that of Li and E.

Macroscale solver. Since the macroscopic model (5.1) is in the form of conservation laws, it is natural to choose as the macroscale solver a finite volume method. Although there are many different finite volume methods that are available for conservation laws (*e.g.*, LeVeque (1990) and Godlewski and Raviart (1996)), many involve the computation of the Jacobian for the flux functions. These are less suited to the present problem since the flux function is not explicitly given to us. An exception is the Lax–Friedrichs central scheme (*e.g.*, Nessyahu and Tadmor (1990)), which is formulated over a staggered grid. As it turns out, this method can easily be coupled with molecular dynamics.

To be more specific, we first rewrite the conservation laws in a generic form,

$$\partial_t u + \partial_x f = 0. \tag{5.6}$$

We will confine our discussion to macroscopically one-dimensional problems since the extension to higher dimensions is straightforward. The first-order central scheme represents the solutions by piecewise constants, which are the average values over each cell:

$$u_k^n = \frac{1}{\Delta x} \int_{x_{k-1/2}}^{x_{k+1/2}} u(x, t^n) \, \mathrm{d}x.$$

Time integration over $[x_j,\ x_{j+1}] \times [t^n,\ t^{n+1})$ leads to the scheme

$$u_{k+1/2}^{n+1} = \frac{u_k^n + u_{k+1}^n}{2} - \frac{\Delta t}{\Delta x} \big(f_{k+1}^n - f_k^n\big), \tag{5.7}$$

with numerical flux

$$f_k^n = \frac{1}{\Delta t}\int_{t^n}^{t^{n+1}} f(x_k, t)\,\mathrm{d}t,$$

which is taken to be $f_k^n = f(x_k, t^n)$ in the central scheme.

Now (5.1) is incomplete since we still do not know the fluxes. Next we describe how this information can be extracted from molecular dynamics (MD) or other atomistic models.

Microscale solver. We first discuss the case when the microscopic model is MD. At each point where numerical fluxes are needed, we perform a local MD simulation to estimate the fluxes. The atomistic simulation will be constrained by the local macroscopic variables, which are the local density, momentum and energy. To initialize the MD simulation, we first arrange a trial configuration of the atoms according to the local density. From the local total energy, momentum and potential energy V of the trial configuration, we can compute the thermal energy K_θ by

$$K_\theta = E - V - \frac{1}{2}\rho v^2.$$

If K_θ is negative, the trial configuration is rejected and another trial configuration with the same number of atoms is generated. Otherwise it is accepted, and a local initial temperature is computed from K_θ. The velocities of the atoms are then given by

$$v_i = v + v_i',$$

where v_i' is thermal velocity given by the Maxwell–Boltzmann distribution with the given local temperature.

The set of ODEs (5.2) can be solved by a standard finite difference scheme such as Verlet's method. For the present analysis, however, we will assume that the ODEs are solved exactly to avoid unnecessary complication. After the MD system equilibrates, we obtain the required quantities by averaging. Specifically, let $\tilde{j}$ be the spatial average over the simulation box of a local flux whose expression was given in (5.4):

$$\tilde{J} = \tilde{J}(X), \quad X = (x_1, x_2, \ldots, x_N, v_1, v_2, \ldots, v_N). \tag{5.8}$$

Then we obtain the corresponding macroscale quantity by time-averaging,

$$J = \frac{1}{T}\int_{\tau}^{\tau+T} \tilde{J}(X(t))\,\mathrm{d}t, \tag{5.9}$$

where τ is the starting point when the time-averaging begins, and T is the duration of the time-averaging. An additional ensemble averaging can also be used if desired. For detailed discussion of the set-up of the MD as well as boundary conditions, see Li and E (2005).

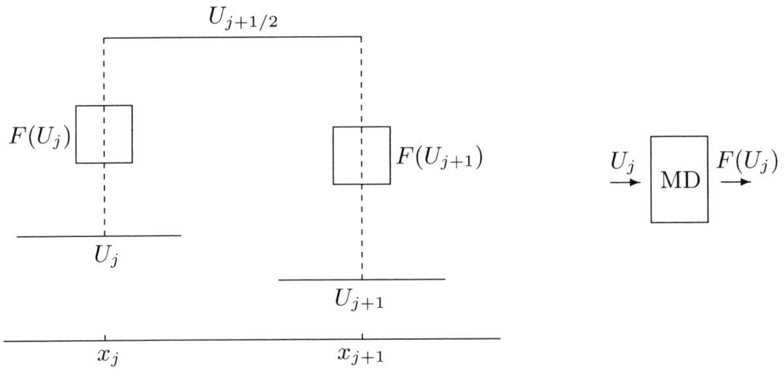

Figure 5.1. A schematic illustration of the numerical procedure. Starting from piecewise constant solutions u_k^n, one integrates (5.6) in time and in the cell x_k, x_{k+1}. The time step Δt is chosen in such a way that the waves coming from $x_{k+1/2}$ will not reach x_k, and thus for $t \in [t^n, t^{n+1}]$, $u(x_k, t) = u_k^n$. If $f(u)$ at x_k is found to be unknown, we perform an MD simulation using u_k^n to invoke and restrict the microscopic process. The required flux is then extracted from the simulation and the integration is completed. Analogously one can embed the MD simulation in higher-order macro schemes or higher dimensions.

The overall numerical procedure is shown schematically in Figure 5.1. At each time step, the scheme (5.7) requires as input the fluxes at grid point x_j to complete the time integration. These flux values are obtained by performing local MD simulations that are consistent with the local macroscale state (ρ, q, E). Once these values are computed, one can advance to the next macro time step using (5.7).

One result from such a method is shown in Figure 5.2. Here the set-up for the macroscale model is a Riemann problem for one-dimensional wave propagation in solids. The result of HMM is compared with that of a direct MD simulation. The microscale model is MD with Lennard–Jones potential.

5.2. Error analysis

As we have seen before, error estimates for HMM proceed in two steps. The first step is to estimate the 'HMM error', that is, the error in the estimated data. The second step is to obtain the overall error estimate.

For the exact solution we will take the solution for the case when the fluxes are given by the averaged quantities for the equilibrium distribution, with parameters given by the local values of the macroscale variables (density, momentum, energy).

First, let us look at the HMM error. Let

$$e(\text{HMM}) = E[|J - \bar{J}|], \tag{5.10}$$

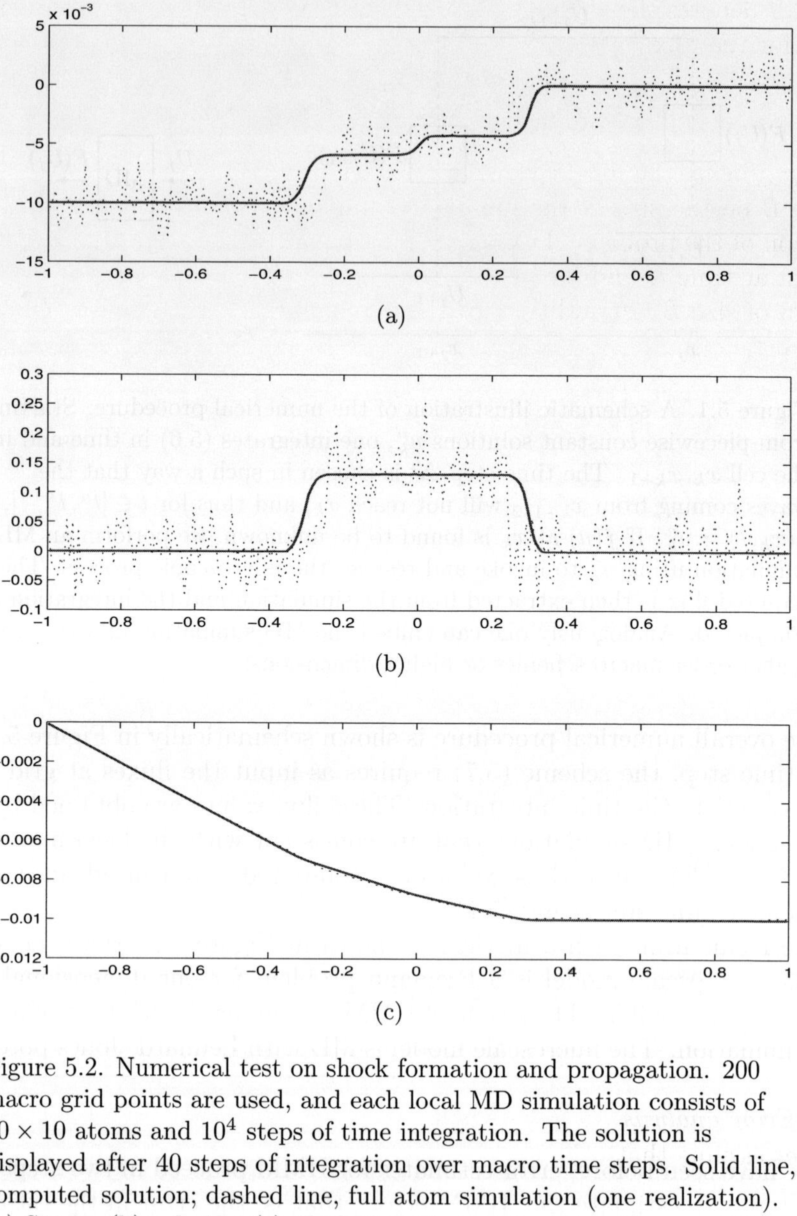

(a)

(b)

(c)

Figure 5.2. Numerical test on shock formation and propagation. 200 macro grid points are used, and each local MD simulation consists of 40×10 atoms and 10^4 steps of time integration. The solution is displayed after 40 steps of integration over macro time steps. Solid line, computed solution; dashed line, full atom simulation (one realization). (a) Strain, (b) velocity, (c) displacement.

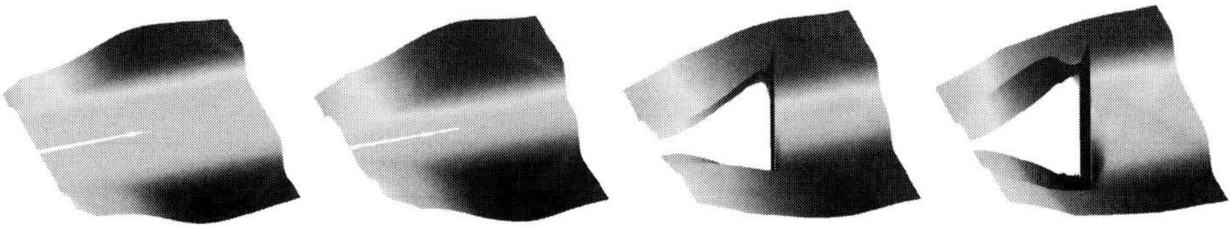

Figure 5.4. An open crack under mode I shock loading.

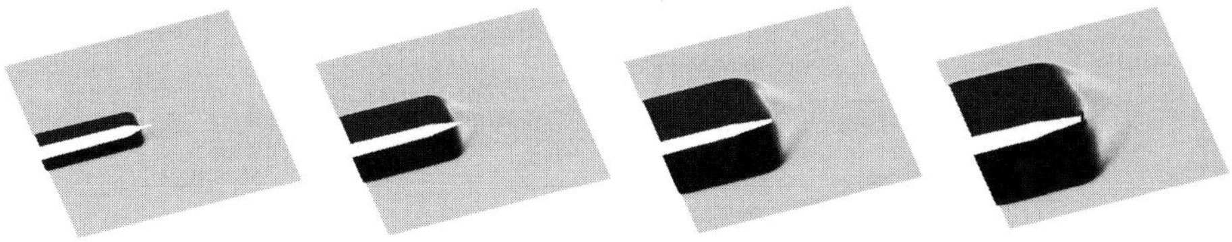

Figure 5.5. An open crack subject to face loading.

Another critical issue is the boundary condition for the atomistic region. This issue has been extensively studied in Li (2008), Li and E (2007) and Li and E (2006).

Numerical results. As the first application, Xiantao Li studied cracks in pure iron at low temperatures. The interatomic potential is the EAM model (Daw and Baskes 1984). The system is a rectangular sample with size 0.78 μm $\times$ 0.78 μm. The plane strain condition was assumed. For the MD model, this is enforced by a periodic boundary condition in the third direction with period equal to one atomic spacing. In the first experiment, tensile stress is rapidly applied from the top and bottom boundaries of the specimen. This generates two shock waves, which later propagate to the centre of the system to interact with a pre-existing crack. Below a certain threshold, no crack growth is observed after the two shocks collide. Instead of shock collisions, reflections off the crack faces are observed behind the crack tip. Above a certain threshold, one begins to observe crack growth, followed by a sequence of high-frequency lattice waves, the first two of which are of elliptical and diamond shapes. Experimental observation shows that spallation does not occur instantaneously when the tension exceeds the spallation strength: it occurs after a brief incubation period. This was also observed in our numerical experiments. A sequence of snapshots is shown in Figure 5.4.

In the second experiment, Li studied the response of a crack to loadings along the crack faces. These results are shown in Figure 5.5. In this case, the shocks are no longer of longitudinal type, and they propagate with a much slower speed. The magnitude required to produce fractures is much higher.

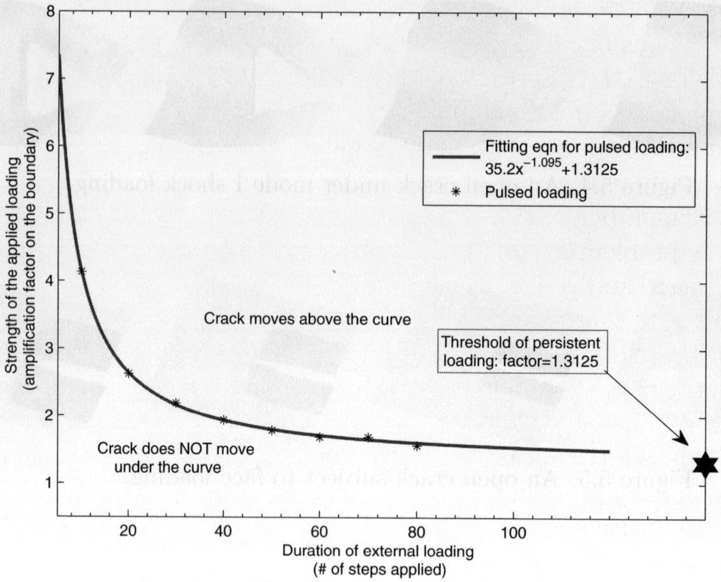

Figure 5.6. The onset of crack initiation.

Figure 5.6 summarizes the results under the impact of shocks of different magnitude. The goal is to investigate the strength of the material under various types of extreme loading conditions. The curve in the figure indicates the onset of the crack growth. In particular, for a given impact duration, the point on the curve indicates the magnitude for which the shock impact leads to fracture. For instantaneous shock loads, only very strong shocks are able to produce a growing crack. For continuing (slow) impacts, the threshold is much lower, and the curve approaches a plateau, which agrees with linear elasticity analysis. The results can be compared to the classical experimental results (Ravi-Chandar and Knauss 1984).

6. Conclusion

When designing multiscale methods, there are two important factors to consider.

1 From a modelling viewpoint, we would like to eliminate *ad hoc* modelling assumptions as much as possible.

2 From a numerical viewpoint, the algorithms have to be practical and must satisfy the standard requirements for numerical algorithms, such as stability and accuracy.

These two aims are not always compatible. Therefore compromises have to be made. This is one lesson we have learned over the past decade in multiscale modelling. HMM provides a general framework that allows us to make such compromises *and* make progress.

Although HMM has shown great promise, much work remains to be done in order to fully realize its potential. One should note, however, that serious obstacles remain before HMM can be applied to more challenging problems. For type A problems, namely problems involving local defects or singularities, the macrostates around the defects or singularities have to be inhomogeneous, and there are difficulties in formulating constrained microscopic models so that they are consistent with inhomogeneous macrostates. In addition, near defects or singularities, the local macroscale solutions are often singular too. Therefore, for numerical efficiency, local mesh refinements are needed. This is a technical problem that needs to be addressed. For type B problems, namely problems for which microscopic models are needed to supplement constitutive relations, the main obstacle is still the cost associated with running the microscopic models.

A major issue for multiscale methods is the effect of fluctuations. This is manifested in several ways. Fluctuation effects are almost always present in microscopic models, such as molecular dynamics or Monte Carlo models. They are also present in detailed models of random media. However, they are absent from most effective macroscale models. In cases when fluctuation effects are important, these effective macroscale models have to be remedied to include the fluctuations (Donev, Bell, Garcia and Alder 2010). How to do this is still very much an open question. In particular, in HMM, to reduce costs, we would like to perform the microscopic simulations on domains as small as possible. But the Central Limit Theorem tells us that fluctuations are inversely proportional to the square root of the volume. Hence computational savings might come at the cost of artificially enlarging the size of the fluctuations (Bal and Jing 2011). These are all issues that will need to be carefully addressed.

Acknowledgement

We are grateful to Xiantao Li for his help in writing this article. The work of Abdulle is supported in part by the Swiss National Science Foundation under Grant 200021 134716/1. The work of E has been supported in part by ONR grant N00014-01-1-0674. The work of Engquist has been supported in part by NSF grant DMS1016577. The work of Vanden-Eijnden has been supported in part by grant DMS-0708140 from NSF and grant N00014-11-1-0345 from ONR.

REFERENCES[6]

A. Abdulle (2005*a*), 'Multiscale methods for advection–diffusion problems', *Discrete Contin. Dyn. Syst.* **suppl.**, 11–21.

A. Abdulle (2005*b*), 'On *a priori* error analysis of fully discrete heterogeneous multiscale FEM', *SIAM Multiscale Model. Simul.* **4**, 447–459.

A. Abdulle (2006), 'Analysis of a heterogeneous multiscale FEM for problems in elasticity', *Math. Models Methods Appl. Sci.* **16**, 615–635.

A. Abdulle (2008), 'Multiscale method based on discontinuous Galerkin methods for homogenization problems', *CR Math. Acad. Sci. Paris* **346**, 97–102.

A. Abdulle (2009), 'The finite element heterogeneous multiscale method: A computational strategy for multiscale PDEs', *GAKUTO Int. Ser. Math. Sci. Appl.* **31**, 135–184.

A. Abdulle (2011), '*A priori* and *a posteriori* error analysis for numerical homogenization: A unified framework', *Ser. Contemp. Appl. Math. CAM* **16**, 280–305.

A. Abdulle (2012), 'Discontinuous Galerkin finite element heterogeneous multiscale method for elliptic problems with multiple scales', *Math. Comp.* **81**, 687–713.

A. Abdulle and Y. Bai (2012), 'Reduced basis finite element heterogeneous multiscale method for high-order discretizations of elliptic homogenization problems', *J. Comput. Phys.*, to appear.

A. Abdulle and W. E (2003), 'Finite difference heterogeneous multi-scale method for homogenization problems', *J. Comput. Phys.* **191**, 18–39.

A. Abdulle and B. Engquist (2007), 'Finite element heterogeneous multiscale methods with near optimal computational complexity', *SIAM Multiscale Model. Simul.* **6**, 1059–1084.

A. Abdulle and M. Grote (2011), 'Finite element heterogeneous multiscale method for the wave equation', *SIAM Multiscale Model. Simul.* **9**, 766–792.

A. Abdulle and A. Nonnenmacher (2009*a*), '*A posteriori* error analysis of the heterogeneous multiscale method for homogenization problems', *CR Math. Acad. Sci. Paris* **347**, 1081–1086.

A. Abdulle and A. Nonnenmacher (2009*b*), 'A short and versatile finite element multiscale code for homogenization problem', *Comput. Methods Appl. Mech. Engrg* **198**, 2839–2859.

A. Abdulle and A. Nonnenmacher (2011*a*), 'Adaptive finite element heterogeneous multiscale method for homogenization problems', *Comput. Methods Appl. Mech. Engrg* **200**, 2710–2726.

A. Abdulle and A. Nonnenmacher (2011*b*), *A posteriori* error estimate in quantities of interest for the finite element heterogeneous multiscale method. Preprint.

A. Abdulle and C. Schwab (2005), 'Heterogeneous multiscale FEM for diffusion problems on rough surfaces', *SIAM Multiscale Model. Simul.* **3**, 195–220.

A. Abdulle and G. Vilmart (2011*a*), Analysis of the finite element heterogeneous multiscale method for nonmonotone elliptic homogenization problems. Preprint.

[6] The URLs cited in this work were correct at the time of going to press, but the publisher and the authors make no undertaking that the citations remain live or are accurate or appropriate.

A. Abdulle and G. Vilmart (2011*b*), 'The effect of numerical integration in the finite element method for nonmonotone nonlinear elliptic problems with application to numerical homogenization methods', *CR Acad. Sci. Paris, Ser. I* **349**, 1041–1046.

A. Abdulle and G. Vilmart (2012*a*), 'Coupling heterogeneous multiscale FEM with Runge–Kutta methods for parabolic homogenization problems: A fully discrete space–time analysis', *Math. Models Methods Appl. Sci.*, to appear.

A. Abdulle and G. Vilmart (2012*b*), '*A priori* error estimates for finite element methods with numerical quadrature for nonmonotone nonlinear elliptic problems', *Numer. Math.*, to appear.

C. Abrams and E. Vanden-Eijnden (2010), 'Large-scale conformational sampling of proteins using temperature-accelerated molecular dynamics', *Proc. Natl Acad. Sci.* **107**, 4961.

M. Ainsworth and R. Rankin (2012), 'Guaranteed computable bounds on quantities of interest in finite element computations', *Internat. J. Numer. Methods Engrg* **89**, 1605–1634.

T. Antoun, L. Seaman, D. Curran, G. Kanel, R. Razorenov and A. Utkin (2003), *Spall Fracture*, Springer.

G. Ariel, B. Engquist and R. Tsai (2009*a*), 'A multiscale method for highly oscillatory ordinary differential equations with resonance', *Math. Comp.* **78**, 929–956.

G. Ariel, B. Engquist and R. Tsai (2009*b*), 'Numerical multiscale methods for coupled oscillators', *Multiscale Model. Simul.* **7**, 1387–1404.

G. Ariel, B. Engquist, H.-O. Kreiss and R. Tsai (2009*c*), Multiscale computations for highly oscillatory problems. In *Multiscale Modeling and Simulation in Science*, Vol. 66 of *Lecture Notes in Computational Science and Engineering*, Springer, pp. 237–287.

G. Ariel, J. M. Sanz-Serna, and R. Tsai (2012), A multiscale technique for finding slow manifolds of stiff mechanical systems. Under review.

D. Arnold, F. Brezzi, B. Cockburn and D. Marini (2001/2002), 'Unified analysis of discontinuous Galerkin methods for elliptic problems', *SIAM J. Numer. Anal.* **39**, 1749–1779.

Z. Artstein, I. Kevrekidis, M. Slemrod and E. Titi (2007*a*), 'Slow observables of singularly perturbed differential equations', *Nonlinearity* **20**, 2463–2481.

Z. Artstein, J. Linshiz and E. Titi (2007*b*), 'Young measure approach to computing slowly advancing fast oscillations', *Multiscale Model. Simul.* **6**, 1085–1097.

I. Babuška (1976), Homogenization and its application: Mathematical and computational problems. In *Numerical Solution of Partial Differential Equations III: Proc. Third Sympos. SYNSPADE 1975*, pp. 89–116.

I. Babuška and J. Osborn (1983), 'Generalized finite element methods: Their performance and their relation to mixed methods', *SIAM J. Numer. Anal.* **20**, 510–536.

G. Bal and W. Jing (2011), 'Corrector theory for MsFEM and HMM in random media', *Multiscale Model. Simul.* **9**, 1549–1587.

R. Becker and R. Rannacher (2001), An optimal control approach to *a posteriori* error estimation in finite element methods. In *Acta Numerica*, Vol. 10, Cambridge University Press, pp. 1–102.

A. Bensoussan, J.-L. Lions and G. Papanicolaou (1978), *Asymptotic Analysis for Periodic Structures*, North-Holland.

L. Bers, F. John and M. Schechter (1957), *Partial Differential Equations*, Vol. 3A of *Lectures in Applied Mathematics*, Wiley.

L. Boccardo and F. Murat (1981), 'Homogénéisation de problèmes quasi-linéaires', *Publ. IRMA, Lille* **3**, 13–51.

S. Boyaval (2008), 'Reduced-basis approach for homogenization beyond the periodic setting', *Multiscale Model. Simul.* **7**, 466–494.

D. Braess (2007), *Finite Elements: Theory, Fast Solvers, and Applications in Elasticity Theory*, third edition, Cambridge University Press.

A. Brandt (2002), 'Multiscale scientific computation: Review 2001', *Multiscale and Multiresolution Methods: Theory and Applications* **20**, 3–96.

M. P. Calvo and J. M. Sanz-Serna (2009), 'Instabilities and inaccuracies in the integration of highly oscillatory problems', *SIAM J. Sci. Comput.* **31**, 1653–1677.

M. P. Calvo and J. M. Sanz-Serna (2010), 'Heterogeneous multiscale methods for mechanical systems with vibrations', *SIAM J. Sci. Comput.* **32**, 2029–2046.

G. Camacho and M. Ortiz (1996), 'Computational modelling of impact damage in brittle materials', *Internat. J. Solids Struct.* **33**, 2899–2938.

Y. Cao, D. Gillespie and L. Petzold (2005), 'Multiscale stochastic simulation algorithm with stochastic partial equilibrium assumption for chemically reacting systems', *J. Comput. Phys.* **206**, 395–411.

R. Car and M. Parrinello (1985), 'Unified approach for molecular dynamics and density-functional theory', *Phys. Rev. Lett.* **55**, 2471–2474.

P. Chartier, A. Murua and J. M. Sanz-Serna (2010), 'Higher-order averaging, formal series and numerical integration I: B-series', *Found. Comput. Math.* **10**, 695–727.

S. Chen, W. E and C.-W. Shu (2005*a*), 'The heterogeneous multiscale method based on the discontinuous Galerkin method for hyperbolic and parabolic problems', *SIAM Multiscale Model. Simul.* **3**, 871–894.

Z. Chen, W. Deng and H. Ye (2005*b*), 'Upscaling of a class of nonlinear parabolic equations for the flow transport in heterogeneous porous media', *Comm. Math. Sci.* **3**, 493–515.

A. Chorin (1967), 'A numerical method for solving incompressible viscous flow problems', *J. Comput. Phys.* **2**, 12–26.

P. Ciarlet and P. Raviart (1972), The combined effect of curved boundaries and numerical integration in isoparametric finite element methods. In *The Mathematical Foundations of the Finite Element Method: With Applications to Partial Differential Equations*, Academic Press, pp. 409–474.

D. Cioranescu and P. Donato (1999), *An Introduction to Homogenization*, Vol. 17 of *Oxford Lecture Series in Mathematics and its Applications*, Oxford University Press.

P. Clément (1975), 'Approximation by finite element functions using local regularization', *Rev. Fran. Automat. Informat. Recherche Opérationnelle Sér. RAIRO Analyse Numérique* **9**, 77–84.

M. Tao, H. Owhadi and J. Marsden (2010), 'Nonintrusive and structure preserving multiscale integration of stiff ODEs, SDEs, and Hamiltonian systems with hidden slow dynamics via flow averaging', *Multiscale Model. Simul.* **8**, 1269–1324.

K. Terada and N. Kikuchi (2001), 'A class of general algorithms for multiscale analyses of heterogeneous media', *Comput. Methods Appl. Mech. Engrg* **190**, 5427–5464.

E. Vanden-Eijnden (2003), 'Numerical techniques for multi-scale dynamical systems with stochastic effects', *Comm. Math. Sci.* **1**, 385–391.

E. Vanden-Eijnden (2007), 'On HMM-like integrators and projective integration methods for systems with multiple time scales', *Comm. Math. Sci.* **5**, 495–505.

R. Verfürth (1996), *A Review of A Posteriori Error Estimation and Adaptive Mesh-Refinement Techniques*, Wiley-Teubner.

T. Wallstrom, S. Hou, M. Christie, L. Durlofsky and D. Sharp (1999), 'Accurate scale up of two phase flow using renormalization and nonuniform coarsening', *Comput. Geosci.* **3**, 67–87.

Y. Xing, W. Grabowski and A. Majda (2009), 'New efficient sparse space–time algorithms for superparameterization on mesoscales', *Monthly Weather Rev.* **137**, 4307–4324.

J. Xu (1996), 'Two-grid discretization techniques for linear and nonlinear PDE', *SIAM J. Numer. Anal.* **33**, 1759–1777.

X.-P. Xu and A. Needleman (1994), 'Numerical simulations of fast crack growth in brittle solids', *J. Mech. Phys. Solids* **42**, 1397–1434.

X. Yue and W. E (2007), 'The local microscale problem in the multiscale modeling of strongly heterogeneous media: Effects of boundary conditions and cell size', *J. Comput. Phys.* **222**, 556–572.

V. Yurinskiĭ (1986), 'Averaging of symmetric diffusion in a random medium', *Sibirsk. Mat. Zh.* **27**, 167–180, 215.

Acta Numerica (2012), pp. 89–305
doi:10.1017/S0962492912000037

Numerical-asymptotic boundary integral methods in high-frequency acoustic scattering*

Simon N. Chandler-Wilde[1]
Ivan G. Graham[2]
Stephen Langdon[1]
Euan A. Spence[2]

[1] *Department of Mathematics and Statistics, University of Reading, Reading RG6 6AX, UK*
E-mail: {S.N.Chandler-Wilde,S.Langdon}@reading.ac.uk

[2] *Department of Mathematical Sciences, University of Bath, Bath BA2 7AY, UK*
E-mail: {I.G.Graham,E.A.Spence}@bath.ac.uk

In this article we describe recent progress on the design, analysis and implementation of hybrid numerical-asymptotic boundary integral methods for boundary value problems for the Helmholtz equation that model time harmonic acoustic wave scattering in domains exterior to impenetrable obstacles. These hybrid methods combine conventional piecewise polynomial approximations with high-frequency asymptotics to build basis functions suitable for representing the oscillatory solutions. They have the potential to solve scattering problems accurately in a computation time that is (almost) independent of frequency and this has been realized for many model problems. The design and analysis of this class of methods requires new results on the analysis and numerical analysis of highly oscillatory boundary integral operators and on the high-frequency asymptotics of scattering problems. The implementation requires the development of appropriate quadrature rules for highly oscillatory integrals. This article contains a historical account of the development of this currently very active field, a detailed account of recent progress and, in addition, a number of original research results on the design, analysis and implementation of these methods.

* Colour online for monochrome figures available at journals.cambridge.org/anu.

CONTENTS

1	Introduction	90
2	BVPs and integral equation formulations	100
3	Asymptotics and hybrid approximation spaces	146
4	Numerical treatment of oscillatory integrals	168
5	Conditioning and coercivity	195
6	Error analysis	227
7	Numerical results	245
	Appendix: Function spaces on Lipschitz domains	269
	References	290

1. Introduction

Acoustic, elastic and electromagnetic wave scattering problems arise in many applications of mathematical, physical and engineering interest, including the modelling of radar, sonar, noise barrier design and atmospheric particle scattering. Often the scattering problem comprises the forward map in the formulation of an inverse problem, for example in non-destructive testing or in methods for detecting hydrocarbon-bearing deposits under the sea bed. While in general the scattered wave has to be found in an inhomogeneous medium, there are a substantial number of applications in which the material is either homogeneous or piecewise homogeneous, at least sufficiently far away from the scatterer. In these cases boundary integral equation (BIE) methods are of considerable interest and form the basis for several commercial scattering codes; see, for example, Chew *et al.* (2004).

This review focuses on the efficient solution of high-frequency acoustic scattering problems in homogeneous media, using integral equation methods. Hence we consider the Helmholtz equation,

$$\Delta u + k^2 u = 0, \tag{1.1}$$

in a domain $\Omega_+ := \mathbb{R}^d \backslash \overline{\Omega_-}$, $d = 2$ or 3, where Ω_- is some bounded open set with surface Γ, and (1.1) is to be solved subject to some suitable boundary condition on Γ and radiation condition in the far field. The Helmholtz equation is of course derived from the linear wave equation under the assumption that all quantities vary harmonically ($\mathrm{e}^{-\mathrm{i}\omega t}$) in time. Here ω is the angular frequency and $k := \omega/c > 0$ is the wavenumber, where c is the speed of sound.

The problem (1.1) has solutions which oscillate in space with wavelength $\lambda = 2\pi/k$. For example the plane waves $u(x) = \exp(\mathrm{i}kx \cdot \hat{a})$, where $\hat{a} \in \mathbb{R}^d$ is a unit vector, are solutions. The number of oscillations grows linearly in

k and so the application of conventional (piecewise polynomial) boundary elements to this problem leads to full matrices of dimension at least $N = O(k^{d-1})$ and a solution time of this order or worse as $k \to \infty$. (Domain finite elements lead to sparse matrices but require even larger N.) Since this lack of robustness with respect to increasing values of k puts high-frequency problems beyond the reach of many standard algorithms, much recent research has been devoted to finding more robust methods.

Of course, saying that k, which has dimension 1/length, is large is meaningless without reference to the size of the scatterer. The dimensionless quantity that *can* meaningfully be thought of as quantifying the oscillatory character of our problem is kL, where L is an appropriate characteristic length of the scattering surface Γ (*e.g.*, the diameter of Γ). Thus a problem can be highly oscillatory even if k is not large provided that L is sufficiently large. But without loss of generality, throughout the review we consider k as the relevant large parameter, equivalently assuming that the unit of length is chosen so that the surface Γ has characteristic length $O(1)$.

The aim of this review is to describe a currently very active area of research which seeks algorithms for scattering problems which (ideally) have bounded error for fixed computational effort as $k \to \infty$, and have computational complexity which is either independent of k, or grows only mildly as k increases. To achieve this aim, the methods we describe explicitly build into the numerical method a certain amount of asymptotic information about the oscillatory nature of the solution as $k \to \infty$, and seek to approximate only the slowly varying components by (piecewise) polynomials. By now these methods have been supported by substantial theoretical justification, as we shall describe.

It is well known that the scattering problem described above is in general a multiscale problem. Figure 1.1 illustrates the resulting total field $u := u^I + u^S$ induced by a scatterer Ω_- composed of the union of a disk and a triangle, when the incident field u^I is a plane wave $u^I(x) = \exp(ikx \cdot \hat{a})$, with unit incident direction $\hat{a}$. The scattered field u^S is found as the solution of the classical 'sound-soft' scattering problem, that is, u^S satisfies (1.1) in Ω_+, the Dirichlet condition $u^S = -u^I$ (and so $u = 0$) on the scattering surface Γ (in this case the union of the circle and the boundary of the triangle) and, in addition, u^S satisfies the usual Sommerfeld radiation condition (given by (2.9)) in the far field. While the incident field u^I oscillates on the single scale k^{-1}, the total field u contains several other scales coming from the scattered field u^S. These include scales of $k^{-1/2}$ (respectively $k^{-1/3}$) associated with widths of zones of transition from illuminated to shadow regions behind diffracting corners (respectively tangency points) These oscillatory and multiscale properties of scattered fields – known for many years in the asymptotics literature – are described in a form useful for numerical analysis in Section 3 of this review.

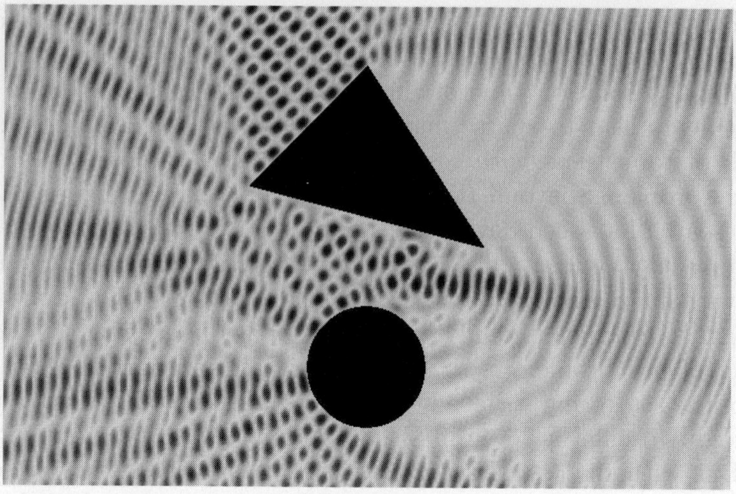

Figure 1.1. The total field when the incident field is a plane wave in the direction $\hat{a} = (\cos\theta, \sin\theta)$, with $\theta = -\pi/18$, and the wavelength is $\lambda = 0.2$ (so $k = 2\pi/\lambda = 10\pi \approx 31.42$). The scatterer has two components, a disk of unit diameter and a triangle.

To formulate BIEs for (1.1), we introduce the standard fundamental solution of the Helmholtz equation, given, in the two-dimensional (2D) and three-dimensional (3D) cases, by

$$\Phi_k(x,y) := \begin{cases} \dfrac{\mathrm{i}}{4} H_0^{(1)}(k|x-y|), & d = 2, \\ \dfrac{\exp(\mathrm{i}k|x-y|)}{4\pi|x-y|}, & d = 3, \end{cases} \tag{1.2}$$

for $x, y \in \mathbb{R}^d$, $x \neq y$, where $H_\nu^{(1)}$ denotes the Hankel function of the first kind of order ν. Using Φ_k, we can build layer potentials that provide solutions to (1.1) in the exterior domain Ω_+, and automatically satisfy the radiation condition at infinity. These potentials conveniently also provide solutions to (1.1) in the interior domain Ω_-. In general all the standard boundary value problems (BVPs) for the Helmholtz equation (1.1) can be formulated as integral equations on Γ using these layer potentials.

For example, in the case of sound-soft scattering, the solution u is fully determined on Ω_+ by its Neumann data $\partial u/\partial n$ on Γ; see Section 2 for details. Moreover the 'far-field' behaviour of u (often of interest in the solution of inverse scattering problems) can be determined by the action of a simple oscillatory linear operator applied to the Neumann data. The required Neumann data can be obtained, for example, by solving the combined potential integral equation

$$\frac{1}{2}\frac{\partial u}{\partial n}(x) + \int_\Gamma \left(\frac{\partial \Phi_k(x,y)}{\partial n(x)} - \mathrm{i}\eta \Phi_k(x,y) \right) \frac{\partial u}{\partial n}(y)\, \mathrm{d}s(y) = f_{k,\eta}(x), \quad x \in \Gamma, \tag{1.3}$$

where, throughout the review, we adopt the convention that the normal derivative is taken outward from Ω_-, and the source term is given by

$$f_{k,\eta}(x) := \frac{\partial u^I}{\partial n}(x) - \mathrm{i}\eta u^I(x), \quad x \in \Gamma. \tag{1.4}$$

The problem (1.3) is well-posed for any fixed choice of the coupling parameter $\eta \in \mathbb{R}\backslash\{0\}$ (see Theorem 2.46). We write (1.3) more compactly as

$$A'_{k,\eta} v := \left(\frac{1}{2} I + D'_k - \mathrm{i}\eta S_k \right) v = f_{k,\eta}, \quad \text{where} \quad v := \partial u / \partial n. \tag{1.5}$$

The operators S_k, D'_k and $A'_{k,\eta}$ will be discussed in detail in Section 2.

Turning to numerical methods, for an operator equation of the general form $Av = f$ posed in $L^2(\Gamma)$, the Galerkin method consists of choosing a finite-dimensional approximating space $\mathcal{V}_N$ and then seeking an approximate solution $v_N \in \mathcal{V}_N$ such that

$$(Av_N, w_N)_{L^2(\Gamma)} = (f, w_N)_{L^2(\Gamma)}, \quad \text{for all } w_N \in \mathcal{V}_N. \tag{1.6}$$

For the discretization of the second-kind integral equation (1.5) there is a classical theory, which holds at least when Γ is sufficiently smooth and when the approximating space consists of piecewise polynomials. Suppose that we solve (1.5) using the Galerkin method (1.6) on a family of N-dimensional spaces $\mathcal{V}_N$ of piecewise polynomial functions of fixed degree on a quasi-uniform sequence of meshes on Γ with diameter $h \to 0$ (so that N is proportional to h^{1-d}). Then it can be shown, using the methods in Atkinson (1997), for example, that there exist constants $C > 0$ and $N_0 > 0$ such that the Galerkin solution v_N satisfies the quasi-optimal error estimate

$$\|v - v_N\| \le C \inf_{w_N \in \mathcal{V}_N} \|v - w_N\|, \tag{1.7}$$

for all $N \ge N_0$, where here and throughout this review $\|\cdot\|$ represents $\|\cdot\|_{L^2(\Gamma)}$, unless otherwise specified. (This theory also extends to some collocation and, with the addition of quadrature, Nyström-like methods.) However, the classical theory does not tell us how C and N_0 depend on k.

These subtle questions are discussed in some detail in Section 6.1 of this review, but we can easily see that the error $\|v - v_N\|$ will be highly k-dependent. This is because v is in general oscillatory, and the jth derivative of v will, roughly speaking, be $O(k^j)$ times bigger than v itself. Thus, using standard estimates for piecewise polynomial approximation of degree p, the error in best approximation appearing on the right-hand side of (1.7) will have an estimate of the form

$$\inf_{w_N \in \mathcal{V}_N} \|v - w_N\| \leq C'(hk)^{p+1}, \tag{1.8}$$

for some constant C' (which may itself grow with k). Thus h will need to decrease with $O(k^{-1})$, and possibly faster, in order to keep the error bounded as $k \to \infty$. Hence, standard (piecewise) polynomial BEMs applied directly to approximate the oscillatory solution v of (1.5) on a surface in $\mathbb{R}^d$ will have complexity at least $O(h^{1-d}) = O(k^{d-1})$ as $k \to \infty$.

By contrast, the *hybrid numerical-asymptotic methods* which are the main subject of this review article exploit more detailed information about the oscillations in v (obtained via asymptotic analysis or exact integral representations) directly in the numerical method. Known highly oscillatory components of v are treated exactly in the algorithm, leaving only more slowly varying components to be approximated by piecewise polynomials. This yields a method which is much more 'robust' as $k \to \infty$.

One of the simplest hybrid numerical-asymptotic methods, and one which we shall discuss at length in Section 3, can be seen as an extension of the 'physical optics' (also called the 'Kirchhoff') approximation. This method assumes that the scattered wave u^S oscillates on the same scale as the incoming plane wave, leading to the ansatz

$$v(x) = kV(x,k)\exp(ikx \cdot \hat{a}), \quad x \in \Gamma. \tag{1.9}$$

For some geometries (*e.g.*, smooth convex scatterers), $V(\cdot,k)$ is much less oscillatory than v, and methods based on approximating $V(\cdot,k)$ using piecewise polynomials have been proposed by a number of authors. This procedure is equivalent to approximating v using a hybrid space $\mathcal{V}_N$ with a basis consisting of products of the plane wave $\exp(ikx \cdot \hat{a})$ and appropriate piecewise polynomial basis functions. Using such a hybrid space $\mathcal{V}_N$, it is possible to show that the best approximation error on the left-hand side of (1.8) increases much more slowly as $k \to \infty$ than the estimate on the right-hand side of (1.8) (which holds for conventional piecewise polynomial spaces). Thus good approximation for large k on relatively coarse meshes (and with relatively few degrees of freedom) can potentially be achieved by hybrid methods.

Unfortunately the benefits of these hybrid methods do not come without cost. When the mesh is fairly coarse and k is large, integration of the kernel of S_k or D'_k over the support of a basis function requires computing an oscillatory integral. The additional oscillations arising via the basis functions further complicate these integrals, which have to be evaluated with a work count growing at worst modestly in k if the overall algorithm is to be successful. Hence this field requires a substantial investment in research into numerical methods for oscillatory integrals; see Section 4 for details.

To do a full numerical analysis of hybrid methods, not only do we require good estimates for the best approximation error (on the right-hand side of (1.7)), but also we need k-explicit estimates for the constants N_0 and C in (1.7). This issue is discussed in Sections 5 and 6.

Summarizing these points, one can think of the numerical analysis of hybrid methods as requiring research on three related questions.

Q1 The design of k-dependent finite-dimensional approximation spaces $\mathcal{V}_N$, with the property that the best approximation error, that is, $\inf_{w_N \in \mathcal{V}_N} \|v - w_N\|$, remains within a given tolerance as $k \to \infty$, with N fixed or growing only slowly with k.

Q2 The design of k-robust methods for computing the oscillatory integrals arising in implementation.

Q3 The proof of sharp estimates for the dependence on k of the 'stability constant' C and the space dimension threshold N_0, ideally showing that these grow only slowly as $k \to \infty$.

This review describes research on these three fundamental issues and related topics, and we summarize its contents here. In Section 2 we describe fundamental solvability results on the relevant BVPs for (1.1) and the reformulation of these BVPs as BIEs, including new BIE formulations that have recently been proposed for high-frequency scattering problems. Given that many scatterers in applications involve corners and edges, we work on general Lipschitz domains. Background results for this section, some of which are not easily found in the literature, are given as an appendix. Section 3 describes the design of good hybrid spaces for a range of scattering problems and the proof of k-robustness of the best approximation error for these spaces (**Q1**). While this section is concerned mostly with 2D problems, hybrid spaces for a class of 3D screen problems are also discussed later in Section 7.6. In Section 4 the question of robust computation of oscillatory integrals which arise in hybrid methods is considered (**Q2**), and rigorous error estimates for these integration schemes, which have been the subject of recent research, are described. Section 5 describes the recently very active field which seeks k-explicit bounds on the conditioning of integral operators such as that in (1.5) (in particular, bounds on the operator, bounds on its inverse, coercivity properties, *etc.*). These results are required for estimating the constants N_0 and C appearing in the formulation of **Q3** above. This is done in Section 6, which in fact gives error estimates for standard as well as hybrid boundary element methods, valid as $k \to \infty$. Finally, Section 7 presents a series of illustratory numerical examples, supporting the estimates of Section 6 and showing that, for a range of scattering problems, arbitrary accuracy is achieved with a computational cost that grows only very mildly with respect to k.

We finish this introduction with some historical remarks on the development of hybrid numerical-asymptotic methods, concluding with some remarks on other methods for the accurate solution of high-frequency scattering problems.

Methods that combine numerical and asymptotic approaches in order to reduce computational cost have been applied within the electromagnetics community for many years. Two papers dating back to 1975 suggest reducing the computational cost at high frequencies by splitting the boundary into different regions, and using numerical methods in some regions and asymptotic approximations in others (Thiele and Newhouse 1975, Burnside, Yu and Marhefka 1975). The BIE method proposed in Burnside *et al.* (1975), for an electromagnetic problem equivalent to plane wave scattering by a sound-hard square, can sensibly be viewed as a prototype of the methods that we develop for scattering by convex polygons in Section 3.3 (and see Sections 7.2–7.4). The related method proposed in Thiele and Newhouse (1975) is a first instance of what has become a popular hybrid BIE-based methodology, whereby high-frequency physical optics approximations (see (3.4) below) are employed on part of the scattering surface, and standard BEM approximations on the remainder, with coupling between these subdomains: see Djordjević and Notaroš (2005), Zhang *et al.* (2010) and the references therein for recent developments.

A hybrid numerical-asymptotic BIE method in exactly the sense of this article was proposed at almost the same time in the acoustics literature by Uncles. In a short proof-of-concept paper, Uncles (1976) proposed the use of essentially the ansatz (1.9), computing scattering by a sound-hard sphere using a piecewise constant approximation for the unknown $V(\cdot, k)$. This hybrid BIE idea (approximating the ratio of scattered to incident field rather than the scattered field itself) was used for the 2D problem of scattering by an impedance half-plane (which we treat with more sophisticated hybrid methods in Section 3.2), by Chandler-Wilde (1988), with numerical results demonstrating the efficiency of this approach at moderate frequencies.

Essentially the same hybrid BIE method to that in Uncles (1976) was proposed independently in the electromagnetics literature in James (1990). Numerical results for the case of a circular scatterer (solving the integral equations (2.109) and (2.111) below) again demonstrated a significant reduction in the number of unknowns required compared to a conventional BEM. A more elaborate method, using hybrid spaces which are close to but less sophisticated than those described below in Sections 3.3.1–3.3.2, was proposed by Wang (1991), whose numerical results demonstrate some accuracy for a number of 2D geometries. The work of Aberegg and Peterson (1995) developed the methods of Uncles and Wang. To tackle 2D problems of electromagnetic scattering by piecewise smooth convex obstacles, solving the Helmholtz equation (1.1), they used the same ansatz (1.9) but introduced a number of aspects that are key to the effective implementation of hybrid methods and which we will study in some detail in Section 3, namely higher-order basis functions, the treatment of corner singularities (via special basis functions in Aberegg and Peterson (1995)), and the use of mesh grading. They report, for many geometries, reasonably accurate results ($\leq 3.5\%$ relative error) with ten times fewer degrees of freedom compared to standard BEMs, and note that the method is robust in that, in cases where the ansatz (1.9) does not capture accurately the oscillation in the solution, the method should in any case be no less accurate for the same number of degrees of freedom than conventional BEMs.

Since 1994 there has been significant numerical analysis interest in hybrid methods for scattering problems, with the majority of investigations focusing (either implicitly or explicitly) on the case of a smooth convex obstacle. This started with the contributions of Abboud *et al.* (1994, 1995), who considered (1.1) subject to an impedance boundary condition on Γ, and formulated this as a first-kind BIE, somewhat different to (1.3) (see (3.10) for details). The ansatz (1.9) was then applied, with the 'slow variable' $V(\cdot, k)$ being approximated using the h-version BEM. This analysis and numerical scheme was subsequently developed by Darrigrand (2002), who proposed fast multipole-based methods to evaluate the oscillatory integrals that arise.

A more advanced approach than that taken by Abboud, Nédélec and Zhou (1995), taking special care to approximate $V(\cdot, k)$ accurately near the shadow boundary (see Section 3 for details), was proposed by Bruno, Geuzaine, Monro and Reitich (2004). They developed a high-frequency Nyström approach, substituting (1.9) directly into (1.5) to obtain a second-kind integral equation for the slow variable $V(\cdot, k)$ (see (3.11)), and then devising a frequency-robust, fast quadrature method for discretizing the corresponding integral operators. This was extended further, with strong emphasis on integration in 3D problems, by Bruno and Geuzaine (2007). At around the same time, Giladi and Keller (2004) – see also Giladi (2007) (following earlier work by Giladi and Keller (2001) on finite element methods) – solved the same equation considered by Bruno *et al.* (2004) using a collocation method, but also took into account the exponentially damped 'creeping waves' behind the shadow boundary. Subsequently Huybrechs and Vandewalle (2006) developed steepest descent-based methods for oscillatory integration and applied these in the implementation of a collocation-type scheme for the BIE (2.109), again using the ansatz (1.9) (Huybrechs and Vandewalle 2007*b*), with the advantage that their approach leads to a sparse linear system.

The same problem of scattering by a sound-soft smooth convex 2D obstacle was considered in Domínguez, Graham and Smyshlyaev (2007), where the ansatz (1.9) was again used. By extending the asymptotic analysis of Melrose and Taylor (1985), Domínguez *et al.* were able to derive rigorous estimates demonstrating that their Galerkin method achieved an error that depended only very mildly on k (see Section 3.1).

For the problem of scattering by an impedance half-plane (see Section 3.2), Chandler-Wilde, Langdon and Ritter (2004), and then Langdon and Chandler-Wilde (2006), proposed and analysed a hybrid Galerkin BEM, proving rigorous error estimates independent of k, as $k \to \infty$. These ideas were extended to scattering by sound-soft convex polygons (Chandler-Wilde and Langdon 2007, Hewett, Langdon and Melenk 2012), curvilinear polygons (Langdon, Mokgolele and Chandler-Wilde 2010), convex polygons with impedance boundary conditions (Chandler-Wilde, Langdon and Mokgolele 2012*b*), and non-convex polygons (Chandler-Wilde, Hewett, Langdon and Twigger 2012*a*), and are described in detail in Sections 3.3 and 3.4.

A Galerkin method in the same spirit, utilizing a high-frequency ansatz similar to those described in Section 3, was applied to the 2D problem of scattering by a flat strip in Davis and Chew (2008), with numerical results suggesting only a very mild dependence of the error on the frequency.

For the case of scattering by a smooth convex obstacle in 3D, Ganesh and Hawkins (2011) again used the ansatz (1.9), and the integral equation for $V(\cdot, k)$ was solved using a discrete Galerkin method with a global polynomial basis and a specially chosen global quadrature rule to handle the oscillatory integrals.

The design of hybrid spaces for the case of multiple scattering (scattering by two or more disjoint convex scatterers) has been considered by Geuzaine, Bruno and Reitich (2005), and a detailed analysis of the phase structure of the solution has been given in Ecevit (2005), Ecevit and Reitich (2009) and Anand, Boubendir, Ecevit and Reitich (2009).

Further discussion of many of the above approaches will be given in Sections 3 and 4. Due to space considerations, this review has not been able to discuss in detail three other techniques which also address the efficient solution of high-frequency scattering problems, and here we give only a few representative references.

The first seeks fast implementations (often specifically tuned to the high-frequency case) of standard discretization methods and includes fast multipole and related fast iterative methods. Research in this direction is still very active and here we mention Rokhlin (1990), Amini and Maines (1998), Christiansen and Nédélec (2000), Farhat, Macedo and Lesoinne (2000), Bruno and Kunyansky (2001), Chandler-Wilde, Rahman and Ross (2002), Donepudi, Jin and Chew (2003), Harris and Chen (2003), Darve and Havé (2004), Livschits and Brandt (2006), Erlangga, Oosterlee and Vuik

(2006), Engquist and Ying (2007), Banjai and Hackbusch (2008), Bruno, Domínguez and Sayas (2012), Ernst and Gander (2012) and Engquist and Ying (2012).

The second is research on partition of unity and related methods in which a number of plane waves is introduced on each element in addition to standard piecewise polynomial boundary elements. These methods are accurate and rapidly convergent for general Helmholtz problems and do not require any prior knowledge of the asymptotics of the solution, but they do not enjoy the k-robustness of the hybrid methods for classes of scattering problems which we describe in this review; in particular, in each of these methods, the degrees of freedom need to increase in proportion to k^{d-1} to maintain accuracy as $k \to \infty$, just as for conventional BEMs, albeit with a lower constant of proportionality. Examples of this approach include de La Bourdonnaye (1994), Perrey-Debain, Trevelyan and Bettess (2003*a*), Perrey-Debain, Trevelyan and Bettess (2003*b*), de La Bourdonnaye and Tolentino (2004), Perrey-Debain, Laghrouche, Bettess and Trevelyan (2004), Perrey-Debain, Trevelyan and Bettess (2005), and Honnor, Trevelyan and Huybrechs (2010). There also exists an extensive literature applying similar ideas within a finite element context; see, for example, Melenk and Babuška (1996), Babuška and Melenk (1997), Cessenat and Després (1998), Monk and Wang (1999), Giladi and Keller (2001), Laghrouche, Bettess and Astley (2002), Huttunen, Monk and Kaipio (2002), Cessenat and Després (2003), Laghrouche, Bettess, Perrey-Debain and Trevelyan (2005), Buffa and Monk (2008), Gittelson, Hiptmair and Perugia (2009), Luostari, Huttunen and Monk (2009), Hiptmair, Moiola and Perugia (2011*a*) and Esterhazy and Melenk (2012).

The third is the solution of appropriate limiting problems, which are valid only if the frequency is sufficiently high (so-called 'ray tracing' methods). Considerable progress has been made on extending these to practical geometries and obtaining error estimates; see, for example, Engquist and Runborg (2003), Benamou (2003), Bleszynski, Bleszynski and Jaroszewicz (2004) and Motamed and Runborg (2009).

Finally we mention that earlier reviews on the subject matter of this article include Bruno and Reitich (2007) and Chandler-Wilde and Graham (2009).

We finish with a brief word on notations and assumed prior knowledge. Throughout we use function space notations that are explained in the Appendix (with cross-referencing from other sections). This use is mild in the more algorithmic Sections 3, 4, and 7, and more pronounced in Sections 2, 5, and 6, where we also need some functional analysis concepts and results which are briefly listed with references at the beginning of the Appendix. We are a little cavalier throughout as to whether we write γu (the trace of u) or $u|_\Gamma$ (the restriction of u to Γ) for the value of the total field on Γ, and likewise whether we write $\partial_n u$ or $\partial u/\partial n$ for the normal derivative. The distinction in these definitions, and that they coincide for the scattering problems that we study, is explained in Section 2.1 (see Theorem 2.12) and Section 2.8.

2. BVPs and integral equation formulations

In this section we formulate the relevant BVPs for the Helmholtz equation, giving details as appropriate of function spaces, radiation conditions, *etc.* We also formulate BIEs, and study properties of the layer potentials and integral operators which these give rise to. We make explicit the relationship between integral equations and BVPs, in particular the conditions under which BIE and BVP formulations are equivalent.

There is much in this section regarding the theory of BIEs for acoustic problems which is not found in other reviews or monographs, in part because many of the results we describe are very recent. In particular, we develop new BIE formulations and new representations for boundary integral operators, which are essential components in the wavenumber-explicit error and stability analysis in Sections 5 and 6.

However, those readers whose primary interest is in the design of hybrid algorithms and their implementation may well prefer to start with Sections 3 and 4, which can be read largely independently of this section.

This article is concerned with scattering problems, so that our focus is naturally on exterior BVPs, that is, problems set in the unbounded exterior of a scatterer. But the theory of BIEs, particularly perhaps BIEs for acoustic problems, depends inextricably on an understanding of the well-posedness of both interior and exterior BVPs, so that necessarily we shall consider interior problems too.

When it comes to the formulation of BVPs and BIEs, there is a degree of choice in the function space setting, in the sense in which the boundary conditions are to be understood, and indeed in the class of domains that one wishes to consider. Overwhelmingly the literature of the modern theory of BIE methods and their numerical solution, especially the solution of these problems on non-smooth domains, uses Sobolev space settings, for which a standard reference is McLean (2000). There is some variation in notations and definitions in respect of Sobolev spaces. We spell out our definitions precisely in the Appendix, mainly following McLean (2000), which has become a standard reference for the theory of BIEs on Lipschitz domains, but indicating explicitly wherever our notations and definitions vary from those of McLean, in which case we usually follow Sauter and Schwab (2011).

For many of the scattering problems we consider, formulations in spaces of continuous functions, or Hölder-continuous functions, are also possible, particularly where the boundary is sufficiently smooth. A standard reference for these is Colton and Kress (1983). We will indicate to what extent and under what conditions these formulations are equivalent in what follows. Some of the results that we wish to use and develop in this article are associated with a third setting for the BVPs, that associated with the harmonic analysis of Calderón–Zygmund operators on the boundaries of Lipschitz domains. There one studies BVPs with data in L^p spaces, understands boundary conditions to hold in a sense of almost everywhere non-tangential convergence, and supplements this with a requirement on the behaviour of maximal functions.

Much of the above material, in particular the deep results using harmonic analysis methods, is not available in an accessible form in the BIE literature, and their implications for BIE methods are not fully understood. At the same time a number of the recent results which we will describe, and which are key to a rigorous error analysis of the new boundary element methods that we will propose, make essential use of these methods of analysis. For this reason the Appendix includes a brief account of these results, which use function spaces specified in terms of maximal functions, and of the relationship of these function spaces to usual Sobolev spaces.

2.1. Acoustic boundary value problems

All the BVPs that we consider are for the Helmholtz equation (1.1) with wavenumber k, which arises in acoustics from the wave equation, satisfied by the perturbation in pressure on an assumption of harmonic ($\mathrm{e}^{-\mathrm{i}\omega t}$) time dependence. We will impose most often the *sound-soft* or Dirichlet boundary condition, namely that u is specified on the boundary Γ, which has been the focus of most of the theory and computation for hybrid numerical-asymptotic methods to date. We will also show numerical results and analysis for problems where an impedance boundary condition (much more widely used in acoustic modelling) is imposed. This is

$$\frac{\partial u}{\partial n} - \mathrm{i}k\beta u = h \tag{2.1}$$

on Γ, where n is the normal directed outwards from the domain of propagation, and β is the *normalized admittance* of the surface. In general β is a function of position (and generally also a function of frequency ω), given by $\beta = \rho c/\zeta$, where ζ is the *surface impedance*, and ρ the density of the fluid in which the acoustic wave propagates; the product ρc is termed the *impedance* of the medium of propagation. The h in (2.1) is given; usually $h = 0$ in scattering problems whenever u denotes the total field. Physical considerations, namely the requirement that the boundary does not emit energy, imply that

$$\operatorname{Re}\beta \geq 0. \tag{2.2}$$

In the special case $\beta = 0$, the impedance boundary condition reduces to the Neumann or *sound-hard* boundary condition. One of many examples where (2.1) is a widely used boundary condition is in outdoor noise propagation, an application we will return to in Section 3.2, where (2.1) with $\beta = 0$ is an appropriate boundary condition on a hard road surface, while (2.1), with $\operatorname{Re}\beta > 0$ and frequency dependent, is a widely used model of a range of sound-absorbing outdoor surfaces; *e.g.*, Taraldsen and Jonasson (2011).

This review contains some new BIE methods which give rise to the study of a generalization of the above boundary condition. Let Z be a bounded vector field defined (at least) on the boundary Γ. The generalization we make replaces the normal derivative in the boundary condition by an oblique derivative, so that the boundary condition is

$$Z \cdot \nabla u - \mathrm{i}k\beta u = h \tag{2.3}$$

on Γ.

We will formulate all of our problems in the Sobolev space setting, and selected problems also in alternative function space settings. In the interior problems below, D is a bounded Lipschitz domain (see Definition A.2 and the paragraph that follows that definition) with boundary Γ, and γ is the trace operator (see Section A.3). We state first the *interior Dirichlet problem*:

$$\begin{aligned}&\text{Given } h \in H^{1/2}(\Gamma), \text{ find } u \in C^2(D) \cap H^1(D)\\ &\text{such that (1.1) holds in } D \text{ and } \gamma u = h \text{ on } \Gamma.\end{aligned} \tag{2.4}$$

The solvability of this standard interior problem is well understood, *e.g.*, McLean (2000, p. 286).

Theorem 2.1. There exists a sequence $0 < k_1 < k_2 < \cdots$ of positive wavenumbers, with $k_m \to \infty$ as $m \to \infty$, such that the interior Dirichlet problem with $h = 0$ has a non-trivial solution. For all other values of $k > 0$ the interior Dirichlet problem has exactly one solution.

The other interior BVPs that are relevant to us are the interior impedance problem and what we will call *the interior oblique impedance problem*, by which we will mean the Helmholtz equation subject to the boundary condition (2.3). Note that, in these Sobolev space formulations, we understand the normal derivative of u in (2.1) as $\partial_n u \in H^{-1/2}(\Gamma)$, where ∂_n is the normal derivative operator defined in equation (A.28). (Of course – see (A.18) – $\partial_n u$ coincides with classical definition of the normal derivative $\partial u/\partial n$ when u is sufficiently regular.) We understand the oblique derivative $Z \cdot \nabla u$ as meaning

$$Z \cdot \nabla u = Z_n \partial_n u + Z \cdot \nabla_\Gamma \gamma u,$$

where $Z_n := Z \cdot n$ is the normal component of Z and ∇_Γ is the surface gradient operator (see (A.14)). Thus the *interior impedance problem* is

$$\begin{aligned}&\text{Given } h \in H^{-1/2}(\Gamma), \text{ and } \beta \in L^\infty(\Gamma), \text{ find } u \in C^2(D) \cap H^1(D)\\ &\text{such that (1.1) holds in } D \text{ and } \partial_n u - \mathrm{i}k\beta\gamma u = h \text{ on } \Gamma;\end{aligned} \tag{2.5}$$

and the *interior oblique impedance problem* is

$$\begin{aligned}&\text{Given } h \in L^2(\Gamma),\ Z \in (L^\infty(\Gamma))^d, \text{ and } \beta \in L^\infty(\Gamma), \text{ find}\\ &u \in C^2(D) \cap H^1(D), \text{ with } \gamma u \in H^1(\Gamma), \text{ such that (1.1)}\\ &\text{holds in } D \text{ and } Z_n\partial_n u + Z \cdot \nabla_\Gamma \gamma u - \mathrm{i}k\beta\,\gamma u = h \text{ on } \Gamma.\end{aligned} \tag{2.6}$$

Note that the requirement, in (2.5), that $\partial_n u = h + \mathrm{i}k\beta\gamma u$ holds, means, in view of the definition of the normal derivative operator ∂_n in (A.28) and (A.29), nothing more or less than the requirement that

$$\int_\Gamma h\,\gamma v\,\mathrm{d}s + \mathrm{i}k\int_\Gamma \beta\,\gamma u\,\gamma v\,\mathrm{d}s = \int_D \left(\nabla u \cdot \nabla \bar{v} + \bar{v}\,\Delta u\right)\mathrm{d}x, \quad v \in H^1(D), \tag{2.7}$$

in the case that $h \in L^2(\Gamma)$. In the case that $h \in H^{-1/2}(\Gamma)$ but $h \notin L^2(\Gamma)$, the integral $\int_\Gamma h\,\gamma v\,\mathrm{d}s$ in (2.7) is understood as the limit

$$\int_\Gamma h\,\gamma v\,\mathrm{d}s = \lim_{j\to\infty}\int_\Gamma h_j\,\gamma v\,\mathrm{d}s,$$

where $(h_j) \subset L^2(\Gamma)$ is any sequence which is convergent to h in the $H^{-1/2}(\Gamma)$ norm. The boundary condition in (2.6) has an analogous interpretation to that of (2.7).

The part of the solvability theory for the interior impedance BVP that we need (Colton and Kress 1983, McLean 2000) is encapsulated in the following theorem. In the statement of this theorem we make the first reference to the following assumption.

Assumption 2.2. Either: (a) $\operatorname{Re}\beta \geq 0$ and $\operatorname{Re}\beta > 0$ on some relatively open subset of Γ; or: (b) $\operatorname{Re}\beta \leq 0$ and $\operatorname{Re}\beta < 0$ on some relatively open subset of Γ.

Theorem 2.3.

(i) Suppose that Assumption 2.2 holds. Then the interior impedance problem has exactly one solution.

(ii) Suppose that $\beta = 0$ (the Neumann boundary case). Then there exists a sequence $0 = k_1 < k_2 < \cdots$ of positive wavenumbers, with $k_m \to \infty$ as $m \to \infty$, such that the interior impedance problem with $h = 0$ has a non-trivial solution. For all other values of $k > 0$ the interior impedance problem with $\beta = 0$ has exactly one solution.

The existence statement in (i) can be obtained by a variety of arguments, not least by BIE methods as a corollary of Theorem 2.30 below. The uniqueness statement in (i) is a consequence of (2.7). In more detail, if u_1 and u_2 are solutions of (2.5) then, defining $u := u_1 - u_2$, it follows from (2.7) applied with $v = u$, since $\Delta u = -k^2 u$, that

$$\mathrm{i}k \int_\Gamma \beta\, |\gamma u|^2\, \mathrm{d}s = \int_D (|\nabla u|^2 - k^2|u|^2)\, \mathrm{d}x. \tag{2.8}$$

Taking imaginary parts, we see that $\int_\Gamma \operatorname{Re}\beta|\gamma u|^2\, \mathrm{d}s = 0$, so that in case (i) we deduce that γu vanishes on some relatively open subset of Γ. Since $\partial_n u = \mathrm{i}k\beta\gamma u$, it follows that $\partial_n u$ vanishes on the same open subset of Γ. By the following version of Holmgren's uniqueness theorem it follows that u is identically zero and so $u_1 = u_2$.

Theorem 2.4. Suppose that D is a bounded Lipschitz open set and that $u \in C^2(D) \cap H^1(D)$ satisfies the Helmholtz equation (1.1) in D and $\gamma u = \partial_n u = 0$ on Γ_0, some non-empty, relatively open subset of $\Gamma = \partial D$. Then $u = 0$ (in D).

Proof. Let $G = (\mathbb{R}^d \setminus \Gamma) \cup \Gamma_0$, so that G is the open set which is the union of Γ_0 and the interior and exterior of Γ. Extend the definition of u from D to G by setting $u(x) = 0$ for $x \in G \setminus D$. Then it follows from Theorem 2.20 below that u in $G \setminus \Gamma_0$ is the difference of two layer potentials, each with density that vanishes on Γ_0. Hence, by Theorem 2.14, $u \in C^2(G)$ and (1.1) holds in G. But (see, *e.g.*, Colton and Kress 1983) solutions of the Helmholtz equation in a domain are real analytic in that domain and, if they vanish in some open subset, vanish identically. Thus $u = 0$ in the component of G that includes $D \cup \Gamma_0$, and in particular $u = 0$ in D. □

Of course, the focus of this article is on solving exterior problems in unbounded domains, for which radiation conditions need to be imposed to ensure uniqueness of solution, expressing mathematically the physical idea that any acoustic field is radiating away from the physical boundary. In the case when the Helmholtz equation (1.1) holds outside some bounded set, the standard radiation condition to impose is the *Sommerfeld radiation condition*, that

$$\frac{\partial u}{\partial r}(x) - \mathrm{i}ku(x) = o(r^{-(d-1)/2}), \tag{2.9}$$

as $r := |x| \to \infty$, uniformly in $\hat{x} := x/r$. The following lemma is an important consequence of this radiation condition; for a proof see, *e.g.*, Colton and Kress (1983). In this lemma, and the remainder of the article, we use the notations $B_R := \{x \in \mathbb{R}^d : |x| < R\}$, $\Gamma_R := \partial B_R = \{x : |x| = R\}$, and $\mathbb{S}^{d-1} := \{x \in \mathbb{R}^d : |x| = 1\}$.

Lemma 2.5. If, for some $R > 0$, u satisfies (1.1) in $\mathbb{R}^d \setminus \overline{B_R}$ and the Sommerfeld radiation condition (2.9), then there exists a function $F \in C^\infty(\mathbb{S}^{d-1})$ (the *far-field pattern*) such that

$$u(x) = \frac{\mathrm{e}^{ikr}}{r^{(d-1)/2}} \big(F(\hat{x}) + O(r^{-1})\big) \tag{2.10}$$

as $r \to \infty$, uniformly in $\hat{x} := x/r$.

A consequence of Lemma 2.5 is the following lemma, derived by applying Green's first identity (A.29) to u in the domain $B_R \cap \Omega_+$ and then letting $R \to \infty$. In this lemma and for the next paragraphs up to and including Theorem 2.10, we suppose that Ω_- is a bounded Lipschitz open set such that $\Omega_+ := \mathbb{R}^d \setminus \overline{\Omega_-}$ is a Lipschitz domain with boundary $\Gamma = \partial\Omega_+$. In this configuration, as noted in Sections 1 and A.5, the normal vector n will always be taken to point out from Ω_- into Ω_+, and $H^1_{\mathrm{loc}}(\Omega_+)$ is as defined in (A.30).

Lemma 2.6. If $u \in C^2(\Omega_+) \cap H^1_{\mathrm{loc}}(\Omega_+)$ and u satisfies the Helmholtz equation (1.1) in Ω_+ and the Sommerfeld radiation condition (2.9), then

$$\int_{\Omega_+} \big(|\nabla u|^2 - k^2|u|^2\big)\,\mathrm{d}x = I - \int_\Gamma \gamma\bar{u}\,\partial_n u\,\mathrm{d}s \tag{2.11}$$

where

$$I := \lim_{R\to\infty} \int_{\Gamma_R} \bar{u}\,\frac{\partial u}{\partial r}\,\mathrm{d}s = \mathrm{i}k \lim_{R\to\infty} \int_{\Gamma_R} |u|^2\,\mathrm{d}s = \mathrm{i}k \int_{\mathbb{S}^{d-1}} |F(\hat{x})|^2\,\mathrm{d}\hat{x},$$

and the integral over Ω_+ in (2.11) is understood as the improper integral $\lim_{R\to\infty} \int_{B_R\cap\Omega_+} \big(|\nabla u|^2 - k^2|u|^2\big)\,\mathrm{d}x$.

We will study, in the unbounded domain Ω_+, the *exterior Dirichlet problem*:

Given $h \in H^{1/2}(\Gamma)$, find $u \in C^2(\Omega_+) \cap H^1_{\mathrm{loc}}(\Omega_+)$ such that (1.1) holds in Ω_+, $\gamma u = h$ on Γ, and u satisfies the radiation condition (2.9). (2.12)

We will also study the *exterior impedance problem*:

Given $h \in H^{-1/2}(\Gamma)$ and $\beta \in L^\infty(\Gamma)$, find $u \in C^2(\Omega_+) \cap H^1_{\mathrm{loc}}(\Omega_+)$ such that (1.1) holds in Ω_+, $\partial_n u + \mathrm{i}k\beta u = h$ on Γ, and u satisfies the radiation condition (2.9). (2.13)

(The sign change in this impedance boundary condition compared to (2.1) is because the normal n here points *into* the domain of propagation.)

In contrast to the interior Dirichlet and Neumann problems, both problems (2.12) and (2.13) have at most one solution for all $k > 0$. This is a consequence of Lemma 2.8 below, which in turn follows from the following key lemma due to Rellich (for a proof see Colton and Kress 1983).

Lemma 2.7. (Rellich) If $u \in C^2(\Omega_+)$ satisfies (1.1) in Ω_+ and the radiation condition (2.9), and $\lim_{R\to\infty} \int_{\Gamma_R} |u|^2 \,\mathrm{d}s = 0$, then u is identically zero.

Rellich's lemma implies the following result (Colton and Kress 1983).

Lemma 2.8. If $u \in C^2(\Omega_+) \cap H^1_{\mathrm{loc}}(\Omega_+)$ satisfies (1.1) in Ω_+, the radiation condition (2.9), and

$$\mathrm{Im} \int_\Gamma \gamma \bar{u} \partial_n u \,\mathrm{d}s \leq 0,$$

then $u = 0$ (in Ω_+).

Proof. The conditions of the lemma imply, by Lemma 2.6, that

$$k \int_{\mathbb{S}^{d-1}} |F(\hat{x})|^2 \,\mathrm{d}\hat{x} = \mathrm{Im} \int_\Gamma \gamma \bar{u} \partial_n u \,\mathrm{d}s \leq 0, \tag{2.14}$$

so that $F = 0$ in $L^2(\mathbb{S}^{d-1})$. It follows from Lemma 2.6 that

$$\lim_{R\to\infty} \int_{\Gamma_R} |u|^2 \,\mathrm{d}s = 0$$

and hence, by Rellich's lemma, that $u = 0$. □

An immediate consequence of Lemma 2.8 is the following uniqueness result for the exterior Dirichlet and impedance problems.

Corollary 2.9. The exterior Dirichlet BVP (2.12) and the exterior impedance BVP (2.13), with $\mathrm{Re}\,\beta \geq 0$, each have at most one solution.

Corollary 2.9 is the uniqueness part of the following theorem.

Theorem 2.10. The exterior Dirichlet BVP and the exterior impedance BVP, with $\mathrm{Re}\,\beta \geq 0$, each have exactly one solution.

Proof. Existence follows from Corollary 2.28 below, which uses results about invertibility of boundary integral operators. □

Our prime interest in the above exterior Dirichlet and impedance problems is that they arise in the context of acoustic scattering. By acoustic scattering we mean the problem of computing the scattered acoustic field u^S produced when an incident field u^I interacts with an obstacle or obstacles (the *scatterer*) occupying some closed set Ω, such that $\Omega_+ := \mathbb{R}^d \setminus \Omega$ is an unbounded domain. By an incident field we mean the following.

Definition 2.11. We call $u^I \in L^1_{\mathrm{loc}}(\mathbb{R}^d)$ an *incident field* if, for some open neighbourhood G of Ω, $u^I|_G \in C^\infty(G)$ and satisfies the Helmholtz equation (1.1) in G. We will refer throughout to the sum $u := u^I + u^S$ as the *total acoustic field* (total field for short).

We deal exclusively with the case when Ω is bounded, except in Section 3.2. Further, except in Section 7.6, we deal exclusively with the case where $\Omega = \overline{\Omega_-}$, with Ω_-, the interior of Ω, a Lipschitz open set, in which case Γ, the surface of the scatterer, is the common boundary of Ω_+ and Ω_-.

We will focus particularly on the case when the incident field is a plane wave, that is, for some $\hat{a} \in \mathbb{R}^d$ with $|\hat{a}| = 1$,

$$u^I(x) = \exp(ikx \cdot \hat{a}), \quad x \in \mathbb{R}^d.$$

However, many of our results apply to general incident fields, for example the field generated by a point source, that is, for some $z \in \mathbb{R}^d \setminus \Omega$,

$$u^I(x) = \Phi_k(x, z), \quad x \in \mathbb{R}^d \setminus \{z\},$$

where Φ_k is the fundamental solution of the Helmholtz equation defined in (1.2).

In this article we will focus on the cases of the *sound-soft scatterer* (where $u = 0$ on Γ), the *impedance scatterer*, where (2.13) with $h = 0$ holds on Γ, and the *sound-hard scatterer*, the special case of the impedance scatterer in which $\partial u / \partial n = 0$ on Γ.

Considering first the *sound-soft scattering problem*, one natural formulation, where our requirement on u^S is continuity rather than membership of a particular Sobolev space, is the following:

$$\begin{aligned}&\text{Find } u^S \in C^2(\Omega_+) \cap C(\overline{\Omega_+}) \text{ such that (1.1) holds in } \Omega_+,\ u = 0 \\ &(\text{so } u^S = -u^I) \text{ on } \Gamma, \text{ and } u^S \text{ satisfies the radiation condition (2.9).}\end{aligned} \tag{2.15}$$

In the case that Ω_+ is a Lipschitz domain, a second natural formulation is to require that the scattered field u^S satisfy the exterior Dirichlet problem (2.12), with boundary data $-u^I|_\Gamma$. In other words u^S satisfies:

$$\begin{aligned}&\text{Find } u^S \in C^2(\Omega_+) \cap H^1_{\text{loc}}(\Omega_+) \text{ such that (1.1) holds in } \Omega_+,\ \gamma u = 0 \\ &(\text{so } \gamma u^S = -u^I|_\Gamma) \text{ on } \Gamma, \text{ and } u^S \text{ satisfies the radiation condition (2.9).}\end{aligned} \tag{2.16}$$

It is shown in Colton and Kress (1998, Theorem 3.7) that problem (2.15) has at most one solution, and the argument in the proof of that theorem shows that if u^S satisfies (2.15) and Ω_+ is Lipschitz, then u^S satisfies (2.16). Further, in the case that Ω_+ is Lipschitz, it is a corollary of Theorem 2.10 that (2.16) has exactly one solution which also satisfies (2.15); to see this it is a matter of showing that if u^S satisfies (2.16) then $u^S \in C(\overline{\Omega_+})$. This does not follow from the Sobolev embedding theorem (Theorem A.1), but is a consequence of elliptic regularity results up to the boundary; *e.g.*, Kenig (1994). Thus the following result holds.

Theorem 2.12. The Dirichlet scattering problem (2.15) has at most one solution. In the case that Ω_+ is a Lipschitz domain, (2.15) and (2.16) share the same unique solution, which satisfies $\partial_n u \in L^2(\Gamma)$.

Proof. In view of the discussion above it remains only to show that $\partial_n u \in L^2(\Gamma)$. But, by Definition 2.11, for some neighbourhood G of Γ, $u^I|_G \in C^\infty(G)$ so that $u^I|_\Gamma \in C^\infty(\Gamma) \subset H^1(\Gamma)$ (see Section A.3). Thus $\gamma u^S \in H^1(\Gamma)$, and the fact that $\partial_n u \in L^2(\Gamma)$ follows from Theorem A.5. □

We remark that, while it is common (following the lead of Colton and Kress 1983, for instance) to make use of the formulation (2.15) in the case when Ω_+ is sufficiently smooth, precisely when Ω_+ is $C^{1,\mu}$ for some $\mu \in (0,1]$, overwhelmingly the recent numerical analysis literature for the case of Lipschitz Ω_+ uses the formulation (2.16). Theorem 2.12 makes clear that the formulation (2.15) is a valid alternative even in the Lipschitz case.

Consider next the *impedance scattering problem* formulated as follows. Restricting attention to the case when Ω_+ is Lipschitz, we require that u^S satisfies the exterior impedance problem (2.13) with

$$h := -\left(\frac{\partial u^I}{\partial n} + ik\beta u^I\right)\bigg|_\Gamma \in L^2(\Gamma). \tag{2.17}$$

In other words, with h given by (2.17), u^S satisfies:

$$\begin{aligned}&\text{Find } u^S \in C^2(\Omega_+) \cap H^1_{\mathrm{loc}}(\Omega_+) \text{ such that (1.1) holds in}\\ &\Omega_+,\ \partial_n u + ik\beta\gamma u = 0 \text{ (so } \partial_n u^S + ik\beta\gamma u^S = h) \text{ on } \Gamma,\\ &\text{and } u^S \text{ satisfies the radiation condition (2.9).}\end{aligned} \tag{2.18}$$

Analogously to Theorem 2.12, an immediate corollary of Theorem 2.10 and Theorem A.5, is the following result.

Corollary 2.13. The impedance scattering problem (2.18) has exactly one solution, which satisfies $\partial_n u \in L^2(\Gamma)$ and $\gamma u \in H^1(\Gamma)$.

2.2. Layer potentials

In Section 2.5 we will formulate the BVPs in the previous subsection as BIEs. Here we introduce the required *layer potentials* and recall some of their properties.

Suppose that Ω_+ is an unbounded Lipschitz open set with boundary Γ, in which case $\Omega_- := \mathbb{R}^d \setminus \overline{\Omega_+}$ is a bounded Lipschitz open set. For $\phi \in L^1(\Gamma)$ and $k > 0$ let

$$\mathcal{S}_k\phi(x) := \int_\Gamma \Phi_k(x,y)\phi(y)\,\mathrm{d}s(y), \quad x \in \mathbb{R}^d \setminus \Gamma, \tag{2.19}$$

and

$$\mathcal{D}_k\phi(x) := \int_\Gamma \frac{\partial \Phi_k(x,y)}{\partial n(y)}\,\phi(y)\,\mathrm{d}s(y), \quad x \in \mathbb{R}^d \setminus \Gamma, \tag{2.20}$$

where the normal n is directed into Ω_+. Note that, in particular, these definitions apply for $\phi \in H^s(\Gamma)$ with $0 \leq s \leq 1$, since of course $H^s(\Gamma) \subset L^2(\Gamma) \subset L^1(\Gamma)$. Further, they extend in a natural way, via the duality pairing $\langle \cdot, \cdot \rangle_\Gamma$ defined by (A.24), to the case that $\phi \in H^s(\Gamma)$ for $-1 \leq s < 0$. In this case, for example, noting that $\Phi_k(x, \cdot) \in C^\infty(\Gamma) \subset H^{-s}(\Gamma)$, $\mathcal{S}_k\phi(x)$ is understood as $\mathcal{S}_k\phi(x) = \langle \Phi_k(x, \cdot), \phi \rangle_\Gamma$, and $\mathcal{D}_k\phi(x)$ is understood similarly. Equivalently, $\mathcal{S}_k\phi(x)$ and $\mathcal{D}_k\phi(x)$ can be understood as the limits

$$\mathcal{S}_k\phi(x) = \lim_{j\to\infty} \mathcal{S}_k\phi_j(x), \quad \mathcal{D}_k\phi(x) = \lim_{j\to\infty} \mathcal{D}_k\phi_j(x), \tag{2.21}$$

where $(\phi_j) \subset L^2(\Gamma)$ is any sequence converging in $H^s(\Gamma)$ to $\phi \in H^s(\Gamma)$.

We will use the above definitions also for $k = 0$ with Φ_0 given by (1.2) in the 3D case ($d = 3$). The definition (1.2) does not make sense when $d = 2$ and here we define Φ_0 by

$$\Phi_0(x, y) := \frac{1}{2\pi} \log\left(\frac{a}{|x - y|}\right), \tag{2.22}$$

for some constant $a > 0$ (the usual choice $a = 1$). In both 2D and 3D, Φ_0 is the standard fundamental solution for the Laplace equation.

Boundary integral equation methods for solving the BVPs of Section 2.1 are based on solutions in terms of layer potentials. This is effective because of the following simple result whose proof we sketch.

Theorem 2.14. For $k \geq 0$ and $\phi \in H^{-1}(\Gamma)$, $\mathcal{S}_k\phi, \mathcal{D}_k\phi \in C^2(\mathbb{R}^d \setminus \Gamma)$ and satisfy (1.1) in $\mathbb{R}^d \setminus \Gamma$. For $k > 0$, $\mathcal{S}_k\phi$ and $\mathcal{D}_k\phi$ satisfy the Sommerfeld radiation condition (2.9).

Proof. For a proof in the case that $\phi \in C(\Gamma)$, see Colton and Kress (1983). For $\mathcal{S}_k\phi$ the result follows for $\phi \in H^{-1}(\Gamma)$ since $C(\Gamma)$ is dense in $H^{-1}(\Gamma)$, so that there exists a sequence $(\phi_j) \subset C(\Gamma)$ with $\|\phi - \phi_j\|_{H^{-1}(\Gamma)} \to 0$ as $j \to \infty$, which implies that $\mathcal{S}_k\phi_j(x) \to \mathcal{S}_k\phi(x)$ uniformly on compact subsets of $\mathbb{R}^d \setminus \Gamma$, from which it follows that $\mathcal{S}_k\phi$ satisfies (1.1). In the case $k > 0$ also, for all sufficiently large R,

$$\max_{|x|\geq R} \bigl(|x|\, |\mathcal{S}_k\phi_j(x) - \mathcal{S}_k\phi(x)|\bigr) \to 0, \quad \text{as } j \to \infty,$$

from which it follows that $\mathcal{S}_k\phi$ satisfies the Sommerfeld radiation condition. The same arguments work for $\mathcal{D}_k\phi$ when $\phi \in H^{-1}(\Gamma)$. □

The above theorem, coupled with Lemma 2.5, implies that, for $k > 0$, $\mathcal{S}_k\phi$ and $\mathcal{D}_k\phi$ both have the representation (2.10) at infinity. In fact (see Colton and Kress 1998, or McLean 2000, p. 294), the far-field patterns of

$\mathcal{S}_k\phi$ and $\mathcal{D}_k\phi$ are, respectively,

$$F_{\mathcal{S}}(\hat{x}) = c_d\, k^{(d-3)/2} \int_\Gamma \exp(-ik\hat{x}\cdot y)\phi(y)\,\mathrm{d}s(y) \qquad (2.23)$$

and

$$F_{\mathcal{D}}(\hat{x}) = -\mathrm{i}\, c_d\, k^{(d-1)/2} \int_\Gamma \hat{x}\cdot n(y)\exp(-ik\hat{x}\cdot y)\phi(y)\,\mathrm{d}s(y), \qquad (2.24)$$

where

$$c_d = \frac{\mathrm{e}^{-\mathrm{i}(d-3)\pi/4}}{2(2\pi)^{(d-1)/2}}.$$

To derive BIEs for scattering problems, we need to supplement the above result with mapping properties of $\mathcal{S}_k$ and $\mathcal{D}_k$. In the following theorem, $\chi \in C_0^\infty(\mathbb{R}^d)$ is any smooth compactly supported function such that $\chi(x) = 1$ in a neighbourhood of $\Omega_- \cup \Gamma$ and, for example, $\chi\mathcal{S}_k$ denotes the composition of the operator $\mathcal{S}_k$ followed by multiplication by χ. A consequence of Theorem 2.15 is that $\mathcal{S}_k\phi \in H^1_{\mathrm{loc}}(\mathbb{R}^d)$ and that $\mathcal{D}_k\phi \in H^1_{\mathrm{loc}}(\Omega_\pm)$. (From (2.23) and (2.24) it follows that $\mathcal{S}_k\phi$ and $\mathcal{D}_k\phi$ decay too slowly at infinity for either to be in $H^1(\Omega_+)$.)

Theorem 2.15. For $-\frac{1}{2} \le s \le \frac{1}{2}$ and $k \ge 0$, the following mappings are bounded:

(i) $\chi\mathcal{S}_k : H^{s-1/2}(\Gamma) \to H^{s+1}(\mathbb{R}^d)$;

(ii) $\chi\mathcal{D}_k : H^{s+1/2}(\Gamma) \to H^{s+1}(\Omega_\pm)$.

A proof of the above theorem for the restricted range $|s| < 1/2$, derived from the key paper by Costabel (1988), is given in McLean (2000). It depends on characterizations of the single- and double-layer potential operators in terms of the adjoints of the trace and normal derivative operators, respectively, and, in particular, depends on (A.17), which holds only for $1/2 < s < 3/2$ and not for $s = 1/2$ (even for smooth boundaries Γ). Thus the proof does not extend to $s = \pm 1/2$. This is of significance for us since, for the most part, it will be precisely the limiting cases $s = \pm 1/2$ that are of interest.

To establish the mapping properties of Theorem 2.15 for $s = \pm 1/2$ (which, as McLean notes, imply the same mapping properties for $|s| < 1/2$ by interpolation), we can use the equivalences of Corollary A.8. These, together with the standard interpolation results, imply that Theorem 2.15 is equivalent to Theorem 2.16 below. In this theorem, the so-called *non-tangential maximal functions*, u^* and $(\nabla u)^*$, are defined as in (A.36) and (A.37), except that now $\Theta(x) := \Theta_+(x) \cup \Theta_-(x)$, where $\{\Theta_\pm(x) : x \in \Gamma\}$ denotes any uniform and sufficient family of non-tangential approach sets in $\Omega_\pm$, so that $\Theta_\pm(x) \subset \Omega_\pm$, $\overline{\Theta_\pm(x)} \cap \Gamma = \{x\}$, and (A.32) holds with the same constant $C > 1$ for every $x \in \Gamma$.

Theorem 2.16. For $\phi, \psi \in H^{-1}(\Gamma)$ let $u = \mathcal{S}_k\phi$ and $v = \mathcal{D}_k\psi$. Then, for some constant $C > 0$ independent of ϕ and ψ:

(i) if $\phi \in L^2(\Gamma)$ then $(\nabla u)^* \in L^2(\Gamma)$ and $\|(\nabla u)^*\|_{L^2(\Gamma)} \leq C\|\phi\|_{L^2(\Gamma)}$;

(ii) if $\psi \in L^2(\Gamma)$ then $v^* \in L^2(\Gamma)$ and $\|v^*\|_{L^2(\Gamma)} \leq C\|\psi\|_{L^2(\Gamma)}$;

(iii) if $\phi \in H^{-1}(\Gamma)$ then $u^* \in L^2(\Gamma)$ and $\|u^*\|_{L^2(\Gamma)} \leq C\|\phi\|_{H^{-1}(\Gamma)}$;

(iv) if $\psi \in H^1(\Gamma)$ then $(\nabla v)^* \in L^2(\Gamma)$ and $\|(\nabla v)^*\|_{L^2(\Gamma)} \leq C\|\psi\|_{H^1(\Gamma)}$.

The proof of this theorem requires deep results from the harmonic analysis literature, specifically that part concerned with the study of Calderón–Zygmund operators, in particular Cauchy integral and layer-potential operators on Lipschitz curves and surfaces.

For a proof of (i) and (ii) we refer the reader to the account in Meyer and Coifman (2000) for the case $k = 0$ and Torres and Welland (1993) and Liu (1995) for the extension to $k > 0$. It is convenient to leave remarks on the proof of (iii) and (iv) (which are corollaries of (i) and (ii)) until we introduce relevant boundary integral operators in the next subsection. Note that by far the main part of the work is to establish results for $k = 0$. Extensions from $k = 0$ to $k > 0$ by perturbation arguments are relatively straightforward because $W_k := \Phi_k - \Phi_0$ is much smoother than Φ_0. Precisely, it follows easily from (1.2) and the power series expansions for Hankel functions that $W_k(x, y) = w_k(x - y)$ where $w_k \in C^\infty(\mathbb{R}^d \setminus \{0\})$ and, for some constant $c_k > 0$,

$$|w_k(x)| + |\nabla w_k(x)| \leq c_k, \quad |\partial^\alpha w_k(x)| \leq \begin{cases} c_k \log\left(|x|^{-1}\right), & d = 2, \\ c_k|x|^{-1}, & d = 3, \end{cases} \tag{2.25}$$

for $|x| \leq 1/2$. Here α is any multi-index with $|\alpha| = 2$ so that $\partial^\alpha w_k$ is any second-order partial derivative (we use here the notation (A.1)). The monograph of Colton and Kress (1983) is a classic example of arguing by perturbation from $k = 0$ to obtain results for $k > 0$; see also Torres and Welland (1993), Liu (1995) and Mitrea (1996).

As in Section A.5, we denote the exterior and interior trace operators by γ_+ and γ_-, and the exterior and interior normal derivative operators by ∂_n^+ and ∂_n^-, respectively, with the normal directed out of Ω_- into the exterior domain Ω_+. Applying Theorem 2.15 with $s = 0$ and Theorem 2.14, we see that, for $\phi \in H^{-1/2}(\Gamma)$ and $\psi \in H^{1/2}(\Gamma)$, both $\chi\mathcal{S}_k\phi$ and $\chi\mathcal{D}_k\psi$ are in $H^1(\Omega_\pm)$, in fact in $H^1(\Omega_\pm; \Delta)$. (This notation is defined below equation (A.26).) It follows from (A.17) that the traces $\gamma_\pm\mathcal{S}_k\phi$ and $\gamma_\pm\mathcal{D}_k\psi$ are well-defined as elements of $H^{1/2}(\Gamma)$, and that (see (A.28)) the normal derivatives $\partial_n^\pm\mathcal{S}_k\phi$ and $\partial_n^\pm\mathcal{D}_k\psi$ are well-defined as elements of $H^{-1/2}(\Gamma)$. Moreover (McLean 2000) the following jump relations hold:

$$\gamma_+\mathcal{S}_k\phi = \gamma_-\mathcal{S}_k\phi, \quad \partial_n^+\mathcal{D}_k\psi = \partial_n^-\mathcal{D}_k\psi. \tag{2.26}$$

A natural question to ask is whether the traces and normal derivatives of the single- and double-layer potentials, $u = \mathcal{S}_k\phi$ and $v = \mathcal{D}_k\psi$, have anything to do with the limiting values of $u(x)$, $v(x)$, $\nabla u(x)$ and $\nabla v(x)$ as x approaches Γ from inside $\Omega_\pm$. Reassuringly, this is the case, at least if the densities ϕ and ψ are smooth enough. As before Theorem 2.16, for $x \in \Gamma$ let $\Theta_\pm(x)$ denote any non-tangential approach set to x from $\Omega_\pm$, so that $\Theta_\pm(x) \subset \Omega_\pm$, $\Theta_\pm(x) \cap \Gamma = \{x\}$, and (A.32) holds. Then, if $\phi \in H^{s-1/2}(\Gamma)$ and $\psi \in H^{s+1/2}(\Gamma)$, for some $s \in (-1/2, 1/2]$, it follows from Lemma A.9 that, for almost every $x \in \Gamma$,

$$\gamma_\pm u(x) = \lim_{y\to x,\, y\in\Theta_\pm(x)} u(y) \quad \text{and} \quad \gamma_\pm v(x) = \lim_{y\to x,\, y\in\Theta_\pm(x)} v(y). \tag{2.27}$$

Further, these non-tangential limits are well-defined, by Theorem 2.16 and Corollary A.8, even for $\phi \in H^{-1}(\Gamma)$ and $\psi \in L^2(\Gamma)$, providing an extension of the notion of the traces of u and v to the cases of densities $\phi \in H^{-1}(\Gamma)$ and $\psi \in L^2(\Gamma)$.

Similarly, if $\phi \in L^2(\Gamma)$ and $\psi \in H^1(\Gamma)$ then, by Theorem 2.16 and Lemma A.10, $\partial_n^\pm u, \partial_n^\pm v \in L^2(\Gamma)$ and, for almost all $x \in \Gamma$,

$$\partial_n^\pm u(x) = \frac{\partial u_\pm}{\partial n}(x) := \lim_{y\to x,\, y\in\Theta_\pm(x)} n(x)\cdot\nabla u(y) \tag{2.28}$$

and

$$\partial_n^\pm v(x) = \frac{\partial v_\pm}{\partial n}(x) := \lim_{y\to x,\, y\in\Theta_\pm(x)} n(x)\cdot\nabla v(y). \tag{2.29}$$

Further, for almost all $x \in \Gamma$, by Theorem 2.16 and Lemma A.10, $\nabla_\Gamma \gamma u \in L^2(\Gamma)$ and

$$\lim_{y\to x, y\in\Theta_\pm(x)} \nabla u(y) = \nabla_\Gamma \gamma_\pm u(x) + n(x)\partial_n^\pm u(x) \tag{2.30}$$

and

$$\lim_{y\to x, y\in\Theta_\pm(x)} \nabla v(y) = \nabla_\Gamma \gamma_\pm v(x) + n(x)\partial_n^\pm v(x). \tag{2.31}$$

2.3. Boundary integral operators

For $k \geq 0$ we define the *acoustic single-* and *double-layer operators*, S_k and D_k, respectively, by

$$\begin{aligned} S_k\phi(x) &:= \int_\Gamma \Phi_k(x,y)\phi(y)\,\mathrm{d}s(y), \\ D_k\phi(x) &:= \int_\Gamma \frac{\partial\Phi_k(x,y)}{\partial n(y)}\phi(y)\,\mathrm{d}s(y), \end{aligned} \tag{2.32}$$

for $x \in \Gamma$, where Φ_k is given by (1.2) in 3D and in 2D by (1.2) for $k > 0$ and by (2.22) for $k = 0$.

When Γ and ϕ are both sufficiently smooth, it is well known that the above integrals are well-defined (the integrands are in $L^1(\Gamma)$) for all $x \in \Gamma$. In particular (Colton and Kress 1983), this is the case if Γ is C^2 and $\phi \in C(\Gamma)$, with both $S_k\phi, D_k\phi \in C(\Gamma)$. In the 2D case, with $\phi \in L^2(\Gamma)$ and Γ Lipschitz, it still holds (by Cauchy–Schwarz, since $\Phi_k(x,\cdot) \in L^2(\Gamma)$ when $d = 2$), that $S_k\phi(x)$ is well-defined for all $x \in \Gamma$ and $S_k\phi \in C(\Gamma)$.

For Lipschitz Γ, the situation is more delicate for the double-layer potential, but for $\phi \in L^2(\Gamma)$, irrespective of the dimension d, $S_k\phi$ and $D_k\phi$ are well-defined by (2.32) for almost all $x \in \Gamma$, with $D_k\phi$ understood as the Cauchy principal value integral

$$D_k\phi(x) := \lim_{\epsilon\to 0} \int_{\Gamma\setminus B_\epsilon(x)} \frac{\partial\Phi_k(x,y)}{\partial n(y)}\phi(y)\,\mathrm{d}s(y), \quad x \in \Gamma, \tag{2.33}$$

where $B_\epsilon(x)$ is the open ball of radius ϵ centred at x and $S_k\phi, D_k\phi \in L^2(\Gamma)$. This result for the double-layer potential is not straightforward and was established first for the case $k = 0$; see, *e.g.*, Meyer and Coifman (2000) and the discussion in the following paragraphs. The extension to $k > 0$ is more straightforward; see Torres and Welland (1993).

As is well known, and will be recalled below, $S_k\phi$ and $D_k\phi$ appear when we take boundary values of the single- and double-layer potentials $\mathcal{S}_k\phi$ and $\mathcal{D}_k\phi$. When we apply the normal derivative operator ∂_n, two additional boundary integral operators, D'_k and H_k, arise, which we will term the (*acoustic*) *adjoint double-layer operator* and the (*acoustic*) *hypersingular operator*, respectively. For $\phi \in L^2(\Gamma)$ and $\psi \in H^1(\Gamma)$ these operators are given explicitly by

$$D'_k\phi(x) := \int_\Gamma \frac{\partial\Phi_k(x,y)}{\partial n(x)}\phi(y)\,\mathrm{d}s(y), \tag{2.34}$$

$$H_k\psi(x) := \frac{\partial}{\partial n(x)}\int_\Gamma \frac{\partial\Phi_k(x,y)}{\partial n(y)}\psi(y)\,\mathrm{d}s(y), \tag{2.35}$$

for almost all $x \in \Gamma$, where the integral defining $D'_k\phi$ is understood as a Cauchy principal value integral (as in (2.33)) while $H_k\psi \in L^2(\Gamma)$ is defined in the sense of (2.29), that is,

$$H_k\psi(x) := \lim_{y\to x,\, y\in\Theta_\pm(x)} n(x)\cdot\nabla\mathcal{D}_k\psi(y). \tag{2.36}$$

Note that for $\psi \in H^{1/2}(\Gamma)$, $\partial_n^+\mathcal{D}_k\psi = \partial_n^-\mathcal{D}_k\psi$ (see (2.26)). It follows from (2.29) that the limits in (2.36) as $y \to x$ from $\Theta_+(x)$ and from $\Theta_-(x)$ coincide.

It is a consequence of Young's inequality for convolutions that S_k is a bounded operator on $L^2(\Gamma)$. A much deeper result is that both D_k and D'_k are bounded operators on $L^2(\Gamma)$; for a clear discussion of the proofs of these results for $k = 0$ see Meyer and Coifman (2000), and see Torres and Welland (1993) for the relatively straightforward extensions to $k > 0$. This boundedness was established only in 1982 as a corollary of the proof of Coifman, McIntosh and Meyer (1982) that the Cauchy integral operator is bounded on $L^2(\Gamma)$ when Γ is the graph of a Lipschitz-continuous function. It was shown soon afterwards, by Verchota (1984), that, in the case $k = 0$, D_k is also a bounded operator on $H^1(\Gamma)$ and S_k a bounded operator from $L^2(\Gamma)$ to $H^1(\Gamma)$; again, for extensions to $k > 0$ see Torres and Welland (1993). The results in Verchota (1984) were achieved through the use of Rellich-type identities, which we will make new use of in Section 5.7 (see the discussion in Section 5.3).

Important identities, which follow by Fubini's theorem for S_k, and by the arguments of Meyer and Coifman (2000) plus Fubini's theorem (to move from $k = 0$ to $k > 0$) for D_k and D_k', are that, for $\phi, \psi \in L^2(\Gamma)$,

$$\int_\Gamma \phi\, S_k \psi \,\mathrm{d}s = \int_\Gamma \psi\, S_k \phi \,\mathrm{d}s, \qquad \int_\Gamma \phi\, D_k \psi \,\mathrm{d}s = \int_\Gamma \psi\, D_k' \phi \,\mathrm{d}s. \tag{2.37}$$

For $k = 0$ (when the kernels of the operators are real) this is precisely a statement that, as operators on the Hilbert space $L^2(\Gamma)$, D_k' is the adjoint of D_k and S_k is self-adjoint. To frame these identities as statements about adjoints for $k > 0$, let $\mathcal{C} : H^s(\Gamma) \to H^s(\Gamma)$ denote the operation of complex conjugation, that is,

$$\mathcal{C}u(x) := \overline{u(x)}, \quad x \in \Gamma,$$

so that $\mathcal{C}$ is an anti-linear bounded operator on $H^s(\Gamma)$ for $|s| \le 1$. Then, if A^* denotes the adjoint of a bounded linear operator A on $L^2(\Gamma)$, it follows from (2.37) that

$$S_k^* = \mathcal{C} S_k \mathcal{C}, \quad D_k^* = \mathcal{C} D_k' \mathcal{C}. \tag{2.38}$$

Combined with standard results on adjoints of operators on Hilbert spaces, this has simple but important consequences, for example that

$$\|D_k\|_{L^2(\Gamma)\leftarrow L^2(\Gamma)} = \|D_k^*\|_{L^2(\Gamma)\leftarrow L^2(\Gamma)} = \|D_k'\|_{L^2(\Gamma)\leftarrow L^2(\Gamma)}. \tag{2.39}$$

A further important consequence follows from writing (2.37) in terms of the duality pairing (A.24), as

$$\langle S_k\phi, \psi\rangle_\Gamma = \langle \phi, S_k^*\psi\rangle_\Gamma, \quad \langle D_k\phi, \psi\rangle_\Gamma = \langle \phi, D_k^*\psi\rangle_\Gamma. \tag{2.40}$$

These identities hold in the first instance just for $\phi, \psi \in L^2(\Gamma)$. But, since S_k is a bounded operator from $L^2(\Gamma)$ to $H^1(\Gamma)$, the first of these identities can be extended to $\phi \in L^2(\Gamma)$, $\psi \in H^{-1}(\Gamma)$, and used to show (together with (2.38)) that S_k extends to a bounded operator from $H^{-1}(\Gamma)$ to $L^2(\Gamma)$. Similarly, since D_k is a bounded operator on $H^1(\Gamma)$ as well as on $L^2(\Gamma)$, the second of these identities implies that D_k' extends to a bounded linear operator on $H^{-1}(\Gamma)$. These remarks sketch the proof of most of the following result.

Theorem 2.17. For $|s| \le 1/2$ and $k \ge 0$ the following mappings are bounded:

$$\begin{aligned} S_k &: H^{s-1/2}(\Gamma) \to H^{s+1/2}(\Gamma), \\ D_k &: H^{s+1/2}(\Gamma) \to H^{s+1/2}(\Gamma), \\ D'_k &: H^{s-1/2}(\Gamma) \to H^{s-1/2}(\Gamma). \end{aligned}$$

Proof. We have shown the above mappings for the limiting cases $s = \pm 1/2$. The results for intermediate values of s follow by interpolation (see, *e.g.*, the introduction and Theorems B.2 and B.11 in Appendix B of McLean (2000)). □

The connection between the above boundary integral operators and the operators $\mathcal{S}_k$ and $\mathcal{D}_k$ is obtained via an extended version of the jump relations (2.26). From McLean (2000) we have, on $H^{-1/2}(\Gamma)$,

$$\gamma_\pm \mathcal{S}_k = S_k, \quad \partial_n^\pm \mathcal{S}_k = \mp\tfrac{1}{2}I + D'_k, \tag{2.41}$$

where I is the identity operator. Similarly, on $H^{1/2}(\Gamma)$, we have

$$\gamma_\pm \mathcal{D}_k = \pm\tfrac{1}{2}I + D_k. \tag{2.42}$$

For $\psi \in H^1(\Gamma)$ we have, from (2.26), (2.36) and (2.29), that

$$H_k\psi = \partial_n^\pm \mathcal{D}_k\psi. \tag{2.43}$$

To see that H_k extends (uniquely) to a bounded operator from $H^{s+1/2}(\Gamma)$ to $H^{s-1/2}(\Gamma)$, for $|s| \le 1/2$, and in particular that (2.43) holds for $\psi \in H^{1/2}(\Gamma)$, our method is again to show this result first for $k = 0$ and then make a perturbation argument. To obtain the result for $k = 0$ a convenient route is to use the result of Verchota (1984) that $S_0 : L^2(\Gamma) \to H^1(\Gamma)$ is a bijection. (There is a subtlety in dimension 2: we have to choose a in (2.22) so that it does not equal the so-called *capacity* of Γ. For example, choosing a larger than the diameter of Γ is sufficient. See Chapter 8 in McLean (2000) and Section 4 of Verchota (1984).) This implies by duality that also $S_0^* = S_0 : H^{-1}(\Gamma) \to L^2(\Gamma)$ is a bijection, and hence, by interpolation, that $S_0 : H^{s-1/2}(\Gamma) \to H^{s+1/2}$ is a bijection for $|s| \le 1/2$. Uniqueness for the interior Dirichlet problem for Laplace's equation and the trace results (2.41) and (2.42) imply that, for $\psi \in H^1(\Gamma)$,

$$\mathcal{D}_0\psi(x) = -\mathcal{S}_0 S_0^{-1}\big(\tfrac{1}{2}I - D_0\big)\psi(x), \quad x \in \Omega_- \tag{2.44}$$

(a similar argument is used in Verchota 1984). By density of $H^1(\Gamma)$ in $L^2(\Gamma)$, and that both the left and right sides of this equation depend continuously on $\psi \in L^2(\Gamma)$, it follows that (2.44) in fact holds for all $\psi \in L^2(\Gamma)$.

The identity (2.44) firstly establishes (iv) in Theorem 2.16, as a corollary of (i) (see the remarks following Theorem 2.16). Then, recalling that Verchota (1984) also tells us that $\frac{1}{2}I - D_0$ is a bijection on $H^s(\Gamma)$ for $s = 0$ and

1, and hence by interpolation for all $0 \leq s \leq 1$, we can rewrite (2.44) as

$$\mathcal{S}_0\phi(x) = -\mathcal{D}_0\left(\tfrac{1}{2}I - D_0\right)^{-1} S_0\phi(x), \quad x \in \Omega_-, \tag{2.45}$$

for $\phi \in H^{-1}(\Gamma)$. Hence we are able to deduce (iii) in Theorem 2.16 as a corollary of (ii). Finally, combining (2.41), (2.42), and (2.44), we see that, on $H^1(\Gamma)$,

$$H_0 = -\partial_n^- \mathcal{S}_0 S_0^{-1}\left(\tfrac{1}{2}I - D_0\right) = -\left(\tfrac{1}{2}I + D_0'\right) S_0^{-1}\left(\tfrac{1}{2}I - D_0\right). \tag{2.46}$$

This identity, combined with the above observation that $S_0 : H^{s-1/2}(\Gamma) \to H^{s+1/2}(\Gamma)$ is a bijection, and Theorem 2.17, allows us to extend the domain of definition of H_0, and implies, together with the bounds (2.25), the following result.

Theorem 2.18. For $|s| \leq 1/2$, the hypersingular operator

$$H_k : H^{s+1/2}(\Gamma) \to H^{s-1/2}(\Gamma),$$

and this mapping is bounded.

2.4. Green's representation theorems

Our main starting point for our numerical schemes will be integral equations obtained from Green's theorems. We begin with the following simple consequence of (A.29).

Theorem 2.19. (Green's second formula) Suppose that D is a Lipschitz open set and that $u, v \in H^1(D; \Delta)$. Then

$$\int_\Gamma \left(\gamma v \partial_n u - \gamma u \partial_n v\right) \mathrm{d}s = \int_D \left(v\Delta u - u\Delta v\right) \mathrm{d}x.$$

From this theorem we deduce *Green's representation theorems.* As in the previous subsection, Ω_- is a bounded Lipschitz open set, and $\Omega_+ = \mathbb{R}^d \setminus \overline{\Omega_-}$ is assumed to be connected, and so is an unbounded Lipschitz domain, and the trace and normal derivative operators, $\gamma_\pm$ and $\partial_n^\pm$, are as defined in Section A.5.

Theorem 2.20. If $u \in H^1(\Omega_-) \cap C^2(\Omega_-)$ and, for some $k \geq 0$, $\Delta u + k^2 u = 0$ in Ω_-, then

$$\mathcal{S}_k \partial_n^- u(x) - \mathcal{D}_k \gamma_- u(x) = \begin{cases} u(x) & x \in \Omega_-, \\ 0 & x \in \Omega_+. \end{cases} \tag{2.47}$$

Proof. For $x \in \Omega_+$ this is an immediate consequence of Theorem 2.19, applied with $D = \Omega_-$ and $v = \Phi_k(\cdot, x)$. For $x \in \Omega_-$, the application of Theorem 2.19 is in Ω_- with a small ball of radius ϵ removed, and the theorem follows on taking the limit $\epsilon \to 0$; see Colton and Kress (1983) for details. □

The following is the version of Green's representation theorem which holds in exterior domains. It is shown by applying Theorem 2.20, with Ω_- replaced by the part of Ω_+ contained in a large ball of radius R, and then letting $R \to \infty$. The integral around the boundary of the large ball vanishes in this limit as a consequence of the Sommerfeld radiation condition (2.9) satisfied by u and by $\Phi_k(\cdot, x)$; see Colton and Kress (1983) for details.

Theorem 2.21. If $u \in H^1_{\mathrm{loc}}(\Omega_+) \cap C^2(\Omega_+)$ and, for some $k > 0$, $\Delta u + k^2 u = 0$ in Ω_+ and u satisfies the Sommerfeld radiation condition (2.9) in Ω_+, then

$$-\mathcal{S}_k \partial_n^+ u(x) + \mathcal{D}_k \gamma_+ u(x) = \begin{cases} u(x) & x \in \Omega_+, \\ 0 & x \in \Omega_-. \end{cases} \tag{2.48}$$

2.5. Boundary integral equation formulations

We have set up the tools that we need to derive BIE formulations of the BVPs in Section 2.1. Our main tool for computation will be so-called *direct* BIE formulations, namely, integral equation formulations derived from the Green's representation theorems in which the solution to the integral equation is either the trace or normal derivative of the solution to the BVP. The starting point to derive these BIEs are the representation formulae in Theorems 2.20 and 2.21.

To describe the integral equations succinctly, we define the matrices of operators

$$P_\pm = \pm \begin{bmatrix} \gamma_\pm \mathcal{D}_k & -\gamma_\pm \mathcal{S}_k \\ \partial_n^\pm \mathcal{D}_k & -\partial_n^\pm \mathcal{S}_k \end{bmatrix}. \tag{2.49}$$

These are the so-called *Calderón projectors*; we will see shortly that these are indeed projection operators (for example on the Hilbert space $H^{1/2}(\Gamma) \times H^{-1/2}(\Gamma)$). Applying the jump relations, (2.41), (2.42) and (2.43), we find that

$$P_\pm = \tfrac{1}{2} I \pm M_k, \tag{2.50}$$

where I is the (2×2 matrix) identity operator and M_k is the matrix of boundary integral operators

$$M_k = \begin{bmatrix} D_k & -S_k \\ H_k & -D'_k \end{bmatrix}. \tag{2.51}$$

To see where P_+ arises, we apply the trace operator γ_+ and then the normal derivative operator ∂_n^+ to (2.48) to obtain two equations which we can write in matrix form as

$$P_+ c_+ u = c_+ u, \tag{2.52}$$

where $c_+ u = [\gamma_+ u, \partial_n^+ u]^T$ is the Cauchy data for u on Γ. Explicitly, using

(2.50), these two equations are

$$(D_k - \tfrac{1}{2}I)\gamma_+ u - S_k\partial_n^+ u = 0 \tag{2.53}$$

and

$$H_k\gamma_+ u - (D_k' + \tfrac{1}{2}I)\partial_n^+ u = 0, \tag{2.54}$$

each a linear relationship between the components $\gamma_+ u$ and $\partial_n^+ u$ of the Cauchy data $c_+ u$. Similar relationships between the components of the Cauchy data $c_- u = [\gamma_- u, \partial_n^- u]^T$ are obtained by applying γ_- and ∂_n^- to (2.47). We summarize these key results in the following lemma.

Lemma 2.22. If $u \in H^1(\Omega_-) \cap C^2(\Omega_-)$ and, for some $k \geq 0$, $\Delta u + k^2 u = 0$ in Ω_-, then $P_-\, c_- u = c_- u$. Similarly, if $u \in H^1_{\mathrm{loc}}(\Omega_+) \cap C^2(\Omega_+)$ and, for some $k > 0$, $\Delta u + k^2 u = 0$ in Ω_+ and u satisfies the Sommerfeld radiation condition (2.9) in Ω_+, then $P_+\, c_+ u = c_+ u$.

Of course, this lemma implies that if, for some $k > 0$, $u \in H^1(\Omega_-) \cap H^1_{\mathrm{loc}}(\Omega_+) \cap C^2(\mathbb{R}^d \backslash \Gamma)$, $\Delta u + k^2 u = 0$ in $\mathbb{R}^d \backslash \Gamma$, and u satisfies the Sommerfeld radiation condition, then $P_\pm\, c_\pm u = c_\pm u$. In particular, by Theorems 2.14 and 2.15, the following lemma holds.

Lemma 2.23. If $u = \mathcal{D}_k\phi_1 - \mathcal{S}_k\phi_2$ with $\phi_1 \in H^{1/2}(\Gamma)$, $\phi_2 \in H^{-1/2}(\Gamma)$, then

$$P_\pm\, c_\pm u = c_\pm u.$$

Further, writing $\phi = [\phi_1, \phi_2]^T$, it follows immediately from the definition (2.49) that

$$c_\pm u = \pm P_\pm \phi,$$

so that

$$P_\pm^2 \phi = \pm P_\pm c_\pm u = \pm c_\pm u = P_\pm \phi.$$

This lemma confirms that $P_\pm$ are projection operators on $H^{1/2}(\Gamma) \times H^{-1/2}(\Gamma)$, *i.e.*, that $P_\pm^2 = P_\pm$. Further, in view of (2.50), this projection property is equivalent to the very useful identity

$$M_k^2 = \tfrac{1}{4}I, \tag{2.55}$$

which, written out in component form, is

$$S_k H_k = D_k^2 - \tfrac{1}{4}I, \quad D_k S_k = S_k D_k', \quad H_k D_k = D_k' H_k, \tag{2.56}$$

plus a further identity, obtained from the first identity by taking adjoints.

Lemma 2.22 is the basis for all the standard direct BIE formulations for interior and exterior acoustic BVPs. For example, if u satisfies the exterior Dirichlet problem (2.12), then it follows immediately from (2.52), in component form (2.53) and (2.54), that $\partial_n^+ u$ satisfies both

$$S_k\partial_n^+ u = (D_k - \tfrac{1}{2}I)h \tag{2.57}$$

and

$$\left(D_k' + \tfrac{1}{2}I\right)\partial_n^+ u = H_k h. \tag{2.58}$$

Similarly, if u satisfies the interior Dirichlet problem (2.4), then, from Lemma 2.22, in particular from the equation $P_- c_- u = c_- u$, it follows that

$$S_k \partial_n^- u = \left(D_k + \tfrac{1}{2}I\right)h \tag{2.59}$$

and

$$\left(D_k' - \tfrac{1}{2}I\right)\partial_n^- u = H_k h. \tag{2.60}$$

Let us make two simple observations here. Firstly, all these equations are BIEs of the form

$$A\phi = \psi, \tag{2.61}$$

where A is a linear combination of boundary integral operators and the identity, ϕ is the solution to be determined and ψ is given data. Secondly we observe that the same operator A can arise from both interior and exterior problems. This has the important implication that, although exterior acoustic problems are generically uniquely solvable (this holds in particular for the exterior Dirichlet and Neumann/impedance problems that we focus on in this article: see Theorem 2.10), the natural BIE formulations of these problems need not be uniquely solvable for all wavenumbers k. For example, Theorem 2.1 shows that the homogeneous interior Dirichlet problem has non-trivial solutions at a sequence k_m of positive wavenumbers. If $k = k_m$ and u is such a solution then $\partial_n^- u$ is a solution of (2.59) with $h = 0$ (a non-trivial solution by Theorem 2.4 which implies that $\partial_n^- u \neq 0$), and so, for $k = k_m$, the BIE (2.57) for the exterior Dirichlet problem has infinitely many solutions (in $H^{-1/2}(\Gamma)$).

In (2.57)–(2.60) we have stated four standard BIEs for the exterior and interior Dirichlet BVPs. Similarly, we can write down BIEs for the exterior and interior Neumann problems (problems (2.5) and (2.13) with β set to zero); indeed these equations are just equations (2.57)–(2.60) re-interpreted as equations where the unknown is the Dirichlet data h and the known function is $\partial_n u$.

Another popular and closely related approach is the *indirect* method where the solution is sought in the form of a layer potential with some unknown density, for example in the form

$$u = \mathcal{S}_k\phi \quad \text{or} \quad u = \mathcal{D}_k\psi, \tag{2.62}$$

for some $\phi \in H^{-1/2}(\Gamma)$ or $\psi \in H^{1/2}(\Gamma)$. By Theorems 2.14 and 2.15 these satisfy each of the BVPs of Section 2.1 provided the relevant boundary condition is satisfied. For example, $u = \mathcal{S}_k\phi$ satisfies the exterior Dirichlet problem (2.12) if and only if $\gamma_+ u = h$, that is,

$$S_k\phi = h; \tag{2.63}$$

similarly $u = \mathcal{D}_k\phi$ satisfies (2.12) if and only if $\gamma_+ u = h$, that is,

$$(\tfrac{1}{2}I + D_k)\psi = h. \tag{2.64}$$

Like (2.57) and (2.58), these equations are of the form (2.61); indeed, with the same operator $A = S_k$ in (2.63) and (2.57), and with closely related operators in (2.58) and (2.64).

We recall (see (2.38)) that D_k' is closely related to the Hilbert space adjoint of D_k. Seeking to generalize this, let us denote by A' the *quasi-adjoint* of an operator A, where we call A' the quasi-adjoint of A if

$$A' = \mathcal{C}A^*\mathcal{C}, \tag{2.65}$$

where A^* is the Hilbert space adjoint of A. We call A *quasi-self-adjoint* if $A' = A$. Then (see (2.38)) D_k' is the quasi-adjoint of D_k and S_k is quasi-self-adjoint, and moreover

$$(\tfrac{1}{2}I + D_k)' = \tfrac{1}{2}I + D_k',$$

that is, the operator in (2.64) is the quasi-adjoint of that in (2.58).

Remark 2.24. An important observation is that A, A^* and (as a consequence of (2.65)) A' all share the same norm. Furthermore, one of the three is invertible if and only if they are all invertible. Moreover, if they are all invertible, then their inverses share the same norm.

The observation that the operator in (2.64) is the quasi-adjoint of that in (2.58) holds more generally: the indirect BIE method gives rise to equations of the form (2.61) where the operator A is the quasi-adjoint of an equation arising from the direct BIE approach. In particular the interior and exterior Dirichlet and Neumann problems all give rise to BIEs of the form (2.61) and the operators that arise are tabulated in Table 2.1. In the column labelled 'Direct', we list the operators in the direct BIEs which follow from Lemma 2.22, while in the column labelled 'Indirect' we show this information for the indirect BIEs which arise from looking for the solution in the form (2.62).

Let us pull out a few points from this table. Observe first that each operator in the third column is the quasi-adjoint of the operator immediately to its left. The message here is, roughly speaking, that the indirect formulation does not give rise to different operators to invert from the direct formulation, in particular all spectral properties (relevant for conditioning and behaviour of iterative solvers) are the same. Secondly, note that the collection of operators arising in the different formulations of the exterior Dirichlet and Neumann problems is precisely the same collection of operators as arises in the formulation of the interior problems. Finally, recall that we argued below equation (2.61) that (2.57) has infinitely many solutions in $H^{-1/2}(\Gamma)$ at wavenumbers k for which the interior Dirichlet problem

for some $\phi \in H^{1/2}(\Gamma)$ and $\eta \in \mathbb{C}$, we see from Theorems 2.14 and 2.15 and from (2.41) and (2.42) that u satisfies (2.12) if and only if

$$A_{k,\eta}\phi = h, \tag{2.71}$$

where $A_{k,\eta}$ is the quasi-adjoint of $A'_{k,\eta}$, that is,

$$A_{k,\eta} = \tfrac{1}{2}I + D_k - \mathrm{i}\eta S_k. \tag{2.72}$$

Equation (2.71) and this method for solving the exterior Dirichlet problem date back to Brakhage and Werner (1965), Leis (1965) and Panič (1965).

Similarly, u given by (2.70) with $\phi \in H^{1/2}(\Gamma)$ satisfies the exterior impedance problem (2.13) if and only if

$$C'_{k,\eta,k\beta}\phi = h \tag{2.73}$$

where, for $\hat{\eta} \in L^\infty(\Gamma)$,

$$C'_{k,\eta,\hat{\eta}} = B'_{k,\eta} + \mathrm{i}\hat{\eta}A_{k,\eta} \quad \text{and} \quad B'_{k,\eta} = H_k + \mathrm{i}\eta\big(\tfrac{1}{2}I - D'_k\big) \tag{2.74}$$

is the quasi-adjoint of $B_{k,\eta}$. Of course, $C'_{k,\eta,k\beta} = B'_{k,\eta}$ in the Neumann BVP case $\beta = 0$.

The corresponding direct formulation for the exterior impedance problem follows essentially from (2.68). If u satisfies (2.13) then, by Lemma 2.22, in particular from (2.53) and (2.54), we see that

$$\big(\tfrac{1}{2}I - D_k\big)\gamma_+ u - \mathrm{i}kS_k(\beta\gamma_+ u) = -S_k h \tag{2.75}$$

and

$$H_k\gamma_+ u + \mathrm{i}k\big(\tfrac{1}{2}I + D'_k\big)(\beta\gamma_+ u) = \big(\tfrac{1}{2}I + D'_k\big)h. \tag{2.76}$$

For the case $\beta = 0$ (the exterior Neumann problem) we have seen in Table 2.1 that equation (2.75) is not uniquely solvable at wavenumbers k for which the homogeneous interior Dirichlet problem has a non-trivial solution, and (2.76) is not uniquely solvable at wavenumbers for which the homogeneous interior Neumann problem has a non-trivial solution. These statements hold equally for $\beta \neq 0$; this can essentially be read off from Table 2.1. In particular, from (2.59) and (2.60) we see that, if u is a non-trivial solution of the homogeneous interior Dirichlet problem ((2.4) with $h = 0$), then $\partial_n^- u$ (which is non-zero by Theorem 2.4) is in the null space of both S_k and $\frac{1}{2}I - D'_k$, and hence in the null space of $\frac{1}{2}I - D'_k - \mathrm{i}k\beta S_k$, for any $\beta \in L^\infty(\Gamma)$. Thus, if (2.4) with $h = 0$ has a non-trivial solution, then $\frac{1}{2}I - D'_k - \mathrm{i}k\beta S_k$ is not invertible. Hence by Remark 2.24 its quasi-adjoint $\frac{1}{2}I - D_k - \mathrm{i}kS_k\beta$, which is the operator on the left-hand side of (2.75), also is not invertible.

We obtain a direct combined potential formulation of the exterior impedance problem by taking a linear combination of equations (2.75) and (2.76), which gives

$$C_{k,\eta,k\beta}\gamma u = A'_{k,\eta}h, \tag{2.77}$$

where $C_{k,\eta,\hat\eta}$ is the quasi-adjoint of $C'_{k,\eta,\hat\eta}$, defined explicitly by

$$C_{k,\eta,\hat\eta}\phi := B_{k,\eta}\phi + \mathrm{i}A'_{k,\eta}(\hat\eta\phi), \quad \phi \in H^{1/2}(\Gamma). \tag{2.78}$$

We can deduce mapping properties of the operators $A_{k,\eta}$, $B_{k,\eta}$, $C_{k,\eta,\hat\eta}$ and their quasi-adjoints straightforwardly from Theorem 2.17. This is the content of the first sentence of the following theorem.

Theorem 2.27. For $k \geq 0$, $\eta \in \mathbb{C}$, $|s| \leq \frac{1}{2}$, and $\hat\eta \in L^\infty(\Gamma)$, the mappings

$$\begin{aligned} A_{k,\eta} &: H^{s+1/2}(\Gamma) \to H^{s+1/2}(\Gamma), & A'_{k,\eta} &: H^{s-1/2}(\Gamma) \to H^{s-1/2}(\Gamma), \\ B_{k,\eta} &: H^{s+1/2}(\Gamma) \to H^{s-1/2}(\Gamma), & B'_{k,\eta} &: H^{s+1/2}(\Gamma) \to H^{s-1/2}(\Gamma), \\ C_{k,\eta,\hat\eta} &: H^{s+1/2}(\Gamma) \to H^{s-1/2}(\Gamma), & C'_{k,\eta,\hat\eta} &: H^{s+1/2}(\Gamma) \to H^{s-1/2}(\Gamma), \end{aligned}$$

are bounded. Further, for $k > 0$ and $\operatorname{Re}\eta \neq 0$ the mappings for $A_{k,\eta}$, $A'_{k,\eta}$, $B_{k,\eta}$ and $B'_{k,\eta}$ are invertible; the mappings for $C_{k,\eta,\hat\eta}$ and $C'_{k,\eta,\hat\eta}$ are also invertible provided also that $\operatorname{Re}\hat\eta \geq 0$.

Proof. That these mappings are bounded follows from Theorems 2.17 and 2.18. It is enough (*cf.* Remark 2.24) to show the invertibility of $A'_{k,\eta}$, $B_{k,\eta}$, and $C_{k,\eta,\hat\eta}$, with the invertibility of the other operators following by passing to adjoints. A main step in proving the invertibility of these three operators is to show that each is Fredholm of index zero for $|s| \leq 1/2$. But this follows from Theorem 2.25 since, for $|s| \leq \frac{1}{2}$, $A'_{k,\eta}$ is a compact perturbation of $\frac{1}{2}I + D'_k$ and $B_{k,\eta}$ and $C_{k,\eta,\hat\eta}$ are compact perturbations of H_k. These compactness results follow from the boundedness of the mappings in Theorem 2.17 and that the operator embedding $H^t(\Gamma)$ into $H^r(\Gamma)$ is compact for $-1 \leq r < t \leq 1$. To establish invertibility, it remains to show injectivity and, arguing as in the proof of Theorem 2.25, it is enough to show this injectivity for the standard case $s = 0$. We give the proof of injectivity for just one of these operators, namely $A'_{k,\eta}$, which is the focus of much of the rest of this article, following Theorem 2.7 in Chandler-Wilde and Langdon (2007). The proofs for the other operators are similar.

So suppose that $\phi \in H^{-1/2}(\Gamma)$ and $A'_{k,\eta}\phi = 0$ and let $u = \mathcal{S}_k\phi$. Then, using (2.41) and Theorems 2.14 and 2.15, we see that u satisfies the interior impedance problem (2.5) with $h = 0$ and $\beta = \eta/k$, in particular $\partial_n^- u - \mathrm{i}\eta\gamma_- u = A'_{k,\eta}\phi = 0$. It follows from Theorem 2.3 that, provided $\operatorname{Re}\eta \neq 0$, $u = 0$ in Ω_-. This (together with Theorems 2.14 and 2.15) implies that u in Ω_+ satisfies the homogeneous exterior Dirichlet problem ((2.12) with $h = 0$), in particular $\gamma_+ u = \gamma_- u = 0$. It follows from Corollary 2.9 that $u = 0$ in Ω_+ so that $\phi = \partial_n^- u - \partial_n^+ u = 0$. □

The implications of this for our combined potential integral equation formulations is spelled out in the following corollary. This corollary provides further the promised proof of the existence part of Theorem 2.10.

Corollary 2.28. Suppose that $k > 0$ and $\eta \in \mathbb{C}$ with $\operatorname{Re}\eta \neq 0$. Then both the following statements hold.

(i) For every $h \in L^2(\Gamma)$ the integral equation (2.71) has exactly one solution $\phi \in L^2(\Gamma)$, and if $h \in H^s(\Gamma)$ for $0 \le s \le 1$ then also $\phi \in H^s(\Gamma)$. If $h \in H^{1/2}(\Gamma)$ and $\phi \in H^{1/2}(\Gamma)$ is the unique solution of (2.71), then u given by (2.70) is the unique solution of (2.12). Conversely, if u is the unique solution of (2.12) then $\partial_n^+ u \in H^{-1/2}(\Gamma)$ is the unique solution of (2.68). Further, if $h = \gamma_+ u \in H^s(\Gamma)$ for $1/2 < s \le 1$ then $\partial_n^+ u \in H^{s-1}(\Gamma)$.

(ii) For every $h \in H^{-1}(\Gamma)$ the integral equation (2.73) has exactly one solution $\phi \in L^2(\Gamma)$, and if $h \in H^s(\Gamma)$ for $-1 \le s \le 0$ then also $\phi \in H^{s+1}(\Gamma)$. If $h \in H^{-1/2}(\Gamma)$ and $\phi \in H^{1/2}(\Gamma)$ is the unique solution of (2.73), then u given by (2.70) is the unique solution of (2.13). Conversely, if u is the unique solution of (2.13) then $\gamma_+ u \in H^{1/2}(\Gamma)$ is the unique solution of (2.77). Further, if $h = \gamma_+ u \in H^s(\Gamma)$ for $-1/2 < s \le 0$ then $\gamma_+ u \in H^{s+1}(\Gamma)$.

The above corollary makes clear that the exterior Dirichlet, Neumann and impedance BVPs can be solved by combined potential integral equation formulations. Essentially the same boundary integral operators arise when we solve the interior impedance BVP (2.5). To see this, extending the notations (2.72) and (2.74), for $\eta \in L^\infty(\Gamma)$, we define the bounded operators $A_{k,\eta} : H^s(\Gamma) \to H^s(\Gamma)$ and $B'_{k,\eta} : H^s(\Gamma) \to H^{s-1}(\Gamma)$, for $0 \le s \le 1$, by

$$A_{k,\eta}\phi = \tfrac{1}{2}\phi + D_k\phi - \mathrm{i}S_k(\eta\phi) \text{ and } B'_{k,\eta}\phi = H_k\phi + \mathrm{i}\big(\tfrac{1}{2}I - D'_k\big)(\eta\phi), \quad (2.79)$$

and, with $A_{k,\eta}$ and $B_{k,\eta}$ defined in this way, define the bounded operator $C_{k,\eta,\hat{\eta}}$, for $\hat{\eta} \in L^\infty(\Gamma)$, by (2.78). Thus the following result contains the invertibility statements about $A_{k,\eta}$ and $B'_{k,\eta}$ in Theorem 2.27 as a special case. Its proof is a straightforward variation on the proof of Theorem 2.27.

Theorem 2.29. If $\eta \in L^\infty(\Gamma)$ and Assumption 2.2 holds with β replaced by η, then the operators $A_{k,\eta}$ and $B'_{k,\eta}$ are invertible as mappings $A_{k,\eta} : H^{s+1/2}(\Gamma) \to H^{s+1/2}(\Gamma)$ and $B'_{k,\eta} : H^{s+1/2}(\Gamma) \to H^{s-1/2}(\Gamma)$, for $|s| \le 1/2$. Further, if also $\hat{\eta} \in L^\infty$ with $\operatorname{Re}\hat{\eta} \ge 0$, then also $C'_{k,\eta,\hat{\eta}} : H^{s+1/2}(\Gamma) \to H^{s-1/2}(\Gamma)$ is invertible for $|s| \le 1/2$.

These generalizations of $A_{k,\eta}$ and $B'_{k,\eta}$ arise when we make a direct formulation of the interior impedance problem (2.5).

Theorem 2.30. If u satisfies the interior impedance problem (2.5), then

$$A_{k,k\beta}\gamma_- u = S_k h, \quad (2.80)$$

$$B'_{k,k\beta}\gamma_- u = -\big(\tfrac{1}{2}I - D'_k\big)h, \quad (2.81)$$

and γu is the unique solution in $H^{1/2}(\Gamma)$ of each of these equations in the case that Assumption 2.2 holds. Conversely, suppose Assumption 2.2 holds. Then, if $\phi \in H^{1/2}(\Gamma)$ is the unique solution of either $A_{k,k\beta}\phi = S_k h$ or $B'_{k,k\beta}\phi = -(\frac{1}{2}I - D'_k)h$, and

$$u(x) = \mathcal{S}_k(h + \mathrm{i}k\beta\phi)(x) - \mathcal{D}_k\phi(x), \quad x \in \Omega_-, \tag{2.82}$$

it follows that u satisfies (2.5).

Proof. Equations (2.80) and (2.81) follow from Lemma 2.22 and (2.50), and the fact that γu is the unique solution of each of these equations follows from Theorem 2.29. Suppose now that Assumption 2.2 holds and $\phi \in H^{1/2}(\Gamma)$ is the unique solution of either (2.80) or (2.81) and u is defined by

$$u(x) := \mathcal{S}_k(h + \mathrm{i}k\beta\phi)(x) - \mathcal{D}_k\phi(x), \quad x \in \mathbb{R}^d \setminus \Gamma. \tag{2.83}$$

Then, by Theorems 2.14 and 2.15, $u \in H^1(\Omega_-) \cap H^1_{\mathrm{loc}}(\Omega_+) \cap C^2(\mathbb{R}^d \setminus \Gamma)$, $\Delta u + k^2 u = 0$ in $\mathbb{R}^d \setminus \Gamma$, and u satisfies the Sommerfeld radiation condition. Further, by Lemma 2.23, (2.50) and (2.80), $\partial_n^+ u - \partial_n^- u = -(h + \mathrm{i}k\beta\phi)$ and $\gamma_- u - \gamma_+ u = \phi$ and moreover $\gamma_+ u = 0$ in the case that $A_{k,k\beta}\phi = S_k h$, while $\partial_n^+ u = 0$ in the case that $B'_{k,k\beta}\phi = -(\frac{1}{2}I - D'_k)h$. Thus, by the uniqueness parts of Theorem 2.10, $u = 0$ in Ω_+ so that $\gamma_+ u = \partial_n^+ u = 0$ and $\partial_n^- u = h + \mathrm{i}\beta k \gamma_- u$. Thus we have shown that $u|_{\Omega_-}$ satisfies (2.5). □

2.7. *Poincaré–Steklov and boundary integral operators*

In this section we explore the connections between boundary integral operators and solution operators for the corresponding BVPs (Dirichlet to Neumann and similar maps). The representations we deduce, several of which appear to be new, will be an important tool in our investigation of the k-dependence of condition numbers of BIE formulations in Section 5.

We noted in Theorem 2.10 that the exterior Dirichlet problem (2.12) is uniquely solvable. Let P^+_{DtN} denote the mapping which takes $h \in H^{1/2}(\Gamma)$ to $\partial_n u$, where u is the solution to (2.12). This mapping $P^+_{\mathrm{DtN}} : H^{1/2}(\Gamma) \to H^{-1/2}(\Gamma)$ is called the exterior *Dirichlet to Neumann map* or exterior *Poincaré–Steklov operator*. Similarly, let $P^+_{\mathrm{NtD}} : H^{-1/2}(\Gamma) \to H^{1/2}(\Gamma)$ denote the mapping which takes $h \in H^{-1/2}(\Gamma)$ to $\gamma u \in H^{1/2}(\Gamma)$, where u is the solution to the exterior Neumann problem ((2.13) with $\beta = 0$). P^+_{NtD} is called the exterior *Neumann to Dirichlet map* or sometimes the exterior *Steklov–Poincaré operator* (Sauter and Schwab 2011). It is immediate from these definitions that these mappings are inverses of each other; what is less obvious, but follows easily from Green's second formula (Theorem 2.19), is that the Steklov–Poincaré and Poincaré–Steklov operators are both quasi-self-adjoint. Thus

$$P^+_{\mathrm{DtN}} P^+_{\mathrm{NtD}} = P^+_{\mathrm{NtD}} P^+_{\mathrm{DtN}} = I, \; (P^+_{\mathrm{DtN}})' = P^+_{\mathrm{DtN}}, \; (P^+_{\mathrm{NtD}})' = P^+_{\mathrm{NtD}}. \tag{2.84}$$

The importance of these mappings for our purposes is their close connection with boundary integral operators. Indeed, we have from Corollary 2.28(i) that, if u satisfies the exterior Dirichlet problem (2.12), then $\partial_n u$ is the unique solution of (2.68). Thus

$$P^+_{\mathrm{DtN}} = \left(A'_{k,\eta}\right)^{-1} B_{k,\eta}. \tag{2.85}$$

We can deduce a further representation for P^+_{DtN} from Corollary 2.28(i) based on solving (2.71) and can deduce representations for P^+_{NtD} from Corollary 2.28(ii), but these representations can equally be read off from (2.84). For example, combining (2.84) and (2.85), we see that

$$P^+_{\mathrm{DtN}} = \left(P^+_{\mathrm{DtN}}\right)' = \left(\left(A'_{k,\eta}\right)^{-1} B_{k,\eta}\right)' = B'_{k,\eta} A^{-1}_{k,\eta}.$$

Let us summarize these representations in a theorem.

Theorem 2.31. It holds that

$$P^+_{\mathrm{DtN}} = \left(A'_{k,\eta}\right)^{-1} B_{k,\eta} = B'_{k,\eta} A^{-1}_{k,\eta} \text{ and } P^+_{\mathrm{NtD}} = B^{-1}_{k,\eta} A'_{k,\eta} = A_{k,\eta}\left(B'_{k,\eta}\right)^{-1}.$$

The operators $P^+_{\mathrm{DtN}} : H^{1/2}(\Gamma) \to H^{-1/2}(\Gamma)$ and $P^+_{\mathrm{NtD}} : H^{-1/2}(\Gamma) \to H^{1/2}(\Gamma)$ are bounded, and have unique extensions to bounded operators $P^+_{\mathrm{DtN}} : L^2(\Gamma) \to H^{-1}(\Gamma)$ and $P^+_{\mathrm{NtD}} : H^{-1}(\Gamma) \to L^2(\Gamma)$. Further,

$$P^+_{\mathrm{DtN}} : H^{1/2+s}(\Gamma) \to H^{-1/2+s}(\Gamma) \quad \text{and} \quad P^+_{\mathrm{NtD}} : H^{-1/2+s}(\Gamma) \to H^{1/2+s}(\Gamma),$$

for $|s| \leq 1/2$, and all these mappings are bounded.

The above representations are only one of many possible boundary integral representations for the Poincaré–Steklov maps; indeed it is most usual in the BIE literature (*e.g.*, Steinbach 2008, Sauter and Schwab 2011) to use representations directly in terms of S_k, D_k, D'_k and H_k, avoiding combined potential operators. One such representation follows from Theorem 2.26 and (2.57); we see that, for values of k for which S_k is invertible,

$$P^+_{\mathrm{DtN}} = S_k^{-1}\left(-\tfrac{1}{2}I + D_k\right).$$

For further such representation see Steinbach (2008) and Sauter and Schwab (2011).

Similar representations can be obtained for interior Dirichlet to Neumann and Neumann to Dirichlet operators, for wavenumbers k for which the Dirichlet and Neumann interior problems are uniquely solvable. But we will focus here on the interior *impedance to Dirichlet* operator $P^{-,\eta}_{\mathrm{ItD}}$ which maps $h \in H^{-1/2}(\Gamma)$ to γu, where u is the solution of the interior impedance problem (2.5) with $\beta = \eta/k$. Provided β satisfies Assumption 2.2, which is the case if and only if $\eta \in L^\infty(\Gamma)$ satisfies Assumption 2.2, it follows from Theorem 2.10 that this is a well-defined mapping from $H^{-1/2}(\Gamma)$ to

$H^{1/2}(\Gamma)$; we see from Theorem 2.30 that

$$P_{\mathrm{ItD}}^{-,\eta} = A_{k,\eta}^{-1} S_k. \tag{2.86}$$

One can also make representations for exterior impedance to Dirichlet operators. For $h \in H^{-1/2}(\Gamma)$ and $\hat{\eta} \in L^\infty(\Gamma)$ with $\mathrm{Re}\,\hat{\eta} \geq 0$, let $P_{\mathrm{ItD}}^{+,\hat{\eta}}$ denote the mapping which takes h to $\gamma u \in H^{1/2}(\Gamma)$, where u denotes what is, by Theorem 2.10, the unique solution of the exterior impedance problem (2.13) with $\beta = \hat{\eta}/k$. From Theorem 2.30 and (2.77) we see that, for every $\eta \in \mathbb{C}$ with $\mathrm{Re}\,\eta \neq 0$,

$$P_{\mathrm{ItD}}^{+,\hat{\eta}} = C_{k,\eta,\hat{\eta}}^{-1} A_{k,\eta}'. \tag{2.87}$$

Just as for the exterior Dirichlet to Neumann map, it follows from Theorem 2.19 that the operators $P_{\mathrm{ItD}}^{-,\eta}$ and $P_{\mathrm{ItD}}^{+,\hat{\eta}}$ are quasi-self-adjoint. Thus, in particular,

$$P_{\mathrm{ItD}}^{-,\eta} = \left(P_{\mathrm{ItD}}^{-,\eta}\right)' = \left(A_{k,\eta}^{-1} S_k\right)' = S_k \left(A_{k,\eta}'\right)^{-1}. \tag{2.88}$$

These representations are one route to establishing the following mapping properties.

Theorem 2.32. Provided $\hat{\eta} \in L^\infty(\Gamma)$ and Assumption 2.2 is satisfied with β replaced by $\hat{\eta}$, the mapping $P_{\mathrm{ItD}}^{-,\hat{\eta}} : H^{-1/2}(\Gamma) \to H^{1/2}(\Gamma)$ is bounded and has a unique extension to a bounded operator $P_{\mathrm{ItD}}^{-,\hat{\eta}} : H^{-1}(\Gamma) \to L^2(\Gamma)$. The same statements hold for $P_{\mathrm{ItD}}^{+,\hat{\eta}}$ for $\hat{\eta} \in L^\infty(\Gamma)$ with $\mathrm{Re}\,\hat{\eta} \geq 0$. Further, provided the respective conditions on $\hat{\eta}$ are satisfied,

$$P_{\mathrm{ItD}}^{\pm,\hat{\eta}} : H^{-1/2+s}(\Gamma) \to H^{1/2+s}(\Gamma),$$

for $|s| \leq 1/2$, and all these mappings are bounded.

The above results represent the exterior DtN and NtD maps and the interior and exterior ItD maps partly in terms of the inverses of $A_{k,\eta}$, $B_{k,\eta}$ and $C_{k,\eta,\hat{\eta}}$. A component of our arguments in Section 5.6, when we study the dependence of the norm of $A_{k,\eta}^{-1}$ on k, will be the following representations in the other direction. While representations, such as those we have presented above, for Poincaré–Steklov operators in terms of boundary integral operators and their inverses are widely used, these representations in the other direction seem less well known.

Theorem 2.33. Suppose that $\eta \in L^\infty(\Gamma)$ and that Assumption 2.2 holds with β replaced by η. Then, for $|s| \leq 1/2$, the mapping $A_{k,\eta}^{-1} : H^{1/2+s}(\Gamma) \to H^{1/2+s}(\Gamma)$ has the representation

$$A_{k,\eta}^{-1} = I - P_{\mathrm{ItD}}^{-,\eta}\left(P_{\mathrm{DtN}}^{+} - \mathrm{i}\eta\right) \tag{2.89}$$

and, provided also that $\hat{\eta} \in L^\infty(\Gamma)$ with $\mathrm{Re}\,\hat{\eta} \geq 0$, the mapping $C_{k,\eta,\hat{\eta}}^{-1}$:

$H^{s-1/2}(\Gamma) \to H^{1/2+s}(\Gamma)$ has the representation

$$\left(C'_{k,\eta,\hat{\eta}}\right)^{-1} = P^{+,\hat{\eta}}_{\mathrm{ItD}} - P^{-,\eta}_{\mathrm{ItD}}\left(I - \mathrm{i}(\eta + \hat{\eta})P^{+,\hat{\eta}}_{\mathrm{ItD}}\right). \tag{2.90}$$

Proof. The left- and right-hand sides of (2.89) are bounded operators on $H^{s+1/2}(\Gamma)$ for $|s| \le 1/2$, by Theorems 2.27, 2.31 and 2.32, so that to show (2.89) it is enough to show that

$$A^{-1}_{k,\eta} h = h - P^{-,\eta}_{\mathrm{ItD}}\left(P^{+}_{\mathrm{DtN}} - \mathrm{i}\eta\right)h \tag{2.91}$$

holds for all h in some dense subset of $L^2(\Gamma)$, *e.g.*, $H^{1/2}(\Gamma)$. So suppose that $h \in H^{1/2}(\Gamma)$ and that $\phi = A^{-1}_{k,\eta} h$, so that

$$A_{k,\eta}\phi = h,$$

and define

$$u = \mathcal{D}_k\phi - \mathrm{i}\mathcal{S}_k(\eta\phi). \tag{2.92}$$

Then, by Lemma 2.23 and (2.50), $\gamma_+ u = h$, $\gamma_- u = h - \phi$ and $\partial_n^+ u - \partial_n^- u = \mathrm{i}\eta\phi$, so that

$$\partial_n^- u - \mathrm{i}\eta\gamma_- u = \partial_n^+ u - \mathrm{i}\eta h.$$

Hence, and by Theorems 2.14 and 2.15, $u|_{\Omega_+}$ satisfies the exterior Dirichlet problem (2.12) with $\gamma_+ u = h$, while $u|_{\Omega_-}$ satisfies the interior impedance problem (2.5) with $\beta = \eta/k$ and impedance data $\partial_n^+ u - \mathrm{i}\eta h$, so that

$$\gamma_- u = P^{-,\eta}_{\mathrm{ItD}}\left(\partial_n^+ u - \mathrm{i}\eta h\right) = P^{-,\eta}_{\mathrm{ItD}}\left(P^{+}_{\mathrm{DtN}} - \mathrm{i}\eta\right)h.$$

and, since $A^{-1}_{k,\eta} h = \phi = h - \gamma_- u$, (2.91) follows.

Arguing similarly, to establish (2.90) it is enough to show that

$$\left(C'_{k,\eta,\hat{\eta}}\right)^{-1} h = P^{+,\hat{\eta}}_{\mathrm{ItD}} h - P^{-,\eta}_{\mathrm{ItD}}\left(I - \mathrm{i}(\eta + \hat{\eta})P^{+,\hat{\eta}}_{\mathrm{ItD}}\right)h \tag{2.93}$$

holds for all h in some dense subset of $H^{-1}(\Gamma)$, *e.g.*, $H^{-1/2}(\Gamma)$. So suppose that $h \in H^{-1/2}(\Gamma)$ and that $\phi = (C'_{k,\eta,\hat{\eta}})^{-1} h$, so that $C'_{k,\eta,\hat{\eta}}\phi = h$, and define u by (2.92). Then, by Lemma 2.23 and (2.50), $\partial_n^+ u + \mathrm{i}\hat{\eta}\gamma_+ u = C'_{k,\eta,\hat{\eta}}\phi = h$, $\gamma_+ u - \gamma_- u = \phi$, and $\partial_n^+ u - \partial_n^- u = \mathrm{i}\eta\phi$, so that

$$\partial_n^- u - \mathrm{i}\eta\gamma_- u = h - \mathrm{i}(\hat{\eta} + \eta)\gamma_+ u.$$

Hence, and by Theorems 2.14 and 2.15, $u|_{\Omega_+}$ satisfies the exterior impedance problem (2.13) with $\beta = \hat{\eta}/k$ and impedance data h, while $u|_{\Omega_-}$ satisfies the interior impedance problem (2.5) with $\beta = \eta/k$ and impedance data $h - \mathrm{i}(\hat{\eta} + \eta)\gamma_+ u$, so that

$$\gamma_- u = P^{-,\eta}_{\mathrm{ItD}}\left(h - \mathrm{i}(\hat{\eta} + \eta)\gamma_+ u\right) = P^{-,\eta}_{\mathrm{ItD}}\left(I - \mathrm{i}(\eta + \hat{\eta})P^{+,\hat{\eta}}_{\mathrm{ItD}}\right)h.$$

and, since $(C'_{k,\eta,\hat{\eta}})^{-1} h = \phi = \gamma_+ u - \gamma_- u$, (2.93) follows. □

We note that, in particular, the above theorem holds for $\hat{\eta} = 0$, in which case (2.90) reduces to

$$\left(B'_{k,\eta}\right)^{-1} = P^{+}_{\mathrm{NtD}} - P^{-,\eta}_{\mathrm{ItD}}\left(I - \mathrm{i}\eta P^{+}_{\mathrm{NtD}}\right). \tag{2.94}$$

It is clear from Theorem 2.31, (2.86), and Theorem 2.33, that $A^{-1}_{k,\eta}$ and the interior and exterior solution operators P^{+}_{DtN} and $P^{-,\eta}_{\mathrm{ItD}}$ are intimately related. In particular, we will use the representation (2.89) to bound the norm of $A^{-1}_{k,\eta}$ as an operator on $L^2(\Gamma)$ in Section 5.6. This method of bounding the norms of inverses of integral operators is not spelt out, but is to some extent implicit in the arguments in Verchota (1984) and Chandler-Wilde and Monk (2008).

Theorem 2.33 is also very useful in shedding light on the physical meanings of the solutions to indirect BIEs. In the direct case, for instance the combined potential equations (2.68) and (2.77), the solution is just the trace or normal derivative of the solution of the BVP that we are solving. The meaning of the solution to an indirect integral equation, for instance (2.71) or (2.73), is more obscure, but is elucidated by Theorem 2.33. For example (2.89) implies that the solution to the direct integral equation formulation (2.71) of the exterior Dirichlet problem is the difference between the Dirichlet data (the trace of the solution of the exterior problem) and the trace of the solution of an interior impedance problem with data depending on the solution to the exterior problem. Thus the solution of (2.71) depends in a complex way on the solutions of both interior and exterior BVPs. This can make direct formulations preferable for numerical computation. As has been well known for many years, the issue is that the singularities in the solution to direct formulations are just those of the boundary traces of the solutions to the original BVPs, while the solutions to indirect integral equations, *e.g.*, (2.71), typically contain singularities related to both solutions of interior and exterior problems, and so require more degrees of freedom to approximate. As pointed out in Bruno *et al.* (2004), there are additional reasons to prefer direct formulations to indirect formulations for high-frequency scattering problems: the oscillatory behaviour of solutions of indirect integral equations is more complex, and depends inextricably on the solutions to both interior and exterior problems in a complicated interacting way.

2.8. Regularity of the solutions of the integral equations and BVPs

In this subsection we make some brief comments on the smoothness of the solutions of the BVPs (2.4), (2.5), (2.12), and (2.13).

Our first result makes clear that, under rather general and mild conditions, the traces γu and $\partial_n u$ are (at least almost everywhere) limiting values of the function and its gradient as Γ is approached from $\Omega_{\pm}$.

Theorem 2.34. Let $\{\Theta_\pm(x) : x \in \Gamma\}$ be families of non-tangential approach sets to x from $\Omega_\pm$ (as defined in Section A.4). (i) If u satisfies the exterior Dirichlet problem (2.12) or the exterior impedance problem (2.13) then $\lim_{y\to x, y\in\Theta_+(x)} u(y) = \gamma^+ u(x)$, for almost all $x \in \Gamma$. (ii) If v satisfies the interior Dirichlet or impedance problems, (2.4) or (2.5), then $\lim_{y\to x, y\in\Theta_-(x)} v(y) = \gamma_- v(x)$. (iii) If u satisfies (2.12) with boundary data $h \in H^1(\Gamma)$ or satisfies (2.13) with data $h \in L^2(\Gamma)$, then also $\nabla_\Gamma \gamma_+ u \in L^2(\Gamma)$ and $\partial_n^+ u \in L^2(\Gamma)$ and

$$\lim_{y\to x, y\in\Theta_+(x)} \nabla u(y) = \nabla_\Gamma \gamma_+ u(x) + n(x)\partial_n^+ u(x),$$

for almost all $x \in \Gamma$. In particular, for almost all $x \in \Gamma$,

$$\partial_n^+ u(x) = \frac{\partial u}{\partial n}(x) := \lim_{y\to x, y\in\Theta_+(x)} n(x)\cdot\nabla u(y). \tag{2.95}$$

(iv) Similarly, if v satisfies (2.13) with data $h \in L^2(\Gamma)$, then also $\nabla_\Gamma \gamma_+ v \in L^2(\Gamma)$ and $\partial_n^+ v \in L^2(\Gamma)$ and

$$\lim_{y\to x, y\in\Theta_+(x)} \nabla v(y) = \nabla_\Gamma \gamma_+ v(x) + n(x)\partial_n^+ v(x),$$

for almost all $x \in \Gamma$.

Proof. By Theorems 2.21 and 2.20 the solutions to all these problems have representations as single- and double-layer potentials. Thus the limits claimed in (i) and (ii) follow from (2.27). Similarly the limits claimed in (iii) and (iv) follow from (2.28)–(2.31), provided that $\gamma_+ u, \gamma_+ v \in H^1(\Gamma)$ and $\partial_n^+ u, \partial_n^+ v \in L^2(\Gamma)$. But for u, this follows from Corollary 2.28(i), while for v this follows from the impedance boundary condition and Corollary 2.28(ii). □

Under stronger smoothness conditions on Γ and on the data, the solution of each BVP and some of its derivatives may be continuous up to the boundary. The following local elliptic regularity result is a result of this sort which is sufficient for our purposes, applying in the important practical case when the boundary Γ is smooth except for corners and edges: for a proof see Theorem 4.18 in McLean (2000) (and see also Theorem 3.27 and its proof in Colton and Kress 1983 for a related result specific to the Helmholtz case).

Lemma 2.35. Suppose that u satisfies the interior or exterior Dirichlet or impedance BVP in $D = \Omega_\pm$ (*i.e.*, u satisfies (2.12), (2.13), (2.4) or (2.5)), and that Γ_0 is a relatively open subset of Γ which is C^∞, and that the boundary data $h \in C^\infty(\overline{\Gamma_0})$ and (in the impedance case) also $\beta \in C^\infty(\overline{\Gamma_0})$. Then u and its partial derivatives of all orders have continuous extensions from D to $D \cup \Gamma_0$.

2.9. The star-combined integral equation

In this subsection we introduce novel generalizations of the combined potential integral equations (2.68) and (2.71), building on recent developments in Spence, Chandler-Wilde, Graham and Smyshlyaev (2011).

Theorem 2.36. If u satisfies the exterior Dirichlet problem (2.12) with $h \in H^1(\Gamma)$, then

$$S_k \partial_n^+ u = \left(-\tfrac{1}{2}I + D_k\right)h \tag{2.96}$$

and

$$\left[n\left(\tfrac{1}{2}I + D_k'\right) + \nabla_\Gamma S_k\right]\partial_n^+ u = \left[nH_k + \nabla_\Gamma D_k - \tfrac{1}{2}\nabla_\Gamma\right]h. \tag{2.97}$$

Hence, for $\eta \in L^\infty(\Gamma)$ and $Z \in (L^\infty(\Gamma))^d$, we have

$$A'_{k,\eta,Z}\partial_n^+ u = B_{k,\eta,Z}h, \tag{2.98}$$

where the operator $A'_{k,\eta,Z} : L^2(\Gamma) \to L^2(\Gamma)$ is defined by

$$A'_{k,\eta,Z} := Z \cdot n\left(\tfrac{1}{2}I + D_k'\right) + Z \cdot \nabla_\Gamma S_k - \mathrm{i}\eta S_k, \tag{2.99}$$

and $B_{k,\eta,Z} : H^1(\Gamma) \to L^2(\Gamma)$ by

$$B_{k,\eta,Z} := Z \cdot \left[nH_k + \nabla_\Gamma D_k - \tfrac{1}{2}\nabla_\Gamma\right] - \mathrm{i}\eta\left(-\tfrac{1}{2}I + D_k\right). \tag{2.100}$$

Proof. We have observed already above in (2.57) that (2.96) holds under the weaker condition that $h \in H^{1/2}(\Gamma)$. If also $h \in H^1(\Gamma)$ then, by Corollary 2.28(i), $\partial_n^+ u \in L^2(\Gamma)$. Then, taking the non-tangential limit of the gradient of (2.66) as x approaches Γ, and using (2.30) and (2.31) and the jump relations (2.41)–(2.43), we obtain (2.97). □

The above derivation is straightforward, but the integral equation formulation (2.98) seems to be very recent (Spence *et al.* 2011). The integral equation can be seen as a generalization of (2.68), reducing to this equation when the choice $Z = n$ is made, since

$$A'_{k,\eta,n} = A'_{k,\eta} \quad \text{and} \quad B_{k,\eta,n} = B_{k,\eta}. \tag{2.101}$$

In the case when Ω_- is star-shaped with respect to an appropriately chosen origin (see Definition 5.5 below), Spence *et al.* (2011) proposed the integral equation formulation (2.97) with $Z(x) := x$, $x \in \Gamma$, and the special choice

$$\eta(x) := k|x| + \mathrm{i}\frac{d-1}{2}. \tag{2.102}$$

We will use, following Spence *et al.* (2011), the notation $\mathscr{A}_k$ for $A_{k,\eta,Z}$ in that case, that is,

$$\mathscr{A}_k := A'_{k,\eta,Z} = Z \cdot n\left(\tfrac{1}{2}I + D_k'\right) + Z \cdot \nabla_\Gamma S_k - \mathrm{i}\eta S_k, \tag{2.103}$$

with η given by (2.102) and $Z(x) := x$, $x \in \Gamma$. Following Spence *et al.* (2011) we will call $\mathscr{A}_k$ the *star-combined* operator.

The following theorem follows from results in Spence *et al.* (2011) discussed in Section 5.7, precisely from Theorem 5.26, which proves the stronger property of coercivity. These results, and their implications for high-frequency scattering computations, are discussed in more detail in Sections 5.7 and 6.4.

Theorem 2.37. If Ω_- is star-shaped with respect to the origin, which implies that, for some $c > 0$,

$$x \cdot n \geq c, \quad \text{for almost all } x \in \Gamma, \tag{2.104}$$

then $\mathscr{A}_k : L^2(\Gamma) \to L^2(\Gamma)$ is invertible with $\|\mathscr{A}_k^{-1}\|_{L^2(\Gamma)\leftarrow L^2(\Gamma)} \leq 2/c$.

The operator $A'_{k,\eta,Z}$ appearing in (2.98) also arises in solving the interior oblique impedance problem (2.6).

Theorem 2.38. The single-layer potential $u = \mathcal{S}_k\phi$, with density $\phi \in L^2(\Gamma)$, satisfies the interior oblique impedance problem (2.6) if

$$A'_{k,k\beta,Z}\phi = h. \tag{2.105}$$

If u satisfies the interior oblique impedance problem, then $u = \mathcal{S}_k\phi$, with $\phi \in L^2(\Gamma)$ satisfying (2.105).

Proof. If $u = \mathcal{S}_k\phi$ with $\phi \in L^2(\Gamma)$ then, by Theorems 2.14, 2.15 and (2.30), u satisfies the interior oblique impedance BVP if and only if

$$Z_n\partial_n^- u + Z \cdot \nabla_\Gamma \gamma_- u - ik\beta\gamma_- u = h.$$

But, using the jump relations (2.41), this equation is precisely (2.105). On the other hand, if u satisfies the interior oblique impedance problem then $\gamma_- u \in H^1(\Gamma)$ and, by Theorem A.5, $\partial_n^- u \in L^2(\Gamma)$ so that $h_0 := \partial_n^- u - ik\gamma_- u \in L^2(\Gamma)$. Now note that, by Theorem 2.3, the interior impedance BVP (2.5) with $\beta = 1$ and $h = h_0$ has exactly one solution. Clearly this solution is u. Now, recall that $A'_{k,k,n} = A'_{k,k}$, and this operator is invertible by Theorem 2.27. Further, the interior impedance problem is the special case of the interior oblique impedance problem in which $Z = n$. So, by the first part of this theorem, we see that $u = \mathcal{S}_k\phi$ with $\phi = \left(A'_{k,k}\right)^{-1}h_0 \in L^2(\Gamma)$. □

From Theorem 2.38 the invertibility of $A'_{k,k\beta,Z}$ is equivalent to the solvability of the interior oblique impedance problem.

Corollary 2.39. For $\beta \in L^\infty(\Gamma)$ and $Z \in (L^\infty(\Gamma))^d$, the operator $A'_{k,k\beta,Z} : L^2(\Gamma) \to L^2(\Gamma)$ is injective if and only if (2.6) with $h = 0$ has only the trivial solution, and is surjective if and only if (2.6) has a solution for every $h \in L^2(\Gamma)$.

Proof. The statement regarding surjectivity is clear from Theorem 2.38. So is the statement regarding injectivity, provided that $u = \mathcal{S}_k\phi = 0$ in Ω_- only if $\phi = 0$. But this is a consequence of Theorems 2.14 and 2.15 and (2.26) and the fact that the exterior Dirichlet problem (2.12) is uniquely solvable. More explicitly, $\phi \in L^2(\Gamma)$ and $u = \mathcal{S}_k\phi$ is zero in Ω_- then u satisfies (2.12) in Ω_+ with $h = 0$ and so $u = 0$ in Ω_+ by Corollary 2.9. Then, by (2.41), $\phi = \partial_n^- u - \partial_n^+ u = 0$. □

The following result is taken from Chandler-Wilde and Spence (2012), and is a consequence of Theorem 2.50 below. It strengthens the above corollary in the case when Z is real-valued and Lipschitz-continuous and Z_n is strictly positive.

Theorem 2.40. If $\beta \in L^\infty(\Gamma)$ and $Z \in (C^{0,1}(\Gamma))^d$ is real-valued, with

$$Z_n(x) := Z(x) \cdot n(x) \geq c, \quad \text{for almost all } x \in \Gamma, \tag{2.106}$$

for some $c > 0$, then the operator $A'_{k,k\beta,Z} : L^2(\Gamma) \to L^2(\Gamma)$ is Fredholm of index zero, and so is invertible if and only if (2.6) with $h = 0$ has only the trivial solution.

The above corollary and theorem show equivalence of questions of solvability of the interior oblique impedance problem to questions of solvability of the integral equation (2.105). One case in which both are clear is the case $Z = n$, when the interior oblique impedance problem reduces to the regular interior impedance problem. Another case is when Ω_- is star-shaped. In that case we have the following corollary of Theorems 2.37 and 2.38.

Corollary 2.41. If Ω_- is star-shaped with respect to the origin, so that (2.104) holds for some $c > 0$, then the interior oblique impedance problem with

$$Z(x) := x \quad \text{and} \quad \beta(x) := |x| + \mathrm{i}\frac{d-1}{2k}, \quad x \in \Gamma,$$

has the unique solution $u = \mathcal{S}_k\phi$ with $\phi = \left(A'_{k,k\beta,Z}\right)^{-1} h \in L^2(\Gamma)$. Moreover, $\|\phi\|_{L^2(\Gamma)} \leq (2/c)\|h\|_{L^2(\Gamma)}$.

The above corollary applies to a special case of the interior oblique impedance BVP: the domain must be star-shaped and particular choices made for Z and β. But one context where this case may have application is in boundary-element-based domain decomposition methods for the Helmholtz equation, for example modifying the method of Steinbach and Windisch (2011). In domain decomposition methods the bounded Lipschitz domain Ω_- is initially divided into a number of subdomains $\Omega_1, \ldots, \Omega_M$, the choice of these up to the user: in particular, it is always possible to arrange that each Ω_j is star-shaped with respect to some point $x_j \in \Omega_j$. In the iterative solution schemes based on domain decomposition methods, a main step is the solution of local problems for the Helmholtz equation on each subdomain Ω_j. In the design of such methods a choice that has to be made is the boundary condition to be used in these local problems. A natural choice in domain decomposition methods for the Helmholtz equation is the impedance boundary condition, so that one solves (2.5) on each subdomain, choosing β (*e.g.*, $\beta = 1$) to ensure that (2.5) is uniquely solvable; a variant on this with additional non-local operators in the impedance boundary condition is used in Steinbach and Windisch (2011). But an interesting alternative may be to use the oblique boundary condition (2.3) on Ω_j, choosing Z and β as in the above corollary, that is,

$$Z(x) := x - x_j \quad \text{and} \quad \beta(x) := |x - x_j| + \mathrm{i}\frac{d-1}{2k}, \qquad x \in \partial\Omega_j.$$

This choice ensures that each local problem is well-posed, as is the integral equation formulation (2.105), but further, as we will see in Section 5.7, this integral equation formulation is coercive in the sense of Section 2.11, with a coercivity constant that is uniform in the wavenumber k.

In Section 2.7 we discussed the relation between boundary integral operators and Poincaré–Steklov operators. These results generalize to the operator $A'_{k,\eta,Z}$ and the interior oblique impedance BVP (2.6). Generalizing the notation $P^{-,\eta}_{\mathrm{ItD}}$ introduced before equation (2.86), let $P^{-,\eta,Z}_{\mathrm{ItD}}$ denote the operator which maps $h \in L^2(\Gamma)$ to $\gamma_- u \in H^1(\Gamma)$, where u is the solution of the interior oblique impedance problem (2.6) with $\beta = \eta/k$. Of course, $P^{-,\eta,Z}_{\mathrm{ItD}}$ is only well-defined for those $\eta \in L^\infty(\Gamma)$ and $Z \in (L^\infty(\Gamma))^d$ for which the interior oblique impedance BVP (2.6) has exactly one solution in the case that $\beta = \eta/k$. By Corollary 2.39 these are precisely the η and Z for which $A'_{k,\eta,Z}$ is invertible, as shown in the following result.

Theorem 2.42. Suppose that $k > 0$ and that $\eta \in L^\infty(\Gamma)$ and $Z \in (L^\infty(\Gamma))^d$ are such that $A'_{k,\eta,Z}$ is invertible. Then

$$P^+_{\mathrm{DtN}} = \big(A'_{k,\eta,Z}\big)^{-1} B_{k,\eta,Z}, \quad P^{-,\eta,Z}_{\mathrm{ItD}} = S_k \big(A'_{k,\eta,Z}\big)^{-1},$$

and, if $\operatorname{ess\,inf}_{x\in\Gamma} |Z_n(x)| > 0$,

$$\big(A'_{k,\eta,Z}\big)^{-1} = \frac{1}{Z_n} I - \left(P^+_{\mathrm{DtN}} - \frac{1}{Z_n}(\mathrm{i}\eta - Z\cdot\nabla_\Gamma)\right) P^{-,\eta,Z}_{\mathrm{ItD}}.$$

Proof. The first two results follow immediately from Theorems 2.36 and 2.38. To see the last equation, suppose that $\phi, h \in L^2(\Gamma)$ and $A'_{k,\eta,Z}\phi = h$ and define $u = \mathcal{S}_k\phi$. Then, by Theorem 2.38 and (2.41),

$$\gamma_+ u = \gamma_- u = P^{-,\eta,Z}_{\mathrm{ItD}} h, \quad \phi = \partial^-_n u - \partial^+_n u.$$

Further,

$$\partial^+_n u = P^+_{\mathrm{DtN}} \gamma_+ u = P^+_{\mathrm{DtN}} P^{-,\eta,Z}_{\mathrm{ItD}} h,$$

Remark 2.45. We have shown below equation (2.76) that (2.75) (and so also (2.111)) is not uniquely solvable at *irregular* wavenumbers k for which the homogeneous interior Dirichlet problem has a non-trivial solution. Further, at these wavenumbers, it does not hold that (2.112) is a solution to the impedance scattering problem for every solution ϕ of (2.111). (This is in contrast to (2.110), which satisfies the sound-soft scattering problem even when (2.109) has more than one solution, which it does for the same irregular wavenumbers.) This 'irregular frequency' property leads, for standard BEM discretizations, to very bad conditioning and inaccurate numerical approximations of u in a neighbourhood of such k. In Section 7.4, through computations using (2.111) *precisely at irregular wavenumbers where the integral equation is not uniquely solvable*, we see that a bonus of the novel discretization methods that we propose is that they, at least in these examples, do not exhibit any conditioning problems at the discrete level, and in fact produce accurate numerical results.

As observed in Remark 2.45, the integral equations (2.109) and (2.111) are not uniquely solvable for all $k > 0$. The standard cure is to use the combined potential integral equations introduced in Section 2.6. The direct versions of the combined potential equations are the integral equation (2.68) for the sound-soft problem (with u replaced by u^S and $h = -u^I|_\Gamma$) and (2.77) for the impedance scattering problem (with u replaced by u^S and h given by (2.17)). Versions of these equations with simplified expressions for the inhomogeneous terms on the right-hand side are stated in the following two theorems. In the first of these theorems we derive a version of the generalized integral equation (2.98) which includes (2.68) as a special case. Note that (2.114) is a restatement of equation (1.5) in Section 1.

Theorem 2.46. Suppose that u^S satisfies the sound-soft scattering problem (2.16) and that $\eta \in L^\infty(\Gamma)$ and $Z \in (L^\infty(\Gamma))^d$. Then $\partial_n^+ u \in L^2(\Gamma)$ satisfies the integral equation

$$A'_{k,\eta,Z}\partial_n^+ u = f_{k,\eta,Z} := [Z \cdot \nabla u^I - \mathrm{i}\eta u^I]|_\Gamma. \tag{2.113}$$

In the case that $Z = n$ and η is constant, this equation simplifies to

$$A'_{k,\eta}\partial_n^+ u = f_{k,\eta} := f_{k,\eta,n} = [\partial_n^+ u^I - \mathrm{i}\eta u^I]|_\Gamma, \tag{2.114}$$

and this equation is uniquely solvable for all $k > 0$ if $\mathrm{Re}\,\eta \neq 0$. In the case that Ω_- is star-shaped with respect to the origin, satisfying (2.104) for some $c > 0$, $Z(x) := x$, for $x \in \Gamma$, and η is given by (2.102), equation (2.113) is

$$\mathscr{A}_k \partial_n^+ u = f_k := [x \cdot \nabla u^I - \mathrm{i}\eta u^I]|_\Gamma, \tag{2.115}$$

with $\mathscr{A}_k$ defined by (2.103), and this equation is uniquely solvable for all $k > 0$.

Proof. That $\partial_n^+ u$ satisfies (2.113) follows from (2.107), (2.30), (2.31) and the jump relations (2.41)–(2.43). That (2.114) is uniquely solvable follows from Theorem 2.27, and that (2.115) is uniquely solvable from Theorem 2.37. □

We will refer to (2.115) as the *star-combined* integral equation.

The second theorem states the standard combined potential integral equation for the impedance scattering problem, dating back to Burton and Miller (1971), who were concerned with the sound-hard case $\beta = 0$.

Theorem 2.47. If u^S satisfies the impedance scattering problem (2.18) and $\eta \in \mathbb{C}$, then $\gamma_+ u \in H^1(\Gamma)$ satisfies the integral equation

$$C_{k,\eta,k\beta}\,\gamma_+ u = -\left[\frac{\partial u^I}{\partial n} - \mathrm{i}\eta u^I\right]\bigg|_\Gamma, \tag{2.116}$$

with $C_{k,\eta,k\beta}$ defined by (2.78). If $\operatorname{Re}\eta \neq 0$ then $\gamma_+ u$ is the unique solution in $H^{1/2}(\Gamma)$ of this equation.

Proof. Equation (2.116) follows from applying the operator $\partial_n^+ - \mathrm{i}\eta\gamma_+$ to (2.108) (*i.e.*, from taking a particular linear combination of Neumann and Dirichlet traces), on using the jump relations (2.41)–(2.43). That $\gamma_+ u$ is the unique solution of this equation follows from Theorem 2.27. □

2.11. Coercivity of boundary integral operators

For the analysis of numerical methods, for example for the combined potential equation (2.68), it is important that the boundary integral operator ($A'_{k,\eta}$ in (2.68)) is invertible, so that the equation we are solving is well-posed. However, additional properties of the operator are needed to prove convergence of numerical schemes and deduce error estimates.

The error analysis of BIE methods is most developed for schemes based on Galerkin methods, and this will be our focus in Section 6. The argument will proceed as outlined in Section 1, that is, we prove a quasi-optimal error estimate of the form (1.7), and then bound the best approximation error. In particular, for the hybrid methods introduced in Section 3, we cannot currently see any prospect of proving estimates of the form (1.7) for anything other than Galerkin schemes. The results that have been obtained in this direction will be described in Section 6.

Our focus in later sections is on high-frequency analysis, proving (1.7) with the dependence of the constant C on k made explicit. However, although the standard combined potential formulations (2.68) and (2.71) date back at least to Brakhage and Werner (1965) and Burton and Miller (1971), the establishment of (1.7) for standard Galerkin methods, even for fixed k (indeed, even for the Laplace case $k = 0$) remains an outstanding open problem for the case of general Lipschitz Γ. We will describe below, and in

Section 5.7, progress in resolving this open problem, indeed in a way which also teases out the dependence on k. We will also describe briefly in this subsection ways that have been proposed in the last five years for modifying equations (2.68) and (2.71), introducing additional operators into these equations which imply additional computational effort, with the benefit that they allow (1.7) to be established (for fixed k) for general Lipschitz Γ.

The Galerkin method starts by writing the integral equation in weak form as follows: find $v \in \mathcal{V}$ such that

$$a(v, w) = F(w), \quad \text{for all } w \in \mathcal{V}. \tag{2.117}$$

Here $\mathcal{V}$ is some Hilbert space, F is an anti-linear continuous functional on $\mathcal{V}$, that is, some element of $\mathcal{V}'$, the dual space of $\mathcal{V}$, and a is a bounded sesquilinear form on $\mathcal{V} \times \mathcal{V}$, so that, for some constant $B \geq 0$,

$$|a(v, w)| \leq B\|u\| \, \|w\|, \quad v, w \in \mathcal{V}. \tag{2.118}$$

(The smallest $B \geq 0$ for which this inequality holds is the *norm* of a.) The Galerkin method for approximating (2.117) then seeks a solution $v_N \in \mathcal{V}_N \subset \mathcal{V}$, where $\mathcal{V}_N$ is a finite-dimensional subspace, requiring that

$$a(v_N, w_N) = F(w_N), \quad \text{for all } w_N \in \mathcal{V}_N. \tag{2.119}$$

In the case when the integral equation takes the form

$$Av = f,$$

where $v \in \mathcal{V}$, $f \in \mathcal{V}'$, and $A : \mathcal{V} \to \mathcal{V}'$ is a bounded linear operator, the standard way to construct the weak form (2.117) is to define

$$a(v, w) := \langle Av, w\rangle \quad \text{and} \quad F(w) := \langle f, w\rangle, \tag{2.120}$$

where, for $f \in \mathcal{V}'$, $w \in \mathcal{V}$, $\langle f, w\rangle$ denotes the action of the functional f on w.

Let us illustrate this general framework by application to three of the integral equation formulations of Section 2.10, using the fact that (see Section A.3) $H^{-s}(\Gamma)$ is (isomorphic to) the dual space of $H^s(\Gamma)$. For (2.109) we take $\mathcal{V} = H^{-1/2}(\Gamma)$, $\mathcal{V}' = H^{1/2}(\Gamma)$, $v = \partial_n^+ u$,

$$a(v, w) := \langle S_k v, w\rangle_\Gamma \quad \text{and} \quad F(w) := \langle u^I|_\Gamma, w\rangle_\Gamma, \tag{2.121}$$

where $\langle \cdot, \cdot\rangle_\Gamma$ is defined in (A.24). For the integral equation (2.113), which will be the main focus in Section 6, the natural weak form is obtained by setting $\mathcal{V} = \mathcal{V}' = L^2(\Gamma)$, $v = \partial_n^+ u$,

$$a(v, w) := (A'_{k,\eta,Z}\, v, w) \quad \text{and} \quad F(w) := (f_{k,\eta,Z}, w). \tag{2.122}$$

Finally, the standard weak formulation of (2.116) is obtained by setting $\mathcal{V} = H^{1/2}(\Gamma)$, $\mathcal{V}' = H^{-1/2}(\Gamma)$, $v = \gamma_+ u$,

$$a(v, w) := \langle C_{k,\eta,k\beta}\, v, w\rangle_\Gamma \quad \text{and} \quad F(w) := \left\langle -\left[\frac{\partial u^I}{\partial n} - i\eta u^I\right]\Big|_\Gamma, w\right\rangle_\Gamma. \tag{2.123}$$

All of these sesquilinear forms are bounded, the norm of the sesquilinear form a being the norm of the corresponding operator,

$$S_k : H^{-1/2}(\Gamma) \to H^{1/2}(\Gamma), \quad A'_{k,\eta,Z} : L^2(\Gamma) \to L^2(\Gamma),$$
$$\text{and} \quad C_{k,\eta,k\beta} : H^{1/2}(\Gamma) \to H^{-1/2}(\Gamma),$$

in the respective cases (2.121), (2.122), and (2.123).

We will call a bounded sesquilinear form a on $\mathcal{V} \times \mathcal{V}$ *coercive* if, for some $\alpha > 0$ (the *coercivity constant*),

$$|a(v, v)| \geq \alpha \|v\|^2, \quad \text{for all } v \in \mathcal{V}. \tag{2.124}$$

Likewise, we will call the bounded linear operator $A : \mathcal{V} \to \mathcal{V}'$ coercive (with coercivity constant α) if the sesquilinear form a given by (2.120) is coercive. It is easy to see that if $A : \mathcal{V} \to \mathcal{V}'$ is coercive then so is its adjoint $A^* : \mathcal{V}' \to \mathcal{V}$ with the same coercivity constant α.

Simple but powerful results hold if a is bounded and coercive: the Lax–Milgram lemma guarantees that (2.117) has exactly one solution $v \in \mathcal{V}$ for every $w \in \mathcal{V}$, with

$$\|v\|_{\mathcal{V}} \leq \alpha^{-1} \|f\|_{\mathcal{V}'}, \tag{2.125}$$

and existence of the Galerkin solution and quasi-optimality is guaranteed by Céa's lemma.

Lemma 2.48. (Céa's lemma) If the sesquilinear form a is bounded and coercive, satisfying (2.118) and (2.124), then (2.119) has exactly one solution $v_N \in \mathcal{V}_N$, which satisfies

$$\|v - v_N\| \leq \frac{B}{\alpha} \inf_{w_N \in \mathcal{V}_N} \|v - w_N\|. \tag{2.126}$$

This version of Céa's lemma will be one of the main analysis tools for our hybrid numerical-asymptotic methods in Section 6. This will be surprising to many readers since variational problems of the form (2.117), arising from BVPs for the Helmholtz equation, are standard examples of indefinite problems where coercivity does not hold, at least for sufficiently large k. It has seemed reasonable to assume that the same should hold true for weak formulations arising via integral equation formulations. However recent results, discussed in Section 5.7, show that coercivity holds for (2.122), for certain choices of Z and a range of geometries, with α bounded away from zero for all sufficiently large k.

In cases where coercivity does not hold (or is not known to hold) analysis of Galerkin methods, at least for fixed k, can be achieved when the operator A takes the form $A = B + C$, where $B : \mathcal{V} \to \mathcal{V}'$ is coercive and $C : \mathcal{V} \to \mathcal{V}'$ is compact. We will say that A is a *compactly perturbed coercive* operator in this case. (Many authors call our 'coercive' and 'compactly perturbed coercive' *elliptic* and *coercive* respectively.) The standard generalization of Céa's lemma to this case is as follows (see, *e.g.*, Theorem 8.11 in Steinbach 2008).

Theorem 2.49. Suppose the sesquilinear form a is given by (2.120), where the bounded linear operator $A : \mathcal{V} \to \mathcal{V}'$ is a compactly perturbed coercive operator and is injective. Suppose moreover that $(\mathcal{V}_N)_{N\in\mathbb{N}}$ is a sequence of approximation spaces converging to $\mathcal{V}$, in the sense that

$$\inf_{w_N\in\mathcal{V}_N} \|w - w_N\| \to 0 \quad \text{as } N \to \infty,$$

for every $w \in \mathcal{V}$. Then there exists $N_0 \in \mathbb{N}$ and $C > 0$ such that, for $N \geq N_0$, (2.119) has exactly one solution $v_N \in \mathcal{V}_N$, which satisfies

$$\|v - v_N\| \leq C \inf_{w_N\in\mathcal{V}_N} \|v - w_N\|.$$

This theorem is relevant to the combined potential integral equation (2.116) for the impedance scattering problem, for general Lipschitz Γ, since $C_{k,\eta,k\beta} : H^{1/2}(\Gamma) \to H^{-1/2}(\Gamma)$ is a compactly perturbed coercive operator (since the same holds for H_k (McLean 2000, Theorem 7.8) and $C_{k,\eta,k\beta} - H_k$ is compact), and is also injective for all $k > 0$ if $\operatorname{Re}\eta \neq 0$ (see Theorem 2.27). In particular, this means that Theorem 2.49 applies to the standard Burton–Miller integral equation (Burton and Miller 1971) for sound-hard scattering (the special case of (2.116) when $\beta = 0$), although the dependence of N_0 and C on k in this case is still unknown.

Perhaps surprisingly, up to now it seemed harder to develop the theory for Galerkin BIE methods in the general Lipschitz case for sound-soft scatterers. Theorem 2.49 does apply to the integral equation (2.109) for the sound-soft scattering problem, but not for every $k > 0$. The reason is that, while, for all $k > 0$, $S_k : H^{-1/2}(\Gamma) \to H^{1/2}(\Gamma)$ is a compactly perturbed coercive operator (see, *e.g.*, Steinbach 2008, Theorem 6.40), S_k is not injective for all $k > 0$: see Theorem 2.25. On the other hand, the operator $A'_{k,\eta} : L^2(\Gamma) \to L^2(\Gamma)$ is injective, indeed invertible, for all $k > 0$ (Theorem 2.27). However, for general Lipschitz Γ, it is an open question whether $A'_{k,\eta}$ is a compactly perturbed coercive operator. Indeed this question is equivalent, since $A'_{k,\eta}$ is compactly perturbed coercive if and only if $A'_{0,0}$ is, as the difference $A'_{k,\eta} - A'_{0,0}$ is compact (*e.g.*, Chandler-Wilde and Langdon 2007), to the long-standing open question: Is

$$A'_{0,0} = \tfrac{1}{2}I + D'_0,$$

considered as an operator on $L^2(\Gamma)$, a compact perturbation of a coercive operator for general Lipschitz Γ?

The answer to this last question is yes if Γ is C^1, for then D'_0 itself is compact (Fabes, Jodeit and Riviere 1978), and so $A'_{0,0}$ and $A'_{k,\eta}$ are compact perturbations of $\frac{1}{2}I$. In the 2D case, when Γ is Lipschitz and a curvilinear polygon (relevant to many of the examples in this article), it has been known since the 1960s (*e.g.*, Shelepov 1969) that the answer is also yes. For general Lipschitz Γ it was shown by Elschner (1992, Appendix A) (*cf.* Steinbach and Wendland 2001, Steinbach 2008, §6.64, Costabel 2007, Section 4.1), that $A'_{0,0} + cS_0$ is coercive, for all sufficiently large $c > 0$, but on $H^{-1/2}(\Gamma)$ rather than $L^2(\Gamma)$, meaning that

$$|((A'_{0,0} + cS_0)v, v)_{H^{-1/2}(\Gamma)}| \geq \alpha \|v\|^2_{H^{-1/2}(\Gamma)},$$

where $(\cdot,\cdot)_{H^{-1/2}(\Gamma)}$ is the inner product on $H^{-1/2}(\Gamma)$. Since S_0 is a compact operator on $H^{-1/2}(\Gamma)$, it follows that $A'_{0,0}$ is compactly perturbed coercive and Theorem 2.49 can be applied to Galerkin methods for (2.113). However, these are not the usual Galerkin methods but non-standard methods, formulated with respect to the non-local $H^{-1/2}(\Gamma)$ inner product, a setting which is less straightforward than $L^2(\Gamma)$ for computation. See Wendland (2009) for a survey that discusses in more detail what is known about these questions for the operator $A'_{0,0}$.

Given these difficulties in the Galerkin analysis for equations involving $A_{k,\eta}$ and $A'_{k,\eta}$, a number of authors (Buffa and Hiptmair 2005, Buffa and Sauter 2006, Engleder and Steinbach 2007, Engleder and Steinbach 2008) have proposed methods involving modified integral operators. Focusing on $A'_{k,\eta}$ for convenience, these modifications have in common that they replace the 'second-kind' part, $\frac{1}{2}I + D'_k$, of the operator by the product of $\frac{1}{2}I + D'_k$ and another non-local operator, so that the resulting product is a compact operator from $H^{-1/2}(\Gamma) \to H^{1/2}(\Gamma)$, and the whole modified operator $A^{\dagger}_{k,\eta}$ is injective (like $A'_{k,\eta}$) and is a compactly perturbed coercive operator from $H^{-1/2}(\Gamma)$ to $H^{1/2}(\Gamma)$. These modifications are attractive theoretically, but have the disadvantage that their implementation is more complicated and expensive.

An alternative and potentially less costly modification is to use $A'_{k,\eta,Z}$ instead of $A'_{k,\eta}$. The following result, which suggests that this may be a promising idea, is shown in Chandler-Wilde and Spence (2012).

Theorem 2.50. If $\eta \in \mathbb{C}$ and $Z \in (C^{0,1}(\Gamma))^d$ is real-valued and satisfies (2.106) for some $c > 0$, then the operator $A'_{k,\eta,Z} : L^2(\Gamma) \to L^2(\Gamma)$ is a compactly perturbed coercive operator.

This result is encouraging but is not quite as helpful as we would like, as we do not know in general whether $A'_{k,\eta,Z}$ is injective. One case where this *is* clear is when $Z = n$ and $\operatorname{Re}\eta \neq 0$, as then $A'_{k,\eta,Z} = A'_{k,\eta}$ is invertible, but the theorem does not apply in this case unless Γ is $C^{1,1}$.

3. Asymptotics and hybrid approximation spaces

Now that we have developed BIE formulations of scattering problems in Section 2 we are ready to discuss hybrid numerical-asymptotic methods for their solution, and in particular to design hybrid approximation spaces informed by asymptotic analysis, addressing question **Q1** posed in Section 1. The methods of interest here will in general approximate the (k-dependent) solution v of a BIE using an ansatz of the form

$$v(x) \approx V_0(x,k) + c(k) \sum_{m=1}^{M} V_m(x,k) \exp(ik\psi_m(x)), \quad x \in \Gamma. \tag{3.1}$$

In this representation, V_0 is a known (generally oscillatory) function, the phases ψ_m are chosen *a priori*, $c(k)$ is a scaling factor, explicitly chosen, and the amplitudes V_m, $m = 1, \ldots, M$, are approximated numerically. The idea (and in many cases this can be rigorously proved) is that, if the phases are carefully chosen, then $V_m(\cdot, k)$, $m = 1, \ldots, M$, will be much less oscillatory than v and so can be better approximated by piecewise polynomials than v itself.

One possible starting point for constructing such hybrid methods is related to the *physical optics* (or *Kirchhoff*) approximation. Consider a plane wave $u^I(x) := \exp(ikx \cdot \hat{a})$ incident on an infinite plane passing through (for simplicity only) the origin, with unit normal direction n. Then, under the assumption of sound-soft scattering, the correct physical scattered field is the reflected plane wave

$$u^R(x) = -\exp(ikx \cdot \hat{a}^R), \quad \text{where} \quad \hat{a}^R = \hat{a} - 2(n \cdot \hat{a})n. \tag{3.2}$$

(The Sommerfeld radiation condition (2.9) is not, of course, relevant in this case as the scatterer is unbounded.) It follows immediately that if $u := u^I + u^R$ is the total wave then its normal derivative $v = \partial u/\partial n$ satisfies

$$v(x) = 2\frac{\partial u^I}{\partial n}(x) = 2ikn \cdot \hat{a} \exp(ikx \cdot \hat{a}), \quad \text{on} \quad \Gamma. \tag{3.3}$$

Based on the observation that in the high frequency case the scatterer acts locally like a plane, a plausible high-frequency approximation for $v = \partial u/\partial n$ in the scattering problem (2.15), at least when the scatterer is convex, is that, on Γ,

$$v \approx \begin{cases} 2\dfrac{\partial u^I}{\partial n} & \text{on the illuminated part,} \\ 0 & \text{on the part in shadow.} \end{cases} \tag{3.4}$$

In line with many other authors, we will refer to this as the *physical optics* approximation; some other authors call this a geometrical optics approximation. Following usual practice, we will also call the approximation to the solution of (2.15) that is obtained by substituting the approximation (3.4) to $\partial u/\partial n$ into the integral representation (2.107), the *physical optics* or *Kirchhoff* approximation. (The analogous approximation in the sound-hard case is that $v = u|_\Gamma$, the restriction of u to Γ, satisfies $v \approx 2u^I|_\Gamma$ on the illuminated part of Γ and $v \approx 0$ on the part in shadow.)

For a general smooth, strictly convex obstacle the scattering surface naturally divides into illuminated and shadow zones,

$$\{x \in \Gamma : n(x) \cdot \hat{a} < 0\} \quad \text{and} \quad \{x \in \Gamma : n(x) \cdot \hat{a} > 0\}, \quad \text{respectively}, \tag{3.5}$$

which are separated by the *shadow boundary*:

$$S := \{x \in \Gamma : n(x) \cdot \hat{a} = 0\}. \tag{3.6}$$

Motivated by the physical optics approximation (3.4), which is valid for high frequency for such scatterers, a plausible ansatz in the case when the incident field is the plane wave $u^I(x) := \exp(ikx \cdot \hat{a})$, as discussed already in Section 1, is that

$$v(x) = \frac{\partial u}{\partial n}(x) \approx kV(x,k)\exp(ikx \cdot \hat{a}), \tag{3.7}$$

where V is a well-behaved non-oscillatory function away from the shadow boundary. Much work has been done on analysing this approximation (see the discussion in Section 3.1). This is our first example of the general ansatz (3.1).

Example 3.1. The ansatz (3.1), with $M = 1$, $V_0 \equiv 0$ and $\psi_1(x) = x \cdot \hat{a}$, motivated by the physical optics approximation (3.4), is sometimes assumed to hold (with V_1 non-oscillatory) on the whole of Γ, even when Γ is not convex. In Section 3.1 we prove its utility, at least for the smooth convex case.

On the other hand, for a convex polygonal scatterer it can be shown (see Section 3.3) that, if V_0 is the physical optics approximation (3.4), that is,

$$V_0 := \begin{cases} 2\dfrac{\partial u^I}{\partial n} & \text{on illuminated sides,} \\ 0 & \text{on sides in shadow,} \end{cases} \tag{3.8}$$

then

$$v = V_0 + \Sigma, \tag{3.9}$$

where Σ denotes a sum of terms (of a similar form to the summation on the right-hand side of (3.1)), arising from diffraction of the incoming wave at the corners of the polygon.

Example 3.2. The ansatz (3.1), with V_0 given by (3.8) and $M = 2n_s$, where n_s is the number of sides of Γ, will be proposed for convex polygons in Section 3.3, with the phases ψ_m determined by the orientation of the sides of Γ.

As we have described already in Section 1, a number of authors in the acoustics and electromagnetics literature, starting with Uncles (1976), have proposed the ansatz of Example 3.1 as the basis for hybrid numerical-asymptotic methods for scattering by convex obstacles. Abboud, Nédélec and Zhou (1994, 1995) developed this idea, apparently independently of earlier work, and took the first steps towards numerical analysis of these methods. These authors considered the impedance scattering problem (2.18) and formulated this as the first-kind BIE

$$-H_k v + k^2\beta S_k(\beta v) - \mathrm{i}kD_k'(\beta v) - \mathrm{i}k\beta D_k v = g_k := -\frac{\partial u^I}{\partial n} - \mathrm{i}k\beta u^I, \quad (3.10)$$

where $v = u|_\Gamma$, the restriction of u to Γ. Then the ansatz of Example 3.1 was used, and the unknown 'slow variable' $V_1(\cdot, k)$ was approximated using the h-version BEM. Although the method in Abboud, Nédélec and Zhou (1994, 1995) lacks a rigorous error analysis, some formal arguments were given which suggested that consistency error for the Galerkin method for (3.10) remains bounded as $k \to \infty$, provided the BEM is employed with $O(k^{(d-1)/3})$ degrees of freedom. In Abboud *et al.* (1995), and also in the thesis Zhou (1995), a range of experiments on smooth and non-smooth convex 2D objects (and some experiments also on the sphere) were presented, illustrating the sharpness of the formal error estimate described above. The authors also discussed there the computation of the Galerkin stiffness matrix, and proposed some methods for computing the matrix entries which (while not being independent of k in cost) do allow a reduction in complexity by making some use of the method of stationary phase.

In terms of key algorithmic ideas, a substantial step forward was taken in the work of Bruno and collaborators (Bruno *et al.* 2004), which again employed the ansatz of Example 3.1 but combined it with a more careful discretization scheme, aimed at avoiding the breakdown of the ansatz near the shadow boundary. Starting from the direct combined potential formulation (2.114), employing the ansatz of Example 3.1, and multiplying each side of (2.114) by $\exp(-\mathrm{i}kx\cdot\hat{a})$, one obtains

$$\left(\frac{1}{2}I + \widetilde{D_k}' - \mathrm{i}\eta\widetilde{S_k}\right)V(\cdot,k) = \mathrm{i}(kn\cdot\hat{a} - \eta), \quad (3.11)$$

where the single-layer and adjoint double-layer operators with modulated kernels, $\widetilde{S_k}$ and $\widetilde{D_k}'$ respectively, are defined by

$$\widetilde{S_k}\phi(x) := \int_\Gamma \Phi_k(x,y)\exp(\mathrm{i}k(y-x)\cdot\hat{a})\phi(y)\,\mathrm{d}s(y), \quad (3.12)$$

and

$$\widetilde{D_k}'\phi(x) := \int_\Gamma \frac{\partial \Phi_k(x,y)}{\partial n(x)} \exp(\mathrm{i}k(y-x)\cdot\hat{a})\phi(y)\,\mathrm{d}s(y). \tag{3.13}$$

These integral operators have oscillatory kernels containing a factor of the form $\exp(\mathrm{i}k\Psi(x,y))$, where $\Psi(x,y) = |x-y| + (y-x)\cdot\hat{a}$. (This will be explained in more detail later: see, *e.g.*, (4.41).) To solve (3.11), Bruno *et al.* (2004) sampled V at points of a suitable mesh on Γ, and applied special integration rules based on partitions of unity and exponentially convergent quadrature, carefully handling the oscillations in the kernel, to obtain a Nyström-type scheme. The result is a fully discrete method which, in the examples computed, turns out to have close to k-independent error for a fixed number of degrees of freedom. To avoid the breakdown of the approximation of Example 3.1 at the shadow boundary S, the mesh is graded in $O(k^{-1/3})$ neighbourhoods of S. The estimates on the derivatives of V in Section 3.1 below could be used to provide consistency estimates for such a procedure. The work of Bruno *et al.* (2004) was continued by Bruno and Geuzaine (2007), concentrating on the formulation of a robust quadrature method for the oscillatory integrals which arise in the discretization of (3.11) in 3D. This is described in more detail in Section 4.5.

Huybrechs and Vandewalle (2007*b*) solved a model 2D problem related to (3.11), by collocation with a piecewise polynomial basis, and the emphasis then was on the construction of efficient quadrature schemes for the entries of the collocation matrix. These were evaluated by a combination of generalized Filon quadrature with a steepest descent approach; see Sections 4.2 and 4.5 for details. By examining the location of the stationary points of the phase Ψ, Huybrechs and Vandewalle were able to demonstrate that their scheme leads to a sparse BEM formulation, with numerical results comparable to those of Bruno *et al.* (2004) being obtained.

Ganesh and Hawkins (2011) solved (3.11) in the case of a 3D smooth convex obstacle, using a discrete Galerkin method with global polynomial basis and a specially chosen global quadrature rule to compute the Galerkin matrix entries (again, see Section 4.5 for details). Following similar lines to the purely formal argument put forward by Abboud, Nédélec and Zhou (1994), one might expect that (because no special action is taken at the shadow boundary) the method would require $O(k^{2/3})$ degrees of freedom to maintain accuracy as $k \to \infty$, and numerical results in Ganesh and Hawkins (2011) support this conjecture. However this method is much more efficient than standard BEM (which requires at least $O(k^2)$ degrees of freedom as $k \to \infty$) and impressive results for rather large k are presented in Ganesh and Hawkins (2011). The method of Ganesh and Hawkins (2011) requires that the scatterer can be mapped to the sphere with an invertible map, but this is always true for convex obstacles.

For the rest of this section we will describe the asymptotic background to the design of various hybrid numerical-asymptotic methods.

3.1. Smooth convex obstacles

Let us consider the sound-soft scattering problem (2.15). Under the assumption that Γ is C^∞ and convex (in 2D or 3D), and has strictly positive curvature, we write the solution $v := \partial u/\partial n$ of (1.3) as

$$v(x) := kV(x,k)\exp(ikx\cdot\hat{a}), \quad x\in\Gamma \tag{3.14}$$

(*i.e.*, we employ the ansatz of Example 3.1). The key question is how $V(x,k)$ behaves for x near the shadow boundary defined in (3.6) (this is essentially equivalent to understanding how the total field u behaves in the transition region between the illuminated and shadow parts of the exterior domain). The first (non-rigorous) results on this question were by Fock (*e.g.*, Fock 1965), with these results valid in a neighbourhood of the shadow boundary whose width shrinks to zero as $k\to\infty$. This was followed by Ludwig (1967), who obtained (non-rigorous) results in a k-independent neighbourhood of the shadow boundary, and subsequently by the rigorous analysis of Buslaev (1964, 1975); see also Babich and Buldyrev (1991, Chapter 13), Babich and Kirpichnikova (1979) and further references therein. The following result is quoted from Melrose and Taylor (1985, Theorem 9.27), but the same result can essentially be found elsewhere in the literature. For example the 'Fundamental Theorem' in Buslaev (1975, §1.9) concerns the case of Neumann boundary conditions and a point source incident wave, although the general method used there can also be directly extended to the Dirichlet boundary conditions and plane wave incidence which are considered here (Buslaev 1962).

Theorem 3.3. There exists $\Delta>0$ such that $V(x,k)$ has the asymptotic expansion

$$V(x,k)\sim\sum_{\ell,m\geq 0}k^{-1/3-2\ell/3-m}b_{\ell,m}(x)\Phi^{(\ell)}(k^{1/3}Z(x)), \quad x\in\Gamma_\Delta, \tag{3.15}$$

as $k\to\infty$, where $\Gamma_\Delta := \{x\in\Gamma : \operatorname{dist}(x,S)\leq\Delta\}$, and S is the *shadow boundary* defined in (3.6).

The precise meaning of the asymptotic expansion in (3.15) can be made clear using the *symbol classes* of Hörmander (1983*a*, p. 236, Definition 7.8.1), also mentioned briefly in Melrose and Taylor (1985, p. 249). In the context of our problem, a function $p = p(x,k)$ (where $x\in\Gamma_\Delta$ and $k\in(0,\infty)$) is said to lie in the symbol class $S^{\mu}_{\rho,\delta}$ if

$$|D^\alpha_k D^n_x p(x,k)|\leq C_{\alpha,n}(1+k)^{\mu-\rho\alpha+\delta n}, \quad \alpha,\ n\in\mathbb{N}\cup\{0\},$$

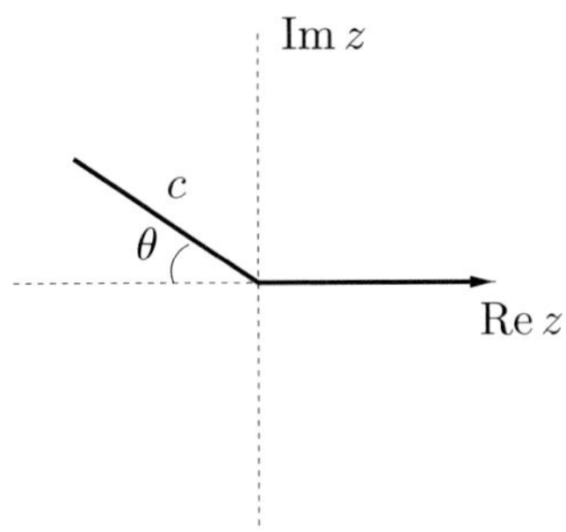

Figure 3.1. The contour c in the complex plane.

where D_x^n denotes a surface derivative of order n on Γ_Δ.

Then, the precise meaning of (3.15) is that, for any $\mu < 0$ and for all $L, M \in \mathbb{N} \cup \{0\}$ sufficiently large (depending on μ), the remainder

$$R_{L,M} := V(x,k) - \sum_{\ell,m=0}^{L,M} k^{-1/3-2\ell/3-m} b_{\ell,m}(x)\Phi^{(\ell)}(k^{1/3}Z(x))$$

satisfies

$$R_{L,M} \in S^{\mu}_{2/3,1/3} \quad \text{for } (x,k) \in I_\Delta \times (0,\infty) \tag{3.16}$$

(*i.e.*, $\rho = 2/3$, $\delta = 1/3$ in this case). This interpretation is equivalent to saying that (3.15) is a conventional asymptotic expansion in the limit as $k \to \infty$, and remains so under term-by-term differentiation on both sides with respect to both x and k. This use of the $\sim$ symbol in (3.15) is prevalent throughout the rigorous asymptotics and microlocal analysis literature, and it always has the same meaning.

In the expansion (3.15), $b_{\ell,m}$ are C^∞ complex-valued functions on Γ_Δ and Z is a C^∞ real-valued function on Γ_Δ, which vanishes to first order on S and is positive-valued on the illuminated zone and negative-valued on the shadow zone (see (3.5)). Moreover $\Phi : \mathbb{C} \to \mathbb{C}$ is an entire function (also known as 'Fock's integral' (*e.g.*, Fock 1965, §7, 12), defined by

$$\Phi(\tau) := \exp(-\mathrm{i}\tau^3/3) \int_c \frac{\exp(-\mathrm{i}z\tau)}{\mathrm{Ai}\,(\exp(2\pi\mathrm{i}/3)z)}\,\mathrm{d}z, \tag{3.17}$$

where Ai is the Airy function and c is the contour depicted in Figure 3.1, with θ chosen to be any sufficiently small positive angle, ensuring the absolute convergence of (3.17). We note that Fock's integral is often denoted $\Psi(\tau)$ in other references (see for example Melrose and Taylor 1985). The asymptotics of $\Phi(\tau)$ are given in Melrose and Taylor (1985, Lemma 9.9) (for $\tau \to +\infty$), whereas for $\tau \to -\infty$ they can be inferred from Babich and Buldyrev (1991, p. 393), Buslaev (1964, Lemma 8); for a summary see Domínguez *et al.* (2007). From these one can obtain the derivative estimates

$$|\Phi(\tau)| \le C_0(1+|\tau|), \qquad |\Phi'(\tau)| \le C_1, \tag{3.18}$$

$$\text{and} \quad |\Phi^{(\ell)}(\tau)| \le C_\ell(1+|\tau|)^{-2-\ell}, \quad \text{for } \ell \ge 2, \tag{3.19}$$

for all $\tau \in \mathbb{R}$, with C_ℓ independent of k.

Using the asymptotic expansion in (3.15) and the interpretation of it in terms of the Hörmander symbol classes, the following result was derived in Domínguez *et al.* (2007).

Corollary 3.4. For all $L, M \in \mathbb{N} \cup \{0\}$, the decomposition

$$V(x,k) = \left[\sum_{\ell,m=0}^{L,M} k^{-1/3-2\ell/3-m} b_{\ell,m}(x)\Phi^{(\ell)}(k^{1/3}Z(x))\right] + R_{L,M}(x,k) \tag{3.20}$$

holds for all $x \in \Gamma_\Delta$, with remainder term satisfying, for all $L, M \in \mathbb{N} \cup \{0\}$,

$$|D_x^n R_{L,M}(x,k)| \le C_{L,M,n}(1+k)^{\mu+n/3}, \tag{3.21}$$

where $C_{L,M,n}$ is independent of k and

$$\mu = -\min\{2(L+1)/3,\ (M+1)\}. \tag{3.22}$$

In numerical analysis we typically need k-explicit estimates on the derivatives of V. These are obtained in the following result, which is a generalization of results in Domínguez *et al.* (2007), to include 3D as well as 2D scattering problems.

Theorem 3.5. For all $n \in \mathbb{N} \cup \{0\}$ there exist constants $C_n > 0$ independent of k and x such that, for all k sufficiently large, and $x \in \Gamma_\Delta$,

$$|D_x^n V(x,k)| \le \begin{cases} C_n & n = 0, 1, \\ C_n\, k^{-1}\,(k^{-1/3} + \operatorname{dist}(x,S))^{-(n+2)} & n \ge 2, \end{cases} \tag{3.23}$$

where D^n denotes any surface derivative operator of order n on Γ_Δ and S is the shadow boundary as defined in (3.6).

Remark 3.6. This theorem quantifies rather precisely the boundary layer effect near the shadow boundary. If $\operatorname{dist}(x,S) \ge C > 0$, then all the derivatives of V decay with $O(k^{-1})$. However if $\operatorname{dist}(x,S) = O(k^{-1/3})$ then for $n \ge 2$, the nth derivative of V grows with $O(k^{(n-1)/3})$. We emphasize also that, up to this point in the theory, this estimate holds only for $x \in \Gamma_\Delta$ (a neighbourhood of the shadow boundary, but independent of k). The extension to all of Γ requires a rigorous application of matched asymptotic expansions: see, *e.g.*, Domínguez *et al.* (2007), where the estimates (3.23) are shown (in the 2D case) to hold for all $x \in \Gamma$.

Proof. Choosing any $n \in \mathbb{N} \cup \{0\}$, we can select L, M so that $-\mu \geq n/3$, with μ defined in (3.22). Then we apply Corollary 3.4 to obtain $V(x,k) = A_{L,M}(x,k) + R_{L,M}(x,k)$, where

$$A_{L,M}(x,k) := k^{-1/3} \sum_{\ell=0}^{L} k^{-2\ell/3} B_{\ell,M}(x) \Phi^{(\ell)}(k^{1/3} Z(x)),$$

$$B_{\ell,M} := \sum_{m=0}^{M} k^{-m} b_{\ell,m}(x),$$

the derivatives of $R_{L,M}$ are bounded as in (3.21), and, by choice of μ,

$$|D_x^n R_{L,M}(x,k)| \leq C_n, \quad \text{for all } k,$$

with C_n denoting a generic constant independent of x and k but possibly depending on n. Now, since all derivatives of $B_{\ell,M}$ are bounded independently of k, we obtain

$$|D_x^n A_{L,M}(x,k)| \leq C_n k^{-1/3} \sum_{j=0}^{n} \sum_{\ell=0}^{L} k^{(j-2\ell)/3} \left| \Phi^{(\ell+j)}(k^{1/3} Z(x)) \right|, \tag{3.24}$$

and the result follows on observing that by (3.18), (3.19), each term in the double sum can be estimated in the required way. □

A hybrid approximation scheme for (1.3) can now be devised by inserting an appropriate piecewise polynomial approximation of V into (3.14). Although methods which do not take any special care near the shadow boundary can still be very effective (Abboud, Nédélec and Zhou 1994, 1995, Aberegg and Peterson 1995, Ganesh and Hawkins 2011), Theorem 3.5 suggests we may do better if we do more work near the shadow boundary. As an example consider the 2D case, where Γ is parametrized by a mapping $\zeta : [0, 2\pi] \to \Gamma$, assumed proportional to arc-length. The shadow boundary consists of two points, here denoted $\zeta(t_i)$, $i = 1, 2$. Following Domínguez *et al.* (2007), we partition the parameter domain into four zones $\Lambda_i : i = 1, \ldots, 4$, where Λ_i, $i = 1, 2$ are small neighbourhoods of t_1, t_2 respectively, chosen as

$$\Lambda_1 := [t_1 - O(k^{-1/3+\delta}),\ t_1 + O(k^{-1/3+\varepsilon})], \tag{3.25}$$

$$\Lambda_2 := [t_2 - O(k^{-1/3+\varepsilon}),\ t_2 + O(k^{-1/3+\delta})], \tag{3.26}$$

with $\delta, \varepsilon \in (0, 1/3)$ parameters to be chosen. For large enough k, Λ_1, Λ_2 are disjoint and the remaining two components of $[0, 2\pi]$ are denoted Λ_3 (contained in the illuminated zone) and Λ_4 (in the shadow zone), with the regions Λ_j touching only at the end points. The hybrid approximation space $\mathcal{V}_N$ is then defined to be the space of functions taking the form (3.14), with V approximated by a polynomial of degree p in each of $\Lambda_i : i = 1, 2, 3$ and

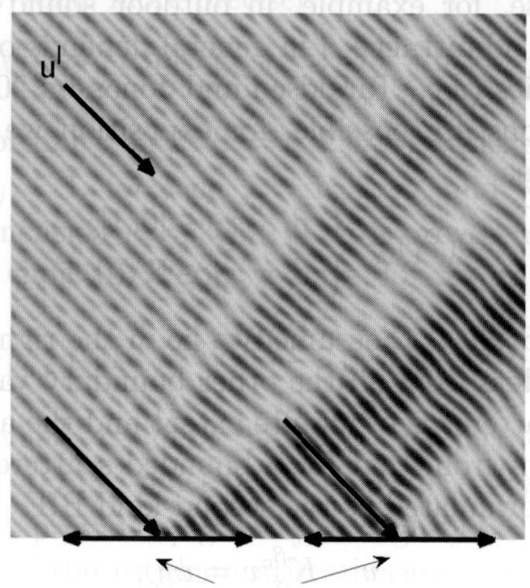

Figure 3.2. Scattered field $u^S = u - u^I$; the arrows represent the incident rays, and those diffracted rays scattered at the discontinuities in the impedance which travel in either direction along the boundary.

Chandler-Wilde (2006, Theorems 2.3, 2.6) have shown that the functions $V^{\pm}$ are not oscillatory, and that their high-order derivatives decay rapidly away from the discontinuities in impedance at $s = 0$ and at $s = 1$.

Theorem 3.7. For $t > 1/k$, $m = 0, 1, \ldots$, there exist constants C_m, dependent only on m, such that

$$\left|V^{\pm(m)}(t)\right| \leq C_m \, k^m (kt)^{-1/2-m} \cos\theta, \tag{3.32}$$

and there exists a constant C, such that

$$\left|V^{\pm}(t)\right| \leq C \, (kt)^{-3/2} (\cos\theta)^{-1}. \tag{3.33}$$

Thus, if θ is bounded away from $-\pi/2$ and $\pi/2$ (the directions of grazing incidence), the estimates (3.33) can be used, while the bounds (3.32) apply for all θ. Numerical experiments in Langdon and Chandler-Wilde (2003*b*, 2006) suggest that, for $m = 0$, these bounds appear to be sharp with respect to both θ and kr. These estimates are crucial to the k-independent convergence rate obtained in Theorem 3.8 below.

Given these estimates the hybrid approximation space $\mathcal{V}_N$ for (3.30) is then chosen to be a space of functions of the form (3.31), with V^+ replaced by certain piecewise polynomials defined on a mesh graded towards 0, and V^- replaced by piecewise polynomials defined on a mesh graded towards 1. A crucial ingredient in the method is the precise design of the mesh, specifically that it is geometric (like (3.47)) on most of the interval $[0, 1]$. Solving (3.30) by the Galerkin method in $\mathcal{V}_N$, the resulting solution $v_N \in \mathcal{V}_N$

has an error estimate given by the following theorem. In this section we shall use the notation $A \lesssim B$ when $A \leq CB$, where C is a constant that does not depend on k or N.

Theorem 3.8. (Langdon and Chandler-Wilde 2006, Theorem 3.5) If $|\beta_1 - \beta_c| < \operatorname{Re}\beta_c$ (*i.e.*, the relative jump in β is not too large), then

$$\|v - v_N\|_{L^2(0,1)} \lesssim \frac{\log^{1/2}(\min(N/\cos\theta, k))}{N^{\nu+1}},$$

where $N \in \mathbb{N}$ is a parameter used in the definition of the graded meshes and ν is the degree of polynomial approximation on each interval of the graded meshes. The space $\mathcal{V}_N$ has dimension $\lesssim N\log(\min(N/\cos\theta, k)$ (in contrast to elsewhere in the review, where we generally use the convention that $\mathcal{V}_N$ has dimension N).

We note that in general the meshes used to approximate V^- and V^+ do not coincide, which ensures the corresponding system matrices are well-conditioned. (Further details are in Langdon and Chandler-Wilde 2006.) This type of approximation scheme arises again below, for example in Section 3.3.

The estimate in Theorem 3.8 is derived by first noting that

$$\|K_\beta^{\beta_c}\|_{L^2(0,1)\leftarrow L^2(0,1)} \leq \frac{|\beta_1 - \beta_c|}{\operatorname{Re}\beta_c} < 1,$$

under the assumption on β. Thus existence and boundedness of $(\mathcal{I} - K_\beta^{\beta_c})^{-1}$ follows from the Banach Lemma. Writing the Galerkin method as

$$(\mathcal{I} - \mathcal{P}_N K_\beta^{\beta_c})v_N = \mathcal{P}_N\psi_{\beta_c}, \tag{3.34}$$

where $\mathcal{P}_N : L^2(0,1) \to \mathcal{V}_N$ is the orthogonal projection onto $\mathcal{V}_N$, with $\|\mathcal{P}_N\|_2 = 1$, it follows that (3.34) is uniquely solvable, and a little algebra (see Section 6.1 for similar arguments) shows that

$$\|v - v_N\|_{L^2(0,1)} \leq \frac{\operatorname{Re}\beta_c}{\operatorname{Re}\beta_c - |\beta_1 - \beta_c|}\|v - \mathcal{P}_N v\|_{L^2(0,1)}.$$

The final result is obtained by using a bound on $\|v - \mathcal{P}_N v\|_{L^2(0,1)}$ proved in Langdon and Chandler-Wilde (2006, Theorem 3.4).

Thus, for fixed N, as k increases the error is bounded *independently* of k. Further results can be found in Langdon and Chandler-Wilde (2003*a*, 2003*b*, 2006), where it is shown, for instance, that the Galerkin method can be implemented with a cost that is independent of k, that the condition number of the Galerkin matrix is bounded independently of k and, via a careful choice of mesh parameters, that the error estimates are uniform with respect to the angle of incidence θ (even as θ approaches grazing angle of incidence).

3.3. Convex polygons

In this subsection we consider the solution of three high-frequency scattering problems, where the scatterers are non-smooth bounded 2D domains. The first is the sound-soft exterior Dirichlet problem (2.16) for a polygonal scatterer ('the sound-soft polygon', with results described in Section 3.3.1), the second is the exterior impedance problem (2.18), again for a polygonal scatterer ('the impedance polygon', Section 3.3.3), and the third is the sound-soft exterior Dirichlet problem for a curvilinear polygon (described briefly in Section 3.3.2). For the two sound-soft problems, (2.16) is reformulated as the BIE (2.114) or (2.115), and the solution $v := \partial u/\partial n$ is written in the form (3.1), using a slightly different ansatz in each case. For the impedance problem, a similar ansatz to (3.1) is used for $v = \gamma u$. Rigorous k-explicit error estimates for approximation in the corresponding hybrid approximation spaces are available for the sound-soft and impedance polygons. For the curvilinear polygon, the validity of the hybrid approximation space is justified by a heuristic argument, and also by numerical results, presented in Section 7.3.

3.3.1. The sound-soft polygon

The chief reference for this subsection is Chandler-Wilde and Langdon (2007), although other references are mentioned below. The treatment begins by establishing that the choice of hybrid space given in Example 3.2 is appropriate for this problem. In fact it is shown that, for $x \in \Gamma$,

$$\begin{aligned} v(x,k) &= V_0(x,k) \\ &\quad + k\sum_{m=1}^{n_s}\big[V_m^+(x,k)\exp(ikx\cdot\hat{d}_m) + V_m^-(x,k)\exp(-ikx\cdot\hat{d}_m)\big], \end{aligned} \tag{3.35}$$

where n_s is the number of sides of the polygon, the unit vector $\hat{d}_m$ is parallel to the mth side Γ_m, the function $V_m^\pm$ is non-zero only on Γ_m, and $V_0(x,k)$ is given by (3.8). It is easily seen that this corresponds to a particular instance of (3.1) with $M = 2n_s$ (note that here v has a different scaling to that used in Chandler-Wilde and Langdon (2007)). Let Γ_m denote the side of the polygon connecting the vertices P_m and P_{m+1}, with length L_m and corresponding exterior angles $\omega_m, \omega_{m+1} \in (\pi, 2\pi)$ at its end-points, and make the convention that $P_{n_s+1} = P_1$ and $\omega_{n_s+1} = \omega_1$. Then, for $x \in \Gamma_m$, $m = 1, \ldots, n_s$, we can rewrite (3.35) as

$$v(s) = V_0(s) + k\big[V_m^+(s)\exp(iks) + V_m^-(L_m - s)\exp(-iks)\big], \quad s \in [0, L_m], \tag{3.36}$$

where s denotes the distance of x from P_m.

To establish the potential of (3.35) as the starting point for a hybrid approximation space, it is necessary to establish the k-explicit regularity of the functions $V_m^\pm(\cdot,k)$. In fact it turns out that the representation (3.35) captures the oscillatory behaviour of v exactly, in the sense that the functions $V_m^\pm$ are non-oscillatory. The key to obtaining this result, as for all of the results in this section, lies in understanding precisely the high-frequency asymptotics of the solution to the scattering problem. Comparing the problems in this subsection with those in Section 3.1, it might at first appear that polygonal scatterers present more of a challenge, due to the difficulties caused by the corners of the polygon. However, it turns out that if x is sufficiently close to a corner, the behaviour of $V_m^\pm$ can be understood using standard techniques for handling elliptic corner singularities (*e.g.*, Grisvard 1985), or by explicitly representing u and $\partial u/\partial n$ near corners via separation of variables in polar coordinates centred on the corner. Using this approach the following theorem was established.

Theorem 3.9. (Chandler-Wilde and Langdon 2007, Corollary 3.4) For $m = 1, \ldots, n_s$, $t > 0$, and for all $n \geq 0$, the bound

$$\left|\frac{\partial^n}{\partial t^n} V_m^+(t)\right| \lesssim M(u)\, k^n (kt)^{-\alpha_m - n} \tag{3.37}$$

holds for $kt \leq 1$, where $\alpha_m := 1 - \pi/\omega_m \in (0, 1/2)$, and

$$M(u) := \sup_{x\in\Omega_+} |u(x)|.$$

The hidden constant in (3.37) is independent of k and t. The same bounds hold for V_m^-, with α_m replaced by α_{m+1}.

The dependence of the constant $M(u)$ on the wavenumber k is not yet fully understood. Hewett *et al.* (2012) showed that when Ω is a star-shaped polygon, $M(u)$ grows at most algebraically with k; specifically,

$$M(u) = O(k^{1/2} \log^{1/2} k) \quad \text{as} \quad k \to \infty. \tag{3.38}$$

However, numerical experiments in Chandler-Wilde and Langdon (2007), Hewett *et al.* (2012) and Chandler-Wilde *et al.* (2012*a*) lead us to conjecture that, in fact, $M(u) = O(1)$ as $k \to \infty$, for both convex and a more general class of non-convex polygons. As yet though, this statement remains to be proved.

Establishing regularity results for $V_m^\pm(x,k)$ when x is further away from the corners turns out to be a somewhat easier task than deriving the analogous estimates for smooth convex scatterers, as in Section 3.1. The key step is the observation (adapted from Chandler-Wilde *et al.* 2004 and related to the ideas in Section 3.2) that one can write down an explicit solution to the Dirichlet boundary value problem for the Helmholtz equation in a half-plane, using the explicit Green's function constructed via the method of images. To this end, let $D_m \subset \Omega_+$ denote the half-plane whose boundary contains Γ_m, and set $G_{m,k}(x,y) = \Phi_k(x,y) - \Phi_k(x,y'_m)$, where y'_m denotes the image of y in the straight line ∂D_m. (This is the Dirichlet Green's function for the half-plane D_m.) Then, for an illuminated side Γ_m (shadow sides are treated similarly), we have the formula

$$u(x) = u^I(x) + u^R(x) + \int_{\partial D_m} \frac{\partial G_{m,k}(x,y)}{\partial n(y)} u(y)\,\mathrm{d}s(y), \quad \text{for } x \in D_m. \quad (3.39)$$

Here u^R is the plane wave reflected from ∂D_m (defined analogously to (3.2)). Taking the normal derivative in (3.39), noting that $\partial G_{m,k}(x,y)/\partial n(y) = 2\partial\Phi_k(x,y)/\partial n(y)$, and recalling that $u = 0$ on Γ, we have

$$\frac{\partial u}{\partial n}(x) = 2\frac{\partial u^I}{\partial n}(x) + 2\int_{\partial D_m \setminus \Gamma_m} \frac{\partial^2 \Phi_k(x,y)}{\partial n(x)\partial n(y)} u(y)\,\mathrm{d}s(y), \quad x \in \Gamma_m. \quad (3.40)$$

The formula (3.40) holds on each illuminated side. The same formula holds also on shadow sides but with the term $2\partial u^I/\partial n$ deleted. Thus, on a side Γ_m, it holds that

$$\frac{\partial u}{\partial n}(x) = V_0(x,k) + k\big[V_m^+(x,k)\,\mathrm{e}^{\mathrm{i}kx\cdot\hat{d}_m} + V_m^-(x,k)\mathrm{e}^{-\mathrm{i}kx\cdot\hat{d}_m}\big], \quad (3.41)$$

where V_0 is defined by (3.8) and the functions $V_m^\pm$ and unit vector $\hat{d}_m$ are defined in (3.42) below. This formula is obtained by decomposing the integral in (3.40) into two parts, corresponding to the two lines either side of Γ_m that extend to infinity in each direction (which we denote by $\Gamma_m^\pm$), so that explicitly

$$V_m^\pm(x,k) := \frac{\mathrm{i}k}{2}\int_{\Gamma_m^\pm} \exp(\mp\mathrm{i}ky\cdot\hat{d}_m)\mu(k|x-y|)u(y)\,\mathrm{d}s(y), \quad (3.42)$$

with $\mu(z) := \exp(-\mathrm{i}z)H_1^{(1)}(z)/z$, for $z > 0$, and with $\hat{d}_m$ the unit vector pointing from Γ_m^+ along Γ_m.

Equation (3.41) is the representation (3.35) promised in Example 3.2, evaluated on Γ_m. Although we cannot evaluate the integrals in (3.42), as they involve the unknown u on $\Gamma_m^\pm$, it can be shown that the $V_m^\pm$ are not oscillatory on Γ_m. This is the content of the following theorem.

Theorem 3.10. (Chandler-Wilde and Langdon 2007, Theorem 3.2) For $m = 1,\ldots,n_s$, $t > 0$, and for all $n \geq 0$, we have

$$\left|\frac{\partial^n}{\partial t^n} V_m^\pm(t)\right| \lesssim M(u)k^n(kt)^{-1/2-n}, \quad (3.43)$$

for $kt \geq 1$. The hidden constant in (3.43) is independent of k and t.

The hybrid approximation space $\mathcal{V}_N$ for this problem is defined as the space of all functions of the form (3.35), with $V_m^\pm$, $m = 1,\ldots,n_s$, each replaced by certain piecewise polynomials of degree p. Specifically, V_m^+ is approximated on Γ_m using piecewise polynomials of degree p, on a mesh which is graded near the corner P_m (to accommodate the blow-up of derivatives near that point) and has points more spread out over the rest of Γ_m. Analogously V_m^- is approximated on Γ_m using piecewise polynomials on a mesh graded towards the other corner, P_{m+1} (recall (3.36)); as for the scheme described in Section 3.2, the meshes used to approximate V_m^+ and V_m^- do not usually coincide and the precise design of the meshes is crucial to the success of the method. Using Theorems 3.9 and 3.10, one can prove the following best approximation estimate for $\mathcal{V}_N$.

Theorem 3.11. (Chandler-Wilde and Langdon 2007, Theorem 4.4) Denote the length of the boundary of the polygon by L, the total number of degrees of freedom by N, and the degree of polynomial approximation on each element by p. Then

$$k^{-1/2} \inf_{w_N \in \mathcal{V}_N} \|v - w_N\| \lesssim M(u)(n_s \log(kL))^{1/2} \left(\frac{n_s \log(kL)}{N} \right)^{p+1}.$$

The hidden constant in this estimate depends only on the corner angles and on p.

Combining with (3.38), this theorem shows that in order to maintain control of the quantity $k^{-1} \inf_{w_N \in \mathcal{V}_N} \|v - w_N\|$, it is sufficient to increase the number of degrees of freedom N slightly faster than $\log k$ as k increases. For a discussion of what are the appropriate quantities to consider when measuring the error, and why, we refer to Section 6.5.

A collocation method based on the identical integral equation formulation and the same approximation space $\mathcal{V}_N$ is implemented in Arden, Chandler-Wilde and Langdon (2007). The numerical results there suggest that there is little difference in accuracy between the Galerkin and the (rather easier to implement) collocation method for this problem, although there is no error analysis for collocation.

3.3.2. Convex curvilinear polygons

The extension of the ideas in Section 3.3.1 to sound-soft convex curvilinear polygons is discussed in Langdon *et al.* (2010). In this case the ansatz for smooth convex obstacles (3.14) is combined with that for convex polygons (3.35), leading to the ansatz (for $v = \partial u/\partial n$)

$$\begin{aligned} v(x,k) &= kV(x,k)\exp(ikx \cdot \hat{a}) \\ &\quad + k\sum_{m=1}^{n_s} \big[V_m^+(x,k)\exp(iks) + V_m^-(x,k)\exp(-iks)\big], \end{aligned} \tag{3.44}$$

for $x = \zeta(s) \in \Gamma$, where ζ again represents arc-length parametrization of Γ, n_s is the number of sides of the curvilinear polygon and each function $V_m^\pm$

is assumed non-zero only on side m. Here, V and $V_m^{\pm}$ are all to be found. Whereas for smooth obstacles the regularity of V can be established, and for convex polygons V is known and $V_m^{\pm}$ can be proved to be slowly oscillating, for convex curvilinear obstacles such results have, to date, remained elusive. Nevertheless, numerical results in Langdon *et al.* (2010) suggest (at least for the examples tested) that, if we use the ansatz (3.44) and approximate V as in Section 3.1 and $V_m^{\pm}$ as in Section 3.3.1, then the number of degrees of freedom required to approximate $(1/k)\partial u/\partial n$ to any given level of accuracy for convex curvilinear polygons grows only logarithmically as k increases. This is a similar result to that described in Theorem 3.11 above for straight-sided convex polygons. Further numerical results in Section 7.3 illustrate how this result appears to extend to the computation of the solution u in the domain, and of the far-field pattern F defined by (2.10) (see also (2.23) and (2.24)).

3.3.3. The impedance polygon

The extension of the method in Section 3.3.1 to scattering from a convex polygon with impedance boundary condition is described in Chandler-Wilde *et al.* (2012*b*). In this case, we have $v = \gamma u$ as the unknown function to be approximated on Γ, rather than $\partial u/\partial n$, and the integral formulation (2.111) is used. Again, v can be expressed as a product of oscillatory and non-oscillatory functions on each side of the polygon, leading to a similar ansatz to that used for the sound-soft convex polygon (3.35), but without the factor k multiplying the summation on the right-hand side, and with a different V_0. Similar techniques to those described in Section 3.3.1 can be applied to establish regularity estimates for $V_m^{\pm}$. In this case, it is proved in Chandler-Wilde *et al.* (2012*b*) that (using the same notation as in Section 3.3.1)

$$k^{-n}\left|\frac{\partial^n}{\partial t^n}V_m^{+}(t)\right| \lesssim \begin{cases} M(u)(kt)^{\beta_m} & \text{for } kt \leq 1, \\ M(u)(kt)^{-1/2-n} & \text{for } kt \geq 1, \end{cases}$$

where the hidden constant is independent of k and t and where $\beta_m = \min\{0, \hat{\alpha}_m - n\}$, with $\hat{\alpha}_m = \pi/\omega_m \in (1/2, 1)$. Similar bounds hold for V_m^{-}, with $\hat{\alpha}_m$ replaced by $\hat{\alpha}_{m+1}$. An order p polynomial approximation space similar to that in Section 3.3.1 is employed, again with geometric mesh grading except within a wavelength of each corner. Letting $\mathcal{V}_N$ denote the resulting approximation space (and using the same notation as in the statement of Theorem 3.11), the best approximation result is

$$k^{1/2} \inf_{w_N \in \mathcal{V}_N} \|v - w_N\| \lesssim M(u)(n_s \log(kL))^{1/2}\left(\frac{n_s \log(kL)}{N}\right)^{p+1}. \qquad (3.45)$$

The hidden constant depends only on the impedance β, the corner angles, and p. This estimate is identical to Theorem 3.11 except that the factor of $k^{-1/2}$ on the left-hand side of the estimate in Theorem 3.11 is now a factor of $k^{1/2}$ (recall that here we have $v = \gamma u$, as opposed to $v = \partial u/\partial n$ in Section 3.3.1). Thus, under the assumption that (3.38) holds also for the impedance case, we can maintain control of $\inf_{w_N \in \mathcal{V}_N} \|v - w_N\|$ by increasing the number of degrees of freedom N slightly faster than $\log k$ as k increases. Results in Section 7.4 (*e.g.*, Figure 7.11) suggest that this may even be pessimistic, with errors decreasing for fixed N as k increases (as would be expected from (3.45) if $M(u) = O(1)$ was known to hold).

3.3.4. Exponential convergence

All of the theoretical and numerical results described in Section 3.3 so far have concerned methods with an algebraic order of convergence. These results have recently been improved upon in Hewett *et al.* (2012), where, for the sound-soft convex polygon, *hp* methods with exponential convergence rate are presented. The analysis of this scheme requires stronger analyticity and regularity estimates on $V_m^{\pm}$, appearing in (3.36), than are provided by Theorems 3.9 and 3.10. In Hewett *et al.* (2012) it is shown that $V_m^{\pm}(t)$, $m = 1, \ldots, n_s$, have an analytic continuation into the right half-plane $\mathrm{Re}\,(t) > 0$, where (using the same notation as in Theorem 3.9)

$$|V_m^+(t)| \lesssim \begin{cases} M(u)|kt|^{-\alpha_m} & 0 < |t| \leq 1/k, \\ M(u)|kt|^{-1/2} & |t| > 1/k, \end{cases} \tag{3.46}$$

and an identical bound holds for $|V_m^-|$, with α_m replaced by α_{m+1}.

We can describe the *hp* space for approximating v in (3.36) by simply describing how V_m^+ and V_m^- are approximated. For V_m^+ we define the geometric mesh

$$x_0 := 0, \qquad x_i := \sigma^{n-i} L_m, \quad i = 1, 2, \ldots, n, \tag{3.47}$$

where L_m is the length of side Γ_m, $0 < \sigma < 1$ is a grading parameter and $n \in \mathbb{N}$. Then we set $\mathcal{V}_m^+$ to be the space of piecewise polynomials of degree p on $[0, L_m]$ with respect to this mesh. Because of the grading, this space approximates V_m^+ well despite the singularity at $s = 0$. We define $\mathcal{V}_m^-$ completely analogously by a mesh graded towards L_m. The approximation space $\mathcal{V}_N$ then consists of functions of the form (3.35), with V_m^+ and V_m^- replaced by functions from $\mathcal{V}_m^+$ and $\mathcal{V}_m^-$ respectively. Typically the number of layers n on each geometric mesh is taken to be proportional to p (alternative choices are discussed in Hewett *et al.* (2012)), in which case the approximation space $\mathcal{V}_N$ has dimension $N = O(p^2)$, and the best approximation result is then (with $v = \partial u/\partial n$)

$$\inf_{w_N \in \mathcal{V}_N} \|v - w_N\| \lesssim M(u) k^{\alpha} \exp(-p\tau), \tag{3.48}$$

where $\tau > 0$ is a constant which represents the rate of exponential decay,

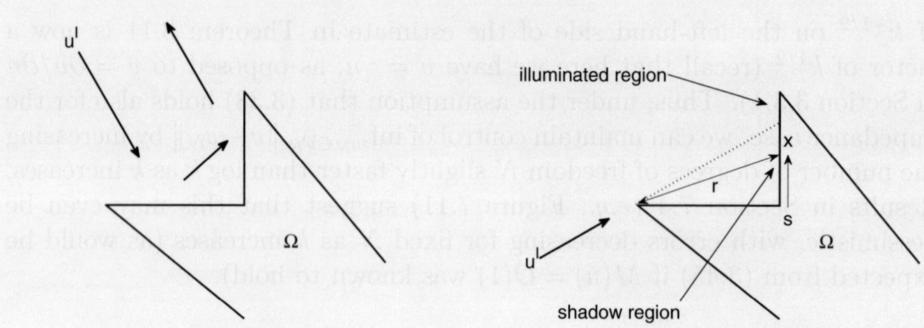

(a) re-reflections (b) partial shadowing

Figure 3.3. New phenomena that arise for non-convex obstacles.

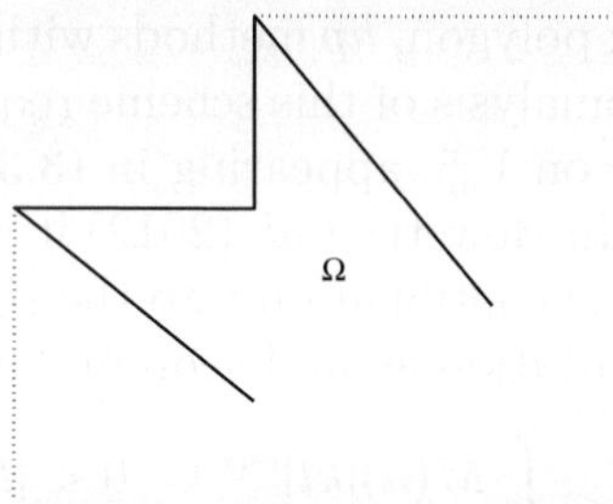

Figure 3.4. Assumption on geometry of non-convex obstacles, that Ω lies entirely within the semi-infinite dashed lines.

and $\alpha = 1 - \min_{m=1,\ldots,n_s}(1 - \pi/\omega_m) \in (1/2, 1)$, with ω_m the external angles at the corners of the polygon. To maintain accuracy as k increases, it is sufficient for p to grow only logarithmically, in which case the algebraically growing k-dependent prefactor in (3.48) (recall (3.38)) will be absorbed into the exponentially decaying factor. Numerical results demonstrating this are provided in Section 7.2 (see Figure 7.5).

3.4. Non-convex polygons

In this subsection we consider the sound-soft scattering problem (2.16) when Γ is the boundary of a non-convex two-dimensional polygon. Non-convexity significantly complicates the behaviour of the solution on Γ, permitting two new phenomena that cannot occur for convex obstacles: re-reflections (where the incident field reflects off one part of Γ onto another part of Γ) and partial illumination (where one part of Γ obscures another, creating a shadow zone and shadow boundary that do not correspond to the definitions (3.5) and (3.6)). These are illustrated in Figure 3.3.

The main reference for the results described in this subsection is Chandler-Wilde *et al.* (2012*a*). Only a restricted class of polygonal geometries is

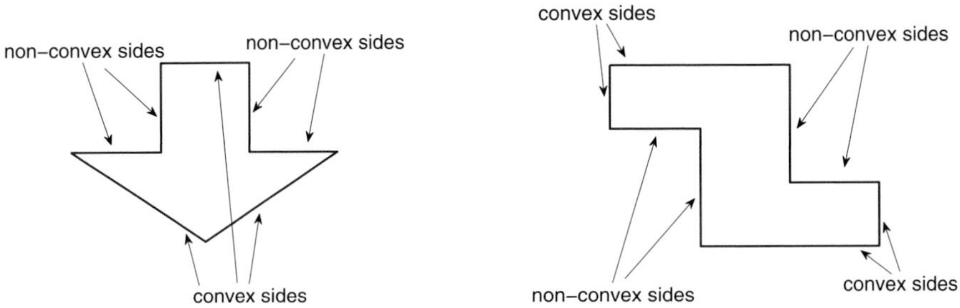

(a) star-shaped polygon (b) non-star-shaped polygon

Figure 3.5. Examples of non-convex polygons satisfying our geometrical assumptions.

described rigorously by the present theory. While there is some evidence that the approximation space we propose will be effective in practice for a wider range of geometries, it will not be appropriate in certain cases, *e.g.*, for the trapping domains described in Section 5.2. The assumptions on the geometry are as follows.

(1) Each external angle is either greater than π (a 'convex corner') or exactly $\pi/2$ (a 'non-convex corner'); this assumption permits the use of a representation formula for the solution in a quarter-plane that leads to a significant simplification of the analysis in Chandler-Wilde *et al.* (2012*a*).

(2) For each non-convex corner, the obstacle Ω must lie within the dashed lines illustrated in Figure 3.4; this assumption ensures that re-reflections and partial shadowing occurring on any side of the polygon can only be caused by a single alternative side of the polygon (and also disallows trapping domains), which eases the analysis considerably.

Such polygons may or may not be star-shaped. Two examples of scatterers satisfying these constraints are shown in Figure 3.5. We call a side Γ_m, $m = 1, \ldots, n_s$, 'convex' if each of its end points is a convex corner and 'non-convex' otherwise. On convex sides, the ansatz (3.36) holds (for $v = \partial u/\partial n$), and the analysis of Sections 3.3.1 and 3.3.4 can be carried over to derive identical best approximation estimates on those sides as for convex polygons.

We now consider the behaviour on a typical non-convex side. For ease of presentation, we consider only the configuration in Figure 3.3(b), and describe a representation for the solution $v = \partial u/\partial n$ on the vertical side in that figure (with r representing the distance from the 'shadowing' corner to a point x on that side, and s representing the distance from the non-convex corner to the point x (thus r can be thought of as a function of s), as shown in the figure). In this case, the side is partially illuminated; we also expect

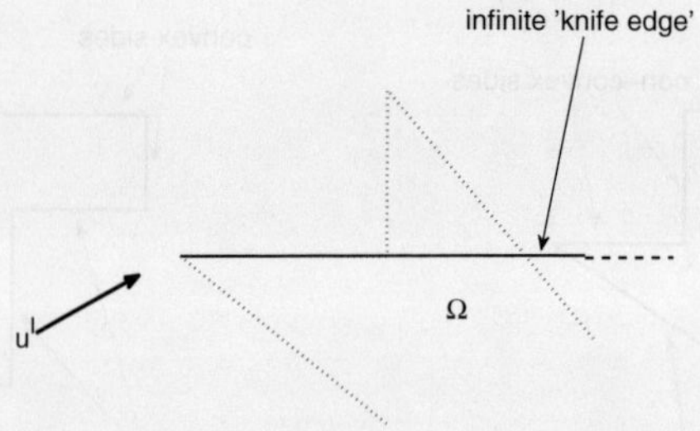

Figure 3.6. The leading-order behaviour u^d on the vertical side of Ω corresponds to the known exact solution to the problem of diffraction by the infinite knife edge in the figure.

diffraction from each corner in the figure. The following theorem describes an appropriate ansatz on such a side (which we denote by Γ_m, and presume to have length L_m).

Theorem 3.12. (Chandler-Wilde *et al.* 2012*a*) On the vertical side Γ_m in Figure 3.3(b), $v = \partial u/\partial n$ satisfies

$$v(s) = V_0(s) + V^+(L_m + s)\exp(iks) + V^-(L_m - s)\exp(-iks) + \tilde{V}(s)\exp(ikr(s)), \qquad (3.49)$$

for $s \in [0, L_m]$, where

$$V_0(s) := 2\frac{\partial u^d}{\partial n}(s),$$

and u^d is the known exact solution to the problem of diffraction of the incident plane wave u^I by the infinite knife edge (with Dirichlet boundary conditions) that would be created were the horizontal side of the polygon to be extended indefinitely to the right, as shown in Figure 3.6, and the rest of Ω ignored. The functions $V^{\pm}$ and $\tilde{V}$ can be shown to be analytic in appropriate regions of the complex plane to allow a similar approximation theory to that described in Section 3.3.4 to be applied; in particular, $V^{\pm}$ satisfy the bounds (3.46), and (for star-shaped polygons)

$$|\tilde{V}(s)| \leq Ck\log^{1/2} k,$$

in a fixed (independent of k) neighbourhood of $[0, L_m]$ in the complex plane.

For alternative configurations such as that corresponding to Figure 3.3(a), very similar results hold. We refer to Chandler-Wilde *et al.* (2012*a*) for details.

There are parallels between the proof of Theorem 3.12 and the proofs of the regularity results for convex polygons described in Section 3.3.1. In particular, whereas the results in Section 3.3.1 are based on a representation of the solution in a half-plane, the proof of Theorem 3.12 is based on the explicit expression for the Green's function in a quarter-plane, with this used in the quarter-plane formed by extending to infinity the two sides subtended by the non-convex corner. This theoretical reliance on the explicit expression for the quarter-plane Green's function has driven the geometrical restrictions imposed here, but we anticipate that an approximation space based on ansatz (3.49) could equally well be applied to more general star-shaped non-convex scatterers, although the analysis does not carry forward to such cases at present.

The hybrid approximation space $\mathcal{V}_N$ for this problem is then defined as the set of functions that on each convex side has the form (3.36), with V_m^+ and V_m^- replaced by functions from $\mathcal{V}_m^+$ and $\mathcal{V}_m^-$ respectively (as described in Section 3.3.4). On non-convex sides we use representations based on the ansatz (3.49); for the configuration described in Theorem 3.12 on Γ_m, V^- is replaced by a function from $\mathcal{V}_m^-$ (to deal with the singularity at the convex corner at the top of the vertical side in Figure 3.3(b)), and V^+ and $\tilde{V}$ are each replaced by a polynomial on the whole of Γ_m, since these functions are analytic in a neighbourhood of $[0, L_m]$.

Assuming for simplicity the same number of layers n on each graded mesh, and the same degree p of polynomial approximation on each element, the result is a scheme with $2n(p+1)$ degrees of freedom per convex side and $(n+2)(p+1)$ per non-convex side. As described in Section 3.3.4, if the number of layers n on each geometric mesh is taken to be proportional to p then the approximation space $\mathcal{V}_N$ has dimension $N = O(p^2)$. Using the regularity results provided by Theorem 3.12 and (3.46) the following best approximation result can be proved (for $v = \partial u/\partial n$).

Theorem 3.13. (Chandler-Wilde *et al.* 2012*a*) For star-shaped non-convex polygons satisfying the geometrical assumptions described above, for some $\tau > 0$,

$$\inf_{w_N \in \mathcal{V}_N} \|v - w_N\| \lesssim M(u) k^{1/2+\alpha} \exp(-p\tau), \tag{3.50}$$

where $\alpha = 1 - \min_{m=1,\ldots,n_s}(1 - \pi/\omega_m) \in (1/2, 1)$, with ω_m the external angles at the corners of the polygon.

To maintain accuracy as k increases, since (3.38) holds, it is again sufficient for p to grow logarithmically as $k \to \infty$. Numerical results illustrating this are provided in Section 7.5 (see Tables 7.6–7.8), where it is also shown that the approximation properties of the best approximation (3.50) carry over to the approximation of the far-field pattern.

3.5. Multiple scattering

A very interesting extension of the method in Bruno *et al.* (2004) and Bruno and Geuzaine (2007) to multiple scattering is given in Geuzaine *et al.* (2005). There it is explained how the integral equation (1.3) may be solved by a

Neumann series approach, where each term in the Neumann series corresponds to the scattering by a single convex obstacle of an incident field consisting of the incident wave combined with previously scattered waves. Each of these single-obstacle scattering problems can be solved by a method similar to the methods described above, except that now the ansatz (3.1) becomes somewhat more complicated. Although there is still only one term in the expansion (*i.e.*, $V_0 = 0$ and $M = 1$), the phase ψ_1 has now to be chosen as a function reflecting the optical distance travelled by rays through all previous reflections (as worked out in detail for the case of two scatterers when each individual scatterer is smooth and convex in Geuzaine *et al.* (2005)). Preliminary numerical tests were provided in Geuzaine *et al.* (2005) which demonstrated the potential for the method. The theory was substantially advanced in the subsequent works of Ecevit (2005) and Ecevit and Reitich (2009), where the implementation of the Neumann series was shown to correspond to a sum over increasing period of a sequence of periodic orbits. Each orbit corresponds to reflections off a fixed set of scatterers, and this allows the convergence rate of the Neumann series to be estimated, for sufficiently high frequency, and permits the formulation of methods for accelerating its convergence. The most recent work in this direction (Anand *et al.* 2009) extended the analysis to the three-dimensional case, where additional considerations on the relative orientation of the scattering bodies come into play.

4. Numerical treatment of oscillatory integrals

In this section we review work on the key implementation issue (see question **Q2** identified in Section 1) of how to assemble the linear systems arising from hybrid methods with a computational time that either remains bounded or else grows only slowly as $k \to \infty$. In Section 4.1 we describe the specific integrals which arise in the numerical treatment of scattering problems by hybrid Galerkin methods. These are particularly challenging examples of the general problem of computing oscillatory integrals, which has a long history and has seen a great deal of research interest in recent years. In Section 4.2 we review classical and modern work on oscillatory integration in general. The oscillatory integrals introduced in Section 4.1 have integrands which are products of explicit oscillatory exponentials and complicated non-oscillatory factors; Filon quadrature rules which require only point-values of the non-oscillatory factors are particularly useful for these. Some classes of such rules are reviewed in Section 4.3, together with recent progress in their error analysis. These are applied to computation of Galerkin integrals in scattering problems in Section 4.4. The section is completed with Section 4.5, in which the integration approaches in a number of other hybrid methods are reviewed.

4.1. Integration problems arising in hybrid Galerkin methods

We will be chiefly concerned with discretization of integral equations arising from the direct and indirect formulations of scattering problems, of the form

$$A_{k,\eta}v := \left(\frac{1}{2}I + D_k - \mathrm{i}\eta S_k\right)v = f, \tag{4.1}$$

$$A'_{k,\eta}v := \left(\frac{1}{2}I + D'_k - \mathrm{i}\eta S_k\right)v = f, \tag{4.2}$$

(recall (2.71) and (2.68) for the general case and (2.114) for the direct formulation in the particular case of sound-soft scattering), where η is usually taken to be a k-dependent constant. We will also be concerned with implementation of the star-combined formulation, which is of the form

$$\mathscr{A}_k v := (x \cdot n)\left(\frac{1}{2}I + D'_k\right)v + x \cdot \nabla_\Gamma S_k v - \mathrm{i}\eta S_k v = f, \tag{4.3}$$

where η is a (smooth) function of $x \in \Gamma$ (see (2.103) and (2.115)). Each equation above is posed in $L^2(\Gamma)$ and in general f is k-dependent.

Let us suppose we take a Galerkin approach based on the ansatz (3.1) and let ϕ, $\widetilde{\phi}$ denote a typical pair of basis functions in this Galerkin method. Disregarding for a moment the surface derivative term (the second term of the middle entry of (4.3)), the Galerkin implementation of all other terms will require computation of various inner products. Firstly inner products which do not involve integral operators appear. These are $(d-1)$-dimensional 'load vector' and 'mass matrix' integrals, of the form

$$(f, \widetilde{\phi}), \quad (bV_0, \widetilde{\phi}) \quad \text{and} \quad (b\phi, \widetilde{\phi}), \tag{4.4}$$

with b being smooth and non-oscillatory, f being the right-hand side in (4.1), (4.2) and (4.3), and V_0 the known, in general oscillatory, leading-order function in (3.1). All inner products involving $V_0(x,k)$ are with respect to x. Secondly there are 'stiffness matrix' inner products involving integral operators. These are in principle $2(d-1)$-dimensional, and in general take the form

$$(D_k\phi, b\widetilde{\phi}), \ (D'_k\phi, b\widetilde{\phi}), \ (S_k\phi, b\widetilde{\phi}), \tag{4.5}$$

or

$$(D_k V_0, b\widetilde{\phi}), \ (D'_k V_0, b\widetilde{\phi}), \ (S_k V_0, b\widetilde{\phi}), \tag{4.6}$$

with b again smooth and non-oscillatory (in some cases $b \equiv 1$). Finally the surface gradient term in (4.3) yields further 'stiffness matrix' inner products of the form

$$(\nabla_\Gamma S_k\phi, \ \widetilde{\phi}x) \quad \text{and} \quad (\nabla_\Gamma S_k V_0, \ \widetilde{\phi}x), \tag{4.7}$$

and these can be reduced to inner products just involving the single-layer

Notation 4.4. If R is any linear integral operator on Γ with kernel function r and if ψ_m are the phases appearing in (4.8), then we denote by $R^{m,m'}$ the integral operator:

$$(R^{m,m'}w)(x) = \int_\Gamma r^{m,m'}(x,y)w(y)\mathrm{d}s(y),$$

with modulated kernel $r^{m,m'}(x,y) := \exp(ik(\psi_m(y) - \psi_{m'}(y))r(x,y)$.

With $\phi, \widetilde{\phi}$ as defined above, it is then easy to see that

$$\begin{aligned}(R\phi, b\widetilde{\phi}) &= (R^{m,m'}(\chi_j P_\ell), b\chi_{j'}P_{\ell'}) \\ &= \int_{\Gamma_{j'}}\int_{\Gamma_j} r^{m,m'}(x,y)\chi_j(y)P_\ell(y)b(x)\chi_{j'}(x)P_{\ell'}(x)\,\mathrm{d}s(y)\,\mathrm{d}s(x),\end{aligned}$$

and the structure of each of the inner products in (4.5) can then be understood to be special cases of this abstract formula. For example the third integral in (4.5) may be written

$$(S_k\phi, b\widetilde{\phi}) = (S_k^{m,m'}(\chi_j P_\ell), b\chi_{j'}P_{\ell'}),$$

where S_k is the integral operator with modulated kernel

$$\Phi_k(x,y)\exp(ik(\psi_m(x) - \psi_{m'}(y))), \quad x, y \in \Gamma. \tag{4.13}$$

Similarly the first and second inner products in (4.5) (involving D_k and D'_k) yield the modulated kernels

$$\frac{\partial\Phi_k(x,y)}{\partial n(y)}\exp(ik(\psi_m(x) - \psi_{m'}(y))), \quad \frac{\partial\Phi_k(x,y)}{\partial n(x)}\exp(ik(\psi_m(x) - \psi_{m'}(y))). \tag{4.14}$$

The inner products in (4.6) can be represented in a similar way, noting that for each the examples discussed in Section 3 we can write

$$V_0(x,k) = \sum_j V_0^j(x,k)\exp(ik\psi_0^j(x)), \tag{4.15}$$

where the known functions V_0^j are not oscillatory. This representation is obvious for the convex obstacles studied in Section 3; that it can also be applied for the non-convex polygons discussed in Section 3.4 is shown in Chandler-Wilde *et al.* (2012*a*).

The construction of robust quadrature rules for the oscillatory integrals (4.5) and (4.6) depends crucially on the explicit extraction of the phase of the oscillatory components, which (once this is done for the leading-order behaviour V_0, as discussed above) is achieved by extracting the phase from the fundamental solution and its normal derivatives. Restricting to (4.5) and to the kernels which arise in the direct boundary integral operator only (*i.e.*, the second and third entries in (4.5); the first entry is analogous), it is easy to see that

$$\Phi_k(x,y) = F(x,y)\exp(\mathrm{i}k|x-y|), \qquad \frac{\partial \Phi_k(x,y)}{\partial n(x)} = G(x,y)\exp(\mathrm{i}k|x-y|), \tag{4.16}$$

where

$$F(x,y) = \begin{cases} \dfrac{\mathrm{i}}{4} h_0(k|x-y|) & d=2, \\ \dfrac{1}{4\pi|x-y|} & d=3, \end{cases} \tag{4.17}$$

with $h_0(z) = \exp(-\mathrm{i}z)H_0^{(1)}(z)$ for $z > 0$, and

$$G(x,y) = \begin{cases} -\dfrac{\mathrm{i}k}{4}\left(\dfrac{n(x)\cdot(x-y)}{|x-y|}\right) h_1(k|x-y|) & d=2, \\ \dfrac{n(x)\cdot(x-y)(\mathrm{i}k|x-y|-1)}{4\pi|x-y|^3} & d=3, \end{cases} \tag{4.18}$$

with $h_1(z) = \exp(-\mathrm{i}z)H_1^{(1)}(z)$ for $z > 0$.

In the 3D case, both F and G are trivially non-oscillatory but of course have derivatives which (in general) blow up with negative powers of $|x-y|$. This statement also holds in 2D, but its proof is a bit more complicated and is postponed to Section 4.1.4.

Collecting together the observations (4.13), (4.14) and (4.16), we see that all of the integrals in (4.5) take the form

$$\int_{\Gamma_j}\int_{\Gamma_{j'}} g(x,y)\exp(\mathrm{i}k\Psi(x,y))\,\mathrm{d}s(y)\,\mathrm{d}s(x) \tag{4.19}$$

with phase

$$\Psi(x,y) = |x-y| + \psi_m(y) - \psi_{m'}(x). \tag{4.20}$$

Here $g(x,y)$ is smooth except at the diagonal $x = y$, and is non-oscillatory in the sense that derivatives of g do not grow any faster as k increases than g itself. The integrals in (4.6) can be represented in a similar way, with different phase functions Ψ varying from problem to problem according to the exact nature of V_0 (specific examples are discussed in Section 4.4).

Remark 4.5. Note that by (4.17) and (4.18), $G(x,y)$ (the non-oscillatory part of the kernel of D'_k) contains a multiplicative factor of k which is absent from $F(x,y)$ (the non-oscillatory part of the kernel of S_k). Moreover in the combined potential formulation (4.1), (4.2), S_k is multiplied by the coupling parameter η, which is usually chosen proportional to k. (The discussion in Remark 5.1 indicates that this is a natural choice.) Thus it may at first appear that $(D'_k\phi,\tilde{\phi})$ and $\mathrm{i}\eta(S_k\phi,\tilde{\phi})$ might be $O(k)$ as $k \to \infty$. However, this is in fact not the case; the oscillatory components of the integral operators lead to decay of $O(1/k)$ as $k \to \infty$ for the corresponding integrals (see

Section 4.2, (4.29) in particular), and hence the matrix entries $(D'_k\phi, \tilde{\phi})$ and $\mathrm{i}\eta(S_k\phi, \tilde{\phi})$ are both, in general, $O(1)$ as $k \to \infty$.

4.1.3. Stiffness matrix integrals for surface gradient of the single-layer potential

It remains to discuss the implementation of the Galerkin integrals (4.7) arising from the star-combined formulation. We do this here only for the 3D case (following the discussion in Spence *et al.* 2011) but we note an analogous procedure can be applied in 2D, and this has been implemented in Kim (2012). Let us denote $\Gamma_0 = \operatorname{supp} \widetilde{\phi}$, so that (4.7) may be written

$$\int_{\Gamma_0} (\widetilde{\phi}(x)x) \cdot \nabla_\Gamma (S_k\phi)(x)\, \mathrm{d}s(x). \tag{4.21}$$

The surface gradient operator may be moved from the right-hand term in (4.21) to the left-hand term by the following integration by parts procedure. We assume that the surface patch Γ^0 is parametrized by a map,

$$\widehat{x} \to \begin{bmatrix} \widehat{x} \\ \xi(\widehat{x}) \end{bmatrix},$$

defined on a 2D parameter domain $\widehat{\Gamma^0}$, where $\xi \in C^2(\widehat{\Gamma^0})$. (This does *not* imply that Γ is globally smooth.) This map is assumed to provide a good parametrization of Γ^0 such that the columns of the Jacobian

$$J(\widehat{x}) := \begin{bmatrix} 1 & 0 \\ 0 & 1 \\ \frac{\partial \xi}{\partial \widehat{x}_1} & \frac{\partial \xi}{\partial \widehat{x}_2} \end{bmatrix}$$

form a basis for the tangent plane at x, the image of $\widehat{x} = (\widehat{x}_1, \widehat{x}_2)$ under this map. Moreover for any vector field $w : \Gamma^0 \to \mathbb{R}^3$, we may also resolve w in the tangent and normal directions via the formula

$$w(x) = J(\widehat{x})\widehat{\omega}(\widehat{x}) + (w(x) \cdot n(x))n(x),$$

for some field $\widehat{\omega} : \widehat{\Gamma^0} \to \mathbb{R}^2$. Then in Spence *et al.* (2011) the formula

$$\begin{aligned} \int_{\Gamma^0} w(x) \cdot \nabla_\Gamma v(x)\, \mathrm{d}s(x) &= \int_{\partial\widehat{\Gamma^0}} (\det J(\widehat{x})^T J(\widehat{x}))^{1/2} \, (\widehat{\omega}(\widehat{x}) \cdot \widehat{n}(\widehat{x}))\, \widehat{v}(\widehat{x})\, \mathrm{d}\zeta(\widehat{x}) \\ &\quad - \int_{\widehat{\Gamma^0}} \widehat{\nabla} \cdot \left[(\det J(\widehat{x})^T J(\widehat{x}))^{1/2} \widehat{\omega}(\widehat{x}) \right] \widehat{v}(\widehat{x})\, \mathrm{d}\widehat{x} \end{aligned} \tag{4.22}$$

is proved for all sufficiently smooth vector fields w and v, where $\widehat{n}(\widehat{x})$ is the outward normal from $\widehat{\Gamma^0}$ at $\widehat{x} \in \partial\widehat{\Gamma^0}$ and $\widehat{\nabla}$ denotes gradient with respect to $\widehat{x}$. Inserting $w = \widetilde{\phi}(x)x$ and $v = S_k\phi$ into (4.22) gives a formula for the integral (4.21), which avoids computing the surface derivative of the single-layer potential.

4.1.4. Phase extraction in the 2D case

We finish this subsection by briefly explaining why the functions F and G appearing in (4.17) and (4.18) are non-oscillatory in the 2D case. This property is proved using the following lemma, which examines the asymptotics of the functions h_0 and h_1 which appear in the formulae (4.17) and (4.18).

Lemma 4.6. For all $n \geq 0$, there exists a constant C_n such that

$$\left|\left(\frac{\mathrm{d}}{\mathrm{d}z}\right)^n h_0(z)\right| \leq C_n \begin{cases} \max\{1+\log(1/z), z^{-n}\} & z \in (0,1], \\ z^{-(n+1/2)} & z \in [1,\infty), \end{cases}$$

and

$$\left|\left(\frac{\mathrm{d}}{\mathrm{d}z}\right)^n h_1(z)\right| \leq C_n \begin{cases} z^{-(n+1)} & z \in (0,1], \\ z^{-(n+1/2)} & z \in [1,\infty). \end{cases}$$

We do not give the proof here, but we note it is obtained by using the following integral formula (see, *e.g.*, Oberhettinger and Badii (1973, (12.31) in part II)):

$$H_0^{(1)}(z) = -\frac{2\mathrm{i}}{\pi}\exp(\mathrm{i}z)\int_0^\infty \frac{\exp(-zt)}{t^{1/2}(t-2\mathrm{i})^{1/2}}\,\mathrm{d}t. \tag{4.23}$$

This immediately gives us a formula for $h_0(z)$ which can be readily differentiated. A formula for $h_1(z)$ is obtained by combining the first derivative of (4.23) with the formula $H_1^{(1)}(z) = -\frac{\mathrm{d}}{\mathrm{d}z}H_0^{(1)}(z)$ (Abramowitz and Stegun 1964, (9.1.27)). More details of the proof of Lemma 4.6 are in Chandler-Wilde and Langdon (2007, Theorem 3.1) and in Kim (2012) .

Now if the 2D contour Γ is parametrized by ζ (say arc length parametrization), we are concerned with estimates for the derivatives of $h_0(kr(s,t))$ and $h_1(kr(s,t))$ where $r(s,t) = |\zeta(s) - \zeta(t)|$. When considering the single-layer potential we can use Lemma 4.6 and the Faá di Bruno formula for multiple application of the chain rule to show (see Kim (2012)) that for each $p_1, p_2 \in \mathbb{N}$, there exists a constant C such that

$$\left|\left(\frac{\partial}{\partial s}\right)^{p_1}\left(\frac{\partial}{\partial t}\right)^{p_2}\Big\{h_0(kr(s,t))\Big\}\right| \leq C\, r(s,t)^{-p_1-p_2},$$

so the function F is not oscillatory. Analogous estimates can be obtained for the function G (Kim 2012).

In Sections 4.3 and 4.4 below, we will describe some oscillatory integration methods which have been used to implement hybrid methods in scattering. Since the computation of oscillatory integrals has a long history and has seen a great deal of recent interest (some of it related to high-frequency scattering and some of it coming from completely different motivations), we insert at this point a (necessarily brief) review of numerical methods for oscillatory integration in general.

4.2. Efficient evaluation of highly oscillatory integrals: a review

In this subsection we describe a range of schemes for computing single integrals of the form

$$I_k^{[a,b]}(g) = \int_a^b g(s) \exp(ik\Psi(s))\, ds, \tag{4.24}$$

and double integrals of the form

$$I_k^{[a,b]\times[a',b']}(g) = \int_a^b \int_{a'}^{b'} g(s,t) \exp(ik\Psi(s,t))\, dt\, ds, \tag{4.25}$$

where $a, b, a', b' \in \mathbb{R}$, $a < b$, $a' < b'$, and $k > 0$. We have in mind here that the function g is integrable on its domain and that it is slowly oscillating compared to $\exp(ik\Psi)$.

Such integrals appear in Galerkin methods for BIEs arising in 2D scattering as described in Section 4.1 (after parametrization of (a part of) the boundary via univariate functions); recall (4.12) and (4.19). Integrals such as (4.25) also appear in collocation methods for 3D scattering problems, as described in Section 4.5, after parametrization of a part of the boundary by a bivariate function. Galerkin methods in 3D lead to integrals such as (4.25) and its four-dimensional analogue (again recalling the formulae (4.12) and (4.19)). We consider specific methods for the 3D case in Section 4.5, but remark that many of the ideas applicable to the efficient evaluation of (4.25) can, in principle at least, be extended to higher dimensions.

Because of the oscillation, conventional (interpolatory) quadrature techniques for evaluating (4.24)–(4.25) perform poorly when k is large, with the computational cost required to achieve a prescribed level of accuracy increasing rapidly as $k \to \infty$. To get around this difficulty, many schemes have been proposed, some of which have been inspired by classical asymptotic integration schemes; see, *e.g.*, Bleistein and Handelsman (1986), Olver (1974) and Wong (1989). Starting with (4.24) and assuming that g and Ψ are sufficiently smooth and that Ψ' does not vanish on $[a, b]$, repeated integration by parts gives

$$I_k^{[a,b]}(g) = -\sum_{j=0}^{n-1} \left(\frac{i}{k}\right)^{j+1} u_j(s) \exp(ik\Psi(s))\Big|_{s=a}^{s=b} + \left(\frac{i}{k}\right)^n \int_a^b g_n(s) \exp(ik\Psi(s))\, ds, \tag{4.26}$$

for $n = 1, 2, \ldots$, where

$$u_0 = \frac{g}{\Psi'}, \quad g_{j+1} = u_j', \quad u_{j+1} = \frac{g_{j+1}}{\Psi'},$$

for $j = 0, 1, 2, \ldots$. Applying the Riemann–Lebesgue lemma (*e.g.*, Wong

1989, p. 200) to the remaining integral, we find that $I_k^{[a,b]}(g)$ is asymptotic to the boundary terms (assuming they do not vanish), that is,

$$I_k^{[a,b]}(g) \sim -\sum_{j=0}^{n-1}\left(\frac{\mathrm{i}}{k}\right)^{j+1} u_j(s)\exp(\mathrm{i}k\Psi(s))\Big|_{s=a}^{s=b} + o(k^{-n}), \quad \text{as } k\to\infty, \tag{4.27}$$

for $n = 1, 2, \ldots$.

For (4.25) on the other hand, making the assumption that $\nabla\Psi(x) \neq 0$ for all $x \in [a,b]\times[a',b']$, applying the divergence theorem and following Wong (1989, p. 425), for instance, we have the similar result that

$$\begin{aligned} I_k^{[a,b]\times[a',b']}(g) = &-\sum_{j=0}^{n-1}\left(\frac{\mathrm{i}}{k}\right)^{j+1}\int_0^L (u_j(\zeta(s))\cdot n(s))\exp(\mathrm{i}k\Psi(\zeta(s)))\,\mathrm{d}s \\ &+\left(\frac{\mathrm{i}}{k}\right)^{n}\int_a^b\int_{a'}^{b'}\tilde{g}_n(s,t)\exp(\mathrm{i}k\Psi(s,t))\,\mathrm{d}t\,\mathrm{d}s, \end{aligned} \tag{4.28}$$

where $\zeta(s)$, $n(s)$ and $L = 2(b-a+b'-a')$ are the arc length parametrization, unit outward normal vector and length, respectively, of the boundary of $[a,b]\times[a',b']$, and

$$u_0 := \frac{\nabla\Psi}{|\nabla\Psi|^2}g, \quad \tilde{g}_{j+1} = (\nabla\cdot u_j), \quad u_{j+1} = \frac{\nabla\Psi}{|\nabla\Psi|^2}\tilde{g}_{j+1},$$

for $j = 0, 1, 2, \ldots$. Under the further assumption that $\frac{\partial}{\partial s}\Psi(\zeta(s)) \neq 0$ on the boundary of $[a,b]\times[a',b']$ (this is equivalent to the non-resonance condition of Iserles and Nørsett (2006)), which ensures that the double integral on the right-hand side of (4.28) is $o(1)$ as $k\to\infty$, we have

$$I_k^{[a,b]\times[a',b']}(g) \sim -\sum_{j=0}^{n-1}\left(\frac{\mathrm{i}}{k}\right)^{j+1}\int_0^L (u_j(\zeta(s))\cdot n(s))\exp(\mathrm{i}k\Psi(\zeta(s)))\,\mathrm{d}s + o(k^{-n}), \tag{4.29}$$

as $k\to\infty$, for $n = 1, 2, \ldots$.

If in (4.27) Ψ' vanishes at any point of $[a,b]$ then one cannot use the integration by parts representation (4.26). Likewise, if $\nabla\Psi$ vanishes at any point of $[a,b]\times[a',b']$ then one cannot use the divergence theorem formula (4.29). In either case, it can easily be shown that as $k\to\infty$, the corresponding oscillatory integral is dominated by contributions obtained by integrating over small neighbourhoods of all such 'stationary phase' points, with the size of the neighbourhoods vanishing as $k\to\infty$.

In particular, in the 1D case it can be shown that if x_s is a stationary point of order $n \geq 1$, that is,

$$\Psi'(x_s) = \Psi''(x_s) = \cdots = \Psi^{(n)}(x_s) = 0, \quad \text{and} \quad \Psi^{(n+1)}(x_s) \neq 0,$$

and if, in addition, $g(x_s) \neq 0$, then $I_k^{[a,b]}(g) = O(k^{-1/(n+1)})$ as $k \to \infty$. (Thus the flatter $\Psi(x)$ is at $x = x_s$, the less rapidly $I_k^{[a,b]}(g)$ vanishes as $k \to \infty$.) Similar results hold for $I_k^{[a,b]\times[a',b']}(g)$: see, *e.g.*, Wong (1989). The classical 'method of stationary phase' then consists of approximating the integrals asymptotically via Taylor series expansions about the stationary points: see, *e.g.*, Olver (1974, p. 96). If a partition of unity is used to isolate the stationary points, then the integrals over the remainder of the domain can be expanded in a similar way to (4.27) and (4.28) above.

Historically, asymptotic formulae such as these have been used to approximate oscillatory integrals arising from scattering problems. For example, the single integrals in (4.29) each take the form (4.24), and thus an obvious way to evaluate (4.25) for large k, under the stated assumptions on Ψ, could be to simply apply (4.27) within (4.29) and to truncate the series. Such an approach may yield accurate results for *sufficiently large* k; however, the asymptotic series (4.27) and (4.29) are not convergent for fixed k, and may diverge very quickly in practice; see, *e.g.*, Ganesh, Langdon and Sloan (2007), where such an approach is analysed within the context of a particular class of 3D scattering problems. Hence it is not possible to obtain the required integrals to a controllable accuracy for any fixed k unless we combine the asymptotic approach with a convergent quadrature rule. On the other hand, a reasonable goal for any quadrature approach is that it should be able to replicate the asymptotic results in the limit as $k \to \infty$, and standard quadrature schemes for which the weights are not explicitly dependent on k will perform poorly.

A simple idea, implemented by Asheim (2008) is to use the series representations (4.26) and (4.27) and to evaluate the remaining single and double integrals numerically (via Filon quadrature, see Section 4.3), rather than integrating the original integral directly. The reduction of order (with respect to k) of the integral being evaluated numerically, compared to more standard approaches, leads to improved overall accuracy. The schemes in Bruno *et al.* (2004) and Bruno and Geuzaine (2007) are also based on the asymptotic expansions above, but applying convergent quadrature rules to evaluate the integrals in the neighbourhood of stationary points; see Section 4.5 for more details.

An alternative asymptotic approach is the 'method of steepest descent' (see, *e.g.*, Jones 1972, Wong 1989, §II.4). Describing this method as it applies to (4.24), the idea is to deform the path of integration into the complex plane in such a way that the oscillatory integral (4.24) over a finite interval is replaced by two or more non-oscillatory 'Laplace-type integrals'. These 'Laplace-type integrals' are integrals over infinite intervals but with non-oscillatory integrands which decay exponentially. The exponential rate of decay, moreover, increases as k increases so that the integrals can be evaluated asymptotically, using, *e.g.*, Watson's Lemma (Jones 1972). An observation dating back at least to Jones (1972) is that, in the limit as $k \to \infty$, very high-order asymptotic approximations to Laplace-type integrals are provided alternatively by generalized Gauss–Laguerre quadrature. Moreover, these approximations are convergent as the number of quadrature points increases. This suggests a 'numerical method of steepest descent', in which the path of integration is deformed to one or more steepest descent paths and generalized Gauss–Laguerre quadrature is used to evaluate the resulting integrals. This amounts to a numerical quadrature scheme for the original oscillatory integral whose performance improves rapidly as k increases. Chandler-Wilde and Hothersall (1995) is a case study of this methodology applied to evaluate numerically a particular Green's function in acoustics. This paper provides an error analysis of the resulting generalized Gauss–Laguerre quadrature approximations, showing moreover how it is possible to modify the approximations to take account of the presence of a simple pole singularity near the steepest descent path of integration, in a way which maintains accuracy uniformly in k and the position of the pole.

The 'numerical method of steepest descent', only hinted at in Jones (1972), and applied only to a particular example in Chandler-Wilde and Hothersall (1995), is developed systematically by Huybrechs and Vandewalle (2006) for the general oscillatory integral (4.24), with further extensions and generalizations in Deaño and Huybrechs (2009). In Huybrechs and Vandewalle (2006) the focus is on single integrals. The methodology has subsequently been extended to double integrals of the form (4.25) (Huybrechs and Vandewalle 2007*a*), and has been used to good effect in a number of schemes (*e.g.*, Honnor *et al.* 2010, Huybrechs and Vandewalle 2007*b*, Asheim and Huybrechs 2010*b*), that we discuss further in Section 4.5. Due in part to the added difficulties arising due to stationary points in the complex plane, particularly for higher-dimensional integration, a complete rigorous error analysis for this approach has proved elusive. However, recent progress by Huybrechs and Olver (2012), who have reinterpreted this 'numerical steepest descent' method as a Filon-type method (see below) with special complex interpolation points, has gone some way to rectifying this, for single integrals at least.

We now turn our focus to Filon quadrature; this procedure turns out to be particularly well suited to integrals of the form (4.24) and (4.25) arising from hybrid Galerkin methods, and is more amenable to the derivation of rigorous error estimates than some of the other schemes described above. For a much more detailed discussion of highly oscillatory quadrature, we refer to the review articles by Iserles, Nørsett and Olver (2005) and Huybrechs and Olver (2009), and also to the more recent literature reviews in Olver (2010), Huybrechs and Olver (2012) and Domínguez, Graham and Smyshlyaev (2011), for example.

4.3. Recent progress on error analysis of Filon rules

The construction of Filon quadrature starts from the simple observation that integrals of the form

$$\int_a^b P_n(s) \exp(iks)\, \mathrm{d}s$$

can be evaluated analytically if P_n is a polynomial. Thus any integral of the form (4.24) with $\Psi(s) = s$ (often referred to as the linear oscillator), can be approximated by replacing g by a suitable polynomial (*e.g.*, the Legendre or Chebyshev interpolant) and then integrating exactly. More generally, if we wish to approximate (4.24), and if Ψ does not have stationary points, the change of variable $\tilde{s} = \Psi(s)$ reduces (4.24) to a new oscillatory integral with a phase which is linear in $\tilde{s}$, at the cost of having to evaluate the inverse function Ψ^{-1}. Even if Ψ has stationary points this procedure can still be applied on a partition of $[a, b]$ with breakpoints chosen to be the stationary points, although now algebraic singularities in the integrand appear at the breakpoints due to the singularity of the inverse function there.

Filon quadrature has a long history starting from Filon (1928) (see also Luke 1954, Bakhvalov and Vasilčeva 1968, and the review in Evans and Webster 1999), and has enjoyed considerable recent interest following a sequence of influential papers by Iserles and co-authors (see, *e.g.*, Iserles 2004, Iserles 2005, Iserles and Nørsett 2004, Iserles *et al.* 2005). In particular, Iserles (2004) showed that for the linear oscillator a Filon rule has an error which will decay with k like $O(k^{-2})$, provided the interpolation points include the end points a, b. This was generalized in Iserles and Nørsett (2005), to obtain faster decay as $k \to \infty$, by using interpolation of derivatives at the end points of the domain of integration. Related methods were proposed by Xiang (2007), and general oscillators Ψ which may have stationary points were treated by Olver (2007, 2010).

In the context of high-frequency scattering, many integrals of the form (4.24) and (4.25) have to be computed, and rules which use only point values of g are particularly attractive, since g is often very complicated (and may itself be an integral involving special functions). In error and complexity analysis it also turns out to be important to control the quadrature error explicitly with respect to both k and N (the number of quadrature points), and to know how this depends on the regularity of g (*e.g.*, in some Sobolev space). In this context, a recent error analysis for Filon–Clenshaw–Curtis (FCC) rules (Domínguez, Graham and Kim 2012, Domínguez *et al.* 2011) has been performed and we now give some details of the results obtained there.

In view of the above remarks we may begin by considering oscillatory integrals of the form (4.24) with $\Psi(s) \equiv s$, that is,

$$I_k^{[a,b]}(g) = \int_a^b g(s) \exp(iks)\, \mathrm{d}s. \tag{4.30}$$

The FCC rule in its simplest form starts with $[a,b] = [-1,1]$ and approximates $I_k^{[-1,1]}(g)$, by replacing g by $Q_N g$, its polynomial interpolant of degree N at the Clenshaw–Curtis points $t_{j,N} := \cos(j\pi/N)$, $j = 0,\ldots,N$, where $N \geq 1$. After writing $Q_N g$ in terms of Chebyshev polynomials $T_n(x) = \cos(n \arccos(x))$, $n = 0,\ldots,N$, and performing some elementary manipulations, the rule may be written as

$$I_{k,N}^{[-1,1]}(g) = \sum_{n=0}^{N}{}'' \alpha_{n,N}(g)\omega_n(k), \tag{4.31}$$

with

$$\alpha_{n,N}(g) = \frac{2}{N}\sum_{j=0}^{N}{}'' \cos\left(\frac{jn\pi}{N}\right) g(t_{j,N}), \quad n = 0,\ldots,N, \tag{4.32}$$

where the notation $\sum''$ means that the first and last terms in the sum are multiplied by $1/2$, and where the weights

$$\omega_n(k) := \int_{-1}^{1} T_n(s)\exp(\mathrm{i}ks)\,\mathrm{d}s, \quad n \geq 0, \tag{4.33}$$

have to be computed. After an initial application of the discrete cosine transform (via fast Fourier transform (FFT), costing $O(N \log N)$ operations), the rule (4.31) can then be applied to any g in an additional $O(N)$ operations; see Domínguez *et al.* (2011) for more detail. In Domínguez *et al.* (2011) a stable and efficient scheme for computing $\{\omega_n(k)\}$ is presented, and a public domain implementation of the rule is provided by Domínguez (2009). Moreover, in Domínguez *et al.* (2011), it is shown that there exists a constant $C > 0$ such that, for $r \in \{0,1\}$ and all integers $m \geq 1$, we have

$$|I_k^{[-1,1]}(g) - I_{k,N}^{[-1,1]}(g)| \leq C \min\left\{1, \left(\frac{1}{k}\right)^r\right\}\left(\frac{1}{N}\right)^{m-r} \|g_c\|_{H^m(-\pi,\pi)}, \tag{4.34}$$

for $k > 0$ and $N \geq 1$. Here $g_c(\theta) = g(\cos\theta)$ is the cosine transform of g and $H^m(-\pi,\pi)$ is the Sobolev space of order m of all 2π-periodic functions. The proof of (4.34) is obtained by noticing that

$$\begin{aligned} I_k^{[-1,1]}(g) - I_{k,N}^{[-1,1]}(g) &= \int_{-1}^{1} (g - Q_N g)(s)\ \exp(\mathrm{i}ks)\,\mathrm{d}s \\ &= \int_0^{\pi} (g - Q_N g)_c(\theta)\exp(\mathrm{i}k\cos\theta)\sin\theta\,\mathrm{d}\theta, \end{aligned}$$

and that $(g - Q_N g)_c$ is the error in the even trigonometric interpolant of degree N to g_c at the points $j\pi/N$, $j = 0,\ldots,N$. Then, integration by parts and standard error estimates for trigonometric interpolation at equally spaced points yield the estimate. It is straightforward to extend (4.34) to all $r \in [0,1]$. In fact a more general version of this theorem with $r = 2$

and m required to be any real number greater than 7/2 was proved in Domínguez *et al.* (2011), making precise the requirements for the $O(k^{-2})$ decay predicted in Iserles (2004).

The convergence rate in (4.34) may be slow if g is not smooth and in this case it is better to apply the rule in a composite fashion with fixed N on meshes graded suitably towards the singular point(s) and obtain convergence by letting the mesh diameter shrink to zero. In order to obtain error estimates for such composite rules, Domínguez *et al.* (2012) prove a variant of (4.34) which shows (emphasizing the case $k \to \infty$), that there exists $\sigma_N > 0$ such that

$$|I_k^{[-1,1]}(g) - I_{k,N}^{[-1,1]}(g)| \leq \sigma_N \left(\frac{1}{k}\right)^r \left\{\int_{-1}^{1} \frac{|g^{(N+1)}(s)|}{\sqrt{1-s^2}}\,\mathrm{d}s\right\}^{1/2}, \tag{4.35}$$

for all $k > 0$ and $N \geq 1$.

Now, to compute (4.30) for any $[a, b]$ we can transplant onto $[-1, 1]$ using the affine change of variables $s = c + h\tilde{s}$, $\tilde{s} \in [-1, 1]$, where and the estimate, analogous to (4.35), is

$$\left|I_k^{[a,b]}(g) - I_{k,N}^{[a,b]}(g)\right| \leq \tilde{\sigma}_N \left(\frac{1}{k}\right)^r h^{N+2-r} \left\{\int_a^b \frac{|g^{(N+1)}(s)|}{\sqrt{(s-a)(b-s)}}\,\mathrm{d}s\right\}^{1/2}. \tag{4.36}$$

If the FCC rule is applied in a composite fashion, the estimate (4.36) may then be used to establish convergence as the subinterval size shrinks to zero, when N is fixed. Optimal convergence for singular integrands g can be obtained by suitable mesh refinement. For example, suppose

$$\|g\|_{N+1,\beta} := \max\left\{\sup_{s\in[0,1]} |g(s)|,\ \sup_{s\in[0,1]} \left|s^{(j-\beta)} D^j g(s)\right|,\ j = 1, \ldots, N+1\right\} < \infty, \tag{4.37}$$

for some $\beta > 0$, and introduce the graded mesh $s_j = (j/M)^q$, $j = 0, \ldots, M$. On the first interval, we use the one-point rule

$$\int_0^{s_1} g(s)\exp(\mathrm{i}ks)\,\mathrm{d}s \approx g(s_1)\int_0^{s_1} g(s)\exp(\mathrm{i}ks)\,\mathrm{d}s,$$

while on all other subintervals we use the FCC rule shifted to that interval in an affine way. Then it was shown (Kim 2012) that the error $E_{k,M,q,N}(g)$ (*i.e.*, the difference between the integral $I_k^{[0,1]}(g)$ and the resulting approximation) may be estimated as

$$|E_{k,M,q,N}(g)| \leq C \left(\frac{1}{k}\right)^r \left(\frac{1}{M}\right)^{N+1-r} \|g\|_{N+1,\beta}, \tag{4.38}$$

provided $q > (N + 1 - r)/(\beta + 1 - r)$, where C depends on N, β and r but not M. The results illustrated here have been extended in Domínguez *et al.* (2012) to the case of logarithmic singularities in g and also the case $g(s) \sim s^{\beta}$, with $\beta \in (-1, 0)$. In the algorithms presented in Domínguez *et al.* (2012) the error decays at least as fast as the exact integral $I_k^{[0,1]}(g)$ as $k \to \infty$. Logarithmic singularities in particular often occur in the integrals arising in hybrid Galerkin methods (recall (4.17) and Lemma 4.6).

Related estimates to those described here may be found in Melenk (2010) (see also Melenk 2008), where general Filon-type rules for (4.24) are considered with g approximated by a general Hermite interpolating polynomial and g, Ψ are considered to be holomorphic in a neighbourhood of $[a, b]$. Error estimates for this case showing exponential convergence with respect to the polynomial degree and allowing sharp estimates of the rate of decay with respect to k are obtained. It is expected that by changing the choice of mesh grading and generalizing the setting to that of piecewise analytic functions, the estimates above could also be extended to show exponential convergence for *hp*-versions of the FCC quadrature, such as is observed (for a related scheme, using Gauss–Legendre rather than Clenshaw–Curtis nodes and weights) in Melenk and Langdon (2007, 2012).

4.4. Galerkin integrals for the 2D case

In this subsection we describe in more detail the robust computation of the oscillatory integrals arising in hybrid Galerkin methods in 2D in both the smooth convex and the polygonal cases. The relevant references are (for the smooth convex case) Domínguez, Graham, Kim and Smyshlyaev (2009), Domínguez *et al.* (2012), Domínguez *et al.* (2011), and Kim (2012), and (for the polygonal case) Melenk and Langdon (2007) and Melenk and Langdon (2012). The central method to be applied is the FCC rule, usually after application of a transformation to render the phase linear. We start in Section 4.4.1 by describing the key ideas in the simplest context of the load vector and mass matrix integrals, as described in Section 4.1.1, which are only oscillatory in the polygonal case. We then proceed in Section 4.4.2 by studying the (in general) more difficult stiffness matrix entries, as introduced in Sections 4.1.2 and 4.1.3.

4.4.1. Load vector and mass matrix entries

Recalling Section 4.1.1, we begin by noting that all load vector and mass matrix entries for the convex polygonal case (as described in Example 4.2) have linear phase, and thus each can be expressed exactly in the form (4.24) with $\Psi(s) = Cs$, where $C \in \mathbb{R}$. The change of variables $\tilde{s} = Cs$ immediately puts each of these integrals in the form (4.30), and then the FCC rules of Section 4.3 can be applied without difficulty.

As a slightly more challenging example we consider one of the integrals described in Example 4.3, which arise in the case of a non-convex polygon with an illuminated non-convex side. Specifically, we consider the construction of an efficient quadrature scheme for evaluating

$$\int_0^1 g(s)\exp(ik(s+\sqrt{s^2+a^2}))\,\mathrm{d}s,$$

where $a>0$ and $g \in L^2(0,1)$ is not oscillatory. We proceed by making the change of variables

$$\tau = s+\sqrt{s^2+a^2},$$

giving $s=(\tau^2-a^2)/2\tau$ and hence $\mathrm{d}s/\mathrm{d}\tau = (1+a^2/\tau^2)/2>0$. Thus the integral can be written as

$$\int_a^{1+\sqrt{1+a^2}} \tilde{g}(\tau)\exp(ik\tau)\,\mathrm{d}\tau, \tag{4.39}$$

where

$$\tilde{g}(\tau) := \frac{1}{2}\left(1+\frac{a^2}{\tau^2}\right) g\left(\frac{\tau^2-a^2}{2\tau}\right)$$

is non-oscillatory. The integral (4.39) thus again takes the form (4.30), allowing us to apply the rules of Section 4.3 directly. The other load vector/mass matrix integrals that arise for convex and non-convex polygons can all be evaluated using a similar procedure.

4.4.2. Stiffness matrix entries

In Section 4.1.2 it is explained that we can cover all the oscillatory integration problems which arise from the 'stiffness matrix' inner products, (4.5) and (4.6), by considering integrals of the form (4.19) with g non-oscillatory but (usually) weakly singular at $x=y$ and with phase Ψ given either by (4.20) (for (4.5)), where $\{\psi_m\}$ are the phases present in the basis functions (see (4.8)), or by a similar formula found by replacing $\psi_m(y)$ in (4.20) with $\psi_0^j(y)$, where ψ_0^j are the phase functions in the representation (4.15) for V_0.

In the computation of (4.19) we will parametrize the boundary Γ by some parametrization ζ. Identifying x with $\zeta(s)$, and y with $\zeta(t)$, we may rewrite (4.19) in the form

$$J_k = \int\!\!\int g(s,t)\exp(ik\Psi(s,t))\,\mathrm{d}t\,\mathrm{d}s. \tag{4.40}$$

(Here we have abused notation slightly: the notation $g(s,t)$ is used to denote the function $g(\zeta(s),\zeta(t))|\zeta'(s)|\,|\zeta'(t)|$, while $\Psi(s,t)$ denotes the function $\Psi(\zeta(s),\zeta(t))$ and the limits on the integrals in (4.40) depend on the choice of the parametrization.)

Recalling the specific phases of the basis functions ψ_m and the leading-order behaviour V_0 for the two examples of a smooth convex obstacle (recall

Example 3.1) and a convex polygon (see (3.35) and (3.8)), we have

$$\Psi(s,t) = \begin{cases} |\zeta(s) - \zeta(t)| + (\zeta(t) - \zeta(s)) \cdot \hat{a} & \text{convex smooth,} \\ |\zeta(s) - \zeta(t)| + \theta s \pm t & \text{convex polygon,} \end{cases} \tag{4.41}$$

for some $\theta \in [-1,1]$ (with $\theta = \pm 1$ corresponding to the inner products (4.5), and $\theta \in (-1,1)$ for the inner products (4.6)).

One approach to computing the integral J_k in (4.40) robustly with respect to k is to introduce a change of variable $\tau = \Psi(s,t)$, thus obtaining a transformed integral with phase which varies linearly in τ (as in Section 4.4.1). Introducing the notation

$$\Psi_{[s]}(t) := \Psi(s,t) \quad \text{and} \quad \Psi^{[t]}(s) := \Psi(s,t),$$

we can choose to interpret this transformation as either $\tau = \Psi_{[s]}(t)$ (defining τ in terms of t, for each s) or $\tau = \Psi^{[t]}(s)$ (defining τ in terms of s, for each t). (The choice of interpretation would be motivated by the desire to avoid, as far as possible, choosing a transformation containing stationary points.) The transformed integral is then suitable for treatment using the Filon rules outlined in Section 4.3 (as described in Section 4.4.1 for the simpler load vector/mass matrix entries), and hence rigorous error estimates are possible. The transformation does however introduce some new difficulties, since the regular domain of integration in (4.40) (usually rectangular or triangular) is transformed to a (usually non-standard) shaped domain and, moreover, inverse functions of Ψ appear in the transformed integrand. However, the method (first suggested by Melenk 2006 and substantially developed by Domínguez 2007; see also Domínguez *et al.* 2009) has enjoyed considerable success in computing the Galerkin integrals.

To give more detail, let us first consider the case of a smooth convex obstacle. Recalling (3.25)–(3.26), Γ is parametrized by $\zeta : [0, 2\pi] \to \Gamma$ and the parameter domain $[0, 2\pi]$ is partitioned into four subintervals $\Lambda_i : i = 1, \ldots, 4$, where Λ_1, Λ_2 correspond to the near-shadow boundary zones (which shrink to zero as $k \to \infty$), Λ_3 corresponds to a subset of the illuminated zone and the remaining interval of parameter space, Λ_4, corresponds to the deep shadow zone. Since the solution is exponentially small in the deep shadow, it is approximated by zero there, and thus only integrals over the domains

$$\Lambda_j \times \Lambda_{j'} \quad \text{for } j, j' = 1, 2, 3 \tag{4.42}$$

have to be considered. These domains are depicted in Figure 4.1(a), where we have written $\Lambda_1 = [a,b]$, $\Lambda_2 = [c,d]$ and $\Lambda_3 = [b,c]$. In the case of the 'diagonal domains' in (4.42) (*i.e.*, those with $j = j'$), the integrand in (4.40) generally has a singularity at $s = t$. These domains are subdivided into two triangles along the diagonal (Figure 4.1(a)), thus confining the singularity to the boundary, as turns out to be important later.

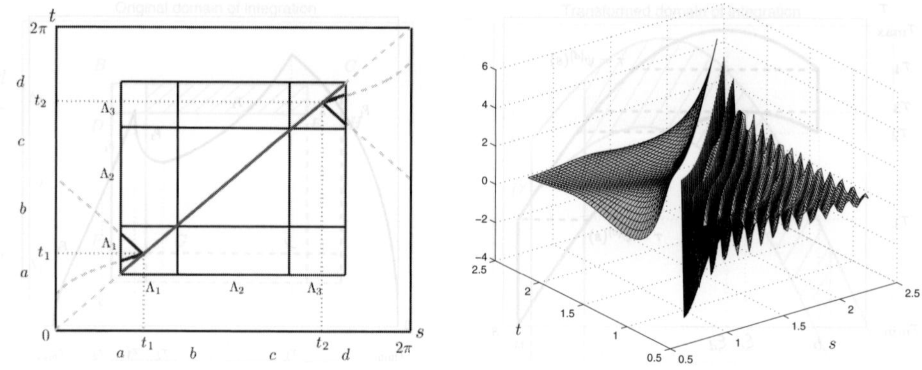

(a) paths of stationary points of the function $\Psi_{[s]}$

(b) detail of the oscillations in the integrand of (4.44) on $\Lambda_1 \times \Lambda_1$ for $k = 30$

Figure 4.3. Locations of stationary points and oscillations of integrand for a smooth convex obstacle.

there.) The composite FCC rules described in Section 4.3 have a convergence rate which remains optimal even in the presence of these singularities, provided the mesh grading is appropriately chosen (see (4.38)).

As mentioned earlier (see also Figure 4.3(a)), the change of variable (4.43) is not valid in $\Lambda_1 \times \Lambda_1$ when $t > s$ or in $\Lambda_3 \times \Lambda_3$ when $t < s$, due to the appearance of stationary points. However, Λ_1 and Λ_3 are both small neighbourhoods of the shadow boundary (recall that $\zeta(t_1)$ and $\zeta(t_2)$ are the shadow boundary points, with t_1 and t_2, marked in Figure 4.3), with length which shrinks to zero as $k \to \infty$. It turns out that in these problematic regions the phase has sufficiently many decaying derivatives as $k \to \infty$ that in fact the integrand in (4.44) is not oscillatory at all, and the integral can be well approximated by standard Clenshaw–Curtis rules, for example. This is explained in some detail in Kim (2012) and is illustrated in Figure 4.3(b). Here the domain above the main diagonal is the one with the stationary points and the integrand is slowly varying. (By contrast the domain below the main diagonal has an oscillatory integrand, but this causes us no difficulty since the change of variable (4.43) has no stationary points here and so this integral can be treated with the transformation (4.43) and then FCC rules, as described above.)

We now consider the case of a convex polygon. For this example, the evaluation of (4.19) can be considered under two distinct scenarios: the case that Γ_j and $\Gamma_{j'}$ are on the same side of the polygon, and the case that Γ_j and $\Gamma_{j'}$ are on different sides of the polygon. For both cases, the basic strategy is identical to that for the smooth convex obstacles described above, namely to use a change of variables in order to transform the integral into a form

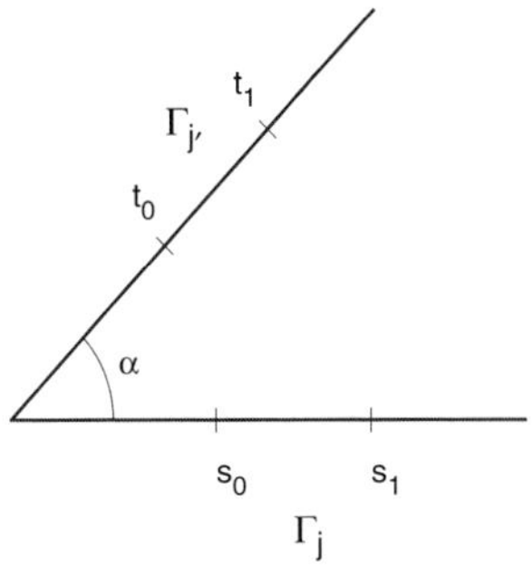

Figure 4.4. Oscillatory integration on the convex polygon: the case where Γ_j and $\Gamma_{j'}$ are supported on different sides.

amenable to evaluation via the FCC quadrature rules of Section 4.3. Further details for this example can be found in Melenk and Langdon (2007, 2012).

First we consider the case for which Γ_j and $\Gamma_{j'}$ each lie on the same side of the polygon. Without loss of generality, we assume that the side on which they lie is parametrized by $\zeta(s) = \{(s,0) : s \in [0,1]\}$, and we suppose further that $\Gamma_j = \{(s,0) : s \in [s_0, s_1]\}$ and $\Gamma_{j'} = \{(t,0) : t \in [t_0, t_1]\}$, where $0 < s_0 < t_0 < s_1 < t_1 < 1$, so that Γ_j and $\Gamma_{j'}$ overlap (the most general situation). In this case the integral (4.40) (with $\Psi(s,t)$ given by (4.41)) is given explicitly by

$$J_k = \int_{s_0}^{s_1} \int_{t_0}^{t_1} g(s,t) \exp(\mathrm{i}k(|s-t| + \theta s + \rho t))\,\mathrm{d}t\,\mathrm{d}s = J_k^{[1]} + J_k^{[2]} + J_k^{[3]},$$

where $\theta \in [-1,1]$, $\rho \in \{-1,1\}$, and

$$\begin{aligned} J_k^{[1]} &= \int_{s_0}^{t_0} \int_{t_0}^{t_1} g(s,t) \exp(\mathrm{i}k((\theta-1)s + (\rho+1)t))\,\mathrm{d}t\,\mathrm{d}s, \\ J_k^{[2]} &= \int_{t_0}^{s_1} \int_{t_0}^{s_1} g(s,t) \exp(\mathrm{i}k(|s-t| + \theta s + \rho t))\,\mathrm{d}t\,\mathrm{d}s, \\ J_k^{[3]} &= \int_{t_0}^{s_1} \int_{s_1}^{t_1} g(s,t) \exp(\mathrm{i}k((\theta+1)s + (\rho-1)t))\,\mathrm{d}t\,\mathrm{d}s. \end{aligned}$$

The integrals $J_k^{[1]}$ and $J_k^{[3]}$ can be trivially put into the form (4.30); supposing for example that $\rho = 1$ and $\theta \neq 1$, we have

$$J_k^{[1]} = \int_{s_0}^{t_0} \left(\int_{t_0}^{t_1} g(s,t) \exp(2\mathrm{i}kt)\,\mathrm{d}t \right) \exp(\mathrm{i}k(\theta-1)s)\,\mathrm{d}s,$$

and both the inner and outer integrals can be evaluated by the FCC rules of Section 4.3 (noting that, in this case, $g(s,t)$ is singular at the point $s = t = t_0$). Evaluation of the integral $J_k^{[2]}$ follows in an identical fashion, splitting the square $(s,t) \in [t_0, s_1] \times [t_0, s_1]$ into the two triangles corresponding to $s > t$ and $s < t$, and evaluating both the inner and outer integrals of each via the FCC rules of Section 4.3, using a Duffy transformation to deal with the singularity on the triangle boundaries corresponding to $s = t$.

Next we consider the case for which Γ_j and $\Gamma_{j'}$ lie on different sides of the polygon. Under the assumption that these sides are not parallel to each other (if the two sides are parallel, similar ideas can be applied, although in that case stationary points may occur; see Melenk and Langdon 2012 for details), we may assume, without loss of generality, the geometrical configuration shown in Figure 4.4, with $\Gamma_j = \{(s,0) : s \in [s_0, s_1]\}$ and $\Gamma_{j'} = \{(t\cos\alpha, t\sin\alpha) : t \in [t_0, t_1]\}$. In this case the integral (4.40) (with $\Psi(s,t)$ given by (4.41)) is given explicitly by

$$J_k = \int_{s_0}^{s_1} \int_{t_0}^{t_1} g(s,t) \exp(ik(\sqrt{s^2 - 2st\cos\alpha + t^2} + \theta s + \rho t))\, dt\, ds,$$

where $\theta \in [-1, 1]$ and $\rho \in \{-1, 1\}$. As above, we will utilize the change of variables (4.43); in this case $t \mapsto \tau$, where

$$\tau = \sqrt{s^2 - 2st\cos\alpha + t^2} + \theta s + \rho t =: \Psi_{[s]}(t) = \Psi^{[t]}(s), \tag{4.47}$$

and we then have

$$\int_{s_0}^{s_1} \int_{\Psi^{[t_0]}(s)}^{\Psi^{[t_1]}(s)} \tilde{g}(s,\tau) \exp(ik\tau)\, d\tau\, ds,$$

with $\tilde{g}$ again given by (4.45). It is shown in Melenk and Langdon (2012) that, for $\theta \in (-1, 1)$,

$$\frac{\partial \tau}{\partial s} = 0 \quad \text{at } s = \left(\cos\alpha - \theta \sqrt{\frac{\sin^2\alpha}{1 - \theta^2}} \right) t,$$

and $\partial^2\tau/\partial s^2 > 0$ for all $t > 0$ (if $\theta = \pm 1$ then $\partial\tau/\partial s \neq 0$). This can be clearly seen in Figure 4.5, where we plot the image of the rectangle $[0, 4] \times [1, 2]$ under the transformation (4.47), for $\theta = 0.5$, $\alpha = \pi/4$ and $\rho = 1$. As for the case of the smooth convex obstacle outlined above, we again proceed by breaking the outer integral up into five pieces, analogous to (4.46) (as shown with the dotted lines in Figure 4.5), and evaluating each of the pieces via a composite FCC rule, grading towards the singularities. We do not give any further details here. The functions corresponding to F_j in (4.46) have a logarithmic singularity (from the fundamental solution appearing in $\tilde{g}$) if $s_0 = t_0 = 0$, that is, if Γ_j and $\Gamma_{j'}$ meet at a corner, and algebraic singularities where $\Psi^{[t_0]}$ and $\Psi^{[t_1]}$ have turning points. The change of variables (4.47) has no stationary points for the geometrical configuration shown in Figure 4.4.

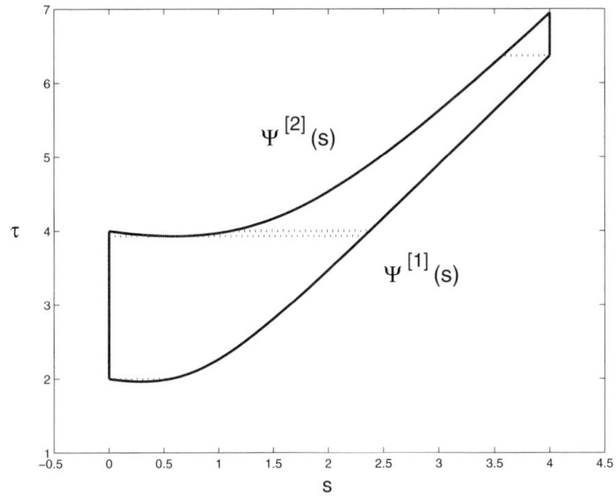

Figure 4.5. The image of the rectangle $[0, 4] \times [1, 2]$ under the transformation (4.47).

4.5. Oscillatory integration in non-variational hybrid schemes

The major advantage of the Galerkin approach in the design of hybrid methods is that the variational framework allows the establishment of error estimates, and most importantly, as we will see in Section 6, these error estimates can be explicit in the wavenumber k. The error estimates permit the development of more efficient methods, for example giving guidance on how to tune the coupling parameter η in the combined potential formulation.

However in practical codes, especially those which are based on second-kind formulations, collocation and Nyström methods often play a prominent role and it is often claimed that these are cheaper to implement than Galerkin methods, due to the apparent reduction in the amount of numerical integration which has to be done.

In collocation methods the solution is approximated by a linear combination of basis functions and the coefficients in the expansion are found by forcing the residual to vanish at a suitable set of points (equivalent to an application of an interpolatory – rather than orthogonal – projection onto the approximation space). Therefore the introduction of hybrid basis functions in collocation methods is straightforward.

The classical 'Nyström scheme' was originally formulated for second-kind integral equations with continuous kernels, and combined a standard quadrature approximation of the integral operator with collocation at the quadrature points: see, *e.g.*, Atkinson (1997) and the (much earlier) references therein. In the modern literature the term 'Nyström' is now often used to describe various methods based on combining quadrature and collocation, even if the collocation points have no relation to the quadrature points.

In this subsection we describe three families of hybrid methods (particularly appropriate for computing scattering by smooth convex obstacles), which are based either on a collocation, Nyström or practical discrete Galerkin approach, focusing particularly on the numerical integration problems which arise and how these are resolved. These are: (i) the Nyström methods of Bruno *et al.* (Bruno *et al.* 2004, Bruno and Geuzaine 2007), which are implemented using a number of innovative ideas including an extension of the method of stationary phase; (ii) the collocation methods of Huybrechs *et al.* (Huybrechs and Vandewalle 2007*b*, Asheim and Huybrechs 2010*b*), which employ a variation on the method of steepest descent to compute the integrals; (iii) the spherical-harmonics-based discrete Galerkin method of Ganesh and Hawkins (2011).

Let us start our discussion by recalling the direct combined potential integral equation (4.2), with solution v expressed using the ansatz (3.14). Then the integral equation to be solved for V is of the form (3.11), that is,

$$\left(\frac{1}{2}I + \widetilde{D_k}' - \mathrm{i}\eta\widetilde{S_k}\right)V \;=\; f, \tag{4.48}$$

where $\widetilde{S_k}$ and $\widetilde{D_k}'$ are the modulated integral operators given by (3.12) and (3.13). Following the discussion in Section 4.1.2, the integral operators $\widetilde{S_k}$ and $\widetilde{D_k}'$ both have an action of the form

$$(RV)(x) = \int_\Gamma g(x,y)\exp(\mathrm{i}k\Psi(x,y))V(y)\,\mathrm{d}s(y), \quad x \in \Gamma, \tag{4.49}$$

where, analogously to (4.19), $g(x,y)$ is smooth, except at the diagonal $x = y$, and non-oscillatory.

In a collocation approach, the action of an operator of the form (4.49) on suitable basis functions has to be computed for some discrete set of collocation points x. (For example, at least in 2D, the FCC methods described in Section 4.3 could be employed for this task.) In a Nyström approach, a quadrature approximation of the integral operator of the form (4.49) is first devised, taking as input a suitable set of point values of V. This yields corresponding approximations of the integral operators in (4.48). The equation (4.48) could then be solved for example by an iterative method which only required the action of these operators. In this way it is possible to devise a Nyström method without needing to adopt the classical collocation method of solving for the Nyström solution.

The work of Bruno *et al.* (2004) and Bruno and Geuzaine (2007) adopts a Nyström approach. We summarize this for the 3D case. The first step is to cover the scattering surface Γ with an atlas of overlapping charts, each assumed to be the image of a rectangle in $\mathbb{R}^2$, under a local invertible map (here assumed smooth). Then introduce a partition of unity subordinate to this covering and note that determining any function V on Γ is equivalent to determining χV for each member χ of the partition of unity. Using the local parametrization, determining χV on Γ is equivalent to determining it on the parameter rectangles. Since χV is smooth and vanishes on the boundary of its parameter rectangle, the FFT can be used to obtain V at any point on Γ quickly from point values on uniform grids in parameter space. (The actual algorithm is quite technical and is described in detail in Bruno and Geuzaine 2007, §3.2.)

Now, turning to computation of the action of the integral operator R in (4.49), the starting point is the observation that as $k \to \infty$, for each $x \in \Gamma$, $RV(x)$ is dominated by contributions from V and its derivatives at the critical points (namely the singular point $y = x =: x_0$ and any points $x_1, \dots, x_n$ where the phase Ψ is stationary, that is, where $\nabla_y \Psi(x, y)|_{y=x_i} = 0$). The stationary points could be determined by Newton's method with starting guesses obtained by geometrical arguments (at least when Γ is convex). Then the integral operator R is approximated by

$$RV(x) \approx \sum_{i=0}^{n} \int_{\Gamma_i} g(x, y) \exp(\mathrm{i}k\Psi(x, y)) c_i(y) V(y) \,\mathrm{d}s(y),$$

where each Γ_i is a neighbourhood of x_i with k-dependent radius chosen to shrink to zero with a rate depending on whether $i = 0$ (singular point) or $i = 1, \dots, n$ (stationary points), and c_i is a smooth 'windowing function' with support in Γ_i.

Finally, given that values of V at all points on Γ are cheap to compute (see above), the integrals over each Γ_i are obtained by integrating in polar coordinates around each x_i. This has the benefit that the Jacobian of the polar coordinate transformation cancels the singularity in the singular integral (over Γ_0) but also oscillations in the transformed integrand are prominent only in the radial direction and so this reduces the complexity of the computation of the contributions to RV. Full details are in Bruno and Geuzaine (2007), and related results using the same ideas are described in Bruno and Reitich (2007). (The observation that such changes of variables can cancel weak singularities in boundary integral operators goes back at least to Wienert (1990); see also Ganesh, Graham and Sivaloganathan (1998) and Graham and Sloan (2002).)

The key novelty of the collocation method employed in Huybrechs and Vandewalle (2007*b*) is the sparsity of the resulting linear system. This is achieved via the use of locally supported basis functions, coupled with the observation that integrals for which the integrand does not contain singularities (those for which the collocation point lies outside the support of the basis function) and for which the phase does not contain stationary points, decay rapidly as k grows. The integrals that remain, which contain complicated phase functions, are evaluated using a special quadrature rule, that combines features of generalized Filon quadrature (Iserles 2005, Iserles and

Nørsett 2005, Olver 2007, Olver 2010) and the numerical method of steepest descent (Huybrechs and Vandewalle 2006, Asheim and Huybrechs 2010*a*). Crucially, this rule requires only local approximation of the amplitude near special points, and this localization is what allows the sparse formulation to hold. The method described in Huybrechs and Vandewalle (2007*b*) is not supported by a rigorous analysis, but the results suggest that this algorithm is comparable in accuracy to others described in this subsection, appearing to offer arbitrary accuracy with a computational cost that is independent of k, and with the added advantage of a sparse rather than a full linear system.

The work in Asheim and Huybrechs (2010*b*) builds upon these ideas, making the observation that, for complicated scattering configurations, if one is only interested in the solution on a particular part of the boundary, then reasonable accuracy can be achieved via a consideration of mainly just local effects, leading to substantial cost savings compared to algorithms for which the solution is sought throughout the domain.

An alternative but closely related approach to the hybrid numerical-asymptotic schemes described in this article is the partition-of-unity-based method with plane wave enrichment, developed by workers such as Bettess and Trevelyan for the solution of general Helmholtz problems for which, in general, identification of the leading-order asymptotic behaviour would be extremely challenging (see, *e.g.*, Perrey-Debain *et al.* 2004 and the references therein). In recent work, Honnor, Trevelyan and Huybrechs have applied the integration rules developed by Huybrechs and Vandewalle (2006, 2007*a*) in the implementation of these methods (Honnor *et al.* 2010, Trevelyan, Honnor and Huybrechs 2007) to good effect.

Another integration scheme for these problems was proposed in Trevelyan and Honnor (2009), with the main idea there being to attempt to line up the direction of the oscillations with the coordinate axes, so as to convert multidimensional oscillatory integrals into oscillatory integrals for which the oscillation is only in one direction. This idea is similar in some ways to the change of variables / FCC approach described in Section 4.4, and has been shown to lead to a signficant reduction in computational cost compared to standard (Gaussian) quadrature.

A fully discrete hybrid Galerkin method for scattering by smooth 3D obstacles (diffeomorphic to the sphere) is introduced in Ganesh and Hawkins (2011). The idea is to approximate V in (4.48) globally using spherical harmonics. This provides an accurate approximation away from the shadow boundary, but is not so effective in the vicinity of the shadow boundary. However, it is argued in Ganesh and Hawkins (2011) that this does not matter in the context of achieving reasonable (global) results, since the amplitude of the oscillations is small near the shadow boundary; numerical results suggest that this amplitude is of the order of a hundredth of the maximum amplitude of V over the entire boundary. A Galerkin scheme is then used to discretize the integral equation, leading to quadruple integrals to be evaluated over the entire surface of the scatterer. The outer Galerkin integrals are discretized by a discrete sum defined using a Gauss-rectangle quadrature rule (similar to the quadrature scheme of Ganesh and Graham (2004), where a fully discrete scheme for low- to mid-range frequency three-dimensional scattering was proposed). The rectangle rule is exponentially accurate for smooth periodic functions (in the azimuthal direction), and the Gauss rule is superalgebraically accurate in the latitudinal direction; indeed this scheme is exact for products of spherical harmonics up to a given order. To maintain global accuracy, the number of quadrature points for these outer integrals must grow with $O(k^{2/3})$.

The inner integrals are then approximated using surface patches, defined for three different types of critical points, namely: the observation point (where the integrand has a removable singularity); the stationary points; and so-called 'steepness points', which correspond to points around which the gradient of the phase is non-zero but small. These points represent, in some sense, the next term in an asymptotic expansion of a method of stationary phase type of approximation. Each of these three types of critical point has a spherical cap placed around it, and the integral is evaluated numerically on this cap, with the radius of the patch depending both on the nature of the critical point and on the distance to the shadow boundary. The singular critical points are evaluated via a rotation and translation procedure, with the Jacobian of the polar coordinate transformation cancelling the singularity in a very similar way to that described above for the scheme of Bruno and Geuzaine (2007). The remaining integrations are carried out via a rectangle rule in the azimuthal direction, and a uniform partition in the non-periodic direction. On each interval of the uniform partition, standard quadrature is applied if the integrand is slowly oscillating, and Filon quadrature (as described in Section 4.3, but using a cubic polynomial interpolatory rule) is applied if the remaining integrand is highly oscillatory. Results are presented for spherical and non-spherical convex scatterers, for a range of wavenumbers $1 \leq k \leq 100000$, and it appears that accuracy (either high- or low-order prescribed tolerance) can be maintained for only a mild growth of $O(k^{2/3})$ in the number of unknowns as k increases.

5. Conditioning and coercivity

In Section 2 we met various standard integral equation formulations of scattering problems and saw that, in many cases, for every $k > 0$, the BVPs were well-posed, with the relevant integral operators bounded as mappings between appropriate Sobolev spaces. In this section we aim to determine *explicitly* the dependence of the integral operators on k, in order to understand better their behaviour for high frequencies. (In many cases the dependence on k is known for the whole frequency range $k > 0$, but we focus on the high-frequency case, which is the subject of the article.) Since the integral operators depend on k in a complicated non-linear way, determining this k-dependence is, in many cases, a challenging problem. We restrict ourselves to discussing mainly the exterior Dirichlet problem (2.12) (which includes the sound-soft scattering problem (2.15)), although some of the bounds we obtain on integral operators will be applicable to other BVPs.

In the context of the three questions posed in Section 1, this section will provide the remaining ingredients to prove k-explicit error estimates of the form (1.7), addressing **Q3**; these ingredients are then assembled in Section 6.

In this section, $(\cdot,\cdot)$ will denote the $L^2(\Gamma)$ inner product and $\|\cdot\|$ will denote either the $L^2(\Gamma)$ norm of an element of $L^2(\Gamma)$ (as in Section 3), or the $L^2(\Gamma) \to L^2(\Gamma)$ norm of an operator. We will mainly be concerned with the combined potential operators $A'_{k,\eta}$ and $A_{k,\eta}$, defined by (2.69) and (2.72) respectively, but we will also give results, where available, for the new class of integral operators introduced in Section 2.9, namely $A'_{k,\eta,Z}$ defined by (2.99), a special case of which is the star-combined operator $\mathscr{A}_k$ (2.103). Recall from Remark 2.24 that $\|A'_{k,\eta}\| = \|A_{k,\eta}\|$, and similarly for their inverses. Thus any results on the $L^2(\Gamma)$ norms of $A'_{k,\eta}$ and $(A'_{k,\eta})^{-1}$ also hold for $A_{k,\eta}$ and $A^{-1}_{k,\eta}$, respectively. For the same reasons, if $A'_{k,\eta}$ is coercive, then $A_{k,\eta}$ is coercive with the same coercivity constant.

The outline of the section is as follows: we begin in Section 5.1 with a summary of the main bounds on $\|A'_{k,\eta}\|$ and $\|(A'_{k,\eta})^{-1}\|$, and thus on the condition number of $A'_{k,\eta}$,

$$\operatorname{cond}(A'_{k,\eta}) = \|A'_{k,\eta}\|\|(A'_{k,\eta})^{-1}\|. \tag{5.1}$$

These bounds immediately suggest how to choose the parameter η to minimize the condition number of $A'_{k,\eta}$ for k large. The next two subsections, Section 5.2 and Section 5.3, give a summary of relevant results and techniques from scattering theory that are key components of the bounds on $\|(A'_{k,\eta})^{-1}\|$ (together with the representation of $(A'_{k,\eta})^{-1}$ in Theorem 2.33), and of the methods used to establish coercivity of $A'_{k,\eta}$ and $\mathscr{A}_k$. There have been several investigations of conditioning and coercivity in the special cases when Γ is the circle or sphere. This is because for these two domains the integral operators diagonalize in the Fourier and spherical harmonic bases respectively, and thus bounds on the quantities of interest, *e.g.*, the norms of the operators, can be obtained by bounding eigenvalues of the operators. We discuss the conditioning and coercivity results obtained in this way for these two special domains in Section 5.4. Sections 5.5 and 5.6 discuss bounds on $\|A'_{k,\eta}\|$ and $\|(A'_{k,\eta})^{-1}\|$, respectively, for more general domains than the circle and sphere. Section 5.7 discusses conditions under which $A'_{k,\eta}$ is coercive, and the corresponding bounds on its coercivity constant.

Establishing coercivity of $A'_{k,\eta}$ and the related star-combined operator $\mathscr{A}_k$ are vital requirements for the error analysis in Section 6. We note that, whereas bounds on $\|A'_{k,\eta}\|$ can be obtained by norm estimation techniques that apply to general oscillatory integral operators, the bounds on $\|(A'_{k,\eta})^{-1}\|$ and the coercivity constant rely heavily on methods and results from classical scattering theory.

Similarly to Section 3, for two quantities D and E which may depend on k we write $D \lesssim E$ if $D \leq C\,E$ for some constant C that is independent of k. Also we write $D \sim E$ if $D \lesssim E$ and $E \lesssim D$.

5.1. *Summary of the main results on the conditioning of $A'_{k,\eta}$ and $A_{k,\eta}$*

Some examples of the behaviour of $\|A'_{k,\eta}\|$ and $\|(A'_{k,\eta})^{-1}\|$ for different domains are shown in Figure 5.1. The following is a short summary of the main bounds on $A'_{k,\eta}$ and $(A'_{k,\eta})^{-1}$ discussed in this section.

- For a general Lipschitz domain (in 2D or 3D),

$$\frac{1}{2} \leq \|A'_{k,\eta}\| = \|A_{k,\eta}\| \lesssim 1 + k^{(d-1)/2}\left(1 + \frac{|\eta|}{k}\right), \tag{5.2}$$

  for all $k > 0$. The upper bound is sharp in 2D if the domain contains a straight line segment. (See Theorem 5.14 and Lemmas 5.17 and 5.18.)

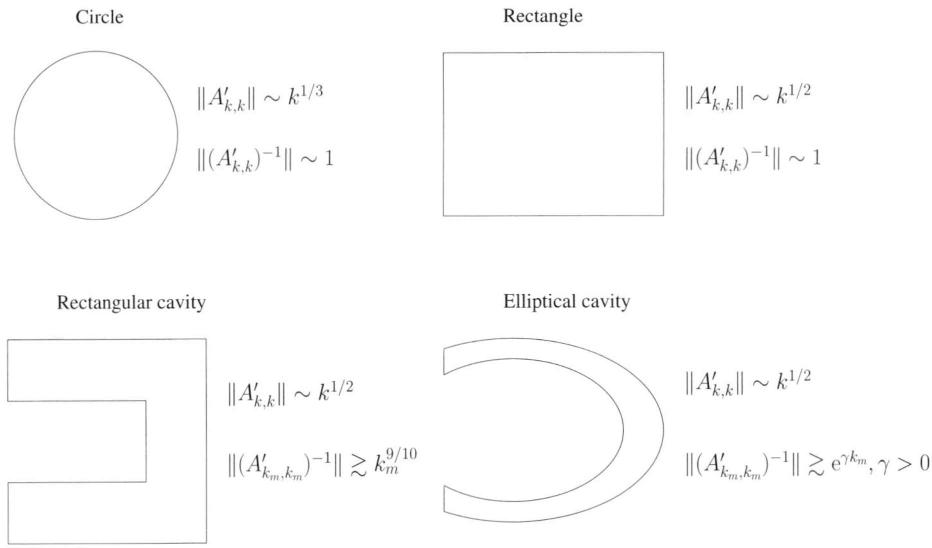

Figure 5.1. Examples of the behaviour of $\|A'_{k,k}\|$ and $\|(A'_{k,k})^{-1}\|$ as $k \to \infty$. Here k_m denotes a specific sequence of wavenumbers, which are different for the rectangular and elliptical cavities (see Section 5.6.2 for more details).

- For a star-shaped (in the sense of Definition 5.5 below) Lipschitz domain (in 2D or 3D),

$$2 \leq \|(A'_{k,\eta})^{-1}\| = \|A^{-1}_{k,\eta}\| \lesssim 1 + \frac{1+k}{|\eta|}, \tag{5.3}$$

  for all $k > 0$ (see Theorem 5.22 and Lemma 5.17). For a general (smooth) non-trapping domain (in the sense of Definition 5.4 below) the upper bound is multiplied by a factor of $k^{2+(d-1)/2}$, although this should be considered a preliminary bound and not a definitive one. Indeed, numerical experiments suggest that the upper bound in (5.3) holds for some non-trapping domains that are not star-shaped. The precise definitions of trapping and non-trapping are given in Definition 5.4, but, roughly speaking, a trapping domain is one that contains an open cavity that can 'trap' high-frequency waves.
- There exist trapping domains such that, for some $\gamma > 0$,

$$\|(A'_{k_m,\eta})^{-1}\| = \|A^{-1}_{k_m,\eta}\| \gtrsim \exp(\gamma k_m)\left(1 + \frac{|\eta|}{k_m}\right)^{-1},$$

  for some sequence $k_m \to \infty$; an example is the so-called elliptical cavity domain in Figure 5.1 (see Theorem 5.24).

Remark 5.1. The commonly recommended choices of η in the literature are to take η proportional to k for all but small values of k, and then η either constant (for $d = 3$) or proportional to $(\log k)^{-1}$ (for $d = 2$) for k small. Up until the last few years, these choices have been based on investigations when Γ is the circle or sphere (see the references in Section 5.4) or computational experience (Bruno and Kunyansky 2001).

The bounds (5.2) and (5.3) show that the choice $\eta \sim k$ is optimal for large k in the sense that it minimizes the rate of growth with respect to k of the upper bound on $\mathrm{cond}(A'_{k,\eta})$. Thus

$$\mathrm{cond}(A'_{k,k}) \lesssim k^{(d-1)/2}, \qquad \text{as } k \to \infty \tag{5.4}$$

for star-shaped Lipschitz domains. This bound is sharp for any star-shaped polygon.

The upper bound in (5.2) is not sharp as $k \to 0$. However, sharp bounds on $\|A'_{k,\eta}\|$ can be obtained in the $k \to 0$ limit which back up the previously recommended choices of η for small k mentioned above (Betcke *et al.* 2011, §2.6, §2.7).

Remark 5.2. A natural question is whether or not the norm bounds given above continue to hold if the operators are approximated by their discrete counterparts, and so to what extent the bounds above are relevant to understanding conditioning at a discrete level. For conventional Galerkin boundary element discretizations this question has been addressed to a large extent in Betcke *et al.* (2011), in particular showing that (5.4) holds at a discrete level for a variety of 2D star-shaped domains.

Remark 5.3. For reasons of space we do not treat in this article the important implications of condition number estimates and coercivity for the behaviour of iterative solvers for the linear systems that arise in Galerkin methods. A recent review of these issues in the context of BIEs for high-frequency scattering is that of Antoine and Darbas (2012).

5.2. Relevant classical scattering results

In this subsection we collect some classical results about the high-frequency Helmholtz equation. Whereas in Section 3 we discussed the high-frequency asymptotics of the solution, here we discuss some classical results that bound the solution in terms of the data (in suitable norms).

These results are taken from a large body of work concerned with both the Helmholtz equation and the wave equation. The problem involving the Helmholtz equation that has received most interest is the question of how the solutions behave as $k \to \infty$. For scattering problems involving the wave equation the most-studied problem is the long-time behaviour of the energy in a neighbourhood of the obstacle. These two problems are obviously related, since when one takes the Fourier transform in time of the wave equation one obtains the Helmholtz equation; however, to understand rigorously the relationship between the two problems one must understand how solutions of the wave equation with singularities in the initial data (such as a delta function) behave when they hit the obstacle. Understanding this so-called *propagation of singularities* was one of the main motivations for the development of the tools of *microlocal analysis.*

Some classic books and monographs describing this body of work in scattering theory from different perspectives are Babich and Buldyrev (2008), Borovikov and Kinber (1994), Hörmander (1985*a*, 1985*b*), Lax and Phillips (1989), Melrose (1995), Morawetz (1975*b*) and Vainberg (1989) (with the Russian texts translations of earlier Russian originals).

A key concept in scattering theory is the geometrical classification of domains as either *trapping* or *non-trapping*. We give the non-technical definition; see, for instance, Lax and Phillips (1989, Epilogue). For a more mathematical, but technical, definition see Melrose and Sjöstrand (1982, Definition 7.20) or Melrose (1979, Definition 1.3). In this definition and the rest of the section, as in Section 2, $B_R := \{x \in \mathbb{R}^d, |x| < R\}$.

Definition 5.4. (Trapping and non-trapping for smooth domains) Assume that Ω_+ is smooth, that is, Γ is C^∞. For any $R > 0$ such that $\overline{\Omega_-} \subset B_R$, consider all possible rays starting in $\Omega_+ \cap B_R$ (*i.e.*, starting from every point and travelling in every possible direction). Whenever a ray hits Γ, continue it according to the law of reflection (angle of incidence equals angle of reflection) until it leaves B_R. We call Ω_+ *trapping* if there are arbitrarily long paths or closed paths of rays; otherwise Ω_+ is called *non-trapping*.

(Note that Definition 5.4 does not cover rays that hit the boundary at a tangent, and there are additional subtleties with these. Indeed, Taylor (1976) showed that, in some cases, one cannot uniquely define the reflection of rays tangent to the boundary.)

We have defined trapping and non-trapping for C^∞ domains since the propagation of singularities of the wave equation on C^∞ domains is fully understood by the results of Melrose and Sjöstrand (1978, 1982). However, when Ω_+ is not C^∞, and especially when it is not C^1, understanding how a ray reflects is more complicated. Following the development of the very effective but non-rigorous Geometrical Theory of Diffraction (Keller 1962), there has been much work on rigorously understanding the propagation of singularities on domains with corners and edges. The recent papers by Melrose, Vasy and Wunsch (2008, 2012) both contain good overviews of this work.

To illustrate the difficulty involved in formulating a definition of trapping for non-smooth domains, consider the particular example of rays hitting a convex polygon. When a ray hits a corner it produces diffracted rays emanating from the corner, and in particular some that travel along the sides of the polygon (understanding these rays on the boundary is implicit in the approximation results for the convex polygon in Section 3.3). This means that there exist 'glancing rays' that travel around the boundary of the polygon (hitting a corner and then either continuing on the next side or travelling back) and do not escape to infinity; thus the exterior of a convex polygon is, in this sense, a trapping domain. At each diffraction from a corner, however, these rays lose energy, and thus the trapping is in a weaker sense than having a closed path of rays (as in Definition 5.4).

We state below some results about solutions of the Helmholtz equation in non-trapping domains (Theorems 5.6 and 5.7) that depend on the propagation of singularities results of Melrose and Sjöstrand (1978, 1982) for C^∞ domains. Using the recent results about propagation of singularities on manifolds with corners and edges from the programme of work by Melrose, Vasy and Wunsch, it is reasonable to believe that analogous (or slightly weaker) results to Theorems 5.6 and 5.7 would hold for the exterior of a non-trapping polygon (in the weaker sense discussed above), for example. However, although the relevant technical tools now exist, such results have not yet been proved. (Actually, we note that, in the special case of the exterior of a polygon, either the results of Gérard and Lebeau (1993), about propagation of singularities from wedges with analytic boundaries, or the results of Hillairet (2005), about propagation of singularities on Euclidean surfaces with conic singularities, may be sufficient to achieve this goal, again with some non-trivial technical work outstanding.)

A slightly different, but related, question is whether there exists the appropriate analogue of the propagation of singularities results of Melrose and Sjöstrand (1978, 1982) for, say, C^2 domains (instead of C^∞ domains), possibly allowing for higher-order diffractive effects. To the authors' knowledge this does not appear to be the case, although we note that Ivrii's work on the Weyl law and spectral asymptotics on domains slightly better than C^1 (Bronstein and Ivrii 2003, Ivrii 2003) contains some results for non-tangent rays.

The preceding paragraphs hopefully give an indication of some of the subtleties associated with the concepts of trapping and non-trapping on non-smooth domains. For the remainder of this section we will consider domains such as the rectangular and elliptical cavities in Figure 5.1 as trapping domains, despite not satisfying Definition 5.4, since they are not C^∞, since for these obstacles there exist closed paths of rays that hit only smooth parts of the boundary.

We have just encountered a slight 'culture clash' between the classic scattering literature and the numerical analysis literature: classical scattering theory is happiest in smooth domains, whereas from the point of view of practical applications we want to consider domains with corners and edges. What then should we aim for in the numerical analysis of scattering problems? An important class of C^∞ domains for which the non-trapping condition holds consists of domains that are star-shaped in the following sense.

Definition 5.5. (Star-shaped) We say that Ω_- is star-shaped with respect to the point $x_0 \in \Omega_-$ if there exists $c > 0$ such that $(x - x_0) \cdot n \geq c$ for almost every $x \in \Gamma$, where n is the normal vector pointing outwards from Ω_-. We say that Ω_- is star-shaped if it is star-shaped with respect to some $x_0 \in \Omega_-$.

If Ω_- is star-shaped and C^∞ then Ω_+ is non-trapping in the sense of Definition 5.4 (Lax and Phillips 1989, Chapter 5, Proposition 3.1) (star-shapedness guarantees that the rays can only be reflected a finite number of times before escaping from the obstacle). Since the normal vector is defined almost everywhere on Lipschitz boundaries, star-shapedness is a well-defined property for Lipschitz domains, and thus a reasonable aim from the point of view of numerical analysis of scattering problems is to prove results for star-shaped Lipschitz domains, and then also for (smooth) non-trapping domains.

Another 'culture clash' is that classical scattering theory is happiest in 3D (essentially because the fundamental solution for the Helmholtz equation $\Phi_k(x, y)$ is an analytic function of k in 3D, but not in 2D). However, from a numerical analysis perspective one often starts in 2D, where computations are easier, and then progresses to 3D.

One way to understand the significance of trapping and non-trapping is the following. Recall that the Helmholtz operator with $k > 0$ in a bounded domain can have non-trivial solutions for certain homogeneous boundary conditions (these are eigenfunctions of the Laplacian). In an unbounded domain the Helmholtz equation with $k > 0$ cannot have any non-trivial solutions satisfying homogeneous boundary conditions, provided an appropriate radiation condition is prescribed (Corollary 2.9). However, as $k \to \infty$, solutions of the Helmholtz equation behave more and more like rays (this is the whole notion of ray theory), and thus if the domain is trapping then, for certain k, there will be solutions of the Helmholtz equation localized in the trapping part of the obstacle and behaving almost like non-trivial solutions satisfying homogeneous boundary conditions. This informal discussion can be made mathematically precise through the concepts of *resonances* and *quasimodes* (see the references in Section 5.6.2).

In Section 5.6.2 we take a closer look at the type of behaviour that solutions of the Helmholtz equation can exhibit in trapping domains (and define the concepts of a resonance and a quasimode); for the remainder of this subsection we look at bounds on the solution operator of the Helmholtz equation in *non-trapping* domains.

Theorem 5.6. (Vainberg 1975) Let Ω_+ be a non-trapping domain (in the sense of Definition 5.4) in 2D or 3D. If $g \in L^2(\Omega_+)$ has compact support, and $v \in H^2_{\text{loc}}(\Omega_+)$ satisfies

$$\Delta v + k^2 v = g,$$

the Sommerfeld radiation condition (2.9), and either

$$\gamma v = 0 \quad \text{or} \quad \partial_n v = 0 \text{ on } \Gamma,$$

then, given any $R > 0$ such that $\overline{\Omega_-} \subset B_R$, for every $k_0 > 0$,

$$k\|v\|_{L^2(\Omega_+\cap B_R)} + \|v\|_{H^1(\Omega_+\cap B_R)} + \frac{1}{k}\|v\|_{H^2(\Omega_+\cap B_R)} \lesssim \|g\|_{L^2(\Omega_+)}, \qquad (5.5)$$

for $k \geq k_0$.

We will not go into detail about how this bound was obtained, referring the reader to the original paper by Vainberg (1975) and also the accounts in Vainberg (1999) and Vainberg (1989) for this information, but we will make two brief remarks: firstly, that the bound is sharp in terms of the powers of k in its k-dependence, and secondly that it relies on the results on propagation of singularities in Melrose and Sjöstrand (1978, 1982). See also Remark 5.9 for more discussion on this type of bound.

Whereas Theorem 5.6 obtained bounds on solutions to the inhomogeneous Helmholtz equation in Ω_+, we now discuss bounds on the Dirichlet to Neumann map for the homogeneous Helmholtz equation (1.1) in Ω_+ (the map P^+_{DtN} introduced in Section 2.7). Via results in Section 2.7 (in particular Theorem 2.33) these will help us obtain bounds on $\|(A'_{k,\eta})^{-1}\|$. The following theorem giving bounds on this map has recently been obtained from Theorem 5.6.

Theorem 5.7. (Lakshtanov and Vainberg 2012) Let Ω_+ be a non-trapping domain (in the sense of Definition 5.4) in 2D or 3D. If u is a solution of (1.1) in Ω_+ satisfying the Sommerfeld radiation condition (2.9), then, for every $k_0 > 0$,

$$\|\partial_n u\|_{H^{-1/2}(\Gamma)} \lesssim k^3 \|\gamma u\|_{H^{1/2}(\Gamma)},$$

and

$$\|\gamma u\|_{H^{1/2}(\Gamma)} \lesssim k \|\partial_n u\|_{H^{-1/2}(\Gamma)},$$

for all $k \geq k_0$.

We note that this paper also contains analogous bounds for $k \in \mathbb{C}$ with $0 < \arg k < \pi/2$ that hold for general Lipschitz obstacles.

Using certain 'multiplier methods', applicable only to bounding $\partial_n u$ in $L^2(\Gamma)$ (to be explained in more detail in Section 5.3), one can obtain the following bound on P^+_{DtN} that is better than the first bound in Theorem 5.7 from the point of view of k-dependence.

Theorem 5.8. Let Ω_- be a Lipschitz star-shaped domain in 2D or 3D. Let u be a solution of (1.1) in Ω_+ satisfying the Sommerfeld radiation condition (2.9). Then, if $\gamma u \in H^1(\Gamma)$,

$$\|\partial_n u\|_{L^2(\Gamma)} \lesssim \|\nabla_\Gamma \gamma u\|_{L^2(\Gamma)} + k\|\gamma u\|_{L^2(\Gamma)}, \tag{5.6}$$

for all $k > 0$ (where ∇_Γ is the surface gradient operator defined in (A.14)).

The bound (5.6) was essentially proved in Morawetz and Ludwig (1968); the authors only considered the smooth star-shaped case, but the arguments are valid for Lipschitz domains with a little extra technical work (for example, (5.6) was proved in the case that Γ is Lipschitz and piecewise smooth in Chandler-Wilde and Monk (2008)).

Using a mixture of the arguments of Lakshtanov and Vainberg (2012) and the multiplier methods used to prove Theorem 5.8, Spence (2012) proved that, when Ω_+ is non-trapping, for every $k_0 > 0$,

$$\|\partial_n u\|_{L^2(\Gamma)} \lesssim k^2 \|\gamma u\|_{H^1(\Gamma)}, \tag{5.7}$$

for all $k \geq k_0$. However, a sharper bound than (5.7) can almost certainly be obtained (at least in 2D) by translating the results of Morawetz (1975*a*) into this setting.

Remark 5.9. (Resolvent estimates) The bound in Theorem 5.6 is a bound on the operator $R(k) := (\Delta + k^2)^{-1}$, that is, the resolvent of the Laplacian. We now briefly discuss bounds of this type because (a) they are a crucial component of scattering theory, and (b) some of the recent developments in the error analysis of BIEs for the Helmholtz equation for large k provide an alternative perspective on these bounds.

(a) Rewriting (5.5) in terms of $R(k)$, it becomes

$$\|\chi R(k)\chi\|_{L^2(\mathbb{R}^d)\to H^j(\mathbb{R}^d)} \lesssim k^{j-1}, \quad j = 0, 1, 2, \tag{5.8}$$

for any $\chi \in C_0^\infty(\mathbb{R}^d)$, and this is the form in which resolvent estimates usually appear in the scattering theory literature. (The bound (5.8) is known as a 'local resolvent estimate' because of the pre- and post-multiplication of $R(k)$ by a cut-off function.) The simplest situation where one encounters a bound of this form is that of the Helmholtz equation in $\mathbb{R}^d$ (the resolvent of the so-called 'free Laplacian'). In this case the estimate (5.8) appears in, for example, Vainberg (1975, Theorem 3) and Burq (2002, §2.1). Generalizations of this estimate (which remove the 'local' nature at the cost of working in more technical function spaces) appear in, for example, Hörmander (1983*b*, Theorem 14.3.7) and Perthame and Vega (1999) (note that Theorem 14.3.7 of Hörmander (1983*b*) is for fixed k, but a rescaling of the independent variable yields the result for arbitrary k). In the case of the Laplacian in the exterior of a bounded obstacle, resolvent estimates were obtained by Vainberg (1975, Theorem 7) (quoted as Theorem 5.6 above) and Morawetz (1975*a*, Theorem I.2D) (although the latter is not quite in the form of (5.8)).

(b) The error analysis for the Helmholtz equation recently developed by Melenk and co-workers, and reviewed in the BIE context in Section 6.3, has at its heart an additive splitting of the solution into a part with finite regularity but k-independent bounds, and a part that is analytic with k-explicit bounds for all its derivatives. In $\mathbb{R}^d$ this splitting leads to an alternative proof of the classical local free resolvent estimate (5.8) (Melenk and Sauter 2010, Lemma 3.5). See also Esterhazy and Melenk (2012, §3) for an overview of this work.

5.3. Rellich- and Morawetz-type identities

In this subsection we discuss the so-called Rellich- and Morawetz-type identities. We do this because many of the results about $\|(A'_{k,\eta})^{-1}\|$ and coercivity of $A'_{k,\eta}$ that we discuss in Sections 5.6 and 5.7 have been proved using these techniques, and several key results that we quoted in Section 2 have been proved using them, notably the regularity result in Theorem 2.12 which follows from Nečas (1967) and the results of Verchota (1984).

The method of multipliers is one of the few truly general techniques in PDE theory. It consists of multiplying the PDE by a judiciously chosen function and integrating (often by parts). In our case we consider the Helmholtz equation, written as $\mathcal{L}u = 0$ with $\mathcal{L} = \Delta + k^2$, and restrict attention to the case where the multiplier is a function of u, namely $\overline{\mathcal{N}u}$, where $\mathcal{N}$ is some linear operator.

In this framework Green's first identity (given in the Appendix as equation (A.26)) with $v = u$ is equivalent to the choice $\mathcal{N}u = u$. Rellich-type identities are obtained by choosing $\mathcal{N}u$ to be some derivative of u: for example, the derivative in the radial direction in the case of star-shaped domains (so $\mathcal{N}u = x \cdot \nabla u$) or a derivative along the vertical coordinate axis in the case of domains that are the hypograph of a function.

The multiplier $x \cdot \nabla u$ was originally introduced by Rellich (1940) to obtain an expression for the eigenvalues of the Laplacian in star-shaped domains as an integral over the boundary (as opposed to the usual expression as an integral over the domain used, for example, in the Rayleigh–Ritz method). Following on from their work using the Rellich multiplier for Laplace's equation (Payne and Weinberger 1955), Payne and Weinberger (1958) generalized the resulting identity to second-order elliptic systems with variable coefficients. This general identity, in which the multiplier is an arbitrary derivative of u (that is $Z \cdot \nabla u$ for some arbitrary vector field Z) was independently introduced by Hörmander (1954). We note that the Rellich identity with multiplier $x \cdot \nabla u$ is often known as the Derrick–Pohozaev identity (see, for example, Evans (1998, §9.4.2)). This is because it was independently introduced by both Pohozaev (1965) and Derrick (1964) in their studies of the possible non-existence of solutions of $\Delta u + \lambda f(u) = 0$. (Actually, Derrick's proof of the identity was not via the multiplier $x \cdot \nabla u$, but via a scaling argument: see Berestycki and Lions (1983, §2.1) for a nice account of both proofs.)

Morawetz-type identities consist of taking certain linear combinations of both derivatives of u and u itself (that is, linear combinations of the Rellich and Green multipliers). These are discussed in more detail for the Helmholtz equation in Section 5.3.1; see, for example, Morawetz (1961, 1968), Strauss (1975) and Morawetz, Ralston and Strauss (1977) for their use in the context of wave equations.

Rellich- and Morawetz-type identities have been used in many different contexts since their introduction; in the following three subsections we discuss three uses relevant to this article.

(1) Bounding the Dirichlet to Neumann map for second-order elliptic PDEs (including the Laplace and Helmholtz equations), Section 5.3.1.

(2) Bounding the energy norm of solutions of the Helmholtz equation, Section 5.3.2.

(3) Proving coercivity of integral operators related to the Helmholtz equation, Section 5.3.3.

Remark 5.10. Here we have understood Rellich- and Morawetz-type identities in the context of multiplier methods. An alternative approach is to view them as consequences of Noether's theorem (relating symmetry groups and conservation laws); see, for example, Olver (1993, Notes to Chapter 4).

5.3.1. Bounding the Dirichlet to Neumann map

For simplicity, let $d = 2$ (the three-dimensional case, $d = 3$, is slightly more complicated) and consider the simplest Rellich-type multiplier: $\mathcal{N}u = x\cdot\nabla u$.

Multiplying $\mathcal{L}u = 0$ by $\overline{\mathcal{N}u}$ and integrating by parts leads to the following identity for solutions of $\mathcal{L}u = 0$:

$$\nabla\cdot\left(2\mathrm{Re}\left(x\cdot\overline{\nabla u}\nabla u\right) + \left(k^2|u|^2 - |\nabla u|^2\right)x\right) = 2k^2|u|^2 \tag{5.9}$$

(for details see, *e.g.*, Spence *et al.* 2011, Lemma 2.1). Assuming $\mathcal{L}u = 0$ in Ω_-, integrating this identity over Ω_- and using the divergence theorem yields

$$\int_\Gamma \left[(x\cdot n)\left(|\partial_n u|^2 + k^2|u|^2 - |\nabla_\Gamma u|^2\right) + 2\mathrm{Re}\left(x\cdot\overline{\nabla_\Gamma u}\partial_n u\right)\right]\mathrm{d}s = 2k^2\int_{\Omega_-}|u|^2\,\mathrm{d}x. \tag{5.10}$$

In the simplest possible case of $k = 0$ (*i.e.*, the Laplace equation) the right-hand side of (5.10) is zero, and so if $\min(x\cdot n) > 0$ then (5.10) can be used to bound $\|\partial_n u\|^2_{L^2(\Gamma)}$ by $\|\nabla_\Gamma u\|^2_{L^2(\Gamma)}$ and *vice versa*. Indeed, from (5.10) it follows that

$$\min_{x\in\Gamma}(x\cdot n)\|\partial_n u\|^2_{L^2(\Gamma)} \le \max_{x\in\Gamma}(x\cdot n)\|\nabla_\Gamma u\|^2_{L^2(\Gamma)} - 2\mathrm{Re}\int_\Gamma x\cdot\overline{\nabla_\Gamma u}\partial_n u\,\mathrm{d}s. \tag{5.11}$$

Using the standard inequality $2ab \le \epsilon a^2 + b^2/\epsilon$ on the final term yields $\|\partial_n u\|^2 \lesssim \|\nabla_\Gamma u\|^2$. All this relies on Ω_- being star-shaped with respect to the origin (recall Definition 5.5). For more general domains one must use a more general vector field Z in the multiplier with $Z\cdot n > 0$ on Γ, and then control the resulting non-divergence terms, but a similar result can be obtained; see, for example, McLean (2000, Theorem 4.2.4).

A key advantage of identities such as (5.10) is that, modulo some non-trivial technical work, they hold when Γ is Lipschitz. Indeed this method of bounding the Dirichlet to Neumann map was used by Nečas (1967) in his proof of Theorem A.5 of the Appendix, by Jerison and Kenig, who independently discovered the Rellich identity (5.9) in their famous work on the Laplacian on Lipschitz domains (see, *e.g.*, Kenig 1994), and implicitly by Verchota (1984) in his proof that $\frac{1}{2}I - D_0$ is invertible in $L^2(\Gamma)$ when Γ is Lipschitz.

Whereas Nečas, and Jerison and Kenig applied the identity (5.9) in an interior domain, Verchota needed to apply (5.9) both in Ω_- and the exterior domain Ω_+. When (5.9) is applied in Ω_+, in order for there to be no contribution from infinity, one requires the surface integral over a large ball to tend to zero as the radius R of the ball tends to infinity. This integral is equal to

$$\int_{\Gamma_R} \left[R\left(|\partial_n u|^2 + k^2|u|^2 - |\nabla_\Gamma u|^2\right) + 2\mathrm{Re}\left(x \cdot \overline{\nabla_\Gamma u}\partial_n u\right)\right] \mathrm{d}s, \tag{5.12}$$

where $\Gamma_R := \partial B_R$. When $k = 0$ and u is a solution of the Laplace equation in Ω_+ that vanishes at infinity, (5.12) tends to zero as $R \to \infty$. However, when u is a solution of the Helmholtz equation that satisfies the radiation condition (2.9), the integral (5.12) does not tend to zero (this is because the derivatives of the Helmholtz solution decay more slowly at infinity than the derivatives of the Laplace solution). Thus, if one seeks to bound the exterior Dirichlet to Neumann map for the Helmholtz equation using (5.9), the sign of the volume terms in (5.10) is such that the terms on Γ bound $\|\partial_n u\|^2_{L^2(\Gamma)}$ in terms of $\|\nabla_\Gamma u\|^2_{L^2(\Gamma)}$ but (5.12) does not tend to zero as $R \to \infty$, and this non-zero contribution cannot be immediately controlled.

Because of this, Morawetz and Ludwig (1968) introduced the multiplier $\mathcal{N}u = r\mathcal{M}u$, where

$$\mathcal{M}u := \frac{x}{r} \cdot \nabla u - \mathrm{i}ku + \frac{d-1}{2r}u, \tag{5.13}$$

and $r = |x|$. This leads to an identity very similar to the Rellich identity (5.9), namely that, for solutions of $\mathcal{L}u = 0$,

$$\nabla \cdot \left(2\mathrm{Re}\left(r\overline{\mathcal{M}u}\nabla u\right) + \left(k^2|u|^2 - |\nabla u|^2\right)x\right) = \left(|\nabla u|^2 - |u_r|^2\right) + |u_r - \mathrm{i}ku|^2, \tag{5.14}$$

where $u_r = (x \cdot \nabla u)/r$. As with the Rellich identity, the non-divergence terms are all positive, and when integrated over Ω_+ the resulting integral over Γ is the same as in (5.10), except that the left-hand side now contains an extra term proportional to $\mathrm{Re}\,(\overline{u}\partial_n u)$ (which can be controlled using the same inequality that we used to deal with the final term in (5.11)). In contrast to the Rellich identity, however, the choice of terms subtracted from $x \cdot \nabla u$ in the multiplier (5.13) means that the analogue of the integral over Γ_R (5.12) tends to zero as $R \to \infty$. This is because, if u satisfies the radiation condition (2.9), the multiplier $r\mathcal{M}u$ is lower order at infinity than $x \cdot \nabla u$: the first two terms in the multiplier $\mathcal{M}u$ defined in (5.13) equal the quantity in the radiation condition (2.9), and the term $\frac{d-1}{2r}u$ subtracts off the next term in the asymptotic expansion, so that

$$\mathcal{M}u(x) = O\left(\frac{1}{r^{(d+3)/2}}\right) \text{ as } r \to \infty$$

(compared to $u_r - iku = O(r^{-(d+1)/2})$ from (2.10)). Thus, the identity (5.14) *can* be used to obtain a bound on the Dirichlet to Neumann map for the Helmholtz equation in Ω_+, ultimately resulting in Theorem 5.8. Following the introduction of the multiplier $\mathcal{M}u$ in Morawetz and Ludwig (1968), more general multipliers were then used by Morawetz (1975*a*) to obtain estimates for the Helmholtz equation (as a stepping stone to estimates for the wave equation).

One area where Rellich identities have been used extensively to treat time harmonic scattering problems is rough surface scattering. A standard model of this phenomenon is the Helmholtz equation posed in the infinite region above the graph of a function $f : \mathbb{R}^{d-1} \to \mathbb{R}$. Here the radiation condition is more subtle than in the bounded obstacle case, and the notion of 'outgoing' must be formulated in terms of the Fourier transform of the solution at a certain height above the surface; see Chandler-Wilde and Monk (2005) and the references therein.

The appropriate Rellich identity to use in this case is that with multiplier $e \cdot \nabla u$, where e is a unit vector in the vertical d-coordinate direction: this ensures that $e \cdot n > 0$ on the boundary (in this case, the graph of the function f). When this identity is applied in the analogue of Ω_+ (the region above the graph of f), we again want there to be no contribution from infinity; this is the case if an integral analogous to (5.12) tends to zero as $R \to \infty$ (where R is now the height above the graph of f). Similar to the bounded obstacle case, this integral does not tend to zero as $R \to \infty$. However, whereas in the bounded obstacle case this problem can be fixed by adding terms to the Rellich multiplier $x \cdot \nabla u$ to mirror the radiation condition, adding terms to the Rellich multiplier $e \cdot \nabla u$ does not appear helpful in the rough surface case (this is related to the more complicated behaviour of the scattered field at infinity when the scatterer is unbounded). Nevertheless, the integral analogous to (5.12) can be controlled using bounds on the Fourier transform of the solution above the surface (Chandler-Wilde and Monk 2005, Lemma 2.2), and thus a Dirichlet to Neumann map estimate can still be obtained (Chandler-Wilde, Heinemeyer and Potthast 2006). It is shown in Spence (2012) that the analogue of these bounds in the bounded obstacle case (Chandler-Wilde and Monk 2008, Lemma 2.1) is almost exactly equivalent to using the Morawetz–Ludwig multiplier (5.13).

5.3.2. Bounding the energy norm of solutions of the Helmholtz equation

By slightly changing the Rellich multiplier $x \cdot \nabla u$ one can ensure that the non-divergence terms in the resulting identity become the energy norm of the solution u, that is, $\|u\|_{1,k,D}$, where

$$\|u\|^2_{1,k,D} := \|\nabla u\|^2_{L^2(D)} + k^2 \|u\|^2_{L^2(D)} \tag{5.15}$$

(where D is either Ω_- or Ω_+). Indeed, the multiplier $\mathcal{N}u = x\cdot\nabla u + \alpha u$ for α a real constant yields

$$\nabla\cdot\big(2\mathrm{Re}\,\big(x\cdot\overline{\nabla u}+\alpha\overline{u}\big)\nabla u + \big(k^2|u|^2-|\nabla u|^2\big)x\big)$$
$$= (d-2-2\alpha)|\nabla u|^2 + (2\alpha - d)k^2|u|^2, \tag{5.16}$$

where d is the dimension. (This identity can also be understood as adding a multiple of the pointwise form of Green's identity to the Rellich identity, and indeed this is the way it has been understood in most of the instances of its use.) Choosing $2\alpha = d-1$ means that the non-divergence terms in (5.16) become $-\|u\|^2_{1,k,D}$. Although of obvious independent interest, bounds on the energy norm can be translated into bounds on the inf-sup constant for the standard variational formulations of the Helmholtz equation that are the starting point for finite element discretizations; see, *e.g.*, Chandler-Wilde and Monk (2008, Lemma 3.4).

The fact that Rellich identities can give bounds on the energy norm was used for the Helmholtz equation in interior domains by Makridakis, Ihlenburg and Babuška (1996, Proposition 2.1), Melenk (1995), Cummings and Feng (2006) and Hetmaniuk (2007). In exterior domains, Chandler-Wilde and Monk (2008) used them to prove the bound (5.5) without the third term on the left-hand side when Ω_- is a star-shaped domain in 2D and 3D, obtaining an explicit value for the hidden constant on the right-hand side of the bound. (In this application the problem that (5.12) does not tend to zero as $R\to\infty$ was overcome via Chandler-Wilde and Monk 2008, Lemma 2.1. An effective alternative is to use the Morawetz–Ludwig multiplier (5.13); see Spence 2012.) For rough surface scattering modelled by the Helmholtz equation, bounds on the energy norm of the solution were obtained via Rellich identities by Chandler-Wilde and Monk (2005), Claeys and Haddar (2007) and Lechleiter and Ritterbusch (2010). Finally, we note that Rellich identities were used to prove analogous results for the equations of linear elasticity by Cummings and Feng (2006), and for the Maxwell equations by Hiptmair, Moiola and Perugia (2011*b*) and Haddar and Lechleiter (2011).

5.3.3. *Proving coercivity of integral operators*

As discussed in Section 5.3.1 above, the Rellich identity was a key component of Verchota's bound on $\|\big(\frac{1}{2}I - D_0\big)^{-1}\|$ (Verchota 1984). In essence, Verchota's argument can be seen as applying the Dirichlet to Neumann map bounds in Ω_- and Ω_+, that follow from (5.11), to the particular solution of the Laplace equation $u=\mathcal{S}_0\phi$. Verchota combines these bounds with the jump relations (2.41) to obtain bounds on $\|(\frac{1}{2}I-D_0')^{-1}\| = \|(\frac{1}{2}I-D_0)^{-1}\|$.

Recently it has become clear that the exact structure of the Rellich-identity terms on Γ when $u=\mathcal{S}_k\phi$ can be used to prove the stronger property (compared to invertibility) of *coercivity* in the sense of (2.124)

(Spence *et al.* 2011). This discovery was motivated by the relatively well-known fact that applying Green's first identity (A.26) with $u = \mathcal{S}_k\phi$ in Ω_- and Ω_+, and then using the jump relations (2.41), yields

$$\operatorname{Re}\left(-\mathrm{i}(S_k\phi, \phi)_{L^2(\Gamma)}\right) \geq 0, \tag{5.17}$$

for all $\phi \in L^2(\Gamma)$; see, *e.g.*, Engleder and Steinbach (2007, Lemma 3.1) or Spence *et al.* (2011, §1.3). (A different proof of this result is given in Nédélec 2001, §3.4.4.) Given that this use of Green's first identity gives information about part of the combined potential operator $A'_{k,\eta}$ (namely the part involving S_k), a natural question is whether repeating the argument with a different identity can obtain information about more, or even all, of $A'_{k,\eta}$.

Repeating the argument leading to (5.17) with Green's identity replaced by the Morawetz–Ludwig identity (5.14) yields the inequality

$$\operatorname{Re}\left(\left(x \cdot n\, D'_k + x \cdot \nabla_\Gamma S_k - \mathrm{i}\eta S\right)\phi, \phi\right)_{L^2(\Gamma)} \geq 0, \tag{5.18}$$

for all $k > 0$ with the particular choice of $\eta(x) = k|x| + \mathrm{i}(d-1)/2$; this shows that the star-combined operator (2.103) is coercive uniformly in k for all star-shaped Lipschitz domains (see Theorem 5.26 below).

Looking at the terms on the left-hand side of (5.18), we see that if the vector field x in the Morawetz–Ludwig identity is replaced by a more general vector field $Z(x)$, and if $Z(x)$ is equal to the normal vector on Γ (so that $Z \cdot \nabla_\Gamma S_k = 0$ and $Z \cdot n = 1$), then the left-hand side becomes $\operatorname{Re}(D'_k - \mathrm{i}\eta S)\phi, \phi)_{L^2(\Gamma)}$. Thus, if $Z(x)$ is also such that the inequality in (5.18) is maintained after replacing x by $Z(x)$, then this inequality shows that the standard combined-potential operator $A'_{k,\eta}$ is coercive, that is, for some $\alpha_{k,\eta} > 0$,

$$|(A'_{k,\eta}\phi, \phi)| \geq \alpha_{k,\eta}\|\phi\|^2, \tag{5.19}$$

for all $\phi \in L^2(\Gamma)$ (and hence $A_{k,\eta}$ is also coercive by the considerations leading to Remark 2.24). Necessary conditions for the inequality analogous to (5.18) to hold are that the non-divergence terms of the analogue of the identity (5.14) are non-negative, and also $Z(x)$ must be proportional to x as $r \to \infty$ for there to be no contribution from infinity (the analogue of (5.12) must tend to zero as $R \to \infty$). In this way, one can prove $A'_{k,\eta}$ is coercive by showing there exists a vector field $Z(x)$ satisfying the constraints outlined above (Spence, Kamotski and Smyshlyaev 2012). This is analogous to Morawetz's reformulation of the energy decay of solutions to the wave equation in terms of the existence of an appropriate multiplier in Morawetz (1975*a*); however, we note that the multipliers that need to be constructed to prove coercivity satisfy more stringent requirements than Morawetz's multipliers; see Spence *et al.* (2012). Coercivity results obtained using this technique are discussed in Section 5.7.

5.4. *Bounds obtained by Fourier analysis*

The majority of this subsection discusses investigations of conditioning and coercivity in the case where Γ is either a circle or a sphere, in which case a complete theory is possible due to the diagonalization of the operators in the Fourier ($d = 2$) or spherical harmonic ($d = 3$) bases. However, we want to emphasize at the beginning that these are not the only domains where these (in some sense) explicit methods can be used. Indeed, the real requirement is that the domain is one in which the Helmholtz equation is separable, in which case classical analytical techniques of applied mathematics can be brought to the problem. For example, boundary integral operators on the ellipse are considered in Kress and Spassov (1983) and Rodin and Steinbach (2003) for the Laplace equation, and in Betcke, Phillips and Spence (2012*b*) for the Helmholtz equation. The separability requirement can be relaxed slightly, but then the analytical techniques become more involved and only yield approximate results; see, *e.g.*, the investigations of the conditioning of first-kind BIEs for the Helmholtz equation in several canonical domains (including the exterior of a crack) in Warnick and Chew (1999, 2001, 2004).

Remark 5.11. One particularly noteworthy investigation using Fourier analysis is Ha-Duong's work on scattering by a flat screen (in 3D) and a crack (in 2D). Coercivity for both S_k and H_k, as mappings between the appropriate trace spaces, was proved in Ha-Duong (1990), but under the assumption that $\operatorname{Im}(k) > 0$ (for an alternative proof, which essentially repeats the argument using Green's identity leading to (5.17) but with $\operatorname{Im} k > 0$, see Ha-Duong 2003). Frequency-explicit coercivity estimates for H_k for $k > 0$ were shown in Ha-Duong (1992), and a non-frequency-explicit coercivity estimate for S_k for $k > 0$ was stated in Costabel (2004, Proposition 2.3) but without proof. These results have recently been sharpened and extended in Chandler-Wilde and Hewett (2012).

The case where Γ is the circle or sphere. As with the rest of this section we focus on the exterior Dirichlet problem (2.12) (which includes the sound-soft scattering problem (2.16)). Partial results for the analogous formulation of the Neumann problem, that is, (2.73) and (2.77) with $\beta = 0$, are given in Kress (1985), Amini (1990) and Amini (1993). The conditioning of the combined-potential integral equations for electromagnetic scattering in these domains was studied for small k in Kress (1985), and the conditioning of certain 'regularized' formulations of the acoustic Dirichlet problem was considered in Buffa and Sauter (2006) (these 'regularized' formulations were discussed briefly in Section 2.11).

When Γ is the circle or sphere, the analysis is simplified by the fact that $D'_k = D_k$ and thus $A'_{k,\eta} = A_{k,\eta}$ (Kress 1985).

Suppose Γ is the unit circle, with parametrization $\zeta(\theta) = (\cos\theta, \sin\theta)$. With this parametrization $L^2(\Gamma)$ is isometrically isomorphic to $L^2[0, 2\pi]$. We can write any $\varphi \in L^2[0, 2\pi] = L^2(\Gamma)$ as

$$\varphi(\theta) = \frac{1}{2\pi}\sum_{m\in\mathbb{Z}} \widehat{\varphi}(m)\exp(im\theta), \quad \text{where } \widehat{\varphi}(m) := \int_0^{2\pi} \varphi(\theta)\exp(-im\theta)\,\mathrm{d}\theta,$$

in which case the L^2-inner product and norm are given by

$$(\varphi, \psi) = \frac{1}{2\pi}\sum_{m\in\mathbb{Z}} \widehat{\varphi}(m)\overline{\widehat{\psi}(m)} \quad \text{and} \quad \|\varphi\|^2 = \frac{1}{2\pi}\sum_{m\in\mathbb{Z}} |\widehat{\varphi}(m)|^2.$$

On the circle the operators S_k, D_k, and hence $A_{k,\eta}$ are convolution operators and all act diagonally on the basis $\{\exp(im\theta)\}_{m\in\mathbb{Z}}$ with the corresponding eigenvalues given by

$$\lambda^{S_k}(m) = \frac{\pi\mathrm{i}}{2} H^{(1)}_{|m|}(k) J_{|m|}(k), \tag{5.20a}$$

$$\lambda^{D_k}(m) = \frac{\pi\mathrm{i}k}{2} H^{(1)}_{|m|}(k) J'_{|m|}(k) - \frac{1}{2}, \tag{5.20b}$$

$$\lambda^{A_{k,\eta}}(m) = \frac{\pi}{2} H^{(1)}_{|m|}(k)\big(\mathrm{i}kJ'_{|m|}(k) + \eta J_{|m|}(k)\big). \tag{5.20c}$$

By this we mean that

$$R\phi(\theta) = \frac{1}{2\pi}\sum_{m=-\infty}^{\infty} \lambda^R(m)\mathrm{e}^{\mathrm{i}m\theta}\widehat{\phi}(m),$$

for $R = S_k, D_k$, or $A_{k,\eta}$; see, *e.g.*, Kress (1985, equation (4.4)) or Domínguez *et al.* (2007, Lemma 1)). Thus $\lambda^R(m)$ is the eigenvalue associated with the eigenfunction $\exp(im\theta)$.

On the sphere the eigenvalues for these operators (in the spherical harmonic basis) are proportional to the expressions (5.20) with m replaced by $m+1/2$ (where the constant of proportionality is independent of k and m); see, *e.g.*, Kress (1985), Domínguez *et al.* (2007).

Since the eigenfunctions $\exp(im\theta)$, $m \in \mathbb{Z}$, are an orthonormal basis of $L^2[0, 2\pi] = L^2(\Gamma)$, bounding norms can be recast as bounding eigenvalues. In particular, with $\mathbb{N}_0 := \mathbb{N} \cup \{0\}$,

$$\|R\| = \sup_{m\in\mathbb{N}_0} |\lambda^R(m)|, \quad \|R^{-1}\| = \Big(\inf_{m\in\mathbb{N}_0} |\lambda^R(m)|\Big)^{-1}. \tag{5.21}$$

Obtaining a bound on the coercivity constant of $A_{k,\eta}$ in (5.19) can also be translated into obtaining a bound on the eigenvalues. Indeed, the variational form obtained when solving $A_{k,\eta}\varphi = g$ with the Galerkin method (see (1.6)) can be rewritten in the Fourier basis as

$$a_{k,\eta}(\varphi, \psi) = (A_{k,\eta}\varphi, \psi) = \frac{1}{2\pi}\sum_{m\in\mathbb{Z}} \lambda^{A_{k,\eta}}(m)\widehat{\varphi}(m)\overline{\widehat{\psi}(m)}. \tag{5.22}$$

If, for example, $\alpha_{k,\eta} > 0$, where

$$\alpha_{k,\eta} := \inf_{m\in\mathbb{N}_0} \mathrm{Re}\,(\lambda^{A_{k,\eta}}(m)), \tag{5.23}$$

then for $\varphi \in L^2(\Gamma)$,

$$\mathrm{Re}\,(a_{k,\eta}(\varphi,\varphi)) = \frac{1}{2\pi}\sum_{m\in\mathbb{Z}} \mathrm{Re}\,(\lambda^{A_{k,\eta}}(m))|\widehat{\varphi}(m)|^2 \geq \alpha_{k,\eta}\|\varphi\|^2; \tag{5.24}$$

so that $\alpha_{k,\eta}$ is a coercivity constant for $A'_{k,\eta}$. (Another potential way of getting a lower bound for the coercivity constant would be to consider the imaginary parts of $\lambda^{A_{k,\eta}}(m)$. However, Figure 5.2 shows that the imaginary parts are not bounded away from zero even in a case where the operator is coercive.)

Using (5.21), and bounding above the appropriate combinations of Bessel and Hankel functions, the following upper bounds on $\|S_k\|$, $\|D_k\|$, and $\|A_{k,\eta}\|$ can be obtained.

Theorem 5.12. For Γ the circle or sphere, for $k > 0$,

$$\|S_k\| \lesssim k^{-2/3}, \quad \|D_k\| = \|D'_k\| \lesssim 1, \tag{5.25}$$

so that

$$\|A_{k,\eta}\| \lesssim 1 + |\eta|k^{-2/3}. \tag{5.26}$$

The bound for $\|S_k\|$ in (5.25) was proved for the circle and sphere by Giebermann (1997), Domínguez *et al.* (2007) and Banjai and Sauter (2007). The bound for $\|D_k\|$ in (5.25) was proved by Banjai and Sauter (2007), with Giebermann (1997) and Domínguez *et al.* (2007) proving the weaker bound $\|D_k\| \lesssim k^{1/3}$. (To be precise we note that Domínguez *et al.* (2007) do not explicitly state the bound for $\|S_k\|$, but obtain the appropriate bound on the eigenvalues (5.20a), and Banjai and Sauter (2007) state the bound for $\|S_k\|$ for the sphere only, but due to the fact that the eigenvalues on the circle are very similar to those on the sphere their result is easily extended to the circle case.)

The bound (5.26) indicates that with the usual choice of $\eta = k$, $\|A_{k,\eta}\| \lesssim 1 + k^{1/3}$. The lower bounds obtained by more general methods in Section 5.5.2 show that this upper bound is sharp. The bound (5.26) also shows that the choice $\eta \sim k^{2/3}$ means that $\|A_{k,\eta}\|$ is bounded uniformly in k. Although this might at first suggest that $\eta = k^{2/3}$ is a better choice than $\eta = k$, the former choice apparently leads to $\|A_{k,\eta}^{-1}\|$ growing like $k^{1/3}$ (this is suggested by Theorem 5.22 and numerical experiments of Betcke *et al.* 2011) and thus the condition number of $A_{k,k^{2/3}}$ in the case of a circle and sphere grows at the same rate as $k \to \infty$ as that of $A_{k,k}$.

Theorem 5.13. (Domínguez *et al.* 2007) When Γ is the circle, with $\alpha_{k,\eta}$ defined by (5.23), there exists a $k_0 > 0$ such that

$$\alpha_{k,k} \geq \frac{1}{2},$$

for all $k \geq k_0$, so that $A_{k,k}$ is coercive with a coercivity constant of 1/2. When Γ is the sphere, (5.19) holds for $\eta = k$ for sufficiently large k, with

$$\alpha_{k,k} \geq \frac{1}{2} - O\left(\frac{1}{k^{2/3}}\right), \quad \text{as } k \to \infty.$$

These lower bounds on the eigenvalues, first conjectured by Giebermann (1997), are much harder to prove than the upper bounds of Theorem 5.12.

Recall that $\alpha_{k,\eta}$ in (5.19) and $\|(A'_{k,\eta})^{-1}\|$ are related by

$$\|(A'_{k,\eta})^{-1}\| \leq 1/\alpha_{k,\eta} \tag{5.27}$$

(this follows immediately from equation (2.125)). Thus the bounds of Theorem 5.13 immediately imply that

$$\|A_{k,k}^{-1}\| \leq 2$$

for all $k \geq k_0$ for the circle, and

$$\|A_{k,k}^{-1}\| \leq 2 + O\left(\frac{1}{k^{2/3}}\right), \quad \text{as } k \to \infty,$$

for the sphere.

5.5. *Bounds on norms of boundary integral operators*

We now discuss bounds on the norms of S_k, D_k, D'_k, $A_{k,\eta}$ and $A'_{k,\eta}$ for much more general domains than just the circle and sphere. (Recall from Remark 2.24 that $\|D_k\| = \|D'_k\|$ and $\|A_{k,\eta}\| = \|A'_{k,\eta}\|$.)

5.5.1. *Upper bounds*

Theorem 5.14. (Chandler-Wilde *et al.* 2009) If Γ is Lipschitz then

$$\|S_k\| \lesssim k^{(d-3)/2}, \quad \|D_k\| \lesssim 1 + k^{(d-1)/2}, \quad \|D'_k\| \lesssim 1 + k^{(d-1)/2}, \tag{5.28}$$

for $k > 0$; thus

$$\|A'_{k,\eta}\| = \|A_{k,\eta}\| \lesssim 1 + k^{(d-1)/2}\left(1 + \frac{|\eta|}{k}\right)$$

for $k > 0$ (uniformly in η).

These bounds were obtained by using the following idea. If T is an integral operator on Γ with kernel $t(x,y)$ then

$$\|T\|_{L^1(\Gamma)\leftarrow L^1(\Gamma)} = \operatorname*{ess\,sup}_{y\in\Gamma} \int_\Gamma |t(x,y)|\,\mathrm{d}s(x),$$

$$\|T\|_{L^\infty(\Gamma)\leftarrow L^\infty(\Gamma)} = \operatorname*{ess\,sup}_{x\in\Gamma} \int_\Gamma |t(x,y)|\,\mathrm{d}s(y)$$

(provided these integrals exist); see, *e.g.*, Jörgens (1982). Thus, by the Riesz–Thorin interpolation theorem,

$$\|T\|_{L^2(\Gamma)\leftarrow L^2(\Gamma)} \le \left(\|T\|_{L^1(\Gamma)\leftarrow L^1(\Gamma)}\|T\|_{L^\infty(\Gamma)\leftarrow L^\infty(\Gamma)}\right)^{1/2};$$

see, *e.g.*, Stein and Weiss (1971, Chapter V, Theorem 1.3). Furthermore, if $|t(x,y)| \le \tilde{t}(x,y)$, where $\tilde{t}$ is such that $\tilde{t}(x,y) = \tilde{t}(y,x)$, then

$$\|T\|_{L^2(\Gamma)\leftarrow L^2(\Gamma)} \le \operatorname*{ess\,sup}_{x\in\Gamma} \int_\Gamma \tilde{t}(x,y)\,\mathrm{d}s(y). \tag{5.29}$$

In the case of S_k the bound (5.29) is applied with $T = S_k$ (with $\tilde{t}(x,y)$ chosen as $|\Phi_k(x,y)|$), whereas for D_k and D'_k it is applied to the perturbation $D_k - D_0$ (since the singularity of these double-layer operators is too strong for the operators themselves to be bounded on $L^1(\Gamma)$ and $L^\infty(\Gamma)$ for general Lipschitz Γ). In the 2D case the bounds on D_k and D'_k in (5.28) are then proved using bounds on (the derivatives of) Hankel functions similar to those in Lemma 4.6.

It is important to note that these bounds ignore the oscillation in k. For example, the method described above gives the bound, for $k > 0$,

$$\begin{aligned} \|S_k\| &\le \operatorname*{ess\,sup}_{y\in\Gamma} \int_\Gamma |\Phi_k(x,y)|\,\mathrm{d}s(y) \\ &= \operatorname*{ess\,sup}_{y\in\Gamma} \int_\Gamma \frac{1}{4\pi|x-y|}\,\mathrm{d}s(y) \lesssim 1, \quad \text{for } d = 3. \end{aligned} \tag{5.30}$$

Despite this apparent crudeness the bound (5.30) is sharp in its k-dependence for many geometries in 2D (and so are the bounds on $\|A'_{k,\eta}\|$ and $\|A_{k,\eta}\|$); we will see this below in Lemma 5.18 when we discuss lower bounds.

The same idea can be used to bound the surface gradients of S_k, $D_k - D_0$, and $D'_k - D'_0$ and this gives bounds on the operators S_k, D_k, and D'_k as mappings from $L^2(\Gamma)$ into $H^1(\Gamma)$ (provided Γ is smooth enough for these operators to be bounded), again using bounds similar to those in Lemma 4.6 in the 2D case.

Theorem 5.15. (Graham *et al.* 2012) If Γ is Lipschitz then

$$\|S_k\|_{H^1(\Gamma)\leftarrow L^2(\Gamma)} \lesssim 1 + k^{(d-1)/2}, \tag{5.31}$$

for all $k > 0$, and if Γ is $C^{2,\mu}$, for some $\mu \in (0,1)$, then

$$\|D_k\|_{H^1(\Gamma)\leftarrow L^2(\Gamma)} \lesssim 1 + k^{(d+1)/2}, \quad \|D_k'\|_{H^1(\Gamma)\leftarrow L^2(\Gamma)} \lesssim 1 + k^{(d+1)/2},$$

for all $k > 0$.

We remark that the requirement that Γ is $C^{2,\mu}$ ensures that the mappings $D_0, D_0' : L^2(\Gamma) \to H^1(\Gamma)$ are bounded; see Kirsch (1989, Theorem 4.2), Colton and Kress (1998, Theorem 3.6).

The bounds of the previous theorem are an important ingredient both in the proof of k-explicit quasi-optimality for the h-version of the BEM discussed in Section 6.1, and in the upper bounds on $\|(A'_{k,\eta})^{-1}\|$ in Section 5.6.1.

In Section 2.9 we introduced the integral operator $A'_{k,\eta,Z}$, defined by (2.99) for $k > 0$, $\eta \in L^\infty(\Gamma)$, and $Z \in (L^\infty(\Gamma))^d$. This definition involves D_k' and S_k (like $A'_{k,\eta}$) but also $\nabla_\Gamma S_k$. Supplementing the bounds (5.28) with (5.31) we obtain that

$$\|A'_{k,\eta,Z}\| \lesssim 1 + k^{(d-1)/2}\left(1 + \frac{\|\eta\|_\infty}{k}\right) \tag{5.32}$$

for all $k > 0$.

Finally, a technique for obtaining upper bounds on $\|S_k\|$ and $\|D_k\|$ that does take into account the oscillatory nature of the kernels is the following standard idea from harmonic analysis (see, *e.g.*, Stein 1993, Chapter 7, §2), whose use in this context was suggested in Chandler-Wilde and Graham (2009). Observe that, for example,

$$\|S_k\| = \|S_k^* S_k\|^{1/2},$$

where S_k^* is the Hilbert space adjoint of S_k (so its kernel is the complex conjugate of the kernel of S_k). Now $S_k^* S_k$ is an integral operator whose norm can be estimated using (5.29) above. The kernel of $S_k^* S_k$ is itself an integral, and this can be estimated using standard techniques for oscillatory integrals. The use of this method of estimating norms is ongoing research, but an initial bound obtained by this technique is the following.

Theorem 5.16. Let Ω_- be a strictly convex C^2 domain with strictly positive curvature in 3D. Then given $\epsilon > 0$ there exists a $k_0 > 0$ such that

$$\|S_k\| \lesssim \frac{1}{k^{1/20-\epsilon}},$$

for $k \geq k_0$.

This final bound should be contrasted with the cruder bound $\|S_k\| \lesssim 1$ for $d = 3$ obtained in Theorem 5.14 above. This shows that taking into account the oscillatory nature of the kernels can, at least for some domains (and especially in 3D), provide better bounds.

5.5.2. Lower bounds

The technique used so far to obtain lower bounds on $\|S_k\|$, $\|D_k\|$, and $\|A'_{k,\eta}\|$ is to design a particular $\phi_k \in L^2(\Gamma)$, depending on k and the geometry of the obstacle, to make $\|S_k\phi_k\|/\|\phi_k\|$ (or the analogous expressions with S_k replaced by D_k or $A'_{k,\eta}$) as large as possible. This technique strongly depends on the geometry of the obstacle, and has almost exclusively to date been used in 2D. The first bound we discuss is an exception in that it holds both in 2D and 3D. It relies on the fact that $A'_{k,\eta}$ is a compact perturbation of the identity on C^1 domains.

Lemma 5.17. (Chander-Wilde *et al.* 2009, Theorem 4.3) In both 2D and 3D, if part of Γ is C^1 then $\|A_{k,\eta}\| \geq 1/2$ and $\|(A'_{k,\eta})^{-1}\| \geq 2$.

A variety of lower bounds on $\|S_k\|$ and $\|D_k\|$ were obtained in Chandler-Wilde, Graham, Langdon and Lindner (2009). We give four examples.

Lemma 5.18. (Chandler-Wilde *et al.* 2009, Theorem 4.2) In the 2D case, if Γ is Lipschitz and contains a straight line section, then there exists $k_0 > 0$ such that

$$\|S_k\| \gtrsim \frac{1}{k^{1/2}} \tag{5.33}$$

for $k \geq k_0$, and

$$\|A'_{k,\eta}\| = \|A_{k,\eta}\| \geq \frac{|\eta|}{k^{1/2}} - 1 + O\left(\frac{|\eta|}{k}\right)$$

for $k \geq k_0$, uniformly in $\eta > 0$.

The lower bounds of Lemma 5.18 shows that the upper bounds on $\|S_k\|$ and $\|A_{k,\eta}\|$ (for $\eta \gtrsim k$) in Theorem 5.14 are sharp in the 2D case if part of Γ contains a straight line segment.

The following two lemmas show that the upper bounds when Γ is a circle, (5.25), are sharp.

Lemma 5.19. (Chandler-Wilde *et al.* 2009, Theorem 4.4) In the 2D case, if Γ is Lipschitz and C^2 in a neighbourhood of some point on the boundary then there exists $k_0 > 0$ such that

$$\|S_k\| \gtrsim \frac{1}{k^{2/3}}$$

for $k \geq k_0$.

Lemma 5.20. (A special case of Chandler-Wilde *et al.* 2009, Theorem 4.7) If Γ is C^1 then there exists $k_0 > 0$ such that

$$\|D_k\| \gtrsim 1$$

for $k \geq k_0$.

Lemma 5.21. (A special case of Chandler-Wilde *et al.* 2009, Theorem 4.6) If Γ is a 2D polygon then there exists $k_0 > 0$ such that

$$\|D_k\| \gtrsim k^{1/4}$$

for $k \geq k_0$.

5.6. *Bounds on inverses*

5.6.1. *Upper bounds on* $\|(A'_{k,\eta})^{-1}\|$

The bounds on $\|A'_{k,\eta}\|$ in Section 5.5 were obtained directly from the definition of the integral operator, with the fact that this operator is used to solve the Helmholtz equation entering only via the fact that its kernel is given in terms of the fundamental solution (1.2). In contrast, the bounds on $\|(A'_{k,\eta})^{-1}\|$ that we discuss in this subsection are obtained by using Theorem 2.33, which expresses $(A'_{k,\eta})^{-1}$ in terms of solution maps to BVPs involving the Helmholtz equation.

The known upper bounds on $\|(A'_{k,\eta})^{-1}\|$ are collected in the following theorem.

Theorem 5.22. If Ω_- is a star-shaped Lipschitz domain in 2D or 3D, then

$$\|(A'_{k,\eta})^{-1}\| \lesssim 1 + \frac{1+k}{|\eta|}, \tag{5.34}$$

for $k > 0$ (Chandler-Wilde and Monk 2008), and if Ω_+ is a non-trapping domain (in the sense of Definition 5.4) in 2D or 3D then, for every $k_0 > 0$,

$$\|(A'_{k,\eta})^{-1}\| \lesssim k^{2+(d-1)/2}\left(1 + \frac{k}{|\eta|}\right) \tag{5.35}$$

for $k \geq k_0$ (Spence 2012).

This theorem is proved using Theorem 2.33, which gives a bound on $\|(A'_{k,\eta})^{-1}\|$ in terms of bounds on the exterior Dirichlet to Neumann map, and on the interior impedance to Dirichlet map. To bound the exterior Dirichlet to Neumann map we use Theorem 5.8 (in the star-shaped case) and equation (5.7) (in the non-trapping case). To bound the interior impedance to Dirichlet map when Ω_- is star-shaped we use the Rellich identity (using the ideas in Section 5.3.1), and for more general domains we use a weaker bound obtained using Green's representation theorem along with the bounds of Theorem 5.15; for details see Chandler-Wilde and Monk (2008) and Spence (2012).

An explicit value for the hidden constant in the bound (5.34) is given in Chandler-Wilde and Monk (2008, Theorem 4.3). For example, for a square of side length $2a$, with the choice $\eta = k$, $\|(A'_{k,\eta})^{-1}\| < 5\frac{1}{4}$ if $ka \geq 1$, for a cube of side $2a$, again with $\eta = k$, $\|(A'_{k,\eta})^{-1}\| < 8$ for $ka \geq 1$. The hidden constant in (5.35) ultimately depends on the hidden constant in (5.5), which is in principle computable.

Regarding the sharpness of the bounds of Theorem 5.22, if $k/|\eta| = O(1)$ as $k \to \infty$, then Lemma 5.17 implies that (5.34) is sharp in its k-dependence. The bound (5.35) for the non-trapping case is almost certainly not sharp, with the numerical experiments of Betcke *et al.* (2011) and Betcke and Spence (2011) indicating that the stronger bound (5.34) holds for several 2D non-trapping and non-star-shaped domains, at least for moderate values of k. (Betcke and Spence (2011) investigate the coercivity constant of $A'_{k,\eta}$, which implies bounds on $\|(A'_{k,\eta})^{-1}\|$ via (5.27).)

5.6.2. Lower bounds on $\|(A'_{k,\eta})^{-1}\|$

From the analysis and numerical experiments described above it is not implausible to conjecture that the upper bound (5.34) on $\|(A'_{k,\eta})^{-1}\|$ holds whenever Ω_+ is non-trapping in both 2D and 3D.

This motivates the question: 'How fast can $\|(A'_{k,\eta})^{-1}\|$ grow when Ω_+ is trapping?' This question was investigated in both 2D and 3D in Chandler-Wilde *et al.* (2009) and Betcke *et al.* (2011), with further work in Betcke, Chandler-Wilde, Graham and Langdon (2012*a*). In this subsection we discuss two theorems taken from the first two of these papers. These theorems give examples of 2D trapping domains for which $\|(A'_{k,\eta})^{-1}\|$ grows through some increasing sequence of wavenumbers; analogous results for certain 3D trapping domains can be found in Betcke *et al.* (2012*a*). In these theorems the $\gtrsim$ notation means that the implied constants in (5.36) and (5.37) below are both independent of m.

Theorem 5.23. (Chandler-Wilde *et al.* 2009, Theorem 5.1) If Ω_+ contains a square of side length $2a$, two parallel sides of which form part of Γ (an example is the rectangular cavity in Figure 5.1), and if $k_m = m\pi/2a$, $m \in \mathbb{N}$, then

$$\|A^{-1}_{k_m,\eta}\| \gtrsim (k_m a)^{9/10}\left(1 + \frac{|\eta|}{k_m}\right)^{-1}. \tag{5.36}$$

Theorem 5.24. (Betcke *et al.* 2011, Theorem 2.8) If, for some $a_1 > a_2 > 0$, Ω_+ contains the ellipse $E := \{(x_1, x_2) : (x_1/a_1)^2 + (x_2/a_2)^2 < 1\}$, and if Γ coincides with the boundary of this ellipse in neighbourhoods of the points $(0, \pm a_2)$, then there exists a sequence $0 < k_0 < k_1 < k_2 < \cdots$, with $k_m \to \infty$ as $m \to \infty$, such that, for some $\gamma > 0$,

$$\|A^{-1}_{k_m,\eta}\| \gtrsim \exp(\gamma k_m)\left(1 + \frac{|\eta|}{k_m}\right)^{-1} \tag{5.37}$$

(an example of such a domain is the elliptical cavity in Figure 5.1).

Numerical evidence for both these estimates was given in Betcke *et al.* (2011), with evidence for (5.36) also given in Löhndorf and Melenk (2011, Example 4.7).

These two theorems are proved by looking at particular solutions of the Helmholtz equation, in the rectangular cavity and elliptical cavity respectively, and translating the properties of the solutions into properties of the integral operator $A'_{k,\eta}$. In order to understand the solutions that we use to prove the theorems, we first need to discuss briefly the concepts of *resonances* and *quasimodes*.

A resonance of the exterior Dirichlet problem (2.12) is a complex wavenumber $k_{\rm res}$ such that there exists a non-trivial solution $u_{\rm res}$ of the Helmholtz equation in Ω_+ satisfying (2.10) and *zero* Dirichlet boundary conditions. From Corollary 2.9 we know that resonances cannot occur for $\operatorname{Im} k \geq 0$ (because in this case if $u = 0$ on Γ then $u = 0$ in Ω_+), and thus any resonances must lie in the lower half complex plane (actually Corollary 2.9 only considers real k, but the result also holds for $\operatorname{Im} k \geq 0$). Note that (2.10) implies that, for $\operatorname{Im} k < 0$, the solution $u_{\rm res}(x)$ must grow exponentially as $r \to \infty$.

The location of resonances in the lower half complex plane and their relationship to trapping are classic questions in scattering theory, with appropriate generalizations of these questions still active research topics today. In the 1967 first edition of Lax and Phillips (1989) it was conjectured that:

(1) for any non-trapping domain there are no resonances in a strip $\{k : -\alpha \leq \operatorname{Im}\{k\} \leq 0\}$ for some constant $\alpha > 0$;

(2) for any trapping domain there is a sequence of resonances $\{k_m\}_{m=1}^{\infty}$ such that $\operatorname{Im}\{k_m\} \to 0$ as $m \to \infty$.

The first conjecture (intimately linked to the question of local energy decay for solutions of the wave equation that was mentioned briefly in Section 5.2) was proved to be true in Vainberg (1975) and Melrose (1979) (using the results of Melrose and Sjöstrand 1978). However, an example of a trapping domain for which there are no resonances in a strip below the real axis was given in Ikawa (1983), and thus the second conjecture is false. The second conjecture is true if one restricts attention to trapping domains that contain a trapped ray that is stable under perturbation – a so-called 'elliptic' trapped ray (this was proved by Stefanov and Vodev using the link with quasimodes; see the references in the discussion of quasimodes below). Thus the elliptical cavity domain in Figure 5.1 contains a sequence of resonances converging to the real axis, but the square cavity domain in the same figure need not. More details about these results can be found in Lax and Phillips (1989, Epilogue), Vainberg (1999), Melrose (1995), for example. (The 2D case is more subtle than the 3D case due to the presence of a branch point in the fundamental solution (1.2) at $k = 0$.) A nice non-technical discussion of resonances is given in Zworski (1999).

A related concept is that of a quasimode. In the context of acoustic scattering, a quasimode is a solution of the inhomogeneous Helmholtz equation for $k > 0$ where the source is small relative to the solution. (We index everything by m at this stage since in what follows we will always consider a family of quasimodes with wavenumbers k_m strictly increasing as $m \to \infty$.) We call v_m a quasimode if, for some compactly supported $g_m \in L^2(\Omega_+)$,

$$\Delta v_m + k_m^2 v_m = g_m,$$

where $k_m > 0$, v_m satisfies zero Dirichlet boundary conditions and the Sommerfeld radiation condition (2.9), and, for some $R > 0$ such that $\Omega_- \subset B_R$, the bound

$$\|v_m\|_{L^2(\Omega_+ \cap B_R)} \gtrsim L(k_m)\|g_m\|_{L^2(\Omega_+)} \tag{5.38}$$

holds where the factor $L(k_m)$ is 'large'. Thus, one can think of k_m^2 as being 'nearly' an eigenvalue of the Laplacian.

How large can we expect $L(k_m)$ to be? Theorem 5.6 tells us that a solution v of the Helmholtz equation in a non-trapping domain with zero Dirichlet boundary conditions and source g satisfies

$$\|v\|_{L^2(\Omega_+ \cap B_R)} \lesssim \frac{1}{k}\|g\|_{L^2(\Omega_+)}.$$

Thus a bound of the form (5.38) cannot hold, even with $L(k_m)$ constant, for non-trapping domains.

One intuitively expects that, if there is a resonance k_{res} in the lower half complex plane close to the positive real axis, then there will be a quasimode with $k_m > 0$ close to k_{res}. This intuition turns out to be correct, and the relationship between resonances and quasimodes was elucidated in Stefanov and Vodev (1995), Stefanov and Vodev (1996), Tang and Zworski (1998) and Stefanov (1999, 2003). For more information on quasimodes in general see, *e.g.*, Lazutkin (1999).

Theorems 5.23 and 5.24 are proved by constructing quasimodes with

$$L(k_m) = \begin{cases} 1 & \text{for the square cavity domain,} \\ \exp(\beta k_m) & \text{for some } \beta > 0, \text{ for the elliptical cavity domain.} \end{cases}$$

The construction of the quasimodes in the square cavity domain given in Chandler-Wilde *et al.* (2009) is based on the fact that $u(x) = \sin(k_m x_1)$ with $k_m = m\pi/a$, $m \in \mathbb{N}$ is an eigenfunction of the 1D Laplacian with eigenvalue k_m^2 on $[0, a]$ under Dirichlet boundary conditions. The quasimode v_m is then defined by $\chi(x)u(x)$, where $\chi(x)$ is a smooth function with compact support in the cavity. In a similar way, the quasimodes in the elliptical cavity domain constructed in Betcke *et al.* (2011) are based on the fact that, for the ellipse

$\{(x_1/a_1)^2 + (x_2/a_2)^2 < 1\}$, with $a_1 > a_2 > 0$, there are eigenfunctions of the Laplacian that become exponentially localized about the stable periodic orbit $\{(0, x_2) : |x_2| \leq a_2\}$ as $k_m \to \infty$.

Given a domain with a family of quasimodes satisfying (5.38), one can show that

$$\|(A'_{k,\eta})^{-1}\| \gtrsim \left(\frac{L(k_m)}{k_m^{d-2}} - \frac{1}{k_m^{(d-1)/2}}\right)\left(1 + \frac{|\eta|}{k_m}\right)^{-1}. \tag{5.39}$$

In the case of the elliptic cavity, this bound combined with the exponential growth of $L(k_m)$ establishes Theorem 5.24, that is, (5.37) holds for any $\gamma < \beta$. Since $L(k_m)$ does not increase for the square cavity, a more subtle argument is needed to obtain the growth in Theorem 5.23.

In the rest of this subsection we briefly sketch how the bound (5.39) is obtained, in large part following Betcke *et al.* (2011, §2.5). We also indicate how to prove Theorem 5.23 given that the bound (5.39) is too crude to achieve this.

Given a quasimode v_m such that both v_m and g_m are compactly supported, first define v_m^I as the Newtonian potential

$$v_m^I(x) := \int_{\Omega_+} \Phi_k(x, y) g_m(y) \,\mathrm{d}y, \quad x \in \mathbb{R}^d.$$

(The I superscript indicates that we will think of v_m^I as an incident field for a Helmholtz scattering problem; in particular note that v_m^I is an incident field in the sense of Definition 2.11.) Now, by standard properties of the Newtonian potential (see, *e.g.*, McLean 2000, Chapter 6, Sauter and Schwab 2011, Chapter 3),

$$\Delta v_m^I + k_m^2 v_m^I = \begin{cases} g_m & \text{in } \operatorname{supp}(g_m) \subset \Omega_+, \\ 0 & \text{otherwise.} \end{cases}$$

Define v_m^S by $v_m^S = v_m - v_m^I$ (the superscript S indicates that we think of v_m^S as the scattered field). Then v_m^S satisfies the sound-soft scattering problem (2.16) for incident field v_m^I. It follows from Theorem 2.46 that

$$A'_{k,\eta}\left(\frac{\partial v_m}{\partial n}\right) = f_m,$$

where

$$f_m := \left(\frac{\partial v_m^I}{\partial n} - \mathrm{i}\eta v_m^I\right)\bigg|_\Gamma.$$

To prove the bound (5.39), we need only show that

$$\left\|\frac{\partial v_m}{\partial n}\right\|_{L^2(\Gamma)} \gtrsim \left(\frac{L(k_m)}{k^{d-2}} - \frac{1}{k_m^{(d-1)/2}}\right)\left(1 + \frac{|\eta|}{k_m}\right)^{-1} \|f_m\|_{L^2(\Gamma)}. \tag{5.40}$$

This bound follows from combining (5.38) with the bounds

$$\left\| \frac{\partial v_m}{\partial n} \right\|_{L^2(\Gamma)} \gtrsim \|v_m\|_{L^2(\Omega_+)} \left(k_m^{(3-d)/2} - \frac{1}{L(k_m)} \right) \tag{5.41}$$

and

$$\|g_m\|_{L^2(\Omega_+)} \gtrsim k_m^{-(d-1)/2} \left(1 + \frac{|\eta|}{k} \right)^{-1} \|f_m\|_{L^2(\Gamma)}. \tag{5.42}$$

The bound (5.41) can be established by using Green's integral representation theorem to write v_m in Ω_+ in terms of the Newtonian potential of g_m and the single-layer potential of $\partial v_m/\partial n$ (the analogue of Theorem 2.21 for the inhomogeneous Helmholtz equation; see, *e.g.*, McLean 2000, Theorems 7.5 and 9.6), using bounds on the norms of the potentials as mappings from Γ to Ω_+ similar to those obtained in Section 5.5.1, and finally using (5.38) to relate the norm of g_m to that of v_m. The bound (5.42) can be established by using the definition of v_m^I in terms of the Newtonian potential, and then results on the norms of the potentials used for the first bound (for details see Betcke *et al.* 2011, §2.5).

As noted above, the lower bound on $\|(A'_{k,\eta})^{-1}\|$ is too crude to prove Theorem 5.23. Instead the result can be obtained by directly obtaining a bound from below on $\|\partial v_m/\partial n\|_{L^2(\Gamma)}$, and a bound from above on $\|f_m\|_{L^2(\Gamma)}$. The former follows trivially from the definition of v_m, whereas the latter requires carefully estimating the oscillatory integrals in the definition of v_m^I (for details see Chandler-Wilde *et al.* 2009, §5).

5.7. Bounds on coercivity constants

The results obtained by Fourier analysis show that $A_{k,k}$ is coercive for sufficiently large k when Γ is the circle and sphere. Given that it is surprising that coercivity holds, even in these special cases (as discussed in Section 2.11), a numerical investigation of the conditions under which coercivity holds was undertaken in Betcke and Spence (2011). Before we report the results of that investigation we discuss the information about $\alpha_{k,\eta}$, the coercivity constant of $A'_{k,\eta}$ in (5.19), that can be obtained through its relationship with $\|(A'_{k,\eta})^{-1}\|$ (5.27). The fact that $\|(A'_{k,\eta})^{-1}\| \geq 2$ from Theorem 5.17 means that $\alpha_{k,\eta} \leq 1/2$. Furthermore, Theorems 5.23 and 5.24 imply that, for the rectangular and elliptical cavities respectively,

$$\alpha_{k_m,\eta} \lesssim k_m^{-9/10} \left(1 + \frac{|\eta|}{k_m} \right), \quad \alpha_{k_m,\eta} \lesssim \exp\left(-\gamma k_m\right) \left(1 + \frac{|\eta|}{k_m} \right) \tag{5.43}$$

(where the k_m are different for each domain). Thus, with the standard choice $\eta \sim k$, even if coercivity holds for some k, it cannot hold uniformly as $k \to \infty$ for these two domains.

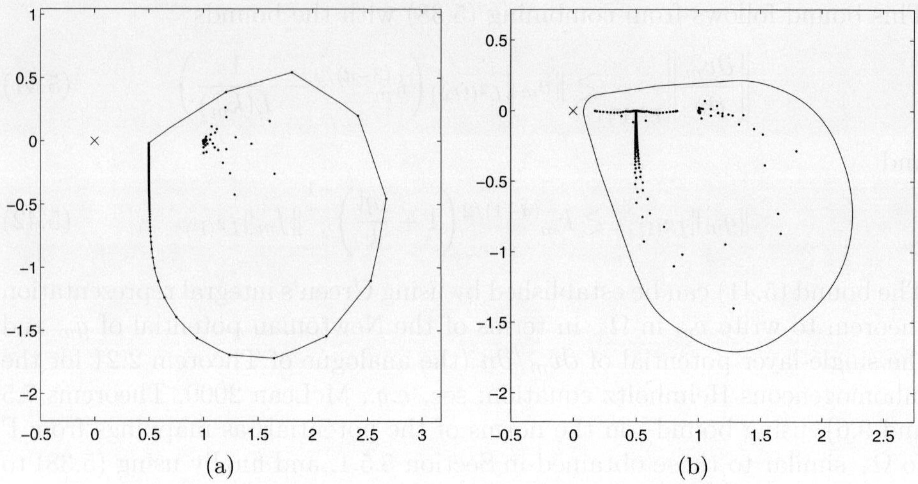

Figure 5.2. Eigenvalues and boundary of the numerical range of the boundary integral operator $A'_{k,k}$ when Γ is the unit circle (a) and the equilateral triangle with side length 1 (b) for $k = 50$, where $\times$ marks the origin.

Coercivity constants for a range of 2D domains were computed using the connection between the coercivity constant of an operator and its numerical range in Betcke and Spence (2011). Recall that the numerical range of a bounded linear operator T on a Hilbert space $\mathcal{V}$ is defined by

$$W(T) := \{(T\phi, \phi) : \phi \in \mathcal{V}, \ \|\phi\| = 1\}.$$

From this definition it is straightforward to see that if T is coercive with coercivity constant α then $\alpha = \inf_{z \in W(T)} |z|$. Thus T is coercive if and only if $0 \notin \overline{W(T)}$.

Standard algorithms from numerical linear algebra exist for computing the numerical range of finite-dimensional operators. Betcke and Spence (2011) used these to compute the numerical range of Galerkin discretizations of $A'_{k,k}$ and proved that these numerical ranges converge to $W(A'_{k,k})$ (in an appropriate sense) as the dimension of the Galerkin approximation space tends to infinity.

Figure 5.2 shows both the boundary of the numerical range and the eigenvalues of $A'_{k,k}$ when Γ is the unit circle and the equilateral triangle for $k = 50$. The circle plot shows that $A_{k,k}$ is coercive with constant $1/2$, confirming Theorem 5.13. The triangle plot shows that $A'_{k,k}$ is also coercive in this case. A striking difference between these two plots is that for the circle the numerical range is the convex hull of the eigenvalues, but for the triangle this is not the case. Recalling the result that the closure of the numerical range of a normal operator is the convex hull of its spectrum (see, *e.g.*, Gustafson

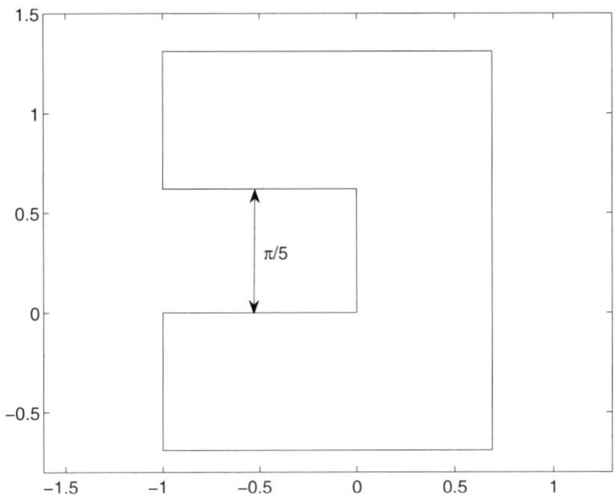

Figure 5.3. A particular rectangular cavity domain. The open cavity has a width of $\pi/5$.

and Rao 1997, Theorem 1.4-4), this indicates that $A'_{k,k}$ is a normal operator on the circle but not on the triangle. From the other computations in Betcke and Spence (2011) it appears that in 2D $A'_{k,\eta}$ is normal if and only if Γ is a circle. Proving one direction is easy: the fact that $A'_{k,\eta}$ and its adjoint both diagonalize in the Fourier basis mean that they commute, and thus $A'_{k,\eta}$ is normal (this is also true for the sphere). Proving the converse is more difficult, and at the time of writing still an open problem. However, the analogous result for S_k has been proved in Betcke *et al.* (2012*b*), and this paper also contains an investigation of the non-normality of $A'_{k,\eta}$ using tools such as pseudospectra (as advocated by Trefethen and Embree 2005).

From the computations in Betcke and Spence (2011) it was conjectured that $A'_{k,k}$ is coercive whenever the domain is non-trapping. Indeed these computations show $A'_{k,k}$ to be coercive uniformly in k for k between 10 and 100 for several non-trapping domains (including non-smooth ones such as an L-shaped polygon), but not coercive for the particular rectangular cavity domain in Figure 5.3. (Computations for higher k were not performed, because of the large matrix dimensions: here the Galerkin discretizations of $A'_{k,k}$ used a piecewise polynomial basis and followed the 'ten points per wavelength' convention.)

The numerical range of $A'_{k,k}$ for $k = 4$ and 5, when Ω_- is the trapping domain shown in Figure 5.3, is shown in Figure 5.4. These wavenumbers were chosen because $k = 5$ is the first of the sequence of wavenumbers for which quasimodes of the exterior Helmholtz problem in this domain can be constructed, with a corresponding resonance in the lower half-plane close to the real axis (recall the discussion in Section 5.6.2). Although $A'_{k,k}$ is invertible for every $k > 0$. Figure 5.4 shows that $A'_{5,5}$ is not coercive,

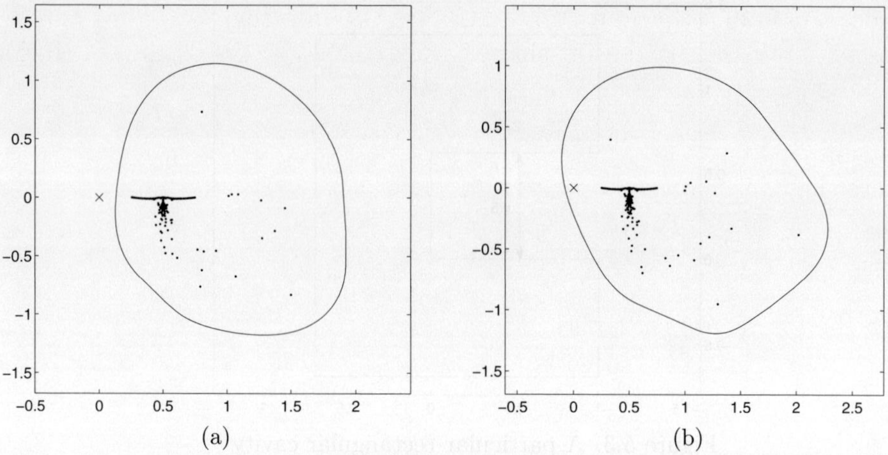

Figure 5.4. The numerical range of $A'_{k,k}$ when Ω_- is the trapping domain from Figure 5.3 in the cases $k = 4$ (a) and $k = 5$ (b), where $\times$ marks the origin.

whereas $A'_{4,4}$ is. Thus, not only is $A'_{k,k}$ not uniformly coercive as $k \to \infty$, as shown by (5.43), but it is not coercive for at least one finite value of k.

Interestingly, the spectra of $A'_{k,k}$ for $k = 4$ and 5 are almost the same, and in particular are both contained in the right half-plane. This illustrates the fact that the spectrum does not determine the behaviour of non-normal operators, as explained in more detail in Trefethen and Embree (2005).

We now discuss two theorems about coercivity; the first regarding $A'_{k,\eta}$, and the second regarding the star-combined operator $\mathscr{A}_k$ defined in (2.103). The first theorem says that, with a choice of η proportional to k, $A'_{k,\eta}$ is coercive on smooth convex domains once k is large enough.

Theorem 5.25. (Spence *et al.* 2012) Let Ω_- be a strictly convex C^3 domain with strictly positive curvature in either 2D or 3D. Then there exists a constant η_0 such that, given $\delta > 0$, there exists $k_0 > 0$ (depending on δ) such that, for $\eta = \eta_0 k$ and $k \geq k_0$,

$$\operatorname{Re}\,(A'_{k,\eta}\phi, \phi)_{L^2(\Gamma)} \geq \left(\frac{1}{2} - \delta\right) \|\phi\|^2_{L^2(\Gamma)}, \tag{5.44}$$

for all $\phi \in L^2(\Gamma)$. This bound also holds with $A'_{k,\eta}$ replaced by $A_{k,\eta}$.

This result essentially includes the earlier results of Domínguez *et al.* (2007) for the circle and sphere, as special cases.

The second theorem shows that, when Ω_- is star-shaped, if we are prepared to use the star-combined operator $\mathscr{A}_k$, which is a slight modification of $A'_{k,\eta}$ (as discussed in Section 2.9), coercivity holds uniformly for all $k > 0$ with only the requirement that Γ is Lipschitz.

Theorem 5.26. (Spence *et al.* 2011) Suppose that Ω_- is a bounded Lipschitz domain that is star-shaped with respect to the origin. Then, for all $\phi \in L^2(\Gamma)$,

$$\mathrm{Re}\,(\mathscr{A}_k\phi, \phi)_{L^2(\Gamma)} \geq \alpha\|\phi\|^2_{L^2(\Gamma)}, \tag{5.45}$$

where the star-combined operator $\mathscr{A}_k$ is given by

$$\mathscr{A}_k = (x \cdot n)\left(\frac{1}{2}I + D'_k\right) + x \cdot \nabla_\Gamma S_k - \mathrm{i}\eta S_k \tag{5.46}$$

with the function η chosen as

$$\eta = k|x| + \mathrm{i}\frac{d-1}{2}, \tag{5.47}$$

and the k-independent coercivity constant α is given by

$$\alpha = \frac{1}{2}\operatorname*{ess\,inf}_{x\in\Gamma}(x \cdot n(x)) > 0. \tag{5.48}$$

Note that Theorem 2.37 (about the invertibility of $\mathscr{A}_k$) follows from (5.45) using the analogue of (5.27) (with $A'_{k,\eta}$ replaced by $\mathscr{A}_k$).

These two theorems were proved using Rellich–Morawetz-type identities and the ideas in Section 5.3.3. Proving that $A'_{k,\eta}$, or some modified version of $A'_{k,\eta}$, is coercive is still open in the general non-trapping case.

6. Error analysis

In this section we return to the problem of proving, explicitly in k, the convergence of various boundary element methods for the Helmholtz equation. The key first step in doing this is to obtain the quasi-optimal error estimate (1.7) with explicit estimates for C and N_0 in terms of k (**Q3** of the Introduction). Then k-explicit convergence rates are obtained by estimating the best approximation error $\inf_{w_n \in \mathcal{V}_N} \|v - w_N\|$ explicitly in k (recall that for a function $w \in L^2(\Gamma)$, $\|w\|$ denotes $\|w\|_{L^2(\Gamma)}$ unless otherwise specified). We discuss this question for the direct and indirect combined potential equations and the star-combined equation, namely

$$A_{k,\eta}v := \left(\frac{1}{2}I + D_k - \mathrm{i}\eta S_k\right)v = f, \tag{6.1}$$

$$A'_{k,\eta}v := \left(\frac{1}{2}I + D'_k - \mathrm{i}\eta S_k\right)v = f, \tag{6.2}$$

(for real $\eta \neq 0$), and

$$\mathscr{A}_k v := (x \cdot n)\left(\frac{1}{2}I + D'_k\right)v + x \cdot \nabla_\Gamma S_k v - \mathrm{i}\eta S_k v = f, \tag{6.3}$$

where η is the smooth function of $x \in \Gamma$ given by (2.102). These equations are well-posed in $L^2(\Gamma)$; see Theorems 2.27 and 2.37. Specifically, this section contains the following.

- For general piecewise polynomial Galerkin methods, k-explicit quasi-optimality results for (6.1) and (6.2) on smooth (but otherwise general) boundaries Γ are given in Sections 6.1 and 6.3. In addition a k-explicit estimate of the best approximation error $\inf_{w_n \in \mathcal{V}_N} \|v - w_N\|$ for a particular class of boundaries Γ in the case of the sound-soft scattering problem (2.16) is given in Section 6.2.
- For a range of hybrid Galerkin methods taken from those discussed in Section 3, k-explicit quasi-optimality and convergence estimates are given in Section 6.4.

For second-kind integral equations such as (6.2) and (6.1), there are several classical approaches to error analysis, all based on viewing the integral operator as a compact k-dependent perturbation of a well-posed k-independent operator. The classical choice for the latter operator is $\frac{1}{2}I$, although a different choice is made in Löhndorf and Melenk (2011): see Section 6.3. The abstract theory of projection methods (*e.g.*, Atkinson 1997) then provides an error analysis, which covers not only the Galerkin method but also collocation and even Nyström methods. Another (closely related) classical approach is via the general variational theory of Galerkin methods, based on the discrete inf-sup condition. Here approximations such as collocation can be treated through an application of the Strang lemma (Ciarlet 2002). Up until recently, neither of these classical approaches has easily lent itself to explicit estimation of C and N_0 in (1.7) in terms of k, so recent research has had to develop new ideas for this task. A recent contribution is Banjai and Sauter (2007), along with the already-mentioned Löhndorf and Melenk (2011). In earlier work, for the particular integral equation (3.30) studied in Section 3.2, a complete k-explicit error analysis was developed in Chandler-Wilde *et al.* (2002) for a simple h-version piecewise-polynomial discrete collocation method, in particular showing that (1.7) holds with C independent of k, provided that kh is below some threshold.

In Section 6.1 we discuss the application of the classical projection analysis for general smooth Γ and for the h-version BEM in 2D and 3D following Graham, Löhndorf, Melenk and Spence (2012). In this analysis, not only must the mesh diameter h decrease with some negative power of k to ensure that the Galerkin equations are solvable ('mesh threshold'), but also C in (1.7) may grow with k. The key ingredients for this theory are bounds on the inverse of $A_{k,\eta}$ (respectively $A'_{k,\eta}$) given in Theorem 5.22 and estimates for the smoothing properties of the compact parts of $A_{k,\eta}$, $A'_{k,\eta}$ as mappings from $L^2(\Gamma)$ to $H^1(\Gamma)$ given in Theorem 5.15. At the end of Section 6.1, we compare the results obtained for the h-version BEM via the projection analysis with those obtained using the Galerkin variational framework in Banjai and Sauter (2007). In Section 6.2 we combine the quasi-optimal error estimate from Section 6.1 with an estimate of the best approximation error to obtain a convergence rate for a special case of (6.2) approximated with a low-order BEM.

The standard piecewise polynomial methods considered in Sections 6.1, 6.2 and 6.3 are not 'high-frequency' methods since they use only conventional piecewise polynomial basis functions, and the particular character of these basis functions is heavily exploited in the analysis. Nevertheless the derivation of k-explicit error estimates for such methods is a topic of recent research interest so it has a natural place in this review. In this case, the highly oscillatory nature of the solution means that the mesh diameter must decrease with increasing k. The general convention is that

$$h \lesssim k^{-1} \tag{6.4}$$

(*i.e.*, a fixed number of grid points per wavelength) should be sufficient for adequate approximation of the solution. While the analysis of Section 6.1 suggests that a somewhat smaller h is required for quasi-optimality (1.7) to hold independently of k, the example in Section 6.2 proves that (6.4) is indeed sufficient for accurate best approximation in the special case of scattering from a smooth convex obstacle, using low-order conventional BEM.

In Section 6.3 we give an overview of the substantial progress made in Melenk (2012) and Löhndorf and Melenk (2011), concerning the analysis of hp-BEM (*i.e.*, boundary elements of order p on meshes of diameter h, with refinement in both h and p allowed). This theory provides sufficient conditions that ensure (1.7) holds, with C independent of k, for general analytic boundaries Γ. These conditions are satisfied if, for example, hk/p is sufficiently small and p grows logarithmically in k.

An alternative and very powerful way to obtain quasi-optimality is to establish the (rather strong) property of coercivity for the relevant boundary integral operator. Despite the fact that the Helmholtz equation with high wavenumber is typically viewed as highly indefinite, some corresponding boundary integral formulations have remarkable coercivity properties, as discussed in Section 5.7. These allow us to establish quasi-optimality for Galerkin methods using *any* approximating subspace without a mesh threshold and would provide alternative proofs of quasi-optimality for h- or hp-BEM. However, more importantly for this article, this approach gives us the only currently known way of establishing quasi-optimality for the hybrid methods introduced in Section 3. This error analysis for hybrid spaces is discussed in Section 6.4.

The error analysis of this section assumes that all (highly oscillatory) Galerkin integrals are computed exactly. Estimates for fully discrete Galerkin methods which combine results of this section with some of the quadrature error estimates from Section 4 to provide rigorous fully discrete error estimates for some hybrid methods are presented in Kim (2012).

Concerning notation in this section, for the h-version BEM (where the polynomial degree p of the basis functions is fixed), the approximating space is denoted by $\mathcal{V}_h$, while for the hp version, we denote it by $\mathcal{V}_{h,p}$. We do not give a formal definition of such families of spaces here, but refer the reader for example to Sauter and Schwab (2011, §4.1) or Löhndorf and Melenk (2011, §3.3). We implicitly assume that the meshes are quasi-uniform and that exact representation of Γ is used. Since we work in $L^2(\Gamma)$, our boundary element spaces may be either continuous or discontinuous piecewise polynomials. In Section 6.4 we will be concerned with hybrid spaces, which in general depend on piecewise polynomials of various degrees and also on k. In this case the relevant spaces, already described in detail in Section 3, will be denoted in the abstract way as $\mathcal{V}_N$. Similar to Sections 3 and 5, for two quantities D and E which may depend on h, p, and k, in this section we write $D \lesssim E$ if $D \leq C\,E$ for some constant C which is independent of h, p, and k (hence C is also independent of N, the dimension of $\mathcal{V}_N$). Also we write $D \sim E$ if $D \lesssim E$ and $E \lesssim D$.

6.1. k-explicit error estimates for the h-version of the BEM

In this subsection we write the combined potential operators on the left-hand side of (6.2) and (6.1) in abstract form as $\lambda I + L_k$, where $\lambda = 1/2$. We also make the assumption that $\eta \in \mathbb{R}$ is chosen so that

$$C^{-1} \leq \eta/k \leq C \tag{6.5}$$

for some constant $C > 0$, and the parameter η does not appear explicitly in the abstract notation L_k. Thus, in abstract form, we are solving the equation

$$(\lambda I + L_k)v = f. \tag{6.6}$$

Recall that since $\lambda I + L_k$ denotes one of the operators $A_{k,\eta}$, $A'_{k,\eta}$ defined by (6.1) and (6.2) this operator is always invertible on $L^2(\Gamma)$ when Γ is Lipschitz; see Theorem 2.27. Moreover, the bounds on $\|A_{k,\eta}^{-1}\|$ presented in Theorem 5.22 are a key ingredient to the theory below.

Now let us consider the h version of the Galerkin method, that is, we seek an approximation $v_h \in \mathcal{V}_h, \subset L^2(\Gamma)$, the space of piecewise polynomials of some fixed degree $p \geq 0$ on shape-regular meshes of diameter h, with h decreasing to zero. The Galerkin equations may be written

$$\big((\lambda I + L_k)v_h, w_h\big) = (f, w_h), \quad \text{for all } w_h \in \mathcal{V}_h. \tag{6.7}$$

With P_h being the orthogonal projection from $L^2(\Gamma)$ onto $\mathcal{V}_h$, the Galerkin equations (6.7) are equivalent to the operator equation

$$\big(\lambda I + P_h L_k\big) v_h = P_h f. \tag{6.8}$$

We begin with a simple classical lemma.

Lemma 6.1. Suppose for some $\delta > 0$,

$$\|(I - P_h)L_k\| \|(\lambda I + L_k)^{-1}\| \leq \frac{\delta}{1+\delta}. \tag{6.9}$$

Then the Galerkin equations have a unique solution and satisfy the quasi-optimal error estimate

$$\|v - v_h\| \leq \lambda(1+\delta)\|(\lambda I + L_k)^{-1}\| \inf_{w_h \in \mathcal{V}_h} \|v - w_h\|. \tag{6.10}$$

Proof. Since $\delta > 0$, the hypothesis implies that

$$\|I - (\lambda I + L_k)^{-1}(\lambda I + P_h L_k)\| \leq \left(\frac{\delta}{1+\delta}\right) < 1.$$

Using the fact that $(I - A)$ is invertible if $\|A\| < 1$ (with $\|(I - A)^{-1}\| \leq (1 - \|A\|)^{-1}$), the previous bound implies that $(\lambda I + L_k)^{-1}(\lambda I + P_h L_k)$ is invertible, with

$$\|(\lambda I + P_h L_k)^{-1}(\lambda I + L_k)\| \leq \frac{1}{1 - \delta/(1+\delta)} = 1 + \delta.$$

Thus $(\lambda I + P_h L_k)$ is invertible and

$$\|(\lambda I + P_h L_k)^{-1}\| \leq (1+\delta)\|(\lambda I + L_k)^{-1}\|.$$

Since we also have

$$\begin{aligned} v - v_h &= v - (\lambda I + P_h L_k)^{-1} P_h f \\ &= (\lambda I + P_h L_k)^{-1}(\lambda v - P_h(f - L_k v)) \\ &= \lambda\big(\lambda I + P_h L_k\big)^{-1}(I - P_h)v, \end{aligned}$$

the required estimate (6.10) follows readily. □

The following corollary is the consequence of Lemma 6.1 when we have at our disposal an estimate of the smoothing power of L_k.

Corollary 6.2. Suppose

$$N(k) := \|L_k\|_{H^1(\Gamma) \leftarrow L^2(\Gamma)} < \infty. \tag{6.11}$$

Then, for all $\delta > 0$ there exists $C_\delta > 0$ such that the condition

$$hN(k)\|(\lambda I + L_k)^{-1}\| \leq C_\delta \tag{6.12}$$

is sufficient to ensure that the quasi-optimal estimate (6.10) holds.

Proof. By standard approximation theory (see, for example Sauter and Schwab (2011, §§4.3.4, 4.3.5)),

$$\|(I - P_h)L_k\| \lesssim h\,N(k), \tag{6.13}$$

and so the result then follows from Lemma 6.1 (with C_δ taken to be $\delta/(1+\delta)$ multiplied by the hidden constant in (6.13)). □

While Lemma 6.1 and Corollary 6.2 provide general criteria which ensure that the quasi-optimality estimate (6.10) holds, a bit more work is needed to obtain k-explicit requirements on h. Combining Theorems 5.14 and 5.15 from Section 5.5 and recalling the choice (6.5) of η, we see that for $s \in [0,1]$, and provided $C^{2,\mu}$, for some $\mu \in (0,1)$, then for all $k_0 > 0$

$$\|L_k v\|_s \lesssim k^{\beta_d+s}\|v\|, \quad \text{where } \beta_d = (d-1)/2, \tag{6.14}$$

for $k \geq k_0$ where $\|\cdot\|_s$ denotes the norm in $H^s(\Gamma)$. (Theorem 5.14 gives this estimate for $s = 0$, Theorem 5.15 gives it for $s = 1$, and the case of $s \in (0,1)$ follows by interpolation.) Using this estimate with $s = 1$, we obtain directly the following result that quantifies the mesh threshold required for quasi-optimality.

Theorem 6.3. Suppose that (6.14) holds for $s = 1$. Then, for each $\delta > 0$ there exists $C_\delta > 0$ (independent of h and k) such that the condition

$$h \leq C_\delta\, k^{-(\beta_d+1)}\|(\lambda I + L_k)^{-1}\|^{-1} \tag{6.15}$$

ensures that the Galerkin equations have a unique solution satisfying the quasi-optimality estimate (6.10).

Remark 6.4. For star-shaped domains, Theorem 5.22 gives us the upper bound $\|(\lambda I+L_k)^{-1}\| \lesssim 1$, and combining this with Theorem 6.3 immediately shows that the quasi-optimality estimate (6.10) holds for star-shaped $C^{2,\mu}$ domains provided $h \lesssim k^{-3/2}$ in 2D, and provided $h \lesssim k^{-2}$ in 3D. While these requirements are somewhat stronger than the convention (6.4), they are sufficient to ensure that quasi-optimality holds with the constant in front of the best approximation error in (6.10) independent of k, at least for these domains.

In fact, as we shall now show, with a bit more work and with an additional assumption, one can sharpen the quasi-optimality estimate (6.10) to show that the Galerkin solution is asymptotically just as good as the best possible approximation to v from $\mathcal{V}_h$, that is, we shall show that

$$\frac{\|v - v_h\|}{\inf_{w_h \in \mathcal{V}_h}\|v - w_h\|} \to 1 \quad \text{as } h \to 0. \tag{6.16}$$

Theorem 6.5. Suppose (6.11) holds, and suppose also that

$$M(k) := \|(\lambda I + L_k^*)^{-1}L_k\|_{H^1(\Gamma)\leftarrow L^2(\Gamma)} < \infty, \tag{6.17}$$

where L_k^* is the adjoint of L_k. Then the condition (6.12) is sufficient to ensure that the Galerkin equations have a unique solution, and furthermore there exists a C_0 independent of h and k such that if

$$C_0\, h\, C(k) \leq 1 \tag{6.18}$$

where

$$C(k) = N(k) + (\lambda + \|L_k\|)M(k),$$

then

$$\inf_{w_h \in \mathcal{V}_h} \|v - w_h\| \leq \|v - v_h\| \leq [1 + C_0\, h\, C(k)] \inf_{w_h \in \mathcal{V}_h} \|v - w_h\|. \tag{6.19}$$

Proof. If we apply P_h to (6.6) and subtract the resulting equation from (6.8), we obtain

$$\lambda(v_h - P_h v) = P_h L_k (v - v_h). \tag{6.20}$$

Then, writing

$$\|v - v_h\|^2 = (v - v_h, v - P_h v) + (v - v_h, P_h v - v_h), \tag{6.21}$$

we essentially now have to show that the second term on the right-hand side goes to zero more quickly than $\|v - v_h\|^2$. This is done by taking the inner product of (6.20) with $v - v_h$ to obtain

$$\begin{aligned} \lambda(v - v_h, P_h v - v_h) &= -(v - v_h, P_h L_k(v - v_h)) \\ &= (v - v_h, (I - P_h) L_k (v - v_h)) \\ &\quad - (v - v_h, L_k(v - v_h)). \end{aligned} \tag{6.22}$$

Using Cauchy–Schwarz and (6.13), the first term on the right-hand side of (6.22) may be estimated as

$$|(v - v_h, (I - P_h) L_k (v - v_h))| \lesssim h\, N(k)\, \|v - v_h\|^2. \tag{6.23}$$

The second term on the right-hand side of (6.22) may be rewritten as

$$\begin{aligned} (v - v_h, L_k(v - v_h)) &= ((\lambda I + L_k)^{-1}(\lambda I + L_k)(v - v_h), L_k(v - v_h)) \\ &= ((\lambda I + L_k)(v - v_h), (\lambda I + L_k^*)^{-1} L_k (v - v_h)) \\ &= ((\lambda I + L_k)(v - v_h), (I - P_h)(\lambda I + L_k^*)^{-1} L_k (v - v_h)), \end{aligned}$$

where the last line uses Galerkin orthogonality (6.20), that is, $P_h(\lambda I + L_k)(v - v_h) = 0$. (This is the classical 'superconvergence argument' for second-kind integral equations; see, *e.g.*, Chandler 1980.) Hence, using Cauchy–Schwarz and (6.13) again, we have

$$|(v - v_h, L_k(v - v_h))| \lesssim h\, (\lambda + \|L_k\|)\, M(k)\, \|v - v_h\|^2. \tag{6.24}$$

Thus, using (6.23) and (6.24) in (6.22) and using the definition of $C(k)$, we obtain

$$|(v - v_h, P_h v - v_h)| \lesssim hC(k)\, \|v - v_h\|^2.$$

Finally, combining this with (6.21) and using Cauchy–Schwarz, we obtain

$$(1 - (C_0/2)C(k)h)\, \|v - v_h\| \lesssim \|v - P_h v\|,$$

for some constant C_0. If the threshold (6.18) holds then this yields the result (6.19). □

We now investigate how $C(k)$ in the condition (6.18) depends on k. For simplicity we assume that Γ is $C^{2,\mu}$ for some $\mu \in (0,1)$, and thus the bound (6.14) holds. In this case the condition (6.12) for existence and uniqueness of a solution to the Galerkin equations is contained within the condition (6.15).

For a bound on the k-dependence of $C(k)$ we need to bound $M(k)$. This is possible using the bounds on $\|(\lambda I + L_k)^{-1}\|$ (and thus on $\|(\lambda I + L_k^*)^{-1}\|$), but the resulting bounds on $M(k)$ are unlikely to be sharp. For example in the case Γ is star-shaped (so $\|(\lambda I + L_k)^{-1}\| \lesssim 1$), consider u and g related by

$$(\lambda I + L_k^*)^{-1} L_k u = g.$$

Then $(\lambda I + L_k^*)g = L_k u$ and $\lambda \|g\|_1 \lesssim k^{\beta_d+1}(\|g\| + \|u\|)$. Moreover, $\|g\| \lesssim \|L_k u\| \lesssim k^{\beta_d}\|u\|$, and so

$$M(k) \lesssim k^{2\beta_d+1}.$$

Thus, using (6.11) and (6.14), $C(k) \lesssim k^{3\beta_d+1}$ and the condition (6.18) becomes

$$hk^{3\beta_d+1} \lesssim 1,$$

that is, $hk^{5/2} \lesssim 1$ in 2D, and $hk^4 \lesssim 1$ in 3D. Finally, we can obtain conditions under which we have asymptotic optimality, *i.e.*, (6.16): from the estimate (6.16) we see this is the case if $hC(k) \to 0$ as $k \to \infty$.

In summary, the standard projection method argument shows that under the mesh threshold (6.15), that is, $hk^{3/2} \lesssim 1$ in 2D and $hk^2 \lesssim 1$ in 3D, we have quasi-optimality with C in (1.7) independent of k for $C^{2,\mu}$ star-shaped domains. The superconvergence argument shows us that, under the stronger conditions that h decreases faster than $k^{-5/2}$ in 2D (*i.e.*, $hk^{5/2} = o(1)$ as $k \to \infty$) and k^{-4} in 3D, (6.16) holds.

We now discuss the relation of the above results to the error analysis of Banjai and Sauter (2007), which is obtained using the variational approach. In particular, Corollary 3.3 of that reference proves that, in the case of piecewise constant BEM (*i.e.*, the approximating space is $\mathcal{V}_h$ with polynomial degree $p = 0$), the Galerkin equations have a unique solution and satisfy the error estimate

$$\|v - v_h\| \lesssim C_c \inf_{w_h \in \mathcal{V}_h} \|v - w_h\|, \tag{6.25}$$

provided that

$$h\, C_c M(k) \text{ is sufficiently small.} \tag{6.26}$$

Here $M(k)$ is defined by (6.17) and C_c is the continuity constant of the sesquilinear form induced by $\lambda I + L_k$, so that

$$|((\lambda I + L_k)u, w)| \le C_c \|u\| \, \|w\|. \tag{6.27}$$

Using (6.14) in the case $s = 0$, to bound C_c, we obtain from the results of Banjai and Sauter (2007) that

$$\|v - v_h\| \lesssim k^{\beta_d} \inf_{w_h \in \mathcal{V}_h} \|v - w_h\|, \tag{6.28}$$

provided

$$hk^{\beta_d} M(k) \text{ is sufficiently small.}$$

This can be compared to the conclusion of Remark 6.4 which says that if Γ is $C^{2,\mu}$ with $\mu \in (0, 1)$ (so that Theorem 6.3 holds) and star-shaped (so that $\|(\lambda I + L_k)^{-1}\| \lesssim 1$) then

$$\|v - v_h\| \lesssim \inf_{w_h \in \mathcal{V}_{h,p}} \|v - w_h\| \quad \text{provided } h \lesssim k^{-(\beta_d+1)}. \tag{6.29}$$

An additional result in Banjai and Sauter (2007) is that, when Γ is the unit sphere, and with the choice $\eta \sim k^{2/3}$, it turns out that $\|L_k\| \lesssim 1$ (see Theorem 5.12 above), and hence $C_c \lesssim 1$ with this choice of η. Banjai and Sauter (2007) then made the additional assumption (backed up by some numerical evidence) that, for the unit sphere and with this choice of η,

$$M(k) \lesssim k \tag{6.30}$$

(where $M(k)$ is defined in (6.17)). Under this assumption, the stronger estimate

$$\|v - v_h\| \lesssim \inf_{w_h \in \mathcal{V}_h} \|v - w_h\| \quad \text{provided } hk \lesssim 1 \tag{6.31}$$

follows from (6.25) and (6.26). However, it is not clear if (6.30) holds.

If (6.31) holds, then the h-version BEM could be described as not suffering from the so-called 'pollution effect', a phenomenon widely studied in a finite element context; see, for example, Babuška and Sauter (2000). For h-version finite element methods the pollution effect has the consequence that the number of degrees of freedom per wavelength must increase as k increases, to maintain accuracy of the finite element solution. However, this effect can disappear when hp refinement is carefully applied; see, for example,

Ainsworth and Wajid (2010) and Melenk and Sauter (2011). Analogously, although the property (6.31) has not yet been proved to hold for the h-BEM, its analogue has recently been proved for the hp-BEM in Löhndorf and Melenk (2011) (that is, they proved that quasi-optimality, with a constant independent of k, holds for the hp-BEM provided the number of degrees of freedom per wavelength is above a certain threshold). We discuss this work in Section 6.3.

6.2. Best approximation error with piecewise polynomials in a particular case

To convert the quasi-optimality property (1.7) into an estimate of the convergence rate for the Galerkin solution, we need to estimate the best approximation error appearing on the right-hand side of (1.7). As discussed above, the convention for ensuring this approximation error is suitably controlled is that h should satisfy (6.4). If $\mathcal{V}_h$ denotes a space of piecewise polynomials of fixed degree p, an estimate which would justify this convention would be

$$\inf_{w_h \in \mathcal{V}_h} \|v - w_h\|_{L^2(\Gamma)} \lesssim (hk)^{p+1} \|v\|_{L^2(\Gamma)}, \tag{6.32}$$

for sufficiently large k. In the following lemma we show that, in a relevant special case, (6.32) does indeed hold.

Lemma 6.6. Suppose Ω_- is a 2D strictly convex domain with C^∞ boundary Γ and strictly positive curvature. Suppose that u is the solution of the sound-soft scattering problem (2.16) and $v = \partial u/\partial n$. Then, for all fixed $\eta \in \mathbb{R}\backslash\{0\}$, (6.32) holds for $p = 0$.

Proof. We use the standard approximation result that

$$\inf_{w \in \mathcal{V}_h} \|v - w\|_{L^2(\Gamma)} \lesssim h\|v\|_{H^1(\Gamma)} \tag{6.33}$$

and seek a k-explicit estimate on $\|v\|_{H^1(\Gamma)}$.

Domínguez *et al.* (2007, Theorem 5.4, Corollary 5.5) have proved that there exists $k_0 > 0$ such that, for all $k \geq k_0$,

$$v(x) = k\, V(x,k) \exp(\mathrm{i}kx \cdot \hat{a}), \tag{6.34}$$

where (see also Theorem 3.5 above), V satisfies

$$\|V(\cdot,k)\|_{L^2(\Gamma)} \lesssim 1 \quad \text{and} \quad \|\nabla_\Gamma V(\cdot,k)\|_{L^2(\Gamma)} \lesssim 1.$$

Now, since $v = \partial u/\partial n$, v also satisfies the integral equation (1.5) with $\eta = 0$ and f given by (1.4), that is,

$$\left(\frac{1}{2}I + D_k'\right) v = \frac{\partial u^I}{\partial n} = \mathrm{i}k\,(n \cdot \hat{a}) \exp(\mathrm{i}kx \cdot \hat{a}).$$

Now, using Theorem 5.14,

$$(1 + k^{1/2})\|v\| \gtrsim \left\| \left(\frac{1}{2} I + D'_k \right) v \right\| = \|f\| = k\|n \cdot \hat{a}\|.$$

Since $n \cdot \hat{a}$ does not vanish identically on Γ, we then have $\|v\| \gtrsim k^{1/2}$. (Note that, for Γ a circle or sphere, this argument gives the stronger bound $\|v\| \gtrsim k$ provided we use the bound on $\|D'_k\|$ in (5.25).)

Now, differentiating (6.34), we obtain

$$\nabla_\Gamma v(x) = k\big(\mathrm{i}v(x)(\hat{a} - (\hat{a} \cdot n)n) + \nabla_\Gamma V(x,k) \exp(\mathrm{i}kx \cdot \hat{a})\big),$$

and thus

$$|v|_{H^1(\Gamma)} \leq k\big(\|v\| + \|\nabla_\Gamma V(\cdot,k)\|\big).$$

Since the first term on the right-hand side grows like $k^{1/2}$ and the second term is bounded in k, it follows that $|v|_{H^1(\Gamma)} \lesssim k\|v\|$ for sufficiently large k, and the result follows on combination of this with (6.33). □

Whereas the previous lemma used asymptotic results about the solution of the sound-soft scattering problem to prove (6.32) for a specific Γ, it is possible to prove a weaker version of (6.32) for much more general Γ using only the fact that $v = \partial u/\partial n$ solves the integral equation (6.2), with f given by (1.4), and the bounds of Theorem 5.15.

Lemma 6.7. (Graham *et al.* 2012) Let $v = \partial u/\partial n$, where u is the solution to the sound-soft scattering problem (2.16) (so v satisfies (6.2) with f given by (1.4); see also (2.114)). If Γ is $C^{2,\mu}$, for some $\mu \in (0,1)$, then for all $k_0 > 0$,

$$\inf_{w_h \in \mathcal{V}_h} \|v - w_h\|_{L^2(\Gamma)} \lesssim hk^{(d+1)/2}\|v\|_{L^2(\Gamma)} \tag{6.35}$$

for all $k \geq k_0$.

The proof of this lemma bounds $\|v\|_{H^1(\Gamma)}$ by $\|v\|_{L^2(\Gamma)}$ using only the second-kind structure of the integral equation (6.2) and the $L^2(\Gamma) \to H^1(\Gamma)$ bounds on D'_k and S_k in Theorem 5.15 (see Graham *et al.* 2012 for the details). Thus this method also gives an analogous result for the *indirect* integral equation (6.1), even though the physical meaning of its solution is much less clear than for the direct equation (as discussed after Theorem 2.33).

Lemma 6.8. (Graham *et al.* 2012) Let v be the solution to the indirect boundary integral equation (6.1) with $f = -u^I$ (thus from (2.71) the scattered field u^S is then equal to $(\mathcal{D}_k - \mathrm{i}\eta\mathcal{S}_k)v$). If Γ is $C^{2,\mu}$, for some $\mu \in (0,1)$, then for all $k_0 > 0$,

$$\inf_{w_h \in \mathcal{V}_h} \|v - w_h\|_{L^2(\Gamma)} \lesssim hk^{(d+1)/2}\|v\|_{L^2(\Gamma)} \tag{6.36}$$

for all $k \geq k_0$.

At first sight it may appear strange that the bounds (6.35) and (6.36) hold for *all* (sufficiently smooth) domains, without any consideration of trapping or non-trapping (which, as we saw in Section 5, can affect the conditioning substantially). However, as mentioned above, all these bounds do is estimate $\|v\|_{H^1(\Gamma)}$ by $\|v\|_{L^2(\Gamma)}$, with no information about how large the latter may be (when Ω_- is star-shaped we know $\|v\|_{L^2(\Gamma)}$ is bounded polynomially in k by Theorem 5.22, whereas for certain trapping domains it is exponentially large by Theorem 5.24).

6.3. *Recent results of Löhndorf and Melenk*

Considerable progress on k-explicit error analysis for (6.1) and (6.2) was recently made by Löhndorf and Melenk (2011), who explore conditions under which the quasi-optimality estimate

$$\|u - u_h\| \leq C \inf_{v_h \in \mathcal{V}_{h,p}} \|u - v_h\| \tag{6.37}$$

holds with C independent of h, p and k, where the approximating space in the Galerkin method is the space $\mathcal{V}_{h,p}$, of piecewise polynomials of degree p on a quasi-uniform mesh of diameter h, and refinement with respect to both h and p is allowed.

To give an example of the results in Löhndorf and Melenk (2011), let us restrict to the indirect boundary integral equation (6.1), and assume that

$$\|(A'^{*}_{k,\eta})^{-1}\| \lesssim 1 \tag{6.38}$$

(as holds, for example, for Lipschitz star-shaped domains by combining Theorem 5.22 with equation (2.65) and the remarks following it). Then a main result (namely Theorem 3.17) in Löhndorf and Melenk (2011) proves that for analytic surfaces Γ, there exist $C, \epsilon, \sigma > 0$ independent of h, p and k such that, provided

$$\left(\frac{h}{\sigma + h}\right)^{p+1} + \left(\frac{kh}{\sigma p}\right)^{p+1} \leq \epsilon k^{-5}, \tag{6.39}$$

then (6.37) holds with C independent of h, p and k.

Because of the superalgebraic convergence with respect to p on the left-hand side of (6.39), it is fairly straightforward to see that condition (6.39) is satisfied and thus quasi-optimality (without pollution) is obtained if

$$\frac{kh}{p} \lesssim 1 \quad \text{and} \quad p \gtrsim \log k. \tag{6.40}$$

Note that this mesh threshold condition is much less stringent than that which had to be imposed in the analysis of the standard h version in the previous subsection. For example, Theorem 6.3 requires $hk^{3/2}$ (respectively hk^2) to be sufficiently small in 2D (respectively 3D) for quasi-optimality to be guaranteed in the case when $\|A_{k,\eta}^{-1}\| \lesssim 1$. Moreover, it is additionally shown in Löhndorf and Melenk (2011) that the modest conditions on h and p in (6.40) remain sufficient for (6.37), even if the right-hand side of (6.38) is allowed to grow in k with polynomial order. (Such growth occurs, for example, in a range of 'trapping' domains: see Section 5.6.2.) Analogous results also hold for the hp-BEM applied to the direct formulation (6.2).

The proof that (6.39) implies (6.37) has required the development in Löhndorf and Melenk (2011) and Melenk (2012) of substantial new technical tools for the Helmholtz equation. We only attempt a brief overview of these tools here.

The first step in the proof of (6.37) is to make a decomposition of the operator $A_{k,\eta}$ in (6.1) in the form

$$A_{k,\eta} = \left(\frac{1}{2}I + D_0 + \mathrm{i}S_0\right) + R_k + k[\mathcal{A}_k]. \tag{6.41}$$

The k-independent first term (in round brackets) has a bounded inverse in $L^2(\Gamma)$. The second term R_k satisfies the estimates

$$\|R_k\|_{H^1(\Gamma)\leftarrow L^2(\Gamma)} \lesssim k \quad \text{and} \quad \|R_k\|_{L^2(\Gamma)\leftarrow L^2(\Gamma)} \le q, \tag{6.42}$$

where $q \in (0,1)$ may be chosen as small as we like. (Recalling (6.14), the norm of the compact part of $A_{k,\eta}$ from $L^2(\Gamma)$ to $H^1(\Gamma)$ generally grows with order k^{β_d+1}, so the left-hand estimate in (6.42) gives a better rate of growth than this for R_k.)

In the third term in (6.41), $\mathcal{A}_k$ is an operator which maps $L^2(\Gamma)$ into functions which are analytic in a neighbourhood of the surface Γ, but may not be continuous across Γ, and $[\cdot]$ denotes the jump of such a function across Γ. (Recall that Γ is assumed analytic in this theory.) Decompositions such as this are discussed in depth in Melenk (2012), where an analogous decomposition is also presented for A_k^{-1}.

The second step in the proof that (6.39) implies (6.37) is to combine these decompositions with the variational theory of the Galerkin method for (6.2) to show (Löhndorf and Melenk 2011, Theorem 3.8) (recalling assumption (6.38)) that there exists a positive constant ϵ so that (6.37) holds provided

$$\xi(h,p,k) \le k^{-7/2}\epsilon,$$

where (with a slight change of the notation in Löhndorf and Melenk 2011), $\xi(h,p,k)$ is a bound on the rate of convergence of hp approximation of $k[u]$ and $[\partial u/\partial n]$ over all u in a certain space of analytic functions defined on a neighbourhood of Γ. Note that the quantity which we have here denoted ξ is closely related to the quantities denoted $\eta_i, i = 1,2$ in Löhndorf and Melenk (2011).

Finally the result is obtained by estimating ξ using the technology for quantifying hp approximation of analytic functions. Löhndorf and Melenk

(2011, Lemma 3.16) have shown that provided kh/p is sufficiently small, then

$$\xi(h,p,k) \lesssim k^{3/2}\left\{\left(\frac{h}{\sigma+h}\right)^{p+1} + \left(\frac{kh}{\sigma p}\right)^{p+1}\right\}.$$

Thus the proof that (6.39) implies (6.37) follows.

To summarize, we note that in Section 6.1 the essential mechanism driving the analysis is that the operator L_k appearing in $\lambda I + L_k$ has a suitable smoothing property (here we just need a k-explicit bound on the norm of L_k, operating from $L^2(\Gamma)$ to $H^1(\Gamma)$). The procedure of Löhndorf and Melenk (2011) is structurally similar: the decomposition in (6.41) is one where $A_{k,\eta}$ is decomposed into an invertible operator plus an (extremely) smoothing one. Ultimately, it is this smoothing property that leads to the condition (6.39).

6.4. Error estimates for hybrid numerical-asymptotic methods

The methods used to prove stability and convergence in the previous two subsections are heavily dependent on the fact that the basis functions in the Galerkin method are piecewise polynomial. For example the standard error estimate for piecewise polynomials in (6.13) plays a crucial role in Corollary 6.2 and also in Theorem 6.5: see (6.24).

In this subsection we are primarily concerned with proving estimates for hybrid methods whose basis functions are not usually piecewise polynomial. To do this we revert to classical Galerkin estimates provided by Céa's lemma; this was stated in a rather general form as Lemma 2.48, and we restate it here for this particular case.

Lemma 6.9. (Céa's lemma) Suppose we have a problem which can be written in weak form as follows. Seek $v \in L^2(\Gamma)$ such that

$$a(v,w) = (f,w), \quad \text{for all } w \in L^2(\Gamma), \tag{6.43}$$

and assume that a satisfies the conditions

$$\text{boundedness} \quad |a(u,w)| \le B_k \|u\|\,\|w\|, \tag{6.44}$$

$$\text{coercivity} \quad |a(u,u)| \ge \alpha_k \|u\|^2, \tag{6.45}$$

for all $u, w \in L^2(\Gamma)$. Let $\mathcal{V}_N$ be any subspace of $L^2(\Gamma)$ (which may depend on discretization parameters such as h or p and also, possibly, on the wave number k). Then consider the following approximate problem. Seek $v_N \in \mathcal{V}_N$ such that

$$a(v_N, w_N) = (f, w_N), \quad \text{for all } w_N \in \mathcal{V}_N.$$

This problem has a unique solution v_N which satisfies the error estimate

$$\|v - v_N\| \le \left(\frac{B_k}{\alpha_k}\right) \inf_{w_N \in \mathcal{V}_N} \|v - w_N\|.$$

We will apply this lemma to the Galerkin formulation in $L^2(\Gamma)$ of the combined potential equations (6.1) and (6.2), that is, we consider the case where the bilinear form a in (6.43) is

$$a_{k,\eta}(u,w) := (A_{k,\eta}u, w), \quad \text{or} \quad a'_{k,\eta}(u,w) := (A'_{k,\eta}u, w). \tag{6.46}$$

Moreover, we shall consider also formulations based on the star-combined formulation (6.3), where a is

$$a_{k,*}(u,w) := (\mathscr{A}_k u, w). \tag{6.47}$$

Using the continuity and coercivity results of Sections 5.5 and 5.7, we can now give some theorems on quasi-optimality for the Galerkin method.

Theorem 6.10. Let Ω_- be a C^3 strictly convex domain with strictly positive curvature in 2D or 3D. Suppose the equation (6.43) is solved with $f \in L^2(\Gamma)$ and the bilinear form a taken to be either $a = a_{k,\eta}$, $a = a'_{k,\eta}$ or $a = a_{k,*}$. Then there exists an $\eta_0 > 0$ (independent of k) and a $k_0 > 0$ such that if $\eta = \eta_0 k$, then

$$\|v - v_N\| \lesssim k^{(d-1)/2} \inf_{w_N \in \mathcal{V}_N} \|v - w_N\|, \quad \text{for all } k \geq k_0 \text{ and for all } N.$$

Proof. By Theorem 5.25 there exists an η_0 independent of k such that the operators $A_{k,\eta}$ and $A'_{k,\eta}$ are coercive for $\eta = \eta_0 k$ and for all k greater than some k_0. In addition, Theorem 5.14 states that with this choice of η, $\|A_{k,\eta}\|$ and $\|A'_{k,\eta}\|$ are $O(k^{(d-1)/2})$; thus the result follows from Céa's lemma. □

Theorem 6.11. Let Ω_- be a Lipschitz star-shaped domain in 2D or 3D. If the equation (6.43) is solved with $a = a_{k,*}$, then for all $k > 0$ and all N

$$\|v - v_N\| \lesssim (1 + k^{(d-1)/2}) \inf_{w_N \in \mathcal{V}_N} \|v - w_N\|.$$

Proof. The proof follows from the uniform coercivity of the star-combined operator, given in Theorem 5.26, and the upper bound of $O(k^{(d-1)/2})$ on its norm given in (5.32). □

Combining the previous two theorems with the best approximation results in Section 3 we can prove a number of results on the convergence rates of the hybrid numerical-asymptotic methods discussed in Section 3. Some examples are given below.

Theorem 6.12. Let Ω_- be a smooth (C^∞) strictly convex domain with strictly positive curvature in 2D. Suppose that the sound-soft scattering problem (2.16) is solved with the Galerkin method using the standard combined potential formulation (2.114) and the hybrid approximation space defined in Section 3.1 (originally introduced in Domínguez *et al.* 2007). Let p denote the degree of the polynomials used in each of the three zones (so

p is proportional to the total number of degrees of freedom of the method), and let n be an integer with $6 \le n \le p+1$. Then there exist η_0, k_0, δ, and c_0, all greater than zero, such that, if the coupling parameter η is chosen to be $\eta = \eta_0 k$, then

$$\|v - v_N\| \lesssim k^{19/18} \left\{ \left(\frac{k^{1/9}}{p} \right)^n + k^{4/9} \exp(-c_0\, k^{\delta}) \right\}, \tag{6.48}$$

for all $k \ge k_0$, where c_0 and δ only depend on Γ. Thus, provided p grows slightly faster than $O(k^{1/9})$, the error is bounded as $k \to \infty$.

In the following three theorems (all concerning polygons),

$$M(u) := \sup_{x \in \Omega_+} |u(x)|,$$

as in Theorem 3.9. As discussed below this theorem, the best currently available bound on $M(u)$ is that it is $O(k^{1/2} \log^{1/2} k)$ as $k \to \infty$, but numerical experiments suggest it is in fact $O(1)$. For this reason, the bounds in the following theorems will be explicit in $M(u)$. Regarding notation, in what follows N denotes the total number of degrees of freedom, p the degree of polynomial approximation on each element, and n_s the number of sides of the polygon (as in Sections 3.2–3.4).

Theorem 6.13. Let Ω_- be a convex polygon. Suppose that the sound-soft scattering problem (2.16) is solved with the Galerkin method using the star-combined potential formulation (6.47) and the hybrid approximation space defined in Section 3.3.1 (originally introduced in Chandler-Wilde and Langdon 2007). Then, with the same notation as in Theorem 3.11, we have that, for all $k_0 > 0$,

$$\|v - v_N\| \lesssim k M(u) (n_s \log(kL))^{1/2} \left(\frac{n_s \log(kL)}{N} \right)^{p+1}, \tag{6.49}$$

for all $k \ge k_0$.

Theorem 6.14. Let Ω_- be a convex polygon. Suppose that the sound-soft scattering problem (2.16) is solved with the Galerkin method using the star-combined potential formulation (6.47) and the hp-type hybrid approximation space defined in Section 3.3.4 (originally introduced in Hewett *et al.* 2012). Then there exists a $\tau > 0$ such that, for all $k_0 > 0$,

$$\|v - v_N\| \lesssim k^{3/2} M(u) \exp(-p\tau) \tag{6.50}$$

for all $k \ge k_0$. Thus, if p grows logarithmically with k then the error decays exponentially. (Note that this is a simplified bound, with the power of k independent of corner angles. A similar philosophy is adopted in the following theorem.)

Theorem 6.15. Let Ω_- be a non-convex polygon of the type considered in Section 3.4. Suppose that the sound-soft scattering problem (2.16) is solved with the Galerkin method using the star-combined potential formulation (6.47) and the *hp*-type hybrid approximation space defined in Section 3.4 (originally introduced in Chandler-Wilde *et al.* 2012*a*). Then there exists a $\tau > 0$ such that, for all $k_0 > 0$,

$$\|v - v_N\| \lesssim k^2 M(u) \exp(-p\tau) \tag{6.51}$$

for all $k \geq k_0$. Thus, if p grows logarithmically with k, then the error decays exponentially.

As mentioned at the beginning of this section, all these error estimates assume that the Galerkin equations are solved exactly for v_N. In practice, the matrix entries of the sesquilinear form are highly oscillatory integrals, as discussed in Section 4. Fully discrete error estimates for hybrid methods in the case of a smooth convex obstacle have been obtained in Kim (2012), by combining Theorem 6.12 with the analysis from Section 4.4 and the quadrature error estimates from Section 4.3 and using the Strang lemma.

6.5. Alternative ways to measure the error

In this section we have concentrated on measuring the accuracy of v_N by considering

$$\|v - v_N\|. \tag{6.52}$$

In this subsection we discuss alternative measures of the error, which basically involve multiplying (6.52) by different powers of k. We consider only the case where $v = \partial u/\partial n$, with u the solution of the sound-soft scattering problem (2.16); the considerations for other Dirichlet boundary conditions, and for impedance boundary conditions, may be different.

Since v itself depends on k, rather than controlling the absolute error (6.52), it might seem more sensible to control relative error measures such as

$$\frac{\|v - v_N\|}{\|v\|} \quad \text{or} \quad \frac{\|v - v_N\|}{\|v^I\|},$$

where $v^I = \partial u^I/\partial n$ (we saw this earlier in Section 6.2). The advantage of the second of these is that the behaviour of $\|v^I\|$ is known, that is, $\|v^I\| \sim k$ as $k \to \infty$ for an arbitrary obstacle. For smooth convex obstacles, the Kirchhoff approximation shows that on the lit side $\|v\|$ grows in proportion to $\|v^I\| \sim k$, and thus it is reasonable to believe (and proved in Lemma 6.6 for the case of a circle) that $\|v\| \sim k$. (Note that the asymptotics from Domínguez *et al.* (2007) used in Lemma 6.6 show that $\|v\| \lesssim k$, and this lemma showed that $\|v\| \gtrsim k^{1/2}$.) Thus, the second measure of error, and also the first in at least the case that Γ is a circle, becomes

$$k^{-1}\|v - v_N\|. \tag{6.53}$$

As a third possibility, in certain cases where we use the ansatz (3.1), we always know the leading-order part of the solution V_0 (this could be zero), thus what is really being computed in the BEM is the difference $v - V_0$. To assess the performance of methods in this case, we can compute the relative L^2 error defined as

$$\frac{\|(v - V_0) - (v_N - V_0)\|}{\|v - V_0\|} = \frac{\|v - v_N\|}{\|v - V_0\|}. \tag{6.54}$$

An alternative is to take the view that the computation of $v = \partial u/\partial n$ is an intermediate step, and that the real goal is to compute u accurately in the domain Ω_+, by substituting the Galerkin approximation v_N to $\partial u/\partial n$ into Green's integral representation (2.107). Denoting the resulting approximation to u by u_N, we may seek to control

$$\frac{\|u - u_N\|_{L^p(G)}}{\|u\|_{L^p(G)}} \quad \text{or} \quad \frac{\|u - u_N\|_{L^p(G)}}{\|u^I\|_{L^p(G)}}, \tag{6.55}$$

where $G \subset \Omega_+$ is some particular region of interest.

Applying the Cauchy–Schwarz inequality to the expression for $u - u_N$ given by Green's integral representation yields the upper bound

$$|u(x) - u_N(x)| \leq c(x)\|v - v_N\|_{L^2(\Gamma)}, \quad c(x) := \left\{\int_\Gamma |\Phi(x, y)|^2 \,\mathrm{d}s(y)\right\}^{1/2}. \tag{6.56}$$

The k-dependence of $\|u\|_{L^p(G)}$ is, in general, unknown, but $\|u^I\|_{L^p(G)}$ is independent of k. In 3D, $c(x)$ is also independent of k, but in 2D $c(x) \sim k^{-1/2}$. Thus, if one seeks to achieve small values for the second measure of relative error in (6.55) by controlling $\|v - v_N\|_{L^2(\Gamma)}$, one needs to ensure that

$$k^{-(3-d)/2}\|v - v_N\|_{L^2(\Gamma)} \tag{6.57}$$

is small. The scaling by $k^{-(3-d)/2}$ in (6.57) is actually rather natural in that it makes the expression (6.57) dimensionless.

In many applications it is not the solution u in Ω_+ that is required, but the far-field pattern of u^S, $F(\hat{x})$, defined in (2.10). In the case of sound-soft scattering, it follows from (2.107) that

$$u^S = -\mathcal{S}_k \frac{\partial u}{\partial n},$$

and an explicit formula for the far-field pattern of this is (2.23), with $\phi = \partial u/\partial n$. If we define F_N to be (2.23) with ϕ replaced by v_N (a Galerkin approximation to $\partial u/\partial n$), then we are again led to the consideration of (6.57), since by the Cauchy–Schwarz inequality we have

$$|F(\hat{x}) - F_N(\hat{x})| \lesssim k^{-(3-d)/2}\|v - v_N\|_{L^2(\Gamma)}. \tag{6.58}$$

We could also measure relative errors of F on $\mathbb{S}^{d-1}$ analogously to (6.55), that is,

$$\frac{\|F - F_N\|_{L^p(\mathbb{S}^{d-1})}}{\|F\|_{L^p(\mathbb{S}^{d-1})}}. \tag{6.59}$$

Remark 6.16. Finally we note that obviously the Hölder inequality could have been used instead of the Cauchy–Schwarz inequality in both (6.56) and (6.58) to give $|u(x) - u_N(x)|$ and $|F(\hat{x}) - F_N(\hat{x})|$ in terms of other L^p norms of $v - v_N$ (with different powers of k). There is currently no theory on how $\|v - v_N\|_{L^p(\Gamma)}$ behaves, except for the case $p = 2$. Note that the numerical results in Section 7.5 (in particular Tables 7.7 and 7.10) show that the error $\|v - v_N\|_{L^1(\Gamma)}$ can be considerably less than $\|v - v_N\|_{L^2(\Gamma)}$ (see also Hewett *et al.* 2012). Regarding the prospects of extending the theory, it is fairly straightforward to modify the best approximation results of Section 3 to apply in $L^p(\Gamma)$ rather than $L^2(\Gamma)$. (Via (A.25), such estimates can moreover be used to get best approximation results in negative Sobolev norms, for example in $H^{-1/2}(\Gamma)$.) However, it is unclear how to prove a version of the k-explicit quasi-optimality results, such as Theorem 6.11, in $L^p(\Gamma)$ rather than $L^2(\Gamma)$.

7. Numerical results

In this section we present numerical results for a range of scattering problems, using the hybrid approximation spaces described in Section 3. In cases where the theory of Section 6 is applicable, the experiments confirm the validity of the error estimates proved in that section. The experiments also show that hybrid methods are effective computational algorithms in cases where a full error analysis is not available. Further, we see that the methods sometimes perform better than the theory predicts, with the number of degrees of freedom required to achieve a prescribed level of accuracy growing, as k increases, more mildly than predicted, and in some cases even decreasing.

These results confirm that hybrid methods achieve, at least for classes of scattering problems, the objectives set out in **Q1** in Section 1. For the examples with smooth scatterers, which we implement making full use of the numerical schemes for oscillatory integrals detailed in Section 4, we also show computation times. Impressively, these remain bounded as k increases at the same time as the accuracy increases, confirming that the objectives set out in **Q2** in Section 1 are also achieved.

7.1. Smooth convex obstacles

We thank T. Kim for the numerical results in this subsection. These illustrate the hybrid method described in Section 3.1, which was implemented

Table 7.1. Error e_p on unit circle with exact solution given by a Fourier series.

p	$k = 125$	$k = 250$	$k = 500$
4	3.49×10^{-2}	3.05×10^{-2}	2.71×10^{-2}
8	8.34×10^{-3}	3.96×10^{-3}	3.04×10^{-3}
12	1.03×10^{-2}	2.99×10^{-3}	1.22×10^{-3}
16	1.33×10^{-2}	4.44×10^{-3}	1.58×10^{-3}
22	1.74×10^{-2}	6.27×10^{-3}	2.42×10^{-3}

Table 7.2. Error e_p on unit circle with exact solution taken to be v_{24}.

p	$k = 1000$	$k = 4000$	$k = 16000$
4	4.41×10^{-2}	2.35×10^{-2}	2.50×10^{-2}
8	2.80×10^{-3}	2.20×10^{-3}	1.82×10^{-3}
12	9.61×10^{-4}	3.43×10^{-4}	2.61×10^{-4}
16	8.26×10^{-4}	1.81×10^{-4}	5.45×10^{-5}
22	5.68×10^{-4}	1.02×10^{-4}	9.68×10^{-6}

using the quadrature scheme described in Section 4.4. Throughout this subsection we compute the sound-soft scattering of an incident plane wave with direction $\hat{a} = (1,0)$ by a smooth convex obstacle. The integral equation we solve is (2.114) with $\eta = k$. The boundary Γ will be parametrized by a mapping $\zeta : [0,2\pi] \to \Gamma$ and the shadow boundary (3.6) consists of two points $x_i = \zeta(t_i)$, $i = 1,2$. With the Fock zones (3.25), (3.26) taken to be

$$\Lambda_1 = [t_1 - ak^{-1/3+\delta},\ t_1 + a'k^{-1/3+\epsilon}], \quad \Lambda_2 = [t_2 - a'k^{-1/3+\epsilon},\ t_2 + ak^{-1/3+\delta}] \tag{7.1}$$

(with parameters described below), the approximating space is $\mathcal{V}_N$, as given by (3.27). Since, in this subsection, the order of the numerical scheme is determined by the polynomial degree p, we denote the space by $\mathcal{V}_p$. The Galerkin approximation of the factor V in (3.14) is then denoted V_p, and the corresponding approximation of v is denoted v_p.

Our first experiment is for the unit circle with $\zeta(t) = (\cos t, \sin t)$. In Table 7.1 we tabulate the proxy for the relative error (recall the discussion around (6.53))

$$e_p = \frac{1}{k}\|v - v_p\| \tag{7.2}$$

for various values of p and k. In this case the solution $v = \partial u/\partial n$ can

Table 7.3. Approximate errors $|V_p((-b,0),k)+2\mathrm{i}|$, for the elliptical scatterer with $b=3$, $c=1$.

p	$k=1000$	$k=4000$	$k=8000$	$k=16000$	Relative time
6	2.94×10^{-3}	8.71×10^{-3}	2.90×10^{-2}	5.26×10^{-2}	1
8	3.44×10^{-3}	2.54×10^{-3}	1.07×10^{-2}	2.20×10^{-2}	1.16
10	3.52×10^{-3}	1.01×10^{-3}	3.98×10^{-3}	9.21×10^{-3}	1.36
12	3.34×10^{-3}	8.34×10^{-4}	1.55×10^{-3}	3.80×10^{-3}	1.65
14	2.65×10^{-3}	7.35×10^{-4}	6.86×10^{-3}	1.49×10^{-3}	1.86
16	3.67×10^{-3}	7.94×10^{-4}	4.50×10^{-4}	5.14×10^{-4}	2.22
18	2.38×10^{-3}	7.05×10^{-4}	3.87×10^{-4}	1.87×10^{-4}	2.54

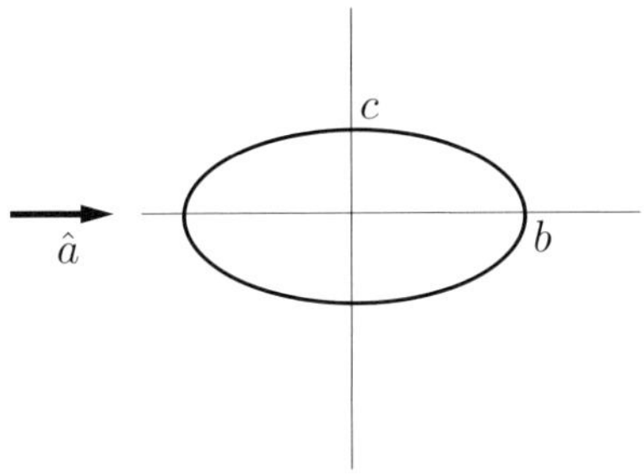

Figure 7.1. Elliptical scatterer with semi-axes b, c.

be found exactly as a Fourier series, and the errors (7.2) can be computed accurately by quadrature. The parameters for the Fock zones are $a=3.52$, $a'=2.20$, $\epsilon=1/9$, $\delta=0$. The table shows convergence as p increases, up to a saturation error which decreases as k increases. This is caused by the exponentially decaying k-dependent (but p-independent) term in the error estimate (6.48). We also notice an improvement in accuracy for fixed p as k increases, although the estimate in Theorem 6.12 does not predict this. The Fourier series providing the exact solution is slow to converge for very large k. We therefore illustrate the same convergence behaviour for larger values of k in Table 7.2, where we give values of e_p computed using v_{24} as an approximation of the exact solution v. Here the exponentially decaying k-dependent part of the error is not visible. In these examples the dimension of the hybrid space and hence the size of the matrix being inverted is $3p+3$.

Our second experiment is for an elliptical scatterer with semi-axes b, c, as shown in Figure 7.1. The parameters for the Fock zones are $a=24$, $a'=16$, $\epsilon=1/9$ and $\delta=0$. Here the exact solution v is unknown, but comparing the physical optics approximation (3.4) with the ansatz (3.7) (recall also (3.3)) it is known that the factor $V((-b,0),k)$ approaches $-2\mathrm{i}$ with rate $O(k^{-1})$ as $k\to\infty$. In Table 7.3 we give the values of the error $|V_p((-b,0),k)+2\mathrm{i}|$

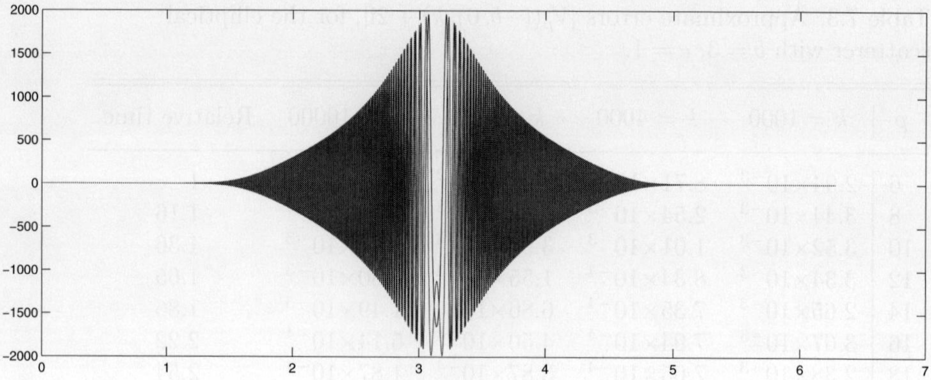

(a) approximation to v

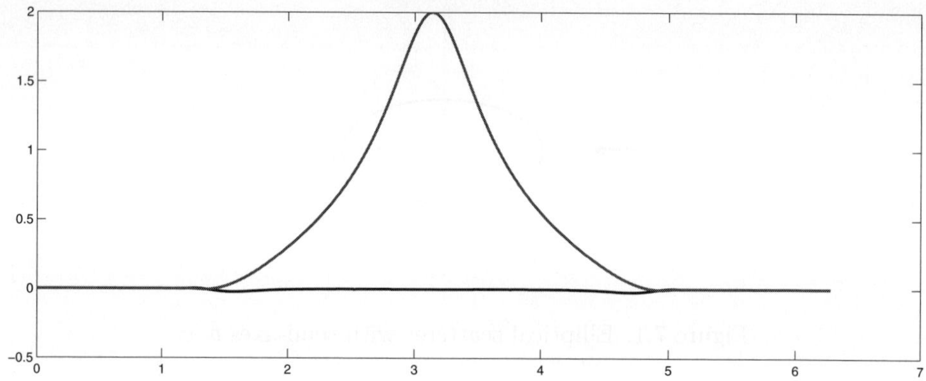

(b) approximation to $V(\cdot, k)$, both real (the bottom curve) and imaginary (the top curve) parts

Figure 7.2. Ellipse with $b = 3$, $c = 1$, $k = 1000$ and $p = 12$.

for the case $b = 3$, $c = 1$ for various values of k and polynomial degree p. In this case the quadrature rules from Section 4.4 are set up once and used for all values of k, so the computational times are fixed with respect to k. (More details of the choice of parameters in the quadrature rules and how to obtain fixed computation times with varying k are given in Kim (2012).) In Table 7.3, for fixed small p, the errors grow slightly with k. For larger p, the errors decrease as $k \to \infty$. The column for $k = 16000$ shows an experimental rate of decay of about $O(\exp(-0.45p))$ as $p \to \infty$. The final column indicates the computation time as a proportion of the time taken for $p = 1$ and shows an optimal growth rate of about $O(p^2)$. (Recall that the stiffness matrix for the Galerkin method has dimension $O(p)$.) For this ellipse, in Figure 7.2(a) we illustrate the computed oscillatory solution v, and in Figure 7.2(b) the corresponding slowly varying factor V (for $k = 1000$, computed with $p = 12$).

Table 7.4. The errors e_p defined in (7.3).

Standard				
p	$k = 200$	$k = 400$	$k = 600$	Time (s)
4	1.47×10^{-2}	3.09×10^{-2}	3.49×10^{-2}	84
8	2.18×10^{-3}	4.16×10^{-3}	4.99×10^{-3}	101
12	1.53×10^{-3}	6.83×10^{-4}	7.87×10^{-4}	135
16	2.25×10^{-3}	6.59×10^{-4}	3.73×10^{-4}	184
Star-combined				
p	$k = 200$	$k = 400$	$k = 600$	Time (s)
4	1.62×10^{-2}	3.54×10^{-2}	4.09×10^{-2}	118
8	2.91×10^{-3}	5.83×10^{-3}	6.24×10^{-3}	136
12	3.04×10^{-3}	9.85×10^{-4}	9.97×10^{-4}	180
16	2.72×10^{-3}	8.36×10^{-4}	6.97×10^{-4}	238

The final example compares the standard combined potential formulation (2.114) (with $\eta = k$) with the star-combined formulation (2.115). The ellipse parameters are $b = 2$, $c = 1$ and the Fock zone parameters are $a = 4.91$, $a' = 3.55$, $\epsilon = 1/9$, $\delta = 0$. The stiffness matrices for the surface gradient of the single-layer potential arising in the star-combined operator were computed using the 2D version of the integration by parts procedure in Section 4.1.3. In Table 7.4 we tabulate the relative error

$$e_p = \frac{\|v - v_p\|}{\|v\|}, \tag{7.3}$$

with the 'exact' solution v obtained (for moderate k only) via the Nyström method of Colton and Kress (1998, §3.5). The relative errors were estimated using Simpson's rule with a sufficient number of quadrature points. Again as k increases, and for low p, the errors may increase slightly. This is consistent with Theorem 6.12. Again, for larger p, the errors decrease as k grows, which is a better result than the theory predicts. The star-combined method performs comparably to the standard method. The computing times are longer (timings are in seconds) due to the appearance of an additional integral operator, namely $\nabla_\Gamma S_k$, but the stronger theory for this method motivates its investigation. The growth rate of the CPU time with increasing p in the star-combined case is of the same order as in the standard case.

Much more detail of these experiments, including an investigation of how the choice of the intervals Λ_1 and Λ_2 should depend on the curvature of Γ at the shadow boundary points, is given in Kim (2012).

7.2. Sound-soft convex polygon

For the case of a sound-soft convex polygon, we present numerical results using the *hp*-version BEM described in Section 3.3.4. We consider scattering by a square of side length 2π (so that the number of wavelengths per side is equal to k), with both grazing and non-grazing incidence angles (showing the robustness of the scheme with respect to angle of incidence). The scatterers, the incident direction vectors and the corresponding total fields for $k = 10$ are all plotted in Figure 7.3. We solve the BIE (2.114), with $\eta = k$, and demonstrate exponential decay of the error on the boundary as the polynomial degree p increases, with only a very mild dependence on the wavenumber k, consistent with the best approximation error estimate (3.48).

Rigorous error estimates for this method are presented in Theorem 6.14 only in the context of the star-combined formulation. However, our numerical results, computed by solving the standard combined potential equation (2.114), are empirically consistent with (6.50). In our experiments we take the same degree p of polynomial approximation on each element, and the same number of layers on each graded mesh (chosen proportional to p; see also Section 3.3.4), using the same parameter choices as given in Hewett *et al.* (2012, §7), to which we refer for details. The total number of degrees of freedom for these examples is $16(p + 1)^2$. Since, as in Section 7.1, the order of the numerical scheme is determined solely by the polynomial degree p, we again denote our approximation of $v = \partial u/\partial n$ by v_p.

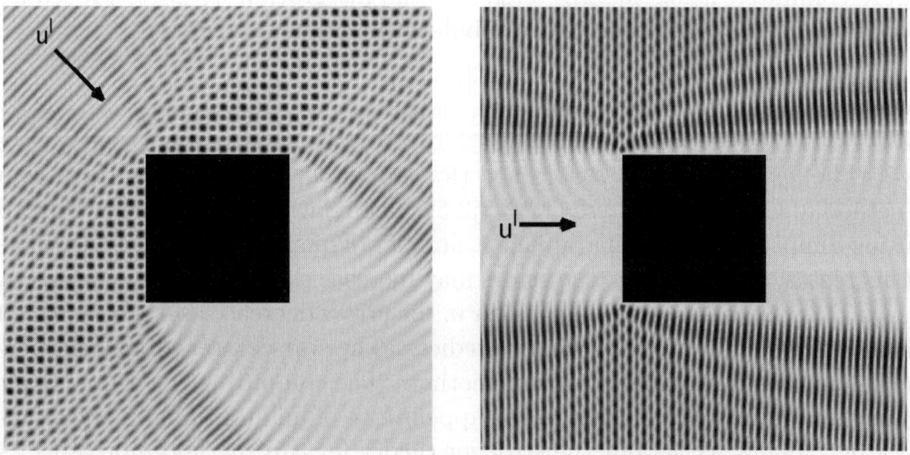

(a) non-grazing incidence (b) grazing incidence

Figure 7.3. Scattering by a sound-soft square of side length 2π, $k = 10$.

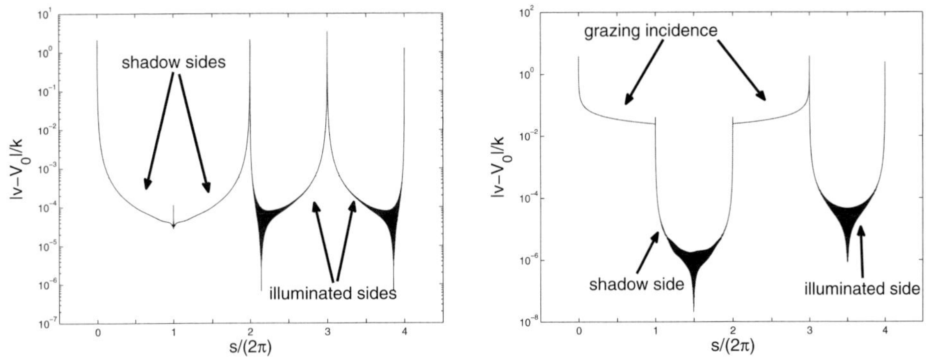

(a) non-grazing incidence (b) grazing incidence

Figure 7.4. Scattering by a sound-soft square of side length 2π, for $k = 160$.

To illustrate the qualitatively different behaviour of the solution on sides with grazing incidence, in Figure 7.4 we plot, for $k = 160$ and for cases (a) and (b) of Figure 7.3, graphs of $|(v - V_0)(s)|/k$ against s, where V_0 is the known leading-order part of the solution (see (3.35) and (3.8)). The 'exact' solution $v = \partial u/\partial n$ is the numerical solution computed with 1024 degrees of freedom (corresponding to $p = 7$). The solution in each case is singular at the corners of the polygon, which are at $s = s_j := 2j\pi$, $j = 0, \ldots, 4$ (note the logarithmic scale on the vertical axis, and also that the solution is plotted for 100000 evenly spaced values of s). As s increases away from the corners, for case (a) $|(v - V_0)(s)|$ decays like $|s - s_j|^{-3/2}$ on all four sides, whereas for case (b) $|(v - V_0)(s)|$ decays like $|s - s_j|^{-3/2}$ on the illuminated and the shadow sides, but like $|s - s_0|^{-1/2}$ and $|s - s_3|^{-1/2}$ on the two sides with grazing incidence, as seen in Figure 7.4(b). This suggests that, for $n = 0$, Theorem 3.10 is sharp in its dependence on t for the case of grazing incidence, but not for non-grazing incidence. (For the impedance half-plane problem we saw this difference between grazing and non-grazing explicitly in (3.32) and (3.33), and we conjecture that similar estimates will hold for the polygonal scatterer considered here.) The function $|v - V_0|$ is highly oscillatory in the (apparently) shaded areas in Figure 7.4.

In Figure 7.5 we plot (on a logarithmic scale) a measure of relative L^2 error different from that in (7.3) (recall the discussion around (6.54)), namely

$$e_p = \frac{\|v - v_p\|}{\|v - V_0\|},$$

against p, for cases (a) and (b) and for a range of values of k. In each case we take the 'exact' reference solution to be that computed with $p = 7$. Full details of how e_p is accurately computed are provided in Hewett *et al.* (2012). The linear plots in Figure 7.5 clearly demonstrate exponential convergence

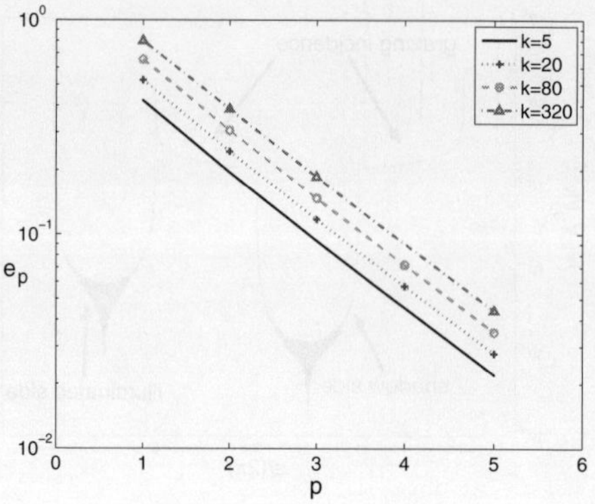

(a) non-grazing incidence

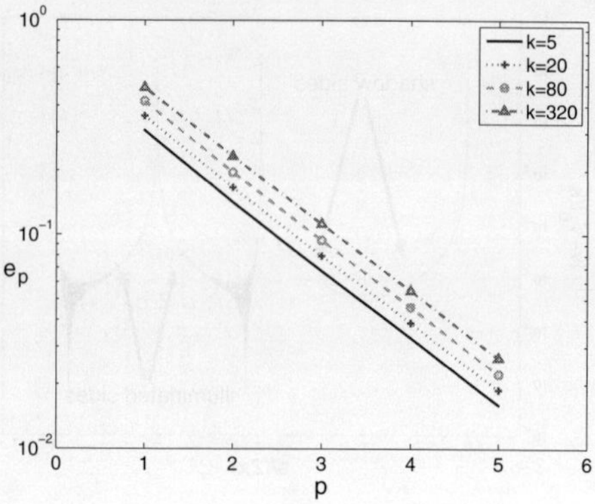

(b) grazing incidence

Figure 7.5. Relative errors e_p for scattering by a sound-soft square of side length 2π.

with respect to the polynomial degree p, with only very mild growth in the relative error e_p as k increases. In particular it is clear from these figures that e_p remains bounded if p is increased by 1 each time k is increased by a factor of 4. Thus increasing p proportional to $\log k$ is enough to maintain accuracy. The scheme clearly copes well with grazing incidence; indeed, because $\|v - V_0\|$ is bigger, if anything the problem with grazing incidence is solved more accurately, as measured by e_p, than the one with non-grazing incidence.

Further numerical results for the same scheme, investigating how the error depends on the geometry of the obstacle (particularly the sharpness of the corners), and also demonstrating its accuracy for computing solutions in the domain Ω_+ and computing far-field patterns, can be found in Hewett *et al.* (2012). These results exhibit computation times that grow approximately logarithmically with k, illustrating the power of the quadrature schemes described in Section 4.4.

7.3. Curvilinear polygon

For the case of a sound-soft curvilinear polygon, we present numerical results using the scheme described in Section 3.3.2. The scatterer we consider is depicted in Figure 7.6, where we show the incident direction and total field, plotted for $k = 20$. The scattering obstacle consists of the intersection of two circles of radius 3, forming a 'lens' shape (with internal corner angles $2\pi/3$), and the incident field is such that the boundary between the shadow and illuminated regions coincides with the corners of the obstacle.

Full details of this example are provided in Langdon *et al.* (2010), where numerical results are presented for the case where the integral equation (2.114), with $\eta = k$, is solved using the hybrid approximation space

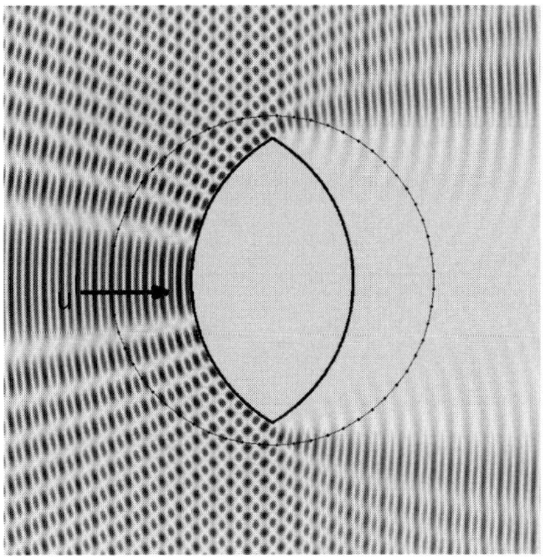

Figure 7.6. Scattering by a curvilinear polygon, $k = 20$; the dotted line indicates the circle on which the total field is computed, for comparison of errors between different discretizations, as described below.

described in Section 3.3.2 and the unknowns V, $V_m^{\pm}$ are approximated with piecewise constant functions. These examples demonstrate that the number of degrees of freedom N required to approximate $v = \partial u/\partial n$ to a prescribed level of relative accuracy (using the measure (6.53)) appears to grow only logarithmically with k as k increases, and that $O(N^{-1})$ convergence is observed for fixed k.

Here, we investigate the accuracy of the approximation to the solution in the domain Ω_+; specifically we compute the total field $u = u^I + u^S$ (by substituting the approximation v_N to $v = \partial u/\partial n = \partial_n u$ into the representation formula (2.107)) on the circle (of radius $r = 3$) surrounding the scatterer, illustrated in Figure 7.6. With a slight abuse of notation, we denote the solution at the point $r(\cos t, \sin t)$ on the circle by $u_N(t)$, $t \in [-\pi, \pi]$, where $t = 0$ corresponds to the point on the circle in the deepest shadow (*i.e.*, furthest to the right in Figure 7.6). Plots of $|u(t)|$ for $k = 20$ and $k = 320$ (computed with $N = 1792$ and $N = 2264$ respectively) are shown in Figure 7.7, with the shadow region indicated.

In Figure 7.8 we plot the relative maximum error on the circle (recall (6.55)),

$$e_N = \frac{\max_{t\in[-\pi,\pi]} |u(t) - u_N(t)|}{\max_{t\in[-\pi,\pi]} |u(t)|},$$

against N for different values of k. The maxima in this formula were approximated by sampling t at $30k$ evenly spaced points on $[-\pi, \pi]$. The algebraic decay of e_N with increasing N is clear in Figure 7.8 (note the logarithmic scale on each axis). Moreover, for fixed N, e_N appears to grow

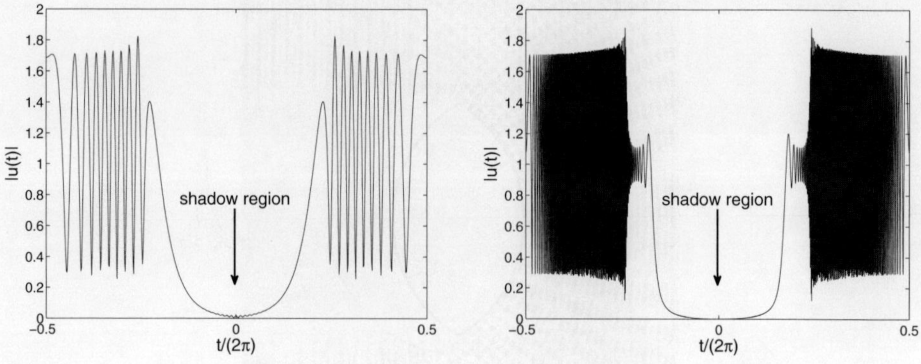

(a) $k = 20$ (b) $k = 320$

Figure 7.7. Total field evaluated on the circle of Figure 7.6.

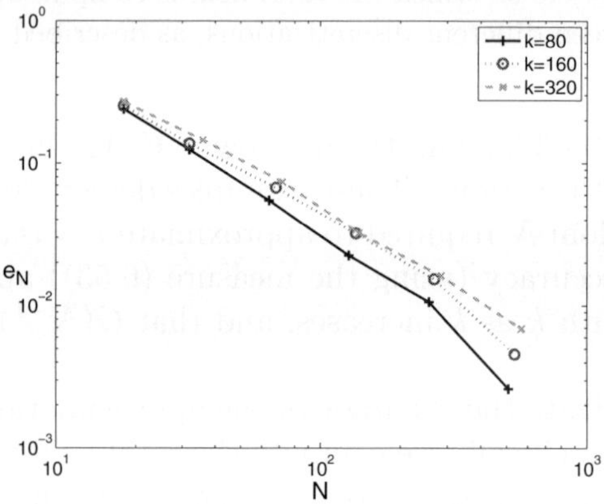

Figure 7.8. Relative maximum errors on the circle of Figure 7.6.

only mildly as k increases. In Table 7.5 (overleaf) we show the gradient of the best fitting (least-squares) linear approximations to the curves, for each fixed value of k; as piecewise constant basis functions have been used, we might expect these gradients to be approximately equal to one. The values vary for different values of k, but the average convergence rate is close to $O(N^{-1})$ in each case.

For this example we also compute an approximation to the far-field pattern F of u^S on the unit circle $\mathbb{S}^1$ (using the formula (2.23), with ϕ replaced by v_N, as discussed in Section 6.5), and the estimated orders of convergence as $N \to \infty$ for various fixed k are given in Table 7.5 (and see also Figure 7.10). With $F(t)$ denoting the far-field pattern in the direction $(\cos t, \sin t)$, plots of $F(t)$, $t \in [-\pi, \pi]$ for $k = 20$ and $k = 320$ (again computed with $N = 1792$ and $N = 2264$ respectively) are shown in

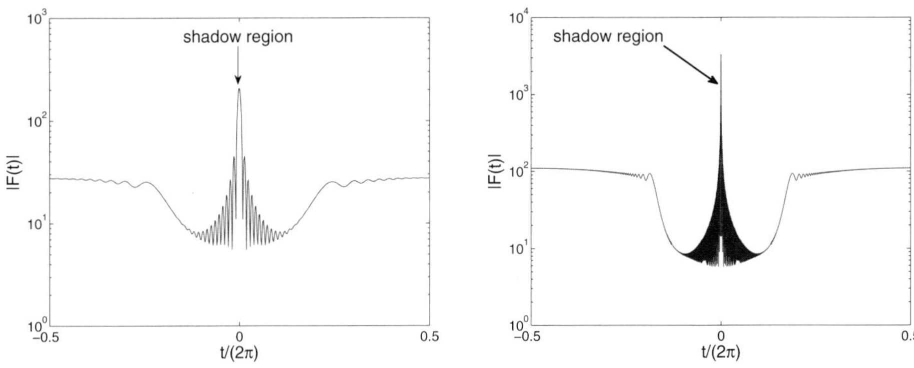

(a) $k = 20$ (b) $k = 320$

Figure 7.9. Far-field patterns, scattering by a curvilinear polygon.

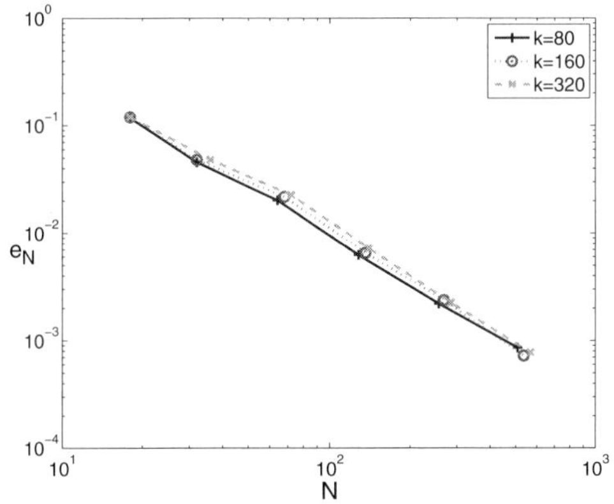

Figure 7.10. Relative absolute maximum errors in the far-field pattern, scattering by a curvilinear polygon.

Figure 7.9, again with the shadow region indicated. In Figure 7.10 we plot approximations to (recall (6.59))

$$e_N = \frac{\|F - F_N\|_{L^\infty(\mathbb{S}^1)}}{\|F\|_{L^\infty(\mathbb{S}^1)}},$$

for various k and N. To approximate the uniform norms, we computed F_N at $30k$ evenly spaced nodes on the unit circle. Again, the algebraic decay with respect to N for fixed k is clear (see Table 7.5), and the error for fixed N remains roughly constant as k increases. We have no convergence theory for this method, but the numerical results indicate that it works well. In particular with $k = 320$ the scatterer contains 640 wavelengths but about 100 degrees of freedom are sufficient to obtain 1% accuracy for the far-field pattern.

Table 7.5. Approximate orders of convergence (with respect to N for fixed k) for the computation of the total field in the domain (Figure 7.8) and the far-field pattern (Figure 7.10).

Figure	$k = 80$	$k = 160$	$k = 320$
Figure 7.8	-1.5	-1.2	-1.1
Figure 7.10	-1.7	-1.5	-1.5

7.4. Impedance convex polygon

Here we present results for scattering by a convex polygon with impedance boundary conditions. The total field u is required to satisfy (2.13) with $h = 0$ and $\beta = 1$, so that the scattered field u^S satisfies (2.18) with h given by (2.17). We use the hybrid method described in Section 3.3.3 (with $p = 0$, that is, piecewise constant approximation to $V_m^{\pm}$) and, as in Section 7.2, we consider scattering by a square of side length 2π, considering incidence fields in both grazing and non-grazing directions. The problem is solved using the BIE (2.111), with unknown $v = \gamma u$. In fact this integral equation is not uniquely solvable for values of k that correspond to irregular frequencies (recall Remark 2.45). All of the examples in this subsection are computed for such values of k, and, as we shall see, the loss of uniqueness at the continuous level does not appear to lead to any problems at the discrete level for the hybrid method. (This is in contrast to conventional boundary element solvers, which are known to exhibit poor accuracy near irregular frequencies.) A comparison of our numerical results with the high-frequency asymptotics for the impedance wedge problem with grazing incidence (contained in Osipov, Hongo and Kobayashi 2002) reveals that our hybrid method provides an accurate solution (see Figure 7.13(b)).

In Figure 7.11(a) we plot the absolute errors $\|v - v_N\|$ (where $v = \gamma u$ is the Dirichlet data of the total field on Γ) for the case of non-grazing incidence, illustrating the convergence with respect to N (the total number of degrees of freedom) for various fixed k. The 'exact' solution v was computed using a large value of N (details are provided in Chandler-Wilde *et al.* 2012*b*). For fixed k, the convergence appears to be $O(N^{-1})$, and for fixed N it appears to be $O(k^{-1/2})$, both of which are consistent with the estimate (3.45) (assuming $M(u) = O(1)$) in the case $p = 0$ (recall piecewise constant approximation to $V_m^{\pm}$ is being used here). In Figure 7.11(b) we also plot the 2-norm condition number of the Galerkin matrix arising from the hybrid method for each value of k and N, showing that the conditioning improves as k increases (for fixed N). Further results and discussion, including discussion of the apparently

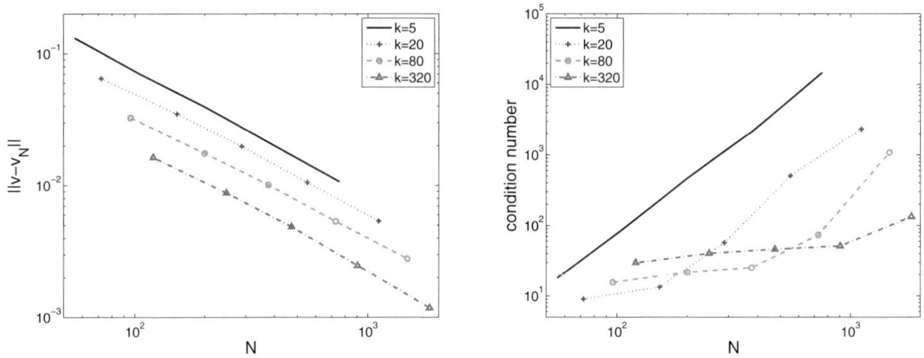

(a) errors (b) condition numbers

Figure 7.11. Scattering by a square with impedance boundary conditions: non-grazing example.

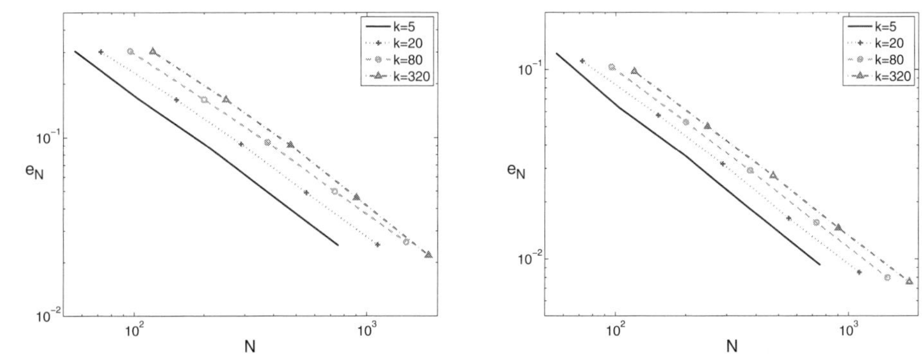

(a) non-grazing incidence (b) grazing incidence

Figure 7.12. Scattering by a square with impedance boundary conditions: relative errors e_N, defined by (7.4).

advantageous aspects of hybrid methods when solving non-uniquely solvable BIEs, are given in Chandler-Wilde *et al.* (2012*b*).

As for the sound-soft problem described in Section 7.2, we again have an exact representation for the leading-order part of the solution V_0 in (3.35) (for details see Chandler-Wilde *et al.* 2012*b*, §2). In Figure 7.12 we show the relative errors, defined (as in (6.54)) by

$$e_N = \frac{\|v - v_N\|}{\|v - V_0\|}, \tag{7.4}$$

for various k and for the cases of incident fields with both grazing and non-grazing angles of incidence. A comparison of the relative errors for the two examples reveals that the results are similar, with the approximation to the

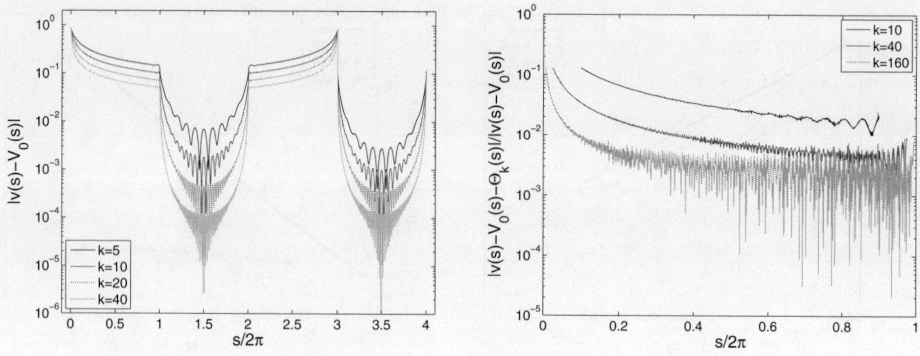

(a) (b)

Figure 7.13. Grazing incidence example: behaviour of solution and comparison with asymptotically derived formula. (a) Boundary solution: $k = 5$ (top), $k = 10$ (second from top), $k = 20$ (second from bottom), $k = 40$ (bottom). (b) Relative difference between boundary solution and asymptotic behaviour for scattering by wedge: $k = 10$ (top), $k = 40$ (middle), $k = 160$ (bottom).

grazing solution, as measured by e_N, slightly more accurate, reflecting the fact that $\|v - V_0\|$ is larger in the grazing incidence case.

Finally, we validate our computations via comparison with an asymptotic approximation, valid in the case of an infinite right-angled wedge with impedance boundary conditions, as derived by Osipov, Hongo and Kobayashi (2002). Specifically, their results predict that, for grazing incidence and when $\beta = 1$, on the edge that runs parallel to the incident direction (with t here representing distance from the corner),

$$v(t) - V_0(t) \sim \frac{\sqrt{2}}{\sqrt{\pi k t}} \exp(\mathrm{i}(kt + \pi/4)) =: \Theta_k(t), \quad \text{as } kt \to \infty. \tag{7.5}$$

In Figure 7.13(a) we show $|v - V_0|$ as a function of arc-length s on the whole boundary, with the sides parallel to the incident direction corresponding to $s \in [0, 2\pi]$ and $s \in [4\pi, 6\pi]$. In Figure 7.13(b) we plot

$$\frac{|v(s) - V_0(s) - \Theta_k(s)|}{|v(s) - V_0(s)|}, \tag{7.6}$$

on the interval $s \in [\lambda, 2\pi - \lambda]$, where (recall Section 1) $\lambda = 2\pi/k$ is the wavelength (*i.e.*, we exclude the region in the immediate vicinity of the corner from this comparison). As k increases, the ratio (7.6) decreases rapidly; for $k = 160$, it is of the order of 1% along almost the entire length of the side of the polygon. Chandler-Wilde *et al.* (2012*b*, Figure 4) reveal that, even for $k = 40$, the numerical and asymptotic solutions are almost indistinguishable except in the immediate vicinity of the corner. These results provide strong support for the claim that the hybrid numerical-asymptotic approximation space has picked out the 'correct' solution to the integral equation, even in the absence of invertibility at the continuous level.

7.5. Non-convex polygon

For the case of a sound-soft non-convex polygon, we present numerical results for the solution of the BIE (2.114), with $\eta = k$, solved using the *hp* scheme of Section 3.4. Rigorous error estimates for this method are presented in Theorem 6.15 only in the context of the star-combined formulation. However, as in Section 7.2, the numerical results below indicate that solving the standard combined potential equation (2.114) yields results consistent with those proved in Theorem 6.15. Specifically, we see exponential decay of the error on the boundary as the polynomial degree p increases, with only a very mild dependence on the wavenumber k, consistent with (and in fact rather better than, in terms of dependence on k) the best approximation error estimate given in Theorem 3.13.

The scatterer is depicted in Figure 7.14, where we show the incident direction and total field for $k = 5$ (a circle, on which we compute the total field below, is also plotted in the figure). Full details of this example are provided in Chandler-Wilde *et al.* (2012*a*). Note that for this example, one (convex) side is entirely illuminated, two sides (one convex and one non-convex) are entirely in shadow, and one (non-convex) side is partially illuminated; the upper section of the vertical side in the figure is illuminated, whilst the lower section is shadowed by the adjacent non-convex (horizontal) side. This type of scenario involving partial illumination of a side is not possible for convex polygonal scatterers, and is one of the reasons for the more complicated asymptotics in this case (recall Figure 3.3).

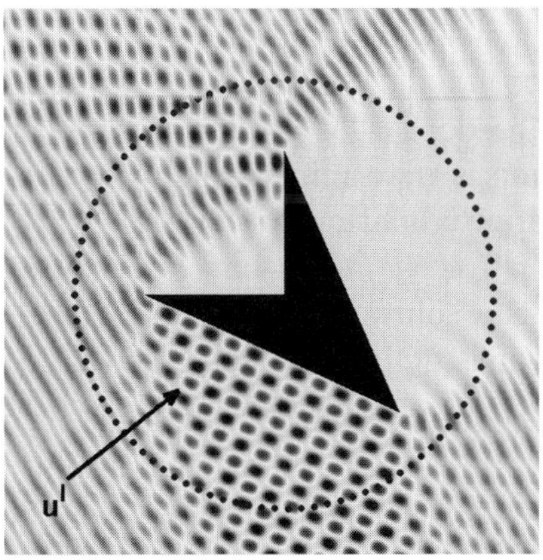

Figure 7.14. Total field, scattering by a non-convex polygon.

Table 7.6. Non-convex polygon: absolute and relative errors, $k = 5, 160$.

	$k=5$					$k=160$			
p	N	$\|v-v_p\|/k$	e_p	e_p/e_{p-1}	p	N	$\|v-v_p\|/k$	e_p	e_p/e_{p-1}
1	56	1.7287	0.8108	–	1	56	0.3336	0.9280	–
2	120	1.3905	0.6522	0.80	2	120	0.2720	0.7564	0.82
3	208	1.1102	0.5207	0.80	3	208	0.2168	0.6029	0.80
4	320	0.8642	0.4053	0.78	4	320	0.1685	0.4686	0.78
5	456	0.6340	0.2974	0.73	5	456	0.1235	0.3434	0.73

Table 7.7. Non-convex polygon: absolute and relative errors (L^2 and L^1), $p = 4$.

		$p=4$			
k	N	$N/(L/\lambda)$	$\|v-v_p\|/k$	e_p	e_p^1
5	320	10.7	0.864	0.405	0.012
10	320	5.3	0.630	0.418	0.016
20	320	2.7	0.432	0.427	0.018
40	320	1.3	0.315	0.440	0.018
80	320	0.7	0.230	0.454	0.019
160	320	0.3	0.169	0.469	0.019

As in Section 7.2 we again take the same degree p of polynomial approximation on each element, and the same number of layers on each mesh (chosen proportional to p). For all examples in this subsection, the total number of degrees of freedom is $N = 12p^2 + 28p + 16$, and since the order of the numerical scheme is determined solely by the polynomial degree p, we again denote our approximation of $v = \partial u/\partial n$ by v_p. Also, as in both Section 7.2 and Section 7.4, we again have an exact representation for the leading-order part of the solution V_0 in (3.49) (see Theorem 3.12). Our measure of relative error is defined in this case (as in (6.54) and (7.4)) by

$$e_p = \frac{\|v - v_p\|}{\|v - V_0\|}, \tag{7.7}$$

where the 'exact' reference solution v is computed with $p = 7$ (and hence $N = 800$) in each case.

In Table 7.6 we show the scaled absolute L^2 errors $\|v - v_p\|/k$, and the relative L^2 errors e_p defined as in (7.7), for $k = 5$ and $k = 160$ and for $p = 1, \ldots, 5$. The last column shows the ratio e_p/e_{p-1}, and the roughly constant value of this ratio demonstrates exponential convergence with respect to p for fixed k. Moreover, for fixed p, the scaled absolute errors decrease as k increases (we return to this point below) and the relative errors remain more or less bounded.

Further, in Table 7.7 we fix $p = 4$ and show the scaled absolute and relative errors for a range of values of k. In this table, $N/(L/\lambda)$ (where L is the length of the boundary) is the number of degrees of freedom per wavelength (which, for conventional boundary element solvers would typically need to be of the order of 10 or so for 'engineering accuracy'). In the final column of Table 7.7 we also show relative $L^1(\Gamma)$ errors, defined by

$$e_p^1 = \frac{\|v - v_p\|_{L^1(\Gamma)}}{\|v - V_0\|_{L^1(\Gamma)}}. \tag{7.8}$$

Our interest in these stems from the fact that the error in the far-field pattern and in solutions in the interior of Ω_+ can be estimated in terms of the error in $L^1(\Gamma)$ (see Remark 6.16). Hence it is interesting to note that in Table 7.7, the relative $L^1(\Gamma)$ error is (over 20 times) smaller in magnitude than the relative $L^2(\Gamma)$ error. (Recall that $\|\cdot\|$ always denotes the L^2 norm on Γ.) The theory in Chandler-Wilde *et al.* (2012*a*) and Hewett *et al.* (2012) only proves $L^2(\Gamma)$ error bounds (and indeed these deteriorate in the polygonal case if an interior corner angle at any corner decreases to zero), and it remains an open question to explain theoretically the observation that the far-field pattern, for example, remains well approximated, even in the case of such extreme angles. (This would follow, as noted in Remark 6.16, if error bounds in $L^1(\Gamma)$ could be established, uniform in the corner angles.)

In Table 7.8, we show the relative error of the solution in the domain,

$$\frac{\|u - u_p\|_{L^2(G)}}{\|u\|_{L^2(G)}} \tag{7.9}$$

(recall (6.55)), where G represents the circle plotted in Figure 7.14 (and $L^2(G)$ is the L^2 norm on the 1D manifold G), for $k = 5$ and $k = 160$, and for $p = 0, \ldots, 5$ (recall that the 'exact' reference solution is that computed with $p = 7$; the L^2 norms in (7.9) are computed using the trapezoidal rule with 30 000 nodes). For fixed k, the relative error appears to decrease exponentially with respect to p, and the errors for fixed p are comparable in magnitude for the two values of k. Moreover, these errors are much smaller than the $L^2(\Gamma)$ errors given in Tables 7.6 and 7.7, and are much closer in size to the relative $L^1(\Gamma)$ errors presented above.

Table 7.8. Relative L^2 error (given by (7.9)) for the solution computed on the circle shown in Figure 7.14.

p	N	$k=5$	$k=160$
0	16	4.7×10^{-2}	1.3×10^{-2}
1	56	3.4×10^{-3}	4.3×10^{-3}
2	120	9.2×10^{-4}	2.6×10^{-4}
3	208	3.0×10^{-4}	9.4×10^{-5}
4	320	1.2×10^{-4}	4.0×10^{-5}
5	456	3.1×10^{-5}	1.6×10^{-5}

Table 7.9. Relative L^2 error for the computation of the far-field pattern, non-convex polygon.

p	Error, $k=5$	Error, $k=160$
0	6.7×10^{-2}	1.9×10^{-2}
1	5.0×10^{-3}	6.1×10^{-3}
2	1.4×10^{-3}	3.6×10^{-4}
3	4.6×10^{-4}	1.3×10^{-4}
4	1.8×10^{-4}	5.6×10^{-5}
5	4.8×10^{-5}	2.2×10^{-5}

In Table 7.9, we show the relative L^2 error in the approximation of the far-field patterns (recall (6.59)),

$$\frac{\|F - F_N\|_{L^2(\mathbb{S}^1)}}{\|F\|_{L^2(\mathbb{S}^1)}}, \tag{7.10}$$

again for $k = 5$ and $k = 160$ (computed, as above, using the trapezoidal rule with 30 000 nodes). The errors are comparable in magnitude and behaviour with those found in Table 7.8 for the domain solution.

The results of Tables 7.6–7.9 suggest that, for fixed p, the accuracy of our approximation to the solution on the boundary (measured by (6.53)), the solution in the domain (measured by (7.9)) and the far-field pattern (measured by (7.10)) all improve as k increases. We conclude this subsection with a closer look at how the results in Table 7.7 compare with the best approximation error estimate given in Theorem 3.13.

To investigate this, we make the hypothesis that $\mathrm{error}(k) := \|v - v_p\|/k \sim k^{\mu}$ as $k\to\infty$, where $\mathrm{error}(k)$ refers to the scaled absolute L^2 error for a particular value of k, and μ denotes the estimated order of growth ($\mu > 0$) or decay ($\mu < 0$). Under this hypothesis, we estimate μ by calculating the gradient of the best fitting least-squares linear approximation to $\log \mathrm{error}(k)$ as a function of $\log k$. The results in Table 7.7 give an estimate $\mu \approx -0.47$.

Now Theorem 3.13 predicts that, for fixed p, the best approximation satisfies

$$\inf_{w_N\in\mathcal{V}_N} \|v - w_N\|/k \lesssim M(u)k^{\alpha-1/2},$$

where $\alpha = 1 - \min_{m=1,\dots,n_s}(1-\pi/\omega_m) \in (1/2, 1)$, with ω_m the external angles at the corners of the polygon. For the geometry of the example considered here, $\min_{m=1,\dots,n_s}(1-\pi/\omega_m) \approx 0.4638$, and hence we expect

$$\inf_{w_N\in\mathcal{V}_N} \|v - w_N\|/k \lesssim M(u)k^{1/2-0.4638}. \tag{7.11}$$

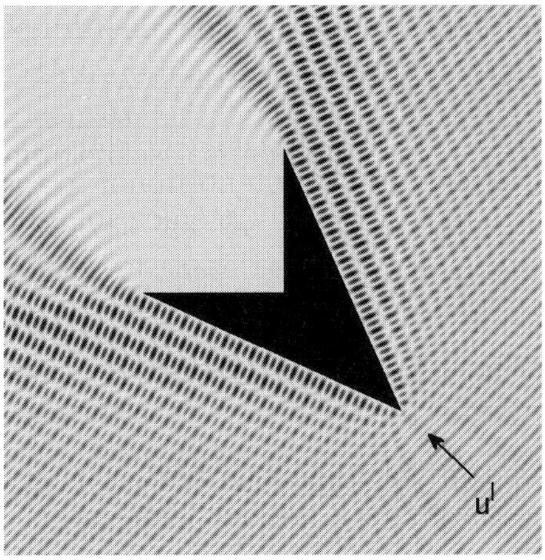

Figure 7.15. Total field, scattering by a non-convex polygon, non-convex sides in shadow.

As discussed in Section 3.3.1, the dependence of the constant $M(u)$ on the wavenumber k is not yet fully understood. If (3.38) were sharp, we would expect $\inf_{w_N \in \mathcal{V}_N} \|v - w_N\|/k \lesssim k^{0.5362} \log^{1/2} k$. However, the results in Table 7.7 suggest that the scaled absolute error $\|v - v_p\|/k$ is decaying, not growing, as $k \to \infty$, and moreover our estimate $\mu \approx -0.47$ tallies remarkably well with the best approximation estimate (7.11) under the assumptions that $M(u) = O(1)$, as $k \to \infty$, and further that the extra factor of $k^{1/2}$ in our best approximation estimate (3.50) for non-convex polygons, compared to our best approximation estimate (3.48) for convex polygons, is spurious.

To investigate this further, we conclude this subsection with a second example on the same geometry but for which the two non-convex sides are both in shadow. The incident direction and total field for $k = 10$ are shown for this example in Figure 7.15. In Figure 7.16 we plot $|(v - V_0)(s)|/k$ for $k = 10$ and $k = 160$. The corner between the two non-convex sides is at $s/(2\pi) = 1$, the corners between convex and non-convex sides are at $s/(2\pi) = 0$ (equivalently $s/(2\pi) = 6$) and $s/(2\pi) = 2$, and the corner between the two convex sides is at $s/(2\pi) = 4$. The singularity in the boundary solution at all corners except for the one between the non-convex sides can be seen in each case (compare with Figure 7.4), as can the increased oscillations for larger k (where there is a shaded region, that represents high oscillation; note also that we are plotting $|(v - V_0)(s)|/k$, with both $\mathrm{Re}\,(v - V_0)$ and $\mathrm{Im}\,(v - V_0)$ oscillating very rapidly). The exact solution is zero at the corner between the non-convex sides, and the apparent spike there is an artefact of the logarithmic scale.

In Table 7.10 we show the scaled absolute and relative errors for this example, computed with $p = 4$ as for the first example above, again for a

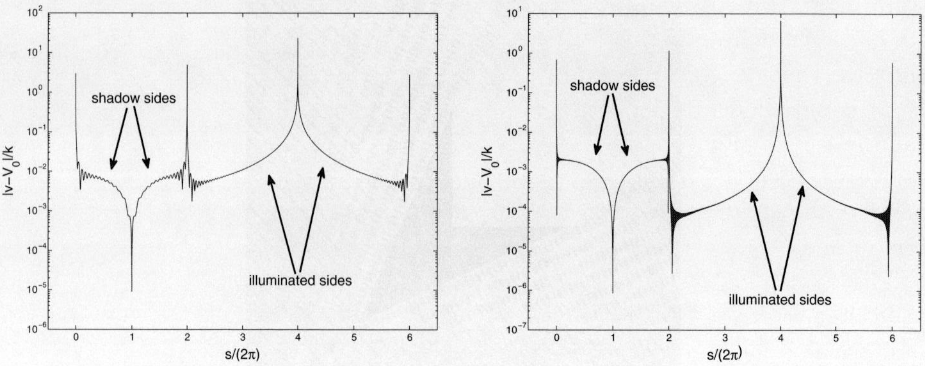

(a) $k = 10$ (b) $k = 160$

Figure 7.16. Scattering by a non-convex polygon, with non-convex sides in shadow.

Table 7.10. Non-convex polygon: scaled absolute and relative errors (L^2 and L^1), $p = 4$; example with non-convex sides in shadow.

	$p = 4$				
k	N	N/λ	$\Vert v - v_p\Vert/k$	e_p	e_p^1
5	320	10.7	0.461	0.279	0.011
10	320	5.3	0.338	0.290	0.013
20	320	2.7	0.249	0.302	0.015
40	320	1.3	0.184	0.316	0.016
80	320	0.7	0.135	0.329	0.017
160	320	0.3	0.100	0.344	0.016

range of values of k. In terms of k-dependence, the results in Table 7.10 closely mimic those in Table 7.7. Making the same hypothesis as above that $\text{error}(k) := \Vert v - v_p\Vert/k \sim k^{\mu}$ as $k \to \infty$, we obtain the estimate $\mu \approx -0.44$. This provides further support for the suggestions that $M(u) = O(1)$ as $k \to \infty$ and that the factor of $k^{1/2}$ on the right-hand side of (3.50) is spurious, as outlined above. These conclusions are further supported by additional numerical results in Chandler-Wilde *et al.* (2012*a*), including an example for which re-reflections occur (as in Figure 3.3(a)).

7.6. Three-dimensional screen

We thank D. P. Hewett for the numerical results in this subsection. Here we consider the problem of scattering of an incident plane wave u^I by a sound-soft three-dimensional screen, $\Gamma = \{(x_1, x_2, 0) : (x_1, x_2) \in (0,1) \times (0,1)\}$. We denote the exterior domain by $\Omega_+ = \mathbb{R}^3 \backslash \overline{\Gamma}$. The precise BVP that we wish to solve is as follows: find $u \in C^2(\Omega_+) \cap H^1_{\mathrm{loc}}(\Omega_+)$ such that (1.1) is satisfied in Ω_+, $u = -u^I$ on Γ, where $u^I(x) = \exp(\mathrm{i}kx \cdot \hat{a})$, and u satisfies the Sommerfeld radiation condition (2.9) with $d = 3$. The precise sense in which the boundary condition holds needs to be expressed in terms of carefully defined trace operators (see Chandler-Wilde and Hewett 2012 for details).

The equivalent 2D problem has been studied by Stephan and Wendland (1984), and more recently by Davis and Chew (2008), who used hybrid numerical-asymptotic methods. Our goal in this subsection is to show how an ansatz of the form (3.1) can be used for this challenging 3D problem to represent the key oscillatory components of the solution. (Some key ingredients for a k-explicit error analysis of this problem are discussed briefly in Remark 5.11.)

The BVP specified above is equivalent to the BIE (see, *e.g.*, Sauter and Schwab 2011, Stephan 1987)

$$u^I(x) = \int_\Gamma \Phi_k(x, y)[\partial_n u](y)\,\mathrm{d}s(y), \quad x \in \Gamma,$$

where $[\partial_n u]$ denotes the jump in $\partial u / \partial n$ across Γ. For the equivalent 2D problem (see Davis and Chew 2008), an appropriate approximation space is that used on each side of a sound-soft convex polygon in the hybrid schemes described in Sections 3.3.1 and 3.3.4 (see also Section 7.2 above). In this case, the correction to the leading-order behaviour (*i.e.*, $v - V_0$ in (3.1)) consists of two 'diffracted' waves, each travelling along the screen from one end to the other, and being rediffracted back from either end *ad infinitum*. For the 3D problem, however, the construction of a hybrid approximation space is rather more difficult. Here, although we can again separate the leading-order behaviour, writing

$$[\partial_n u](x) = 2[\partial_n u^I](x) + \sum_{m=1}^{M} V_m(x,k)\exp(\mathrm{i}k\psi_m(x)), \quad x \in \Gamma, \tag{7.12}$$

it turns out to be much more difficult to identify M and the phase functions $\psi_m(x)$, $m = 1, \ldots, M$, such that the corresponding amplitudes $V_m(x,k)$ are relatively non-oscillatory. The reason for this is that 'edge waves' and 'corner waves', diffracted by the edges and corners of the screen respectively, are rediffracted infinitely often by the other edges and corners of the screen, taking a different direction of travel after each rediffraction. Although the waves lose energy on each rediffraction, the difficulties in representing all phases of this series of rediffractions in such a way that the amplitudes are (completely) non-oscillatory is significantly harder than in 2D.

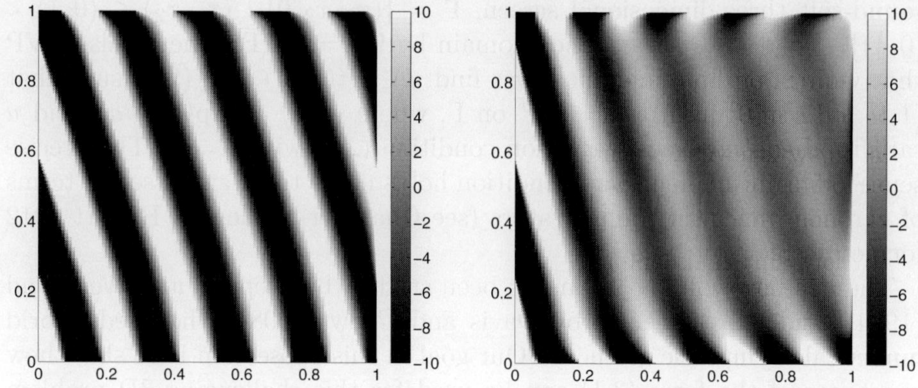

(a) Re $[\partial_n u]$ (b) Re $[\partial_n u] - 2$Re $[\partial u^I/\partial n]$

Figure 7.17. Solution Re $[\partial_n u]$ to (7.12), computed with standard BEM, for $\lambda = 0.2$ and $\hat{a} = (3,1,1)/\sqrt{11}$.

To illustrate this, in Figure 7.17(a) we plot the real part of $[\partial_n u]$ (away from the edges of the screen), calculated using a standard BEM, with piecewise constant basis functions on a rectangular mesh with at least ten grid points per wavelength, graded towards the edges. The solution is singular at the edges, hence we exclude a narrow region close to the edge of the screen from our plot, so that the detail elsewhere can be more clearly seen. For this example $\hat{a} = (3,1,1)/\sqrt{11}$, and $\lambda = 0.2$ (and hence $k = 2\pi/\lambda = 10\pi$); the screen is thus five wavelengths long in each direction, and our standard boundary element space has more than 2500 degrees of freedom. In Figure 7.17(b) we plot the same quantity, but with the leading-order behaviour $2[\partial_n u^I]$ subtracted off. This remainder is what we seek to approximate numerically, and the question is how to choose functions ψ_m in (7.12) such that amplitude functions $V_m(x,k)$ are slowly oscillating. Figure 7.17(b) makes clear that this will not be a simple task.

We now demonstrate that, with a judicious choice of ψ_m in (7.12), the solution $v := [\partial_n u]$ can be represented to a high degree of accuracy away from the edges of the screen using only a small M. To investigate this, we consider a similar example to that plotted in Figure 7.17, but now with $\lambda = 0.05$, so $k = 40\pi$, in which case the screen occupies twenty wavelengths in each direction. We compute the solution v_h to this problem by piecewise constant BEM on a fine mesh with at least ten grid points per wavelength, graded towards the boundary. Using v_h as an 'exact' solution, we consider how to achieve a good approximation to v_h in the interior of Γ by an ansatz of the form (7.12). Our choice of M and ψ_m, $m = 1, \ldots, M$, is discussed below. Moreover, $V_m(x,k)$ are chosen to be constant on each element of a coarse mesh (consisting, for this example, of sixteen squares, arranged in

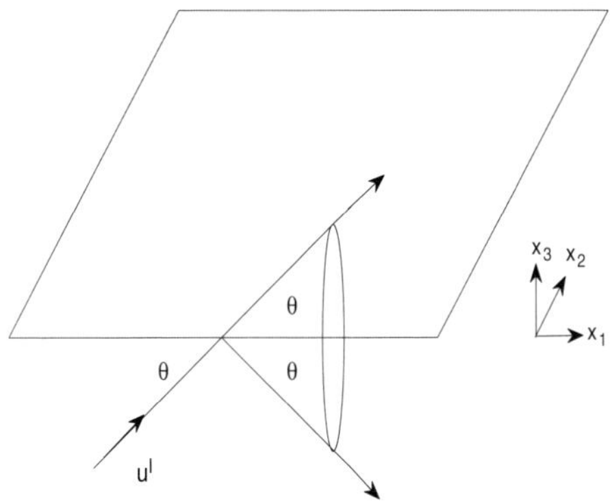

Figure 7.18. Incident wave u^I, diffracted by an edge of the screen; the angle the incident wave makes with the edge is θ, and this is the angle that the waves described in (A) and (E), which travel across the surface of the screen, make with the edge.

a four-by-four grid, each occupying approximately five wavelengths in each direction). More precisely, on each element of the coarse mesh we compute constants ν_m so that

$$v_H := 2[\partial_n u^I](x) + \sum_{m=1}^{M} \nu_m \exp(ik\psi_m(x)) \tag{7.13}$$

is the best approximation to v_h with respect to a discrete L^2 norm on the fine mesh over all choices of constants $\{\nu_m\}_{m=1}^M$. Although this is not a numerical method for the 3D screen problem, it demonstrates the potential of hybrid methods for 3D problems like this.

In Table 7.11 we investigate the following possible ways of choosing ψ_m (some of which can be sensibly used in combination with each other).

(A) Plane waves propagating in the direction of the singly edge-diffracted rays predicted by the Geometrical Theory of Diffraction (*e.g.*, Keller 1962, Borovikov and Kinber 1994); see Figure 7.18. There is one such wave associated with each of the four edges.

(B) Radial waves (emanating from the corners), where $\psi_m(x) = |x - x_m|$, with x_m, $m = 1, \ldots, 4$, the four corners of the screen.

(C) Rediffractions of the plane waves described in (A). These represent the first 'reflection' of each plane wave in the first edge it hits as it travels across the surface of the screen.

(D) Uniformly distributed plane waves; in each case a uniform distribution of directions was used, but the results shown in Table 7.11 represent the average over a large number of experiments, with in each case the evenly spaced directions being rotated by the same random perturbation, different in each experiment.

Table 7.11. Comparison between various approximation schemes, for scattering by a 3D screen of size twenty wavelengths in each direction, using only sixteen coarse elements (away from the edges).

Approximation space	M	e_2	e_∞
(D)	4	15%	28%
(D)	16	13%	23%
(A)	4	8.0%	16%
(A) and (C)	8	7.9%	16%
(E)	4	1.4%	5.0%
(E) and (B)	8	1.2%	4.4%

In addition we also investigated a variation on the above, which we label (E) in Table 7.11, where we tried the same plane waves described in (A), but with a slightly more sophisticated approximation to the singly diffracted component (incorporating decay and the edge singularity), so that instead of approximating $V_m(x,k)$ on each element of the coarse mesh by a constant, we instead approximate $V_m(x,k)$ by a constant multiplied by $1/\sqrt{kR}$, where R represents the distance from the edge in the direction of propagation. These 'Sommerfeld waves' represent the exact solution for scattering by an infinite half-plane (again, see, *e.g.*, Keller 1962, Borovikov and Kinber 1994), and have singularities at the edges, which are not apparent here as we have excluded in the optimization procedure that part of the screen that lies within a wavelength of the edges.

For any suitable $D \subset \Gamma$, we denote by $\|f\|_{2,h,D}$ the approximation to $\|f\|_{L^2(D)}$ obtained by replacing f by its piecewise constant interpolant on the fine mesh at the element centroids. Denoting the elements of the coarse mesh by Γ_m, $m = 1,\ldots,16$ (for this example), and letting $\tilde{\Gamma} := \cup_{m=1}^{16}\Gamma_m$, in Table 7.11 we present values of

$$e_2 := \frac{\|v_h - v_H\|_{2,h,\tilde{\Gamma}}}{\|v_h\|_{2,h,\tilde{\Gamma}}} \quad \text{and} \quad e_\infty := \max_{m=1,\ldots,16} \frac{\|v_h - v_H\|_{2,h,\Gamma_m}}{\|v_h\|_{2,h,\Gamma_m}}.$$

The column M in Table 7.11 refers to the number of terms in the summation on the right-hand side of (7.13). Note that the first row of the table indicates a relative least-squares error of 15% and a relative maximum error of 28%, and this turns out to be no better than taking $M = 0$ in (7.13). However, we see that the edge plane waves (A) and particularly the

Sommerfeld waves (E) are far more effective. Thus in this case, while standard BEM approximation of $[\partial_n u]$ would require approximately 2500 degrees of freedom in each coarse element in order to achieve 'engineering accuracy' (approximately 1%), applying the ansatz (7.12) with a judicious choice of Sommerfeld waves (E) can lead to an overall error of approximately 1.4% with only four degrees of freedom per coarse element, so 64 degrees of freedom overall, a very significant reduction indeed.

Appendix: Function spaces on Lipschitz domains

Large sections of this review, namely those where we give details of algorithms and show numerical results, can be understood for the most part without any ideas from functional analysis or many notions of function spaces. But both of these play an essential role in the mathematical analysis of the algorithms that we propose. In this appendix we collect some technical details concerning function spaces on Lipschitz domains. There is an emphasis on Sobolev spaces, for which our notations and definitions follow McLean (2000), but we also provide what appears to be the only account in the boundary element literature of spaces defined in terms of maximal functions, and their relationship to Sobolev spaces. These results are relevant to Sections 2 and 5.

Large parts of Sections 2, 5, and 6 use standard notions and results of linear functional analysis (one relevant reference is Chapter 2 of McLean (2000)). In these sections we assume that the reader is familiar with the idea of a Banach space, a linear space equipped with a norm $\|\cdot\|$, which is complete with respect to convergence in this norm, and with the basic theory of bounded and compact linear operators on Banach spaces. If $A : X \to Y$ is a bounded linear operator from a Banach space X to a Banach space Y we will say that A is invertible if A is bijective: by the Banach theorem this is the case if and only if the inverse operator A^{-1} is bounded. We will sometimes classify $A : X \to Y$ as Fredholm, and talk about the index of a Fredholm operator (see, *e.g.*, McLean 2000). The main facts from this theory that we will need are that A is Fredholm of index zero if and only if A is the sum of an invertible and a compact operator, and that if A is Fredholm of index zero then it is injective if and only if it is surjective.

We shall use the notion of a dual space. Following Kato (1995), for example, we shall define the dual space of a Banach space V to be the set of bounded *anti-linear* functionals on V. We will assume also that the reader is familiar with the idea of a Hilbert space (a linear space V equipped with an inner product $(\cdot,\cdot)$, which induces a norm on the linear space defined by $\|v\| := (v,v)^{1/2}$, and which is a Banach space with respect to this norm), and with the idea of the adjoint of a bounded linear operator on a Hilbert space and with some of its elementary properties.

A.1. Function spaces

For $D \subset \mathbb{R}^d$ (we shall suppose that $d = 2$ or 3 throughout, though of course most of our definitions work for arbitrary space dimensions), let $C(D)$ denote the set of continuous functions $\phi : D \to \mathbb{C}$, and let $BC(D)$ denote those $\phi \in C(D)$ that are also bounded (of course, $C(D) = BC(D)$ if D is compact). $BC(D)$ is a Banach space with norm $\|\phi\|_{BC(D)} = \sup_{x\in D} |\phi(x)|$. We will frequently use the shorter notation $\|\cdot\|_\infty$ for $\|\cdot\|_{BC(D)}$ when the set D is clear from the context.

In the case that D is measurable (with respect to d-dimensional Lebesgue measure), with non-zero measure, which in particular is the case if D is an open set or its closure $\overline{D}$, we will denote by $L^p(D)$, for $1 \le p < \infty$, the set of functions $\phi : D \to \mathbb{C}$ that are Lebesgue-measurable with $\int_D |\phi|^p \, \mathrm{d}x < \infty$. $L^p(D)$ is a Banach space with the norm $\|\cdot\|_{L^p(D)}$ defined by $\|\phi\|_{L^p(D)} = (\int_D |\phi|^p \, \mathrm{d}x)^{1/p}$, with the understanding that elements of $L^p(D)$ are regarded as equal if they are equal almost everywhere. Of course, equipped with the inner product $(\cdot,\cdot)_{L^2(D)}$, defined by

$$(u, v)_{L^2(D)} := \int_D u\bar{v} \, \mathrm{d}x,$$

which we will frequently abbreviate as $(\cdot,\cdot)$, $L^2(D)$ is a Hilbert space, and $\|u\|_{L^2(D)} = (u, u)^{1/2}_{L^2(D)}$.

Similarly, $L^\infty(D)$ denotes the set of functions $\phi : D \to \mathbb{C}$ that are Lebesgue-measurable and essentially bounded, in which case $\|\phi\|_{L^\infty(D)} := \|\phi\|_\infty := \operatorname{ess\,sup}_{x\in D} |\phi(x)| < \infty$. By *essentially bounded* we mean here that there is some bounded function ψ such that $\phi = \psi$ almost everywhere, and $\|\phi\|_\infty = \operatorname{ess\,sup}_{x\in D} |\phi(x)|$ denotes the infimum of $\sup_{x\in D} |\psi(x)| < \infty$ over all $\psi : D \to \mathbb{C}$ with $\phi = \psi$ almost everywhere. Note that $BC(D)$ is a closed subspace of $L^\infty(D)$ and our two definitions of $\|\cdot\|_\infty$ coincide on $BC(D)$.

For $1 \le p \le \infty$, and a measurable $D \subset \mathbb{R}^d$, $L^p_{\mathrm{loc}}(D)$ denotes the set of $u : D \to \mathbb{C}$ which are locally L^p: this means that, for every bounded measurable $G \subset D$, $u|_G \in L^p(G)$,

Given an open set $D \subset \mathbb{R}^d$ and $u : D \to \mathbb{C}$, we denote the partial derivatives of u (when they exist) by

$$(\partial^\alpha u)(x) := \left(\frac{\partial}{\partial x_1}\right)^{\alpha_1} \cdots \left(\frac{\partial}{\partial x_d}\right)^{\alpha_d} u(x), \tag{A.1}$$

for every *multi-index* $\alpha = (\alpha_1, \ldots, \alpha_d) \in \mathbb{N}_0^d$. (Here $\mathbb{N}_0 = \mathbb{N} \cup \{0\}$.) The *order* of the partial derivative $\partial^\alpha u$ is the number $|\alpha| = \alpha_1 + \cdots + \alpha_d$. For any integer $n \in \mathbb{N}_0$, let

$$C^n(D) := \{u : \partial^\alpha u \text{ exists and is continuous on } D \text{ for } |\alpha| \le n\},$$

let $C^n(\overline{D}) := \{v|_{\overline{D}} : v \in C^n(\mathbb{R}^d)\}$, and let $C^\infty(D) := \cap_{n\in\mathbb{N}_0} C^n(D)$. Complementing this last notation, for any *closed* set $G \subset \mathbb{R}^d$, let

$$C^\infty(G) := \{v|_G : v \in C^\infty(\mathbb{R}^d)\}.$$

Of course, $C^0(D)$ is the same as $C(D)$. For $u \in C(D)$ the *support* of u, $\operatorname{supp} u$, is the closure in D of $\{x \in D : u(x) \neq 0\}$, and

$$C_0^\infty(D) := \{u \in C^\infty(D) : \operatorname{supp} u \text{ is a compact subset of } D\}.$$

As usual, for $u \in C^1(D)$, $\nabla u \in (C(D))^n$ will denote the vector whose jth component is $\frac{\partial u(x)}{\partial x_j}$.

Given a set $D \subset \mathbb{R}^d$ we will say that $u \in C(D)$ is Hölder-continuous with index $\mu \in (0, 1]$ if there exists an $L \geq 0$ such that

$$|u(x) - u(y)| \leq L|x - y|^\mu, \quad x, y \in D. \tag{A.2}$$

Note that, by this definition, every Hölder-continuous function is uniformly continuous. If (A.2) holds with $\mu = 1$ then we say that u is Lipschitz-continuous and will call L a *Lipschitz constant* for u. We let $C^{0,\mu}(D) \subset BC(D)$ denote the set of functions that are Hölder-continuous with index μ and also bounded. This set is a Banach space with norm $\|\cdot\|_{C^{0,\mu}(D)}$ defined by

$$\|u\|_{C^{0,\mu}(D)} := \|u\|_\infty + \sup_{x,y \in D,\, x \neq y} \frac{|u(x) - u(y)|}{|x - y|^\mu}.$$

Given an open set $D \subset \mathbb{R}^d$, $u \in L^1_{\text{loc}}(D)$, and a multi-index $\alpha \in \mathbb{N}_0^d$, we will say that $\partial^\alpha u$ exists in a weak sense if there exists $w \in L^1_{\text{loc}}(D)$ such that

$$\int_D wv \, \mathrm{d}x = (-1)^{|\alpha|} \int_D u \partial^\alpha v \, \mathrm{d}x, \quad v \in C_0^\infty(D),$$

in which case we define $\partial^\alpha u := w$. This notation coincides with (A.1) in the case that $u \in C^n(D)$ for some $n \geq |\alpha|$.

This leads to our first definition of Sobolev spaces. For every non-empty open subset $D \subset \mathbb{R}^d$, $p \in [1, \infty]$, and $n \in \mathbb{N}_0$, we define the Sobolev space $W_p^n(D)$ of order n by

$$W_p^n(D) := \{u \in L^p(D) : \text{the weak derivative } \partial^\alpha u \text{ exists and is in } L^p(D) \text{ for } |\alpha| \leq n\}.$$

$W_p^n(D)$ is a Banach space when equipped with the norm $\|\cdot\|_{W_p^n(D)}$ defined by

$$\|u\|_{W_p^n(D)} := \begin{cases} \left(\sum_{|\alpha| \leq n} \int_D |\partial^\alpha u(x)|^p \, \mathrm{d}x\right)^{1/p} & 1 \leq p < \infty, \\ \max_{|\alpha| \leq n} \operatorname{ess\,sup}_{x \in D} |\partial^\alpha u(x)| & p = \infty. \end{cases}$$

A standard result is that $W_\infty^1(\mathbb{R}^d) = C^{0,1}(\mathbb{R}^d)$, with equivalence of norms.

In fact, for any open set D, if $u \in C^{0,1}(D)$ with Lipschitz constant L, then $\nabla u(x)$ exists for almost every $x \in D$, with $|\nabla u(x)| \leq L$; this is Rademacher's theorem (Rademacher 1919). Further, ∇u is the gradient of u in the sense of weak derivatives, so that $u \in W^1_\infty(D)$.

The following is a second standard definition of Sobolev spaces. For $u \in L^1(\mathbb{R}^d)$, let $\hat{u} = \mathcal{F}u$ denote the Fourier transform of u, defined by

$$\hat{u}(\xi) = (2\pi)^{-d/2} \int_{\mathbb{R}^d} \exp(-\mathrm{i}x \cdot \xi) u(x)\, \mathrm{d}x, \quad \xi \in \mathbb{R}^d. \tag{A.3}$$

For $u, v \in C_0^\infty(\mathbb{R}^d)$ and $s \in \mathbb{R}$ define the inner product $(u, v)_{H^s(\mathbb{R}^d)}$ by

$$(u, v)_{H^s(\mathbb{R}^d)} := \int_{\mathbb{R}^d} (1 + |\xi|^2)^s \hat{u}(\xi) \bar{\hat{v}}(\xi)\, \mathrm{d}\xi.$$

Equipped with $(\cdot, \cdot)_{H^s(\mathbb{R}^d)}$, $C_0^\infty(\mathbb{R}^d)$ is a pre-Hilbert space. The completion of this space is the Hilbert space $H^s(\mathbb{R}^d)$, with norm

$$\|u\|_{H^s(\mathbb{R}^d)} := (u, u)^{1/2}_{H^s(\mathbb{R}^d)} = \left(\int_{\mathbb{R}^d} (1 + |\xi|^2)^s |\hat{u}(\xi)|^2\, \mathrm{d}\xi \right)^{1/2}.$$

Clearly, $H^s(\mathbb{R}^d) \subset H^t(\mathbb{R}^d)$ for $s > t$. Our choice of normalization in the definition of the Fourier transform is such as to ensure that the extension of $\mathcal{F}$ to $L^2(\mathbb{R}^d)$ is a unitary operator, in particular

$$(\mathcal{F}u, \mathcal{F}v)_{L^2(\mathbb{R}^d)} = (u, v)_{L^2(\mathbb{R}^d)}, \quad u, v \in L^2(\mathbb{R}^d). \tag{A.4}$$

It follows immediately from (A.3) that $\mathcal{F}$ maps $L^1(\mathbb{R}^d)$ to $L^\infty(\mathbb{R}^d)$ and is bounded with norm $\leq (2\pi)^{-d/2}$, and from (A.4) that $\mathcal{F}$ is an isometrice isomorphism on $L^2(\mathbb{R}^d)$. It is a standard application of the Riesz–Thorin interpolation theorem (Stein and Weiss 1971, Chapter V, Theorem 1.3) that $\mathcal{F}$ is a bounded operator from $L^p(\mathbb{R}^d)$ to $L^q(\mathbb{R}^d)$, for $1 \leq p \leq 2$, where $p^{-1} + q^{-1} = 1$, with

$$\|\mathcal{F}\|_{L^q(\mathbb{R}^d) \leftarrow L^p(\mathbb{R}^d)} \leq (2\pi)^{-\theta d}, \tag{A.5}$$

where

$$\theta = \frac{1}{p} - \frac{1}{2} = \frac{1}{2} - \frac{1}{q}. \tag{A.6}$$

Note that (A.4) implies that, for $u, v \in C_0^\infty(\mathbb{R}^d)$,

$$(u, v)_{H^0(\mathbb{R}^d)} = (u, v)_{L^2(\mathbb{R}^d)},$$

so that $H^0(\mathbb{R}^d) = L^2(\mathbb{R}^d) = W_2^0(\mathbb{R}^d)$ and $H^s(\mathbb{R}^d) \subset L^2(\mathbb{R}^d)$ for $s > 0$. Indeed,

$$H^n(\mathbb{R}^d) = W_2^n(\mathbb{R}^d), \quad n \in \mathbb{N}_0,$$

with equivalence of norms (McLean 2000). If s is large enough then we can think of the elements of $H^s(\mathbb{R}^d)$ also as continuous functions, as the following theorem (see, *e.g.*, McLean 2000) shows.

Theorem A.1. (Sobolev embedding theorem) If $0 < \mu < 1$ and $s = d/2 + \mu$, then $H^s(\mathbb{R}^d) \subset C^{0,\mu}(\mathbb{R}^d)$ (in the sense that if $u \in H^s(\mathbb{R}^d)$ then u is equal almost everywhere to some $v \in C^{0,\mu}(\mathbb{R}^d)$). Further, for some $C > 0$ depending only on d and s, $\|u\|_{C^{0,\mu}(\mathbb{R}^d)} \leq C \|u\|_{H^s(\mathbb{R}^d)}$, for all $u \in H^s(\mathbb{R}^d)$.

Another important embedding theorem makes a connection between the spaces $L^p(\mathbb{R}^d)$ and $H^s(\mathbb{R}^d)$. An easy consequence of (A.5), the definition of the Sobolev norm, and Hölder's inequality is that, for $1 \leq p \leq 2$,

$$L^p(\mathbb{R}^d) \subset H^s(\mathbb{R}^d), \quad \text{for } s < -\theta d, \tag{A.7}$$

where θ is given by (A.6). Further, for some constant $C > 0$ depending only on d, s, and p, $\|u\|_{H^s(\mathbb{R}^d)} \leq C \|u\|_{L^p(\mathbb{R}^d)}$, for $s < -\theta d$.

For every $s \in \mathbb{R}$ we can extend the inner product $(\cdot,\cdot)_{L^2(\mathbb{R}^d)}$ in a natural way to a sesquilinear form $\langle \cdot,\cdot \rangle$ on $H^{-s}(\mathbb{R}^d) \times H^s(\mathbb{R}^d)$. If $u \in H^{-s}(\mathbb{R}^d)$, $v \in H^s(\mathbb{R}^d)$, and (u_j) and (v_j) are sequences in $C_0^\infty(\mathbb{R}^d)$, converging to u and v, respectively, with respect to the norms on $H^{-s}(\mathbb{R}^d)$ and $H^s(\mathbb{R}^d)$, then

$$\langle u, v \rangle := \lim_{j\to\infty} (u_j, v_j)_{L^2(\mathbb{R}^d)}.$$

The mapping u^* given by $v \mapsto u^*(v) := \langle u, v \rangle$ is a bounded anti-linear functional on $H^s(\mathbb{R}^d)$, and so an element of the dual space $(H^s(\mathbb{R}^d))^*$ of $H^s(\mathbb{R}^d)$. In fact, as shown in McLean (2000), this mapping $u \mapsto u^*$ from $H^{-s}(\mathbb{R}^d)$ to $(H^s(\mathbb{R}^d))^*$ is an isometric isomorphism, so that $H^{-s}(\mathbb{R}^d)$ is an isometric realization of the dual space of $H^s(\mathbb{R}^d)$.

For every $s \in \mathbb{R}$, $u \in H^s(\mathbb{R}^d)$, and open set $G \subset \mathbb{R}^d$ we say that $u|_G = 0$ if

$$\langle u, v \rangle = 0, \quad \text{for all } v \in C_0^\infty(\mathbb{R}^d) \text{ with } \operatorname{supp} v \subset G.$$

We define the support of $u \in H^s(\mathbb{R}^d)$ to be the largest closed subset G of $\mathbb{R}^d$ such that $u|_{\mathbb{R}^d \setminus G} = 0$.

Equip $C^\infty_{\text{comp}}(\overline{D}) := \{U|_D : U \in C_0^\infty(\mathbb{R}^d)\}$ with a norm $\|\cdot\|_{H^s(D)}$ defined by

$$\|u\|_{H^s(D)} := \inf_{U \in C_0^\infty(\mathbb{R}^d),\, U|_D = u} \|U\|_{H^s(\mathbb{R}^d)}. \tag{A.8}$$

The completion of this normed space is the Sobolev space $H^s(D)$. More succinctly,

$$H^s(D) = \{U|_D : U \in H^s(\mathbb{R}^d)\},$$

where $U|_D$ denotes the restriction of U to D in the sense defined in McLean (2000). (For $s \geq 0$, $U|_D$ is just the restriction of the function U to D in the usual sense, *i.e.*, $u = U|_D$ means $u(x) = U(x)$, $x \in D$.)

A.2. Lipschitz domains

Following McLean (2000, §3), for example, we will say that an open set D is *Lipschitz* if its boundary ∂D is bounded and, roughly speaking, ∂D is everywhere locally the graph of a function which is Lipschitz-continuous.

Definition A.2. We say that the open set D is Lipschitz if ∂D is compact and there exist finite families $\{W_i\}$, $\{M_i\}$, $\{f_i\}$ and $\{D_i\}$ such that:

(i) the family $\{W_i \subset \mathbb{R}^d\}$ is a finite open cover of the boundary ∂D;

(ii) every $f_i : \mathbb{R}^{d-1} \to \mathbb{R}$ is Lipschitz-continuous with Lipschitz constant $M_i \geq 0$;

(iii) the family $\{D_i \subset \mathbb{R}^d\}$ is such that D and D_i have the same intersection with W_i, for each i;

(iii) for each i there exists a rigid motion $r_i : \mathbb{R}^d \to \mathbb{R}^d$ such that $r_i(D_i) = \{(x', x_d) \in \mathbb{R}^{d-1} \times \mathbb{R} : x_d > f_i(x')\}$, the hypograph of the function f_i;

As usual, by a *rigid motion* in (iii) we mean some composition of a rotation and translation. We note that we can, if we wish, without loss of generality, impose that the rigid motion r_i in (iii) be chosen to be a pure rotation.

We note that our Definition A.2 of a Lipschitz open set coincides with the definition in McLean (2000) of a Lipschitz domain. But we will, as is standard, use the word *domain* exclusively to mean a *connected* open set. So, for us (as in Sauter and Schwab 2011, for example), a Lipschitz domain will mean a Lipschitz open set which is connected.

It is easy to see that D is a Lipschitz open set as defined above if and only D is the union of a finite family of Lipschitz domains whose closures are pairwise disjoint. The concept of a Lipschitz open set is slightly more general, and has the advantage over the concept of a Lipschitz domain that D is a Lipschitz open set if and only if $\mathbb{R}^d \setminus \overline{D}$ is a Lipschitz open set, the point being that D can be connected without $\mathbb{R}^d \setminus \overline{D}$ being connected.

Throughout this review we will say that ∂D is Lipschitz if D is a Lipschitz open set. If the conditions in Definition A.2 hold and, additionally, each $f_i \in C^m(\mathbb{R}^{d-1})$, for some $m \in \mathbb{N}$, then we will say that D is a boundary of class C^m (or that ∂D is C^m for short). If the above conditions hold with each $f_i \in C^m$ and, additionally, for some $\mu \in (0, 1]$, all the partial derivatives of f_i of order m are in $C^{0,\mu}(\mathbb{R}^{d-1})$, then we will say that D is of class $C^{m,\mu}$ (or that ∂D is $C^{m,\mu}$, for short). We will say that ∂D is C^∞ if ∂D is C^m for every $m \in \mathbb{N}$.

The following theorem (McLean 2000, Theorem 3.30) connects the Sobolev spaces that we introduced in the previous subsection.

Theorem A.3. If $D \subset \mathbb{R}^d$ is a Lipschitz open set, then $W_2^n(D) = H^n(D)$ for $n \in \mathbb{N}_0$.

A.3. Sobolev spaces on the boundary and trace operators

We will need to work with Sobolev spaces on the boundary of Lipschitz open sets. Let us consider first the case when $f : \mathbb{R}^{d-1} \to \mathbb{R}$ is in $C^{0,1}(\mathbb{R}^{d-1}) = W^{1,\infty}(\mathbb{R}^{d-1})$ and $D = \{(x', x_d) \subset \mathbb{R}^d : x' \in \mathbb{R}^{d-1}, x_d > f(x')\}$ (we say that D is a *Lipschitz hypograph* in this case), and let $\Gamma = \partial D = \{(x', f(x')) : x' \in \mathbb{R}^{d-1}\}$. In the obvious way, we say that $\Gamma_S = \{(x', f(x')) : x' \in S\} \subset \Gamma$ is measurable if $S \subset \mathbb{R}^{d-1}$ is Lebesgue-measurable, and define a surface measure $\mathrm{d}s$ and surface integrals of functions $\phi \in L^1(\Gamma)$ that are Lebesgue-integrable with respect to this surface measure, by

$$\int_\Gamma \phi \, \mathrm{d}s := \int_{\mathbb{R}^{d-1}} \phi((x', f(x')))\sqrt{1 + |\nabla f(x')|^2} \, \mathrm{d}x'. \tag{A.9}$$

In the natural way, $L^p(\Gamma)$ is defined to be the set of those measurable $\phi : \Gamma \to \mathbb{C}$ for which $\int_\Gamma |\phi|^p \, \mathrm{d}s < \infty$, with $\|\phi\|_{L^p(\Gamma)} = (\int_\Gamma |\phi|^p \, \mathrm{d}s)^{1/p}$, for $1 \le p < \infty$. $L^\infty(\Gamma)$ is the set of essentially bounded functions, with $\|\phi\|_{L^\infty(\Gamma)} := \operatorname{ess\,sup}_{x \in \Gamma} |\phi(x)|$. We note that, with respect to the surface measure, for almost every $x \in \Gamma$, there exists a unit normal $n(x)$ to Γ, given by

$$n(x) = \frac{(-\nabla f(x'), 1)}{\sqrt{1 + |\nabla f(x')|^2}}. \tag{A.10}$$

For every $\psi \in L^2(\Gamma)$ define $\psi_f \in L^2(\mathbb{R}^{d-1})$ by

$$\psi_f(x') = \psi((x', f(x'))), \quad x' \in \mathbb{R}^{d-1}.$$

For $u \in C^\infty_{\mathrm{comp}}(\overline{D})$ define $\phi = \gamma u \in C^\infty(\Gamma) \cap L^2(\Gamma)$, the *trace* of u, by $\gamma u = u|_\Gamma$. Note that $\phi_f \in W^1_2(\mathbb{R}^{d-1}) = H^1(\mathbb{R}^{d-1})$ since, for almost all $x' \in \mathbb{R}^{d-1}$, namely all x' for which $\nabla f(x')$ exists, it holds that

$$\nabla \phi_f(x') = \nabla_{x'} u(x) + \frac{\partial u(x)}{\partial x_d} \nabla f(x'), \tag{A.11}$$

where $\nabla_{x'} := \sum_{m=1}^{d-1} e_m \frac{\partial}{\partial x_m}$ and e_m is the unit vector in the direction x_m. It is then easy to see that $\nabla \phi_f$ given by (A.11) is also the weak derivative of ϕ_f. Unless Γ is smoother than Lipschitz, one cannot expect that $\phi_f \in H^s(\mathbb{R}^{d-1})$ for any $s > 1$.

These observations motivate our definition of boundary Sobolev spaces. For $0 \le s \le 1$, let

$$H^s(\Gamma) := \{\phi \in L^2(\Gamma) : \phi_f \in H^s(\mathbb{R}^{d-1})\}.$$

$H^s(\Gamma)$ is a Hilbert space equipped with the inner product defined by

$$(\phi, \psi)_{H^s(\Gamma)} = (\phi_f, \psi_f)_{H^s(\mathbb{R}^{d-1})}, \tag{A.12}$$

and the above observations imply that $C^\infty_{\mathrm{comp}}(\Gamma) := \{u|_\Gamma : u \in C^\infty_0(\mathbb{R}^d)\} \subset$

$H^1(\Gamma) \subset H^s(\Gamma)$, $0 \le s \le 1$; a little more work shows that $C^\infty_{\text{comp}}(\Gamma)$ is dense in $H^s(\Gamma)$, for $0 \le s \le 1$.

If $u \in C^\infty_{\text{comp}}(\overline{D})$ then, for almost all $x \in \Gamma$, namely all $x = (x', f(x')) \in \Gamma$ for which $\nabla f(x')$ exists, we have

$$\nabla u(x) = n(x)\,(n(x) \cdot \nabla u(x)) + \nabla_\Gamma \phi(x), \tag{A.13}$$

where $\phi = \gamma u \in H^1(\Gamma)$ and $\nabla_\Gamma \phi$ is the *surface gradient* of ϕ. This surface gradient is defined implicitly by (A.13) and explicitly by

$$\nabla_\Gamma \phi(x) = (\nabla_\Gamma \phi)((x', f(x'))) = \sum_{m=1}^{d-1} t_m(e_m + f_m e_d), \tag{A.14}$$

where, for $m = 1, \ldots, d-1$, f_m and t_m are the mth components of $\nabla f(x')$ and $\nabla_\Gamma \phi(x)$, respectively, and $e_m + f_m e_d$ is a surface tangent vector. The vector of coefficients $t = (t_1, \ldots, t_{d-1})^T$ is the unique solution of the linear system $At = r$, where $r = (r_1, \ldots, r_{d-1})^T$, with

$$r_m = \frac{\partial \phi_f(x')}{\partial x_m},$$

and the matrix A has entry $\delta_{mn} + f_m f_n$ in row m column n, where δ_{mn} is the Kronecker delta. A is symmetric and positive definite, with $a^H A a \ge |a|^2$, for every $a \in \mathbb{C}^{d-1}$, where a^H is the conjugate transpose of a. This last inequality implies that $|t| \le |r|$. Explicitly, in 2D, $t_1 = r_1/(1 + f_1^2)$. In 3D,

$$t_1 = \frac{(1 + f_2^2) r_1 - f_1 f_2 r_2}{1 + |\nabla f|^2} \quad \text{and} \quad t_2 = \frac{-f_1 f_2 r_1 + (1 + f_1)^2 r_2}{1 + |\nabla f|^2}. \tag{A.15}$$

It is clear that, if $\phi \in H^1(\Gamma)$, $\nabla_\Gamma \phi(x)$ is well-defined by (A.14) for almost all $x \in \Gamma$, and $\nabla_\Gamma \phi \in (L^2(\Gamma))^d$. Indeed, $\phi \in H^1(\Gamma)$ if and only if $\phi \in L^2(\Gamma)$ and $\nabla_\Gamma \phi \in (L^2(\Gamma))^d$, and an equivalent norm on $H^1(\Gamma)$ is

$$\|\phi\|' := \left(\int_\Gamma \left(|\nabla_\Gamma \phi|^2 + |\phi|^2 \right) \mathrm{d}s \right)^{1/2}, \tag{A.16}$$

so that ∇_Γ is a bounded operator from $H^1(\Gamma)$ to $(L^2(\Gamma))^d$.

For $u \in C^\infty_{\text{comp}}(\overline{D})$ we have defined above the trace of u, $\gamma u \in C^\infty_{\text{comp}}(\Gamma) \cap L^2(\Gamma)$. It is a standard result, due in full to Costabel (1988), that $\gamma u \in H^{s-1/2}(\Gamma)$, for $1/2 < s < 3/2$. Further, for some constant $C_s > 0$ depending only on f and s,

$$\|\gamma u\|_{H^{s-1/2}(\Gamma)} \le C_s \|u\|_{H^s(D)}, \quad \frac{1}{2} < s < \frac{3}{2}. \tag{A.17}$$

Since, by definition, $C^\infty_{\text{comp}}(\overline{D})$ is dense in $H^s(D)$, it follows that γ extends to a bounded operator $\gamma : H^s(D) \to H^{s-1/2}(\Gamma)$ and (A.17) holds for $1/2 < s < 3/2$ and all $u \in H^s(D)$.

Since ∇ is a bounded operator from $u \in H^2(D) = W_2^2(D)$ to $(H^1(D))^d$, the composition $\gamma \circ \nabla$ is a bounded operator from $H^2(D)$ to $(H^{1/2}(\Gamma))^d$, with $\gamma \circ \nabla u = \nabla u|_\Gamma$ for $u \in C^\infty_{\text{comp}}(\overline{D})$. For $u \in C^\infty_{\text{comp}}(\overline{D})$, define $\partial_n u \in L^2(\Gamma)$ by

$$\partial_n u(x) = \frac{\partial u}{\partial n}(x) := n(x) \cdot \nabla u(x), \tag{A.18}$$

for all $x = (x', f(x')) \in \Gamma$ for which $\nabla f(x')$ exists, which is almost all $x \in \Gamma$. Clearly, for some constants $C > 0$ and $C' > 0$ depending only on f,

$$\|\partial_n u\|_{L^2(\Gamma)} \leq C \|\gamma \nabla u\|_{L^2(\Gamma)} \leq C \|\gamma \nabla u\|_{H^{1/2}(\Gamma)} \leq C' \|u\|_{H^2(\Gamma)}.$$

Since $C^\infty_{\text{comp}}(\overline{D})$ is dense in $H^2(D)$ it follows that ∂_n can be extended to a bounded operator $\partial_n : H^2(D) \to L^2(\Gamma)$. We will extend the definition of ∂_n further to certain other subspaces of $H^1(D)$ below. Note that it follows from (A.13) and (A.18) and the density of $C^\infty_{\text{comp}}(\overline{D})$ in $H^2(D)$ that

$$\gamma \nabla u = n\, \partial_n u + \nabla_\Gamma \gamma u, \quad u \in H^2(D). \tag{A.19}$$

Standard Sobolev space formulations of Neumann and impedance boundary value problems (see (2.5) and (2.13)) use negative order Sobolev spaces. For $-1 \leq s < 0$ define an inner product $(\cdot,\cdot)_{H^s(\Gamma)}$ on $L^2(\Gamma)$ by (A.12), and let $H^s(\Gamma)$ denote the completion of $L^2(\Gamma)$ with respect to this inner product. This definition, and the results above for $0 \leq s \leq 1$, imply that $C^\infty_{\text{comp}}(\Gamma)$ is dense in $H^s(\Gamma)$ for $|s| \leq 1$. Since $H^{-s}(\mathbb{R}^{d-1})$ is an isometric realization of the dual space $(H^s(\mathbb{R}^{d-1}))^*$, for $s \in \mathbb{R}$, it follows that the mapping $\phi \mapsto \phi^*$ from $H^{-s}(\Gamma)$ to $(H^s(\Gamma))^*$, given by $\phi^*(\psi) = \langle \phi, \psi \rangle_\Gamma$, $\psi \in H^s(\Gamma)$, where

$$\langle \phi, \psi \rangle_\Gamma := \langle \phi_f, \psi_f \rangle,$$

is an isometric isomorphism, for $|s| \leq 1$, so that $H^{-s}(\Gamma)$ is an isometric realization of $(H^s(\Gamma))^*$ for $|s| \leq 1$. Note that, for $\phi, \psi \in L^2(\Gamma)$,

$$\langle \phi, \psi \rangle_\Gamma = \int_{\mathbb{R}^{d-1}} \phi_f(x') \overline{\psi_f(x')}\, \mathrm{d}x' = \int_\Gamma \frac{\phi(x)\overline{\psi(x)}}{\sqrt{1 + |\nabla f(x')|^2}}\, \mathrm{d}s(x).$$

Clearly, for $\phi \in H^1(\Gamma)$, $\nabla_\Gamma \phi \in L^2_t(\Gamma)$, where $L^2_t(\Gamma) := \{\psi \in (L^2(\Gamma))^d : n \cdot \psi = 0\}$ is the closed subspace of $(L^2(\Gamma))^d$ consisting of tangential vector fields. $L^2_t(\Gamma)$ is a Hilbert space with inner product defined by

$$(\phi, \psi)_{L^2_t(\Gamma)} = \int_\Gamma \phi \cdot \bar{\psi}\, \mathrm{d}s.$$

Let $\nabla^*_\Gamma : L^2_t(\Gamma) \to (H^1(\Gamma))^* = H^{-1}(\Gamma)$ denote the adjoint of $\nabla_\Gamma : H^1(\Gamma) \to L^2_t(\Gamma)$, so that

$$(\nabla_\Gamma \phi, \psi)_{L^2_t(\Gamma)} = \langle \phi, \nabla^*_\Gamma \psi \rangle_\Gamma, \quad \phi \in H^1(\Gamma),\ \psi \in L^2_t(\Gamma). \tag{A.20}$$

(The operator $-\nabla_\Gamma^*$ is closely related to the surface divergence operator; see, *e.g.*, §3.4 in Monk 2003.) Like ∇_Γ, ∇_Γ^* is a local operator, in the sense that $\mathrm{supp}(\nabla_\Gamma^*\psi) \subset \mathrm{supp}(\psi)$, for $\psi \in L_t^2(\Gamma)$ (which is clear from (A.20)). A further important observation, which we justify for $d = 2$ and 3, is that there exists a dense subspace $X_t \subset L_t^2(\Gamma)$ such that $\nabla_\Gamma^*(X_t) \subset L^2(\Gamma)$. In 2D, since $L_t^2(\Gamma) = \{\phi s_t : \phi \in L^2(\Gamma)\}$, where $s_t := (e_1 + f_1 e_2)/\sqrt{1+f_1^2}$ is a particular tangential vector field in $(L^\infty(\Gamma))^2$, such a subspace is $X_t := \{\phi s_t : \phi \in C^\infty_{\mathrm{comp}}(\Gamma)\}$. For, if $\phi \in C^\infty_{\mathrm{comp}}(\Gamma)$, then

$$\nabla_\Gamma^*(\phi s_t)(x) = -\phi_f'(x_1),$$

for almost all $x \in \Gamma$. In 3D, a suitable choice is $X_t = \{u \times n : u \in (C^\infty_{\mathrm{comp}}(\Gamma))^3\}$ (where $\times$ here is the standard vector product). This is easily seen to be dense in $L_t^2(\Gamma)$. Further, by Stokes' theorem, which holds on Lipschitz surfaces as a consequence of the version of the divergence theorem, Theorem A.4 below (see, *e.g.*, Proposition 1.4 in Buffa and Christiansen (2003)), if $\phi \in C^\infty_{\mathrm{comp}}(\Gamma)$ and $\psi = U|_\Gamma \times n \in X_t$, for some $U \in C_0^\infty(\mathbb{R}^d)$, it follows that

$$\int_\Gamma \nabla_\Gamma\phi \cdot \psi \,\mathrm{d}s = -\int_\Gamma \phi\, (\nabla \times U)|_\Gamma \cdot n \,\mathrm{d}s. \qquad \text{(A.21)}$$

Further, this equation extends to the case $\phi \in H^1(\Gamma)$ by density of $C^\infty_{\mathrm{comp}}(\Gamma)$ in $H^1(\Gamma)$. Thus $\nabla_\Gamma^*\psi = -\sqrt{1+|\nabla f|^2}\,(\nabla \times U)|_\Gamma \cdot n \in L^2(\Gamma)$.

The above definitions concern the case when D is a Lipschitz hypograph. If $r(D)$ is the Lipschitz hypograph of some $f \in C^{0,1}(\mathbb{R}^{d-1})$, for some rigid motion $r : \mathbb{R}^d \to \mathbb{R}^d$, then we define $H^s(\Gamma)$ in the same way as when D is a Lipschitz hypograph, except that $\phi_f(x') := \phi[r^{-1}((x', f(x')))]$, for $\phi \in L^2(\Gamma)$.

Now we describe the extension of this procedure to the case when D is a Lipschitz open set with boundary $\partial D = \Gamma$, so that D satisfies the conditions in Definition A.2. We first choose a partition of unity $\{\chi_i\}$ subordinate to the open cover $\{W_i\}$ in that definition. This means that we choose $\chi_i \in C_0^\infty(W_i)$ for each i in such a way that each $\chi_i \ge 0$ and $\sum_i \chi_i(x) = 1$ for all $x \in \Gamma$. (As a first step it is enough to find positive functions $\psi_i \in C_0^\infty(W_i)$ so that $\sum_i \psi_i(x) \neq 0$ for each $x \in \Gamma$, and then define $\chi_i(x) = \psi_i(x)/\sum_m \psi_m(x)$.) For each i let $\Gamma_i := \partial D_i$, where D_i is as in Definition A.2. In terms of these notations we can define a surface measure $\mathrm{d}s$ and surface integrals of functions $\phi \in L^1(\Gamma)$ that are Lebesgue-integrable with respect to this surface measure, by

$$\int_\Gamma \phi \,\mathrm{d}s := \sum_i \int_{\Gamma_i} \chi_i \phi \,\mathrm{d}s. \qquad \text{(A.22)}$$

(The integrals on the right-hand side make sense since the support of $\chi_i\phi_i$ is a subset of $\Gamma \cap \Gamma_i$ and should be interpreted as in (A.9).) Both the set

$L^1(\Gamma)$ and the value of the integral (A.22) are independent of the choice of the families in Definition A.2 and independent of the choice of the partition of unity.

Define an inner product on $C^\infty(\Gamma)$ by

$$(\phi, \psi)_{H^s(\Gamma)} := \sum_i (\chi_i \phi, \chi_i \psi)_{H^s(\Gamma_i)}. \tag{A.23}$$

For $|s| \leq 1$ the Sobolev space $H^s(\Gamma)$ is defined to be the completion of $C^\infty(\Gamma)$ with respect to the norm this inner product induces. For $0 \leq s \leq 1$, equivalently, $H^s(\Gamma)$ is the set of those $\phi \in L^2(\Gamma) = H^0(\Gamma)$ such that $\chi_i \phi \in H^s(\Gamma_i)$ for each i. Define a mapping $\phi \mapsto \phi^*$, from $C^\infty(\Gamma)$ to $(H^s(\Gamma))^*$, for $0 \leq s \leq 1$, by $\phi^*(\psi) = \langle \phi, \psi \rangle_\Gamma$, $\psi \in H^s(\Gamma)$, where

$$\langle \phi, \psi \rangle_\Gamma := \int_\Gamma \phi \bar{\psi} \, \mathrm{d}s = (\phi, \psi)_{L^2(\Gamma)}. \tag{A.24}$$

It is not difficult to see that this mapping extends to a bounded linear map from $H^{-s}(\Gamma)$ to $(H^s(\Gamma))^*$; indeed this map is an isomorphism since $H^{-s}(\Gamma)$ is a realization of $(H^s(\Gamma))^*$, for $|s| \leq 1$, when Γ is a Lipschitz graph. Thus $H^{-s}(\Gamma)$ is a realization of $(H^s(\Gamma))^*$ for $0 \leq s \leq 1$ also in the general Lipschitz case; indeed we see this holds for $|s| \leq 1$ by passing to duals. Considered as a mapping from $H^{-s}(\Gamma) \times H^s(\Gamma) \to \mathbb{C}$, $\langle \cdot, \cdot \rangle_\Gamma$ is a sesquilinear form which we will refer to as the *duality pairing*. We will sometimes make the usual slight abuse of notation and write $\langle \phi, \psi \rangle_\Gamma$ as the integral $\int_\Gamma \phi \bar{\psi} \, \mathrm{d}s$ in cases where, for some $-1 \leq s < 0$, $\phi \in H^{-s}(\Gamma)$, $\psi \in H^s(\Gamma) \subset L^2(\Gamma)$ but $\phi \notin L^2(\Gamma)$.

As discussed in McLean (2000), the space $H^s(\Gamma)$ does not depend on the particular choice of families in Definition A.2 or on the choice of the partition of unity. Further, while these choices affect the definition of the inner product (A.23), the norms arising from distinct choices are equivalent. In particular, the norms on $L^2(\Gamma)$ and $H^0(\Gamma)$ are equivalent.

For general Lipschitz Γ, at each point $x \in \Gamma \cap W_i = \Gamma_i \cap W_i$, the formulae for the unit normal and the surface gradient are given in terms of the definitions (A.10) and (A.14), for $\phi \in H^1(\Gamma)$, in the obvious way, and (A.16) is again an equivalent norm on $H^1(\Gamma)$ and $\nabla_\Gamma : H^1(\Gamma) \to (L^2(\Gamma))^d$ a bounded operator. The normal derivative operator ∂_n is extended to the general Lipschitz case, as a bounded operator $H^2(D) \to L^2(\Gamma)$ through (A.18), and (A.19) holds. The trace operator $\gamma : H^s(D) \to H^{s-1/2}(\Gamma)$ remains a bounded operator in the general Lipschitz case for $1/2 < s < 3/2$ (*i.e.*, (A.17) holds for $u \in H^s(D)$), and has a continuous right inverse for $1/2 < s \leq 1$: in particular there exists a bounded linear operator

$$\gamma^\dagger : H^{1/2}(\Gamma) \to H^1(D)$$

such that $\gamma \gamma^\dagger \phi = \phi$, $\phi \in H^{1/2}(\Gamma)$ (McLean 2000, Theorem 3.37). It follows from (A.7) that, for $1 \leq p \leq 2$, $L^p(\Gamma) \subset H^s(\Gamma)$, for $s < -\theta(d-1)$, where θ

is given by (A.6), and that, for some constant $C > 0$ depending only on p, s, and Γ,

$$\|\phi\|_{H^s(\Gamma)} \leq C\|\phi\|_{L^p(\Gamma)}, \quad \phi \in L^p(\Gamma). \tag{A.25}$$

For example, this inequality holds for $2(d-1)/d < p \leq 2$ if $s = -1/2$.

An essential tool throughout is the following version of the divergence theorem (*e.g.*, McLean 2000, Theorem 3.34).

Theorem A.4. (Divergence theorem) If D is a Lipschitz open set and $F \in (C^\infty_{\mathrm{comp}}(\overline{D}))^d$ then

$$\int_D \nabla \cdot F \,\mathrm{d}x = \int_\Gamma n \cdot \gamma F \,\mathrm{d}s,$$

where n is the unit normal outwards from D.

An important consequence of the divergence theorem, of the density of $C^\infty_{\mathrm{comp}}(\overline{D})$ in $H^s(D)$, and of the continuity of the trace and normal derivative operators, is Green's first identity:

$$\int_D \nabla u \cdot \nabla \bar{v} \,\mathrm{d}x = -\int_D \bar{v}\,\Delta u \,\mathrm{d}x + \int_\Gamma \gamma\bar{v}\,\partial_n u \,\mathrm{d}s, \quad u \in H^2(D),\ v \in H^1(D). \tag{A.26}$$

For every Lipschitz open set $D \subset \mathbb{R}^d$, let $H^1(D;\Delta)$ denote the set of $v \in H^2(D) = W^1_2(D)$ such that $\Delta v \in L^2(D)$ exists in a weak sense, that is, there exists $w \in L^2(D)$ such that $\int_D wu \,\mathrm{d}x = \int_D v\Delta u \,\mathrm{d}x$, for all $u \in C^\infty_0(D)$, in which case we define $\Delta v := w$. $H^1(D;\Delta)$ is a Hilbert space with the inner product

$$(u,v)_{H^1(D;\Delta)} := \int_D \big(\Delta u\,\Delta\bar{v} + \nabla u \cdot \nabla\bar{v} + u\,\bar{v}\big)\,\mathrm{d}x.$$

A standard result (Theorem 3.40 in McLean 2000) is that $C^\infty_0(D)$ is dense in $H^1_0(D) := \{u \in H^1(D) : \gamma u = 0\}$. This, together with (A.26) and the weak definition of Δv, imply the following version of Green's first identity:

$$\int_D \nabla u \cdot \nabla\bar{v}\,\mathrm{d}x = -\int_D \bar{v}\,\Delta u\,\mathrm{d}x, \quad u \in H^1(D;\Delta),\ v \in H^1_0(D). \tag{A.27}$$

In the case that D is Lipschitz, the above formula (A.26) provides a means of extending the definition of ∂_n to a bounded operator $\partial_n : H^1(D;\Delta) \to H^{-1/2}(\Gamma)$. Recalling that $H^{-1/2}(\Gamma)$ is a realization of the dual space of $H^{1/2}(\Gamma)$ and recalling the isomorphism $\phi \mapsto \phi^*$ which takes $H^{-1/2}(\Gamma)$ to $(H^{1/2}(\Gamma))^*$, one defines $\phi = \partial_n u \in H^{-1/2}(\Gamma)$, for every $u \in H^1(D;\Delta)$, and an associated $\phi^* \in (H^{1/2}(\Gamma))^*$, by

$$\langle \phi, \psi\rangle_\Gamma = \phi^*(\psi) := \int_D (\nabla u \cdot \nabla\bar{v} + \bar{v}\,\Delta u)\,\mathrm{d}x, \quad \psi \in H^{1/2}(\Gamma), \tag{A.28}$$

where $v := \gamma^\dagger\psi \in H^1(D)$ satisfies $\gamma v = \psi$. We will use the definition (A.28) of $\partial_n u \in H^{-1/2}(\Gamma)$ particularly with $\psi = \gamma w$, for some $w \in H^1(D)$, when $v = \gamma^\dagger \gamma w$ on the right-hand side of this equation. In general $\gamma^\dagger \gamma w \neq w$, but $w - \gamma^\dagger \gamma w \in H_0^1(D)$ so that it follows from the definition (A.28) and (A.27) that

$$\langle \partial_n u, \gamma w \rangle_\Gamma = \int_D \big(\nabla u \cdot \nabla \bar{w} + \bar{w}\, \Delta u\big)\, \mathrm{d}x, \quad u \in H^1(D;\Delta),\ w \in H^1(D), \tag{A.29}$$

another version of Green's first identity. When $u \in H^2(D) \subset H^1(D;\Delta)$, it follows, comparing (A.26) and (A.29), that the two definitions we have made for $\partial_n u$ coincide.

The following regularity theorem (see McLean 2000, Theorem 4.24, or Nečas 1967) will be very useful in this article. This result is proved using Rellich-identity-type arguments, of the sort that are central to many of our results in Section 5.

Theorem A.5. Suppose D is a Lipschitz open set and $u \in H^1(D;\Delta)$. Then $\partial_n u \in L^2(\Gamma)$ if and only if $\gamma u \in H^1(\Gamma)$. Further, if $\partial_n u \in L^2(\Gamma)$ then, for some $C > 0$ depending only on Γ,

$$\|\gamma u\|_{H^1(\Gamma)} \leq C\big(\|\partial_n u\|_{L^2(\Gamma)} + \|u\|_{H^1(D;\Delta)}\big)$$

and

$$\|\partial_n u\|_{L^2(\Gamma)} \leq C\big(\|\gamma u\|_{H^1(\Gamma)} + \|u\|_{H^1(D;\Delta)}\big).$$

As an important application of this theorem, note that $\partial_n u \in L^2(\Gamma)$ if $u \in H^1(D,\Delta)$ and $\gamma u = 0$.

Roughly speaking, for $u \in H^1(D)$, γu is the value of u on Γ and, for $u \in H^1(D;\Delta)$, $\partial_n u$ is the value of $\frac{\partial u}{\partial n}$ on Γ: these statements are true precisely as they stand when $u \in C^\infty_{\mathrm{comp}}(\overline{D})$. It is thus not surprising that, if $v \in C^\infty_{\mathrm{comp}}(\overline{D})$ and $v = 1$ in a neighbourhood of Γ, then $\gamma(uv) = \gamma u$, for $u \in H^1(D)$, and $\partial_n(uv) = \partial_n u$ for $u \in H^1(D;\Delta)$.

In the scattering problems that we study in this article, when D is an unbounded open set, the solution is not in $H^1(D)$ but is in the linear spaces

$$H^1_{\mathrm{loc}}(D) := \{u \in L^2_{\mathrm{loc}}(D) : vu \in H^1(D),\ v \in C^\infty_{\mathrm{comp}}(\overline{D})\} \tag{A.30}$$

and

$$H^1_{\mathrm{loc}}(D;\Delta) := \{u \in L^2_{\mathrm{loc}}(D) : \nabla u \in (L^2_{\mathrm{loc}}(D))^d, \Delta u \in L^2_{\mathrm{loc}}(D)\}. \tag{A.31}$$

The trace and normal derivative operators, γ and ∂_n, extend in a natural way as linear operators $\gamma : H^1_{\mathrm{loc}}(D) \to H^{1/2}(\Gamma)$ and $\gamma : H^1_{\mathrm{loc}}(D;\Delta) \to H^{-1/2}(\Gamma)$, defined by $\gamma u = \gamma(uv)$ and $\partial_n u = \partial_n(uv)$, where v is an arbitrary $v \in C^\infty_{\mathrm{comp}}(\overline{D})$ with $v = 1$ in a neighbourhood of Γ.

A.4. Maximal functions

The Sobolev spaces and trace operators introduced in the subsections above provide powerful and widely used tools to formulate and study elliptic boundary value problems in Lipschitz domains. Indeed, we use these tools for precisely this purpose in Section 2. But, to develop some of our key theoretical results, we need to supplement these methods with results associated with the study, in the harmonic analysis community, of Calderón–Zygmund operators and singular integral operators on the boundaries of Lipschitz domains.

These additional results provide other benefits which we briefly exhibit. In particular they enable us define a sense in which the trace operator γ, defined as a mapping from $H^s(D)$ to $H^{s-1/2}(\Gamma)$, for $1/2 < s < 3/2$, can be extended to the limiting values $s = 1/2, 3/2$, provided that we restrict the domain of the operator to a subspace of $H^s(D)$. They also enable us to explain in what sense $\partial_n u$, for $u \in H^1(D;\Delta)$, is connected to limiting values of $n(y) \cdot \nabla u(x)$ as $x \in D$ approaches $y \in \Gamma$.

Key notions that we will need are those of non-tangential convergence and non-tangential maximal functions. Let D be a Lipschitz open set with boundary Γ. Given $x \in \Gamma$, we call the set $\Theta(x)$ a *non-tangential approach set (to x from D)* if $\Theta(x) \subset D$, $\overline{\Theta(x)} \cap \Gamma = \{x\}$, and, for some constant $C > 1$,

$$|x - y| \leq C\text{dist}(y, \Gamma), \quad y \in \Theta(x). \tag{A.32}$$

Now suppose that we are given families specifying the Lipschitz open set as in Definition A.2, in particular an open cover $\{W_i\}$ and an associated family of rigid motions $\{r_i\}$. We choose $\{r_i\}$, as we note we are able to do immediately below the definition, to be a family of pure rotations. Further, suppose that we are given a partition of unity $\{\chi_i\}$ subordinate to this open cover, as defined before equation (A.22). We will call a family $\{\Theta(x) : x \in \Gamma\}$ of non-tangential approach sets *sufficient (with respect to these families $\{W_i\}$, $\{r_i\}$, and $\{\chi_i\}$)* if, for some $\varepsilon > 0$,

$$\ell_{i,\varepsilon}(x) := \{x + t\hat{b}_i : 0 < t \leq \varepsilon\} \subset \Theta(x), \quad \text{for all } x \in \text{supp}(\chi_i) \cap \Gamma. \tag{A.33}$$

Here $\hat{b}_i$ is the unit vector that satisfies $\hat{b}_i = r_i^{-1}(e_d)$, so that, for every $x \in W_i \cap \Gamma$, $x + t\hat{b}_i \in D$ for all sufficiently small $t > 0$. Further, we will call the family $\{\Theta(x) : x \in \Gamma\}$ *uniform* if the condition (A.32) holds with the same constant $C > 1$ for all $x \in \Gamma$, and if, for some constant $c > 0$, the diameter of $\Theta(x)$ is $\leq c$, for every $x \in \Gamma$.

It is easy to see that there exists $\varepsilon > 0$ such that (A.33) holds, that is, there exists, given any Lipschitz open set defined as in Definition A.2 and an associated subordinate partition of unity $\{\chi_i\}$, a sufficient family of non-tangential approach sets. Indeed, choosing $\varepsilon > 0$ such that $\ell_{i,\varepsilon}(x) \subset D$ for all i and $x \in \text{supp}(\chi_i) \cap \Gamma$, an example of such a sufficient family is

$$\Theta(x) := \bigcup_{i,\, x \in \text{supp}(\chi_i)} \ell_{i,\varepsilon}(x). \tag{A.34}$$

If $\varepsilon > 0$ is chosen sufficiently small then this choice is also uniform. On the other hand, an obvious example of a uniform family of non-tangential approach sets is obtained by choosing $C > 1$ and $c > 0$ and defining

$$\Theta(x) := \{y \in D : |x - y| < \max(c, C\text{dist}(y, \Gamma)\}. \tag{A.35}$$

If C is chosen sufficiently large then this choice is also sufficient.

Given a family of non-tangential approach sets and a function $u \in C(D)$, we define a *maximal function* $u^* : \Gamma \to [0, \infty]$ by

$$u^*(x) := \sup_{y \in \Theta(x)} |u(y)|, \quad x \in \Gamma. \tag{A.36}$$

In the case that $u \in C^1(D)$, one can define also a maximal function for the gradient by

$$(\nabla u)^*(x) := \sup_{y \in \Theta(x)} |\nabla u(y)|, \quad x \in \Gamma. \tag{A.37}$$

Commonly, in these definitions, the non-tangential approach sets are chosen to be open finite cones, the finite cone $\Theta(x)$ having its vertex at x (see, *e.g.*, Verchota 1984).

Recall that the trace operator γ is well-defined as a mapping from $H^s(D)$ to $H^{s-1/2}(D)$ for $1/2 < s < 3/2$. Formulations of the trace operator in terms of non-tangential convergence enable us to extend the trace operator as a mapping from certain subspaces of $H^{1/2}(D)$ to $L^2(\Gamma)$. In particular, we will see that we can choose this subspace as a subspace of solutions of the Helmholtz equation. This result, which is contained in Corollary A.8 and Lemma A.9 below, is based on the following equivalence result for harmonic functions ($u \in C^2(D)$ is said to be harmonic if $\Delta u = 0$ in D) (see Jerison and Kenig 1995, §5). In this result we introduce the notation

$$\delta(x) := \text{dist}(x, \Gamma), \quad \text{for } x \in D.$$

Theorem A.6. (Jerison and Kenig 1995, Corollaries 5.5 and 5.7) Suppose that D is a bounded Lipschitz open set, $u \in C^2(D)$ is harmonic and let $\{\Theta(x) : x \in \Gamma\}$ be any uniform and sufficient family of tangential approach sets. Then the following statements are equivalent:

(i) $u \in H^{1/2}(D)$;

(ii) $u^* \in L^2(\Gamma)$;

(iii) $\|u\|_{1,\delta} := \left(\int_D \left(|u|^2 + \delta \, |\nabla u|^2 \right) \mathrm{d}x \right)^{1/2} < \infty.$

Further, if $u^* \in L^2(\Gamma)$ then

$$\phi(x) := \lim_{y\to x, y\in\Theta(x)} u(y) \tag{A.38}$$

exists for almost all $x \in \Gamma$, with $\phi \in L^2(\Gamma)$. Moreover, $\|\phi\|_{L^2(\Gamma)}$, $\|u^*\|_{L^2(\Gamma)}$, $\|u\|_{H^{1/2}(D)}$, and $\|u\|_{1,\delta}$ are equivalent norms on the space of harmonic functions u in D for which $u \in H^{1/2}(D)$.

Similarly, the following statements are equivalent:

(a) $u \in H^{3/2}(D)$;

(b) $(\nabla u)^* \in L^2(\Gamma)$;

(c) $\|u\|_{2,\delta} := \left(\int_D \left(|u|^2 + |\nabla u|^2 + \delta \sum_{|\alpha|=2} |\partial^\alpha u|^2\right) \mathrm{d}x\right)^{1/2} < \infty.$

Further, if $(\nabla u)^* \in L^2(\Gamma)$ then

$$\psi(x) := \lim_{y\to x, y\in\Theta(x)} \nabla u(y) \tag{A.39}$$

exists for almost all $x \in \Gamma$, with $\psi \in (L^2(\Gamma))^d$. Moreover, $\|\phi\|^2_{L^2(\Gamma)} + \|\psi\|^2_{L^2(\Gamma)}$, $\|u^*\|_{L^2(\Gamma)} + \|(\nabla u)^*\|_{L^2(\Gamma)}$, $\|u\|_{H^{3/2}(D)}$, and $\|u\|_{2,\delta}$ are equivalent norms on the space of harmonic functions u in D for which $u \in H^{3/2}(D)$.

The above result enables us to extend the definition of the trace operator γ to harmonic functions in $H^{1/2}(D)$, defining γu to be the non-tangential limit ϕ. This holds also for solutions of the Helmholtz equation in D. In dimensions $d = 2$ and 3 this is a consequence of the next result, which follows from the embedding theorem, Theorem A.1, and from properties of the Newtonian potential.

Theorem A.7. Suppose that D is a bounded Lipschitz open set, let $u \in C^2(D)$, and let $\{\Theta(x) : x \in \Gamma\}$ be any uniform and sufficient family of tangential approach sets.

(A) If $\Delta u \in L^2(D)$, then (i)–(iii) of Theorem A.6 are equivalent and, if $u^* \in L^2(\Gamma)$, then the limit (A.38) exists for almost all $x \in \Gamma$, with $\|\phi\|_{L^2(\Gamma)} \le \|u^*\|_{L^2(\Gamma)}$. Moreover, $\|u^*\|_{L^2(\Gamma)} + \|\Delta u\|_{L^2(D)}$, $\|u\|_{H^{1/2}(D)} + \|\Delta u\|_{L^2(D)}$, and $\|u\|_{1,\delta} + \|\Delta u\|_{L^2(D)}$ are all equivalent norms for u.

(B) If $\Delta u \in H^1(D)$, then (a)–(c) of Theorem A.6 are also equivalent and, if $(\nabla u)^* \in L^2(\Gamma)$, then the limit (A.39) exists for almost all $x \in \Gamma$, with $\|\psi\|_{L^2(\Gamma)} \le \|(\nabla u)^*\|_{L^2(\Gamma)}$. Moreover, $\|u^*\|_{L^2(\Gamma)} + \|(\nabla u)^*\|_{L^2(\Gamma)} + \|\Delta u\|_{H^1(D)}$, $\|u\|_{H^{3/2}(D)} + \|\Delta u\|_{H^1(D)}$, and $\|u\|_{2,\delta} + \|\Delta u\|_{H^1(D)}$ are all equivalent norms for u.

Proof. **(A)** Let $g = \Delta u$ and write u as $u = v + w$, where $w = W|_D$ and W is the Newtonian potential

$$W(x) = -\int_{\mathbb{R}^d} \Phi_0(x, y)\, \tilde{g}(y)\, \mathrm{d}y, \quad x \in \mathbb{R}^d,$$

where $\tilde{g}$ is the extension of g by zero from D to $\mathbb{R}^d$. Then, where $\chi \in C_0^\infty(\mathbb{R}^d)$, with $\chi(x) = 1$ for $x \in D$, it holds that $\chi W \in H^2(\mathbb{R}^d)$ with $\|\chi W\|_{H^2(\mathbb{R}^d)} \leq C\|g\|_{L^2(D)}$ (McLean 2000, Theorem 6.1), and that $\Delta W = \tilde{g}$, so that $\Delta w = g$. Further, by the Sobolev embedding theorem, Theorem A.1, since $d \leq 3$, $\chi W \in BC(\mathbb{R}^d)$, with $\|\chi W\|_{BC(\mathbb{R}^d)} \leq C\|g\|_{L^2(D)}$. Thus $w \in C(\overline{D}) \cap H^2(D)$, with $\Delta w = g$. Further, $\|w\|_{C(\overline{D})} + \|w\|_{H^2(D)} \leq C\|g\|_{L^2(D)}$, so that (i)–(iii) hold for w with $\|w^*\|_{L^2(\Gamma)} + \|w\|_{H^{1/2}(D)} + \|w\|_{1,\delta} \leq C\|g\|_{L^2(D)}$. It follows that $v \in C(D)$ satisfies $\Delta v = 0$, so hence that $v \in C^2(D)$ by elliptic regularity. So, by Theorem A.6, (i)–(iii) are equivalent for v and hence also for u, and the equivalence of the norms on u stated in (A) follows.

(B) This follows similarly, except that now $\tilde{g}$ is chosen to be any compactly supported function in $H^1(\mathbb{R}^d)$ such that $\tilde{g}|_D = g$ and $\|\tilde{g}\|_{H^1(\mathbb{R}^d)} \leq 2\|g\|_{H^1(D)}$. (That such a choice is possible is a consequence of (A.8).) Then $\chi W \in H^3(\mathbb{R}^d)$ with $\|\chi W\|_{H^3(\mathbb{R}^d)} \leq C\|g\|_{H^1(D)}$ (McLean 2000, Theorem 6.1) so that $\|\nabla(\chi W)\|_{H^2(\mathbb{R}^d)} \leq C\|g\|_{H^1(D)}$ and, by the Sobolev embedding theorem, Theorem A.1, $w \in C^1(\overline{D})$ with $\|w\|_{C^1(\overline{D})} \leq C\|g\|_{H^1(D)}$. □

An immediate corollary is that Theorem A.6 extends to solution of the Helmholtz equation.

Corollary A.8. Suppose that D is a bounded Lipschitz open set, $u \in C^2(D)$ and, for some constant $k \in \mathbb{C}$, $\Delta u + k^2 u = 0$ in D, and let $\{\Theta(x) : x \in \Gamma\}$ be any uniform and sufficient family of tangential approach sets. Then (i)–(iii) in Theorem A.6 are equivalent, as are (a)–(c). If $u^* \in L^2(\Gamma)$, then the limit (A.38) exists for almost all $x \in \Gamma$, with $\|\phi\|_{L^2(\Gamma)} \leq \|u^*\|_{L^2(\Gamma)}$, while if $(\nabla u)^* \in L^2(\Gamma)$, then the limit (A.39) exists for almost all $x \in \Gamma$, with $\|\psi\|_{L^2(\Gamma)} \leq \|(\nabla u)^*\|_{L^2(\Gamma)}$. Moreover, defining, for $s \geq 0$,

$$H^s_{\mathrm{HE}}(D) := \{v \in H^s(D) \cap C^2(D) : \Delta v + k^2 v = 0\}, \tag{A.40}$$

which is a closed subspace of $H^s(D)$, it holds that $\|u^*\|_{L^2(\Gamma)} + \|u\|_{L^2(D)}$, $\|u\|_{H^{1/2}(D)}$, and $\|u\|_{1,\delta}$ are equivalent norms on $H^{1/2}_{\mathrm{HE}}(D)$ and $\|(\nabla u)^*\|_{L^2(\Gamma)} + \|u\|_{H^1(D)}$, $\|u\|_{H^{3/2}(D)}$, and $\|u\|_{2,\delta}$ are equivalent norms on $H^{3/2}_{\mathrm{HE}}(D)$.

Proof. This follows immediately from Theorem A.7: in the case that any one of (i)–(iii) holds it follows that $g := \Delta u = -k^2 u \in L^2(D)$; in the case that any one of (a)–(c) holds it follows that $g \in H^1(D)$. □

Define a trace operator $\tilde{\gamma} : H^{1/2}_{\mathrm{HE}}(D) \to L^2(\Gamma)$ by $\tilde{\gamma} u = \phi$, where ϕ is defined by (A.38). By Corollary A.8, $\tilde{\gamma}$ is a bounded linear operator, and the next lemma shows that this trace operator extends the trace operator γ to a subspace of $H^{1/2}(\Gamma)$.

Lemma A.9. If D is a bounded Lipschitz open set and $u \in H^s_{\mathrm{HE}}(D)$, with $s > 1/2$, then $\tilde{\gamma} u = \gamma u$.

Proof. Suppose that $s > 1/2$ and $u \in H^s_{\mathrm{HE}}(D)$. Since $\Delta u + k^2 u = 0$ in D, $u \in C^\infty(D)$ by standard interior elliptic regularity results (*e.g.*, Theorem 3.5 in Colton and Kress 1983).

Let $\{\Theta(x) : x \in \Gamma\}$ be a family of non-tangential approach sets that is uniform and sufficient with respect to particular families $\{W_i\}$, $\{r_i\}$, and $\{\chi_i\}$ (see the definition just above (A.33)). It is enough that we show that $\int_\Gamma (\tilde{\gamma} u - \gamma u)\phi \, \mathrm{d}s = 0$ for all $\phi \in C^\infty(\Gamma)$. Indeed, it is enough to show that this holds for all $\phi \in C^\infty(\Gamma)$ with $\mathrm{supp}(\phi) \subset \mathrm{supp}(\chi_i)$, for some i. Let ϕ be such a function, and let $\Gamma^* = \mathrm{supp}(\phi) \subset \mathrm{supp}(\chi_i) \cap \Gamma$. Now, as in (A.33), let $\hat{b}_i = r_i^{-1}(e_d)$, and define $U_t \in C(\Gamma^*)$, for all sufficiently small $t > 0$, by $U_t(x) = u(x + \hat{b}_i t)$, for $x \in \Gamma^*$. (Thus U_t is essentially u evaluated on a parallel surface to Γ^*, distance t away.) Since $\{\Theta(x) : x \in \Gamma\}$ is sufficient, for all sufficiently small $\epsilon > 0$ we have $\ell_{i,\epsilon}(x) \subset \Theta(x)$ for all $x \in \Gamma^*$ (our notation here is that of (A.33)). Since $u \in H^{1/2}_{\mathrm{HE}}(D)$, Corollary A.8 implies that $u^* \in L^2(\Gamma)$, from which it follows that

$$\int_{\Gamma^*} U_t \phi \, \mathrm{d}s \to \int_{\Gamma^*} \tilde{\gamma} u \, \phi \, \mathrm{d}s = \int_\Gamma \tilde{\gamma} u \, \phi \, \mathrm{d}s,$$

as $t \to 0$, by the dominated convergence theorem.

On the other hand, choose $v \in H^s(\mathbb{R}^d)$ such that $u = v|_D$ and, for $t > 0$, define $v_t \in H^s(\mathbb{R}^d)$ by $v_t(x) = v(x + t\hat{b}_i)$, $x \in \mathbb{R}^d$, and let $u_t := v_t|_D \in H^s(D)$. Since $u \in C^\infty(D)$, u_t is C^∞ in a neighbourhood of Γ^*, for all sufficiently small $t > 0$. Thus, for all sufficiently small $t > 0$, $\gamma u_t(x) = u_t(x) = U_t(x)$, $x \in \Gamma^*$. So, for some constant $C > 0$,

$$\left| \int_{\Gamma^*} (U_t - \gamma u)\phi \, \mathrm{d}s \right| = \left| \int_\Gamma \gamma(u_t - u)\phi \, \mathrm{d}s \right| \le C \|u_t - u\|_{H^s(D)} \le C \|v_t - v\|_{H^s(\mathbb{R}^d)},$$

since γ is a bounded map from $H^s(D)$ to $H^{s-1/2}(\Gamma)$. But that $\|v_t - v\|_{H^s(\mathbb{R}^d)} \to 0$ as $t \to 0$ is clear if $v \in C_0^\infty(\mathbb{R}^d)$, and follows for all $v \in H^s(\mathbb{R}^d)$ by density of $C_0^\infty(\mathbb{R}^d)$ in $H^s(\mathbb{R}^d)$. □

Similarly, the mapping $u \mapsto \psi$, where ψ is given by (A.39), from $H^{3/2}_{\mathrm{HE}}(D)$ to $(L^2(\Gamma))^d$, is a bounded linear operator, connected to the operators we have already introduced by the following lemma.

Lemma A.10. If D is a bounded Lipschitz open set, $u \in H^{3/2}_{\mathrm{HE}}(D)$ and ψ is defined by (A.39), then $\partial_n u \in L^2(\Gamma)$, $\gamma u \in H^1(\Gamma)$ and, for almost every $x \in \Gamma$,

$$\psi(x) = n(x)\partial_n u + \nabla_\Gamma \gamma u(x).$$

Proof. We note that $H^{3/2}_{\mathrm{HE}}(D) \subset H^1(D;\Delta)$, and recall that $\partial_n u \in H^{-1/2}(\Gamma)$ is defined by (A.28). We first prove that $\partial_n u \in L^2(\Gamma)$ and that

$$\partial_n u(x) = \frac{\partial u}{\partial n}(x) := n(x) \cdot \lim_{y \to x, y \in \Theta(x)} \nabla u(y), \qquad \text{(A.41)}$$

for almost all $x \in \Gamma$. (Here – recall the definition of sufficient above (A.33) – $\{\Theta(x) : x \in \Gamma\}$ is any family of non-tangential approach sets that is uniform and sufficient with respect to particular families $\{W_i\}$, $\{r_i\}$, and $\{\chi_i\}$, with $\{\chi_i\}$ a partition of unity on Γ subordinate to the finite open cover $\{W_i\}$.) Since, by Corollary A.8, $\psi \in (L^2(\Gamma))^d$, it holds that $\partial u/\partial n \in L^2(\Gamma)$. Recalling the definition (A.28), we see that, if we can show the following version of Green's identity, that

$$\int_\Gamma \frac{\partial u}{\partial n} \gamma \bar{v} \, \mathrm{d}s = \int_D \big(\nabla u \cdot \nabla \bar{v} + \bar{v}\, \Delta u\big)\, \mathrm{d}x, \quad v \in H^1(D), \qquad \text{(A.42)}$$

then (A.41) and $\partial_n u \in L^2(\Gamma)$ will follow. This version of Green's identity, proved under an assumption that $(\nabla u)^* \in L^2(\Gamma)$, is standard in the harmonic analysis literature (*e.g.*, §4.4 in Taylor 2000, §15.5 in Meyer and Coifman 2000). To see how it is obtained, making clear how the Lipschitz character of D and the condition $(\nabla u)^* \in L^2(\Gamma)$ are relevant, we give a proof different from, and arguably simpler than, that in, *e.g.*, Taylor (2000).

Let $\{1, 2, \ldots, N\}$ be the finite index set for the family $\{\chi_i\}$, *i.e.*, $\{\chi_i\} = \{\chi_i : i = 1, \ldots, N\}$. Define $\chi_0 \in C^\infty(\bar{D})$ by

$$\chi_0(x) = 1 - \sum_{i=1}^{N} \chi_i(x), \quad x \in \bar{D}.$$

Since $C^\infty(\bar{D})$ is dense in $H^1(D)$ and $\Delta u = -k^2 u \in L^2(D)$, it is enough to show (A.42) for $v \in C^\infty(\bar{D})$. But to show (A.42) for $v \in C^\infty(\bar{D})$ it is enough to show that (A.42) holds for v replaced by $v_i := \chi_i v$, $i = 0, \ldots, N$. For if this holds then, since $\sum_{i=0}^{N} \nabla \chi_i = 0$ in D,

$$\begin{aligned}\int_\Gamma \frac{\partial u}{\partial n} \gamma \bar{v} \, \mathrm{d}s &= \sum_{i=0}^{N} \int_\Gamma \frac{\partial u}{\partial n} \gamma \bar{v}_i \, \mathrm{d}s \\ &= \sum_{i=0}^{N} \int_D \big(\nabla u \cdot \nabla \bar{v}_i + \bar{v}_i\, \Delta u\big)\, \mathrm{d}x\end{aligned}$$

$$= \sum_{i=0}^{N} \int_D \chi_i (\nabla u \cdot \nabla \bar{v} + \bar{v}\,\Delta u)\,\mathrm{d}x$$

$$= \int_D (\nabla u \cdot \nabla \bar{v} + \bar{v}\,\Delta u)\,\mathrm{d}x.$$

So it remains to show (A.42) with v replaced by v_i, for $i = 0, \ldots, N$. For $i = 0$, since $\gamma\chi_0 = 0$ so that $\chi \in H_0^1(D)$, this follows from (A.29). For $i = 1, \ldots, N$, write $u = U|_D$, for some $U \in H^{3/2}(\mathbb{R}^d)$, and $v = V|_D$, for some $V \in C^\infty(\mathbb{R}^d)$, and, where D_i is as in Definition A.2, let $\Gamma_i := \partial D_i$, and $\hat{b}_i := r_i^{-1}(e_d)$. Further, for $t > 0$, let $D_{i,t} := D_i + t\hat{b}_i$ and $\Gamma_{i,t} := \partial D_{i,t} = \Gamma_i + t\hat{b}_i$. Then, for $t > 0$, where $G := \operatorname{supp}(\chi_i) \cap \bar{D}_{i,t}$, $U|_G = u|_G \in C^\infty(G)$, so that, by (A.26),

$$\int_{\Gamma_{i,t}} \frac{\partial U}{\partial n} V_i\,\mathrm{d}s = \int_{D_{i,t}} (\nabla U \cdot \nabla \bar{V}_i + \bar{V}_i\,\Delta U)\,\mathrm{d}x, \qquad \text{(A.43)}$$

for $t > 0$, where $V_i = \chi_i V$. Noting that, for $x \in \Gamma_{i,t} \cap W_i$ and all sufficiently small $t > 0$, we have, for some constant $C > 0$ independent of x and t, that

$$\left| V_i(x) \frac{\partial U}{\partial n}(x) \right| = \left| v_i(x) \frac{\partial u}{\partial n}(x) \right| \le C(\nabla u)^*(x - t\hat{b}_i),$$

taking the limit as $t \to 0$ in (A.43) and using the dominated convergence theorem we see that

$$\int_{\Gamma_i} \frac{\partial U}{\partial n} \gamma \bar{V}_i\,\mathrm{d}s = \int_{D_i} (\nabla U \cdot \nabla \bar{V}_i + \bar{V}_i\,\Delta U)\,\mathrm{d}x.$$

But this is just (A.42) with v replaced by v_i, since $\operatorname{supp}(\chi_i) \subset W_i$, $D_i \cap W_i = D \cap W_i$, and $U = u$ and $V_i = v_i$ in D.

We have shown that $\partial_n u \in L^2(\Gamma)$ and that (A.41) is true. It follows from Theorem A.5 that $\gamma u \in H^1(\Gamma)$. It remains to show that, for almost every $x \in \Gamma$,

$$\psi_T(x) := \psi(x) - n(x)(n(x) \cdot \psi(x)) = \nabla_\Gamma \gamma u(x).$$

Noting that $\gamma u = \tilde{\gamma} u$ by Lemma A.9, and arguing as in the proof of that lemma, we see that it is enough to show that $\int_\Gamma (\psi_T - \nabla_\Gamma \tilde{\gamma} u) \cdot \phi\,\mathrm{d}s = 0$, for all $\phi \in (L^2(\Gamma))^d$ with $\operatorname{supp}(\phi) \subset \operatorname{supp}(\chi_i)$, for some i. But, for such a ϕ, since $\operatorname{supp}(\phi) \subset \Gamma_i \cap \Gamma$, with a slight abuse of notation we have

$$\int_\Gamma (\psi_T - \nabla_\Gamma \tilde{\gamma} u) \cdot \phi\,\mathrm{d}s = \int_{\Gamma_i} (\psi_T - \nabla_\Gamma \tilde{\gamma} u) \cdot \phi_T\,\mathrm{d}s, \qquad \text{(A.44)}$$

where $\phi_T \in L_t^2(\Gamma_i)$ is the tangential part of ϕ, *i.e.*, $\phi_T := \phi - n(n \cdot \phi)$. It is sufficient to show that the integral on the right-hand side of (A.44) is zero for all $\phi_T \in X_t$ with $\operatorname{supp}(\phi_T) \subset \operatorname{supp}(\chi_i)$, where X_t is the dense subset of $L_t^2(\Gamma_i)$ that satisfies $\nabla_\Gamma^*(X_t) \subset L^2(\Gamma_i)$ introduced just below (A.20). But, for such a ϕ_T, and using the notations Γ^* and U_t introduced in the proof of Lemma A.9, it follows from (A.20), $(\nabla u)^* \in L^2(\Gamma)$, and the dominated convergence theorem, that

$$\begin{aligned}
\int_{\Gamma_i} \nabla_\Gamma \tilde{\gamma} u \cdot \phi_T \, \mathrm{d}s &= \int_{\Gamma_i} \frac{\tilde{\gamma} u \nabla_\Gamma^* \phi_T}{\sqrt{1+|\nabla f|^2}} \, \mathrm{d}s \\
&= \int_{\Gamma^*} \frac{\tilde{\gamma} u \nabla_\Gamma^* \phi_T}{\sqrt{1+|\nabla f|^2}} \, \mathrm{d}s \\
&= \lim_{t\to 0} \int_{\Gamma^*} \frac{U_t \nabla_\Gamma^* \phi_T}{\sqrt{1+|\nabla f|^2}} \, \mathrm{d}s \\
&= \lim_{t\to 0} \int_{\Gamma_i} \frac{U_t \nabla_\Gamma^* \phi_T}{\sqrt{1+|\nabla f|^2}} \, \mathrm{d}s \\
&= \lim_{t\to 0} \int_{\Gamma_i} \nabla_\Gamma U_t \cdot \phi_T \, \mathrm{d}s \\
&= \int_{\Gamma_i} \psi_T \cdot \phi_T \, \mathrm{d}s. \qquad \square
\end{aligned}$$

In this subsection, and in the related material in Section 2.2, we have provided an introduction, aimed specifically at the Helmholtz equation, to the large body of research in the harmonic analysis literature on boundary value problems in Lipschitz domains with L^2 boundary data and on associated properties of layer potentials. For more detail we refer the reader to Coifman *et al.* (1982), Verchota (1984), Torres and Welland (1993), Liu (1995), Jerison and Kenig (1995), Mitrea (1996), Chandler-Wilde and Langdon (2007), Spence *et al.* (2011) and to the monographs by Kenig (1994), Meyer and Coifman (2000), Taylor (2000), and Mitrea, Mitrea and Taylor (2001).

A.5. Exterior and interior traces

To study the properties of boundary integral representations we will need to consider functions defined on both sides of a Lipschitz boundary. So suppose that Ω_- is a bounded Lipschitz open set and let $\Omega_+ := \mathbb{R}^d \setminus \overline{\Omega_-}$ and $\Gamma = \partial\Omega_- = \partial\Omega_+$. Then Ω_+ is an unbounded Lipschitz open set, and Γ is the Lipschitz boundary of both Ω_- and Ω_+, with Ω_- the interior and Ω_+ the exterior of Γ. In this configuration, we will use the notation $\gamma_\pm$ to denote the trace operator γ from $H^s(\Omega_\pm) \to H^{s-1/2}(\Gamma)$, for $1/2 < s < 3/2$, and $\partial_n^\pm$ will be our notation for the normal derivative operator ∂_n from $H^1(\Omega_\pm;\Delta)$ to $H^{-1/2}(\Gamma)$, with the normal n directed out of Ω_- into Ω_+. Thus γ_+ and ∂_n^+ are trace and normal derivatives operators from the exterior side of Γ and γ_- and ∂_n^- those from the interior.

Acknowledgements

This review has benefited greatly from the contributions of our collaborators and also from the comments and suggestions of a number of people. We would like to acknowledge the contributions of the following (listed in alphabetical order): L. Banjai, T. Betcke, V. Domínguez, M. Ganesh, D. P. Hewett, R. Hiptmair, D. Huybrechs, V. Ivrii, I. V. Kamotski, T. Kim, E. Lakshtanov, M. Lindner, J. M. Melenk, A. Moiola, M. Mokgolele, V. P. Smyshlyaev, A. Twigger, B. R. Vainberg, A. Vasy and J. Wunsch.

REFERENCES[1]

T. Abboud, J.-C. Nédélec and B. Zhou (1994), 'Méthode des équations intégrales pour les hautes fréquences', *CR Acad. Sci.* I **318**, 165–170.

T. Abboud, J.-C. Nédélec and B. Zhou (1995), Improvement of the integral equation method for high frequency problems. In *Mathematical and Numerical Aspects of Wave Propagation: Mandelieu–La Napoule, 1995*, SIAM, pp. 178–187.

K. R. Aberegg and A. F. Peterson (1995), 'Application of the integral equation-asymptotic phase method to two-dimensional scattering', *IEEE Trans. Antennas Propagat.* **43**, 534–537.

M. Abramowitz and I. A. Stegun (1964), *Handbook of Mathematical Functions with Formulas, Graphs, and Mathematical Tables*, Dover.

M. Ainsworth and H. A. Wajid (2010), 'Optimally blended spectral-finite element scheme for wave propagation and nonstandard reduced integration', *SIAM J. Numer. Anal.* **48**, 346–371.

S. Amini (1990), 'On the choice of the coupling parameter in boundary integral formulations of the exterior acoustic problem', *Appl. Anal.* **35**, 75–92.

S. Amini (1993), 'Boundary integral solution of the exterior acoustic problem', *Comput. Mech.* **13**, 2–11.

S. Amini and N. D. Maines (1998), 'Preconditioned Krylov subspace methods for boundary element solution of the Helmholtz equation', *Internat. J. Numer. Methods Engng* **41**, 875–898.

A. Anand, Y. Boubendir, F. Ecevit and F. Reitich (2009), 'Analysis of multiple scattering iterations for high-frequency scattering problems II: The three-dimensional scalar case', *Numer. Math.* **114**, 373–427.

X. Antoine and M. Darbas (2012), Integral equations and iterative schemes for acoustic scattering problems. Chapter 2 of *Numerical Methods for Acoustics Problems* (F. Magoules, ed.), Saxe-Coburg Publications.

S. Arden, S. N. Chandler-Wilde and S. Langdon (2007), 'A collocation method for high-frequency scattering by convex polygons', *J. Comput. Appl. Math.* **204**, 334–343.

[1] The URLs cited in this work were correct at the time of going to press, but the publisher and the authors make no undertaking that the citations remain live or are accurate or appropriate.

A. Asheim (2008), 'A combined Filon/asymptotic quadrature method for highly oscillatory problems', *BIT Numer. Math.* **48**, 425–448.

A. Asheim and D. Huybrechs (2010*a*), 'Asymptotic analysis of numerical steepest descent with path approximations', *Found. Comput. Math.* **10**, 647–671.

A. Asheim and D. Huybrechs (2010*b*), 'Local solutions to high frequency 2D scattering problems', *J. Comput. Phys.* **229**, 5357–5372.

K. E. Atkinson (1997), *The Numerical Solution of Integral Equations of the Second Kind*, Cambridge Monographs on Applied and Computational Mathematics, Cambridge University Press.

V. M. Babich and V. S. Buldyrev (1991), *Short-Wavelength Diffraction Theory*, Springer.

V. M. Babich and V. S. Buldyrev (2008), *Asymptotic Methods in Short-Wavelength Diffraction Theory*, Alpha Science Series on Wave Phenomena.

V. M. Babich and N. Y. Kirpichnikova (1979), *The Boundary-Layer Method in Diffraction Problems*, Springer.

I. Babuška and J. M. Melenk (1997), 'The partition of unity method', *Internat. J. Numer. Methods Engng* **40**, 727–758.

I. M. Babuška and S. A. Sauter (2000), 'Is the pollution effect of the FEM avoidable for the Helmholtz equation considering high wave numbers?', *SIAM Review* **42**, 451–484.

N. S. Bakhvalov and L. G. Vasilčeva (1968), 'Evaluation of the integrals of oscillating functions by interpolation at nodes of Gaussian quadrature', *USSR Comp. Math. Phys.* **8**, 241–249.

L. Banjai and W. Hackbusch (2008), 'Hierarchical matrix techniques for low and high frequency Helmholtz equation', *IMA J. Numer. Anal.* **28**, 46–79.

L. Banjai and S. Sauter (2007), 'A refined Galerkin error and stability analysis for highly indefinite variational problems', *SIAM J. Numer. Anal.* **45**, 37–53.

J.-D. Benamou (2003), 'An introduction to Eulerian geometrical optics (1992–2002)', *J. Sci. Comput.* **19**, 63–93.

H. Berestycki and P. L. Lions (1983), 'Nonlinear scalar field equations I: Existence of a ground state', *Arch. Rat. Mech. Anal.* **82**, 313–345.

T. Betcke and E. A. Spence (2011), 'Numerical estimation of coercivity constants for boundary integral operators in acoustic scattering', *SIAM J. Numer. Anal.* **49**, 1572–1601.

T. Betcke, S. N. Chandler-Wilde, I. G. Graham, S. Langdon and M. Lindner (2011), 'Condition number estimates for combined potential boundary integral operators in acoustics and their boundary element discretisation', *Numer. Methods Partial Diff. Equations* **27**, 31–69.

T. Betcke, S. N. Chandler-Wilde, I. G. Graham and S. Langdon (2012*a*), On trapping domains and exponential ill-conditioning in acoustic scattering. In preparation.

T. Betcke, J. Phillips and E. A. Spence (2012*b*), Spectral decompositions and nonnormality of boundary integral operators in acoustic scattering. Submitted for publication.

N. Bleistein and R. A. Handelsman (1986), *Asymptotic Expansions of Integrals*, Dover.

E. Bleszynski, M. Bleszynski and T. Jaroszewicz (2004), 'Development of new algorithms for high frequency electromagnetic scattering', *Comput. Model. Engng Sci.* **5**, 295–317.

V. A. Borovikov and B. Y. Kinber (1994), *Geometrical Theory of Diffraction*, Vol. 37 of *IEE Electromagnetic Waves Series*, IEE.

H. Brakhage and P. Werner (1965), 'Über das dirichletsche Aussenraumproblem für die helmholtzsche Schwingungsgleichung', *Archiv Math.* **16**, 325–329.

M. Bronstein and V. Ivrii (2003), 'Sharp spectral asymptotics for operators with irregular coefficients I: Pushing the limits', *Comm. Partial Diff. Equations* **28**, 83–102.

O. P. Bruno and C. A. Geuzaine (2007), 'An $O(1)$ integration scheme for three-dimensional surface scattering problems', *J. Comput. Appl. Math.* **204**, 463–476.

O. P. Bruno and L. A. Kunyansky (2001), 'Surface scattering in three dimensions: an accelerated high-order solver', *Proc. Royal Soc. London* A **457**, 2921–2934.

O. P. Bruno and F. Reitich (2007), High order methods for high-frequency scattering applications. In *Modeling and Computations in Electromagnetics* (H. Ammari, ed.), Springer, pp. 129–164.

O. P. Bruno, V. Domínguez and F. Sayas (2012), Convergence analysis of a high-order Nyström integral-equation method for surface scattering problems. Submitted for publication. Arxiv: 1109.6352.

O. P. Bruno, C. A. Geuzaine, J. A. Monro and F. Reitich (2004), 'Prescribed error tolerances within fixed computational times for scattering problems of arbitrarily high frequency: The convex case', *Phil. Trans. Royal Soc. London* A **362**, 629.

A. Buffa and S. H. Christiansen (2003), 'The electric field integral equation on Lipschitz screens: Definitions and numerical approximation', *Numer. Math.* **94**, 229–267.

A. Buffa and R. Hiptmair (2005), 'A coercive combined field integral equation for electromagnetic scattering', *SIAM J. Numer. Anal.* **42**, 621–640.

A. Buffa and P. Monk (2008), 'Error estimates for the ultra weak variational formulation of the Helmholtz equation', *M2AN: Math. Model. Numer. Anal.* **42**, 925–940.

A. Buffa and S. Sauter (2006), 'On the acoustic single layer potential: Stabilization and Fourier analysis', *SIAM J. Sci. Comput.* **28**, 1974–1999.

W. D. Burnside, C. L. Yu and R. J. Marhefka (1975), 'A technique to combine the geometrical theory of diffraction and the moment method', *IEEE Trans. Antennas Propagat.* **23**, 551–557.

N. Burq (2002), 'Semi-classical estimates for the resolvent in nontrapping geometries', *Internat. Math. Res. Not.* **2002**, 221–241.

A. J. Burton and G. F. Miller (1971), 'The application of integral equation methods to the numerical solution of some exterior boundary-value problems', *Proc. Royal Soc. London* A **323**, 201–210.

V. S. Buslaev (1962), 'Formulas for the short-wave asymptotic behavior in the diffraction problem by convex bodies' (in Russian), *Vestnik Leningrad. Univ.* **17**, 5–21.

V. S. Buslaev (1964), 'Short wave asymptotic behavior in the problem of diffraction by smooth convex contours' (in Russian), *Trudy Matematicheskogo Instituta imeni Steklova* **73**, 14–117.

V. S. Buslaev (1975), 'On the asymptotic behavior of the spectral characteristics of exterior problems for the Schrödinger operator', *Izvestiya Rossiiskoi Akademii Nauk. Seriya Matematicheskaya* **39**, 149–235.

O. Cessenat and B. Després (1998), 'Application of an ultra weak variational formulation of elliptic PDEs to the two-dimensional Helmholtz problem', *SIAM J. Numer. Anal.* **35**, 255–299.

O. Cessenat and B. Després (2003), 'Using plane waves as base functions for solving time harmonic equations with the ultra weak variational formulation', *J. Comput. Acoust.* **11**, 227–238.

G. A. Chandler (1980), Superconvergence for second kind integral equations. In *The Application and Numerical Solution of Integral Equations*, Sijthoff and Nordhoff, pp. 103–117.

S. N. Chandler-Wilde (1988), Ground effects in environmental sound propagation. PhD thesis, University of Bradford.

S. N. Chandler-Wilde and I. G. Graham (2009), Boundary integral methods in high frequency scattering. In *Highly Oscillatory Problems* (B. Engquist, A. Fokas, E. Hairer and A. Iserles, eds), Cambridge University Press.

S. N. Chandler-Wilde and D. P. Hewett (2012), High frequency scattering by screens: Frequency-explicit coercivity estimates. University of Reading preprint MPS-2012-03. Submitted for publication.

S. N. Chandler-Wilde and D. C. Hothersall (1995), 'Efficient calculation of the Green function for acoustic propagation above a homogeneous impedance plane', *J. Sound Vib.* **180**, 705–724.

S. N. Chandler-Wilde and S. Langdon (2007), 'A Galerkin boundary element method for high frequency scattering by convex polygons', *SIAM J. Numer. Anal.* **45**, 610–640.

S. N. Chandler-Wilde and P. Monk (2005), 'Existence, uniqueness, and variational methods for scattering by unbounded rough surfaces', *SIAM J. Math. Anal.* **37**, 598–618.

S. N. Chandler-Wilde and P. Monk (2008), 'Wave-number-explicit bounds in time-harmonic scattering', *SIAM J. Math. Anal.* **39**, 1428–1455.

S. N. Chandler-Wilde and E. A. Spence (2012), Coercive modifications of the double-layer potential operator on Lipschitz domains. In preparation.

S. N. Chandler-Wilde, I. G. Graham, S. Langdon and M. Lindner (2009), 'Condition number estimates for combined potential boundary integral operators in acoustic scattering', *J. Integral Equations Appl.* **21**, 229–279.

S. N. Chandler-Wilde, E. Heinemeyer and R. Potthast (2006), 'A well-posed integral equation formulation for three-dimensional rough surface scattering', *Proc. Royal Soc.* A **462**, 3683–3705.

S. N. Chandler-Wilde, D. P. Hewett, S. Langdon and A. Twigger (2012*a*), A high frequency boundary element method for scattering by a class of non-convex obstacles. University of Reading preprint MPS-2012-04. Submitted for publication.

S. N. Chandler-Wilde, S. Langdon and M. Mokgolele (2012*b*), 'A high frequency boundary element method for scattering by convex polygons with impedance boundary conditions', *Commun. Comput. Phys.* **11**, 573–593.

S. N. Chandler-Wilde, S. Langdon and L. Ritter (2004), 'A high-wavenumber boundary-element method for an acoustic scattering problem', *Phil. Trans. Royal Soc. London* A **362**, 647–671.

S. N. Chandler-Wilde, M. Rahman and C. R. Ross (2002), 'A fast two-grid and finite section method for a class of integral equations on the real line with application to an acoustic scattering problem in the half-plane', *Numer. Math.* **93**, 1–51.

W. C. Chew, J. M. Song, T. J. Cui, S. Velamparambil, M. L. Hastriter and B. Hu (2004), 'Review of large scale computing in electromagnetics with fast integral equation solvers', *Comput. Model. Engng Sci.* **5**, 361–372.

S. H. Christiansen and J. C. Nédélec (2000), 'Preconditioners for the numerical solution of boundary integral equations from electromagnetism', *CR Acad. Sci.* I **331**, 733–738.

P. G. Ciarlet (2002), *The Finite Element Method For Elliptic Problems*, Classics in Applied Mathematics, SIAM.

X. Claeys and H. Haddar (2007), 'Scattering from infinite rough tubular surfaces', *Math. Methods Appl. Sci.* **30**, 389–414.

F. Cobos, D. E. Edmunds and A. J. B. Potter (1990), 'Real interpolation and compact linear operators', *J. Funct. Anal.* **88**, 351–365.

R. R. Coifman, A. McIntosh and Y. Meyer (1982), 'L'intégrale de Cauchy définit un opérateur borné sur L^2 pour les courbes lipschitziennes', *Ann. of Math.* **116**, 361–387.

D. Colton and R. Kress (1998), *Inverse Acoustic and Electromagnetic Scattering Theory*, Springer.

D. L. Colton and R. Kress (1983), *Integral Equation Methods in Scattering Theory*, Wiley.

M. Costabel (1988), 'Boundary integral operators on Lipschitz domains: Elementary results', *SIAM J. Math. Anal.* **19**, 613–626.

M. Costabel (2004), Time-dependent problems with the boundary integral equation method, Chapter 22 of *Encyclopedia of Computational Mechanics*, Wiley.

M. Costabel (2007), Some historical remarks on the positivity of boundary integral operators. In *Boundary Element Analysis* (M. Schanz and O. Steinbach, eds), Vol. 29 of *Lecture Notes in Applied and Computational Mechanics*, Springer, pp. 1–27.

P. Cummings and X. Feng (2006), 'Sharp regularity coefficient estimates for complex-valued acoustic and elastic Helmholtz equations', *Math. Models Methods Appl. Sci.* **16**, 139.

E. Darrigrand (2002), 'Coupling of fast multipole method and microlocal discretization for the 3-D Helmholtz equation', *J. Comput. Phys.* **181**, 126–154.

E. Darve and P. Havé (2004), 'A fast multipole method for Maxwell equations stable at all frequencies', *Phil. Trans. Royal Soc. London* A **362**, 603–628.

C. P. Davis and W. C. Chew (2008), 'Frequency-independent scattering from a flat strip with TE_z-polarized fields', *IEEE Trans. Antennas Propagat.* **56**, 1008–1016.

A. de La Bourdonnaye (1994), 'A microlocal discretization method and its utilization for a scattering problem', *CR Acad. Sci.* I **318**, 385–388.

A. de La Bourdonnaye and M. Tolentino (2004), 'Reducing the condition number for microlocal discretization problems', *Phil. Trans. Royal Soc. London* A **362**, 541–559.

A. Deaño and D. Huybrechs (2009), 'Complex Gaussian quadrature of oscillatory integrals', *Numer. Math.* **112**, 197–219.

G. H. Derrick (1964), 'Comments on nonlinear wave equations as models for elementary particles', *J. Math. Phys.* **5**, 1252–1254.

M. Djordjević and B. M. Notaroš (2005), 'Higher order hybrid method of moments-physical optics modeling technique for radiation and scattering from large perfectly conducting surfaces', *IEEE Trans. Antennas Propagat.* **53**, 800–813.

V. Domínguez (2007), private communication.

V. Domínguez (2009), Modified Clenshaw Curtis rule package for a class of highly oscillatory integrals.
http://www.unavarra.es/personal/victor_dominguez/clenshawcurtisrule

V. Domínguez, I. G. Graham and T. Kim (2012), Filon–Clenshaw–Curtis rules for highly oscillatory integrals with algebraic singularities and stationary points. In preparation.

V. Domínguez, I. G. Graham and V. P. Smyshlyaev (2007), 'A hybrid numerical-asymptotic boundary integral method for high-frequency acoustic scattering', *Numer. Math.* **106**, 471–510.

V. Domínguez, I. G. Graham and V. P. Smyshlyaev (2011), 'Stability and error estimates for Filon–Clenshaw–Curtis rules for highly-oscillatory integrals', *IMA J. Numer. Anal.* **31**, 1253–1280.

V. Domínguez, I. G. Graham, T. Kim and V. P. Smyshlyaev (2009), 'Recent progress on hybrid numerical asymptotic boundary integral methods for high-frequency scattering problems', *Proc. UKBIM7* (H. Power, A. La Rocca and S. T. Baker, eds), Faculty of Engineering, University of Nottingham, pp. 15–23.

K. C. Donepudi, J. M. Jin and W. C. Chew (2003), 'A grid-robust higher-order multilevel fast multipole algorithm for analysis of 3-D scatterers', *Electromagnetics* **23**, 315–330.

F. Ecevit (2005), Integral equation formulations of electromagnetic and acoustic scattering problems: High-frequency asymptotic expansions and convergence of multiple scattering iterations. PhD thesis, University of Minnesota.

F. Ecevit and F. Reitich (2009), 'Analysis of multiple scattering iterations for high-frequency scattering problems I: The two-dimensional case', *Numer. Math.* **114**, 271–354.

J. Elschner (1992), 'The double layer potential operator over polyhedral domains I: Solvability in weighted Sobolev spaces', *Appl. Anal.* **45**, 117–134.

S. Engleder and O. Steinbach (2007), 'Modified boundary integral formulations for the Helmholtz equation', *J. Math. Anal. Appl.* **331**, 396–407.

S. Engleder and O. Steinbach (2008), 'Stabilized boundary element methods for exterior Helmholtz problems', *Numer. Math.* **110**, 145–160.

B. Engquist and O. Runborg (2003), Computational high frequency wave propagation. In *Acta Numerica*, Vol. 12, Cambridge University Press, pp. 181–266.

B. Engquist and L. Ying (2007), 'Fast directional multilevel algorithms for oscillatory kernels', *SIAM J. Sci. Comput.* **29**, 1710–1737.

B. Engquist and L. Ying (2012), Fast algorithms for high frequency wave propagation. In *Numerical Analysis of Multiscale Problems* (I. G. Graham, T. Y. Hou, O. Lakkis and R. Scheichl, eds), Vol. 83 of *Lecture Notes in Computational Science and Engineering*, Springer, pp. 127–161.

Y. A. Erlangga, C. W. Oosterlee and C. Vuik (2006), 'A novel multigrid based precconditioner for heterogeneous Helmholtz problems', *SIAM J. Sci. Comput.* **27**, 1471–1492.

O. G. Ernst and M. J. Gander (2012), Why it is difficult to solve Helmholtz problems with classical iterative methods. In *Numerical Analysis of Multiscale Problems* (I. G. Graham, T. Y. Hou, O. Lakkis and R. Scheichl, eds), Vol. 83 of *Lecture Notes in Computational Science and Engineering*, Springer, pp. 325–363.

S. Esterhazy and J. M. Melenk (2012), On stability of discretizations of the Helmholtz equation. In *Numerical Analysis of Multiscale Problems* (I. G. Graham, T. Y. Hou, O. Lakkis and R. Scheichl, eds), Vol. 83 of *Lecture Notes in Computational Science and Engineering*, Springer, pp. 285–324.

G. A. Evans and J. R. Webster (1999), 'A comparison of some methods for the evaluation of highly oscillatory integrals', *J. Comput. Appl. Math.* **112**, 55–69.

L. C. Evans (1998), *Partial Differential Equations*, AMS.

E. B. Fabes, M. Jodeit and N. M. Riviere (1978), 'Potential techniques for boundary value problems on C^1 domains', *Acta Mathematica* **141**, 165–186.

C. Farhat, A. Macedo and M. Lesoinne (2000), 'A two-level domain decomposition method for the iterative solution of high frequency exterior Helmholtz problems', *Numer. Math.* **85**, 283–308.

V. Filippov (1976), 'Rigorous justification of the shortwave asymptotic theory of diffraction in the shadow zone', *J. Sov. Math.* **6**, 577–626.

L. N. G. Filon (1928), 'On a quadrature formula for trigonometric integrals', *Proc. Royal Soc. Edinburgh* **49**, 38–47.

V. A. Fock (1965), *Electromagnetic Diffraction and Propagation Problems*, Vol. 1, Pergamon Press.

M. Ganesh and I. G. Graham (2004), 'A high-order algorithm for obstacle scattering in three dimensions', *J. Comput. Phys.* **198**, 211–242.

M. Ganesh and S. Hawkins (2011), 'A fully discrete Galerkin method for high frequency exterior acoustic scattering in three dimensions', *J. Comput. Phys.* **230**, 104–125.

M. Ganesh, I. G. Graham and J. Sivaloganathan (1998), 'A new spectral boundary integral collocation method for three-dimensional potential problems', *SIAM J. Numer. Anal.* **35**, 778–805.

M. Ganesh, S. Langdon and I. H. Sloan (2007), 'Efficient evaluation of highly oscillatory acoustic scattering surface integrals', *J. Comput. Appl. Math.* **204**, 363–374.

P. Gérard and G. Lebeau (1993), 'Diffusion d'une onde par un coin', *J. Amer. Math. Soc.* **6**, 341–424.

C. Geuzaine, O. Bruno and F. Reitich (2005), 'On the $O(1)$ solution of multiple-scattering problems', *IEEE Trans. Magnetics* **41**, 1488–1491.

K. Giebermann (1997), Schnelle Summationsverfahren zur numerischen Lösung von Integralgleichungen für Streuprobleme im $\mathbb{R}^3$. PhD thesis, University of Karlsruhe.

E. Giladi (2007), 'Asymptotically derived boundary elements for the Helmholtz equation in high frequencies', *J. Comput. Appl. Math.* **198**, 52–74.

E. Giladi and J. B. Keller (2001), 'A hybrid numerical asymptotic method for scattering problems', *J. Comput. Phys.* **174**, 226–247.

E. Giladi and J. B. Keller (2004), An asymptotically derived boundary element method for the Helmholtz equations. In *Proc. 20th Annual Review of Progress in Applied Computational Electromagnetics*, Applied Computational Electromagnetics Society, USA.

C. J. Gittelson, R. Hiptmair and I. Perugia (2009), 'Plane wave discontinuous Galerkin methods: Analysis of the h-version', *M2AN: Math. Model. Numer. Anal.* **43**, 297–331.

I. G. Graham and I. H. Sloan (2002), 'Fully discrete spectral boundary integral methods for Helmholtz problems on smooth closed surfaces in $\mathbb{R}^3$', *Numer. Math.* **92**, 289–323.

I. G. Graham, M. Löhndorf, J. M. Melenk and E. A. Spence (2012), Wavenumber explicit h-BEM for high frequency scattering. In preparation.

P. Grisvard (1985), *Elliptic Problems in Nonsmooth Domains*, Vol. 24, Pitman.

K. E. Gustafson and D. K. M. Rao (1997), *Numerical Range: The Field of Values of Linear Operators and Matrices*, Universitext, Springer.

T. Ha-Duong (1990), 'On the transient acoustic scattering by a flat object', *Japan J. Appl. Math.* **7**, 489–513.

T. Ha-Duong (1992), 'On the boundary integral equations for the crack opening displacement of flat cracks', *Integral Equations and Operator Theory* **15**, 427–453.

T. Ha-Duong (2003), On retarded potential boundary integral equations and their discretisation. In *Topics in Computational Wave Propagation*, Vol. 31 of *Lecture Notes in Computational Science and Engineering*, Springer, pp. 301–336.

H. Haddar and A. Lechleiter (2011), 'Electromagnetic wave scattering from rough penetrable layers', *SIAM J. Math. Anal.* **43**, 2418–2443.

T. Harge and G. Lebeau (1994), 'Diffraction par un convexe', *Invent. Math.* **118**, 161–196.

P. J. Harris and K. Chen (2003), 'On efficient preconditioners for iterative solution of a Galerkin boundary element equation for the three-dimensional exterior Helmholtz problem', *J. Comput. Appl. Math.* **156**, 303–318.

U. Hetmaniuk (2007), 'Stability estimates for a class of Helmholtz problems', *Commun. Math. Sci.* **5**, 665–678.

D. P. Hewett, S. Langdon and J. M. Melenk (2012), A high frequency hp boundary element method for scattering by convex polygons. University of Reading preprint MPS-2011-18. Submitted for publication.

L. Hillairet (2005), 'Contribution of periodic diffractive geodesics', *J. Funct. Anal.* **226**, 48–89.

R. Hiptmair, A. Moiola and I. Perugia (2011*a*), 'Plane wave discontinuous Galerkin methods for the 2D Helmholtz equation: Analysis of the p-version', *SIAM J. Numer. Anal.* **49**, 264–284.

R. Hiptmair, A. Moiola and I. Perugia (2011*b*), 'Stability results for the time-harmonic Maxwell equations with impedance boundary conditions', *Math. Mod. Methods Appl. Sci.* **21**, 2263–2287.

M. E. Honnor, J. Trevelyan and D. Huybrechs (2010), 'Numerical evaluation of the two-dimensional partition of unity boundary integrals for Helmholtz problems', *J. Comput. Appl. Math.* **234**, 1656–1662.

L. Hörmander (1954), Uniqueness theorems and estimates for normally hyperbolic partial differential equations of the second order. In *Tolfte Skandinaviska Matematikerkongressen*, Lunds Universitets Matematiska Institution, pp. 105–115.

L. Hörmander (1983*a*), *The Analysis of Linear Partial Differential Operators I: Distribution Theory and Fourier Analysis*, Springer.

L. Hörmander (1983*b*), *The Analysis of Linear Partial Differential Operators II: Differential Operators With Constant Coefficients*, Springer.

L. Hörmander (1985*a*), *The Analysis of Linear Partial Differential Operators III: Pseudo-Differential Operators*, Springer.

L. Hörmander (1985*b*), *The Analysis of Linear Partial Differential Operators IV: Fourier Integral Operators*, Springer.

T. Huttunen, P. Monk and J. P. Kaipio (2002), 'Computational aspects of the ultra-weak variational formulation', *J. Comput. Phys.* **182**, 27–46.

D. Huybrechs and S. Olver (2009), Highly oscillatory quadrature. In *Highly Oscillatory Problems* (B. Engquist, A. Fokas, E. Hairer and A. Iserles, eds), Cambridge University Press.

D. Huybrechs and S. Olver (2012), 'Superinterpolation in highly oscillatory quadrature', *Found. Comput. Math.* **12**, 203–228.

D. Huybrechs and S. Vandewalle (2006), 'On the evaluation of highly oscillatory integrals by analytic continuation', *SIAM J. Numer. Anal.* **44**, 1026–1048.

D. Huybrechs and S. Vandewalle (2007*a*), 'The construction of cubature rules for multivariate highly oscillatory integrals', *Math. Comp.* **76**, 1955–1980.

D. Huybrechs and S. Vandewalle (2007*b*), 'A sparse discretization for integral equation formulations of high frequency scattering problems', *SIAM J. Sci. Comput.* **29**, 2305–2328.

M. Ikawa (1983), 'On the poles of the scattering matrix for two strictly convex obstacles', *J. Math. Kyoto Univ* **23**, 127–194.

A. Iserles (2004), 'On the numerical quadrature of highly-oscillating integrals I: Fourier transforms', *IMA J. Numer. Anal.* **24**, 365–391.

A. Iserles (2005), 'On the numerical quadrature of highly-oscillating integrals II: Irregular oscillators', *IMA J. Numer. Anal.* **25**, 25–44.

A. Iserles and S. Nørsett (2004), 'On quadrature methods for highly oscillatory integrals and their implementation', *BIT* **44**, 755–772.

A. Iserles and S. P. Nørsett (2005), 'Efficient quadrature of highly oscillatory integrals using derivatives', *Proc. Royal Soc. London* A **461**, 1383–1399.

A. Iserles and S. P. Nørsett (2006), 'Quadrature methods for multivariate highly oscillatory integrals using derivatives', *Math. Comp.* **75**, 1233–1258.

A. Iserles, S. P. Nørsett and S. Olver (2005), Highly-oscillatory quadrature: The story so far. In *Proc. ENuMath, Santiago de Compostela* (A. Bermudez de Castro *et al.*, eds), Springer, pp. 97–118.

V. Ivrii (2003), 'Sharp spectral asymptotics for operators with irregular coefficients II: Domains with boundaries and degenerations', *Comm. Partial Diff. Equations* **28**, 103–128.

R. M. James (1990), 'A contribution to scattering calculation for small wavelengths: The high frequency panel method', *IEEE Trans. Antennas Propagat.* **38**, 1625–1630.

D. Jerison and C. E. Kenig (1995), 'The inhomogeneous Dirichlet problem in Lipschitz domains', *J. Funct. Anal.* **130**, 161–219.

D. S. Jones (1972), 'Asymptotic behavior of integrals', *SIAM Rev.* **14**, 286–317.

K. Jörgens (1982), *Linear Integral Operators*, Pitman

T. Kato (1995), *Perturbation Theory for Linear Operators*, corrected printing of second edition, Springer.

J. B. Keller (1962), 'Geometrical theory of diffraction', *J. Optical Soc. Amer.* A **52**, 116–130.

C. E. Kenig (1994), *Harmonic Analysis Techniques for Second Order Elliptic Boundary Value Problems*, AMS.

T. Kim (2012), Asymptotic and numerical methods in high frequency scattering. PhD thesis, University of Bath.

A. Kirsch (1989), 'Surface gradients and continuity properties for some integral operators in classical scattering theory', *Math. Methods Appl. Sci.* **11**, 789–804.

R. Kress (1985), 'Minimizing the condition number of boundary integral operators in acoustic and electromagnetic scattering', *Quart. J. Mech. Appl. Math.* **38**, 323–341.

R. Kress and W. T. Spassov (1983), 'On the condition number of boundary integral operators in acoustic and electromagnetic scattering', *Numer. Math.* **42**, 77–95.

O. Laghrouche, P. Bettess and R. J. Astley (2002), 'Modelling of short wave diffraction problems using approximating systems of plane waves', *Internat. J. Numer. Meth. Engng* **54**, 1501–1533.

O. Laghrouche, P. Bettess, E. Perrey-Debain and J. Trevelyan (2005), 'Wave interpolation finite elements for Helmholtz problems with jumps in the wave speed', *Comput. Meth. Appl. Mech. Engng* **194**, 367–381.

E. Lakshtanov and B. Vainberg (2012), 'A priori estimates for high frequency scattering by obstacles of arbitrary shape', *Comm. Partial Diff. Equations*, to appear.

S. Langdon and S. N. Chandler-Wilde (2003*a*), A Galerkin boundary element method for an acoustic scattering problem, with convergence rate independent of frequency. In *Proc. 4th UK Conference on Boundary Integral Methods* (S. Amini, ed.), Salford University Press, pp. 67–76.

S. Langdon and S. N. Chandler-Wilde (2003*b*), 'A GTD-based boundary element method for a surface scattering problem', *Proc. Institute Acoustics* **25**, 224–233.

S. Langdon and S. N. Chandler-Wilde (2006), 'A wavenumber independent boundary element method for an acoustic scattering problem', *SIAM J. Numer. Anal.* **43**, 2450–2477.

S. Langdon, M. Mokgolele and S. N. Chandler-Wilde (2010), 'High frequency scattering by convex curvilinear polygons', *J. Comput. Appl. Math.* **234**, 2020–2026.

P. D. Lax and R. S. Phillips (1989), *Scattering Theory*, Vol. 26 of *Pure and Applied Mathematics*, second edition, Academic Press.

V. Lazutkin (1999), Semiclassical asymptotics of eigenfunctions. In *Partial Differential Equations V*, Vol. 34 of *Encyclopedia of Mathematical Sciences*, Springer.

G. Lebeau (1984), 'Régularité Gevrey 3 pour la diffraction', *Comm. Partial Diff. Equations* **9**, 1437–1494.

A. Lechleiter and S. Ritterbusch (2010), 'A variational method for wave scattering from penetrable rough layers', *IMA J. Appl. Math.* **75**, 366–391.

R. Leis (1965), 'Zur dirichletschen Randwertaufgabe des Außenraumes der Schwingungsgleichung', *Math. Z.* **90**, 205– 211.

C. Liu (1995), The Helmholtz equation on Lipschitz domains. IMA Preprint Series No. 1356, Institute of Mathematics and its Applications, Minnesota.

I. Livschits and A. Brandt (2006), 'Accuracy properties of the wave-ray multigrid algorithm for Helmholtz problems', *SIAM J. Sci. Comput.* **28**, 1228–1251.

M. Löhndorf and J. M. Melenk (2011), 'Wavenumber-explicit *hp*-BEM for high frequency scattering', *SIAM J. Numer. Anal.* **49**, 2340–2363.

D. Ludwig (1967), 'Uniform asymptotic expansion of the field scattered by a convex object at high frequencies', *Comm. Pure Appl. Math.* **20**, 103–138.

Y. L. Luke (1954), 'On the computation of oscillatory integrals', *Math. Proc. Cambridge Phil. Soc.* **50**, 269–277.

T. Luostari, T. Huttunen and P. Monk (2009), Plane wave methods for approximating the time harmonic wave equation. In *Highly Oscillatory Problems: Computation, Theory and Applications* (B. Engquist, A. Fokas, E. Hairer and A. Iserles, eds), Cambridge University Press.

C. H. Makridakis, F. Ihlenburg and I. Babuška (1996), 'Analysis and finite element methods for a fluid–solid interaction problem in one dimension', *Math. Models Methods Appl. Sci.* **6**, 1119–1141.

W. C. H. McLean (2000), *Strongly Elliptic Systems and Boundary Integral Equations*, Cambridge University Press.

J. M. Melenk (1995), On generalized finite element methods. PhD thesis, The University of Maryland.

J. M. Melenk (2006), private communication.

J. M. Melenk (2008), On the convergence of Filon quadrature. ASC Report 08/2008, Institute for Analysis and Scientific Computing, TU Wien.

J. M. Melenk (2010), 'On the convergence of Filon quadrature', *J. Comput. Appl. Math.* **234**, 1692–1701.

J. M. Melenk (2012), 'Mapping properties of combined field Helmholtz boundary integral operators', *SIAM J. Math. Anal.*, to appear.

J. M. Melenk and I. Babuška (1996), 'The partition of unity finite element method: Basic theory and applications', *Comput. Meth. Appl. Math.* **139**, 289–314.

J. M. Melenk and S. Langdon (2007), An *hp*-BEM for high frequency scattering by convex polygons. In *Proc. WAVES2007*, University of Reading, pp. 93–95.

J. M. Melenk and S. Langdon (2012), A fully discrete *hp* boundary element method for high frequency scattering by convex polygons. In preparation.
J. M. Melenk and S. Sauter (2010), 'Convergence analysis for finite element discretizations of the Helmholtz equation with Dirichlet-to-Neumann boundary conditions', *Math. Comp.* **79**, 1871–1914.
J. M. Melenk and S. Sauter (2011), 'Wavenumber explicit convergence analysis for Galerkin discretizations of the Helmholtz equation', *SIAM J. Numer. Anal.* **49**, 1210–1243.
R. Melrose (1979), 'Singularities and energy decay in acoustical scattering', *Duke Math. J.* **46**, 43–59.
R. B. Melrose (1995), *Geometric Scattering Theory*, Stanford Lectures, Cambridge University Press.
R. Melrose and J. Sjöstrand (1978), 'Singularities of boundary value problems I', *Comm. Pure Appl. Math.* **31**, 593–617.
R. Melrose and J. Sjöstrand (1982), 'Singularities of boundary value problems II', *Comm. Pure Appl. Math.* **35**, 129–168.
R. B. Melrose and M. E. Taylor (1985), 'Near peak scattering and the corrected Kirchhoff approximation for a convex obstacle', *Adv. Math.* **55**, 242–315.
R. Melrose, A. Vasy and J. Wunsch (2008), 'Propagation of singularities for the wave equation on edge manifolds', *Duke Math. J.* **144**, 109–193.
R. Melrose, A. Vasy and J. Wunsch (2012), 'Diffraction of singularities for the wave equation on manifolds with corners', *Astérisque*, to appear.
Y. Meyer and R. Coifman (2000), *Wavelets: Calderón–Zygmund and Multilinear Operators*, Cambridge University Press.
M. Mitrea (1996), 'Boundary value problems and Hardy spaces associated to the Helmholtz equation in Lipschitz domains', *J. Math. Anal. Appl.* **202**, 819–842.
D. Mitrea, M. Mitrea and M. Taylor (2001), *Layer Potentials, the Hodge Laplacian, and Global Boundary Problems in Nonsmooth Riemannian Manifolds*, Vol. 150, no. 713 of *Memoirs of the American Mathematical Society.*
P. Monk (2003), *Finite Element Methods for Maxwell's Equations*, Clarendon Press.
P. Monk and D.-Q. Wang (1999), 'A least-squares method for the Helmholtz equation', *Comput. Meth. Appl. Mech. Engng* **175**, 121–136.
C. S. Morawetz (1961), 'The decay of solutions of the exterior initial-boundary value problem for the wave equation', *Comm. Pure Appl. Math.* **14**, 561–568.
C. S. Morawetz (1968), 'Time decay for the nonlinear Klein–Gordon equation', *Proc. Royal Soc. London* A **306**, 291–296.
C. S. Morawetz (1975*a*), 'Decay for solutions of the exterior problem for the wave equation', *Comm. Pure Appl. Math.* **28**, 229–264.
C. S. Morawetz (1975*b*), *Notes on Time Decay and Scattering for Some Hyperbolic Problems*, Vol. 19 of *Regional Conference Series in Applied Mathematics*, SIAM.
C. S. Morawetz and D. Ludwig (1968), 'An inequality for the reduced wave operator and the justification of geometrical optics', *Comm. Pure Appl. Math.* **21**, 187–203.
C. S. Morawetz, J. V. Ralston and W. A. Strauss (1977), 'Decay of solutions of the wave equation outside nontrapping obstacles', *Comm. Pure Appl. Math.* **30**, 447–508.

M. Motamed and O. Runborg (2009), Approximation of high frequency wave propagation problems. In *Highly Oscillatory Problems: Computation, Theory and Applications* (B. Engquist, A. Fokas, E. Hairer and A. Iserles, eds), Cambridge University Press.

J. Nečas (1967), *Les Méthodes Directes en Théorie des Équations Elliptiques*, Masson.

J. C. Nédélec (2001), *Acoustic and Electromagnetic Equations: Integral Representations for Harmonic Problems*, Springer.

F. Oberhettinger and L. Badii (1973), *Tables of Laplace Transforms*, Springer.

F. J. W. Olver (1974), *Asymptotics and Special Functions*, Computer Science and Applied Mathematics, Academic Press.

P. J. Olver (1993), *Applications of Lie Groups to Differential Equations*, second edition, Springer.

S. Olver (2007), 'Moment-free numerical approximation of highly oscillatory integrals with stationary points', *Eur. J. Appl. Math.* **18**, 435–447.

S. Olver (2010), 'Fast, numerically stable computation of oscillatory integrals with stationary points', *BIT Numer. Math.* **50**, 149–171.

O. Osipov, K. Hongo and H. Kobayashi (2002), 'High-frequency approximations for electromagnetic field near a face of an impedance wedge', *IEEE Trans. Antennas Propagat.* **50**, 930–940.

O. I. Panič (1965), 'On the solubility of exterior boundary-value problems for the wave equation and for a system of Maxwell's equations' (in Russian), *Uspehi Mat. Nauk* **20**, 221–226.

L. E. Payne and H. F. Weinberger (1955), 'New bounds in harmonic and biharmonic problems', *J. Math. Phys.* **33**, 291–307.

L. E. Payne and H. F. Weinberger (1958), 'New bounds for solutions of second order elliptic partial differential equations', *Pacific J. Math.* **8**, 551–573.

E. Perrey-Debain, O. Laghrouche, P. Bettess and J. Trevelyan (2004), 'Plane-wave basis finite elements and boundary elements for three-dimensional wave scattering', *Phil. Trans. Royal Soc. London* A **362**, 561–577.

E. Perrey-Debain, J. Trevelyan and P. Bettess (2003*a*), 'Plane wave interpolation in direct collocation boundary element method for radiation and wave scattering: numerical aspects and applications', *J. Sound Vib.* **261**, 839–858.

E. Perrey-Debain, J. Trevelyan and P. Bettess (2003*b*), 'Use of wave boundary elements for acoustic computations', *J. Comput. Acoust.* **11**, 305–321.

E. Perrey-Debain, J. Trevelyan and P. Bettess (2005), 'On wave boundary elements for radiation and scattering problems with piecewise constant impedance', *IEEE Trans. Antennas Propagat.* **53**, 876–879.

B. Perthame and L. Vega (1999), 'Morrey–Campanato estimates for Helmholtz equations', *J. Funct. Anal.* **164**, 340–355.

S. I. Pohozaev (1965), 'Eigenfunctions of the equation $\Delta u + \lambda f(u) = 0$', *Soviet Math. Dokl.* **6**, 1408–1411.

G. Popov (1987), 'Some estimates of Green's functions in the shadow', *Osaka J. Math.* **24**, 1–12.

S. Prössdorf and B. Silbermann (1991), *Numerical Analysis for Integral and Related Operator Equations*, Birkhäuser.

H. Rademacher (1919), 'Über partielle und totale Differenzierbarkeit von Funktionen mehrer Variabeln und über die Transformation der Doppelintegrale (in German)', *Math. Ann.* **79**, 340–359.

F. Rellich (1940), 'Darstellung der Eigenwerte von $\Delta u + \lambda u = 0$ durch ein Randintegral', *Math. Z.* **46**, 635–636.

G. J. Rodin and O. Steinbach (2003), 'Boundary element preconditioners for problems defined on slender domains', *SIAM J. Sci. Comput.* **24**, 1450–1464.

V. Rokhlin (1990), 'Rapid solution of integral equations of scattering theory in two dimensions', *J. Comput. Phys.* **86**, 414–439.

S. A. Sauter and C. Schwab (2011), *Boundary Element Methods*, Springer.

V. Y. Shelepov (1969), 'The index of an integral operator of potential type in the space L_p', *Soviet Math. Dokl.* **10**, 754–757.

E. A. Spence (2012), Wave-number-explicit bounds in time harmonic scattering. In preparation.

E. A. Spence, S. N. Chandler-Wilde, I. G. Graham and V. P. Smyshlyaev (2011), 'A new frequency-uniform coercive boundary integral equation for acoustic scattering', *Comm. Pure Appl. Math.* **64**, 1384–1415.

E. A. Spence, I. V. Kamotski and V. P. Smyshlyaev (2012), Coercivity of combined boundary integral equations in high frequency scattering. In preparation.

P. Stefanov (1999), 'Quasimodes and resonances: Sharp lower bounds', *Duke Math. J.* **99**, 75–92.

P. Stefanov (2003), 'Sharp upper bounds on the number of resonances near the real axis for trapping systems', *Amer. J. Math.* **125**, 183–224.

P. Stefanov and G. Vodev (1995), 'Distribution of resonances for the Neumann problem in linear elasticity outside a strictly convex body', *Duke Math. J.* **78**, 677–714.

P. Stefanov and G. Vodev (1996), 'Neumann resonances in linear elasticity for an arbitrary body', *Commun. Math. Phys.* **176**, 645–659.

E. M. Stein (1993), *Harmonic Analysis: Real-Variable Methods, Orthogonality, and Oscillatory Integrals*, Princeton University Press.

E. M. Stein and G. Weiss (1971), *Introduction to Fourier Analysis on Euclidean Spaces*, Princeton University Press.

O. Steinbach (2008), *Numerical Approximation Methods for Elliptic Boundary Value Problems: Finite and Boundary Elements*, Springer.

O. Steinbach and W. L. Wendland (2001), 'On C. Neumann's method for second-order elliptic systems in domains with non-smooth boundaries', *J. Math. Anal. Appl.* **262**, 733–748.

O. Steinbach and M. Windisch (2011), 'Stable boundary element domain decomposition methods for the Helmholtz equation', *Numer. Math.* **118**, 171–195.

E. P. Stephan (1987), 'Boundary integral equations for screen problems in R^3', *Integral Equations and Operator Theory* **10**, 236–257.

E. P. Stephan and W. L. Wendland (1984), 'An augmented Galerkin procedure for the boundary integral method applied to two-dimensional screen and crack problems', *Appl. Anal.* **18**, 183–219.

W. A. Strauss (1975), 'Dispersal of waves vanishing on the boundary of an exterior domain', *Comm. Pure Appl. Math.* **28**, 265–278.

S. H. Tang and M. Zworski (1998), 'From quasimodes to resonances', *Math. Res. Lett.* **5**, 261–272.

G. Taraldsen and H. Jonasson (2011), 'Aspects of ground effect modelling', *J. Acoust. Soc. Amer.* **129**, 47–53.

M. E. Taylor (1976), 'Grazing rays and reflection of singularities of solutions to wave equations', *Comm. Pure Appl. Math.* **29**, 1–38.

M. E. Taylor (2000), *Tools for PDE: Pseudodifferential Operators, Paradifferential Operators, and Layer Potentials*, AMS.

G. A. Thiele and T. H. Newhouse (1975), 'A hybrid technique for combining moment methods with the geometrical theory of diffraction', *IEEE Trans. Antennas Propagat.* **23**, 62–69.

R. H. Torres and G. V. Welland (1993), 'The Helmholtz equation and transmission problems with Lipschitz interfaces', *Indiana Univ. Math. J.* **42**, 1457–1485.

L. N. Trefethen and M. Embree (2005), *Spectra and Pseudospectra*, Princeton University Press.

J. Trevelyan and M. E. Honnor (2009), 'A numerical coordinate transformation for efficient evaluation of oscillatory integrals over wave boundary elements', *J. Integral Equations Appl.* **21**, 447–468.

J. Trevelyan, M. Honnor and D. Huybrechs (2007), 'Numerical steepest descent evaluation of 2D partition of unity boundary integrals for Helmholtz problems', *Oberwolfach Reports* **4**, 354–357.

R. J. Uncles (1976), 'Numerical solution of acoustic scattering problems at intermediate frequencies', *J. Acoust. Soc. Amer.* **60**, 266–267.

F. Ursell (1968), 'Creeping modes in a shadow', *Proc. Cambridge Philos. Soc.* **64**, 171–191.

B. R. Vainberg (1975), 'On the short wave asymptotic behaviour of solutions of stationary problems and the asymptotic behaviour as $t \to \infty$ of solutions of non-stationary problems', *Russian Math. Surveys* **30**, 1–58.

B. R. Vainberg (1989), *Asymptotic Methods in Equations of Mathematical Physics* (E. Primrose, trans.), Gordon & Breach Science Publishers.

B. R. Vainberg (1999), Asymptotic expansion as $t \to \infty$ of the solutions of exterior boundary value problems for hyperbolic equations and quasiclassical approximations. In *Partial Differential Equations V*, Vol. 34 of *Encyclopedia of Mathematical Sciences*, Springer.

G. Verchota (1984), 'Layer potentials and regularity for the Dirichlet problem for Laplace's equation in Lipschitz domains', *J. Funct. Anal.* **59**, 572–611.

D.-S. Wang (1991), 'Current-based hybrid analysis for surface-wave effects on large scatterers', *IEEE Trans. Antennas Propagat.* **39**, 839–850.

K. F. Warnick and W. C. Chew (1999), 'Convergence of moment-method solutions of the electric field integral equation for a 2-D open cavity', *Microwave and Optical Technology Lett.* **23**, 212–218.

K. F. Warnick and W. C. Chew (2001), 'On the spectrum of the electric field integral equation and the convergence of the moment method', *Internat. J. Numer. Methods Engng* **51**, 31–56.

K. F. Warnick and W. C. Chew (2004), 'Error analysis of the moment method', *IEEE Antennas and Propagation Magazine* **46**, 38–53.

W. L. Wendland (2009), On the double layer potential. In *Analysis, Partial Differential Equations and Applications* (A. Cialdea *et al.*, eds), Birkhäuser, pp. 319–334.

L. Wienert (1990), Die Numerische Approximation von Randintegraloperatoren für die Helmholtzgleichung im $\mathbb{R}^3$. PhD thesis, University of Göttingen.

R. Wong (1989), *Asymptotic Approximation of Integrals*, Vol. 34 of *Classics in Applied Mathematics*, SIAM.

S. Xiang (2007), 'Efficient Filon-type methods for $\int_a^b f(x)\,e^{i\omega g(x)}\,dx$', *Numer. Math.* **105**, 633–658.

A. Zayaev and V. Filippov (1985), 'Rigorous justification of the asymptotic solutions of sliding wave type', *J. Math. Sci.* **30**, 2395–2406.

A. Zayaev and V. Filippov (1986), 'Rigorous justification of the Friedlander–Keller formulas', *J. Math. Sci.* **32**, 134–143.

Y. Zhang, X.-W. Zhao, D. G. Doñoro, S.-W. Ting and T. K. Sarkar (2010), 'Parallelized hybrid method with higher-order MoM and PO for analysis of phased array antennas on electrically large platforms', *IEEE Trans. Antennas Propagat.* **58**, 4110–4115.

B. Zhou (1995), Méthode des équations intégrales pour la résolution des problèmes de diffraction à haut fréquence. PhD thesis, Université de Paris XI.

M. Zworski (1999), 'Resonances in physics and geometry', *Notices Amer. Math. Soc.* **46**, 319–328.

Acta Numerica (2012), pp. 307–378
doi:10.1017/S0962492912000049

Mathematical cancer therapy planning in deep regional hyperthermia*

Peter Deuflhard[‡,†] Anton Schiela[‡] and Martin Weiser[‡]
Zuse Institute Berlin,
Takustrasse 7,
D-14195 Berlin, Germany
E-mail: deuflhard@zib.de

This paper surveys the mathematics required for a typically challenging problem from computational medicine: cancer therapy planning in deep regional hyperthermia. In the course of many years of close cooperation with clinics, the medical problem has given rise to many subtle mathematical problems, some of which were unsolved when the project started. Efficiency of numerical algorithms, *i.e.*, computational speed and monitored reliability, plays a decisive role in the medical treatment. Off-the-shelf software had turned out to be insufficient to meet the requirements of medicine. Instead, new mathematical theory as well as new numerical algorithms had to be developed. In order to make our algorithms useful in the clinical environment, new visualization software, *i.e.*, a 'virtual lab', including three-dimensional geometry processing of individual virtual patients, had to be designed and implemented. Moreover, before the problems could be attacked by numerical algorithms, careful mathematical modelling had to be done. Finally, parameter identification and constrained optimization for the PDEs had to be newly analysed and realized over the individual patient's geometry. Our new techniques had an impact on the specificity of the treatment of individual patients and on the construction of an improved hyperthermia applicator.

* Colour online for monochrome figures available at journals.cambridge.org/anu.
† Institute of Mathematics, Freie Universität Berlin, Arnimallee 6, D-14195 Berlin, Germany
‡ Supported by the DFG Research Center MATHEON 'Mathematics for key technologies', Berlin.

CONTENTS

1	Introduction	308
2	Patient-specific virtual lab	311
3	Mathematical modelling	316
4	Efficient simulation	337
5	PDE-constrained optimization	349
6	Impact on medical technology	366
References		371

1. Introduction

The term *hyperthermia* in cancer medicine means a heat treatment of tumour tissue. In 1961, Scheid had already detected 'functional specialities in the microcirculation of carcinomas' under the influence of heating: see Scheid (1961). In an early attempt, Manfred von Ardenne (1907–1997) had experimented with what is today called *full-body* hyperthermia: he immersed cancer patients in a hot bath or exposed them to infrared radiation. However, his medical results were unclear: in rare cases, tumour growth was reduced, whereas in most cases there was no effect at all. Today such a procedure is only considered for cancer patients with widespread micro-metastases that cannot be detected by medical imaging. Instead, *regional* hyperthermia has evolved as an option. Here, tumours are only locally heated, while healthy tissue is generally saved. With increasing understanding about the generation and function of heat shock proteins, a lower temperature bound of roughly 42.5° appeared to be advisable. An upper temperature bound in the tumour should be 'not too high' to ensure that healthy tissue is kept close enough to the body temperature of 37°, and that the systemic body stress remains tolerable. At present, hyperthermia is applied via two typical modes.

- *Downstaging.* Hyperthermia treatment by itself can cause the tumour to have sharper boundaries, thus facilitating operative tumour extinction by the oncosurgeon – an effect that helps to reduce the probability of relapse.
- *Combination therapy.* In principle, one might think of a strategy where heat alone simply 'burns' the tumour, which would then be secreted by the kidneys as necrotic tissue. However, as long as healthy tissue cannot be guaranteed to be preserved unburnt in such a process, the therapeutic temperature level must remain moderate. In a temperature window between 41°C and 45°C, so-called *deep* regional hyperthermia (as opposed to surface hyperthermia, *e.g.*, for extremities) is usually combined with chemo- and radiotherapy: see, *e.g.*, Issels *et al.* (1987). In fact, testing

> hyperthermia alone in a research situation, where other therapies such as chemo- or radiotherapy might help to cure the patient, is not ethically justified. This is why the actual cases open to our simulation studies were only deep-seated non-operable tumours.

As early as 1984, it had become clear that careful hyperthermia treatment planning required computer simulations: see, *e.g.*, Roemer and Cetas (1984). In the late 1980s, the physician (and physicist) Wust came to ZIB seeking advice about the fast numerical solution of a PDE called the bio-heat transfer equation, which turned out to be of Helmholtz type in the uncritical case. His interest was in two-dimensional computation, since the input from individual patients comprised two-dimensional cross-sections taken from a stack of images. Initially (see Wust *et al.* 1989, 1991), he was recommended to use the 1985 version of the two-dimensional elliptic solver PLTMG by Bank (1985). However, the two-dimensional calculations appeared to be in strong contradiction to the practical experience of medical radiologists: various patients complained about 'hot spots' quite distant from the tumour being heated, whereas the simulations did not exhibit any such phenomena. It was the mathematicians who suggested that patients should be regarded as three-dimensional! As a consequence, the new version of the three-dimensional elliptic solver Kaskade (Deuflhard, Leinen and Yserentant 1989, Bornemann, Erdmann and Kornhuber 1993, Götschel, Weiser and Schiela 2012) came into play. This started a close collaboration between the numerical analysis group at ZIB and the radiology department at Charité, a collaboration that continued over more than a decade: see, *e.g.*, Wust *et al.* (1999) and Sreenivasa *et al.* (2003). In the course of time, nearly all mathematical and computer science tools necessary to tackle the challenging hyperthermia problems were tested thoroughly. As a result, all pieces were redesigned, until finally the whole computational line appeared to be fast, robust, and reliable enough to be used in a clinical environment. For this purpose, integrated software, the *virtual lab* HyperPlan, was elaborated (with eventually about half a million lines of code). It included a selection of both visualization and numerical algorithms for the construction of individual three-dimensional *virtual patients*.

On the numerical side, it contained adaptive multilevel finite element methods for the parabolic bio-heat transfer (BHT) equation coupled with the time-harmonic heterogeneous Maxwell's equation, the latter in the then still underrated intermediate regime where the wavelength is of the size of the obstacles. For the BHT equation, Yserentant's hierarchical basis preconditioner for two dimensions (Yserentant 1986, Deuflhard *et al.* 1989), as well as Xu's BPX preconditioner for three dimensions (Xu 1989, Bramble, Pasciak and Xu 1990; in the variant due to Yserentant 1993 and Bornemann *et al.* 1993), appeared at just the right time. For the time-harmonic

Maxwell's equation in the radiowave regime, the key turned out to be the use of (a) Whitney forms (Whitney 1957), the lowest-order three-dimensional finite edge elements (Nédélec 1980, Bossavit 1988, Beck 1993) to avoid spurious discrete solutions, and (b) the discrete Helmholtz decomposition due to Hiptmair (see Hiptmair 1996, Beck and Hiptmair 1999, Beck *et al.* 1999 and Hiptmair 2002*b*) within multigrid methods. In 2000, a first survey on this kind of work was given in Deuflhard (2000). These PDE solvers had been coupled to compute an optimal tuning of the antenna parameters. As a first step, this has been modelled by an unconstrained minimization problem, where the medically necessary heating constraints were taken into account via some penalty term.

Unfortunately, the DFG funding of the ZIB–Charité cooperation (within the collaborative research unit SFB 273) terminated at the end of 2002. Fortunately, the topic of regional hyperthermia could be further pursued within the DFG Research Center Matheon, now on a more mathematical basis. The above unconstrained minimization problem with penalty term was substituted by the medically more reasonable constrained minimization problem, where the heating constraints were modelled as strict inequalities. When this substitution was envisaged, the associated mathematical topic (adaptive multilevel methods for PDE-constrained optimization) was far from mature enough to be used with medical reliability. It is the purpose of the present survey to report on progress made since then in the context of past achievements.

Outline

The paper is organized as follows. In Section 2 we give the basics of the virtual lab HyperPlan. This software environment fixes all patient-specific details necessary to pose the hyperthermia problem: individual geometry, material coefficients, electromagnetic antennas at the applicator, and the water bolus. After that, in Section 3, the medical problem is transformed into a mathematical model in terms of partial differential equations (PDEs), a Helmholtz equation in the uncritical case for the distribution of heat in the body, and the time-harmonic Maxwell equation, which is structurally similar to a Helmholtz equation, in the critical case. The therapeutic goal is transformed into a variety of mathematical optimization problems of increasing difficulty, starting from a traditional functional with penalized temperature constraints and continuing to a medically preferable functional with PDE equality constraints and strict temperature inequality constraints. On this basis, Section 4 discusses efficient numerical techniques to solve the PDEs. Section 5 presents the recent theoretical analysis and algorithmic progress for PDE-constrained optimization. An interior point method is derived and worked out as a numerical path-following method in terms of a barrier parameter. As it turns out, the log-barrier function, well-known from finite-dimensional optimization, needs to be replaced by a hierarchy of rational barrier functions in function space optimization. Finally, in Section 6, the impact of the mathematical developments over the years with respect to clinical application is surveyed.

2. Patient-specific virtual lab

The many years of common work between ZIB and Charité in the hyperthermia project taught much to both sides, and this has paid off in all later collaborations between mathematics and medicine. This is why we start with a description of the basic pattern (paradigm) and patient-specific three-dimensional grid generation.

2.1. Paradigm of virtual medicine

For the cooperation between medicine and mathematics the following three-step procedure has evolved (see Figure 2.1).

(I) Construct an individual *virtual patient*. The construction of a patient-specific geometrical three-dimensional model starts from a stack of two-dimensional images originating from CT, MRT or ultrasound, for example. In a first computational step, three-dimensional details of the patient's body need to be identified, a procedure called *segmentation*. In order to obtain reasonable *starting grids* for a multigrid PDE solution, surface triangulation of interfaces, coarsening of surface grids, and finally tetrahedral grid generation are successively performed.

(II) Model and simulate the treatment in a *virtual lab*. The modelling of the therapy in question consists of a description of the essential physiological and physical processes, as accurate as is needed for the medical problems to be tackled. Usually, the description consists of a set of partial differential equations (PDEs) and, in the best case, a functional to be minimized under constraints. In order to suggest an 'optimal' treatment planning, these PDEs need to be solved quickly and reliably. Moreover, in order to be of use as a basis for medical decision, any suggested therapy requires careful visualization. All this must be realized on a hardware and software platform that can be made available in a clinical environment.

(III) Transfer the simulation results *back to the clinical environment*. In some prephase, the model and its simulation will be validated against the clinical situation. Any discrepancies between model and reality need to be discussed and, if necessary, taken into account by remodelling. Once this adaptation has been done, the computationally suggested 'optimal' therapy is transferred back to patient-specific treatment in the clinical environment.

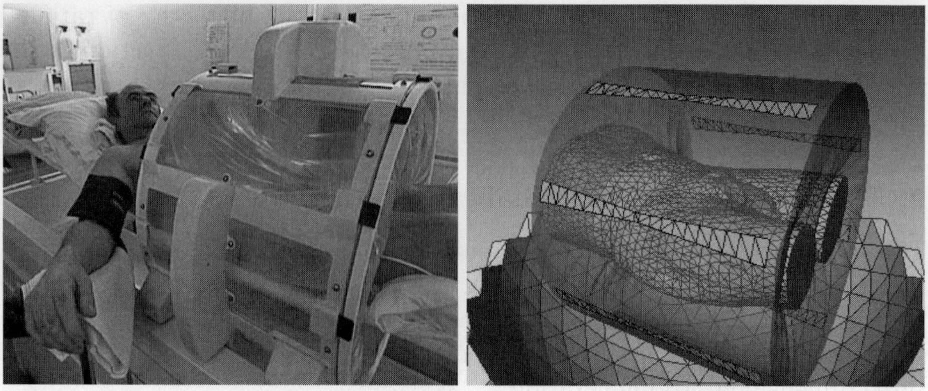

(a) (b)

Figure 2.1. Hyperthermia treatment planning. (a) Real patient in hospital.[1] (b) Virtual patient in the virtual lab.

Step (I) above is essentially independent of the particular medical problem, but only dependent on the body region treated: see Section 2.2 below. Step (II) depends on the medical problem in question, which in this case is deep regional hyperthermia: see Sections 3, 4.1 and 5. Step (III) will be illustrated by many examples of virtual patients inserted in the text throughout the paper.

2.2. Construction of geometrical three-dimensional model

The above paradigm has been implemented in the virtual lab HyperPlan (Stalling *et al.* 1996, Deuflhard 2000), an integrated software system that has been especially developed to support Hyperthermia treatment Planning in a clinical environment. The implementation of roughly half a million lines of code made heavy use of C++ software tools such as OpenGL (graphic library) and OpenInventor (object-oriented class library) which are themselves of high algorithmic complexity.

HyperPlan combines the following features.

- Efficient three-dimensional grid generation techniques, which we will describe immediately below.
- Fast and reliable simulation techniques, *e.g.*, adaptive multilevel finite element methods: see Section 4.1 below.
- Efficient visualization techniques, which are beyond the scope of the present article (for reference see, *e.g.*, Hege and Stalling 1998 and Stalling 1998).

[1] The patient agreed to be shown recognizably within a scientific context.

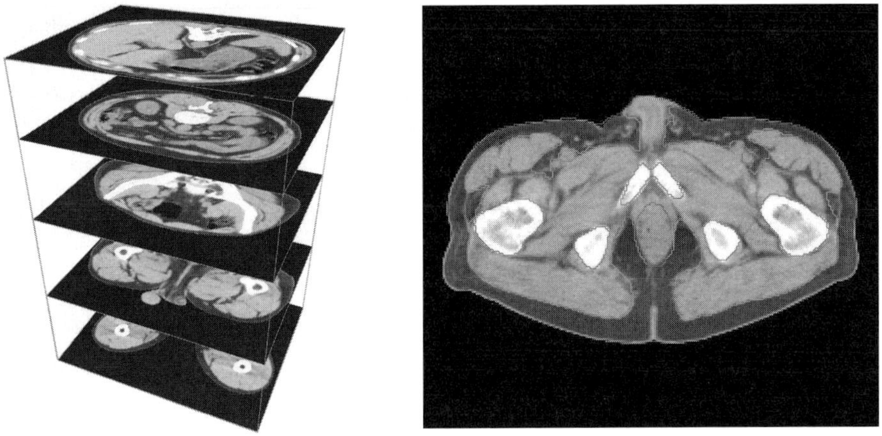

(a) (b)

Figure 2.2. Medical imaging input. (a) Stack of two-dimensional computer tomograms (CTs). (b) Segmented raw image data.

Medical imaging input
Patient-specific geometrical input usually comes from medical imaging techniques such as CT, MRT, ultrasound or mixtures thereof. We skip the case where different sources of input have to be matched (the so-called registration problem: see, *e.g.*, Zitov and Flusser 2003). As shown for CT in Figure 2.2, this input comes as a stack of two-dimensional images that merely give mass density information.[2] In view of the solution of PDEs on the specific patient data, we need more than just density information: see below under segmentation. Moreover, for the purpose of presenting numerical results to medical doctors, a three-dimensional grid generation with clear material identification needs to be realized. This is done in several steps, which are described next.

Segmentation
As can be seen from Figure 2.2, the original CT input contains only information about the mass density distribution in planar cross-sections of the body. What we need, however, to solve PDEs on an individual patient, is a clear correspondence of the various body parts with material properties such as electromagnetic or thermal constants. This task is called segmentation. From computer science, several segmentation techniques such as region growing (Adams and Bischof 1994) or watershed methods (Vincent and Soille 1991) are quite popular. They typically work on the stack of two-dimensional cross-sections and, in the best case, exploit some connection between neighbouring two-dimensional images. Such techniques

[2] More precisely, it gives information about the attentuation factor in tissue.

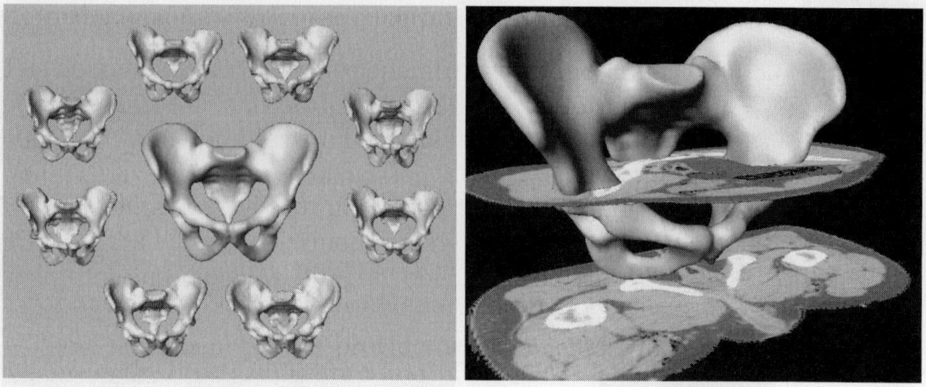

(a) (b)

Figure 2.3. Segmentation of a pelvic region by statistical shape analysis. (a) Representation of known patient data by their average and eight individual data (construction of essential coordinate frame). (b) Simultaneous adaptation of a shape model for the full stack of two-dimensional images of new patient data (representation in the frame of essential coordinates). From Lamecker, Seebass, Hege and Deuflhard (2004).

usually require a significant amount of user interaction, which is not desirable. Therefore, at ZIB, we followed a rather different approach called *statistical shape analysis* (Cootes, Hill, Taylor and Haslam 1994), which moves through three dimensions. In principle, this method performs the following steps.

(1) Start from a sample set of carefully segmented patient data, a time-consuming interactive job done by experts. For illustration, see the eight pelvic regions arranged in a circle in Figure 2.3(a).

(2) Compute an 'average' shape. See the centre shape in Figure 2.3(a). In order to compute a theoretically satisfactory average, one must solve the associated correspondence problem between two surfaces (see, *e.g.*, Lamecker 2008) wherein the way is paved towards a 'variational shape analysis'.

(3) Compute the differences 'individual–average' as fluctuations.

(4) Select 'essential coordinates' by selecting the fluctuations with the help of PCA (principal components analysis).

(5) Whenever new patient data are to be segmented, try to simply compute the essential coordinates for the new data: see Figure 2.3(b). If this works to sufficient accuracy, then exit. Otherwise enrich the patient database by these new data and go to step (1).

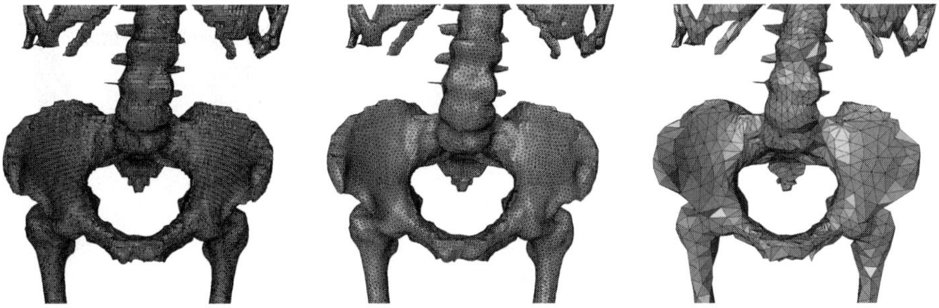

Figure 2.4. Grid coarsening in the pelvic region. From 478 000 nodes via 98 000 nodes to 11 000 nodes.

At the beginning of the above procedure, the number of essential coordinates will grow with all new patient data, but after a while it will get saturated, thus defining the 'essential dimension' of the patient data set at hand. Of course, in order to keep this dimension low, patient data should be clustered according to criteria such as sex, age, sports activities, biological group, *etc.* Figure 2.3 illustrates such a three-dimensional segmentation for some pelvic region. All in all, the techniques summarized here have reached a state of maturity that makes them equivalent in quality to a 'second human radiologist': see the award-winning paper by Kainmueller, Lange and Lamecker (2007), from an international contest on liver segmentation.

Surface triangulation coarsening

Typically, meshes generated after segmentation are much too fine: for an illustration see Figure 2.4(a). For the purpose of visualization, this large number can be handled, but in view of the numerical solution of PDEs on the virtual patients (*e.g.*, by multigrid methods), this number is far too large. For this reason, some precoarsening of the grids is needed. The associated iterative process is based on minimizing local approximation errors of coarser triangulated surfaces, which is closely related to local curvature: wherever the local curvature is 'low', the number of local nodes is reduced, whereas it is left unaltered where the local curvature is 'high' (see, *e.g.*, Heckbert and Garland (1997)). In Figure 2.4, the coarsening procedure (including mesh quality control) due to Zilske, Lamecker and Zachow (2008) is illustrated for a pelvic bone region. For multigrid methods to be discussed below in Section 4.1, this kind of grid then serves as the initial grid.

Tetrahedral grid generation

As a final step, the interface triangular grids are used as a starting point to construct spatial tetrahedral grids in the full volume. This is done by a special adaptation of the 'advancing front' algorithm of Löhner and Parikh (1988). A more recent improvement is the remeshing technique due to Zachow, Zilske and Hege (2007). This final step is illustrated in Figure 2.5(c).

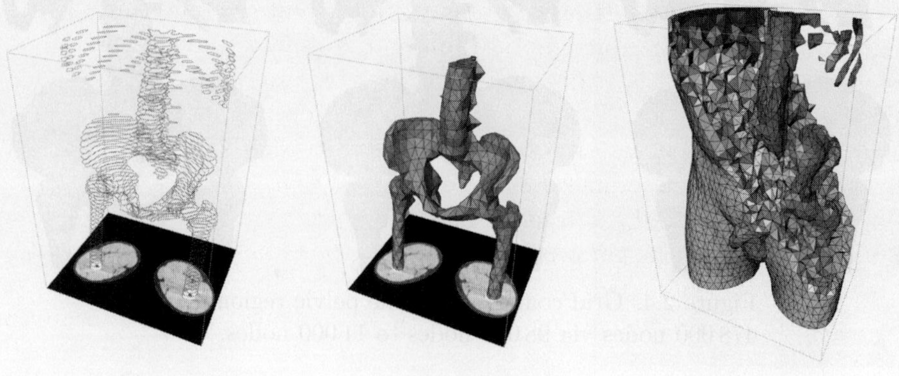

Figure 2.5. Stepwise construction of adaptive three-dimensional grid from a stack of cross-sectional images as shown in Figure 2.2. (a) Segmented surface and interface lines between different material regions. (b) Coarsened triangulation of surface and interfaces, as illustrated in Figure 2.4. (c) Generation of tetrahedral volume grid.

Remark 2.1. HyperPlan was the predecessor of the popular system Amira (Stalling, Westerhoff and Hege 2005), which applies to general systems in science and medicine.

3. Mathematical modelling

The purpose of this section is to describe the mathematical modelling necessary to translate the medical treatment planning problem into a mathematical formulation accessible to numerical methods. It covers the physics of heat generation by radio-wave emission in the antennas of the applicator, which is modelled by the time-harmonic Maxwell equation, and the transport of heat inside the human body, which is modelled by some bioheat transfer equation. Moreover, it includes the identification of unknown parameters in the mathematical models and several choices of quantitative formulations of the therapeutic goal, which leads to optimal control problems that are dealt with in Section 5.

Hyperthermia treatment of a real hospitalized patient is depicted in Figure 2.1(a). As the patient has quite a large rectal carcinoma, his abdomen is lying inside a so-called applicator, which contains eight antennas (pairwise coupled). The antennas emit radiowaves of frequency 100 MHz, which are absorbed in the human body, thus generating heat. The associated mathematical model, the time-harmonic Maxwell equation, is discussed in Section 3.1. Between the applicator and the body is a so-called water bolus to ease the entry of the radiowaves into the body and, at the same time, to cool the skin – thus setting a temperature value at the body surface. The absorption of the electromagnetic radiation is different for different body compartments (essentially depending on the portion of water in the tissue). By virtue of blood flow as well as heat conduction, the heat is distributed in the body. This process is modelled mathematically by a parabolic PDE, the so-called bio-heat transfer equation: see Section 3.2. A solution of the coupled system of PDEs is a necessary prerequisite to compute an optimal tuning of the antenna parameters such that the tumour is heated (within a temperature window), but not healthy tissue. This leads to a formulation in terms of mathematical optimization problems, to be elaborated in Section 3.4.

3.1. Emission of radiowaves

From first principles, the radiowaves emitted by the applicator antennas are known to be governed by Maxwell's equations. Simplified models are available for the case when the wavelength is much larger or much smaller than the geometry of the objects. In the hyperthermia case, however, the wavelength in water (which is a good approximation for the material of the body) is about 30 cm, which is the order of magnitude of the human body (the 'obstacle'), *i.e.*, just in the intermediate regime. As a consequence, there is no way to avoid solving the full Maxwell equations for the heterogeneous medium 'water bolus–human body'.

Time-harmonic Maxwell equation: strong formulation

Fortunately, only a single angular frequency ω is emitted from the antennas, which means that we can make the time-harmonic ansatz

$$E(x,t) = \mathrm{Re}\,(E(x)\mathrm{e}^{-\mathrm{i}\omega t}), H(x,t) = \mathrm{Re}\,(H(x)\mathrm{e}^{-\mathrm{i}\omega t}),$$

for the electromagnetic fields $E, H \in \mathbb{C}^3$. Insertion into the time-dependent Maxwell equations for linear isotropic dielectric media,

$$\mathrm{div}(\mu H) = 0, \tag{3.1a}$$

$$\mathrm{div}(\epsilon E) = 0, \tag{3.1b}$$

$$\mathrm{curl}\, H = \sigma E + \epsilon \frac{\partial E}{\partial t}, \tag{3.1c}$$

$$\mathrm{curl}\, E = -\mu \frac{\partial H}{\partial t}, \tag{3.1d}$$

applying the curl operator to (3.1d), and inserting (3.1c) yields the so-called *double-curl equation* or *time-harmonic Maxwell equation*

$$\mathrm{curl}\, \frac{1}{\mu}\, \mathrm{curl}\, E = \epsilon' \omega^2 E \tag{3.2}$$

for the complex electrical field E. Here μ is the magnetic permeability and $\epsilon' = \epsilon + \mathrm{i}\sigma/\omega$ the complex dielectric constant related to Ohm's law $j = \sigma E$, with electric conductivity $\sigma \geq 0$ for any current density j and the generic dielectric constant $\epsilon > 0$. The material constants μ, ϵ, σ depend, as the name indicates, on the local materials (*e.g.*, muscle, fat, bone, *etc.*) and should therefore, more precisely, be written as $\mu(x), \epsilon(x), \sigma(x)$. They are allocated on the basis of the geometrical patient model: see Section 2.2 above.

For spatially constant material parameters (homogeneous material), the term on the left-hand side may be reformulated to lead to

$$-\Delta E + \nabla \underbrace{\operatorname{div} E}_{=0} = \mu\epsilon'\omega^2 E. \tag{3.3}$$

Here the Laplace operator Δ is applied componentwise, such that (3.3) represents three decoupled Helmholtz-type equations in the critical case. Otherwise, with material jumps, the three PDEs are coupled: see Section 4.1.

Boundary conditions

The computational domain of interest (see Figure 2.1(b)) consists of an artificially chosen sphere coating the water bolus, the antenna array, and the virtual patient model. Hence we have three types of boundary conditions to be imposed. First, the metallic antennas on the physical boundary of the applicator are regarded as perfect conductors, which implies *homogeneous* Dirichlet boundary conditions for the tangential components $E_t = (n \times E) \times n$ of the electric field E, *i.e.*,

$$E_t = 0 \quad \Leftrightarrow \quad n \times E = 0 \quad \text{on} \quad \Gamma_{\text{ant}}.$$

Here Γ_{ant} denotes the thin metallic antennas shown in Figure 3.1 and n denotes the outer unit normal of the surface Γ_{ant}. Second, the small gap between the trapezoidal antenna sheets is modelled by an artificial boundary Γ_{gap} on which the tangential component of E is prescribed, so as to yield the voltage of the power generator when integrated across the gap:

$$E_t = E_t^0 \quad \text{on} \quad \Gamma_{\text{gap}}.$$

Third, on the external sphere Γ_{ext}, the Sommerfeld radiation condition (also called the Silver–Müller condition in the electrical engineering community)

$$n \times \left(\frac{1}{\mu} \operatorname{curl} E\right) - n \times \left(\frac{\mathrm{i}\omega}{\mu c} n \times E\right) = 0 \quad \text{on} \quad \Gamma_{\text{ext}} \tag{3.4}$$

is prescribed as an *approximate transparent* boundary condition in order to avoid reflection at the artificial boundary Γ_{ext}. Here $c = 1/\sqrt{\epsilon\mu}$, the electromagnetic wave velocity in the dielectric material, has been inserted.

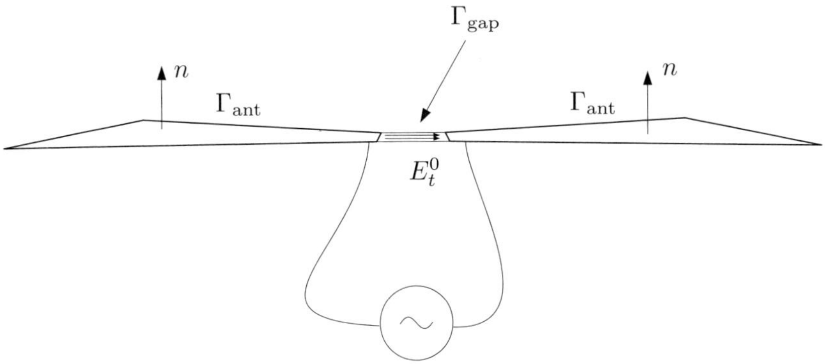

Figure 3.1. Antenna modelling with Dirichlet boundary conditions.

Time-harmonic Maxwell equation: weak formulation

The weak formulation starts from the double-curl equation (3.2) directly. For *lossy* materials (*i.e.*, for $\sigma > 0$), the material constant ϵ' is complex. As a consequence, the above PDE with boundary conditions represents a *non-self-adjoint* operator. Then, following Beck and Hiptmair (1999), we introduce the appropriate Hilbert space for the operator equation as

$$\begin{aligned} H_\Gamma(\text{curl};\Omega) := \{v \in L_2(\Omega)^3; \text{curl}\, v \in L_2(\Omega)^3,\ n \times v = 0 \text{ on } \Gamma_{\text{ant}} \cup \Gamma_{\text{gap}}, \\ n \times v \in L_2(\Gamma_{\text{ext}})^3 \text{ on } \Gamma_{\text{ext}}\}, \end{aligned}$$

equipped with the scalar product

$$\langle v, w\rangle_{H(\text{curl})} := \langle \text{curl}\, v, \text{curl}\, w\rangle_{L_2(\Omega)} + \langle v, w\rangle_{L_2(\Omega\cup\Gamma\text{ext})},$$

and a sesquilinear form $a : H_\Gamma(\text{curl};\Omega) \times H_\Gamma(\text{curl};\Omega) \to \mathbb{C}$ (conjugate linear in the first component):

$$\begin{aligned} a(E, v) := \int_\Omega \left(\frac{1}{\mu}(\text{curl}\, E)^* \,\text{curl}\, v - \omega^2(\epsilon' E)^* v\right) \mathrm{d}x \\ - \int_{\Gamma_{\text{ext}}} \frac{\mathrm{i}\omega}{\mu c}(n \times E)^*(n \times v)\, \mathrm{d}s. \end{aligned} \tag{3.5}$$

We end up with the following variational problem: find $E \in H_\Gamma(\text{curl};\Omega)$ such that

$$a(E, v) = f(v) \quad \forall\, v \in H_\Gamma(\text{curl};\Omega), \tag{3.6}$$

where $f : H_\Gamma(\text{curl};\Omega) \to \mathbb{C}$ is a linear form containing the inhomogeneity of the Dirichlet conditions.

Properties of (3.6)

In order to study the question of *uniqueness* of a solution, we follow the usual procedure: see Monk (2003, Theorem 4.12). Suppose there exists some $\bar{E}$ such that $a(\bar{E}, \bar{E}) = 0$, which implies that both its real and its imaginary part vanish. We assume homogeneous, possibly lossy material, *i.e.*, $\sigma \geq 0$. For the imaginary part we get

$$-\mathrm{Im}\, a(\bar{E}, \bar{E}) = \omega\left(\sigma \int_{\Omega} |\bar{E}|^2_{\mathbb{C}^3}\, \mathrm{d}x + \frac{1}{\mu c}\int_{\Gamma_{\mathrm{ext}}} |n \times \bar{E}|^2_{\mathbb{C}^3}\, \mathrm{d}s\right).$$

Both integrals are positive. For lossless material ($\sigma = 0$), the first term vanishes. The second one is positive, unless the external boundary integral vanishes, which would be the case for homogeneous Dirichlet or Neumann conditions. For the radiation condition (3.4), however, this is not the case. For lossy material ($\sigma > 0$), both terms are positive, which leads to a contradiction to the assumption that they should vanish. If a solution of (3.6) exists at all, it is unique whenever $\sigma \geq 0$.

Two further features of (3.5) deserve special attention: (a) the first term in the volume integral, representing the principal part of the PDE, has an 'ample' nullspace

$$H^0_\Gamma(\mathrm{curl}; \Omega) = \ker \mathrm{curl}$$

containing all gradient functions, in particular, functions of high frequency, (b) in this nullspace, the second term in the volume integral, representing the zero-order term in the PDE, is negative definite, since the factor $-\omega^2 \mathrm{Re}\, \epsilon' = -\omega^2 \epsilon$ is negative. Thus, the functional a is highly *indefinite*. The indefiniteness is the more marked the higher the angular frequency ω, a fact clearly given in hyperthermia (with 100 MHz).

On simply connected Lipschitz domains $H^0_\Gamma(\mathrm{curl}; \Omega)$ gives rise to a regular decomposition of $H_\Gamma(\mathrm{curl}; \Omega)$ (Hiptmair 2002*a*):

$$v = \phi + \mathrm{grad}\, p, \quad \phi \in H^1(\Omega)^3,\ p \in H^1(\Omega),\ \mathrm{grad}\, p \in H^0_\Gamma(\mathrm{curl}; \Omega), \tag{3.7}$$

such that

$$\|\phi\|_{H^1} + \|p\|_{H^1} \leq C\|v\|_{H(\mathrm{curl})}. \tag{3.8}$$

In this way, our problem is split into a vectorial H^1-part and a negative definite H^1-component for the potentials. This decomposition is a modification of the classical *Helmholtz* decomposition.

Interface conditions at material jumps

Both quantities μ and ϵ have jumps at material interfaces that have been determined by segmentation: see Section 2.2 above. At these interfaces special conditions must be observed, which in weak formulation are automatically satisfied (via the incorporated vanishing of the remainder terms stemming from an integration by parts). These conditions imply a kink in the electric field vector E at the boundaries; for an illustration in FastLIC representation (Hege and Stalling 1998) see Figure 3.2.

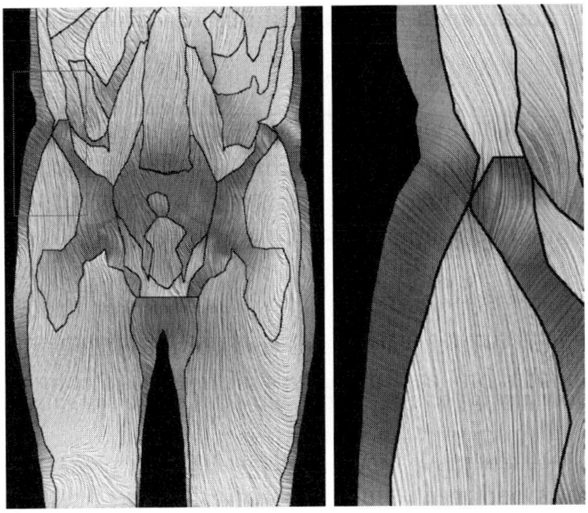

(a) (b)

Figure 3.2. FastLIC representation of electrical field lines (Hege and Stalling 1998). (a) Abdominal body part. (b) Zoom.

Superposition
The radiowaves (100 MHz) are emitted from k_E channels of pairwise coupled $2k_E$ antennas. Since the Maxwell equations are linear, superposition can be applied. Let $E_k(x)$ denote the electric field (the antenna profile) originating from channel k for $k = 1, \ldots, k_E$. Then the total electrical field can be written as a linear combination

$$E(x) = \sum_{k=1}^{k_E} \alpha_k E_k(x), \quad \alpha_k \in \mathbb{C}, \tag{3.9}$$

in terms of complex amplitudes $\alpha_1, \ldots \alpha_{k_E}$. They may be parametrized via

$$\alpha_k = a_k \exp(-\mathrm{i}\theta_k) \quad k = 1, \ldots, k_E, \tag{3.10}$$

with real amplitudes a_k and phase delays θ_k that can be controlled independently. Where appropriate, the k_E complex parameters α_k will be interpreted as $2k_E$ real parameters $\{\operatorname{Re} \alpha_k, \operatorname{Im} \alpha_k\}$.

Remark 3.1. Kremer and Louis (1990) treated the problem of approximating a given electrical field (obtained, *e.g.*, from optimization) via excitation through n antennas. For their theoretical analysis, they define a hyperthermia operator that maps arbitrary antenna excitations at a boundary circle of a cylinder to any electrical fields in that cylinder. This operator is shown to be compact in L_2. By means of singular value decomposition they can conclude that hot spots outside the tumour volume can only be suppressed for sufficiently large n.

3.2. Heat distribution inside the human body

Compared to the *generation* of heat as discussed in the preceding section, the *distribution* of heat inside a specific patient's body is much harder in terms of modelling.

Heat absorption
Inside the body, the electromagnetic waves generate heat by absorption to be characterized via the *absorption rate density*

$$S = \frac{1}{2}\sigma |E|^2_{\mathbb{C}^3}, \tag{3.11}$$

where σ is the electric conductivity of the specific tissue. If one exploits the superposition (3.9), one ends up with the quadratic term

$$S(x) = \frac{1}{2}\sigma(x)|E(x)|^2_{\mathbb{C}^3} = \frac{1}{2}\sigma(x)\sum_{j,l=1}^{k_E} \alpha_j^* \alpha_l \, E_j^*(x)\, E_l(x), \tag{3.12}$$

where σ varies among different tissues. From the definition (3.12) it is apparent that simultaneous phase shifts of the controls α_k lead to a corresponding phase shift of the electrical field E and thus have no thermal impact at all. Therefore, one of the phases can be fixed *a priori*, *e.g.*, by setting $\operatorname{Im}(\alpha_{k_E}) = 0$. We will therefore think of α containing only $2k_E - 1$ real parameters, which will be important in Section 5.1 to get rid of a trivial local non-uniqueness of optimal solutions.

Bio-heat-transfer (BHT) equation: strong formulation
The standard heat transfer model used in our simulations has been the so-called bio-heat-transfer (BHT) equation over the domain $\Omega \subset \mathbb{R}^3$ covered by the patient's body:

$$\rho c \frac{\partial T}{\partial t} = \operatorname{div}(\kappa \operatorname{grad} T) - c_b w\,(T - T_{\text{blood}}) + S \quad \text{in } \Omega, \tag{3.13}$$

where ρ is the density, c the specific heat capacity of tissue, κ the thermal conductivity, c_b the specific heat capacity of blood, w the mass flow rate of blood per unit volume of tissue, the so-called *perfusion*, T_{blood} the blood temperature in the unheated body (*e.g.*, 37°C), and S the absorption rate density as defined in (3.12). All material constants are understood to be local constants, different in different tissues. Already in 1948, this Helmholtz-type equation in the uncritical case was proposed by Pennes (1948) on the basis of theoretical consideration *and* experimental measurements – a rare scientific combination!

The boundary conditions are of Robin type:

$$-\kappa \frac{\partial T}{\partial n} = h(T - T_{\text{bolus}}) \quad \text{on } \partial\Omega. \tag{3.14}$$

For the time-dependent case initial conditions

$$T(0,x) = T_a$$

are prescribed. In hyperthermia, the steady state is typically reached after about 15–20 minutes and maintained for 40–60 minutes. For this reason, we may well be content with solving the stationary elliptic problem rather than the time-dependent parabolic problem.

As with the other material coefficients, the discontinuity of the thermal conductivity here also yields interface conditions across the tissue interfaces. Again, these conditions are automatically satisfied if the PDE is solved in its weak form.

BHT equation: weak formulation

Here we interpret the stationary BHT equation (with notation as in (3.13)) as an operator equation:

$$\langle A(T) - B(\alpha), \varphi\rangle = 0 \quad \forall \varphi \in C^\infty(\Omega), \tag{3.15}$$

where A and B are defined by

$$\begin{aligned}\langle A(T), \varphi\rangle &:= \int_\Omega \kappa\,(\operatorname{grad} T)^{\mathrm{T}} \operatorname{grad}\varphi + c_b w\,(T - T_{\text{blood}})\varphi \,\mathrm{d}x \\ &\quad + \int_{\partial\Omega} h(T - T_{\text{bolus}})\varphi\,\mathrm{d}S, \\ \langle B(\alpha), \varphi\rangle &:= \int_\Omega \frac{\sigma}{2}|E(\alpha)|^2_{\mathbb{C}^3}\varphi\,\mathrm{d}x.\end{aligned} \tag{3.16}$$

The domain Ω consists of a number of subdomains Ω_i, corresponding to various types of tissue. All coefficients depend on the tissue type and are chosen piecewise constant on each subdomain associated with a tissue type. They may vary significantly from tissue to tissue. For a more detailed description of the parameters we refer to Deuflhard, Weiser and Seebass (2000).

Multiscale heat transport

Inside the patient's body, heat is distributed only by diffusion in tissue and by direct transport in blood vessels (with velocity $v(x)$, say), but not by any process described by the zero-order term in the BHT equation (3.13) above. Therefore, from first principles, a PDE of *diffusion–advection* type including the source term seems to be more appropriate:

$$\rho c\,\frac{\partial T}{\partial t} = \operatorname{div}(\kappa \operatorname{grad} T) - \rho c\, v^{\mathrm{T}} \operatorname{grad} T + S, \tag{3.17}$$

However, due to the multiscale structure of the velocity field v, both from the modelling aspect and from the point of view of computational complexity,

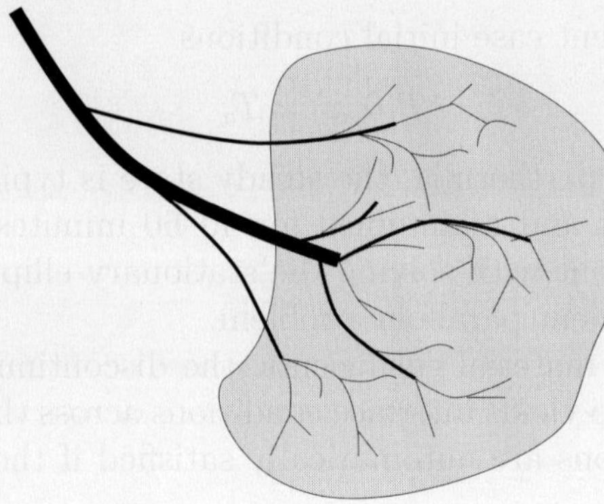

Figure 3.3. Three-scale pattern of heat distribution in the human body: large blood vessels (macroscale), medium-size blood vessels (mesoscale), capillaries (microscale). Homogenization of capillaries leads to a porous medium macroscale model.

such a PDE cannot be used to successfully describe the situation within an individual patient.

The multiscale vascular situation is schematically represented in Figure 3.3. Three scales can be distinguished.

(1) On the macroscale, large blood vessels should be treated as full three-dimensional objects (*e.g.*, as objects in a finite element setting; see also Lagendijk, Schellekens, Schipper and van der Linden 1984).

(2) On some mesoscales, smaller vessels could be modelled as one-dimensional objects (but with a finite width; see D'Angelo and Quarteroni 2007, Quarteroni, Tuveri and Veneziani 2000 and Massing 2007).

(3) On the microscale, detailed phenomena within the capillaries cannot be resolved, neither from modelling nor computational perspectives. That is why some kind of averaging has to take place, which results in a macroscale model for a porous medium region.

Wust *et al.* (1989) had suggested the above BHT equation as the averaged microscale model. It had been derived by a potential flow assumption in view of Darcy's law – known to arise from homogenization techniques in reservoir simulation. That assumption was certainly not self-explanatory in the present context. Therefore we turn to the mathematical *homogenization* approach (see, *e.g.*, Bensoussan, Lions and Papanicolau 1978), which was worked out by Hochmuth and Deuflhard (Deuflhard and Hochmuth 2004, Hochmuth and Deuflhard 2004) for the case of hyperthermia.

In the homogenization framework, we consider a periodic arrangement of microscopic cubic cells εY, periodic along the three spatial coordinate axes y_1, y_2, y_3, with insertions of blood εQ, the capillaries, assumed to have constant temperature T_{blood}: see Figure 3.4. Let Ω_ε denote the solid tissue

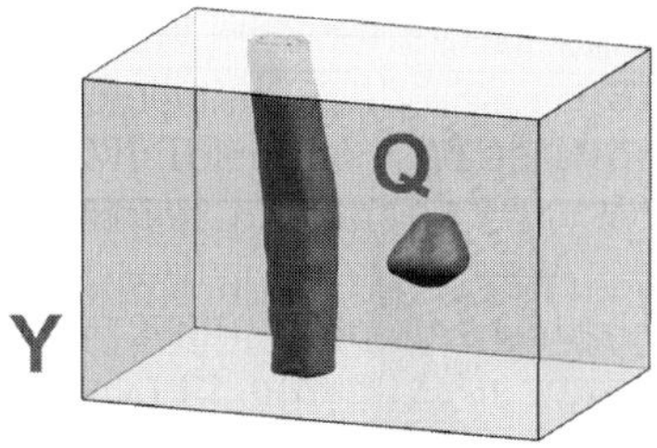

Figure 3.4. Normalized cells Y with inserted blood regions Q: non-isolated as well as isolated case. Surface of insertions $\partial Q \setminus \partial Y$.

volume without the capillaries. Therein heat transfer takes place only by diffusion. For simplicity, Deuflhard and Hochmuth (2004) scaled the thermal conductivity to $\kappa = 1$. As a consequence, the temperature T_ε is defined by the simple model equation

$$-\Delta T_\varepsilon = S_\varepsilon \quad \text{in } \Omega_\varepsilon,$$

with thermal source term $S_\varepsilon = \frac{1}{2}\sigma_\varepsilon |E_\varepsilon|^2$, exterior Dirichlet boundary conditions

$$T_\varepsilon = T_{\text{bas}} \quad \text{on } \partial\Omega_\varepsilon,$$

in terms of some basic temperature T_{bas}, and interior Robin boundary conditions

$$\frac{\partial T_\varepsilon}{\partial n} = \varepsilon\alpha(T^\varepsilon_{\text{blood}} - T_\varepsilon) \quad \text{on } \partial(\varepsilon Q).$$

Due to Deuflhard and Hochmuth (2004), the following asymptotic result can be shown to hold.

Theorem 3.2. Notation as just introduced. Let $|\cdot|$ denote the Lebesgue measure and define relative surfaces and volumes with and without blood by

$$\theta_{\text{surface}} := \frac{|\partial Q \setminus \partial Y|}{|Y \setminus Q|}, \qquad \theta_{\text{volume}} := \frac{|Y \setminus Q|}{|Y|} < 1.$$

Assume that, in weak*-convergence, we have

$$S_\epsilon \rightharpoonup \theta_{\text{volume}} S, \quad T^\epsilon_{\text{blood}} \rightharpoonup \theta_{\text{volume}} T_{\text{blood}}.$$

Then, for $\varepsilon \to 0$, the temperature converges weakly as

$$T_\epsilon - T^\varepsilon_{\text{blood}} \rightharpoonup \theta_{\text{volume}}(T_0 - T_{\text{blood}}) \quad \text{in } L_2,$$

where the limit temperature T_0 satisfies the PDE

$$-\operatorname{div}(\mathcal{A} \operatorname{grad} T_0) + \alpha\theta_{\text{surface}}(T_0 - T_{\text{blood}}) = S \quad \text{in } \Omega$$

with Dirichlet boundary conditions

$$T_0 = T_{\text{bas}} \quad \text{on } \partial\Omega$$

and perturbed conductivity

$$\mathcal{A} = a_{ij}, \quad a_{ij} = \delta_{ij} - \frac{1}{|Y \setminus Q|} \int_{Y \setminus Q} \frac{\partial \chi^j}{\partial y_i} \, \mathrm{d}y,$$

where the functions χ^j, $j = 1, 2, 3$ are understood to be solutions of

$$-\Delta \chi^j = 0 \quad \text{in } Y \setminus Q, \quad \frac{\partial(\chi^j - y_j)}{\partial n}, \quad \chi^j \quad Y\text{-periodic.} \tag{3.18}$$

The proof is a bit technical, mainly because of the occurrence of Robin boundary conditions (for $T^{\varepsilon}_{\text{blood}} = 0$ a proof was given earlier; see, *e.g.*, the book by Allaire, Damlamian and Hornung 1995, and references therein). It starts from the above diffusion–advection equation (3.17), assumed locally in a microscale periodic cell structure which is spatially averaged. Details are left to Deuflhard and Hochmuth (2004). Summarizing, the above theorem confirms the validity of the BHT equation in the microvascular regime – apart from some perturbation factors $\theta_{\text{surface}}, \theta_{\text{volume}} \approx 1$ that take care of the relative blood volume versus the full volume.

Remark 3.3. An approach aiming at directed advective heat transport in tumours has been developed by Gerber (2007), but not validated.

Temperature-dependent perfusion

In a more detailed investigation, temperature-induced vasodilation leads to an increased perfusion of heated tissue, and thus gives rise to a more realistic, *nonlinear* version of the BHT equation. Experiments (Song *et al.* 1984) have shown that the blood flow in normal tissues, *e.g.*, skin and muscle, increases significantly when heated up to 41–43°C, whereas in the tumour zone the blood flow may increase or decrease with temperature depending on the type of tumour. On this experimental basis, the perfusion $w = w(x, T)$ can be chosen monotonically increasing with respect to T in the muscle and fatty tissue, but monotonically decreasing in tumour tissue.

Simulations show significant qualitative differences between the temperature distributions predicted by the linear and nonlinear heat transfer model, as illustrated in Figure 3.5. Generally speaking, self-regulation of healthy tissue reflected by the nonlinear model reduces 'hot spots' caused by local maxima of the absorbed electromagnetic fields and hence allows the total energy deposition to rise. This is one reason for slightly better tumour heating (*c.* 0.5°C) predicted by the nonlinear model. An analogous result is reported in Tompkins *et al.* (1994*b*) for ferromagnetic thermoseed hyperthermia. An important finding is that the nonlinear model has an impact on the optimal treatment parameters as well. Maximal discrepancies turned out to be 22° for the phases θ_k and 0.22 for the relative amplitudes a_k; for notation see (3.10). For a more detailed discussion see Deuflhard *et al.* (2000) and Lang, Erdmann and Seebass (1999).

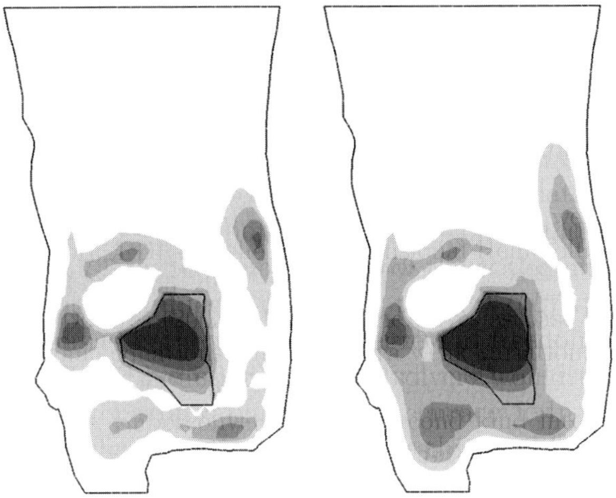

(a) (b)

Figure 3.5. Optimized temperature distributions in a frontal section of the pelvic region. *Black lines*, body outline and tumour contour; *light grey to dark grey shading*, regions heated from 39°C to 43°C. (a) Linear model. (b) Nonlinear model.

Besides local vasodilation, systemic thermoregulation can play an important role, *e.g.*, via centrally controlled vasodilation of the skin and a corresponding shift of perfusion from the body core to the extremities, or an increased heart rate. A feedback mechanism observed rather recently is the so-called *steal effect*, where different regions compete for a joint blood supply. An increased perfusion in one region, *e.g.*, a heated muscle that actively reduces its peripheral resistance by vasodilation, will lead to a larger blood flow and hence a greater decrease of effective blood pressure along the vessels feeding both regions. In this case, another region with less capability for thermoregulation, *e.g.*, a tumour, will experience reduced perfusion and a smaller cooling effect. Such a non-local interaction may open up new possibilities for therapy optimization. It might lead to higher tumour temperatures if, instead of focusing the power on the tumour, one were to heat up muscles supplied by the same vessels in order to cut off the tumour from cooling perfusion.

3.3. Parameter identification

All of the above thermoregulation models contain a number of parameters, most of which are unknown. Such parameters must be determined either from direct measurements or from computations comparing indirect measurements with the applied model. The most relevant parameters are as follows.

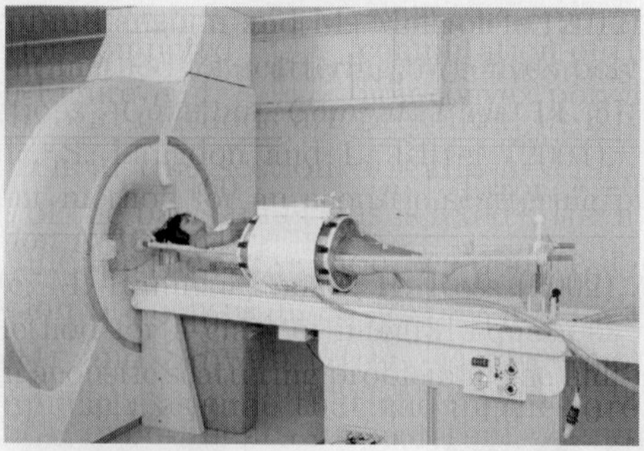

Figure 3.6. Hyperthermia applicator system in open MRT.

- *Material coefficients.* Thermal, electrical, and physiological properties of tissue depend on the specific patient under consideration. Several techniques are available to get hold of reasonable values. Parts of these data are obtained from measurements: see, *e.g.*, Wust *et al.* (1991, Table 1) for electromagnetic and bioheat data, or Seebass *et al.* (2000) with values from Paulsen, Geimer, Tang and Boyse (1999), where data for vessels, lungs, spleen were taken from Gabriel, Lau and Gabriel (1996). In Wust *et al.* (1999, Figure 1), only slight differences between different patients for the absorption rate density are reported. One of the most important parameters is the *perfusion* w, both due to its dominant effect on the temperature and its high inter-individual variability. It can be measured by MRI using a contrast agent, but this is not generally indicated.

- *Geometry.* Since therapy planning and, in particular, segmentation takes a significant amount of time, the patient's geometry is acquired some time before the therapy starts. Moreover, several therapy sessions are performed based on this single geometry. Positioning the patient in exactly the same location every time is a major practical difficulty. Even with the utmost care, spatial deviations of roughly 15 mm are unavoidable. Movements of the patient during therapy lead to additional offsets. Changes in body mass and intestine or bladder contents also affect the actual geometry.

- *Technical parameters.* Clinically available HF power generators are less stable in maintaining requested amplitudes and phase shifts than would be desirable. Moreover, part of the HF energy is reflected at contact resistances in coaxial cable connectors between the power generator and the applicator. These resistances depend on the mechanical force with which the connectors have been put together. While the transmission properties of the cable network can, in principle, be measured, this is not always possible in clinical practice.

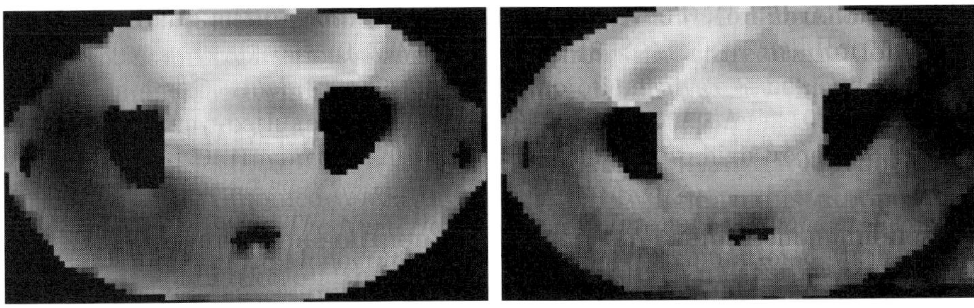

(a) (b)

Figure 3.7. Cross-section of absorption rate density in simulation (a) and MR measurement (b) for a phantom (Weihrauch *et al.* 2007).

Computational identification
A validation of the models describing the electrical field and the heat transport requires *in vivo* measurements of relevant quantities, in particular the temperature. If sufficient data are acquired, the parameters entering the models can be identified, which allows improvement of the therapy. Three-dimensional distributed data sets can be obtained from magnetic resonance tomography (MR) measurements (Rieke and Pauly 2008) in a hybrid MR–hyperthermia applicator: see Figure 3.6. At a rather limited number of discrete times, temperature probes can be placed inside the patient's body. By MR thermometry based on positron resonance frequency shift (PRFS), a combination of tissue temperature increase $T - T_0$ and perfusion w can be measured. Due to lack of better understanding of the relevant processes, a simple measurement model of the form

$$m \approx a(T - T_0) + bw$$

has been used, where the coefficients a and b depend on the tissue type. In fatty tissue in particular, the perfusion influence vanishes almost completely ($b \approx 0$). Resolution, signal-to-noise ratio, and measurement time are mutually dependent. For useful thermometry data, a measurement takes at least 30 s with current MR scanners.

Adaptation of electric fields
As mentioned above, several model parameters are not exactly known in clinical practice. These modelling errors lead to predictions significantly deviating from measurements (see Figure 3.7).

A radical remedy would be to omit the numerical solution of Maxwell's equation completely and to identify the self-adjoint heating effect $M(x) \in \mathbb{C}^{k_E, k_E}$ of applicator control settings using only the k_m MR thermometry measurements (Kowalski, Behnia, Webb and Jin 2002). Assuming the correctness of the computed absorption rate density and the validity of the

Identification of perfusion

Among the heat transfer model parameters perfusion is the most influential one. Hence, measurement or identification of *individual* perfusion values would be very valuable. Quite accurate values can be obtained from MR data using contrast agents (Lüdemann, Wust and Gellermann 2008). The application of contrast agents may incur an undesirable risk for the patient, which is why they are only used if absolutely necessary. Moreover, due to the required injection of the contrast agent, only very few of these measurements can be obtained, such that a continuous monitoring of thermoregulatory feedback is impossible. Consequently, such data are not generally available.

As a non-invasive alternative, the MR thermometry data m mentioned above can be considered. This leads to an identification problem for temperature and perfusion (Gänzler, Volkwein and Weiser 2006):

$$\min_{T,w} \|m - aT - bw\|^2_{L_2(\Omega)} + \gamma\|w - w_0(T)\|^2_H + b(w, w_0(T)),$$

subject to the equality constraints (3.13) and (3.14). Here, $w_0(T)$ is the *a priori* perfusion model. The norm

$$\|w\|^2_H := \|w\|^2_{L_2(\Omega)} + \sum_{i=1}^{\sharp\text{tissues}} \|\nabla w\|^2_{L_2(\Omega_i)}$$

for the regularization term is constructed based on the assumption that the perfusion depends smoothly on the temperature in every tissue region, but may jump across tissue boundaries. The barrier function b enforces non-negativity of the perfusion, but has to be skewed such that its minimizer is the reference perfusion $w_0(T)$, in order not to introduce a systematic bias.

Clinical application of that concept is prevented by the fact that there can be two different value pairs of temperature and perfusion which are compatible with both the BHT equation and the measurements, which is a consequence of the bilinearity of the BHT equation. Hence, while local consistency is fostered by regularization, global consistency cannot be guaranteed. In fact, numerical experiments indicate that the occurrence of both compatible values must be expected (Weiser 2009).

The consequence is that more data are required to provide the clinical staff with reliable identified temperature and perfusion values. Such data might come from additional measurements taken with MR sequences, which exhibit impact coefficients a and b different from PRFS.

3.4. *Therapy planning problem*

Our first step is to formulate the medical problem of therapy planning, which then has to be translated into mathematical terms.

Medical problem setting

Due to the production of so-called heat shock proteins, the temperature should preferably be above 42.5°C. Up to now the heating could not be safely restricted to tumour cells alone, not least since it is impossible to distinguish tumour cells reliably from healthy ones. Therefore, in order to be on the safe side for healthy tissue, local temperatures in tumours cannot be allowed to be too high. Thus the therapeutic aim is in general not to burn tumour cells, but to make them more sensitive to other treatments such as radiotherapy or chemotherapy (cytostatica). *In vitro*, experiments have confirmed an increased sensitivity of tumour cells by a factor of 7 over healthy ones, an effect also seen in a milder form *in vivo*: see, *e.g.*, Brizel *et al.* (1996). The physiological mechanism behind this effect is captured in the so-called oxygen enhancement ratio (OER). Consequently, the general aim can be specified as follows:

Control the antenna parameters such that tumour tissue is heated, while healthy and uncertain tissue is not heated above a prescribed threshold temperature.

After these preparations, we are now ready to translate the above medical problem into a mathematical one. Let T denote the temperature inside the body. Following the medical intention, an optimal temperature distribution should satisfy the following items.

- In the region V_{tumour} populated by tumour cells, the temperature should be sufficiently high to incur a therapeutic benefit. In view of the metabolism of heat shock proteins, a temperature level $T_t \geq 42.5$°C should be maintained.
- In the region V_{healthy} populated by a significant portion of healthy tissue, the temperature should not exceed a tissue-type-dependent threshold $T_h \approx 42$–44°C, in order to avoid subcutaneous burns.

In addition, there is a region where both tumour and healthy cells are present, which can be treated in a special way (see Section 6.2).

Apart from the medical intention, technical restrictions are imposed by the applicator. The amplitude of each antenna is limited, as is the total power emitted. Moreover, the antenna phases are only reliable as long as the amplitudes are not too different.

These aims and restrictions, partially conflicting, have been modelled and formulated as mathematical optimization problems in a variety of ways, which we will present in the following.

Cost functionals

The first approaches to the quantification of therapeutic benefit to be maximized were formulated in terms of the absorption rate density (3.11) alone

(Paulsen *et al.* 1999, Kremer and Louis 1990, Köhler, Maass, Wust and Seebass 2001), based on the idea that high power deposition leads to high temperature. The associated cost functional to be minimized is

$$f(E) = -\frac{1}{2}\int_{V_{\text{tumour}}} \sigma|E(\alpha)|^2_{\mathbb{C}^3}\,\mathrm{d}x. \tag{3.21}$$

An advantage is that this functional can be easily minimized by solving a low-dimensional eigenvalue problem (Kremer and Louis 1990, Köhler *et al.* 2001), as long as no temperature constraints are prescribed.

For medical reasons, a temperature-based objective functional is preferable. Therefore, in Nikita, Maratos and Uzunoglu (1993) and Paulsen *et al.* (1999), such a functional has been suggested. The idea is to penalize 'cold spots' in the tumour below a therapeutic temperature $T_t \approx 43°C$, since the impact of hyperthermia on cancer cells drops quickly below this threshold. This leads to the functional

$$f(T) = \int_{V_{\text{tumour}}} (T_t - T)^2_+\,\mathrm{d}x \tag{3.22}$$

with $(\cdot)_+ = \max(\cdot, 0)$. This approach can be extended to the concept of thermal dose, with the aim of minimizing the fraction of surviving tumour cells. The thermal damage rate inflicted on cancer cells follows the *Arrhenius law* (Dewey 1994, Szasz and Vincze 2006), with a rate constant

$$k(T) = A\exp\left(\frac{\Delta E}{RT}\right).$$

Here, the frequency factor A and the activation energy ΔE depend on the tissue type, and R is the universal gas constant. An estimate s of the fraction of surviving cells is given by the simple differential equation

$$\dot{s}(t,T) = k(T)s(t,T), \quad s(0,T) = 1,$$

which gives rise to

$$s(t,T) = \exp(tk(T)). \tag{3.23}$$

Thus the objective functional to be minimized is the fraction of cancer cells surviving after a prescribed duration t of a therapy session:

$$f(T) = \int_{V_{\text{tumour}}} \exp\left(-tA\exp\left(\frac{\Delta E}{RT}\right)\right)\mathrm{d}x. \tag{3.24}$$

A schematic representation of such a function is given in Figure 3.9. The thermal isoeffect dose is an established quantity for assessing the therapeutic benefit of a treatment (Dewey 1994, Rosner, Clegg, Prescott and Dewhirst 1996). Treatment planning based on the tumour cell survival has been proposed for thermoseed placement (Tompkins *et al.* 1994*a*), but up to

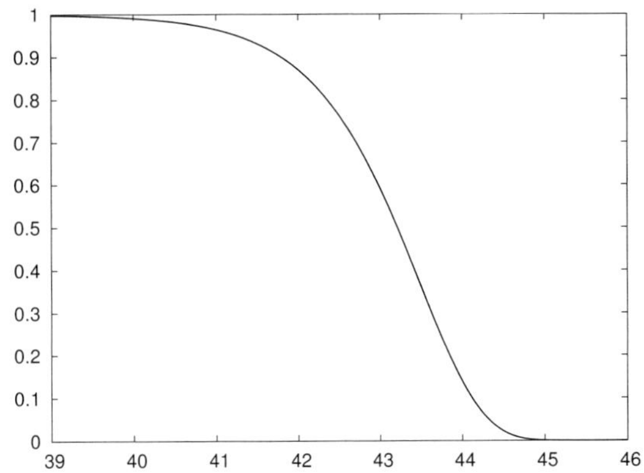

Figure 3.9. Dependence of cancer cell survival rate on tissue temperature according to (3.23).

now rather *ad hoc* cost functionals like (3.22) have been used in regional hyperthermia planning.

Temperature constraints

At the time when mathematical therapy planning started, a sufficient mathematical understanding of state constraints and an efficient algorithmic treatment were lacking. A common approach was therefore an external penalization of 'hot spots' arising in healthy tissue with temperatures above a tissue-dependent threshold T_{lim}, occasionally augmented by an additional, weaker penalization of heating healthy tissue above a comfortable temperature $T_h < T_{\mathrm{lim}}$ (Erdmann, Lang and Seebass 1997). This leads to extended cost functionals

$$J(T) = f(T) + \int_{V_{\mathrm{healthy}}} \left(\gamma_1 (T - T_h)_+^2 + \gamma_2 (T - T_{\mathrm{lim}})_+^2\right) \mathrm{d}x,$$

with fixed penalty parameters $0 \leq \gamma_1 \ll \gamma_2$. Due to external penalization, however, 'hot spots' cannot be completely prevented. In fact, they are bound to arise unless $\gamma_2 \to \infty$. An analysis of the asymptotic behaviour performed in Schiela and Hintermüller (2011) suggests that even under favourable circumstances the maximum constraint violation only decreases by a rate of $O(\gamma_2^{-2/5})$. We would like to point out that in hot spots with less regularity of the temperature distribution, in particular due to discontinuous temperature bound or singularities of the absorption rate density, the maximum violation decreases even more slowly.

A strict *prevention of hot spots* in healthy tissue will require the explicit satisfaction of state constraints of the kind

$$T(x, \alpha) \leq T_{\mathrm{lim}}, \quad x \in V_{\mathrm{healthy}}. \tag{3.25}$$

One way to guarantee feasibility of the temperature distribution is to use a barrier method. In Section 5 below, we will work out results in terms of both theory and algorithms.

Control constraints

Additional constraints directly on the control α are imposed to model the limited power of the microwave applicator:

$$\begin{aligned} |\alpha_k|_{\mathbb{C}} &\le \alpha_{\max}, \quad k = 1, \ldots, k_E, \\ \sum_{k=1}^{k_E} |\alpha_k|_{\mathbb{C}}^2 &\le \alpha_{\max}^{\text{total}}. \end{aligned} \tag{3.26}$$

If desired, stability of antenna phases can be ensured by confining the amplitude ratios to

$$\alpha_{\text{ratio}}^{-1} \le \frac{|\alpha_k|_{\mathbb{C}}}{|\alpha_j|_{\mathbb{C}}} \le \alpha_{\text{ratio}} \quad j, k = 1, \ldots, k_E, \quad \text{with } \alpha_{\text{ratio}} > 1,$$

or by requesting all antennas to have the same amplitude:

$$|\alpha_j|_{\mathbb{C}} = |\alpha_k|_{\mathbb{C}}, \quad j, k = 1, \ldots, k_E.$$

Optimal control problems

Putting all the pieces together, different optimization problems can be formulated. Omitting the control constraints, we state two variants that have been used for therapy planning.

Penalized problem:

$$\min_{\alpha \in \mathbb{C}^{k_E}, T \in H^1(\Omega)} \int_{V_{\text{tumour}}} (T_t - T)_+^2 \, \mathrm{d}x + \int_{V_{\text{healthy}}} \left(\gamma_1 (T - T_h)_+^2 + \gamma_2 (T - T_{\text{lim}})_+^2\right) \mathrm{d}x \tag{3.27a}$$

subject to the BHT equation (3.15),

$$A(T) - B(\alpha) = 0. \tag{3.27b}$$

Constrained problem:

$$\min_{\alpha \in \mathbb{C}^{k_E}, T \in H^1(\Omega)} \int_{V_{\text{tumour}}} \exp\left(-tA \exp\left(\frac{\Delta E}{RT}\right)\right) \mathrm{d}x \tag{3.28a}$$

subject to the BHT equation (3.15) and temperature constraints,

$$A(T) - B(\alpha) = 0, \quad T \le T_{\text{lim}}. \tag{3.28b}$$

This above set of goal functionals and constraints supplies the starting point for the actual computation of optimal therapies for individual patients.

4. Efficient simulation

In the numerical simulation of the whole hyperthermia procedure, *computational efficiency* is of utmost importance, since clinical decision-making requires numerical algorithms that run on a local PC *both fast and reliably.* The simulation involves solving both the time-harmonic Maxwell equation and the BHT equation. The latter, a Helmholtz equation in the non-critical case, can be readily solved by algorithms which are now standard textbook material: see, *e.g.*, Deuflhard and Weiser (2012). That is why we skip this part here. Rather, we concentrate on the two main sources of computational cost, the adaptive multilevel finite element methods for the time-harmonic Maxwell equation in Section 4.1 and their unidirectional coupling with the BHT equation in Section 4.2. An initial survey of the decisive role of these methods in hyperthermia treatment planning has been given in Deuflhard and Seebass (1999).

4.1. Adaptive FEM for the time-harmonic Maxwell equation

The first algorithmic attempts to solve the time-harmonic Maxwell equation were based on the rather popular FDTD method: see Sullivan (1990).

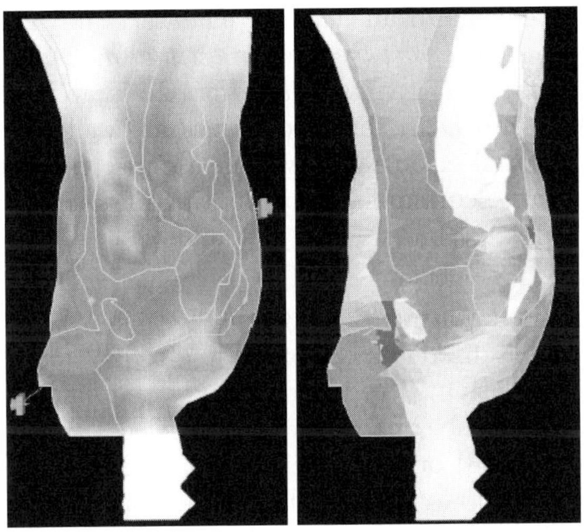

(a) (b)

Figure 4.1. Comparison of electrical field computations. (a) Popular FDTD method (Sullivan 1990) based on the strong formulation (without correct interface conditions). (b) Adaptive finite element method based on the weak formulation (with automatically correct interface conditions). See Section 4.1.

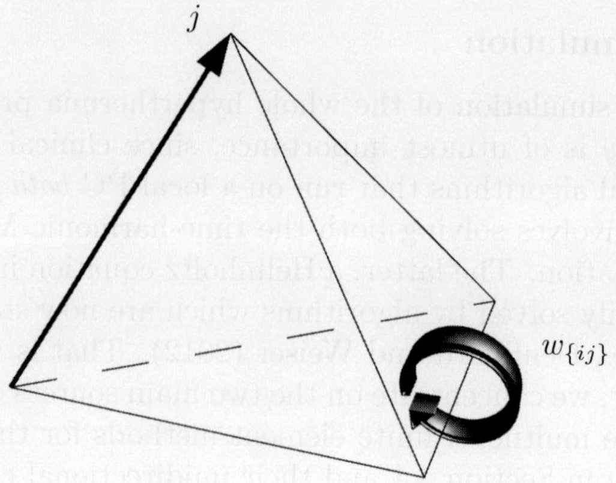

Figure 4.2. The tetrahedral edge element shape function $w_{\{ij\}}$ associated to edge $\{ij\}$ is a circular field rotating around the opposite edge.

These methods started directly from the voxel discretization, that is, the raw data, as input for numerical integration, certainly an intriguing feature. In addition, the methods appeared to be very fast; but, after testing them in a hyperthermia environment, it turned out that they were unable to treat the interface conditions correctly, which in an FEM would be automatically satisfied. A comparison of results for a special virtual patient is given in Figure 4.1.

The typical ingredients of an *adaptive multilevel* finite element method (FEM) are (a) an *a posteriori local error estimator*, (b) an *adaptive mesh refinement strategy*, and (c) a *fast solver* for the large linear systems arising. For Poisson-type PDEs, such techniques are well developed. For the time-harmonic Maxwell equation, however, a deeper understanding of the indefinite structure is necessary to reach optimal multigrid complexity.

Edge elements

We return to the weak formulation (3.5) in Section 3.1 above. In order to obtain a finite element discretization of the weak equation (3.6), we generate a *tetrahedral* triangulation $\mathcal{T}_h$ of the domain Ω and employ Nédélec's curl-conforming finite elements of first order (Nédélec 1980), also called *edge elements* (Bossavit 1988) or Whitney 1-forms. Let us denote the space of linear Lagrange elements by S_h and the nodal basis functions associated to vertex i by λ_i. Then the edge element basis function associated with the edge $e = \{ij\}$ is defined by

$$w_{\{ij\}} = \lambda_i \operatorname{grad} \lambda_j - \lambda_j \operatorname{grad} \lambda_i \tag{4.1}$$

(see Figure 4.2). The corresponding FE space is

$$ND_h := \operatorname{span}\{w_{\{ij\}}\}.$$

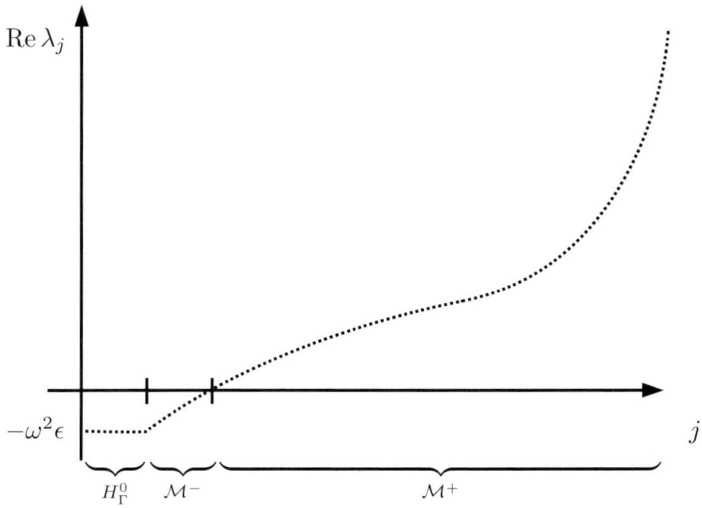

Figure 4.3. Eigenvalue structure for the time-harmonic Maxwell operator from (3.5).

Edge elements possess several distinct advantages compared to vectorial Lagrange elements. They guarantee the desired continuity of the tangential components of the electric field while preserving jumps of the normal components. In particular, concave metallic edges and internal boundaries between materials with jumps in the coefficients can be naturally incorporated in agreement with physical continuity relations. Moreover, they avoid the occurrence of spurious modes, whereas nodal basis functions may give rise to wrong solutions (see Bossavit 1990). The convergence behaviour of edge element discretizations of (3.6) is studied in Monk (1992). For the actual realization we note that for a vector field E the degree of freedom $E_e \in \mathbb{C}$ associated with each edge e in $\mathcal{T}_h$ is given by the path integral

$$E_e = \int_e E \cdot t_e \,\mathrm{d}s \tag{4.2}$$

along the edge e with direction vector t_e.

Spectral properties

Before we dive into algorithmic details, let us have a closer look at the spectral structure of the time-harmonic Maxwell operator from (3.5) that occurs in the weak equation (3.6). As illustrated in Figure 4.3, its spectrum divides into three distinct parts: (i) eigenvalues with negative real part originating from the nullspace $H^0_\Gamma(\mathrm{curl}, \Omega)$ of the curl-operator, (ii) eigenvalues with negative real part to be associated with some space $\mathcal{M}^-$ spanned by the corresponding eigenfunctions, and (iii) eigenvalues with positive real part, which go with some space $\mathcal{M}^+$.

Coarse mesh

Suppose we formally expand both sides of the weak equation (3.6) in terms of eigenfunctions of the Maxwell operator. Then the system would decouple in terms of these eigenfunctions. Those terms on the right-hand side with *smallest* coefficients will dominate the solution structure in the generic case – assuming, of course, that all eigenfunctions are present in the actual solution. Recall that, in the Laplace operator, the *smooth* eigenfunctions have the smallest eigenvalues and thus dominate the solution structure. The situation is very different for the indefinite time-harmonic Maxwell operator. To see this, a simple plane wave analysis has been given in Beck *et al.* (1999), which we repeat here. For simplicity, let the material be homogeneous, so that the coefficients μ, ϵ are constants. Let $E(x) := \exp(ikx)e$ denote a plane spatial wave in a polarization direction e. Upon ignoring the boundary terms in A, eigenvalues $|k|^2 - \omega^2\mu\epsilon$ will come up. Obviously, wavenumbers k close to the critical wavenumber $k_C = \omega\sqrt{\mu\epsilon}$ give rise to the smallest eigenvalues of the time-harmonic operator A. The associated eigenfunctions are bound to dominate the solution generically. Let the critical wavelength be defined by

$$\lambda_C = \frac{2\pi}{k_C} = \frac{2\pi}{\omega\sqrt{\mu\epsilon}}.$$

Then an appropriate coarse mesh able to resolve solutions in the space $\mathcal{M}^-$ (see Figure 4.3) should obey a mesh size restriction of the kind

$$h \leq \frac{\lambda_C}{2}, \tag{4.3}$$

which is often called the *Nyquist condition*. It will again show up below in the context of multigrid methods. In passing we note that the above restriction depends, of course, on the material. In the hyperthermia setting, it is only restrictive in the water bolus, producing a nearly uniform mesh therein.

Large linear system solvers

Replacing $H_\Gamma(\text{curl};\Omega)$ by ND_h in the variational formulation (3.6) leads to a large, very sparse linear equation system

$$Au = b \tag{4.4}$$

with complex symmetric and indefinite matrix A. The solution of (4.4) can be obtained by either direct or iterative solvers. *Direct* solvers have experienced significant progress in recent years (Davis 2006). Despite a still considerable memory requirement due to fill-in, they are very effective on coarse up to moderately fine meshes. It is only for very fine grids, or if low memory requirement is important, that iterative solvers have to be used.

Iterative solvers face two difficulties, as can be seen from the eigenvalue distribution of the operator A in Figure 4.3, which carries over to the matrix A. First, the large eigenvalues grow with decreasing mesh size h. The growing condition number of the matrix A renders both matrix decomposition methods and Krylov methods inefficient on fine meshes. Second, due to the eigenvalues with negative real part, usual positive smoothers such as Jacobi or Gauss–Seidel methods do not converge: the spectral radius of their iteration matrix $I - B^{-1}A$ exceeds 1, so the corresponding modes in $H^0_\Gamma \cup \mathcal{M}^-$ are amplified rather than reduced.

Adaptive mesh refinement

Due to the complex geometrical structure and strongly varying material properties of virtual patients, corner singularities and other local solution features are bound to arise. Hence we consider adaptive mesh refinement to be of crucial importance to ensure both efficiency of the solver and reliability of its computational results. The foundation of adaptive mesh refinement are *a posteriori error estimators*, which are most often based on localized operations. As we are dealing with equations of a partly hyperbolic nature in this context, it is clear that any local estimator will be unable to capture far-field contributions. Accordingly we prefer to use the name *error indicator* in the sense of Babuška and Aziz (1972).

Among the different flavours of error indicators, we consider hierarchical error indicators in the spirit of Deuflhard *et al.* (1989), which have been observed to yield very favourable results with efficiency quite close to 1 on simple test problems (Beck *et al.* 1999, Section 8.2). Quantitative reliability is, of course, of utmost importance in medical applications.

Conceptually, the ansatz space $V^L = ND_h$ is extended by a hierarchical surplus V^H to give a higher-order ansatz space $V^+ = V^L \oplus V^H$, in which (4.4) assumes the block structure

$$\begin{pmatrix} A^{LL} & A^{LH} \\ A^{HL} & A^{HH} \end{pmatrix} \begin{pmatrix} u^L \\ u^H \end{pmatrix} = \begin{pmatrix} b^L \\ b^H \end{pmatrix}. \tag{4.5}$$

For the mere purpose of error indication, (4.5) only needs to be solved approximately via a defect equation for the hierarchical components:

$$A^{HH} u^H = b^H - A^{HL} u^L. \tag{4.6}$$

Hierarchical extensions of the lowest-order edge elements have been described in Savage and Peterson (1996), where the basis functions are given in terms of Lagrangian nodal functions λ_i associated to the vertices i of the tetrahedral mesh. With each edge $\{ij\}$, one basis function

$$w^H_{\{ij\}} = \lambda_i \operatorname{grad} \lambda_j + \lambda_j \operatorname{grad} \lambda_i$$

Table 4.1. Estimated discretization errors versus refinement level j with node numbers N_j. The 'coarse' mesh ($j = 0$) is taken from Figure 4.4.

j	N_j	η
0	128 365	10.1 %
1	373 084	4.96 %
2	1 085 269	2.54 %

and with each face $\{ijk\}$, two basis functions

$$w^{H,1}_{\{ijk\}} = \lambda_i\lambda_j \operatorname{grad}\lambda_k - \lambda_i\lambda_k \operatorname{grad}\lambda_j, \quad w^{H,2}_{\{ijk\}} = \lambda_i\lambda_j \operatorname{grad}\lambda_k - \lambda_j\lambda_k \operatorname{grad}\lambda_i$$

are associated. An approximate solution of (4.6) may be obtained via one block-Jacobi step only (!), where it is essential to keep the (2×2)-block-entries of both functions $w^{H,l}_{\{ijk\}}$, $l = 1, 2$, attached to each face of the triangulation (see Beck *et al.* 1999 for details). At first glance, this block-Jacobi sweep seems to be critical, as the matrix contains negative eigenvalues and the related eigenmodes may be amplified. However, as the modulus of all negative eigenvalues is comparatively small, these modes can be expected to give only minor contributions. Let $\tilde{u}^H$ denote the first Jacobi iterate of the exact u^H. For the definition of a *relative* discretization error estimator we set

$$\eta^2 = \frac{\|\tilde{u}^H\|_E^2}{\|u^L\|_E^2},$$

where, in the absence of a better idea, we employ the norm

$$\|v\|_E^2 := \int_\Omega \left(\frac{1}{\mu} \operatorname{curl} v^* \operatorname{curl} v + \omega^2 \epsilon v^* v \right) \mathrm{d}x + \int_{\Gamma_{\text{ext}}} \beta (n \times v^*)(n \times v) \, \mathrm{d}s. \tag{4.7}$$

This error estimator is the key ingredient in an *adaptive refinement strategy* in the spirit of a technique suggested by Babuška and Rheinboldt (1978) for purely elliptic problems. As an illustration of an adaptive grid constructed along these lines, results for a virtual patient mesh are shown in Figure 4.4.

Table 4.1 gives a list of error indicators as a function of the refinement levels starting from the mesh of Figure 4.4.

Multilevel preconditioning

In the hyperthermia problem, a multigrid V-cycle preconditioner is used within a conjugate residual iterative method (CR): see Freund (1992). On

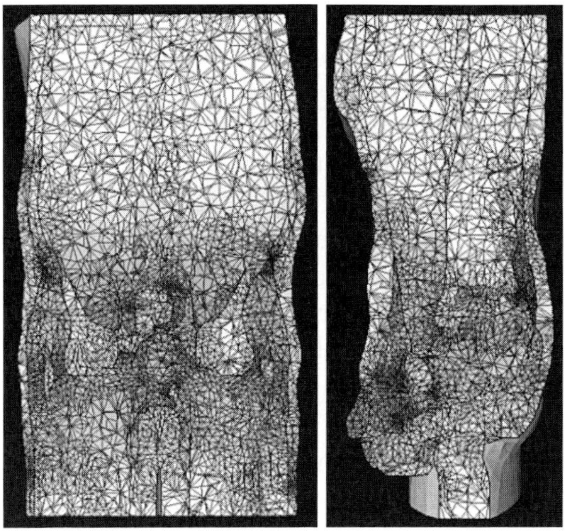

(a) (b)

Figure 4.4. Adaptive mesh of a virtual patient consisting of about 128 000 nodes. Note that a uniform grid would have required an estimated 16 000 000 nodes. (a) Frontal cut. (b) Sagittal cut.

Table 4.2. Convergence history for multilevel solvers with standard Gauss–Seidel smoothing (Std) compared to multiplicative (M-H) and additive (A-H) versions using hybrid Gauss–Seidel smoothing on each level. The iteration is terminated if the ratio of the Euclidean norms of residuals and right-hand sides is below 10^{-5}. From Deuflhard and Seebass (1999).

Ref. depth	Nodes	No. of iterations Std	M-H	A-H	CPU (min) Std	M-H	A-H
0	128 365	4250	354	413	150	24	20
1	373 084	4832	265	277	800	76	60
2	1 085 269	> 10000	186	194	> 2000	215	160

the subspace $\mathcal{M}^+$, *standard multigrid* methods are effective in reducing the error – however, at the expense of amplifying error components in $H^0_\Gamma \cup \mathcal{M}^-$. This leads to a deterioration of the whole multigrid procedure in the present case: see Table 4.2 under columns Std. Therefore, the standard multigrid approach needs to be modified.

The general applicability of multigrid methods to indefinite problems has been investigated in Bramble, Kwak and Pasciak (1994). In order to capture the error part in $\mathcal{M}^-$, certain restrictions on the mesh size of the coarse grid have to be observed – unfortunately, given there in terms of unknown theoretical quantities. Fortunately, both local mode analysis and numerical experience confirmed that it is enough to observe the above Nyquist condition (4.3). The coarse grid correction is then computed by a direct solver. Treating the part originating from $H_\Gamma^0(\mathrm{curl},\Omega)$ is more delicate, since it contains functions with all spatial frequencies and therefore cannot be captured on the coarse mesh.

For Nédélec finite elements there exists a discrete counterpart of the representation of $H_\Gamma^0(\mathrm{curl},\Omega)$ as a space of gradients of potentials. Thus, on simply connected domains each curl-free element of ND_h can be represented as the gradient of a scalar potential in the space S_h of nodal finite elements:

$$ND_h \cap H_\Gamma^0(\mathrm{curl},\Omega) = \mathrm{grad}\, S_h, \tag{4.8}$$

Moreover, the stable decomposition (3.7) has a (perturbed) discrete counterpart, too. To this end, let $I^{\mathrm{curl}} : (S_h)^3 \to ND_h$ denote the edge interpolation operator, defined via (4.2). Then

$$v = \tilde{v} + I^{\mathrm{curl}}\phi + \mathrm{grad}\, p, \quad \phi \in (S_h)^3,\ p \in S_h,\ \tilde{v} \in ND_h \tag{4.9}$$

such that

$$\|h^{-1}\tilde{v}\|_{L_2} + \|\phi\|_{H^1} + \|p\|_{H^1} \le C\|v\|_{H(\mathrm{curl})}. \tag{4.10}$$

This outstanding feature of Nédélec elements (see, *e.g.*, Beck and Hiptmair 1999, Hiptmair 2002*a*) is of crucial importance for the construction of efficient multigrid solvers for the linear systems arising, as we will illustrate for the simpler example of *positive definite problems.*

A multigrid method, applied to a linear system $Au = b$ on a space V, consists of a (non-direct) splitting

$$V = \sum V_i$$

and preconditioners B_i, defined on each V_i. Here we will choose V_i as one-dimensional spaces that come from a multilevel hierarchy and $B_i = A$. The convergence theory then depends on the stability condition

$$\inf_{v=\sum v_i} \sum \|v_i\|_A^2 \le K_1\|v\|_A^2, \tag{4.11}$$

and the strengthened Cauchy–Schwarz inequality

$$\langle v_k, v_l\rangle_A \le \gamma_{kl}\|v_k\|_A\|v_l\|_A \quad K_2 := \|\gamma\|_2, \tag{4.12}$$

which must be established for the given splitting (Xu 1992). In the additive case (*e.g.*, Jacobi-type smoothing) the product K_1K_2 gives an estimate for the condition number of the preconditioned system. If K_1K_2 admits an upper bound that is independent of h, we obtain an optimal preconditioner, *i.e.*, a bounded condition number. Similar results hold in the multiplicative case, *e.g.*, for Gauss–Seidel-type smoothing.

To understand why standard multigrid smoothers are inefficient in an $H(\mathrm{curl})$-setting, consider $\|\cdot\|_A = \|\cdot\|_{H(\mathrm{curl})}$ and a multilevel splitting

$$ND_h = \sum \Phi_i,$$

where Φ_i are one-dimensional spaces, spanned by the edge element basis functions on different levels. Now let $v \in ND_h \cap H^0_\Gamma(\mathrm{curl}, \Omega)$ (recall that $\mathrm{curl}\, v = 0$), such that

$$v = \sum_{\phi_i \in \Phi_i} \phi_i.$$

The basis functions satisfy $\|\phi_i\|_{L_2} \leq c h_i \|\phi_i\|_{H(\mathrm{curl})}$. Thus we obtain

$$\|v\|^2_{H(\mathrm{curl})} = \|v\|^2_{L_2} \leq C \inf_{v=\sum \phi_i} \sum \|\phi_i\|^2_{L_2} \leq C \inf_{v=\sum \phi_i} \sum h_i^2 \|\phi_i\|^2_{H(\mathrm{curl})}. \tag{4.13}$$

Whenever v bears high spatial frequencies, this implies

$$K_1 \geq C h^{-2},$$

and thus leads to mesh-dependent condition numbers and contraction rates.

Returning again to (4.8), we may introduce a potential $p \in S_h$, such that $v = \mathrm{grad}\, p$. Then, for a multilevel decomposition

$$S_h = \sum P_i,$$

where P_i are one-dimensional subspaces spanned by the nodal basis functions of S_h on different levels, standard H^1-multigrid theory yields

$$\inf_{p=\sum p_i} \sum \|p_i\|^2_{H^1} \leq K_{1,H^1} |p|^2_{H^1} = K_{1,H^1} \|\mathrm{grad}\, p\|^2_{L_2} = K_{1,H^1} \|v\|^2_{H(\mathrm{curl})}. \tag{4.14}$$

For general $v \in ND_h$, we can apply the stable discrete decomposition (4.9), which finally yields a stable multigrid decomposition of the form

$$V = \sum \Phi_i + \sum \mathrm{grad}\, P_i, \tag{4.15}$$

where Φ_i and P_i are taken from a multilevel hierarchy of ND_h and S_h, respectively. Then a multigrid estimate similar to (4.14) for the rotational part of v, together with (4.14), yields an h-independent constant K_1. Note that the mere existence of such a splitting is sufficient: it need not be computed explicitly. More details can be found in Beck and Hiptmair (1999), Beck *et al.* (1997, 1999), Hiptmair (1999), Hiptmair and Xu (2007) and Xu, Chen and Nochetto (2009).

On the basis of the splitting (4.15), a hybrid smoothing procedure was constructed in Beck and Hiptmair (1999) and Beck *et al.* (1999). There, as a basic solver the conjugate residual (CR) method was used, which is similar to the well-known conjugate gradient algorithm, but adjusted to symmetric indefinite systems. Its basic operations are Gauss–Seidel sweeps, both in the Nédélec space ND_h, coping with the elliptic part of A, and in the space S_h of potentials. Within this framework, efficient transfer operators between field representations in S_h and ND_h are essential.

If we represent a vector field $v \in ND_h \cap H^0_\Gamma(\text{curl}, \Omega)$ by $v = \text{grad}\, p$, then the representation in ND_h can be obtained easily from (4.2):

$$v_e = \int_{P_1}^{P_2} \text{grad}\, p \cdot t \, \text{d}s = p(P_2) - p(P_1).$$

Here P_1 and P_2 denote the positions of the endpoints of the edge e.

In the following, P_{S_h} denotes the transfer operator from the potential space S_h into the Nédélec space ND_h on $\mathcal{T}_h$. $P^*_{S_h}$ denotes the adjoint operator, defining the canonical restriction. Then the smoother A_S for the nullspace is constructed by the Galerkin product $A_S = P^*_{S_h} A P_{S_h}$. The hybrid smoothing algorithm for a given right-hand side vector $b \in ND_h$ will provide an updated solution vector $u \in ND_h$ via the following steps:

One **Gauss–Seidel** step for $Au = b$
$\tilde{b} \leftarrow b - Au$
$b_S \leftarrow P^*_{S_h} \tilde{b}$
$u_S \leftarrow 0$
One **Gauss–Seidel** step for $A_S u_S = b_S$
$u \leftarrow u + P_{S_h} u_S$

In order to be on safer theoretical grounds, we realize symmetric preconditioning, which means an immediately following step of the adjoint procedure. Note that in this approach both spaces are treated in a *multiplicative* fashion in the sense of domain decomposition methods.

As an alternative, we propose a symmetric *additive* version:

One **symmetric Gauss–Seidel** step for $Au = b$
$b_S \leftarrow P^*_{S_h} b$
$u_S \leftarrow 0$
One **symmetric Gauss–Seidel** step for $A_S u_S = b_S$
$u \leftarrow u + P_{S_h} u_S$

If we employ adaptive mesh refinement, thus creating a sequence of nested triangulations $\mathcal{T}_0 \subset \mathcal{T}_1 \subset \ldots \subset \mathcal{T}_h$, then the extension to a multilevel solver is quite straightforward. Taking into consideration that the associated sequence of Nédélec spaces is nested, *i.e.*, $ND_0 \subset ND_1 \subset \cdots \subset ND_h$, we may adopt the classical multigrid idea (Hackbusch 1985) by using canonical grid transfer operations between these spaces, but employing a hybrid smoother on each level.

Finally, in order to demonstrate the efficiency of this hybrid algorithm, we present comparative results for both the multiplicative and the additive hybrid algorithm using the discrete Helmholtz decomposition in Table 4.2 above. As can be seen, the hybrid versions exhibit a superior performance and do not deteriorate with increasing refinement level. Note that we have no direct factorization available on the coarse grid, thus facing comparatively large iteration counts. A proof of optimal multigrid complexity of this type of algorithm can be found in Beck, Deuflhard, Hege and Stalling (1997) and Hiptmair (1996).

4.2. Coupling of the BHT and the Maxwell equations

In the covered temperature range of 30–50°C, the electromagnetic material parameters are essentially constant, such that the coupling of the time-harmonic Maxwell equation and the BHT equation is unidirectional. Thus, we can first solve the time-harmonic Maxwell equation and insert the results into the BHT equation. Since the Maxwell equation is linear, we can compute the antenna profiles E_k for each antenna channel $k = 1, \ldots, k_E$ and use superposition afterwards. For the insertion into the BHT equation, however, a distinction between the linear and the nonlinear case is necessary.

Linear BHT equation

As can be seen from equation (3.12), the superposition of the electric field E from k_E antenna profiles gives rise to a superposition of k_E^2 source modes to the source term S:

$$\sum_{j,l=1}^{k_E} \alpha_j^* \alpha_l E_j^* E_l = \\ \sum_{j=1}^{k_E} |\alpha_j|_{\mathbb{C}}^2 |E_j|_{\mathbb{C}}^2 + 2 \sum_{j=2}^{k_E} \sum_{l=1}^{j-1} \big(\mathrm{Re}\,(\alpha_j^* \alpha_l) \mathrm{Re}\,(E_j^* E_l) - \mathrm{Im}\,(\alpha_j^* \alpha_l) \mathrm{Im}\,(E_j^* E_l) \big).$$

The source modes,

$$S_{jl} = \begin{cases} \mathrm{Re}\,(E_j^* E_l) & j \le l, \\ \mathrm{Im}\,(E_j^* E_l) & j > l, \end{cases}$$

correspond to temperature elevations T_{jl} above the basic tissue temperature T_{bas} for $\alpha = 0$ and can be computed in advance by solving the homogeneous BHT equation

$$\begin{aligned} -\mathrm{div}(\kappa\, \mathrm{grad}\, T_{jl}) &= -c_b w T_{jl} + \frac{1}{2} \sigma S_{jl} && \text{in } \Omega, \\ -\kappa \frac{\partial T_{jl}}{\partial n} &= h T_{jl} && \text{on } \partial\Omega. \end{aligned}$$

With coefficients $\gamma_{jl} = 2\,\mathrm{Re}\,(\alpha_j^* \alpha_l)$ for $j < l$, $\gamma_{jj} = |\alpha_j|^2_{\mathbb{C}}$, and $\gamma_{jl} = -2\,\mathrm{Im}\,(\alpha_j^* \alpha_l)$ for $j > l$, the temperature is obtained as

$$T(\alpha) = T_{\mathrm{bas}} + \sum_{j,l=1}^{k_E} \gamma_{jl} T_{jl}.$$

Let $\mathrm{cost}_{\mathrm{linBHT}}$ denote the computational cost for one solution of the BHT equation. Then the total computational cost sums up as

$$\mathrm{cost}_{\mathrm{total}} = k_E\,\mathrm{cost}_{\mathrm{Maxwell}} + (k_E^2 + 1)\,\mathrm{cost}_{\mathrm{linBHT}} + \mathrm{cost}_{\mathrm{Opt}}. \tag{4.16}$$

As an illustration, see Table 6.1 below. The cost for the optimization as a whole depends, of course, on the number of iterations.

Nonlinear BHT equation

In the case of a temperature-dependent perfusion $w(T)$, some kind of iteration is needed to take care of this nonlinearity. Deuflhard and Seebass (1999) have suggested the following fixed point iteration:

k_E **Maxwell solves:** E_k

let $\alpha^0 := 0$

for $m = 0, \ldots, n_{\mathrm{iter}}$:

  one *nonlinear* BHT equation solve for $S(\alpha^m)$
    supplies temperature T^m and perfusion $w^m := w(T^m)$

  $k_E^2 + 1$ ***linear* BHT equation solves** for S_{jl} using w^m
    supplies T^m_{bas} and temperature modes T^m_{jl}

  optimization $\min_\alpha f(T^m_{\mathrm{bas}} + \sum_{j,l=1}^{k_E} \gamma_{jl}(\alpha) T^m_{jl})$
    supplies α^{m+1}

The idea behind this algorithm has been that the Maxwell solves are considerably more expensive than the BHT solves. The iteration converges linearly and quite fast at an observed contraction rate $\theta \approx 0.3$. So the total computational cost sums up as

$$\begin{aligned}\mathrm{cost}_{\mathrm{total}} = {}& k_E\,\mathrm{cost}_{\mathrm{Maxwell}} \\ & + n_{\mathrm{iter}}\,(\mathrm{cost}_{\mathrm{nonlinBHT}} + (k_E^2 + 1)\,\mathrm{cost}_{\mathrm{linBHT}} + \mathrm{cost}_{\mathrm{Opt}}).\end{aligned} \tag{4.17}$$

In typical computations, $n_{\mathrm{iter}} \approx 6$ optimization iterations are required. The nonlinear BHT equation has been solved by a Newton iteration in function space (Deuflhard *et al.* 2000), typically requiring no more than two steps due to the fast local convergence.

5. PDE-constrained optimization

In Section 3.4, we had already worked out the medically preferable optimization problem with *strict temperature limits in healthy tissue*. When this aspect had come into the focus of interest, there were no efficient algorithms to tackle such problems. That is why thorough theoretical investigations then began: we present these in Section 5.1, and postpone algorithmic details to Section 5.2. In order to derive efficient *adaptive* algorithms, we follow a multistage computational paradigm. In the first stage, we develop the necessary numerical analysis background in a function space setting, which covers both optimality conditions and formal algorithms for the infinite-dimensional problem. In the second stage, we construct an iterative method, again in the infinite-dimensional setting, for the nonlinear problems arising. Only in the last stage, the innermost loop, where linear operator equations are to be solved, do we discretize the infinite-dimensional problem up to a prescribed error tolerance. This kind of scheme has emerged as the general basis for any adaptive PDE algorithm.

5.1. Theoretical background

The hyperthermia treatment planning problem can be formulated as an optimization problem *in function space*, subject to equality constraints and inequality constraints, in the following form:

$$\min f(T, \alpha), \tag{5.1a}$$

subject to

$$A(T) - B(\alpha) = 0, \tag{5.1b}$$

$$|\alpha_k|_{\mathbb{R}^2} \leq \alpha_{\max} \quad T \leq T_{\text{lim}}. \tag{5.1c}$$

Here A and B represent the BHT equation in weak form (3.15). From Section 3.2 we have α as a set of $2k_E - 1$ *real* parameters.

Various cost functionals f and inequality constraints have been discussed in Section 3.4. The most challenging parts are the constraints on the temperature $T \leq T_{\text{lim}}$. They can be considered as an infinite number of inequality constraints in function space. In PDE-constrained optimization and optimal control they are classified as pointwise *state constraints*, and have been subject to intense research in recent years; see, *e.g.*, Casas (1986), Meyer, Tröltzsch and Rösch (2006), Hintermüller and Kunisch (2006) and Deckelnick and Hinze (2007).

The following derivations are based on Schiela and Weiser (2010) and Schiela (2009). More details and proofs can be found therein.

Barrier regularization and strict feasibility

In *finite* dimensions, barrier regularizations give rise to the highly popular class of interior point methods (Terlaky 1996, Wright 1987). With $y = (y_1, \ldots, y_n)$, the principal idea is to replace an inequality-constrained problem of the kind

$$\min_{y \in \mathbb{R}^n} f(y) \text{ subject to } y_i \geq 0 \quad i = 1 \ldots n$$

by an unconstrained problem, depending on a parameter $\mu > 0$,

$$\min_{y \in \mathbb{R}^n} f(y) - \mu \sum_{i=1}^{n} \log(y_i),$$

where the functional f has been extended by a sum of *logarithmic barrier functions.* This new minimization problem with minimizer y^μ can then be solved approximately. By driving the barrier parameter to its limit $\mu = 0$ we can approximate the solution of the original problem. The logarithmic terms ensure strict feasibility $y_i^\mu > 0$ for all $\mu > 0$, so Newton's method can be applied to the corresponding first-order optimality conditions in a *feasible* neighbourhood of y^μ.

If we want to extend this idea to *infinite*-dimensional function spaces, *e.g.*, for the problem

$$\min_{y \in Y} f(y) \text{ subject to } y(x) \geq 0 \quad \forall x \in K,$$

then a straightforward idea might be to substitute the above finite sum by an integral. Following this approach, the associated barrier regularization would then read

$$\min_{y \in Y} f(y) - \mu \int_K \log(y(x))\, dx.$$

However, we have to take into account two additional issues.

(i) We have to achieve strict feasibility of y^μ in the sense that $y^\mu(x) \geq \varepsilon(\mu) > 0$ for $x \in K$ and for all $\mu > 0$.

(ii) The neighbourhood of y^μ that contains the Newton iterates has to be taken with respect to the $\|\cdot\|_\infty$-norm (or stronger). Any norm that is finite for a pointwise unbounded function yields neighbourhoods that are not contained in the feasible set.

Unfortunately, the logarithmic barrier functional guarantees strict feasibility of y^μ only *almost everywhere* in K, and possibly allows touch points where $y^\mu(x) = 0$ for some $\mu > 0$, even when y^μ is a smooth function. A simple example of such a case has been constructed in Schiela (2009). Of course, in such a case no feasible neighbourhood exists, even in the $\|\cdot\|_\infty$-norm.

A way out of this difficulty is shown by the following lemma.

Lemma 5.1. Let $K \subset \mathbb{R}^d$ be a compact set, satisfying a cone property. Let $y \geq 0$ be in the space $C^\beta(K)$ of Hölder-continuous functions, $1/y^m \in L_1(K)$, $0 < \beta \leq 1$. Assume that $m \geq d/\beta$. Then $1/y \in C(K)$.

Proof. Assume that without loss of generality $0 \in K$ and $y(0) = 0$. We will show that this contradicts the assumption $1/y^m \in L_1(K)$. By the Hölder-continuity of y we infer that $y(x) < cr^\beta$ for all $x \in B(0,r) \cap K$. Using the cone property of K, there is a cone $C_0 \subset K$, and we can compute for sufficiently small $R < 1$

$$\|y^{-m}\|_{L_1(K)} \geq \int_{C_0} y(x)^{-m}\,\mathrm{d}x \geq c\int_{[0,R]} r^{-\beta m} r^{d-1}\,\mathrm{d}r \geq c\int_{[0,R]} r^{-1}\,\mathrm{d}r = \infty,$$

which shows $1/y^m \notin L_1(K)$. Hence, by contradiction, $y > 0$ in K, which implies by compactness of K that there exists $\psi > 0$ with $y \geq \psi$ and thus $1/y \in C(K)$. □

Hence, in order to obtain strict feasibility of barrier minimizers in Hölder spaces, we may use a hierarchy of *higher-order barrier functions*

$$l_m(y) := \begin{cases} -\mu\log(y) & m = 1, \\ \dfrac{\mu^m}{(m-1)y^{m-1}} & m > 1, \end{cases}$$

and solve

$$\min_{y \in C^\beta(K)} f(y) + \int_K l_m(y(x))\,\mathrm{d}x.$$

Then, for sufficiently large m and Hölder-continuous y, the existence of a touch point $y(x) = 0$ would imply

$$\int_K l_m(y(x))\,\mathrm{d}x = \infty,$$

in contradiction to the optimality of y.

BHT equation as operator equation in function space

In order to study the weak form of the BHT equation (3.15) in a functional analytic setting, we have to impose some regularity assumptions. Our set of assumptions has to cover the case of discontinuous coefficients, which arise due to the different types of tissues in the human body.

The presence of state constraints forces us to impose an $\|\cdot\|_\infty$-topology on the space of temperature distributions. For this reason, we have to depart from the usual H^1-framework for elliptic equations, since $H^1(\Omega) \not\subset C(\overline{\Omega})$ for spatial dimension greater than one.

To proceed, let $q > 3$, and define its dual exponent q' via $1/q + 1/q' = 1$. Let $D_q \subset H^1(\Omega)$ be the set of all T, such that $A(T) \in W^{1,q'}(\Omega)^*$, *i.e.*,

$$\langle A(T), \varphi\rangle \leq M\|\varphi\|_{W^{1,q'}} \quad \forall \varphi \in C^\infty(\overline{\Omega}).$$

We have to achieve that the operator

$$A(\cdot) - B(\cdot) : D_q \times \mathbb{R}^{2k_E-1} \to W^{1,q'}(\Omega)^*$$

is well defined, and D_q can be embedded into $C(\overline{\Omega})$. For a strictly monotone perfusion term $w(T,x)(T - T_{\text{blood}})$, general results from elliptic regularity theory (Haller-Dintelmann, Rehberg, Meyer and Schiela 2009) show that A has a continuous inverse $A^{-1} : W^{1,q'}(\Omega)^* \to C(\overline{\Omega})$, *i.e.*, $D_q \subset C(\overline{\Omega})$, and even $\|T\|_{C^\beta(\overline{\Omega})} \le c\|A(T)\|_{W^{1,q'}(\Omega)^*}$ for some $\beta > 0$, where $C^\beta(\overline{\Omega})$ is the space of Hölder-continuous functions. Moreover, D_q only depends on the principal part of A, and is thus independent of T.

To ensure that B maps into $W^{1,q'}(\Omega)^*$, we have to impose a technical regularity assumption on the antenna profiles, which requires slightly more regularity than the usual one for Maxwell's equation, namely $E_k \in H(\text{curl}, \Omega)$. We assume that each E_k is contained in $L_q(\Omega, \mathbb{C}^3)$. Then $B(\alpha)$ is contained in $L_{q/2}(\Omega)$, which, in turn, is contained in $W^{1,q'}(\Omega)^*$ by the Sobolev embedding theorem. In this way singularities in the temperature distribution T can be excluded. Such regularity has been shown under certain smoothness assumptions on the tissue interfaces in Druet *et al.* (2011) and Druet (2007). In any case, practical experience shows that such singular temperature distributions do not occur in the hyperthermia setting. In this framework A and B are then twice continuously Fréchet-differentiable as long as the perfusion term is twice differentiable with respect to T. The derivative of A with respect to T,

$$A' : C(\overline{\Omega}) \supset D_q \to W^{1,q'}(\Omega)^*,$$

is not continuous, but it has a continuous inverse. Moreover, since the principal part of A is linear, $A'(T) - A'(\tilde{T})$ does not contain a differential operator. These facts allow us to apply results such as the inverse function theorem to A, and thus to establish continuous differentiability of the control-to-state mapping

$$\alpha \mapsto T(\alpha) := (A^{-1} \circ B)(\alpha).$$

Application of the barrier regularization to the pointwise state constraints

$$T(x) \le T_{\text{lim}}(x) \quad \forall x \in \Omega$$

immediately leads us to identify the above variable y with $T_{\text{lim}}(x) - T(x)$. Consequently, we arrive at the following barrier minimization problem:

$$\begin{aligned} &\min f(T,\alpha) + \int_\Omega l_m(T_{\text{lim}} - T)\,\mathrm{d}x, \\ &\qquad \text{subject to} \quad A(T) - B(\alpha) = 0. \end{aligned} \tag{5.2}$$

Our regularity considerations yield $T \in C^\beta(\Omega)$, and since T_{lim} is constant on each type of tissue, Lemma 5.1 applies separately on each tissue. With these considerations a generalization of the Weierstrass theorem on existence of minimizers yields the following result (for details of the proof see Schiela and Weiser 2010).

Theorem 5.2. There exists $m < \infty$, such that for every $\mu > 0$ the barrier problem (5.2) has an optimal solution, which is strictly feasible with respect to the inequality constraints.

First- and second-order optimality conditions

For a strictly feasible barrier minimizer (T, α), first- and second-order optimality conditions were derived in Schiela and Weiser (2010). To this end, it is convenient to define the Lagrangian function of the problem:

$$L(T, \alpha, \lambda) := f(T, \alpha) + \int_\Omega l_m(T_{\text{lim}} - T)\,\mathrm{d}x + \langle \lambda, A(T) - B(\alpha)\rangle.$$

Here, as $A(T) - B(\alpha) \in W^{1,q'}(\Omega)^*$, λ is a Lagrangian multiplier in $W^{1,q'}(\Omega)$.

Theorem 5.3. (KKT conditions) If (T, α) is a locally optimal strictly feasible solution of (5.2), then there exists a unique Lagrange multiplier $\lambda \in W^{1,q'}(\Omega)$, such that

$$0 = F(T, \alpha, \lambda) := \begin{cases} f_T(T, \alpha) + \int_\Omega l'_m(T_{\text{lim}} - T)\,\mathrm{d}x + A'(T)^*\lambda, \\ f_\alpha(T, \alpha) - B'(\alpha)^*\lambda, \\ A(T) - B(\alpha). \end{cases} \tag{5.3}$$

Of course, F also depends on the parameter μ, since l'_m does so. Since $A'(T)^*$ is a differential operator, the first line of this system is another (linear elliptic) partial differential equation, the *adjoint equation*, and λ is called the *adjoint state*.

Our algorithm will be based on this version of the KKT conditions and can therefore be considered as a *primal* interior point method. Alternatively one could also introduce Lagrangian multipliers for the state constraints, which would lead to *primal–dual* interior point methods. These are known to be more efficient than their primal counterparts for *finite-dimensional* problems. However, the analysis in function space yields that these multipliers can only be represented as Borel measures (Casas 1986), which are very hard to treat numerically. Below, a specific pointwise damping step is presented that recovers some of the benefits of primal–dual methods.

Theorem 5.4. (second-order optimality conditions) Let (T, α, λ) be a solution of (5.3). Moreover, let

$$\mathcal{K} := \ker\left(A'(T), -B'(\alpha)\right) := \{(\delta T, \delta\alpha) : A'(T)\delta T - B'(\alpha)\delta\alpha = 0\}$$

denote the kernel of the linearized constraints, and let

$$H(T,\alpha,\lambda)(\delta T,\delta\alpha)^2 := L_{TT}(T,\alpha,\lambda)\delta T^2 + L_{\alpha\alpha}(T,\alpha,\lambda)\delta\alpha^2 \tag{5.4}$$

denote the Hessian of the Lagrangian.

(i) If (T,α) is a minimizer of (5.2), then $H(T,\alpha,\lambda)$ is positive semi-definite on $\mathcal{K}$.

(ii) $H(T,\alpha,\lambda)$ is positive definite on $\mathcal{K}$ if and only if (T,α) is a local minimizer of (5.2) and J satisfies a local quadratic growth condition. Then, for each $(r_1,r_2,r_3) \in H^1(\Omega)^* \times \mathbb{R}^{2k_E-1} \times W^{1,q'}(\Omega)^*$, the linear system

$$\begin{pmatrix} L_{TT}(T,\alpha,\lambda) & 0 & A'(T)^* \\ 0 & L_{\alpha\alpha}(T,\alpha,\lambda) & -B'(\alpha)^* \\ A'(T) & -B'(\alpha) & 0 \end{pmatrix} \begin{pmatrix} \delta T \\ \delta\alpha \\ \delta\lambda \end{pmatrix} = \begin{pmatrix} r_1 \\ r_2 \\ r_3 \end{pmatrix} \tag{5.5}$$

has a unique solution $(\delta T,\delta\alpha,\delta\lambda) \in D_q \times \mathbb{R}^{2k_E-1} \times H^1(\Omega)$, depending continuously on (r_1,r_2,r_3).

Finally, in

$$L_{TT}(T,\alpha,\lambda)\delta T^2 = f_{TT}(T,\alpha)\delta T^2 + \int_\Omega l''_m(T_{\lim} - T)\delta T^2\,\mathrm{d}x + \langle\lambda, A''(T)\delta T^2\rangle \tag{5.6}$$

we encounter the second derivative l''_m of the barrier term. This factor can be very large if T is close to its upper bound, and thus induces a strong scaling to be taken into account in solution algorithms.

Central path

The solvability of (5.5) in a neighbourhood of a locally unique solution according to Theorem 5.4 allows us to compute Newton steps for the solution of the KKT system (5.3). Here it is important to obtain $\delta T \in D_q \subset C(\overline{\Omega})$, instead of the weaker result $\delta T \in H^1(\Omega)$. This ensures that Newton's method stays in an L_∞-neighbourhood of the minimizer and yields feasible iterates.

Introducing the shorthand notation $z := (T,\alpha,\lambda)$, we can formulate the following result.

Corollary 5.5. If $H(z)$ is positive definite on $\ker(A',-B')(T,\alpha)$, Newton's method, applied to $F(z)$, converges locally superlinearly to a zero z^μ of F.

Proof. Since F' depends continuously on z, we may just use a qualitative local convergence result for Newton's method (see, *e.g.*, Ortega and Rheinboldt 1970, Theorem 10.2.2). □

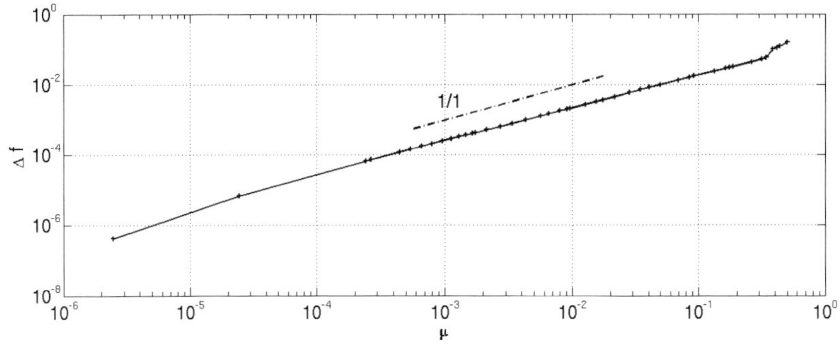

Figure 5.1. Experimental results on the rate of convergence of the function values (see (5.7)) for a hyperthermia planning problem.

In particular, the implicit function theorem yields local existence and differentiability of the so-called *central path* of barrier minimizers.

Corollary 5.6. If $H(z)$ is positive definite on $\ker(A', -B')(T, \alpha)$, then, locally, there is a differentiable path $\mu \mapsto z^\mu$ of local minimizers of the barrier problems, defined in some open interval $]\overline{\mu}, \underline{\mu}[\supset \mu$.

A convergence theory for the path of barrier minimizers has been established for the case of a linear partial differential equation as an equality constraint and a convex functional. In this case, estimates of the form

$$f(T^\mu, \alpha^\mu) - f(T_{\text{opt}}, \alpha_{\text{opt}}) \le C\mu \tag{5.7}$$

have been established, which implies under a quadratic growth condition on α the estimate

$$\|T^\mu - T_{\text{opt}}\|_{L_\infty(\Omega)} + |\alpha^\mu - \alpha_{\text{opt}}|_{\mathbb{R}^{2k_E-1}} \le C\sqrt{\mu}.$$

Furthermore, the qualitative result of Corollary 5.5 can be quantified in this case, giving a μ-dependent lower bound on the radius of convergence. Then a global convergence result can be established (Schiela 2008) for a barrier path-following method.

In hyperthermia, the equality constraints are nonlinear, and so our problem is generically non-convex. However, in a neighbourhood of the central path, under second-order sufficient optimality conditions, our barrier problems behave like convex ones, so we can expect these results to carry over. Numerical results give clear evidence for this conjecture, as can be seen in Figure 5.1.

From an algorithmic point of view, the second-order optimality conditions yield well definedness of a Newton path-following method, as presented in the following section, once an initial guess close to the central path is found.

To find such a point, algorithms have to be used that are robust with respect to non-convexity. For our numerical experiments we employ a new algorithm (see the forthcoming report by Schiela and Weiser 2012) that is based on the idea of a composite step method (see, *e.g.*, Conn, Gould and Toint 2000) with a cubic regularization (Weiser, Deuflhard and Erdmann 2007) of the functional and an affine covariant treatment of the constraints.

5.2. Adaptive central path-following in function space

Section 5.1 laid the groundwork for a Newton path-following method with respect to a barrier parameter μ. The aim of this section is to describe the development of an efficient algorithm based on this principle. We base our algorithmic approach on an affine covariant adaptive path-following method, as described in Deuflhard (2006, Chapter 5). However, compared to the standard case, additional difficulties appear.

- Our problem is posed in function space, and thus Newton steps cannot be computed exactly, which means that an inexact Newton corrector has to be used.
- The radius of convergence of Newton's method tends to zero as $\mu \to 0$. Thus, the subproblems become more and more difficult to solve. This has to be taken into account in an adaptive update strategy for μ.

An algorithm that considers these issues will consist of three loops.

(i) In the outer path-following loop, the barrier parameter μ is reduced successively. This requires an adaptive strategy.

(ii) Each step of the outer loop invokes an inexact Newton method in function space. Here we have to design termination criteria, and criteria on the success of the Newton iteration. An estimate of the *local contraction quantity* $\Theta(z)$ is crucial in this respect.

(iii) In the inner loop each Newton step is computed up to a desired relative accuracy. Typical sources of errors are truncation errors of iterative solvers, but also discretization errors which arise during the approximate computation of Newton steps in function space by finite element discretizations. We have to devise error estimators that are suited to the particular problem structure.

Local convergence of an inexact Newton method

A local affine covariant convergence theory of inexact Newton methods lays the groundwork for our algorithmic considerations. A detailed discussion of this topic can be found in Deuflhard (2006). For the convenience of the reader we give a simplified presentation, based on the analysis of Schiela and Günther (2011).

The starting point is the following very simple local convergence result for Newton's method. It is a generalization of the affine covariant refined Newton–Mysovskikh theorem due to Deuflhard and Potra (1992), which can be applied to semi-smooth systems F as well. It is formulated exclusively in terms of quantities in the domain space of the problem, *i.e.*, affine covariant, and covers the inexact computation of Newton steps.

Theorem 5.7. Let R be a linear space, Z a normed space, $D \subset Z$ and $F : D \to R$. Assume there exists $z_* \in D$ with $F(z_*) = 0$. For given $z \in D$ assume that there exists an invertible linear mapping $F'(z)(\cdot) : Z \to R$ such that

$$\Theta(z) := \frac{\left\|F'(z)^{-1}\left[F'(z)(z - z_*) - (F(z) - F(z_*))\right]\right\|}{\|z - z_*\|} \tag{5.8}$$

is well defined. Assume that an inexact Newton step results in

$$z_+ := z - F'(z)^{-1}F(z) + e,$$

where the relative error $\gamma(z) := \|e\|/\|z - z_*\|$ is bounded by

$$\gamma(z) + \Theta(z) \le \beta < 1. \tag{5.9}$$

Then

$$\|z_+ - z_*\| \le \beta\|z - z_*\|. \tag{5.10}$$

Proof. We compute for one inexact Newton step

$$\begin{aligned}\|z - z_*\| &= \|z - F'(z)^{-1}F(z) + e - z_*\| \\ &\le \left\|F'(z)^{-1}\left[F'(z)(z - z_*) - (F(z) - F(z_*))\right]\right\| + \|e\|.\end{aligned}$$

Inserting the definition of $\Theta(z)$ and assumption (5.9), we obtain

$$\|z - z_*\| \le (\Theta(z) + \gamma(z))\|z - z_*\| \le \beta\|z - z_*\|. \qquad \square$$

Equations (5.9) and (5.10) show that the local convergence of Newton's method is governed by the local contraction rate Θ and the relative accuracy γ. As we will show, in an adaptive Newton path-following algorithm, both quantities can be controlled algorithmically.

To be able to react adaptively to the nonlinearity of the problem we have to provide computable estimates for Θ and γ. We will construct an estimate for γ in the paragraph below about adaptive refinement.

Next we derive a computable estimate for $\Theta(z)$. From (5.8) we observe that $\Theta(z)$ cannot be computed exactly unless z_* is known. However, with the help of a simplified Newton step,

$$\overline{\delta z} := -F'(z)^{-1}F(z_+),$$

we can replace the solution z_* in (5.8) by the new iterate z_+, and compute

$$[\Theta(z)] := \frac{\|F'(z)^{-1}[F'(z)(z-z_+) - (F(z) - F(z_+))]\|}{\|z - z_+\|} = \frac{\|\overline{\delta z}\|}{\|\delta z\|}. \quad (5.11)$$

In view of local quadratic convergence of Newton's method, we may define a parametrized model for $\Theta(z)$ as follows:

$$[\widetilde{\Theta}(z)] := \frac{[\omega]}{2}\|z - z_*\|.$$

The constant $[\omega]$ can be computed for given z via (5.11):

$$[\omega] := \frac{2[\Theta(z)]}{\|z - z_+\|} = \frac{2\|\overline{\delta z}\|}{\|\delta z\|^2}.$$

Then $[\omega]$ can be interpreted as an estimate from below for an affine covariant Lipschitz constant ω, as used in Deuflhard (2006). This gives an estimate for the radius of contraction, *i.e.*, the radius of the largest ball around x_*, where a contraction $\Theta(z) \leq \Theta$ takes place:

$$r_\Theta \geq \frac{2\Theta}{\omega}. \quad (5.12)$$

Replacing ω by our estimate $[\omega]$, we get a quantitative estimate of the size of the region of convergence. Moreover, the remaining error via Newton's method satisfies the triangle inequality,

$$\|z_+ - z_*\| = \Theta(z)\|z - z_*\| \leq \Theta(z)(\|z - z_+\| + \|z_+ - z_*\|),$$

and thus, solving for $\|z_+ - z_*\|$ and replacing $\Theta(z)$ by $[\Theta(z)]$, we obtain the estimate

$$[\|z_+ - z_*\|] := \frac{[\Theta(z)]}{1 - [\Theta(z)]}\|z - z_+\| \quad (5.13)$$

for the remaining error.

Adaptive reduction of barrier parameter

The geometric picture of a successful path-following method is sketched in Figure 5.2. For some prescribed desired contraction $\Theta_d < 1$ (in practice one chooses $\Theta_d \in [0.1, 0.5]$), each iterate z_j should remain in the region of convergence around a target solution z^{μ_j}, which lies on the central path, *i.e.*,

$$\|z_j - z^{\mu_j}\| \leq r_{\Theta_d}(\mu_j). \quad (5.14)$$

We want to achieve this inductively by a suitable adaptive choice of μ_j. First of all, r_{Θ_d} can be estimated via (5.12), if we assume a relation between ω and μ. Numerical experiments, and estimates for related problem classes (Weiser, Gänzler and Schiela 2008), suggest the relation $\omega \sim \mu^{-1/2}$.

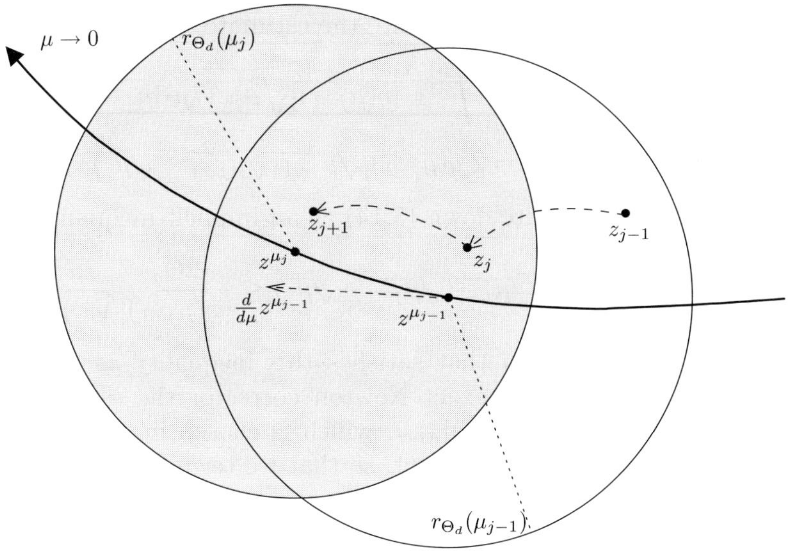

Figure 5.2. Sketch of a Newton path-following method. In each corrector step the iterates approach a target point on the central path and eventually enter the region of convergence of the next target.

To obtain a choice of μ_j that yields (5.14), we also have to provide an estimate for

$$\|z_j - z^{\mu_j}\| \leq \|z_j - z^{\mu_{j-1}}\| + \|z^{\mu_{j-1}} - z^{\mu_j}\|.$$

The first term of the right-hand side is the remaining error of the last Newton corrector and can be estimated via (5.13). The second term depends on the *sensitivity* of z^μ with respect to μ, and we have the estimate

$$\|z^{\mu_{j-1}} - z^{\mu_j}\| = \left\| \int_{\mu_j}^{\mu_{j-1}} \frac{\mathrm{d}}{\mathrm{d}\mu} z^\mu \,\mathrm{d}\mu \right\| \leq \int_{\mu_j}^{\mu_{j-1}} \underbrace{\left\| \frac{\mathrm{d}}{\mathrm{d}\mu} z^\mu \right\|}_{\eta(\mu)} \mathrm{d}\mu. \tag{5.15}$$

For finite-dimensional problems or for fixed discretizations, we may estimate this slope simply by finite differences in the iterates. If, as in our approach, adaptive grid refinement is performed during the corrector, finite differences are not reliable due to the presence of discretization errors. Then, at the cost of one system solve, an estimate $[\eta(\mu_{j-1})]$ can be computed via the formula

$$\frac{\partial}{\partial z} F(z^{\mu_{j-1}}) \frac{\mathrm{d}}{\mathrm{d}\mu} z^{\mu_{j-1}} + \frac{\partial}{\partial \mu} F(z^{\mu_{j-1}}),$$

where we have to replace $z^{\mu_{j-1}}$ by its approximation z_j. Then, using the relation $\eta(\mu) \sim \mu^{-1/2}$, we may compute the estimate

$$[\|z^{\mu_{j-1}} - z^{\mu_j}\|] := \int_{\mu_j}^{\mu_{j-1}} [\eta(\mu_{j-1})]\sqrt{\mu_{j-1}/\mu}\,\mathrm{d}\mu$$
$$= 2[\eta(\mu_{j-1})]\sqrt{\mu_{j-1}}(\sqrt{\mu_{j-1}} - \sqrt{\mu_j}).$$

To summarize, we may write down (5.14) as an implicit inequality in μ_j:

$$[\|z_j - z^{\mu_{j-1}}\|] + 2[\eta(\mu_{j-1})]\sqrt{\mu_{j-1}}(\sqrt{\mu_{j-1}} - \sqrt{\mu_j}) \leq \frac{2\Theta_d}{[\omega(\mu_{j-1})]}\sqrt{\frac{\mu_{j-1}}{\mu_j}}, \quad (5.16)$$

and compute the smallest μ that satisfies this inequality as a prediction for μ_j. In the subsequent inexact Newton corrector the estimated error is reduced up to a tolerance $\mathrm{TOL}_{\mathrm{corr}}$, which is chosen in such a way that usually one corrector step is sufficient, so that we terminate if

$$[\|z_j - z^{\mu_{j-1}}\|] \leq \mathrm{TOL}_{\mathrm{corr}} := \rho\frac{2\Theta_d}{[\omega(\mu_j)]}, \quad \rho \in [1, 1.2].$$

This rule suggests a *multiplicative* update strategy. Define $\sigma_j := \mu_j/\mu_{j-1}$. Then (5.16) can be written as an implicit inequality in σ_j:

$$[\|z_j - z^{\mu_{j-1}}\|][\omega(\mu_{j-1})] + 2[\eta(\mu_{j-1})][\omega(\mu_{j-1})]\mu_{j-1}(1 - \sqrt{\sigma_j}) \leq \frac{2\Theta_d}{\sqrt{\sigma_j}}. \quad (5.17)$$

Under our model assumptions $[\omega] \sim \mu^{-1/2}$, $[\eta] \sim \mu^{-1/2}$, which imply $\mathrm{TOL}_{\mathrm{corr}} \sim \sqrt{\mu}$, we observe that μ_{j-1} cancels out in (5.17), and thus σ is constant. Of course our model assumptions only hold approximately in computational practice, but their use leads to an unbiased and strictly positive update $\mu_{j-1} \to \mu_j > 0$.

If it terminates successfully, we accept the result of the corrector as the new iterate z_j, and compute μ_{j+1} as the new homotopy parameter, just as before. In the case of unsuccessful termination (for example, if $[\Theta(z)] > 1$ for some iterate z), with the additional knowledge of $[\omega]$ gained in this correction step, a more conservative choice for μ_j can be computed and the corrector is restarted at z_{j-1}.

Local norms for barrier subproblems

Up to now our considerations have not included any special structure of our problem. One way to do this is to choose a norm specific to the problem. In the context of barrier methods the strong scaling property of the Hessian, as observed in (5.6), should be taken into account. Thus (for fixed μ) a suitable local norm can be defined as follows:

$$\|\delta z\|_z^2 := \|\delta T\|_{L_2(\Omega)}^2 + \int_\Omega l_m''(T_{\mathrm{lim}} - T)\delta T^2\,\mathrm{d}x + |\delta\alpha|_{\mathbb{R}^{2k_E - 1}}^2 + \|\delta\lambda\|_{L_2(\Omega)}^2. \quad (5.18)$$

Qualitatively speaking, the inclusion of the barrier function l_m into the norm has the effect that errors are given more weight in regions of Ω where T is close to its upper bound $T_{\lim}$. Thereby it is taken into account that the problem is more nonlinear and thus more sensitive to errors in these regions.

In practical computations, the multiplier term $\|\delta\lambda\|^2_{L_2(\Omega)}$ can be omitted. The reason is that λ can be reconstructed approximately from the primal variables T, α by suitable multiplier update rules (Conn, Gould and Toint 2000) and plays the role of an auxiliary variable.

In a neighbourhood of the central path, where positive definiteness of the Hessian H of the Lagrangian from (5.4) can be assumed on $\ker(A', -B')$, the generic L_2-norms of δT and $\delta\alpha$ in (5.18) can be substituted by the norm induced by H for the tangential components, which dominate during the homotopy due to the mild nonlinearity of the state equation. This leads to the affine conjugate norm

$$\|\delta z\|^2_{H(z)} := H(z)(\delta T, \delta\alpha)^2. \tag{5.19}$$

The normal components can also be taken into account if desired: see, *e.g.*, Volkwein and Weiser (2002).

As the local norm changes in every Newton step, contraction $\Theta < 1$ for a fixed norm need not imply contraction between different local norms. Whether this is guaranteed depends on the actual impact of z on $\|\cdot\|_z$. For details, we refer to Deuflhard (2006) and Volkwein and Weiser (2002).

Pointwise damping

Barrier methods rely on iterates that are feasible with respect to the inequality constraints. Since in the barrier context Newton's method approximates a rational function by a linear one, Newton steps tend to violate the constraints. On the other hand, the nonlinearity, introduced by the barrier functions, is given in a pointwise fashion and highly structured. This should be exploited.

In the following we propose a modification of Newton's method, which addresses these issues and may be considered as a pointwise damping strategy. The idea exploits the pointwise structure of the problem and guarantees strict feasibility of the iterates. In the whole discussion $\mu > 0$ is fixed.

For what follows, define the Lagrangian function of the unconstrained problem:

$$\tilde{L}(z) := f(T, \alpha) + \langle \lambda, A(T) - B(\alpha) \rangle.$$

Further, let $z_+ = (T_+, \alpha_+, \lambda_+)$ be the possibly infeasible next Newton iterate, which satisfies

$$F(z) + F'(z)(z_+ - z) = 0. \tag{5.20}$$

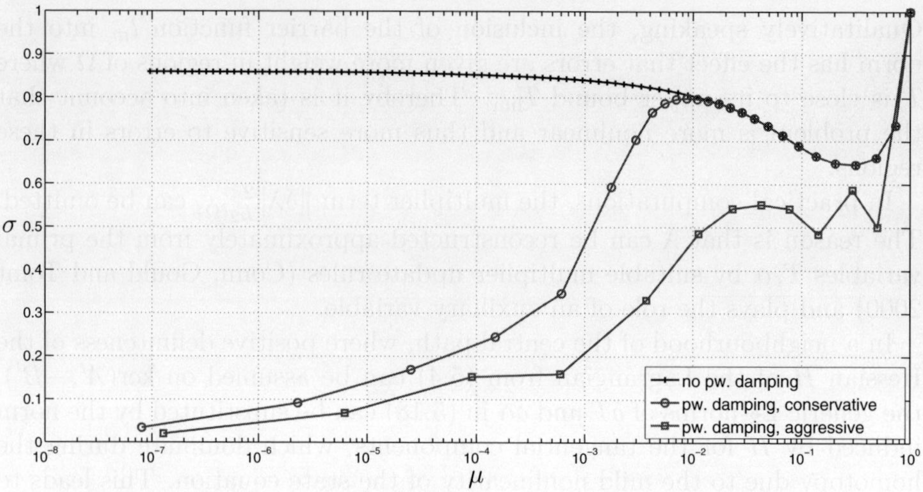

Figure 5.3. Comparison of path-following algorithm with and without pointwise damping. Horizontal axis: barrier parameter μ. Vertical axis: reduction factor $\sigma := \mu_+/\mu$. For pointwise damping one can observe $\sigma \to 0$ for $\mu \to 0$.

The main idea is to construct a modified *feasible* iterate $z_C = (T_C, \alpha_+, \lambda_+)$ that satisfies the first row of (5.3), *i.e.*,

$$\tilde{L}_T(z_C) + l'_m(T_{\lim} - T_C) + A'(T_C)^*\lambda_+ = 0. \tag{5.21}$$

This is an equation in $(H^1)^*$ which seems to be difficult to solve for T_C. However, if we subtract the first row of (5.20) from (5.21) we observe that the temperature-independent differential operators in $A'(\cdot)^*$ cancel out, and we obtain a *pointwise* equation in T_C that depends on z and z_+. If we neglect those higher-order terms that are independent of l_m, we end up with a pointwise equation of the form

$$\tilde{L}_{TT}(z)T_C + l'_m(T_{\lim} - T_C) = r(z, z_+).$$

This equation can be solved for T_C at each node of the discretization. For scalar equations there are efficient and robust algorithms available (*e.g.*, Brent 1971), and the computational cost of performing this step is marginal compared to the overall computation time.

To obtain a pointwise damping step we use only those $T_C = T_C(x)$ for which $|T_C - T| \le |T_+ - T|$; these are exactly those for which $T_+ \ge T_C \ge T$ holds. Hence, we are able to compute a *strictly* feasible corrected iterate T_C from a possibly infeasible iterate T_+ in a *natural way*. It can be shown that

$$\lim_{\|T - T^\mu\|_\infty \to 0} \frac{\|T_C - T_+\|_\infty}{\|T - T_+\|_\infty} = 0.$$

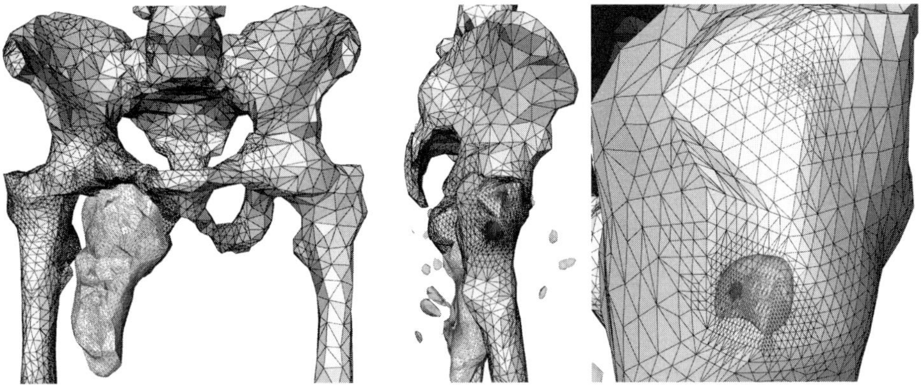

(a) (b) (c)

Figure 5.4. Bone and tumour surfaces with adaptively refined grid. (a) Frontal view. (b) Sagittal view. (c) Zoom into upper femur with 43°C isosurface; adaptive refinement occurs near hot spots.

This implies that undamped steps are recovered asymptotically close to a strictly feasible solution. An estimate for the Newton contraction $\Theta(z)$ can be computed in analogy to the case described above.

Our pointwise damping strategy has a marked effect on the numerical performance of the path-following method. As Figure 5.3 indicates, an algorithm without pointwise damping shows relatively slow linear convergence behaviour for an academic test problem. In contrast, with pointwise damping the reduction of μ often takes place at a superlinear rate. This is particularly beneficial in the context of our hyperthermia problem.

Discretization and adaptive grid refinement

So far we have assumed that our algorithm can compute Newton steps in function space exactly. However, this is not realistic because the Newton steps have to be discretized for computation. Discretization of the temperature T and the Lagrange multiplier λ is done straightforwardly by linear finite elements, and the precomputed electric fields are discretized using edge elements of first order, as described in Section 4.

As an additional numerical difficulty the barrier integrals and their derivatives cannot be evaluated exactly. However, it can be shown that the use of the *trapezoidal rule* is sufficient to obtain satisfactory *a priori* error estimates (Hinze and Schiela 2011).

For *a posteriori* error estimation and adaptive grid refinement we rely once more on hierarchical error estimators, as presented in Section 4.1.

Discretizing (5.5) with an extended ansatz space, we obtain a block linear system of the form (4.5)

$$\begin{pmatrix} F'(z)^{LL} & F'(z)^{LH} \\ F'(z)^{HL} & F'(z)^{HH} \end{pmatrix} \begin{pmatrix} \delta z^L \\ \delta z^H \end{pmatrix} = - \begin{pmatrix} F(z)^L \\ F(z)^H \end{pmatrix}. \tag{5.22}$$

The only difference is that each block of this system is in turn a block matrix of the form (5.5). Finally, as in Section 4.1, we end up with the problem of solving the block linear system

$$F'(z)^{HH}\delta z^H = r^H := -(F(z)^H + F'(z)^{HL}\delta z^L). \tag{5.23}$$

Once δz^H is computed, one can compute cheaply an estimate in the affine conjugate norm defined in (5.19). If $\delta z^H = (\delta T^H, \delta\alpha^H, \delta\lambda^H)$, define $\delta\tilde{z}^H := (\delta T^H, \delta\alpha^H, -\delta\lambda^H)$. Then one can compute via (5.5) that

$$[\|e\|_{H(z)}]^2 := \langle \delta\tilde{z}^H, r^H\rangle = \langle \delta\tilde{z}^H, F'(z)^{HH}\delta z^H\rangle = \|\delta z^H\|^2_{H(z)},$$

because the off-diagonal blocks of $F'(z)^{HH}$ cancel out. Hence, we can estimate the error in the same norm that is used to measure the length of the Newton steps.

Now we can use these estimates to *control* the relative error

$$[\gamma(z)] := \frac{\langle \delta z^H, \tilde{r}^H\rangle^{1/2}}{\|\delta z^L\|_{H(z)}}$$

by adaptive mesh refinement. The refinement loop is stopped as soon as the estimate for $[\gamma(z)]$ drops below a given relative tolerance $\mathrm{TOL_{disc}} \in [0.1, 0.5]$. A closely related adaptive refinement scheme results from goal-oriented error estimation for optimal control problems, considering the error in the cost functional (Becker, Kapp and Rannacher 2000, Weiser 2012).

In the context of Theorem 5.7 it can be seen that the corresponding inexact Newton iteration will converge asymptotically linearly in function space. A superlinear or quadratic convergence rate is not aimed for, since this usually leads to unnecessarily high accuracy and a corresponding dramatic increase of unknowns, impeding the efficiency of the algorithm.

Solution of linear systems

Finally, we have to solve discretized versions of the system (5.5) in order to compute the discretized Newton steps, and also to solve the defect equation (5.23).

This can be done by reducing it to a linear system in the control $\alpha \in \mathbb{R}^{2k_E-1}$ via a Schur complement approach. To obtain the reduced linear system, each of the discretized PDEs $A'(T)v = r$ and $A'(T)^*w = s$ has to be solved for $2k_E - 1$ right-hand sides. For a wide range of discretizations a sparse direct solver (*e.g.*, MUMPS: Amestoy *et al.* 2011) performs quite efficiently. Alternatively, one can solve the PDEs iteratively up to a certain accuracy by a preconditioned conjugate gradient method and perform an outer iteration to obtain a solution of the total system.

In particular, when solving (5.23) approximately, the solution of the PDEs $A'(x)v = r$ and its adjoint can be found iteratively via a Jacobi preconditioned conjugate gradient method.

Putting together all the pieces described above, we now arrive at the following pseudo-code.

algorithm `AdaptivePathfollowing`(z_0, μ_0) *(Initial Guess)*

for $j = 1, \ldots$ *(Path-following Loop)*
    $(\widetilde{z}; \text{success}) \leftarrow$ `InexactNewtonCorrector`$(z_{j-1}; \mu_{j-1})$
    if success
        compute new $\mu_j < \mu_{j-1}$
        $z_j \leftarrow \widetilde{z}$
    else
        compute corrected $\mu_j > \mu_{j-1}$
        $z_j \leftarrow z_{j-1}$

subroutine `InexactNewtonCorrector`$(z; \mu)$:

do *(Newton Iteration Loop)*
    do *(Inexact Newton Step)*
        RefineMarkedElements
        $z_+ := z - F_h'(z)^{-1} F_h(z)$
        $[\|e\|_{H(z)}] :=$ ErrorEstimator:Discretization(z)
        MarkElements
    while $[\|e\|_{H(z)}] > \mathrm{TOL}_{\mathrm{disc}} \|z_+ - z\|$
    compute a pointwise damped iterate z_C
    compute $[\Theta(z)]$
    compute $[\|z_C - z^\mu\|_z]$
    success $= ([\|z_C - z^\mu\|_z] \leq \mathrm{TOL}_{\mathrm{corr}})$
    failure $= ([\Theta(z)] > \text{RequiredContraction})$
    $z \leftarrow z_C$
while not(success $\vee$ failure)
return$(z; \text{success})$

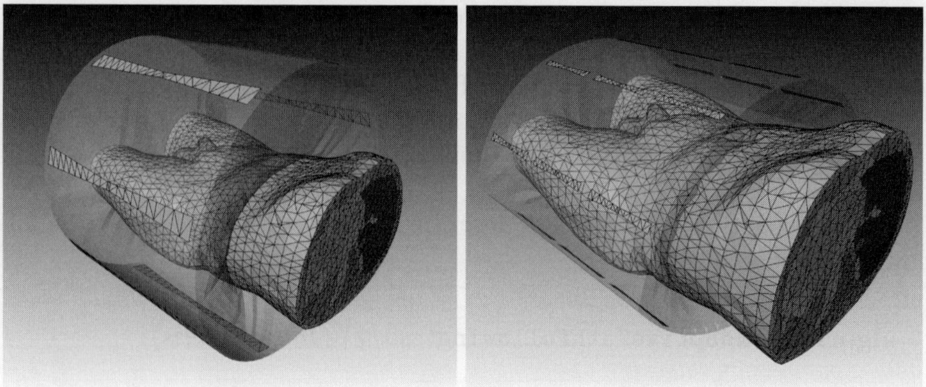

(a) (b)

Figure 6.1. Applicator design. (a) Old applicator, Sigma-60, with 8 antennas ($k_E = 4$) and circular cross-section; see also Figure 2.1(b). (b) New applicator, Sigma-Eye, with 24 antennas ($k_E = 12$) and eye-shaped cross-section, designed on the basis of our simulations.

6. Impact on medical technology

The earlier versions of our virtual lab HyperPlan have already had significant impact on treatment via deep regional hyperthermia, part of which we will present here.

6.1. Redesign of applicator

During the past ten years the applicator most frequently used for regional hyperthermia had been the Sigma-60 applicator of BSD Medical Corporation, Salt Lake City, Utah:[3] see Figure 6.1(a). This applicator consists of 8 antennas (*i.e.*, $k_E = 4$ channels) arranged on a ring with a diameter of 60 cm, hence its name. The space between the antenna ring and the patient's body (see Figure 2.1(a) for an actual photo and Figure 6.1(a) for a computer image) is filled with a *water bolus* containing de-ionized water. In early 1988, based on the results of our first simulations, the new Sigma-Eye applicator (see Figure 6.1(b)), was introduced. It has 24 antennas (*i.e.*, $k_E = 12$ channels) arranged in three parallel rings, thus allowing for an additional power steering along the patient's axis. The much smaller bolus volume has an 'eye'-shaped cross-section, hence the name.

In order to illustrate the progress, let us document some comparative results. Table 6.1 lists the computation times for the whole simulation process (on a SUN UltraSparc), based on the linear BHT as heat transfer model.

[3] Now in Munich, Germany.

Table 6.1. Coarse grid computation times (SUN UltraSparc) for the whole simulation process with linear BHT equation. From Deuflhard and Seebass (1999).

	Sigma-60 $k_E = 4$	Sigma-Eye $k_E = 12$
Grid generation	15 min	
Field calculation	80 min	120 min
Temperature calculation	2 min	20 min
Optimization	6 sec	1 min

Table 6.2. Comparison of old applicator Sigma-60 and the new applicator Sigma-Eye for three patients with different tumour locations. From Deuflhard and Seebass (1999).

	Part of tumour volume heated to above 43°C	
	Sigma-60	Sigma-Eye
Distal (supra-anal) rectal carcinoma	17.5%	62.5%
Highly presacral rectal carcinoma	0.7%	18.4%
Cervical carcinoma at pelvic wall	24.8%	49.1%

Note that, essentially due to its smaller bolus volume, the field calculation time per channel of the Sigma-Eye is only about half that of the Sigma-60. Moreover, recall from the cost model (4.16) that the temperature calculation times roughly increase with k_E^2, whereas the Maxwell solves need $O(k_E)$.

In Table 6.2, the progress in hyperthermia is documented at three patients with typical tumour locations. The new applicator has significantly improved tumour heating, keeping 'hot spots' in healthy tissue at a constant level or slightly reducing them – and not merely for these cases. Incidentally, the second location is the most difficult one, since the tumour is to a large extent enclosed by bony structures.

In Figure 6.2, we illustrate the progress in the resulting therapy with visualizations of the optimal solution in both cases.

Frequency study

Initial frequency studies were performed by Paulsen *et al.* (1999) for 100, 150, and 200 MHz, but only for power functionals such as (3.21); for their applicator test cases, they came up with an 'optimal' frequency of about

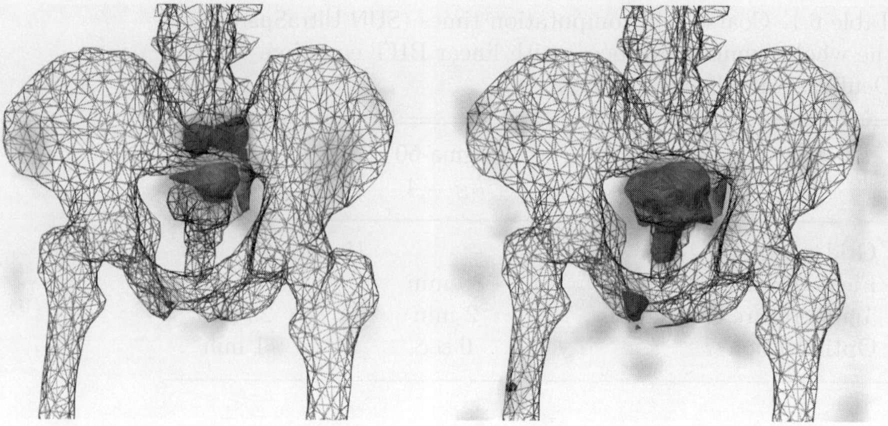

(a) (b)

Figure 6.2. Optimal temperature distribution for a rectal carcinoma at a frequency of 100 MHz. (a) Old applicator (Sigma-60). (b) New applicator (Sigma-Eye).

150 MHz. In Figure 6.3 we show our results for the new applicator in Figure 6.2(b). For all simulations, in various patient cases, a frequency of around 150 MHz also appeared to be preferable.

6.2. *Case study: femoral tumour*

Finally, in order to give an impression of the computational impact of the new optimization techniques, we present a case study. The anonymous patient data refer to a femoral tumour: see Figure 6.4(a). Of the various objective functions from Section 3.4 we selected the Arrhenius functional (3.24) inside the tumour. Outside the tumour, in healthy tissue, we imposed the strict restriction $T \leq 44°C$, and a mild quadratic penalty term, if tissue were heated above 41°C. To treat the uncertain tissue around the tumour, we surrounded the tumour with a region of width just one coarse mesh size: for this region we replaced the constraints with a strong quadratic penalty term, thus smoothing the temperature profiles to be computed. The $2k_E$ electrical fields were computed off-line once; they are a property of the given applicator. In order to satisfy the temperature constraints in healthy tissue, we applied the barrier algorithm from Section 5. The intermediate values of the barrier parameter μ are automatically determined from adaptive control, as described in Section 5.2. In the present example, the path-following method started at $\mu = 0.5$ and took the number of μ-steps documented in Table 6.3 down to $\mu = 10^{-6}$. The state shown in Figure 6.4(e) was reached after 7 μ-steps. The remaining 39 steps have been computed merely for

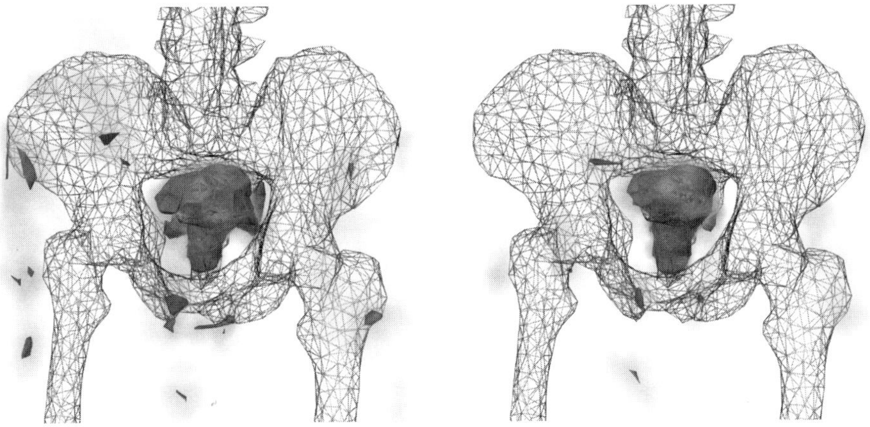

(a) (b)

Figure 6.3. Optimal temperature distribution in new applicator for a rectal carcinoma at frequencies different from the one in Figure 6.2(b). (a) 150 Mhz. (b) 200 MHz.

reference purposes. In practice one would of course employ an automatic termination criterion.

Numerical central path-following

The computational homotopy chain for $\mu \to 0$ started at $\mu = 0.5$ and a conservative solution (no 43°C isothermal surface active): see again Figure 6.4(a). Upon examination of the other images in Figure 6.4, one observes that intermediate steps may well supply interesting planning options, depending on additional medical information to be taken into account, which opens up new possibilities for the medical staff in charge of the hyperthermia treatment. For large μ, the thermal dose is lower than for smaller values. For ever smaller μ, the appearance of hot spots becomes clearly visible. However, due to the barrier's regularization, these regions are still below the prescribed upper bounds of 44°C. Hence, some individual weighting of the occurrence of hot spots versus a better covering of the tumour by the temperature isosurface may be employed.

Mesh history

In contrast to the old penalty approach, the new approach via the PDE-constrained optimization solves the full nonlinear problem including adaptive mesh refinement within one algorithmic sweep. The geometry of the tissue regions is adjusted during mesh refinement. For practical reasons, the maximal number of degrees of freedom had to be restricted. Table 6.3 gives details of three refinement levels at different barrier parameters μ.

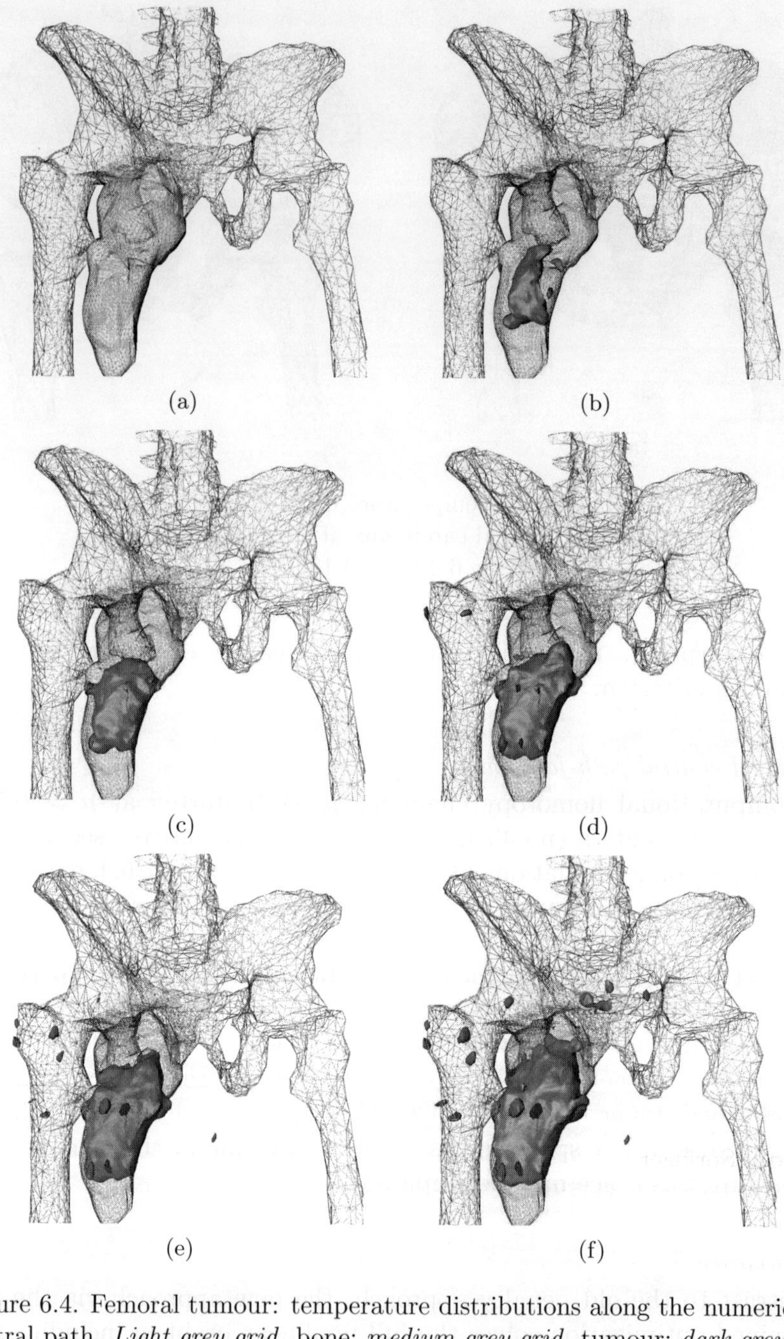

Figure 6.4. Femoral tumour: temperature distributions along the numerical central path. *Light grey grid*, bone; *medium grey grid*, tumour; *dark grey surface*, isothermal surface for 43°C. (a) Just bone and tumour (identical to $\mu = 0.5$), (b) $\mu = 0.44$, (c) $\mu = 0.42$, (d) $\mu = 0.38$, (e) $\mu = 0.26$, (f) $\mu = 10^{-6}$. Obviously, the difference between (f) and (e) is marginal.

Table 6.3. Computational effort spent on the therapy planning problem for the femoral tumour. The first line represents a computation on the coarse grid.

Tetrahedra	Unknowns	μ-steps
74 457	28 691	39
617 178	222 972	46
3 658 357	1 275 057	55

Acknowledgements

The authors gratefully acknowledge many years of funding by the Deutsche Forschungsgemeinschaft: in the collaborative research unit SFB 273 (1996–2002), in a series of individual projects (WU 235/1-1, WU 235/1-2, WE 2937/2-2, WU 235/3-1) and finally within the DFG Research Center MATHEON (in the first two periods, 2002–2010). For the present paper, they wish to thank the ZIB members Stefan Zachow and Hans Lamecker for helpful discussions and support concerning the mathematical visualization, Lin Zschiedrich for intensive discussions on the electrodynamical part, and Lars Lubkoll for computational assistance. Finally, the authors wish to thank Charlie Elliott, who happened to be on our campus as an Alexander-von-Humboldt awardee, for his careful reading (which he called a 'superficial reading').

REFERENCES

R. Adams and L. Bischof (1994), 'Seeded region growing', *IEEE Trans. Pattern Analysis and Machine Intelligence* **16**, 641–647.

G. Allaire, A. Damlamian and U. Hornung (1995), Two-scale convergence on periodic surfaces and applications. In *Mathematical Modelling of Flow through Porous Media* (A. Bourgeat *et al.*, eds), World Scientific, chapter 38.

P. R. Amestoy, A. Buttari, I. S. Duff, A. Guermouche, J.-Y. L'Excellent and B. Uçar (2011), MUMPS. In *Encyclopedia of Parallel Computing* (D. Padua, ed.), Springer.

I. Babuška and A. Aziz (1972), Survey lectures on the mathematical foundations of the finite element method. In *Mathematical Foundations of the Finite Element Method with Applications to Partial Differential Equations* (A. Aziz, ed.), Academic Press, pp. 1–359.

I. Babuška and W. Rheinboldt (1978), 'Error estimates for adaptive finite element computations', *SIAM J. Numer. Anal.* **15**, 736–754.

R. Bank (1985), PLTMG User's Guide. University of California, San Diego.

R. Beck (1993), Feldberechnung in dreidimensionalen Leitungsstrukturen der Mikroelektronik mittels p-adaptiver Finite-Elemente-Methoden. PhD thesis, Fachbereich Elektrotechnik, Technische Universität Berlin.

R. Beck and R. Hiptmair (1999), 'Multilevel solution of the time-harmonic Maxwell's equations based on edge elements', *Int. J. Numer. Methods Engng* **45**, 901–920.

R. Beck, P. Deuflhard, H.-C. Hege and D. Stalling (1997), Numerical algorithms and visualization in medical treatment planning. In *Mathematical Visualization: Experiments, Simulations and Environments* (H.-C. Hege and K. Polthier, eds), Springer International.

R. Beck, P. Deuflhard, R. Hiptmair, R. Hoppe and B. Wohlmuth (1999), 'Adaptive multilevel methods for edge element discretizations of Maxwell's equations', *Surveys Math. Indust.* **8**, 271–312.

R. Becker, H. Kapp and R. Rannacher (2000), 'Adaptive finite element methods for optimal control of partial differential equations: Basic concepts', *SIAM J. Control Optim.* **39**, 113–132.

A. Bensoussan, J.-L. Lions and G. Papanicolau (1978), *Asymptotic Analysis for Periodic Structures*, Vol. 5 of *Studies in Mathematics and Applications*, North-Holland.

F. A. Bornemann, B. Erdmann and R. Kornhuber, R. (1993), 'Adaptive multilevel methods in three space dimensions', *Int. J. Numer. Meth. Engng* **36**, 3187–3203.

A. Bossavit (1988), 'Whitney forms: A class of finite elements for three-dimensional computation in electromagnetism', *Inst. Electr. Eng. Processing* A **135**, 493–500.

A. Bossavit (1990), 'Solving Maxwell's equation in a closed cavity and the question of spurious modes', *IEEE Trans. Magn.* **26**, 702–705.

J. Bramble, D. Kwak and J. Pasciak (1994), 'Uniform convergence of multigrid V-cycle iterations for indefinite and nonsymmetric problems', *SIAM J. Numer. Anal.* **31**, 1746–1763.

J. H. Bramble, J. E. Pasciak and J. Xu (1990), 'Parallel multilevel preconditioners', *Math. Comput.* **55**, 1–22.

R. P. Brent (1971), 'An algorithm with guaranteed convergence for finding a zero of a function', *Comput. J.* **14**, 422–425.

D. Brizel, S. Scully, J. Harrelson, L. Layfield *et al.* (1996), 'Radiation therapy and hyperthermia improve the oxygenation of human soft tissue sarcomas', *Cancer Res.* **56**, 5347–5350.

E. Casas (1986), 'Control of an elliptic problem with pointwise state constraints.', *SIAM J. Control Optim.* **24**, 1309–1318.

A. Conn, N. Gould and P. Toint (2000), *Trust-Region Methods*, SIAM.

T. Cootes, A. Hill, C. Taylor and J. Haslam (1994), 'Use of active shape models for locating structures in medical images', *Image and Vision Computing* **12**, 355–365.

C. D'Angelo and A. Quarteroni (2007), 'On the coupling of 1D and 3D diffusion–reaction equations: Application to tissue perfusion problems', *Math. Mod. Meth. Appl. Sci.* **18**, 1481–1504.

T. Davis (2006), *Direct Methods for Sparse Linear Systems*, SIAM.

K. Deckelnick and M. Hinze (2007), 'Convergence of a finite element approximation to a state-constrained elliptic control problem', *SIAM J. Numer. Anal.* **45**, 1937–1953.

P. Deuflhard (2000), Differential equations in technology and medicine: Computational concepts, adaptive algorithms, and virtual labs. In *Computational Mathematics Driven by Industrial Problems* (R. Burkard, P. Deuflhard, A. Jameson, J.-L. Lions and G. Strang, eds), Springer, pp. 69–125.

P. Deuflhard (2006), *Newton Methods for Nonlinear Problems: Affine Invariance and Adaptive Algorithms*, second edn, Springer.

P. Deuflhard and R. Hochmuth (2004), 'Multiscale analysis of thermoregulation in the human microvascular system', *Math. Meth. Appl. Sci.* **27**, 971–989.

P. Deuflhard and F. Potra (1992), 'Asymptotic mesh independence of Newton–Galerkin methods via a refined Mysovskii theorem', *SIAM J. Numer. Anal.* **29**, 1395–1412.

P. Deuflhard and M. Seebass (1999), Adaptive multilevel FEM as decisive tools in the clinical cancer therapy hyperthermia. In *Domain Decomposition Methods in Sciences and Engineering* (C.-H. Lai, P. E. Bjørstad, M. Cross and O. B. Widlund, eds), Domain Decomposition Press, Bergen, pp. 403–414.

P. Deuflhard and M. Weiser (2012), *Adaptive Numerical Solution of Partial Differential Equations*, de Gruyter textbook.

P. Deuflhard, P. Leinen and H. Yserentant (1989), 'Concepts of an adaptive hierarchical finite element code', *Impact Comp. Sci. Engng* **1**, 3–35.

P. Deuflhard, M. Weiser and M. Seebass (2000), 'A new nonlinear elliptic multilevel FEM applied to regional hyperthermia', *Comput. Vis. Sci.* **3**, 115–120.

W. Dewey (1994), 'Arrhenius relationships from the molecule and cell to the clinic', *Int. J. Hyperthermia* **10**, 457–483.

P.-É. Druet (2007), Higher integrability of the Lorentz force for weak solutions to Maxwell's equations in complex geometries. WIAS Preprint 1270, Weierstraß-Institut für Angewandte Analysis und Stochastik (WIAS).

P.-E. Druet, O. Klein, J. Sprekels, F. Tröltzsch and I. Yousept (2011), 'Optimal control of three-dimensional state-constrained induction heating problems with nonlocal radiation effects', *SIAM J. Control Optim.* **49**, 1707–1736.

B. Erdmann, J. Lang and M. Seebass (1997), Optimization of temperature distributions for regional hyperthermia based on a nonlinear heat transfer model. Technical report SC-97-59, Zuse Institute Berlin.

R. W. Freund (1992), 'Conjugate gradient-type methods for linear systems with complex symmetric coefficient matrices', *SIAM J. Sci. Statist. Comput.* **13**, 425–448.

S. Gabriel, R. Lau and C. Gabriel (1996), 'The dielectric properties of biological tissues III: Parametric models for the dielectric spectrum of tissues', *Phys. Med. Biol.* **41**, 2271–2293.

S. Gerber (2007), Perfusionsmodellierung in menschlichen Tumoren. Master's thesis, Freie Universität Berlin.

T. Gänzler, S. Volkwein and M. Weiser (2006), 'SQP methods for parameter identification problem arising in hyperthermia', *Optim. Meth. Software* **21**, 869–887.

S. Götschel, M. Weiser and A. Schiela (2012), Solving optimal control problems with the Kaskade 7 finite element toolbox. In *Advances in DUNE* (A. Dedner, B. Flemisch and R. Klöfkorn, eds), Springer, to appear 2012.

W. Hackbusch (1985), *Multi-Grid Methods and Applications*, Vol. 4 of *Computational Mathematics*, Springer.

R. Haller-Dintelmann, J. Rehberg, C. Meyer and A. Schiela (2009), 'Hölder continuity and optimal control for nonsmooth elliptic problems', *Appl. Math. Optim.* **60**, 397–428.

P. Heckbert and M. Garland (1997), Survey of polygonal surface simplification algorithms. In *Siggraph 97 Course Notes*, Vol. 25, ACM Press.

H.-C. Hege and D. Stalling (1998), Fast LIC with piecewise polynomial filter kernels. In *Mathematical Visualization: Algorithms and Applications* (H.-C. Hege and K. Polthier, eds), Springer International.

M. Hintermüller and K. Kunisch (2006), 'Feasible and non-interior path-following in constrained minimization with low multiplier regularity', *SIAM J. Control Optim.* **45**, 1198–1221.

M. Hinze and A. Schiela (2011), 'Discretization of interior point methods for state constrained elliptic optimal control problems: Optimal error estimates and parameter adjustment', *Comput. Optim. Appl.* **48**, 581–600.

R. Hiptmair (1996), Multilevel preconditioning for mixed problems in three dimensions. Master's thesis, University of Augsburg, Mathematics Institute.

R. Hiptmair (1999), 'Multigrid method for Maxwell's equations', *SIAM J. Numer. Anal.* **36**, 204–225.

R. Hiptmair (2002*a*), Finite elements in computational electromagnetism. In *Acta Numerica*, Vol. 11, Cambridge University Press, pp. 237–341.

R. Hiptmair (2002*b*), Numerical methods for computational electromagnetism. Master's thesis, University of Augsburg, Mathematics Institute.

R. Hiptmair and J. Xu (2007), 'Nodal auxiliary space preconditioning in $H(\mathrm{curl})$ and $H(\mathrm{div})$ spaces', *SIAM J. Numer. Anal.* **45**, 2483–2509.

R. Hochmuth and P. Deuflhard (2004), 'Multiscale analysis for the bio-heat transfer equation: The nonisolated case', *Models Meth. Appl. Sci.* **14**, 1621–1634.

R. D. Issels, M. Wadepohl, K. Tiling, M. Müller, H. Sauer and W. Wilmanns (1987), 'Regional hyperthermia combined with systemic chemotherapy in advanced abdominal and pelvic tumors', *Recent Results Cancer Res.* **107**, 236–243.

D. Kainmueller, T. Lange and H. Lamecker (2007), Shape constrained automatic segmentation of the liver based on a heuristic intensity model. In *Proc. MICCAI Workshop 3D Segmentation in the Clinic: A Grand Challenge*, pp. 109–116.

T. Köhler, P. Maass, P. Wust and M. Seebass (2001), 'A fast algorithm to find optimal controls of multiantenna applicators in regional hyperthermia', *Phys. Med. Biol.* **46**, 2503–2514.

M. E. Kowalski, B. Behnia, A. G. Webb and J.-M. Jin (2002), 'Optimization of electromagnetic phased-arrays for hyperthermia via magnetic resonance temperature estimation', *IEEE Trans. Biomed. Engng* **49**, 1229–1241.

J. Kremer and A. K. Louis (1990), 'On the mathematical foundations of hyperthermia therapy', *Math. Meth. Appl. Sci.* **13**, 467–479.

J. Lagendijk, M. Schellekens, J. Schipper and P. van der Linden (a984), 'A three-dimensional description of heating patterns in vascularized tissues during hyperthermia treatment', *Phys. Med. Biol.* **29**, 495–507.

H. Lamecker (2008), Variational and statistical shape modeling for 3D geometry reconstruction. PhD thesis, Fachbereich Mathematik und Informatik, Freie Universität Berlin.

H. Lamecker, M. Seebass, H.-C. Hege and P. Deuflhard (2004), A 3D statistical shape model of the pelvic bone for segmentation. In *Medical Imaging 2004: Image Processing* (J. M. Fitzpatrick and M. Sonka, eds), Vol. 5370 of *Proceedings of SPIE*, pp. 1341–1351.

J. Lang, B. Erdmann and M. Seebass (1999), 'Impact of nonlinear heat transfer on temperature control in regional hyperthermia', *IEEE Trans. Biomed. Engrg* **46**, 1129–1138.

R. Löhner and P. Parikh (1988), 'Generation of three-dimensional unstructured grids by the advancing-front method', *Int. J. Numer. Methods Fluids* **8**, 1135–1149.

L. Lüdemann, P. Wust and J. Gellermann (2008), 'Perfusion measurement using DCE-MRI: Implications for hyperthermia', *Int. J. Hyperthermia* **24**, 91–96.

A. Massing (2007), 1D-reduktion thermal signifikanter Aderstränge in der Hyperthermie-Modellierung. Master's thesis, Freie Universität Berlin.

C. Meyer, F. Tröltzsch and A. Rösch (2006), 'Optimal control problems of PDEs with regularized pointwise state constraints', *COAP* **33**, 206–228.

P. Monk (1992), 'A finite element method for approximating the time-harmonic Maxwell equations', *Numer. Math.* **63**, 243–261.

P. Monk (2003), *Finite Element Methods for Maxwell's Equations*, Numerical Mathematics and Scientific Computation, Oxford Science Publications.

J. C. Nédélec (1980), 'Mixed finite elements in $\mathbb{R}^3$', *Numer. Math.* **35**, 315–341.

K. Nikita, N. Maratos and N. Uzunoglu (1993), 'Optimal steady-state temperature distribution for a phased array hyperthermia system', *IEEE Trans. Biomed. Engng* **40**, 1299–1306.

J. Ortega and W. Rheinboldt (1970), *Iterative Solution of Nonlinear Equations in Several Variables*, Academic Press.

K. Paulsen, S. Geimer, J. Tang and W. Boyse (1999), 'Optimization of pelvic heating rate distributions with electromagnetic phased arrays', *Int. J Hyperthermia* **15**, 157–186.

H. H. Pennes (1948), 'Analysis of tissue and arterial blood temperatures in the resting human forearm', *J. Applied Physiology* **1**, 93–122.

A. Quarteroni, M. Tuveri and A. Veneziani (2000), 'Computational vascular fluid dynamics: Problems, models, and methods', *Comp. Vis. Sci.* **2**, 163–197.

M. Ranneberg, M. Weiser, M. Weihrauch, V. Budach, J. Gellermann and P. Wust (2010), 'Regularized antenna profile adaptation in online hyperthermia treatment', *Medical Physics* **37**, 5382–5394.

V. Rieke and K.-B. Pauly (2008), 'MR thermometry', *J. Magn. Reson. Imaging* **27**, 376–390.

R. B. Roemer and T. C. Cetas (1984), 'Applications of bioheat transfer simulations in hyperthermia', *Cancer Research* (*Suppl.*) **44**, 4788–4798.

G. Rosner, S. Clegg, D. Prescott and M. Dewhirst (1996), 'Estimation of cell survival in tumours heated to nonuniform temperature distributions', *Int. J. Hyperthermia* **12**, 223–239.

J. Savage and A. Peterson (1996), 'High-order vector finite elements for tetrahedral cells', *IEEE Trans Microwave Theory Techn.* **44**, 874–879.

P. Scheid (1961), 'Funktionelle Besonderheiten der Mikrozirkulation im Karzinom', *Bibliotheca Anatomica* **1**, 327–335.

A. Schiela (2008), An interior point method in function space for the efficient solution of state constrained optimal control problems. ZIB report 07-44, Zuse Institute Berlin.

A. Schiela (2009), 'Barrier methods for optimal control problems with state constraints', *SIAM J. Optim.* **20**, 1002–1031.

A. Schiela and A. Günther (2011), 'An interior point algorithm with inexact step computation in function space for state constrained optimal control', *Numer. Math.* **119**, 373–407.

A. Schiela and M. Hintermüller (2011), On the length of the primal–dual path in Moreau–Yosida-based path-following for state constrained optimal control: Analysis and numerics. ZIB report 11-37, Zuse Institute Berlin.

A. Schiela and M. Weiser (2010), Barrier methods for a control problem from hyperthermia treatment planning. In *Recent Advantages in Optimization and its Applications in Engineering*, Springer, pp. 419–428.

A. Schiela and M. Weiser (2012), An affine invariant composite-step method for non-convex optimal control problems. Technical report, Zuse Institute Berlin. In preparation.

M. Seebass, R. Beck, J. Gellermann, J. Nadobny and P. Wust (2000), Electromagnetic phased arrays for regional hyperthermia: Optimal frequency and antenna arrangement. Technical report, Zuse Institute Berlin.

C. Song, A. Lokshina, J. Rhee, M. Patten and S. Levitt (1984), 'Implication of blood flow in hyperthermic treatment of tumors', *IEEE Trans. Biomed. Engrg.* **31**, 9–16.

G. Sreenivasa, J. Gellermann, B. Rau, J. Nadobny, P. Schlag, P. Deuflhard, R. Felix and P. Wust (2003), 'Clinical use of the hyperthermia treatment planning system HyperPlan to predict effectiveness and toxity', *Int. J. Radiation Oncology Biol. Phys.* **55**, 407–419.

D. Stalling (1998), Fast texture-based algorithms for vector field visualization. PhD thesis, Fachbereich Mathematik und Informatik, Freie Universität Berlin.

D. Stalling, M. Seebass, H.-C. Hege, P. Wust, P. Deuflhard and R. Felix (1996), HyperPlan: An integrated system for treatment planning in regional hyperthermia. In *Congress on Hyperthermic Oncology*, Vol. II (C. Franconi, G. Arcangeli and R. Cavaliere, eds), University Tor Vergata, Rome, pp. 552–554.

D. Stalling, M. Westerhoff and H.-C. Hege (2005), Amira: A highly interactive system for visual data analysis. In *The Visualization Handbook* (C. Hansen and C. Johnson, eds), Elsevier, chapter 38, pp. 749–767.

D. M. Sullivan (1990), 'Three-dimensional computer simulation in deep regional hyperthermia using the FDTD method', *IEEE Trans. Microwave Theory and Techniques* **38**, 204–211.

A. Szasz and G. Vincze (2006), 'Dose concept of oncological hyperthermia: Heat-equation considering the cell destruction', *J. Cancer Research Therapeutics* **2**, 171–181.

T. Terlaky (1996), *Interior Point Methods of Mathematical Programming*, Springer.

D. Tompkins, R. Vanderby, S. Klein, W. Beckman, R. Steeves and B. Paliwal (1994*a*), 'The use of generalized cell-survival data in a physiologically based objective function for hyperthermia treatment planning: A sensitivity study with a simple tissue model implanted with an array of ferromagnetic thermoseeds', *Int. J. Radiation Oncology Biol. Phys.* **30**, 929–943.

D. Tompkins, R. Vanderby, S. Klein, W. Beckman, R. Steeves, D. Frey and B. Palival (1994*b*), 'Temperature-dependent versus constant-rate blood perfusion modelling in ferromagnetic thermoseed hyperthermia: Results with a model of the human prostate', *Int. J. Hyperthermia.*

L. Vincent and P. Soille (1991), 'Watersheds in digital spaces: An efficient algorithm based on immersion simulations', *IEEE Trans. Pattern Analysis and Machine Intelligence* **13**, 583–598.

S. Volkwein and M. Weiser (2002), 'Affine invariant convergence analysis for inexact augmented Lagrangian–SQP methods', *SIAM J. Control Optim.* **41**, 875–899.

M. Weihrauch, P. Wust, M. Weiser, J. Nadobny, S. Eisenhardt, V. Budach and J. Gellermann (2007), 'Adaptation of antenna profiles for control of MR guided hyperthermia (HT) in a hybrid MR–HT system', *Medical Physics* **34**, 4717–4725.

M. Weiser (2009), 'Optimization and identification in regional hyperthermia', *Int. J. Appl. Electromagn. and Mech.* **30**, 265–275.

M. Weiser (2012), 'On goal-oriented adaptivity for elliptic optimal control problems', *Optim. Meth. Software*, to appear.

M. Weiser, P. Deuflhard and B. Erdmann (2007), 'Affine conjugate adaptive Newton methods for nonlinear elastomechanics', *Optim. Meth. Software* **22**, 413–431.

M. Weiser, T. Gänzler and A. Schiela (2008), 'A control reduced primal interior point method for PDE constrained optimization', *Comput. Optim. Appl.* **41**, 127–145.

H. Whitney (1957), *Geometric Integration Theory*, Princeton University Press.

S. Wright (1987), *Primal–Dual Interior-Point Methods*, SIAM.

P. Wust, J. Nadobny, R. Felix, P. Deuflhard, W. John and A. Louis (1989), 'Numerical approaches to treatment planning in deep RF-hyperthermia', *Strahlenther. Onkol.* **165**, 751–757.

P. Wust, J. Nadobny, R. Felix, P. Deuflhard, A. Louis and W. John (1991), 'Numerical approaches to treatment planning in deep RF-hyperthermia', *Int. J. Hyperthermia* **7**, 157–173.

P. Wust, J. Nadobny, M. Seebass, D. Stalling, J. Gellermann, H.-C. Hege, P. Deuflhard and R. Felix (1999), 'Influence of patient models and numerical methods on predicted power deposition patterns', *Int. J. Hyperthermia* **15**, 519–540.

J. Xu (1989), Theory of multilevel methods. Dissertation, Pennsylvania State University, University Park, USA.

J. Xu (1992), 'Iterative methods by space decomposition and subspace correction', *SIAM Review* **34**, 581–613.

J. Xu and L. Chen and R. H. Nochetto (2009), Optimal multilevel methods for $H(\text{grad})$, $H(\text{curl})$, and $H(\text{div})$ systems on graded and unstructured grids. In *Multiscale, Nonlinear and Adaptive Approximation*, Springer, pp. 599–659.

H. Yserentant (1986), 'On the multi-level splitting of finite element spaces', *Numer. Math.* **49**, 379–412.

H. Yserentant (1993), Old and new convergence proofs for multigrid methods. In *Acta Numerica*, Vol. 2, Cambridge University Press, pp. 285–326.

S. Zachow, M. Zilske and H.-C. Hege (2007), 3D reconstruction of individual anatomy from medical image data: Segmentation and geometry processing. ZIB report 07-41, Zuse Institute Berlin.

M. Zilske, H. Lamecker and S. Zachow (2008), Adaptive remeshing of non-manifold surfaces. In *Eurographics 2008 Annex Conf. Proc.*, pp. 207–211.

B. Zitov and J. Flusser (2003), 'Image registration methods: A survey', *Image Vision Comput.* **21**, 977–1000.

Acta Numerica (2012), pp. 379–474
doi:10.1017/S0962492912000050

High-performance computing systems: Status and outlook*

J. J. Dongarra
University of Tennessee
and
Oak Ridge National Laboratory
and
University of Manchester
E-mail: dongarra@eecs.utk.edu

A. J. van der Steen
NCF/HPC Research, L. J. Costerstraat 5,
6827 AR Arnhem, The Netherlands
E-mail: steen@hpcresearch.nl

This article describes the current state of the art of high-performance computing systems, and attempts to shed light on near-future developments that might prolong the steady growth in speed of such systems, which has been one of their most remarkable characteristics. We review the different ways devised to speed them up, both with regard to components and their architecture. In addition, we discuss the requirements for software that can take advantage of existing and future architectures.

* Colour online for monochrome figures available at journals.cambridge.org/anu.

CONTENTS

1	Introduction	380
2	The main architectural classes	381
3	Shared-memory SIMD machines	384
4	Distributed-memory SIMD machines	386
5	Shared-memory MIMD machines	389
6	Distributed-memory MIMD machines	391
7	ccNUMA machines	395
8	Clusters	397
9	Processors	399
10	Computational accelerators	418
11	Networks	434
12	Recent trends in high-performance computing	439
13	HPC challenges	452
	References	471

1. Introduction

High-performance computer systems can be regarded as the most powerful and flexible research instruments today. They are employed to model phenomena in fields as varied as climatology, quantum chemistry, computational medicine, high-energy physics and many others. In this article we present some of the architectural properties and computer components that make up high-performance computers at present, and also give an outlook on systems to come. Even though the speed of computers has increased tremendously over the years (often a doubling in speed every two or three years), the need for ever faster computers is still there and will not disappear in the foreseeable future.

Before going on to descriptions of the machines themselves, it is useful to consider some mechanisms that are or have been used to increase performance. The hardware structure or *architecture* determines to a large extent the possibilities and impossibilities of speeding up a computer system beyond the performance of a single CPU core. Another important factor that is considered in combination with the hardware is the capability of compilers to generate efficient code to be executed on the given hardware platform. In many cases it is hard to distinguish between hardware and software influences, and one has to be careful in the interpretation of results when ascribing certain effects to hardware or software peculiarities or both. In this article we will place most emphasis on hardware architecture. For a description of machines that can be classified as 'high-performance' we refer readers to Culler, Singh and Gupta (1998) or van der Steen (1995).

The rest of the paper is organized as follows. Section 2 discusses the main architectural classification of high-performance computers. Section 3 presents shared-memory vector SIMD machines. Section 4 discusses distributed-memory SIMD machines. Section 5 looks at shared-memory MIMD machines. Section 6 considers distributed-memory MIMD machines. Section 7 discusses ccNUMA machines, which are closely related to shared-memory systems. Section 8 presents clusters. Section 9 overviews processors, and looks at what is currently available. Section 10 presents computational accelerators, GPUs, and FPGAs. Section 11 discusses networks and what is commercially available. Section 12 overviews recent trends in high-performance computing. Section 13 concludes with an examination of some of the challenges we face in the effective use of high-performance computers.

2. The main architectural classes

For many years, the taxonomy of Flynn (1972) has proved to be useful in the classification of high-performance computers. This classification is based on the ways of manipulating instruction and data streams and comprises four main architectural classes. We will first briefly sketch these classes; afterwards we will fill in some details and describe each class separately.

SISD (single instruction, single data) machines. These are the conventional systems that contain one CPU and hence can accommodate one instruction stream that is executed serially. Today almost all large servers have more than one CPU, but each of these executes instruction streams that are unrelated. Therefore, such systems should still be regarded as (several) SISD machines acting on different data spaces. Examples of SISD machines are workstations, offered by many vendors. The definition of SISD machines is given here for the sake of completeness, but we will not discuss this type of machine in our report.

SIMD (single instruction, multiple data) machines. Such systems often have a large number of processing units, so that all may execute the same instruction on different data in lockstep. Thus, a single instruction manipulates many data items in parallel. Examples of SIMD machines in this class are the CPP Gamma II and the Quadrics Apemille, which are no longer marketed. Nevertheless, the concept is still interesting and is returning in the form of a co-processor in HPC systems, albeit in a somewhat restricted form in some computational accelerators such as graphical processing units (GPUs).

Another subclass of SIMD systems comprises vector processors, which act on arrays of similar data, rather than on single data items, using specially structured CPUs. When data can be manipulated by these vector units, results can be delivered with a rate of one, two and – in special cases –

three per clock cycle (a clock cycle being defined as the basic internal unit of time for the system). So, vector processors execute in an almost parallel way, but only when executing in vector mode on their data. In this case they are several times faster than when executing in conventional scalar mode. For practical purposes, therefore, vector processors are usually regarded as SIMD machines. Examples of such systems are the NEC SX-9B and the Cray X2.

MISD (multiple instruction, single data) machines. Theoretically, in this type of machine multiple instructions should act on a single stream of data. As yet no practical machine in this class has been constructed, nor are such systems easy to conceive. We will disregard them in the following discussions.

MIMD (multiple instruction, multiple data) machines. These machines execute several instruction streams in parallel on different data. The difference from the multiprocessor SIMD machines mentioned above lies in the fact that the instructions and data are related, because they represent different parts of the same task to be executed. Thus, MIMD systems may run many sub-tasks in parallel in order to shorten the time-to-solution for the main task to be executed. There is a large variety of MIMD systems, and, particularly for this class, the Flynn taxonomy does not quite suffice for the classification of systems. Systems that behave very differently, such as a 4-processor NEC SX-9 vector system and a 100 000-processor IBM BlueGene/P, fall into this class. In the following we will make another important distinction between classes of systems and treat them accordingly.

Shared-memory systems. Shared-memory systems have multiple CPUs, all of which share the same address space. This means that the knowledge of where data are stored is of no concern to the user as there is only one memory accessed by all CPUs on an equal basis. Shared-memory systems can be either SIMD or MIMD. Single-CPU vector processors can be regarded as an example of the former, while the multi-CPU models of these machines are examples of the latter. We will sometimes use the abbreviations SM-SIMD and SM-MIMD for the two subclasses.

Distributed-memory systems. In this case each CPU has its own associated memory. The CPUs are connected by some network and may exchange data between their respective memories when required. In contrast to shared-memory machines the user must be aware of the location of the data in the local memories and will have to move or distribute these data explicitly when needed. Again, distributed-memory systems may be either SIMD or MIMD. The first class of SIMD systems mentioned, which operate in lockstep, all have distributed memories associated with the processors. As we will see, distributed-memory MIMD systems exhibit a large variety of topologies in their interconnection network. The details of this topology are largely hidden from the user, which is quite helpful with respect to portability of applications but may have an impact on performance. For distributed-memory systems we will sometimes use the terms DM-SIMD and DM-MIMD to indicate the two subclasses.

As already mentioned, although the difference between shared and distributed-memory machines seems clear-cut, this is not always the case from the user's point of view. For instance, the late Kendall Square Research systems employed the idea of 'virtual shared memory' on a hardware level. Virtual shared memory can also be simulated at the programming level: a specification of High Performance Fortran (HPF) was published in 1993 (High Performance Fortran Forum 1993) which, by means of compiler directives, distributes the data over the available processors. Therefore, the system on which HPF is implemented in this case will look like a shared-memory machine to the user. Other vendors of massively parallel processing (MPP) systems, such as SGI, also support proprietary virtual shared-memory programming models, due to the fact that these physically distributed memory systems are able to address the whole collective address space. So, for the user, such systems have one *global address space* spanning all of the memory in the system. We will say a little more about the structure of such systems in Section 7. In addition, packages such as TreadMarks (Amza *et al.* 1995) provide a 'distributed-shared-memory' environment for networks of workstations.[1] Since 2006, Intel has marketed its 'Cluster OpenMP' (based on TreadMarks) as a commercial product. It allows the use of the shared-memory OpenMP parallel model to be used on distributed-memory clusters. For the last few years companies such as ScaleMP and 3Leaf have provided products to aggregate physical distributed memory into virtual shared memory.

Lastly, so-called partitioned global address space (PGAS) languages such as Co-Array Fortran (CAF) and Unified Parallel C (UPC) are gaining popularity due to the recent emergence of multi-core processors. With proper implementation this allows a global view of the data, and language facilities make it possible to specify the processing of data associated with a (set of) processor(s) without the need to explicitly move data around.

Distributed processing takes the DM-MIMD concept one step further: instead of many integrated processors in one or several boxes, workstations, mainframes, *etc.*, are connected by (Gigabit) Ethernet, or other, faster networks, and set to work concurrently on tasks in the same program. Conceptually, this is no different from DM-MIMD computing, but the communication between processors can be much slower. Packages that initially were

[1] A comprehensive bibliography of distributed shared memory systems can found at www.cs.umd.edu/~keleher/dsmbiblio/dsmbiblio.html.

made to realize distributed computing, such as PVM (standing for parallel virtual machine: Geist *et al.* 1994), and MPI (message-passing interface: Snir *et al.* 1998, Gropp *et al.* 1998) have become *de facto* standards for the 'message passing' programming model. MPI and PVM have become so widely accepted that they have been adopted by all vendors of distributed-memory MIMD systems and even on shared-memory MIMD systems, for compatibility reasons. In addition, there is a tendency to cluster shared-memory systems by a fast communication network to obtain systems with a very high computational power. For example, the NEC SX-9 and the Cray X2 both have this structure. Thus, within the clustered nodes a shared-memory programming style can be used, while between clusters message passing should be used. It must be said that PVM is no longer used very much and the development has stopped. MPI has now more or less become the *de facto* standard.

For SM-MIMD systems we mention OpenMP (Chandra *et al.* 2001, Chapman, Jost and van der Pas 2007),[2] which can be used to parallelize Fortran and C/C++ programs by inserting comment directives (Fortran 77/90/95) or pragmas (C/C++) into the code. OpenMP has quickly been adopted by all major vendors and has become a well-established standard for shared-memory systems.

Note, however, that for both MPI-2 and OpenMP 2.5 many systems/compilers only implement part of the latest standards. Therefore one has to enquire carefully whether a particular system enjoys the full functionality of these standards. The standard vendor documentation will almost never be clear on this point.

3. Shared-memory SIMD machines

This subclass of machines is practically equivalent to single-processor vector processors, although other interesting machines in this subclass have existed (*i.e.*, VLIW machines: van der Steen 1990), and may emerge again in the near future. In the block diagram in Figure 3.1 we depict a generic model of a vector architecture. The single-processor vector machine will have only one of the vector processors shown here and the system may even have its scalar floating-point capability shared with the vector processor (as was the case in some Cray systems). It may be noted that the VPU does not show a cache. Vector processors may have a cache but in many cases the vector unit cannot take advantage of it, and execution speed may in some cases be unfavourably affected because of frequent cache overflow. Of late, however, this tendency has been reversed because of the increasing gap in

[2] OpenMP Application Interface, version 2.5: www.openmp.org/.

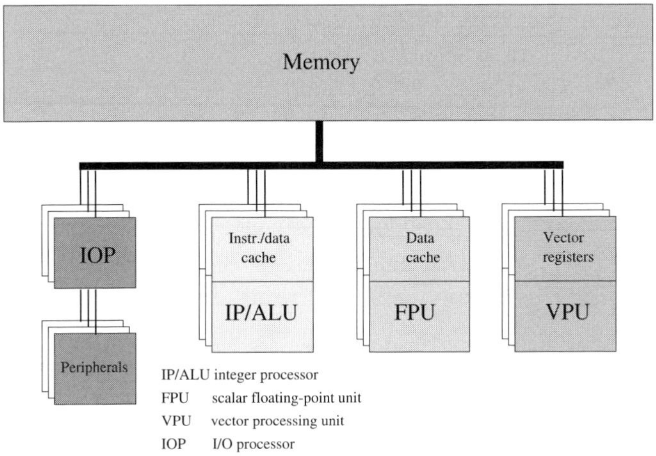

Figure 3.1. Block diagram of a vector processor.

speed between the memory and the processors: the Cray X2 has a cache, and NEC's SX-9 vector system has a facility that is somewhat like a cache.

Although vector processors have existed that loaded their operands directly from memory, and stored the results again immediately in memory (CDC Cyber 205, ETA-10), present-day vector processors use vector registers. This impairs the speed of operations while providing much more flexibility in gathering operands and manipulation with intermediate results.

Because of the generic nature of Figure 3.1 no details of the interconnection between the VPU and the memory are shown. Still, these details are very important for the effective speed of a vector operation: when the bandwidth between memory and the VPU is too small it is not possible to take full advantage of the VPU because it has to wait for operands or has to wait before it can store results. When the ratio of arithmetic to load/store operations is not high enough to compensate for such situations, severe performance losses may be incurred. The influence of the number of load/store paths for the dyadic vector operation $c = a + b$ (a, b, and c vectors) is depicted in Figure 3.2. Because of the high costs of implementing these data paths between memory and the VPU, compromises are often sought and the full required bandwidth (*i.e.*, two load operations and one store operation at the *same* time) is seldom realized. Only Cray Inc., in its former Y-MP, C-series and T-series, has employed this very high bandwidth. Vendors now rely on additional caches and other tricks to hide the lack of bandwidth.

The VPUs are shown as a single block in Figure 3.1. However, there is considerable diversity in the structure of VPUs. Every VPU consists of a number of vector functional units, or 'pipes', that fulfil one or several functions in the VPU. Every VPU will have pipes that are designated

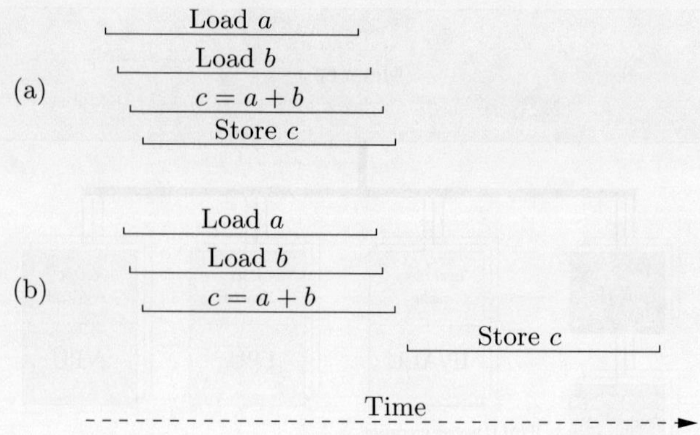

Figure 3.2. Schematic diagram of a vector addition. Case (a) when two load pipes and one store pipe are available; case (b) when two load/store pipes are available.

to perform memory access functions, thus ensuring the timely delivery of operands to the arithmetic pipes and storage of results in memory. Usually there will be several arithmetic functional units for integer/logical arithmetic, for floating-point addition, for multiplication and sometimes a combination of both, a so-called compound operation. Division is performed by an iterative procedure, table look-up, or a combination of both using the add-and-multiply pipe. In addition, there will almost always be a mask pipe to enable operation on a selected subset of elements in a vector of operands. Lastly, such sets of vector pipes can be replicated within one VPU (2 to 16-fold replication occurs). Ideally, this will increase performance per VPU by the same factor, provided the bandwidth to memory is adequate.

Lastly, it must be remarked that vector processors as described here are no longer considered a viable economic option, and both the Cray X2 and the NEC SX-9 will disappear in the near future: vector units within standard processor cores and computational accelerators have invaded the vector processing area. Although they are less efficient and have bandwidth limitations, they are so much cheaper that they out-compete classical vector processors.

4. Distributed-memory SIMD machines

Machines of the DM-SIMD type are sometimes also known as *processor array* machines (Hockney and Jesshope 1988). Because the processors of these machines operate in lockstep, *i.e.*, all processors execute the same instruction at the same time (but on different data items), no synchronization between processors is required. This greatly simplifies the design of such systems. A *control processor* issues the instructions that are to be executed by the processors in the processor array. At present, no commercially avail-

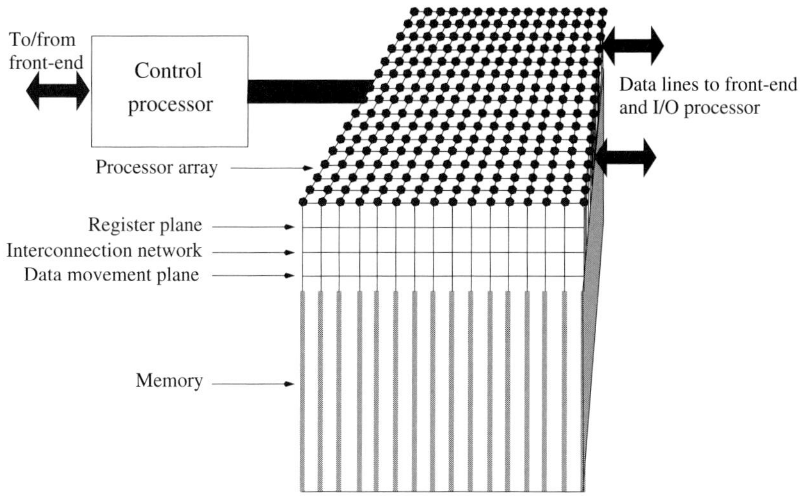

Figure 4.1. A generic block diagram of a distributed-memory SIMD machine.

able machines of the processor array type are marketed. However, because of the shrinking size of devices on a chip, it may be worthwhile locating a simple processor with its network components on a single chip, thus making processor array systems economically viable again. In fact, common graphical processing units (GPUs) share many characteristics with processor array systems. This is why we still discuss this type of system.

DM-SIMD machines use a front-end processor to which they are connected by a data path to the control processor. Operations that cannot be executed by the processor array or by the control processor are offloaded to the front-end system. For instance, I/O may be through the front-end system, by the processor array machine itself, or by both. Figure 4.1 shows a generic model of a DM-SIMD machine, from which actual models will deviate to some degree. Figure 4.1 might suggest that all processors in such systems are connected in a 2D grid, and indeed the interconnection topology of this type of machine always includes the 2D grid. As opposing ends of each grid line are also always connected, the topology is rather that of a torus. This is not the only interconnection scheme: they might also be connected in 3D, diagonally, or in more complex structures.

It is possible to exclude processors in the array from executing an instruction on certain logical conditions, but this means that during the time of this instruction these processors are idle (a direct consequence of the SIMD-type operation), which immediately lowers the performance. Another factor that may adversely affect the speed occurs when data required by processor i resides in the memory of processor j; in fact, as this occurs for all processors at the same time, this effectively means that data will have to be permuted across the processors. To access the data in processor j, the data will have to be fetched by this processor and then sent through the routing network to processor i. This may be fairly time-consuming. For both reasons mentioned, DM-SIMD machines are rather specialized in their use when one wants to employ their full parallelism. Generally, they perform excellently in digital signal and image processing, and on certain types of Monte Carlo simulation where virtually no data exchange between processors is required, and exactly the same types of operation are done on massive data sets with a size that can be made to fit comfortably into these machines. They will also perform well in gene-matching applications.

The control processor as depicted in Figure 4.1 may be more or less intelligent. It issues the instruction sequence that will be executed by the processor array. In the worst case (*i.e.*, a less autonomous control processor), when an instruction is not fit for execution on the processor array (*e.g.*, a simple print instruction), it might be offloaded to the front-end processor, which may be much slower than execution on the control processor. In the case of a more autonomous control processor this can be avoided, thus saving processing interrupts both on the front-end and on the control processor. Most DM-SIMD systems have the ability to handle I/O independently from the front-end processors. This is favourable because the communication between the front-end and back-end systems is avoided. A (specialized) I/O device for the processor array system is generally much more efficient in providing the necessary data directly to the memory of the processor array. These I/O systems are especially important for very data-intensive applications such as radar and image processing.

A feature that is peculiar to this type of machine is that the processors are sometimes of a very simple bit-serial type, *i.e.*, the processors operate on the data items bit-wise, irrespective of their type. Therefore, operations on integers, for example, are produced by software routines on these simple bit-serial processors, which takes at least as many cycles as the operands are long. So, a 32-bit integer result will be produced twice as fast as a 64-bit result. For floating-point operations a similar situation holds, although the number of cycles required is a multiple of that needed for an integer operation. As the number of processors in this type of system is mostly large (1024 or larger, though the Quadrics Apemille was a notable exception), the slower operation on floating-point numbers can be often compensated for by their number, while the cost per processor is quite low compared to full floating-point processors. In some cases, however, floating-point co-processors were added to the processor array. Their number was 8–16 times lower than that of the bit-serial processors because of the cost argument. An advantage of bit-serial processors is that they may operate on operands of any length. This is particularly advantageous for random number generation (which often boils down to logical manipulation of bits) and for signal processing, since in both cases operands of only 1–8 bits are abundant. Because, as mentioned, the execution time for bit-serial machines is proportional to the length of the operands, this may result in significant speed-ups.

At present there are no DM-SIMD systems on the market, but some types of computational accelerator (see Section 10) share many characteristics with DM-SIMD systems that existed until recently. We will briefly discuss properties of these accelerators later.

5. Shared-memory MIMD machines

One subclass of this type of machine was shown in Figure 3.1. In fact, the single-processor vector machine discussed there was a special case of a more general type. The figure shows that more than one FPU and/or VPU may be possible in one system.

The main problem one faces in shared-memory systems is that of the connection of the CPUs to each other and to the memory. As more CPUs are added, the collective bandwidth to the memory should ideally increase linearly with the number of processors, while each processor should preferably communicate directly with all others without the much slower alternative of having to use the memory in an intermediate stage. Unfortunately, full interconnection is quite costly, growing with $O(n^2)$ while increasing the

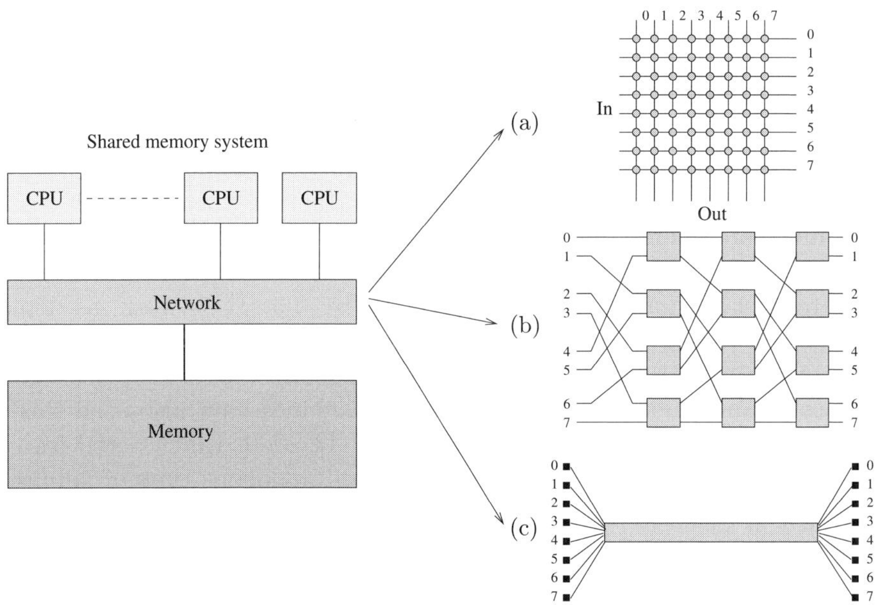

Figure 5.1. Some examples of interconnection structures used in shared-memory MIMD systems. (a) Crossbar, (b) Ω-network, (c) central data bus.

number of processors with $O(n)$. So, various alternatives have been tried. Figure 5.1 shows some of the interconnection structures that are (and have been) used.

As can be seen from the figure, a crossbar uses n^2 connections, an Ω-network uses $n \log_2 n$ connections, while with the central bus there is only one connection. This is reflected in the use of each connection path for the different types of interconnections. For a crossbar each data path is direct and does not have to be shared with other elements. In the case of the Ω-network there are $\log_2 n$ switching stages, and as many data items may have to compete for any path. For the central data bus all data have to share the same bus, so n data items may compete at any time.

The bus connection is the least expensive solution, but it has the obvious drawback that bus contention may occur, thus slowing down computations. Various intricate strategies have been devised using caches associated with the CPUs to minimize the bus traffic. This leads, however, to a more complicated bus structure, which raises the costs. In practice it has proved very hard to design buses that are fast enough, especially where processor speed has been increasing very quickly, imposing an upper bound on the number of processors thus connected, which in practice appears not to exceed 10–20 processors. In 1992, a new standard (IEEE P896) for a fast bus was defined, to connect either internal system components or to connect external systems. This bus, called the Scalable Coherent Interface (SCI), provides a point-to-point bandwidth of 200–1000 MB/s. It has been used in the HP Exemplar systems, but also within a cluster of workstations as offered by SCALI. The SCI is much more than a simple bus and it can act as the hardware network framework for distributed computing: see James, Laundrie, Gjessing and Sohi (1990). However, it has now been effectively superseded by InfiniBand (see Section 11).

A multi-stage crossbar is a network with a logarithmic complexity, and has a structure which is situated somewhere between a bus and a crossbar with respect to potential capacity and costs. The Ω-network, as depicted in Figure 5.1, is an example. Commercially available machines such as the IBM eServer p575, the SGI Altix UV, and many others use(d) a network structure, but a number of experimental machines have also used this or a similar kind of interconnection. The BBN TC2000, which acted as a virtual shared-memory MIMD system, used an analogous type of network (a butterfly network), and it is likely that new machines will also use it, especially as the number of processors grows. For a large number of processors the $n \log_2 n$ connections quickly become more attractive than the n^2 used in crossbars. Of course, the switches at the intermediate levels should be sufficiently fast to cope with the bandwidth required. Obviously, not only the *structure* but also the *width* of the links between the processors is important: a network using 16-bit parallel links will have a bandwidth which is 16 times higher than a network with the same topology implemented with serial links.

Until recently multiprocessor vector processors used crossbars. This was feasible because the maximum number of processors within a system node was small (16 at most). In the late Cray X2 the number of processors had increased so much, however, that it had to change to a logarithmic network topology (see Section 11). Not only does it become harder to build a crossbar of sufficient speed for the larger numbers of processors, but the processors themselves generally also increase in speed individually, doubling the problems of making the speed of the crossbar match that of the bandwidth required by the processors.

Whichever network is used, the type of processors could in principle be arbitrary for any topology. In practice, however, bus-structured machines cannot support vector processors as their speeds would grossly mismatch with any bus that could be constructed at reasonable cost. All available bus-oriented systems use RISC processors, as far as they still exist. The local caches of the processors can sometimes alleviate the bandwidth problem if data access can be satisfied by the caches, thus avoiding references to the memory.

The systems discussed in this subsection are of the MIMD type and therefore different tasks may run on different processors simultaneously. In many cases synchronization between tasks is required, and again the interconnection structure is very important here. Some Cray vector processors in the past employed special communication registers within the CPUs (the X-MP and Y-MP/C series) by which they could communicate directly with the other CPUs they have to synchronize with. This is, however, no longer practised as it is viewed as too costly a feature. The systems may also synchronize via the shared memory. Generally, this is much slower but it can still be acceptable when the synchronization occurs relatively seldom. Of course, in bus-based systems communication also has to be done via a bus. This bus is mostly separated from the data bus to ensure a maximum speed for the synchronization.

6. Distributed-memory MIMD machines

The class of DM-MIMD machines represents undoubtedly the largest fraction in the family of high-performance computers. A generic diagram is given in Figure 6.1. The figure shows that within a computational node A, B, *etc.*, a number of processors (four in this case) draw on the same local memory, and the nodes are connected by some network. Consequently, when a processor in node A needs data present in node B, this has to be accessed through the network – hence the characterization of the system

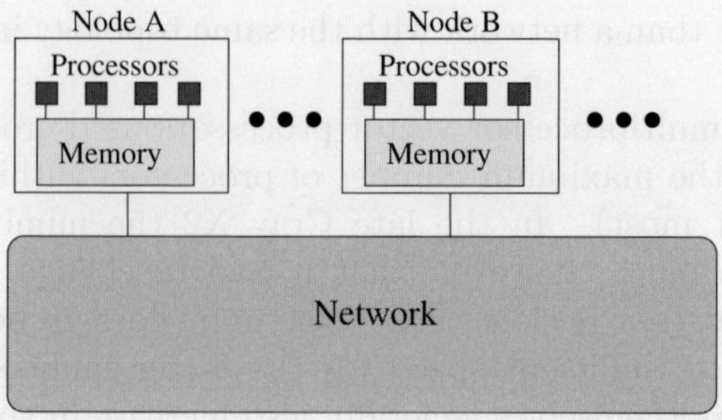

Figure 6.1. Generic diagram of a DM-MIMD machine.

as being of the distributed memory type. The vast majority of all HPC systems today are a variation of the model shown in Figure 6.1.

This type of machine is more difficult to deal with than shared-memory machines and DM-SIMD machines. The latter machines are processor array systems, in which the data structures that are candidates for parallelization are vectors and multi-dimensional arrays laid out automatically on the processor array by system software. For shared-memory systems the data distribution is completely transparent to the user. This is generally quite different for DM-MIMD systems, where the user has to distribute the data over the processors and the data exchange between processors has to be performed explicitly when using the so-called message-passing parallelization model (which is the case in the vast majority of programs). The initial reluctance to use DM-MIMD machines seems to have decreased. This is partly due to the now existing standard for communication software (Geist *et al.* 1994, Snir *et al.* 1998, Gropp *et al.* 1998) and partly because, at least theoretically, this class of system can outperform all other types of machines.

Alternatively, instead of message passing, a partitioned global address space parallelization model may be used with a programming language such as Unified Parallel C (UPC) or Co-Array Fortran.[3] In this case one still has to know where the relevant data are, but no explicit sending/receiving between processors is necessary. This greatly simplifies the programming but the compilers are still either fairly immature or even in an experimental stage, which does not always guarantee great performance, to say the least.

The advantages of DM-MIMD systems are clear. The bandwidth problem that haunts shared-memory systems is avoided because the bandwidth scales up automatically with the number of processors. Furthermore, the speed of the memory, which is another critical issue with shared-memory systems (to get a peak performance that is comparable to that of DM-MIMD systems, the processors of the shared-memory machines should be very fast

[3] Unified Parallel C home page: upc.gwu.edu/.
Co-Array Fortran home page: www.co-array.org/.

and the speed of the memory should match it), is less important for DM-MIMD machines, because more processors can be configured without the aforementioned bandwidth problems.

Of course, DM-MIMD systems also have their disadvantages. The communication between processors is slower than in SM-MIMD systems, and so the synchronization overhead, in the case of communicating tasks, is generally orders of magnitude higher than in shared-memory machines. Moreover, the access to data that are not in the local memory belonging to a particular processor have to be obtained from non-local memory (or memories). This again is very slow compared to local data access. When the structure of a problem dictates a frequent exchange of data between processors and/or requires many processor synchronizations, it may well be that only a very small fraction of the theoretical peak speed can be obtained. As already mentioned, the data and task decomposition are factors that mostly have to be dealt with explicitly, which may be far from trivial.

It will be clear from the paragraph above that both the topology and the speed of the data paths are also crucial for the practical usefulness of DM-MIMD machines. Again, as in Section 5 on SM-MIMD systems, the richness of the connection structure has to be balanced against cost. Of the many conceivable interconnection structures, only a few are popular in practice. One of these is the so-called hypercube topology, as depicted in Figure 6.2(a).

A nice feature of the hypercube topology is that for a hypercube with 2^d nodes, the number of steps to be taken between any two nodes is at most d. Thus, the dimension of the network grows only logarithmically with the number of nodes. In addition, it is theoretically possible to simulate any other topology on a hypercube: trees, rings, 2D and 3D meshes, *etc.* In practice, the exact topology for hypercubes is of secondary importance, because all systems in the market today employ what is called 'wormhole routing', or variants thereof. This means that when a message is to be sent from node i to node j, a header message is sent from i to j, resulting in a direct connection between these nodes. As soon as this connection is established, the proper data are sent through this connection without disturbing the operation of the intermediate nodes. Except for a small amount of time in setting up the connection between nodes, the communication time has become fairly independent of the distance between the nodes. Of course, when several messages in a busy network have to compete for the same paths, waiting times are incurred, as in any network that does not directly connect any processor to all others, and often rerouting strategies are employed to circumvent busy links if the connecting network supports it. Further, the network nodes themselves have become quite powerful and, depending on the type of network hardware, may send and reroute message packages in a way that minimizes contention.

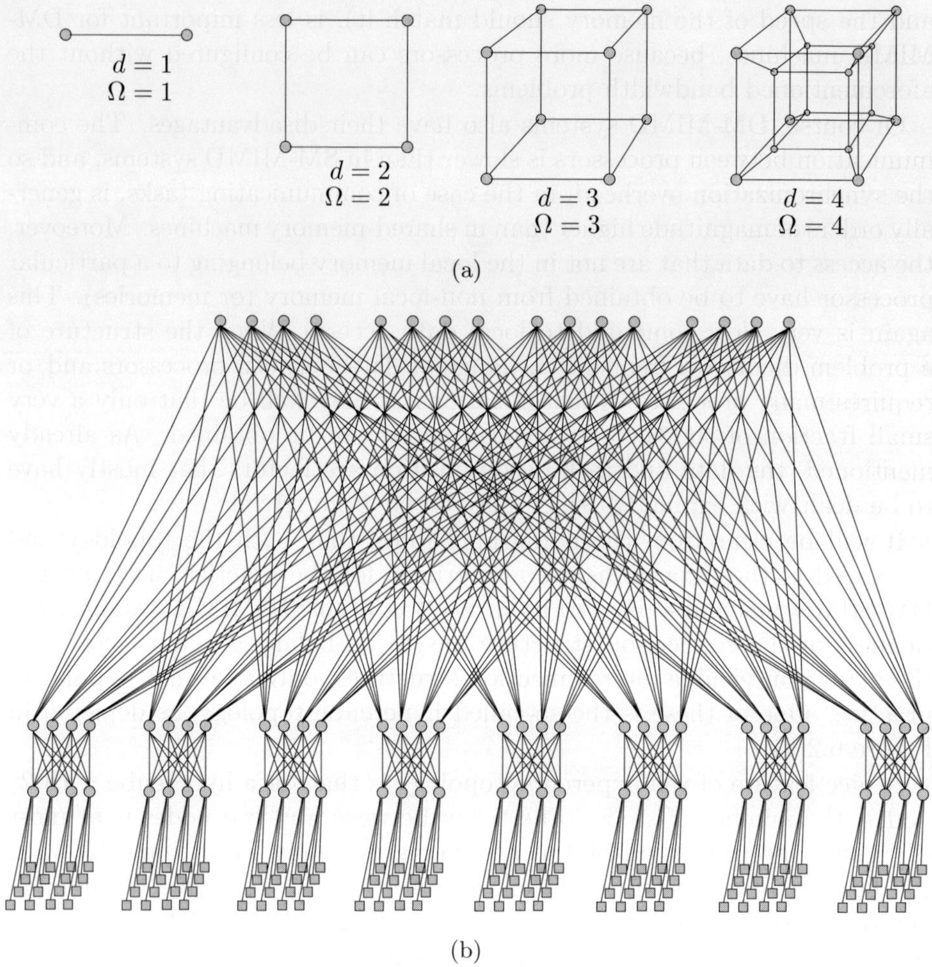

Figure 6.2. Some frequently used networks for DM machine types. (a) Hypercubes, dimension 1–4. (b) A 128-way fat tree.

Another cost-effective way to connect a large number of processors is by means of a *fat tree*. In principle a simple tree structure for a network is sufficient to connect all nodes in a computer system. However, in practice it turns out that, near the root of the tree, congestion occurs because of the concentration of messages that first have to traverse the higher levels in the tree structure before they can descend again to their target nodes. The fat tree amends this shortcoming by providing more bandwidth (mostly in the form of multiple connections) in the higher levels of the tree. One speaks of an N-ary fat tree when the levels towards the roots are N times the number of connections in the level below it. Figure 6.2(b) shows an example of a quaternary fat tree with a bandwidth in the highest level that is four times that of the lower levels.

A number of massively parallel DM-MIMD systems seem to favour a 2D or 3D mesh (torus) structure. The rationale for this seems to be that most large-scale physical simulations can be mapped efficiently on this topology and that a richer interconnection structure hardly pays off. However, some systems maintain (an) additional network(s) besides the mesh to handle certain bottlenecks in data distribution and retrieval (Horie *et al.* 1991). This philosophy has also been followed on IBM's BlueGene systems.

A large fraction of systems in the DM-MIMD class employ crossbars. For relatively small numbers of processors (in the order of 64) this may be a direct or 1-stage crossbar, while to connect larger numbers of nodes multi-stage crossbars are used, *i.e.*, the connections of a crossbar at level 1 connect to a crossbar at level 2, *etc.*, instead of directly to nodes at more remote distances in the topology. In this way it is possible to connect many thousands of nodes through only a few switching stages. In addition to the hypercube structure, other logarithmic complexity networks such as butterfly, Ω, or shuffle-exchange networks and fat trees are often employed in such systems.

As with SM-MIMD machines, a node may in principle consist of any type of processor (scalar or vector) for computation or transaction processing together with local memory (with or without cache) and, in almost all cases, a separate communication processor with links to connect the node to its neighbours. Today, the node processors are mostly off-the-shelf RISC processors, sometimes enhanced by vector processors. A problem that is peculiar to DM-MIMD systems is the mismatch of communication versus computation speed that may occur when the node processors are upgraded without also speeding up the intercommunication. In many cases this may result in turning computation-limited problems into communication-limited problems.

7. ccNUMA machines

As already mentioned in the Introduction, there is a trend to build systems that have a rather small (up to 16) number of processors, tightly integrated in a cluster: a symmetric multiprocessing (SMP) node. The processors in such a node are virtually always connected by a 1-stage crossbar, while these clusters are connected by a less costly network. Such a system may look as depicted in Figure 7.1. Note that in Figure 7.1 all CPUs in a cluster are connected to a common part of the memory. (Figure 7.1 looks functionally identical to Figure 6.1, but there is a difference that cannot be expressed in the figure: all memory is directly accessible by all processors without the necessity to transfer the data explicitly.)

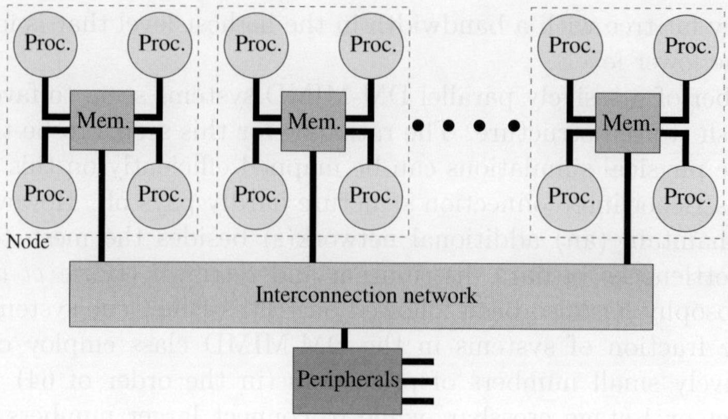

Figure 7.1. Block diagram of a system with a 'hybrid' network: clusters of four CPUs are connected by a crossbar. The clusters are connected by a less expensive network, *e.g.*, a butterfly network.

The most important ways to let the SMP nodes share their memory are S-COMA, which stands for simple cache-only memory architecture, and ccNUMA, or cache coherent non-uniform memory access. Therefore, such systems can be considered as SM-MIMD machines. On the other hand, because the memory is physically distributed, it cannot be guaranteed that a data access operation always will be satisfied within the same time. In S-COMA systems the cache hierarchy of the local nodes is extended to the memory of the other nodes. So, when data are required that do not reside in the local node's memory, they are retrieved from the memory of the node where they are stored. In ccNUMA this concept is further extended in that all memory in the system is regarded (and addressed) globally. So, a data item may not be physically local but logically it belongs to one shared address space. Because the data can be physically dispersed over many nodes, the access time for different data items may well be different, which explains the term 'non-uniform data access'. The term 'cache coherent' refers to the fact that, for all CPUs, any variable that is to be used must have a consistent value. Therefore, it must be ensured that the caches that provide these variables are also consistent in this respect. There are various ways to ensure that the caches of the CPUs are coherent. One is the *snoopy bus protocol*, in which the caches listen in on transport of variables to any of the CPUs and update their own copies of these variables if they have them and they are requested by a local CPU. Another way is the *directory memory*, a special part of memory which enables the caches to keep track of all the copies of variables and of their validity.

At present, no commercially available machine uses the S-COMA scheme. In contrast, there are several popular ccNUMA systems (Bull's bullx R422 series, HP Superdome, and SGI Ultraviolet) that are commercially available.

An important characteristic of NUMA machines is the *NUMA factor*. This factor shows the difference in latency for accessing data from a local memory location as opposed to a non-local one. Depending on the connection structure of the system, the NUMA factor for various parts of a system can differ from part to part: accessing data from a neighbouring node will be faster than from a distant node, for which a number of stages of a crossbar might possibly be traversed. So, when a NUMA factor is mentioned, this is mostly for the largest network cross-section, *i.e.*, the maximal distance between processors.

Since the appearance of multi-core processors, the ccNUMA phenomenon has also manifested itself in processors with multiple cores: the first-level and second-level caches belong to a particular core and therefore, when another core needs data residing in another core's cache, retrieval is via the complete memory hierarchy of the processor chip. This is typically orders of magnitude slower than when retrieval is from the local cache.

For all practical purposes we can classify these systems as being SM-MIMD machines, also because special assisting hardware/software (such as a directory memory) has been incorporated to establish a single system image although the memory is physically distributed.

8. Clusters

The adoption of clusters, *i.e.*, collections of workstations/PCs connected by a local network, has metaphorically exploded since the introduction of the first Beowulf cluster in 1994. The attraction lies in the (potentially) low cost of both hardware and software and the control that builders/users have over their system. The interest in clusters is shown by the creation of the IEEE Task Force on Cluster Computing (TFCC),[4] which reviews the field of cluster computing on a regular basis. Further, books describing how to build and maintain clusters have greatly added to their popularity (Sterling, Salmon, Becker and Savaresse 1999, Spector 2000). As the cluster scene has become a mature and attractive market, large HPC vendors and many start-up companies have entered the field, and offer more-or-less out-of-the-box cluster solutions for those groups who do not want to build their cluster from scratch (hardly anyone builds from scratch these days).

The number of vendors selling cluster configurations has become so large that it is not possible to include all their products in this report. In addition, there is generally great variation in the usage of clusters and they are more often used for *capability computing* (*i.e.*, the system is employed for one or a few programs for which no alternative is readily available in terms of computational capability), while the integrated machines are primarily used for

[4] TFCC home page: www.clustercomputing.org

capacity computing (*i.e.*, the system is employed to the full by use of most of its available cycles by many, often very demanding, applications and users). Traditionally, vendors of large supercomputer systems have learned to provide for capacity computing, as the precious resources of their systems were required to be used as effectively as possible. In contrast, Beowulf clusters typically use the Linux operating system, although a small minority use Microsoft Windows. These operating systems either lack the tools to make use of clusters for capacity computing, or the tools are immature. However, as clusters become on average both larger and more stable, there is a trend to use them as computational capacity servers too, particularly because there is now a plethora of cluster management and monitoring tools. The article by van der Steen (2000) considers some of the aspects that are necessary conditions for this kind of use, such as available cluster management tools and batch systems. The systems assessed then are now quite obsolete, but many of the conclusions are still valid. An important but not very surprising conclusion was that the speed of the network is crucial in all but the most computation-limited applications. Another notable observation was that using compute nodes with more than one CPU may be attractive from the point of view of compactness and (possibly) energy and cooling aspects, but that performance can be severely damaged by the fact that more CPUs have to draw on a common node memory. The bandwidth of the nodes is in this case not up to the demands of memory-intensive applications.

As cluster nodes have become available with 4–8 processors, where each processor may also have up to 12 processor cores, this issue has become all the more important, and one might have to choose between capacity-optimized nodes with more processors but less bandwidth per processor core, or capability-optimized nodes with fewer processors per node but higher bandwidth available for the processors in the node. This choice is not particular to clusters (although the phenomenon is relatively new for them): it also occurs in the integrated ccNUMA systems. Interestingly, as remarked in the previous section, in clusters the ccNUMA memory access model is turning up now in the cluster nodes; as for the larger nodes, it is no longer possible to guarantee symmetric access to all data items for all processor cores (evidently, for a core, a data item in its own local cache will be available more quickly than for a core in another processor).

Fortunately, there is now a fair choice of communication networks available in clusters. Of course Gigabit Ethernet or 10 Gigabit Ethernet are always possible, which are attractive for economic reasons, but have the drawback of a high latency ($\approx$ 10–40 μs). Alternatively, there are networks that operate from user space at high speed and with a latency that approaches those of the networks in integrated systems. These will be discussed in Section 11.

9. Processors

In comparison to 10 years ago, the processor scene has become drastically different. In the period 1980–1990, proprietary processors and, in particular, vector processors were the driving force of supercomputers, but today that role has been taken by common off-the-shelf processors. In fact there are only two companies left that produce vector systems, while all other systems on offer are based on RISC CPUs or x86-like ones. Therefore it is useful to give a brief description of the main processors that populate present supercomputers, and look ahead to the processors that will follow in the coming year. Nonetheless, we will be more conservative in this section than in the description of systems in general, because processors are being turned out at a tremendous pace whereas planning ahead for new generations takes years. We will therefore stick to actually existing components or β versions of a processor.

The RISC processor scene has shrunk significantly in the last few years. The Alpha and PA-RISC processors have disappeared in favour of the Itanium processor product line and, interestingly, the MIPS processor line, which appeared and then disappeared, as they were used in the highly interesting SiCortex systems. Unfortunately SiCortex has had to close down recently, and with it the MIPS processors. In addition, the Itanium processor is no longer used in HPC.

The disappearance of RISC processor families demonstrates a trend that is both worrying and interesting: worrying because diversity in the processor field is decreasing severely and, with it, the choice of systems in this sector. On the other hand there is a trend to enhance systems based on run-of-the-mill processors with special-purpose add-on processors in the form of FPGAs (field programmable gate arrays), or other computational accelerators, because their performance, price level, power consumption, and ease of use have improved to such a degree that they offer attractive alternatives for certain applications.

The very notion of a 'RISC processor' has been eroded somewhat in the sense that processors that execute the Intel x86 (CISC) instruction set now have most of the characteristics of a RISC processor. Both the AMD and Intel x86 processors in fact decode the CISC instructions almost entirely into a set of RISC-like fixed-length instructions. Furthermore, both processor lines feature out-of-order execution, both are able to address and deliver results natively in 64-bit length, and the bandwidth from memory to the processor core(s) has become comparable to those of RISC/EPIC processors. A distinguishing factor is still the mostly much larger set of registers in RISC processors.

Another notable development of the last few years is the placement of multiple processor cores on a processor chip and the introduction of various

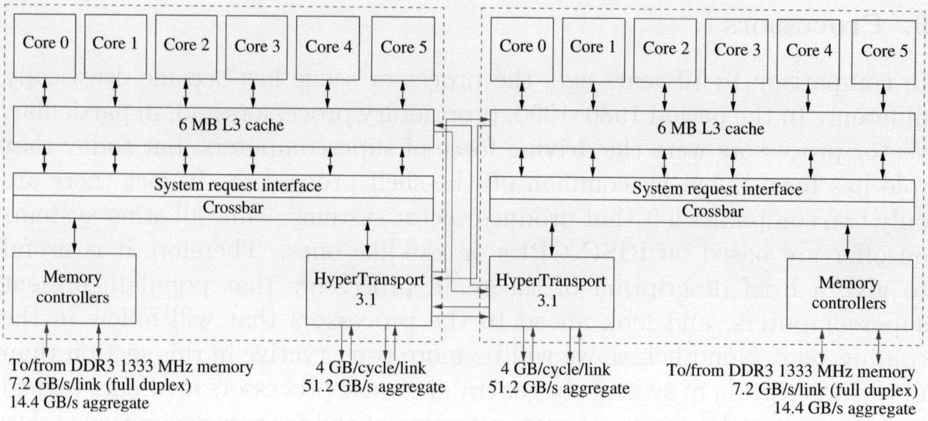

Figure 9.1. Block diagram of an AMD Opteron Magny-Cours processor.

forms of multi-threading. We will discuss these developments for each of the processors separately.

There are two processors one would perhaps expect in this section but are nevertheless not discussed: the Godson 3A and the Itanium Tukwila processors. The first processor, a Chinese one, based on the MIPS architecture, is not available in any machine marketed now or to be marketed in the near future (it is to be succeeded by the Godson 3B early next year). The newest Itanium processor no longer plays a role in the HPC scene and is therefore also omitted.

9.1. AMD Magny-Cours

All AMD processors are clones with respect to Intel's x86 instruction set architecture. The 12-core Opteron variant called 'Magny-Cours', made available in March 2010, is no exception. It is built with a feature size of 45 nm, and in fact the chip is a package containing two modified 6-core Istanbul chips running at a maximum of 2.3 GHz in the 6176 SE variant. The two chips are connected through 16-bit HyperTransport 3.1 links to each other's L3 caches with a single-channel speed of 12.8 GB/s, as shown in Figure 9.1.

The clock frequencies of the various parts of the chip are independent and different: while the processor operates at a speed of of 2.3 GHz, the HyperTransport links run at 3.2 GHz and the four memory buses (two per 6-core chip) run at only 1.8 GHz, thus limiting the maximum bandwidth between memory and the chip to only 28.8 GB/s. AMD has made this choice to limit the power consumption, although the new chips accommodate DDR3 memory at a speed of 1333 MHz, which means that the bandwidth could potentially have been 42.7 GB/s. As in the Istanbul processor, the Magny-Cours processor exploits the 'HT Assist' function. HT Assist sets aside 1 MB in the L3 cache, which contains the position and status of the cache

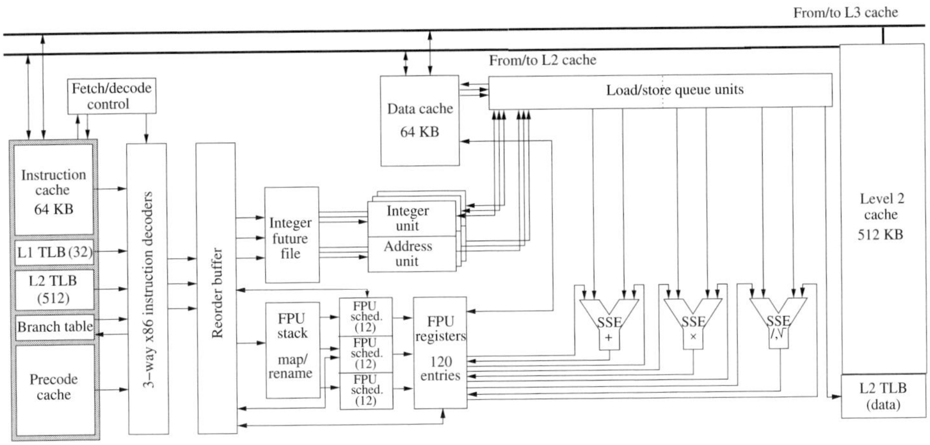

Figure 9.2. Block diagram of an AMD Magny-Cours processor core.

lines in use on the chip. In this way the change in status of cache variables does not have to be broadcast to all cores, but can simply be read from this part of the L3 cache, thus lowering the traffic in the interconnection fabric significantly. This set-up is in fact an implementation of cache coherence via directory memory, as explained in Section 7. Comparison experiments with the earlier Shanghai processor have shown that HT Assist can be highly beneficial thanks to more bandwidth being available for operand transfer. Because the number of cores has doubled with regard to the Istanbul processor, the HT Assist function has become all the more important.

Although they use the x86 instruction set, the AMD processors can be regarded as fully fledged RISC processors: they support out-of-order execution, have multiple floating-point units, and can issue up to nine instructions simultaneously. A block diagram of a processor core is shown in Figure 9.2. It is in effect identical to the Istanbul processor core. The six cores on the chip are connected by an on-chip crossbar. It also connects to the memory controller and, as stated earlier, to its companion chip and other processors on the board via HyperTransport.

The figure shows that a core has three pairs of integer execution units and address generation units that, via an 32-entry integer scheduler, take care of the integer computations and address calculations. Both the integer future file and the floating-point scheduler are fed by the 72-entry reorder buffer, which receives the decoded instructions from the instruction decoders. The decoding in the Opteron core has become more efficient than in the earlier processors: SSE instructions now decode into 1 micro-operation (μ-op) as do most integer and floating-point instructions. In addition, a piece of hardware called the 'sideband stack optimizer' has been added (not shown in the figure), that takes care of the stack manipulations in the instruction stream, thus making instruction reordering more efficient, and thereby increasing the effective number of instructions per cycle.

The floating-point units allow out-of-order execution of instructions via the FPU stack map and rename unit. It receives floating-point instructions from the reorder buffer and reorders them if necessary, before handing them over to the FPU scheduler. The floating-point register file is 120 elements deep, on a par with the number of registers available in RISC processors.[5]

The floating-point part of the processor contains three units: floating add and multiply units that can work in superscalar mode, resulting in two floating-point results per clock cycle and a unit handling 'miscellaneous' operations, such as division and square-root. Because of the compatibility with Intel's processors, the floating-point units are also able to execute Intel SSE2/3 instructions and AMD's own 3DNow! instructions. However, there is the general problem that such instructions are not directly accessible from higher-level languages such as Fortran 90 or C/C++. Both instruction sets were originally meant for massive processing of visualization data, but are increasingly used for standard dense linear algebra operations.

Due to the shrinkage of technology to 45 nm, each core can harbour a secondary cache of 512 KB. Because of the accommodation of DDR3 memory at a bus speed of 1333 MHz, the total bandwidth (but with the limitation of the 1.8 GHz memory interface) a channel transports is 7.2 GB/s or 14.4 GB/s per 6-core chip.

AMD's HyperTransport is derived from licensed Compaq technology and similar to that employed in HP/Compaq's former EV7 processors. It allows for 'glueless' connection of several processors to form multiprocessor systems with very low memory latencies. The Magny-Cours processor uses fourth-generation HyperTransport 3.1, which transfers 12.8 GB/s at 16-bit wide per unidirectional link. The HyperTransport interconnection possibility makes it highly attractive for building SMP-type clusters or to couple computational accelerators (see Section 10) directly to the same memory as the standard processor.

9.2. IBM POWER6

In the systems that feature IBM's supercomputer line, the p575 series, the nodes contain the POWER6 chip as the computational engine. This will change shortly and therefore we will also discuss the POWER7 processor in Section 9.3, but at the time of writing, the POWER6 is still the processor for IBM's high-end HPC systems. As compared to its predecessor, the POWER5+, there are quite some differences, both in the chip lay-out and in the two cores that reside on a chip. Figure 9.3 shows the layout of the

[5] For x86 instructions, 16 registers in a flat register file are present instead of the register stack typical of Intel architectures.

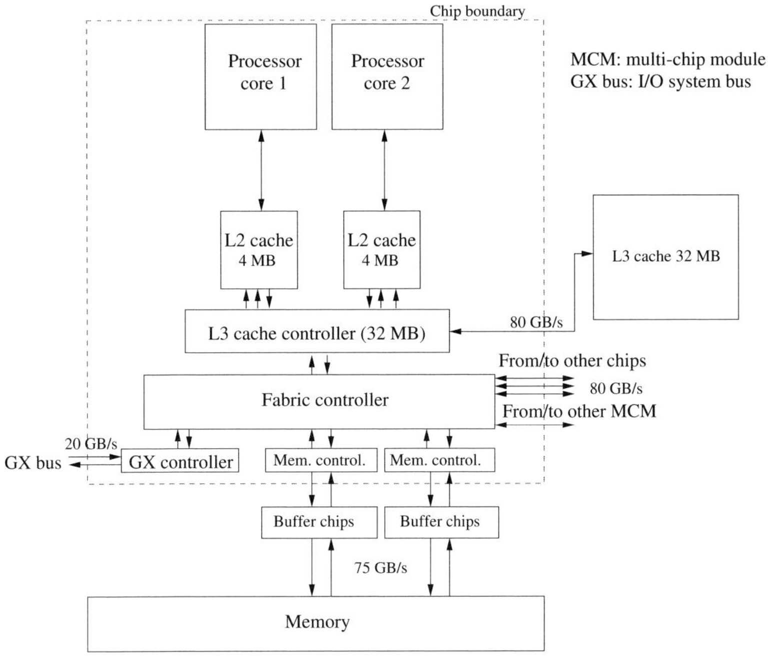

Figure 9.3. Diagram of the IBM POWER6 chip layout.

cores, caches, and controllers on the chip. Already there are significant changes: instead of a 1.875 MB shared L2 cache, each core now has its own 4 MB 8-way set-associative L2 cache that operates at half the core frequency. In addition, there are two memory controllers that connect via buffer chips to the memory and, depending on the number of buffer chips and data widths (both are variable), can have a data read speed ≤ 51.7 GB/s and a write speed of ≤ 25.85 GB/s, *i.e.*, with a core clock cycle of 4.7 GHz up to 11 B/cycle for a memory read and half of that for a memory write. Furthermore, the separate buses for data and coherence control between chips are now unified with a choice of both kinds of traffic, occupying 50 % of the bandwidth, or 67 % for data and 33 % for coherence control. The off-chip L3 cache has shrunk from 36 to 32 MB. It is a 16-way set-associative victim cache that operates at 1/4 of the clock speed.

Further, the core has changed considerably: it is depicted in Figure 9.4. The clock frequency has increased from 1.9 GHz in the POWER5+ to 4.7 GHz for the POWER6 (water-cooled version), an increase of almost a factor of 2.5, while the power consumption has stayed in the same range as that of the POWER5+. This has partly come about by a technology shrink from a 90 nm to 65 nm feature size. It also means that some features of the POWER5+ have disappeared. For instance, the POWER6 largely employs static instruction scheduling, except for a limited amount

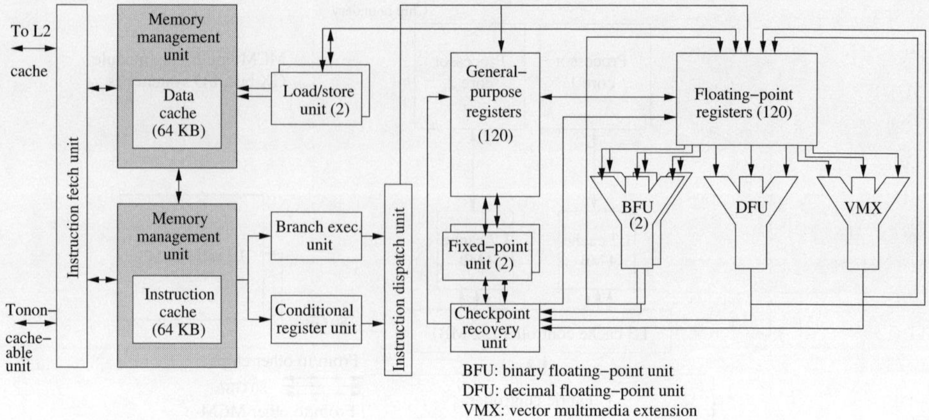

Figure 9.4. Block diagram of the IBM POWER6 core.

of floating-point instruction scheduling, because some of these can be fitted into empty slots left by division and square-root operations. The circuitry required for dynamic instruction scheduling that could thus be removed has, however, been replaced by new units. Besides the two fixed-point units (FXUs) and the two binary floating-point units (BFUs) that were already present in the POWER5+, there is now a decimal floating-point unit (DFU) and a vector multimedia extension (VMX) unit, akin to Intel's SSE units for handling multimedia instructions. In fact, the VMX unit is inherited from the IBM PowerPC's AltiVec unit. The DFU is IEEE 754R compliant. It is obviously for financial calculations and is hardly of consequence for HPC use. Counting only the operations of the BPUs both executing fused multiply–adds (FMAs), the theoretical peak performance in 64-bit precision is 4 flop/cycle or 18.8 Gflop/s/core. A checkpoint recovery unit has been added that is able to catch faulty FXU and FPU (both binary and decimal) instruction executions and reschedule them for retrial. Because of the large variety of functional units, a separate instruction dispatch unit ships the instructions that are ready for execution to the appropriate units, while a significant part of instruction decoding has been pushed into the instruction fetch unit, including updating the branch history tables.

The BFUs not only execute the usual floating-point instructions such as add, multiply and FMA, but also take care of division and square-root operations. A new phenomenon is that integer divide and multiply operations are also executed by the BFUs, again saving on circuitry and therefore power consumption. In addition, these operations can be pipelined in this way and yield a result every two clock cycles.

The L1 data cache has been doubled in comparison to the POWER5+ and is now 64 KB like the L1 instruction cache. Both caches are 4-way set-associative.

The simultaneous multi-threading (SMT) that was already present in the POWER5+ has been retained in the POWER6 processor, and has been improved by a higher associativity of the L1 I and D caches and a larger dedicated L2 cache. Also, instruction decoding and dispatch are dedicated for each thread. By using SMT the cores are able to keep two process threads at work at the same time. The functional units get instructions for the functional units from any of the two threads, whichever is able to fill a slot in an instruction word that will be issued to the functional units. In this way a larger fraction of the functional units can be kept busy, improving the overall efficiency. For very regular computations a single thread (ST) mode may be better, because in SMT mode the two threads compete for entries in the caches, which may lead to trashing in the case of regular data access. Note that SMT is somewhat different from the 'normal' way of multi-threading. In this case a thread that stalls for some reason is stopped and replaced by another process thread that is awoken at that time. Of course this takes some time, which must be compensated for by the thread that has taken over. This means that the second thread must be active for a fair number of cycles (preferably a few hundred cycles at least). SMT does not have this drawback but scheduling the instructions of both threads is quite complicated, especially where only very limited dynamic scheduling is possible.

Because of the much higher clock cycle, and the fact that the memory DIMMs are attached to each chip, it is no longer possible to maintain perfect SMP behaviour within a 4-chip node, *i.e.*, it matters whether data are accessed from a chip's own memory or from the memory of a neighbouring chip. Although the data are only one hop away, there is a ccNUMA effect that one has to be aware of in multi-threaded applications.

9.3. IBM POWER7

As remarked earlier, at this moment IBM is not yet offering HPC systems with the POWER7 inside. Hitachi is already offering a variant of its SR16000 system with the POWER7 processor and IBM is expected to follow shortly. So, it is appropriate to discuss this chip. Figure 9.5 shows the layout of the cores, caches, and memory controllers on the chip. The technology from which the chips are built is identical to that of the POWER6 45 nm Silicon-On-Insulator, but in all other respects the differences from the former generation are large. Firstly, the number of cores has quadrupled. Further, the memory speed has increased from DDR2 to DDR3 via two on-chip memory controllers. As in earlier POWER versions, the inbound and outbound bandwidths from memory to chip are different: 2 B/cycle in and 1.5 B/cycle out. With a bus frequency of 6.4 GHz and 4 in/out channels per controller, this amounts to 51.2 GB/s inward and 38.4 GB/s outward.

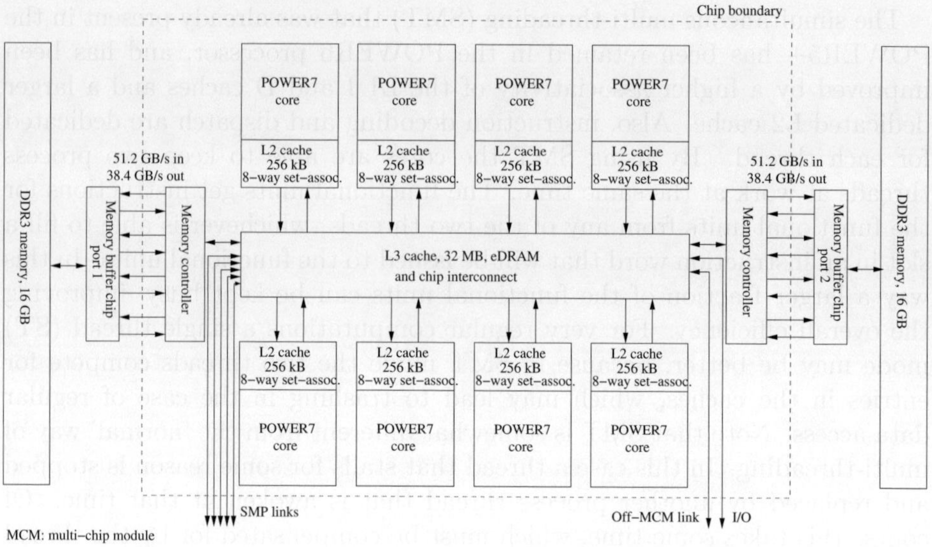

Figure 9.5. Diagram of the IBM POWER7 chip layout.

IBM asserts that an aggregate sustained bandwidth of ≈ 100 GB/s can be reached. Although this is very high in absolute terms, with a clock frequency of 3.5–3.86 GHz for the processors this is no luxury. Therefore it is possible to run the chip in so-called TurboCore mode. In this case four of the 8 cores are turned off and the clock frequency is raised to 4.14 GHz, thus almost doubling the bandwidth for the active cores. As one core is capable of absorbing/producing 16 B/cycle, when executing a fused floating multiply–add operation the bandwidth requirement of one core at 4 GHz is already 64 GB/s. So, the cache hierarchy and possible prefetching are extremely important for a reasonable occupation of the many functional units.

Another new feature of the POWER7 is that the L3 cache has been moved onto the chip. To be able to do this IBM chose to implement the 32 MB L3 cache in embedded DRAM (eDRAM) instead of SRAM as is usual. eDRAM is slower than SRAM but much less bulky, and because the cache is now on-chip, the latency is considerably lower (about a factor of 6). The L3 cache communicates with the L2 caches, which are private to each core. The L3 cache is partitioned in that it contains 8 regions of 4 MB, one region per core. Each partition serves as a victim cache for the L2 cache to which it is dedicated, and in addition to the other 7 L3 cache partitions.

Each chip features 5 10-B SMP links that support SMP operations of up to 32 sockets.

Further, at the core level there are many differences from its predecessor. A single core is depicted in Figure 9.6. To begin with, the number of floating-point units is doubled to four, each capable of a fused multiply–add

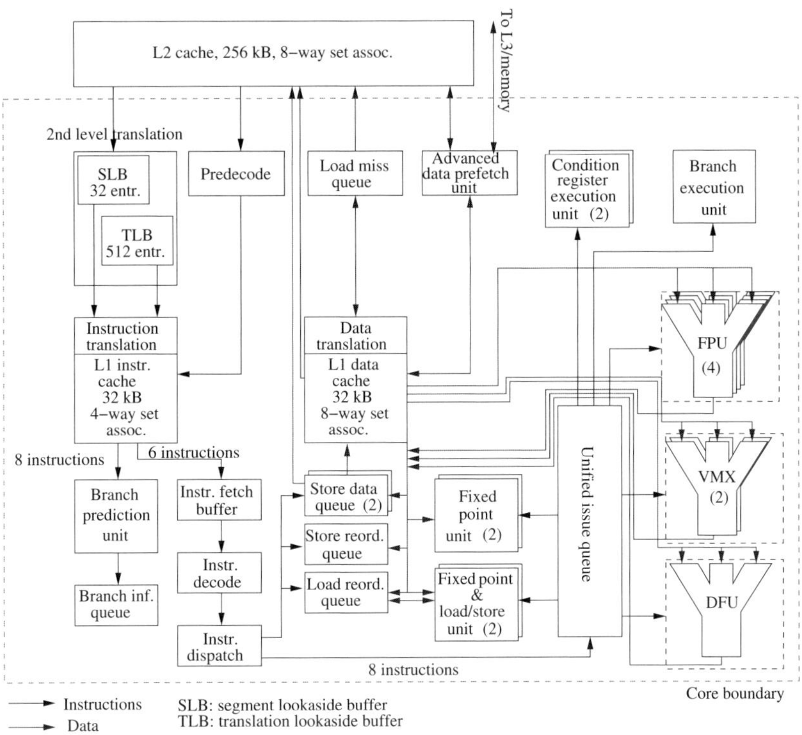

Figure 9.6. Block diagram of the IBM POWER7 core.

operation per cycle. Assuming a clock frequency of 3.86 GHz, this means that a peak speed of 30.88 Gflop/s can be attained with these units. A feature that was omitted from the POWER6 core has been re-implemented in the POWER7 core: dynamic instruction scheduling assisted by the load and load reorder queues. As shown in Figure 9.6, there are two 128-bit VMX units. One of them executes vector instructions akin to the x86 SSE instructions. However, there is also a VMX permute unit that can order non-contiguous operands such that the VMX execute unit can handle them. The instruction set for the VMX unit is an implementation of the AltiVec instruction set, which is also employed in the PowerPC processors. There are also similarities to the POWER6 processor: the core contains a DFU and a checkpoint recovery unit that can re-schedule operations that have failed for some reason.

Another difference that cannot be shown is that the cores now support four SMT threads instead of two. This will be very helpful for the large numbers of functional units to be kept busy. Eight instructions can be taken from the L1 instruction cache. The instruction decode unit can handle six instructions simultaneously while eight instructions can be dispatched every cycle to the various functional units.

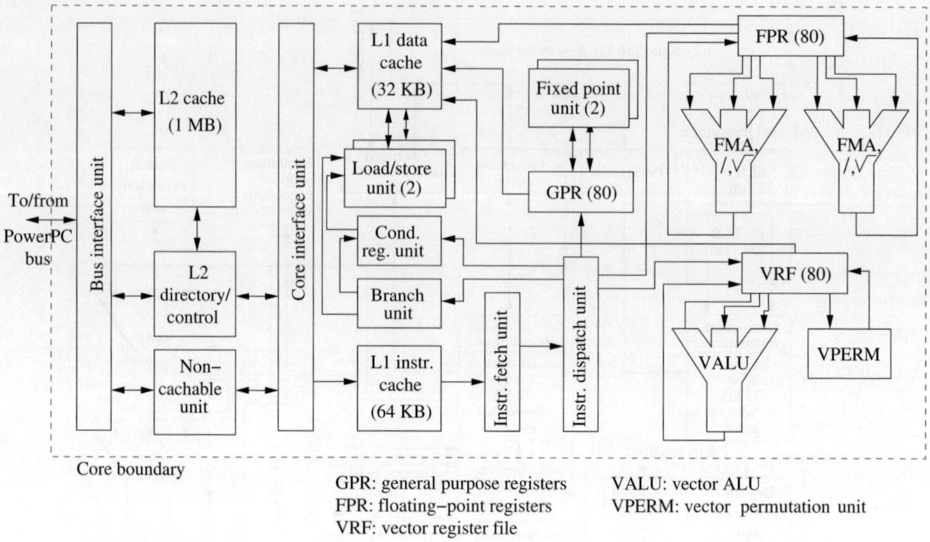

Figure 9.7. Block diagram of the IBM PowerPC 970MP core.

The POWER7 core has elaborate power management features that reduce the power usage for parts that are idle for some time. There are two power-saving modes: *nap* mode and *sleep* mode. In the former the caches and translation lookaside buffers (TLBs) stay coherent to re-activate quickly. In sleep mode, however, the caches are purged and the clock turned off. Only the minimum voltage to maintain the memory contents is applied. Obviously the wake-up time is longer in this case but the power saving can be significant.

9.4. IBM PowerPC 970MP processor

A number of IBM systems are built from JS21 blades, the largest being the Mare Nostrum system at the Barcelona Supercomputing Centre. On these blades a variant of the large IBM PowerPC processor family is used, the dual core PowerPC 970MP. It is a series of dual-core processors, the fastest of which has a clock cycle of 2.2 GHz. A block diagram of a processor core is given in Figure 9.7.

A peculiar trait of the processor is that the L1 instruction cache is twice as large as the L1 data cache: 64 KB versus 32 KB. This is explained partly by the fact that up to 10 instructions can be issued every cycle to the various execution units in the core. Apart from two floating-point units that perform the usual dyadic operations, there is an AltiVec vector facility with a separate 80-entry vector register file, a vector ALU that performs (fused) multiply–add operations, and a vector permutation unit that attempts to order operands such that the vector ALU is used optimally. The vector unit was designed for graphics-like operations, but works quite nicely on data for other purposes as long as access is regular and the operand type agrees. Theoretically, the speed of a core can be 13.2 Gflop/s/core when both FPUs turn out the results of a fused multiply–add and the vector ALU does the same. One PowerPC 970MP should therefore have a theoretical peak performance of 26.4 Gflop/s. The floating-point units also perform square-root and division operations.

Apart from the floating-point and vector functional units, two integer fixed-point units and two load/store units are present in addition to a conditional register unit and a branch unit. The latter uses two algorithms for branch prediction that are applied according to the type of branch to be taken (or not). The success rate of the algorithms is constantly monitored. Correct branch prediction is very important for this processor as the pipelines of the functional units are quite deep: from 16 for the simplest integer operations to 25 stages in the vector ALU. So, a branch miss can be very costly. The L2 cache is integrated and has a size of 1 MB. To keep the load/store latency low, hardware-initiated prefetching from the L2 cache is possible and 8 outstanding L1 cache misses can be tolerated. The operations are dynamically scheduled and may be out of order. In total 215 operations may be in flight simultaneously in the various functional units, also due to the deep pipelines.

The two cores on a chip have common arbitration logic to regulate the data traffic from and to the chip. There is no third-level cache between the memory and the chip on the board housing them. This is possible because of the moderate clock cycle and the rather large L2 cache.

9.5. IBM BlueGene processors

In the last few years two BlueGene types of systems have become available: the BlueGene/L and the BlueGene/P, the successor of the former. Both feature processors based on the PowerPC 400 processor family.

BlueGene/L processor

This processor is in fact a modified PowerPC 440 processor, which is made especially for the IBM BlueGene family. It runs at a speed of 700 MHz. The modification lies in tacking on floating-point units (FPUs) that are not part of the standard processor but can be connected to the 440's APU bus. Each FPU contains two floating-point functional units capable of performing 64-bit multiply–adds, divisions and square-roots. Consequently, the theoretical peak performance of a processor core is 2.8 Gflop/s. Figure 9.8 shows the embedding of two processor cores on a chip. As can be seen from the figure, the L2 cache is very small: only 2 KB divided in a read and a write part. In fact it is a prefetch and store buffer for the rather large L3 cache.

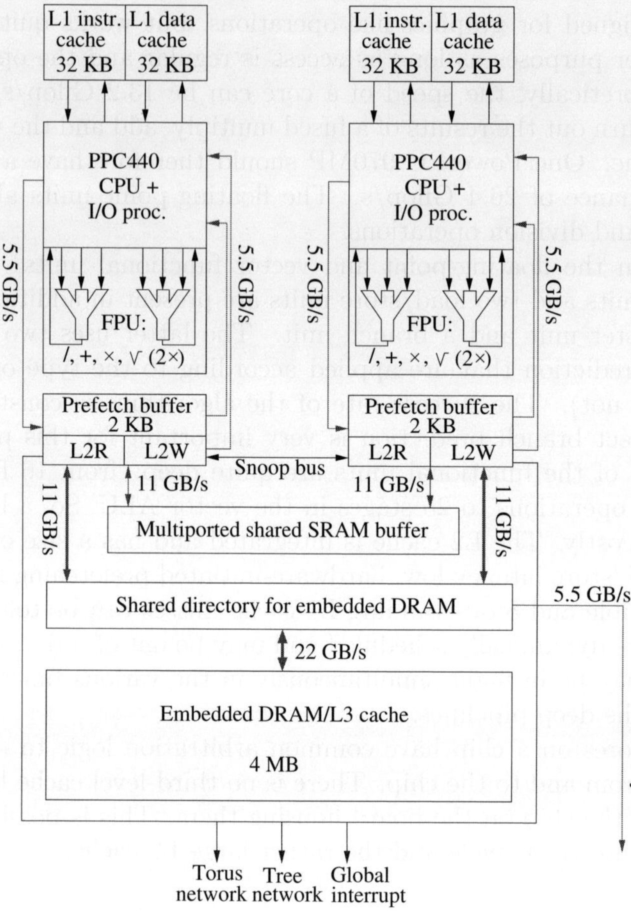

Figure 9.8. Block diagram of an IBM BlueGene/L processor chip.

The bandwidth to and from the prefetch buffer is high, 16 B/cycle to the CPU and 8 B/cycle to the L2 buffer. The memory resides off-chip with a maximum size of 512 MB. The data from other nodes are transported through the L2 buffer, bypassing the L3 cache in the first instance.

BlueGene/P processor

Like the BlueGene/L processor, the BlueGene/P processor is based on the PowerPC core – the PowerPC 450 in this case – at a clock frequency of 850 MHz and with floating-point enhancements similar to those applied to the PPC 440 in the BlueGene/L. The BlueGene/P node contains 4 processor cores, which brings the peak speed to 13.6 Gflop/s/node. The block diagram in Figure 9.9 shows some details. As can be seen from the figure, the structure of the core has not changed much with respect to the BlueGene/L. The relative bandwidth from the L2 cache has been maintained: 16 B/cycle

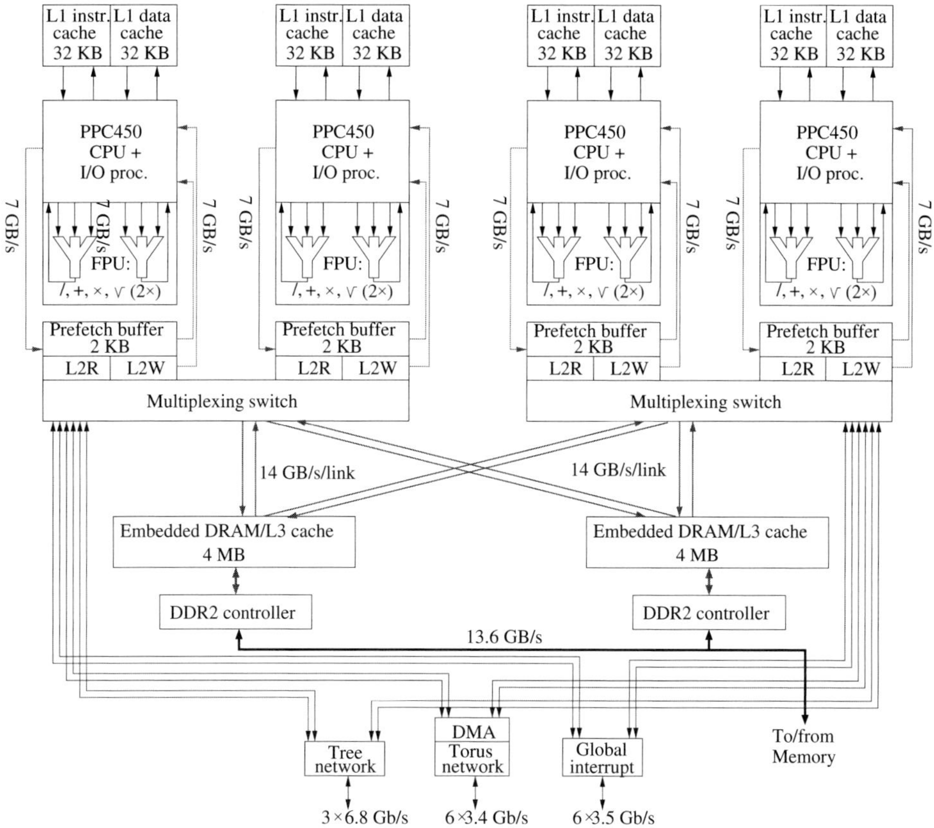

Figure 9.9. Block diagram of an IBM BlueGene/P processor chip.

for reading and 8 B/cycle for writing. In contrast to the BlueGene/L, the cores operate in SMP mode through multiplexing switches that connect pairs of cores to the two 4 MB L3 embedded DRAM chips. So, the L3 size has doubled. Also, the memory per node has increased to 2 GB from 512 MB.

9.6. Intel Xeon

Two variants of Intel's Xeon processors are employed at present in HPC systems (clusters as well as integrated systems): the Nehalem EX, officially the X7500 chip series, and the Westmere EP, officially the X5600 series. Although there is a great deal of common ground, they are sufficiently different that we can discuss each processor separately.

Nehalem EX

The Nehalem EX became available in March 2010 or, more formally, the X7500 series of processors can be regarded as a heavy-duty server extension

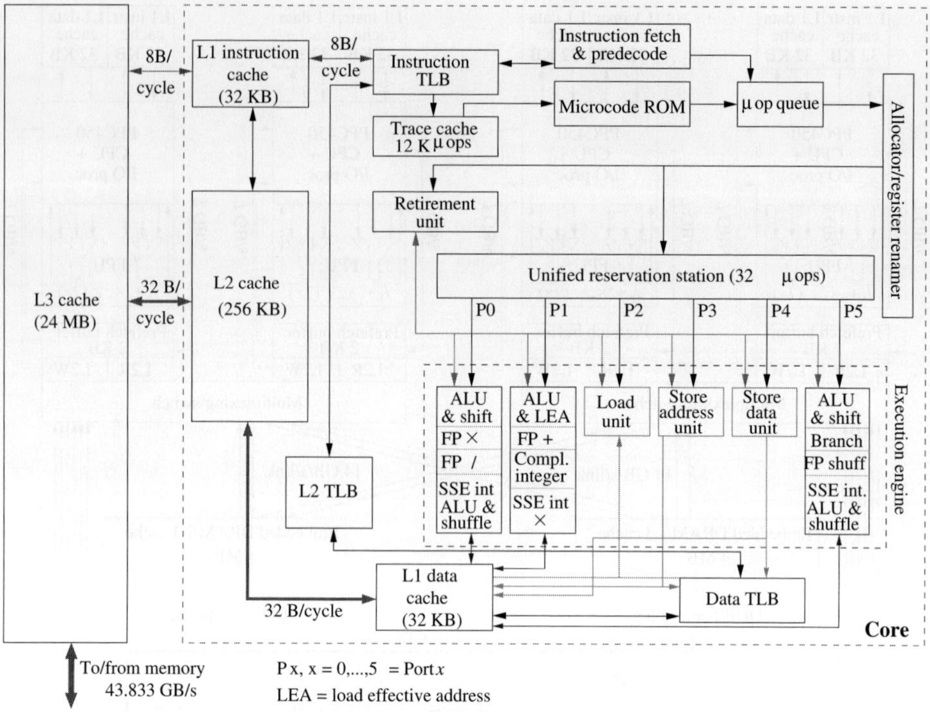

Figure 9.10. Block diagram of an Intel Nehalem processor core.

of the earlier Nehalem EP (X5500) processor. As such, it has twice the number of cores – eight – and the number of QPI links has also doubled to four, to enable the connection to other processors on a server board. In addition the bandwidth is raised, because the number of memory interfaces has increased from three to four. As the bandwidth per memory channel is 10.833 GB/s, the aggregate bandwidth of the chip is 43.3 MB/s. The Nehalem EX is employed in SGI's Altix UV systems and a variety of products from cluster vendors. Below we show the block diagram of the processor core, which is in fact identical to that of the Nehalem EP and built in the same 45 nm technology. A block diagram of the core is given in Figure 9.10, while the layout of an 8-core chip is shown in Figure 9.11.

To stay backwards-compatible with the x86 (IA-32) instruction set architecture, which comprises a CISC instruction set, Intel developed a mode in which these instructions are split into so-called micro-operations (μ-ops) of fixed length that can be treated in the way RISC processors do. In fact the μ-ops constitute a RISC operation set. The price to be paid for this much more efficient instruction set is an extra decoding stage. Branch prediction has been improved and a second-level TLB cache has been added.

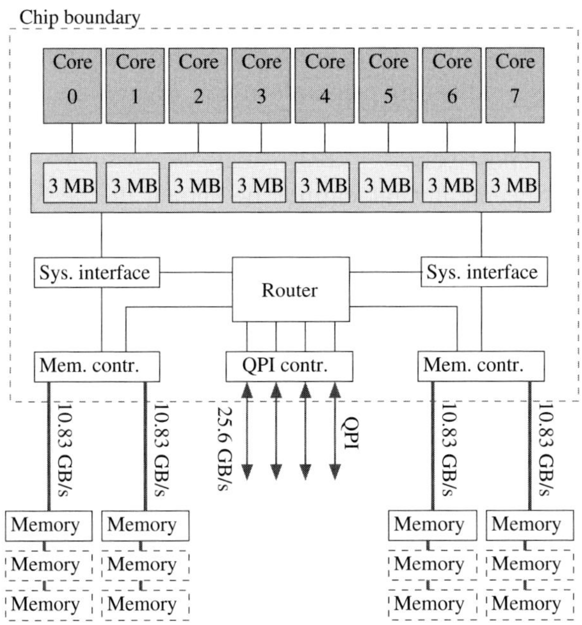

Figure 9.11. Diagram of a Nehalem EX processor.

As in the earlier core architecture, four μ-ops/cycle and some macro-instructions as well as some μ-ops can be fused, resulting in less instruction handling, easier scheduling and better instruction throughput, because these fused operations can be executed in a single cycle. In the Nehalem, two additional μ-ops can be fused in comparison to the core architecture.

As can be seen in Figure 9.10, the processor cores have an execution trace cache which holds partly decoded instructions of former execution traces that can be drawn upon, thus forgoing the instruction decode phase, which might produce holes in the instruction pipeline. The allocator dispatches the decoded instructions, the μ-ops, to the unified reservation station that can issue up to 6 μ-ops/cycle to the execution units, collectively called the execution engine. Up to 128 μ-ops can be in flight at any time. Figure 9.10 shows that ports 0 and 5 drive two identical integer ALUs as well as integer SSE units. Ports 0, 1 and 5 take care of the various floating-point operations.

The two integer arithmetic logic units (ALUs) at ports 0 and 5 are kept simple in order to be able to run them at twice the clock speed. In addition there is an ALU at port 1 for complex integer operations that cannot be executed within one cycle. The floating-point units also contain additional units that execute the Streaming SIMD Extensions 4 (SSE4) repertoire of instructions, an instruction set of more than 190 instructions, that was initially meant for vector-oriented operations such as those in multimedia, and 3D visualization applications, but is also an advantage for regular vector operations as occur in dense linear algebra. The length of the operands for these units is 128 bits. The Intel compilers have the ability to address the SSE4 units. This enables in principle much higher floating-point performance. Ports 2, 3 and 4 serve the load unit, the store address unit, and the store data unit, respectively.

A notable enhancement that cannot be shown in the figures is that the Nehalem (again) supports multi-threading, much in the style of IBM's simultaneous multi-threading, and is called 'hyperthreading' by Intel. Hyperthreading was earlier introduced in the Pentium 4, but disappeared in later Intel processors because the performance gain was very low. Now with a much higher bandwidth and larger caches, speed-ups of more than 30 % for some codes have been observed with hyperthreading. Another feature that cannot be shown is the so-called Turbo mode. This means that the clock cycle can be raised from its nominal speed (2.91 GHz for the fastest variant) by steps of 133 MHz to over 3 GHz as long as the thermal envelope of the chip is not exceeded. So, when some cores are relatively idle other cores can take advantage by operating at a higher clock speed.

The L1 caches have the same size as in the Nehalem's predecessor, but the L2 cache is much smaller: 256 KB instead of 6 MB. It is much faster, however, and able to deliver requested data in 10 cycles or less. The Nehalems feature a common L3 cache that is used by all eight cores in the EX version. Each core has its own section of 3 MB, but when data are not found in the section of a core the other sections can be searched for the missing data item(s). The L3 cache is inclusive, which means that it contains all data that are in the L2 and L1 cache. The consequence is that when a data item cannot be found in the L3 cache it is also not in any of the caches of the other cores, and therefore one need not search them.

In Figure 9.11 it can be noticed that, as well as the first bank of memory of $\leq$ 32 GB, a second and third bank are also depicted, represented by dashed boxes. This means that it is indeed possible to have up to 96 GB of memory/processor. However, this can only be done at the expense of the memory bus speed: for one bank it is 1333 MB/s, for two banks 1066 MB/s, and for three banks only 800 MB/s. So, the latter two options may be chosen, for instance, for database systems that benefit from a large memory that need not be at the very highest speed. For HPC purposes, however, configurations with only one memory bank/processor will usually be offered.

Westmere EP

The Westmere EP (X5600 series) is a 32 nm technology shrink of the Nehalem EP chip. The smaller feature size is used to place 6 cores on a die. The fastest variant, the X5690, has a clock cycle of 3.46 GHz at 130 W. The structure of the core is the same as in the Nehalem processors (see Figure 9.10) but there are slight differences in the instruction set for the

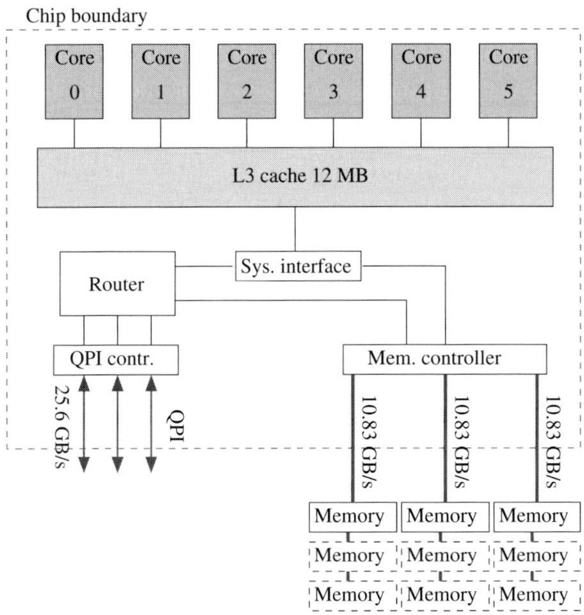

Figure 9.12. Diagram of a Westmere EP processor.

Advanced Encryption Standard (AES). The new instructions, among which is a carry-less multiplication, are said to speed up the en-/decryption rate by a factor of three. Also, the Westmere EP supports the use of 1 GB pages. The packaging on the chip is, apart from the number of cores, identical to that of the Nehalem EP chip with exception of the shared L3 cache. The size of this is halved from 24 MB to 12 MB. The chip layout is depicted in Figure 9.12.

9.7. The SPARC processors

Since Sun was taken over by Oracle all processor development has been shelved. The development of the SPARC processor architecture is now in the hands of Fujitsu, now proceeding with its own SPARC64 implementation. Fujitsu/Siemens markets its HPC servers based on the latter processor. Below we discuss the current SPARC chip, which is commercially available in the Fujitsu machines. Although a follow-on processor, the SPARC64 VIII, seems ready for incorporation in Japan's 10 petaflop/s system, which is currently being built, we only discuss the SPARC64 VII here as this is the one that is commercially available. At present it is not known when its successor will appear on the open market.

The SPARC64 VII is, obviously, Fujitsu's seventh generation of the processor. Of course, the processor must be able to execute the SPARC instruction set but the processor internals are rather different from Sun's late

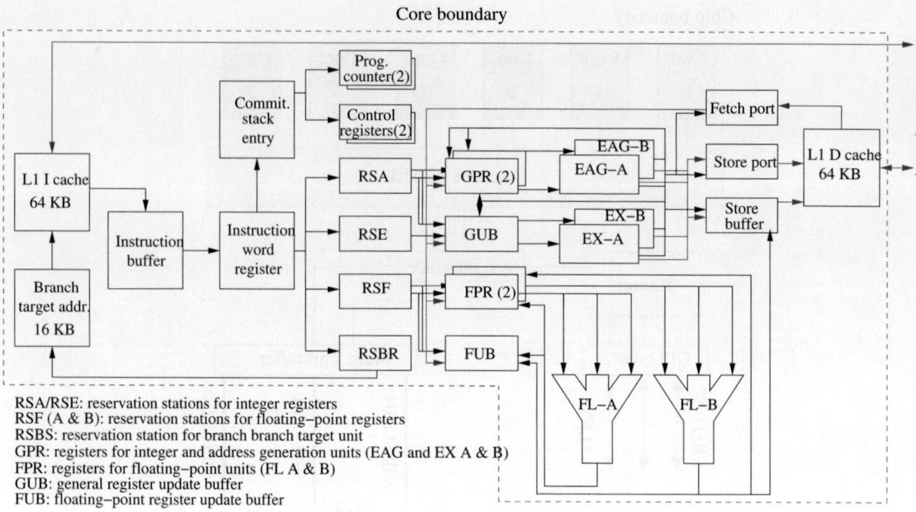

Figure 9.13. Block diagram of a Fujitsu SPARC64 VII processor core.

implementations. Figure 9.13 shows a block diagram of one core of the quad-core SPARC64 VII.

In fact, the core architecture has not changed from the SPARC64 VI, but thanks to the decrease in the feature size from 90 nm to 65 nm, four cores can now be placed on a chip while the highest available clock frequency is raised from 2.4 GHz to 2.52 GHz.

The L1 instruction and data caches are 64 KB, twice as small as in the SPARC64 VI core, and both 2-way set-associative. This decrease in size is somewhat surprising and probably due to the technology shrink to 65 nm feature size. There is also an instruction buffer (IBF), that contains up to 48 4-byte instructions and continues to feed the registers through the instruction word register when an L1 I-cache miss has occurred. A maximum of four instructions can be scheduled each cycle and find their way via the reservation stations for address generation (RSA), integer execution units (RSE), and floating-point units (RSF) to the registers. The two general register files serve both the two address generation units EAG-A, and EAG-B, and the integer execution units EX-A and EX-B. The latter two are not equivalent: only EX-A can execute multiply-and-divide instructions. There are also two floating-point register files (FPR), that feed the two floating-point units FL-A and FL-B. These units are different from those of Sun in that they are able to execute fused multiply–add instructions, as is also the case in the POWER and Itanium processors. Consequently, a maximum of four floating-point results/cycle can be generated. In addition, FL-A and FL-B also perform divide and square-root operations, in contrast to the SPARC4+, which has a separate unit for these operations. Because of their

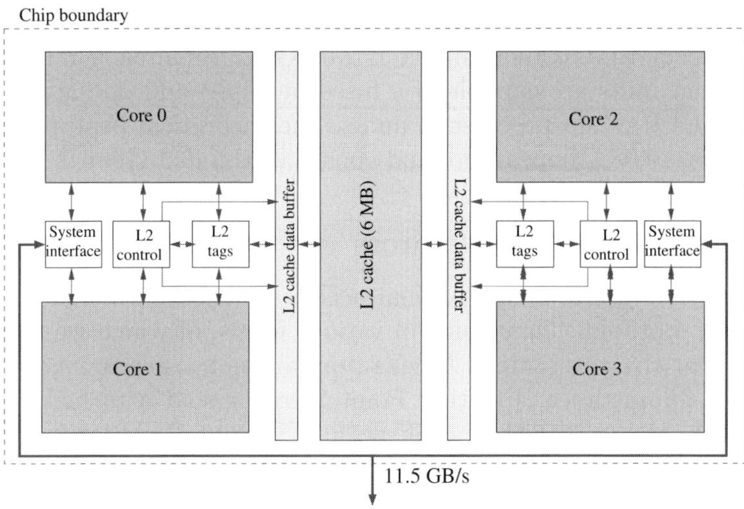

Figure 9.14. Block diagram of a Fujitsu SPARC64 VII processor chip. Four cores share the L2 cache.

iterative nature, the divide and square-root operations are not pipelined. The feedback from the execution units to the registers is decoupled by update buffers: GUB for the general registers and FUB for the floating-point registers.

The dispatch of instructions via reservation stations, each of which can hold 10 instructions, gives the opportunity for speculative dispatch: *i.e.*, dispatching instructions for which the operands are not yet ready at the moment of dispatch but will be by the time the instruction is actually executed. The assumption is that it results in a more even flow of instructions to the execution units.

The SPARC64 VII does not have a third-level cache, but on chip there is a large (6 MB) unified L2 12-way set-associative write-through cache that is shared by the 4 cores in a processor, as can be seen in Figure 9.14. Note that the system bandwidth is the highest available. For the lower-end systems this bandwidth is about 8 GB/s.

The memory management unit (not shown in Figure 9.13) contains separate sets of translation lookaside buffers (TLBs) for instructions and for data. Each set is composed of a 32-entry μTLB and a 1024-entry main TLB. The μTLBs are accessed by high-speed pipelines by their respective caches.

What cannot be shown in the diagrams is that, like the IBM and Intel processors, the SPARC VII is dual-threaded per core. The type of multi-threading is similar to that found in the Intel processors, and is called 'simultaneous multi-threading', differing from the type of multi-threading present in the IBM processors but with the same name. At this moment the highest clock frequency SPARC64 available is 2.52 GHz. As already remarked, the floating-point units are capable of a fused multiply–add operation, like the POWER and Itanium processors, and so the theoretical peak performance is at present 10.08 Gflop/s/core and consequently 40.3 Gflop/s/processor.

10. Computational accelerators

In the last few years computational accelerators have emerged, and now have a firm foothold. They come in various forms, of which we will discuss some general characteristics. Accelerators are not a new phenomenon: in the 1980s, for instance, Floating Point Systems sold attached processors such as the AP120-B with a peak performance of 12 Mflop/s, easily ten times faster than the general-purpose systems they were connected to. Further, the processor array machines described in Section 4 could be regarded as accelerators for matrix-oriented computations in their time. A similar phenomenon is recurring now. HPC users are never content with the performance of the machines they have at their disposal and are continuously looking for ways to speed up their calculations, or parts of them. Accelerator vendors are complying with this wish, and at present there is a fair number of products that, when properly deployed, can deliver significant performance gains.

The scene is roughly divided into three unequal parts:

(1) graphical cards or graphical processing units (GPUs as opposed to the general CPUs),
(2) general floating-point accelerators,
(3) field programmable gate arrays.

The appearance of accelerators is believed to have set a trend in high-performance computing, namely that the processing units should be diversified according to their abilities – not unlike the occurrence of different functional units within a CPU core.[6] In a few years this will lead to hybrid systems incorporating different processors for different computational tasks. Of course, processor vendors can choose to (attempt to) integrate such special-purpose processing units within their main processor line, but for now it is uncertain if or how this will happen.

When speaking of special-purpose processors, *i.e.*, computational accelerators, one should realize that they are indeed good at some specialized computations but totally unable to perform others. So, not all applications can benefit from them, and of those that can, not all can benefit to the

[6] In principle it is entirely possible to perform floating-point computations with integer functional units, but the costs are so high that no one will attempt it.

same degree. Furthermore, using accelerators effectively is not at all trivial. Although the software development kits (SDKs) for accelerators have recently improved enormously, for many applications it is still a challenge to obtain significant speed-up. An important factor in this is that data must be shipped in and out of the accelerator and the bandwidth of the connecting bus is in most cases a severe bottleneck. One generally tries to overcome this by overlapping data transport to/from the accelerator with processing. Tuning the computation and data transport task can be cumbersome. This hurdle has been recognized by several software companies, such as Acceleware, CAPS, and RapidMind (now absorbed by Intel). They offer products that automatically transform standard C/C++ programs into a form that integrates the functionality of GPUs, multi-core CPUs (which are often not used optimally), and, in the case of RapidMind, of Cell processors.

There is one other important consideration that makes accelerators popular: in comparison to general-purpose CPUs they are all very power-efficient, sometimes by orders of magnitude when expressed in flop/Watt. Of course, they will do only part of the work in a complete system, but the power savings can still be considerable, which is very attractive today.

We will now proceed to discuss the three classes of accelerators mentioned above. It must be realized, though, that developments in this field are extremely rapid, and therefore the information given here will become obsolete very fast and hence could be of an approximate nature.

10.1. Graphical processing units

Graphics processing is characterized by doing the same (floating-point) operation on massive amounts of data. To accommodate this way of processing, graphical processing units (GPUs) consist of a large number of relatively simple processors, fast but limited local memory, and fast internal buses to transport the operands and results. Until recently all calculations, and hence results, were in 32-bit precision. This is hardly of consequence for graphics processing as the colour of a pixel in a scene may be a shade off without anyone noticing. HPC users often have computational demands similar to those in the graphical world: the same operation on very many data items. So, it was natural to look into GPUs with their many integrated parallel processors and fast memory. The first adopters of GPUs from the HPC community therefore disguised their numerical program fragments as graphical code (*e.g.*, by using the graphical language OpenGL) to get fast results, often with remarkable speed-ups. Another advantage is that GPUs are relatively cheap because of the enormous numbers sold for graphical use in virtually every PC. One drawback is the 32-bit precision of the typical GPU and, in some cases more importantly, there is no error correction available. By carefully considering which computation really needs 64-bit precision and which does not, and adjusting algorithms accordingly the use of a GPU can be entirely satisfactory, however. GPU vendors have been quick to focus on the HPC community. They have tended to rename their graphics cards GPGPUs, *i.e.*, general-purpose GPUs, although the product is largely identical to the graphics cards sold in every shop. But there have also been real improvements to attract HPC users: 64-bit GPUs have come onto the market. In addition, it is no longer necessary to reformulate a computational problem into a piece of graphics code. Both ATI/AMD and NVIDIA claim IEEE 754 compatibility (being the floating-point computation standard) but neither of them support it to the full. Error correction, as usual for general-purpose CPUs, is becoming available (see p. 421). There are C-like languages and runtime environments available that make the life of a developer of GPUs much easier: for NVIDIA this is CUDA, which has become quite popular with users of these systems. AMD/ATI is concentrating on the newly defined standard OpenCL (see below), which is somewhat more cumbersome but still provides a much better alternative to emulating graphics code.

When one develops a code for a particular GPU platform it cannot be transferred to another without considerable effort in rewriting the code. This drawback is taken up by the GPU vendors (and not only them). Recently OpenCL has become available, which in principle is platform-independent, thus protecting the development effort put into the acceleration of a program. At present, Apple, ATI/AMD, Intel, NVIDIA, and PetaPath are members of the consortium that are willing to provide an OpenCL language interface. First experiences with OpenCL version 1.0, as provided by the Khronos Group, showed generally low performance, but with the new enhanced release of OpenCL 1.1 as of June 2010 this situation is improving.

Another way to be (relatively) independent of the platform is to employ a language transformer. For instance, CAPS provides such transforming tools that can target different types of accelerators or multi-core CPUs. With CAPS' product HMPP, the transformation is brought about by inserting pragmas in the C code or comment directives in Fortran code. HMPP is the only code that has the ability to accelerate Fortran code on general GPU accelerators. The Portland Group sells a CUDA/Fortran compiler that only targets NVIDIA GPUs.

In the following we describe some high-end GPUs that are more or less targeting the HPC community.

ATI/AMD

In June 2010 the latest product from ATI (now wholly owned by AMD) was announced: the ATI FireStream 9370 card. The actual delivery was scheduled by AMD in the third quarter of 2010. Information about the

Table 10.1. Some specifications for the ATI/AMD FireStream 9370 GPU.

Number of processors	1600
Memory (GDDR5)	4 GB
Clock cycle	825 MHz
Internal memory bandwidth	$\leq$ 147.2 GB/s
Peak performance (32-bit)	2.64 Tflop/s
Peak performance (64-bit)	528 Gflop/s
Power requirement, typical	170 W
Power requirement, peak	225 W
Interconnect (PCIe Gen2)	$\times$16, 8 GB/s
ECC, error correction	No
Floating-point support	Partial (32/64-bit)

card, however, is still scant. There is not enough information available for a block diagram but we list in Table 10.1 some of the most important features of the processor. The specifications given indicate that, per core, two floating-point results per cycle can be generated, presumably the result of an add and a multiply operation. Whether these results can be produced independently or result from linked operations is not known, because of the lack of information. Unlike NVIDIA's Fermi card, discussed below, the FireStream 9370 does not support error correction yet. So, one has to be careful in assessing the outcomes of numerically unstable calculations.

Like its direct competitor, NVIDIA, ATI offers a free software development kit, SDK v.2.01, which supports OpenCL 1.1, Direct X11 and ComputeX. The earlier languages, such as BROOK+ and the very low-level Close-To-Metal software development vehicles, are no longer supported.

NVIDIA

NVIDIA is the other big player in the GPU field with regard to HPC. Its latest product is the Tesla C2050/C2070, also known as the 'Fermi' card. A simplified block diagram is shown in Figure 10.1. The GigaThread Engine is able to schedule different tasks in the streaming multiprocessors (SMs) in parallel. This greatly improves the occupation rate of the SMs and thus the throughput. As shown in Figure 10.1, three (or four) SMs per graphics processor cluster (GPC) are present. At the moment no more than a total of 14 SMs are available, although 16 were planned. When the 40 nm production process has sufficiently improved, the number of SMs may increase from 14 to the originally planned 16. A newly introduced feature is the L2 cache, shared by all SMs. Also, there is DMA support to get data from the host's memory without having to interfere with the host CPU. The

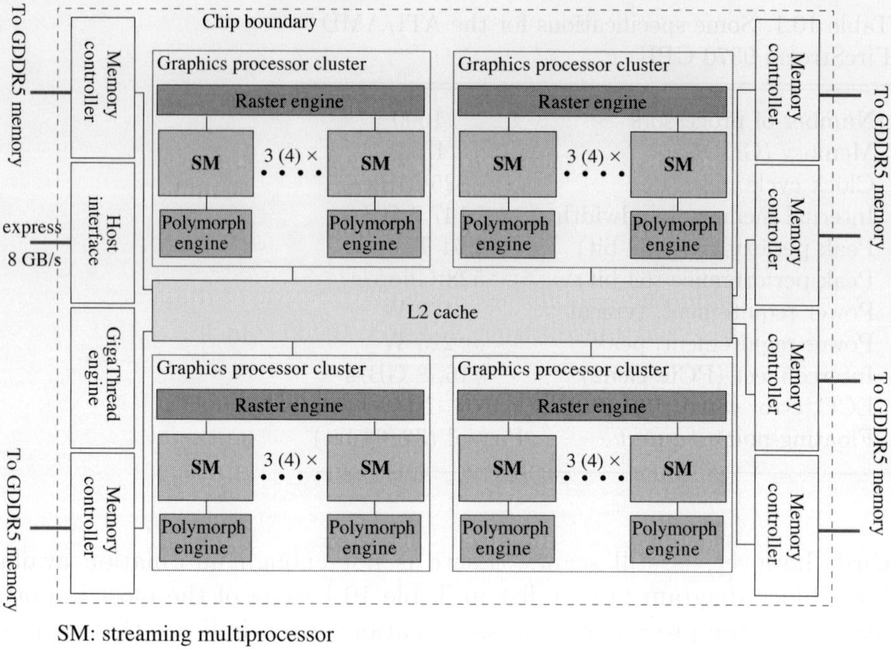

Figure 10.1. Simplified block diagram of the NVIDIA Tesla C2050/C2070 GPU.

GPU memory is GDDR5 and is connected to the card via six 64-bit wide memory interfaces for a bandwidth of about 150 GB/s.

Each SM in turn harbours 32 cores, which used to be named streaming processors (SPs) but now are called CUDA cores by NVIDIA. A diagram of an SM with some internals is given in Figure 10.2. Via the instruction cache's two warp schedulers (a warp is a bundle of 32 threads), the program threads are pushed onto the SPs. In addition each SM has four special function units, which take care of the evaluation of functions that are more complicated than can be profitably computed by the simple floating-point units in the SPs. Lastly, we list some properties of the Tesla C2050/70 in Table 10.2.

From these specifications, it can be inferred that two 32-bit floating-point results per core per cycle can be delivered. The peak power requirement given will probably be an appropriate measure for HPC workloads. A large proportion of the work being done will be from the BLAS library, provided by NVIDIA, or more specifically, its dense matrix–matrix multiplication. This operation occupies any computational core to the fullest and will therefore consume close to the maximum of the power. As can be seen from the table, the only difference between the C2050 and the C2070 is the amount

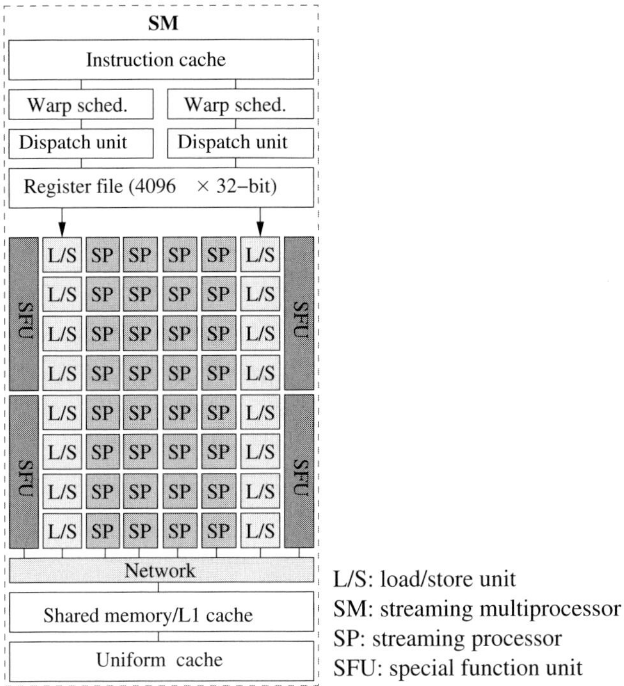

Figure 10.2. Diagram of a streaming processor of the NVIDIA Tesla C2050/C2070.

of memory: the C2050 features 3 GB of GDDR5 memory while the C2070 has double that amount.

Like ATI, NVIDIA provides an SDK comprising a compiler named CUDA, libraries that include BLAS and FFT routines, and a runtime system that accommodates both Linux (Red Hat and SUSE) and Windows. CUDA is a C/C++-like language with extensions and primitives that cause operations to be executed on the card instead of on the CPU core that initiates the operations. Transport to and from the card is done via library routines, and many threads can be initiated and placed in appropriate positions in the card memory so as not to cause memory congestion on the card. This means that for good performance one needs knowledge of the memory structure on the card to exploit it accordingly. This is not unique to the C2050 GPU: it pertains to the ATI FireStream GPU and other accelerators as well.

NVIDIA also supports OpenCL, though CUDA is at present much more popular among developers. For Windows users the NVIDIA Parallel Nsight for Visual Studio is available, which should ease the optimization of the program parts run on the cards.

Table 10.2. Some specifications for the NVIDIA Tesla C2050/70 GPU.

Number of processors	448
Memory (GDDR5), C2050	3 GB
Memory (GDDR5), C2070	6 GB
Internal bandwidth	$\leq$ 153 GB/s
Clock cycle	1.15 GHz
Peak performance (32-bit)	1.03 Tflop/s
Peak performance (64-bit)	515 Gflop/s
Power requirement, peak	238 W
Interconnect (PCIe Gen2)	$\times$8, 4 GB/s; $\times$16, 8 GB/s
ECC, error correction	Yes
Floating-point support	Full (32/64-bit)

10.2. General computational accelerators

Although we have so far looked at the GPUs primarily from the perspective of computational accelerators, they are of course originally full-blown high-end graphical processors. Several vendors have developed accelerators that did not have graphical processing in mind as the foremost application to be served (although they might not be bad in this respect when compared to general CPUs). The future of general computational accelerators is problematic: in principle it is entirely possible to make accelerators that can compete with GPUs, or with the FPGA-based accelerators discussed in Section 10.3, but the volume will always be much lower than that of the other two accelerator variants, which is reflected in the production cost.

Below we discuss two of these general accelerators for the sake of completeness, but it is doubtful that they will survive as marketable products.

PetaPath

PetaPath is a spin-off of ClearSpeed, to position ClearSpeed products in the HPC market. ClearSpeed works in the embedded processor sector but a main product, namely the CSX700 processor, is well equipped for HPC work. We discuss this processor in some detail below.

The ClearSpeed products are in their third generation. Unlike GPUs, ClearSpeed processors were made to operate on 64-bit floating-point data from the start and have full error correction incorporated. The latest processor is the CSX700 chip, which is packaged in a number of products. The most common is the e710 card, which fits into a PCIe slot on any PC or server unit. A variant with a different form factor but with the same functionality is the e720 card, which can be put into blade servers. PetaPath also markets, apart from the separate cards, its Feynman e740 and e780

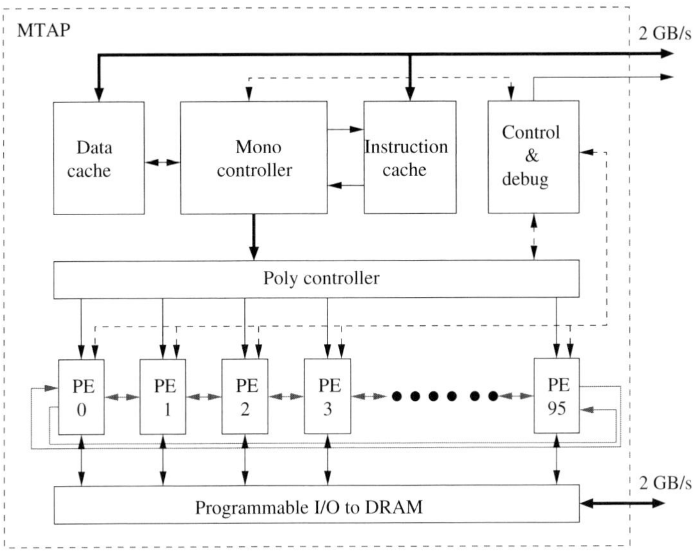

Figure 10.3. Block diagram of a ClearSpeed MTAP unit. Two of these units reside on a CSX700 chip.

units, which house 4 and 8 e720 cards and which connect to a host server by PCIe Gen. 2, 16×, *i.e.*, at 8 GB/s. The bandwidth for the individual cards is 2 GB/s, however. As the peak performance of a single e720 card is 96 Gflop/s, the peak performances of the Feynman e740 and e780 are 384 and 768 Gflop/s, respectively.

The power consumption of the e710/e720 card is extremely low: 25 W maximal, 15 W typical. This is partly due to the low clock frequency of 250 MHz. The e710 card contains, aside from the CSX700 processor, 2 GB DDR2 SDRAM, and an FPGA that manages the data traffic to and from the card. As stated earlier, the interconnect to the host system is compliant with PCIe 8×, amounting to a bandwidth of 2 GB/s. ClearSpeed is quite complete in giving technical details. So, we are able to show a block diagram of the CSX processor in Figure 10.3. Two so-called multi-threaded array processor (MTAP) units are located on one CSX700 chip. As can be seen, an MTAP contains 96 processors (with 4 redundant ones per MTAP). They are controlled via the poly controller, 'poly' being the indication for the data types that can be processed in parallel. The processing elements themselves are able to communicate fast between themselves via a dedicated ring network. Every cycle, a 64-bit data item can be shifted to the right or to the left through the ring. In Figure 10.4 we show the details of a processing element. A maximum of two 64-bit floating-point results can be generated per cycle. As one MTAP contains 96 PEs and there are 2 MTAPs

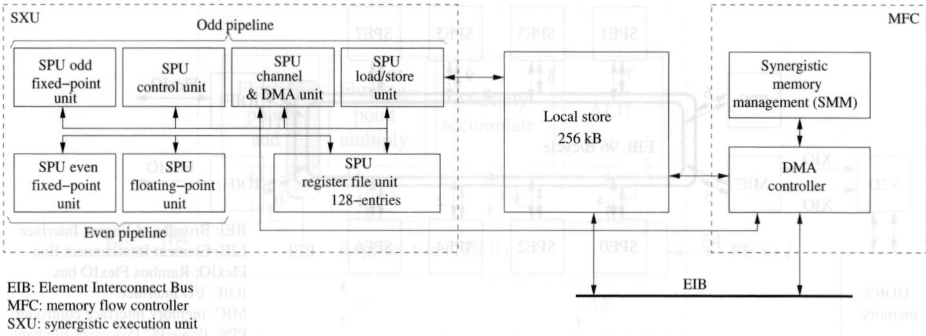

Figure 10.6. Block diagram of an IBM PowerXCell synergistic processing element (SPE).

the PPE takes care of operating system tasks and coordinates the work to be done by the SPEs. All devices in the processor are connected by the Element Interconnect Bus (EIB). The EIB in fact consists of four 16-B-wide rings that transport data in opposite directions to minimize the distance between the devices in the processor. The devices connected to the EIB are numbered, to allow data to be transferred from one device to another. Up to 96 B/cycle can be transferred, amounting to 307.2 GB/s. Although the PowerXCell uses DDR2 memory, the processor proper is designed for use with Rambus memory. This has been taken care of by including the X2D device, which translates the DDR memory requests into Rambus requests and *vice versa*. The two I/O interfaces are controlled through the Broadband Engine Interface (BEI). They have different functions: IOIF1 takes care of the usual external I/O devices via the IOIF protocol, while IOIF0 is able to use the internal I/O protocol, BIF, which is also used on the EIB rings. In this way it is possible to connect to other Cell processors.

The SPEs are the computational workhorses of the Cell processor. We show the internals of an SPE in Figure 10.6. Roughly, there are three important parts in an SPE: the synergistic execution unit (SXU), which contains the functional units for computation, load/store, and DMA control; the local store, which contains the local data to be operated on; and the memory flow controller (MFC), which in turn contains the DMA controller and the memory management unit. As shown in Figure 10.6, in the SXU, the functional units are organized into an odd and an even pipeline. Two instructions can be issued every cycle, one for each of these pipelines. This also implies that one floating-point instruction can be issued per cycle. Depending on the type of operands, this can yield four 32-bit results or two 64-bit results per cycle (in the PowerXCell, in the original Cell processor a 64-bit result can be delivered every 13 cycles, hence the much lower double-precision performance). Note that an SPE does not have any form of cache.

Rather, data are brought in from external memory by DMA instructions via the EIB. This leads to much lower memory latency when a data item is not in the local store. Up to 16 DMA requests can be outstanding for any of the SPEs. As all SPEs are independent, up to 128 DMA requests can be in flight. Of course, this explicit memory management does not make for easy programming. So, one must be careful in managing the data to get (close to) optimal performance.

IBM has put much effort into a software development kit for the Cell processor. It is freely available and, apart from the necessary compilers, there is an extensive library for managing the data transport both from the PPE to the SPEs, between SPEs, initiating the processes on the SPEs, retrieving the results, and managing program overlays. As the local stores in the SPEs are small, the old concept of overlays has been revived. The program is divided into units that depend on each other but do not constitute the whole program. By loading and unloading the units in the correct sequence one can still execute the total program. In addition, there are debugging and performance analysis tools. The total program development can be done using IBM's IDE, Eclipse.

The PowerXCell 8i won its share of fame for its use in the Roadrunner system at Los Alamos National Laboratory. In this system 3240 so-called triblades are connected by InfiniBand. A triblade consists of two QS22 blades, each containing two PowerXCell processors, and an LS21 blade with two Opteron processors. This configuration was the first to break the LINPACK petaflop barrier. This fact certainly helped to increase interest in the Cell processor as an accelerator platform. At present, there are many research projects under way to assess the applicability of Cell BE accelerators and to make their learning curve less steep.

10.3. FPGA-based accelerators

An FPGA (field programmable gate array) is an array of logic gates that can be hardware-programmed to fulfil user-specified tasks. In this way one can devise special-purpose functional units that may be very efficient for this limited task. Moreover, it is possible to configure a multiple of these units on an FPGA that work in parallel. So, potentially, FPGAs may be good candidates for the acceleration of certain applications. Because of their versatility it is difficult to specify where they will be most useful. In general, though, they are not used for heavy 64-bit precision floating-point arithmetic. Excellent results have been reported in searching, pattern matching, signal- and image-processing, encryption, *etc.* The clock cycle of FPGAs is low compared to that of present CPUs: 100–550 MHz, which means that they are very power-efficient. Vendors provide runtime environments and drivers that work with Linux as well as Windows.

Traditionally, FPGAs are configured by means of a hardware description language (HDL), such as VHDL or Verilog. This is very cumbersome for the average programmer as one has to explicitly define not only details such as the placement of the configured devices but also the width of the operands to be be operated on, *etc.* This problem has been recognized by FPGA-based vendors and a large variety of programming tools and SDKs have come into existence. Unfortunately, they differ enormously in approach, and the resulting programs are far from compatible. Further, for FPGA-based accelerators, as for GPUs, there is an initiative to develop a unified API that will ensure compatibility between platforms. The non-profit OpenFPGA consortium is heading this effort. Various working groups are concentrating on, for instance, a core library, an application library, and an API definition. There is no unified way to program FPGAs platform independently, however, and it may take a long time to get there.

The two big players on the FPGA market are Altera and Xilinx. However, in the accelerator business one will seldom find these names mentioned, because the FPGAs they produce are packaged in a form that makes them unusable for accelerator purposes.

It is not possible to fully discuss all vendors that offer FPGA-based products. One reason is that there is a very large variety of products, ranging from complete systems to small appliances housing one FPGA and the appropriate I/O logic to communicate with the outside world. To complicate matters further, the FPGAs themselves come in many variants, *e.g.*, with I/O channels, memory blocks, multipliers, or DSPs already configured (or even fixed) and one can choose FPGAs that have, for instance, a PowerPC405 embedded. Therefore we present FPGA accelerators only in the most global way, and our treatment is necessarily incomplete.

In the following we will discuss products of vendors that have gone to great lengths to not expose their users to the use of HDLs, although for the highest benefits this cannot always be avoided. Necessarily, our choice of topics is again somewhat arbitrary, because this area is changing extremely rapidly.

Convey

The Convey HC-1 was announced in November 2008. It is an example of the hybrid solutions that have arisen to avoid the unwieldy HDL programming of FPGAs, while still benefiting from their potential acceleration capabilities. The HC-1 comprises a familiar x86 front-end with a modified Centos Linux distribution under the name of Convey Linux. Furthermore, there is a co-processor part that contains four Xilinx V5 FPGAs that can be configured into a variety of 'personalities' that would accommodate users from different application areas. The personalities offered include the oil and gas industry, the financial analytics market, and the life sciences.

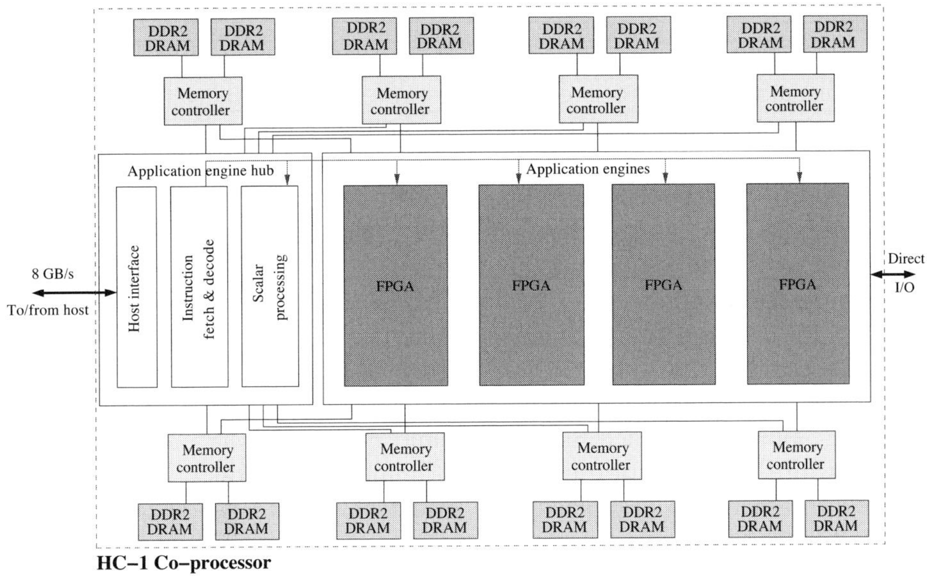

Figure 10.7. Block diagram of the Convey HC-1.

In Figure 10.7 we give a diagram of the HC-1 co-processor's structure. A personality that will be used often for scientific and technical work is the vector personality. Thanks to the compilers provided by Convey standard, code in Fortran and C/C++ can be automatically vectorized and execute the vector units configured in the four FPGAs, for a total of 32 function pipes. Each of these contains a vector register file: four pipes that can execute floating multiply–add instructions, a pipe for integer, logical, divide, and miscellaneous instructions and a load/store pipe. For other selected personalities, the compilers will generate code that is optimal for the instruction mix generated for the appropriately configured FPGAs in the application engine.

The application engine hub shown in Figure 10.7 contains the interface to the x86 host, but also the part that maps the instructions onto the application engine. In addition, it will perform some scalar processing that is not readily passed on to the application engine.

Because the system has many different faces, it is hard to speak about *the* peak performance of the system. As yet there is too little experience with the HC-1 to compare it one-to-one with other systems in terms of performance. However, it is clear that the potential speed-up for many applications can be large.

Kuberre

Since May 2009 Kuberre has marketed its FPGA-based HANSA system. The information provided is extremely scant. The company has traditionally been involved in financial computing, and with the rising need for HPC in this sector Kuberre has built a system that houses 1–16 boards, each with 4 Altera Stratix II FPGAs and 16 GB of memory, in addition to one dual core x86-based board that acts as a front-end. The host board runs the Linux or Windows OS and the compilers.

For programming, a C/C++ or Java API is available. Although Kuberre is naturally oriented to the financial analytics market, the little material that is accessible shows that libraries such as ScaLAPACK, Monte Carlo algorithms, FFTs and wavelet transforms are available. For the life sciences, standard applications such as BLAST and Smith–Waterman are present. The standard GNU C libraries can also be linked seamlessly.

The processors are organized in a grid fashion and use a 256 GB distributed shared cache to combat data access latency. The system comes configured with 768 RISC CPUs for what are called 'generic C/C++ programs', or as 1536 double-precision cores for heavy numerical work. It is possible to split the system to run up to 16 different 'contexts' (reminiscent of Convey's personalities: see p. 430). Part of the machine may be dedicated to a life sciences application, while other parts work on encryption and numerical applications.

As for the Convey HC-1, it is hardly possible to give performance figures, but a fully configured machine with 16 boards should be able to obtain 250 Gflop/s on the LINPACK benchmark.

The material publicly available does not allow us to show a reliable block diagram, but this may come about later when the system is installed at sites that want to evaluate it.

SRC

Until two years ago SRC was the only company that sold a full stand-alone FPGA accelerated system, named the SRC-7. Now it has to share this space with Convey and Kuberre. In addition the so-called SRC-7 MAP station is now marketed, MAP being the processing unit that contains 2 Altera Stratix II FPGAs. Furthermore, SRC has the IMAP card as a product that can be plugged into a PCIe slot on any PC.

SRC has gone to great lengths to ban the term 'FPGA' from its documentation. Instead, it talks about implicit versus explicit computing. In SRC terms implicit computing is performed on standard CPUs, while explicit computing is done on its (reconfigurable) MAP processor. The SRC-7 systems have been designed with the integration of both types of processors in mind, and in this sense it is a hybrid architecture as well, because shared extended memory can be put into the system that is equally accessible by

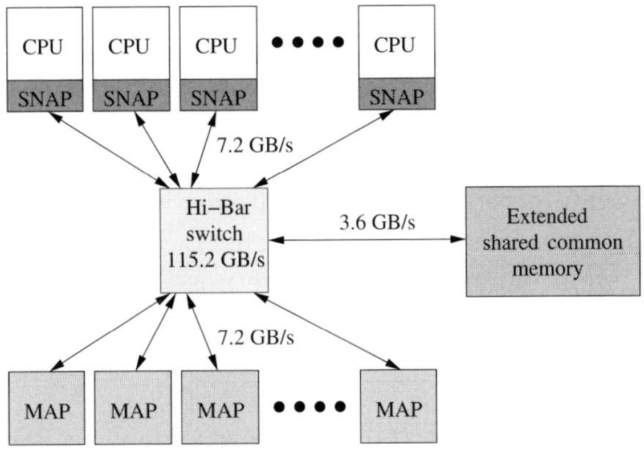

Figure 10.8. Approximate machine structure of the SRC-7.

both the CPUs and the MAP processors. We show a sketch of the machine structure in Figure 10.8. It shows that CPUs and MAP processors are connected by a 16×16 so-called Hi-Bar crossbar switch with a link speed of 7.2 GB/s. The maximum aggregate bandwidth in the switch is 115.2 GB/s, enough to route all 16 independent data streams. The CPUs must be of the x86 or x86_64 type. So, both Intel and AMD processors are possible. As can be seen in the figure, the connection to the CPUs is made through SRC's proprietary SNAP interface. This accommodates the 7.2 GB/s bandwidth but isolates it from the vendor-specific connection to memory. Instead of configuring a MAP processor, common extended memory can also be configured. This allows for shared-memory parallelism in the system across CPUs and MAP processors.

The MAP station is a shrunken version of the SRC-7: it contains an x86(_64) CPU, a MAP processor, and a 4×4 Hi-Bar crossbar that allows common extended memory to be configured.

SRC and Convey are the only accelerator vendors that support Fortran. SRC does this through its development environment Carte. As with Convey and Kuberre, C/C++ is also available. The parallelization and acceleration are largely done by putting comment directives in Fortran code and pragmas in C/C++ code. Also, explicit memory management and prefetching can be done in this way. The directives/pragmas cause a bitstream to be loaded onto the FPGAs in one or more MAP processors that configure them and execute the target code. Furthermore, there is an extensive library of functions, a debugger and a performance analyser. When one wants to employ specific non-standard functionality, *e.g.*, computing with arithmetic of non-standard length, one can create a so-called 'application-specific functional unit'. In fact, one then configures one or more of the FPGAs directly and one has to fall back on VHDL or Verilog for this configuration.

11. Networks

Fast interprocessor networks are, together with fast processors, decisive factors for good integrated parallel systems and clusters. In the early days of clusters the interprocessor communication, and hence the scalability of applications, was hampered by the high latency and the lack of bandwidth of the network that was used (mostly Ethernet). This situation has changed greatly, and to give a balanced view of the possibilities opened by the improved networks, a discussion of some of these networks is in order. Networks have been employed as an important component of 'integrated' parallel systems.

Of course, Gigabit Ethernet (GbE) is now widely available, and with a maximum theoretical bandwidth of 125 MB/s would be able to fulfil a useful role for some applications that are not latency-bound. Furthermore, 10 Gigabit Ethernet (10 GbE) is increasingly available. The adoption of Ethernet is hampered by the latencies that are incurred when the TCP/IP protocol is used for the message transmission. In fact, the transmission latencies without this protocol are much lower: about 5 μs for GbE and 0.5 μs for 10 GbE. Using the TCP/IP protocol, however, gives rise to latencies of somewhat less than 40 μs and in-switch latencies of 30–40 μs for GbE and a 4–10 μs latency for 10 GbE. As such it is not quite on a par with the ubiquitous InfiniBand interconnects with regard to latency and bandwidth. However, the costs are lower and may compensate for a somewhat lower performance in many cases. Various vendors, such as Myrinet and SCS, have circumvented the problem with TCP/IP by implementing their own protocol, thus using standard 10 GbE equipment but with their own network interface cards (NICs) to handle the proprietary protocol. In this way

Table 11.1. Some bandwidths and latencies for various networks as measured with an MPI ping-pong test.

Network	Bandwidth GB/s	Latency μs
Arista 10 GbE (stated)	1.2	4.0
BLADE 10 GbE (measured)	1.0	4.0
Cray SeaStar2+ (measured)	6.0	4.5
Cray Gemini (measured)	6.1	1.0
IBM (InfiniBand) (measured)	1.2	4.5
SGI NumaLink 5 (measured)	5.9	0.4
InfiniBand (measured)	1.3	4.0
InfiniPath (measured)	0.9	1.5
Myrinet 10-G (measured)	1.2	2.1

latencies of 2–4 μs can be achieved: well within the range of other network solutions. Very recently Mellanox came out with 40 GbE on an InfiniBand fabric. It is too early, however, to give characteristics of this new medium.

We restrict ourselves here to networks that are independently marketed as the proprietary networks for systems such as those of Cray and SGI, and are discussed together with the systems in which they are incorporated. We do not pretend to be complete, because in this new field, players enter and leave the scene at a high rate. Rather, we present the main developments which one is likely to meet when one scans the high-performance computing arena. Unfortunately, the spectrum of network types is narrowed by the demise of Quadrics. Quadrics' QsNet^{II} was rather expensive but it had excellent characteristics. The next generation, QsNet^{III}, was on the brink of deployment when the Italian mother company Alinea terminated Quadrics – much to the regret of HPC users and vendors.

A complication with the fast networks offered for clusters is the connection with the nodes. Where in integrated parallel machines the access to the nodes is customized and can be made such that the bandwidth of the network matches the internal bandwidth in a node, in clusters one has to make do with the PCI bus connection that comes with the PC-based node. The type of PCI bus which ranges from 32-bit wide at 33 MHz to 64-bit wide at 66 MHz determines how fast the data from the network can be shipped in and out of the node, and therefore the maximum bandwidth that can be attained in internode communication. In practice, the available bandwidths are in the range 110–480 MB/s. Since 1999 PCI-X has been available, initially at 1 GB/s, in PCI-X 2.0 also at 2 and 4 GB/s. Coupling with PCI-X is at present mostly superseded by its successor PCI-Express 1.1 (PCIe). This provides a 200 MB/s bandwidth per data lane where 1×, 2×, 4×, 8×, 12×, 16×, and 32× multiple data lanes are supported: this makes it fast enough for the host bus adapters of any communication network vendor so far. So, for the networks discussed below, often different bandwidths are quoted, depending on the PCI bus type and the supporting chipset. Therefore, when speeds are quoted, it is always with the proviso that the PCI bus of the host node is sufficiently wide/fast.

Lately, PCIe 2, commonly known as PCIe Gen2, has emerged, with twice the bandwidth. Currently PCIe Gen2 is mostly used within servers to connect to high-end graphics cards (including GPUs used as computational accelerators) at speeds of 4–8 GB/s, but evidently it could also be used to connect to either other computational accelerators or network interface cards that are designed to work at these speeds.

An idea of network bandwidths and latencies for some networks, both proprietary and vendor-independent, is given in Table 11.1. Warning: The entries are only approximate because they also depend on the exact switch and host bus adapter characteristics as well as on the internal bus speeds of the systems. The circumstances under which these values were obtained was very diverse. So, there is no guarantee that these are the optimum attainable results.

11.1. *InfiniBand*

InfiniBand has rapidly become a widely accepted medium for internode networks. The specification was finished in June 2001. Since 2002, a number of vendors have started to offer their products based on the InfiniBand standard. A very complete description (1200 pages) can be found in Shanley (2002). InfiniBand is employed to connect various system components within a system. Via host channel adapters (HCAs), the InfiniBand fabric can be used for interprocessor networks, attaching I/O subsystems, or to multi-protocol switches like Gigabit Ethernet switches, *etc.* Because of this versatility, the market is not limited just to the interprocessor network segment, and so InfiniBand has become relatively inexpensive due to the high volume of sales at present. The characteristics of InfiniBand are rather nice. There are product definitions both for copper and glass fibre connections, switch and router properties are defined, and multiple connections can be employed for high bandwidth. Also, the way messages are broken up into packets and reassembled, as well as routing, prioritizing, and error handling, are all described in the standard. This makes InfiniBand independent of one particular technology and, because of its completeness, it is a good basis upon which to implement a communication library (such as MPI).

Conceptually, InfiniBand knows of two types of connectors to the system components: host channel adapters (HCAs), already mentioned, and target channel adapters (TCAs). The latter are typically used to connect to I/O subsystems, while HCAs concern us more as they are the connectors used in interprocessor communication. InfiniBand defines a basic link speed of 2.5 Gb/s (312.5 MB/s) but also a 4× and 12× speed of 1.25 GB/s and 3.75 GB/s, respectively. Moreover, HCAs and TCAs can have multiple ports that are independent and allow for higher reliability and speed.

Messages can be sent on the basis of remote direct memory access (RDMA) from one HCA/TCA to another: an HCA/TCA is permitted to read/write the memory of another HCA/TCA. This enables very fast transfer once permission and a write/read location are given. A port, together with its HCA/TCA, provides a message with a 128-bit header which is IPv6 compliant and which is used to direct it to its destination via cut-through wormhole routing. In each switching stage the routing to the next stage is decoded and sent on. Short messages of 32 B can be embedded in control messages, which cuts down on the negotiation time for control messages.

InfiniBand switches for HPC are normally offered with 8–864 ports and now mostly at a speed of 1.25 GB/s. However, Sun is now providing a

3456-port switch for its Constellation cluster systems. Switches and HCAs accommodating twice this speed (double data rate, DDR) are now common, but are being replaced more and more by quad data rate (QDR), which became available in late 2008. Obviously, to take advantage of this speed at least PCI Express must be present at the nodes to which the HCAs are connected. The switches can be configured in any desired topology, but in practice a fat tree topology is almost always preferred (see Figure 6.2(b)). How much of the raw speed can be realized depends, of course, on the quality of the MPI implementation imposed on the InfiniBand specifications. A ping-pong experiment on InfiniBand-based clusters with different MPI implementations has shown bandwidths of 1.3 GB/s, and an MPI latency of 4 μs for small messages is quoted by Mellanox, one of the large InfiniBand vendors. The in-switch latency is typically about 200 ns. For the QDR 2.5 GB/s products, the MPI bandwidth indeed nearly doubles while the latency stays approximately the same. At the time of writing, QDR InfiniBand products are available from Mellanox and QLogic. A nice feature of QDR InfiniBand is that it provides dynamic routing, which is not possible with earlier generations. In complicated communication schemes this feature should alleviate contention on some data paths by letting the message take an alternative route.

Because of the recent profusion of InfiniBand vendors, the price is now on a par with, or lower than, those of other fast network vendors such as Myrinet (Section 11.3) and 10 GbE.

11.2. InfiniPath

InfiniPath only provides HCAs with a 4-wide (1.25 GB/s) InfiniBand link on the network side and connecting to a HyperTransport bus or PCI-Express on the computer side. For systems with AMD processors on board, the HyperTransport option is particularly attractive because of the direct connection to the host's processors. This results in very low latencies for small messages. PathScale, the vendor of the InfiniPath HCAs, quotes latencies as low as 1.29 μs. Obviously, this type of HCA cannot be used with systems based on non-AMD processors. For these systems the HCAs with PCI-Express can be used. They have a slightly higher, but still low, latency of 1.6 μs. The effective bandwidth is also high: a uni-directional bandwidth of $\approx$ 950 MB/s can be obtained using MPI for both types of HCA.

The InfiniPath HBAs do not contain processing power themselves. Any processing associated with the communication is done by the host processor. According to PathScale this is an advantage because the host processor is usually much faster than the processors employed in switches. An evaluation report from Sandia National Lab (Doerfler 2005) seems to corroborate this assertion.

PathScale only offers HCAs (and the software stack coming with it) and these can be used by any InfiniBand switch vendor that adheres to the OpenIB protocol standard, which includes pretty much all of them.

11.3. Myrinet

Until recently Myrinet was the market leader in fast cluster networks and it is still one of the largest. The Myricom company which sells Myrinet started in 1994 with its first Myrinet implementation (Boden *et al.* 1995), as an alternative to Ethernet for connecting the nodes in a cluster. Apart from the higher bandwidth, around 100 MB/s at that time, the main advantage was that it entirely operated in user space, thus avoiding operating system interference and the delays that come with it. This meant that the latency for small messages was around 10–15 μs. Latency and bandwidth compared nicely with the proprietary networks of integrated parallel systems of Convex, IBM, and SGI at the time. Although such a network came at a non-negligible cost, in many cases it proved a valuable alternative to either an Ethernet-connected system or an even costlier integrated parallel system.

Since then hardware upgrades and software improvements have made Myrinet the network of choice for many cluster builders, and until a few years ago there was hardly an alternative when a fast, low-latency network was required.

Like InfiniBand, Myrinet uses cut-through routing for an efficient utilization of the network. Also, RDMA is used to write to/read from the remote memory of other host adapter cards, called Lanai cards. These cards interface with the PCI-X of PCI Express bus of the host they are attached to. Myrinet allows copper cables or fibres as signal carriers. The latter form gives a high flexibility in the connection and much headroom in the speed of signals, but the fibre cables and connectors are rather delicate, which can lead to damage when cluster nodes have to be serviced.

Myrinet offers ready-made 8–256 port switches (8–128 for its newest product: see below). The 8 and 16 port switches are full crossbars. In principle all larger networks are built from these using a Clos network topology. An example for a 64-port systems is shown in Figure 11.1. A Clos network is another example of a logarithmic network with the maximum bi-sectional bandwidth of the endpoints. Note that 4 ports of the 16 $\times$ 16 crossbar switches are unused, but other configurations need either more switches or connections or both.

Since the start of 2006 Myricom, like many InfiniBand switch vendors, has provided a multi-protocol switch (and adapters), the Myri-10G. Apart from Myricom's own MX protocol it also supports 10 Gigabit Ethernet, which makes it easy to connect to external nodes/clusters – an ideal start-

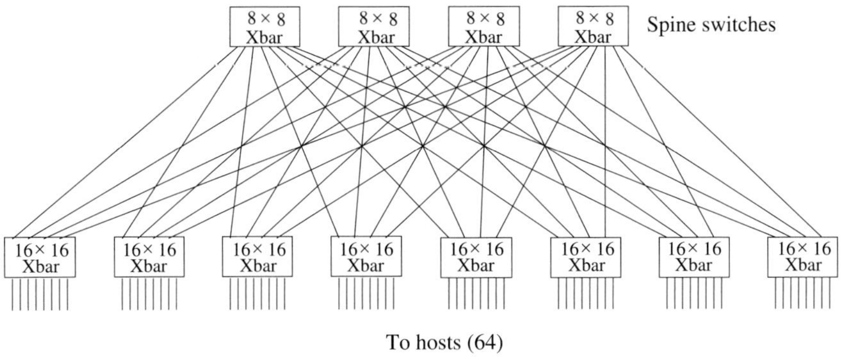

Figure 11.1. An 8×16 Clos network using 8 and 16 port crossbar switches to connect 64 processors.

ing point for building grids from a variety of systems. The specifications as given by Myricom are quite good: ≈ 1.2 GB/s for the uni-directional theoretical bandwidth for both its MX protocol and about the same for the MX emulation of TCP/IP on Gigabit Ethernet. According to Myricom, there is no difference in bandwidth between MX and MPI, and the latencies are claimed to be the same: just over 2 μs.

12. Recent trends in high-performance computing

In this section we analyse major recent trends and changes in high-performance computing. Massively parallel processors (MPPs) became successful in the early 1990s due to their better price/performance ratios, which was enabled by advances in microprocessors. The success of microprocessor-based symmetric multiprocessor (SMP) concepts, even for very high-end systems, was the basis for the emergence of cluster concepts in the early 2000s. During the first half of the decade clusters of PCs and workstations became the prevalent architecture for many HPC application areas on all ranges of performance. However, the Japanese Earth Simulator vector system demonstrated that many scientific applications could benefit greatly from other computer architectures. At the same time there has been renewed broad interest within the scientific HPC community concerning new hardware architectures and new programming paradigms. The IBM BlueGene system is one early example of a shifting design focus for large-scale systems.

12.1. Introduction

Looking back on the last four decades, the HPC market has always been characterized by rapid change in vendors, architectures, technologies and system usage. Despite all these changes, the evolution of performance on a

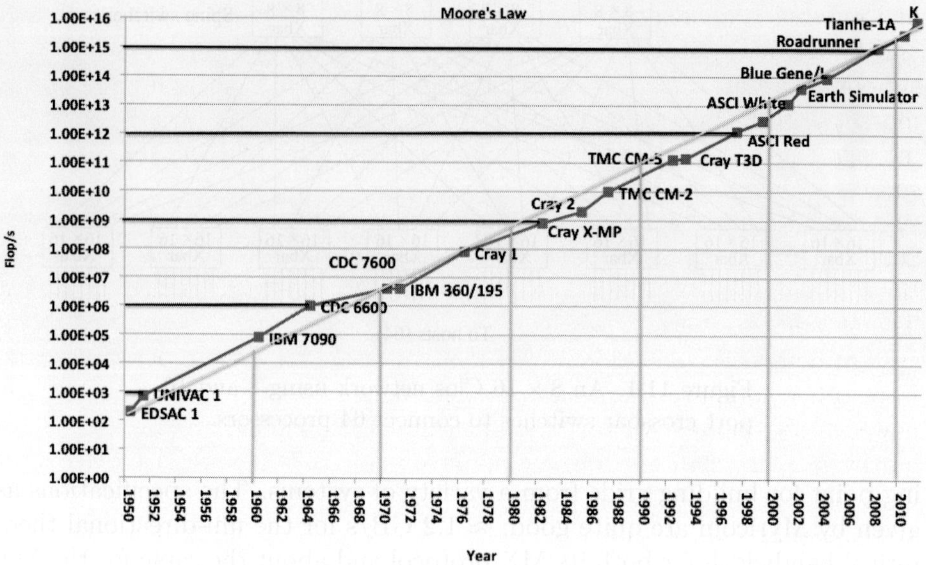

Figure 12.1. Performance of the fastest computer systems for the last six decades.

large scale seems to be a very steady and continuous process. Moore's Law is often cited in this context. Figure 12.1 plots the peak performance of various computers of the last six decades, all 'supercomputers' of their time (Hockney and Jesshope 1988, Meuer 1994), and demonstrates how well this law holds for nearly the entire lifespan of modern computing. On average we see an increase in performance of two orders of magnitude every decade. In this section we analyse recent major trends and changes in the HPC market. For this, we focus on systems which had at least some commercial relevance. This paper extends a previous analysis of the HPC market in Meuer, Strohmaier, Dongarra and Simon (2011). Historical overviews with different focuses can be found in Wilson (1994) and Woodward (1996). Section 12.3 analyses the trend in the first half of the 2000s, and Section 12.4 looks to the future.

The initial success of vector computers in the 1970s was driven by raw performance. The introduction of this type of computer system started the era of 'supercomputing'. In the 1980s the availability of standard development environments and application software packages became more important. Next to performance, these criteria determined the success of MP vector systems, especially with industrial customers. MPPs became successful in the early 1990s due to their better price/performance ratios, enabled by the attack of the 'killer micros'. In the lower and medium market segments, the MPPs were replaced by microprocessor-based SMP systems in the middle of the 1990s. Towards the end of the 1990s, only the companies which had entered the emerging markets for massively parallel database servers and financial applications attracted enough business volume to be able to support the hardware development for the numerical high-end computing market as well. Success in the traditional floating-point intensive engineering applications was no longer sufficient for survival in the market. The success of microprocessor-based SMP concepts, even for the very high-end systems, was the basis for the emergence of cluster concepts in the early 2000s. During the first half of the decade clusters of PCs and workstations became the prevalent architecture for many application areas in the TOP500 on all ranges of performance. However, the Earth Simulator vector system demonstrated that many scientific applications can benefit greatly from other computer architectures. At the same time there has been renewed broad interest within the scientific HPC community concerning new hardware architectures and new programming paradigms. The IBM BlueGene/L system is one early example of a shifting design focus for large-scale systems. The IBM Roadrunner system at Los Alamos National Laboratory broke the petaflops threshold in June 2008. And in November 2011 the Japanese K computer reached 10.5 petaflop/s using 705 024 cores of conventional design.

12.2. A short history of supercomputers

In the second half of the 1970s the introduction of vector computer systems marked the beginning of modern supercomputing. These systems offered a performance advantage of at least one order of magnitude over conventional systems of that time. Raw performance was the main if not the only selling argument. In the first half of the 1980s the integration of vector systems in conventional computing environments became more important. Only those manufacturers which provided standard programming environments, operating systems and key applications were successful in getting industrial customers, and survived. Performance was mainly increased by improved chip technologies and by producing shared-memory multiprocessor systems.

Fostered by several US government programmes, massively parallel computing with scalable systems using distributed memory became the centre of interest at the end of the 1980s. The main goal for their development was overcoming the hardware scalability limitations of shared-memory systems. The increase in performance of standard microprocessors after the RISC revolution, together with the cost advantage of large-scale productions, formed the basis for the 'attack of the killer micros'. The consequence was a transition from ECL to CMOS chip technology and the use of 'off-the-shelf' microprocessors instead of custom-designed processors for MPPs.

The traditional design focus for MPP systems was the very high end of performance. In the early 1990s the SMP systems of various workstation manufacturers, as well as the IBM SP series, which targeted the lower and medium market segments, gained great popularity. Their price/performance ratios were better due to not including support for very large configurations and due to economies of scale. Due to the vertical integration of production, it was no longer economically feasible to produce and focus on the highest end of computing power alone. The design focus for new systems shifted to the market of medium-performance systems.

The acceptance of MPP systems not only for engineering applications but also for new commercial applications, especially for database applications, emphasized different criteria for market success, such as stability of systems, continuity of manufacturer and price/performance. Success in commercial environments became a new important requirement for a successful supercomputer business towards the end of the 1990s. Due to these factors and the consolidation in the number of vendors in the market, hierarchical systems built with components designed for the broader commercial market replaced homogeneous systems at the very high end of scientific computing. The marketplace readily adopted clusters of SMPs, while academic research focused on clusters of workstations and PCs.

12.3. 2000–2005: clusters, Intel processors, and the Earth Simulator

In the early 2000s, clusters built with off-the-shelf components gained more and more attention, not only as academic research objects, but also as computing platforms with end-users of HPC systems. By 2004, this group of clusters represented the majority of new systems on the Top500 in a broad range of application areas. One major consequence of this trend was the rapid rise in the use of Intel processors in HPC systems. Though virtually absent at the high end at the beginning of the decade, Intel processors are now used in the majority of HPC systems. Clusters in the 1990s were mostly self-made systems designed and built by small groups of dedicated scientist or application experts. This changed rapidly as soon as the market for clusters based on PC technology matured. Now the large majority of Top500-class clusters are manufactured and integrated by either a few traditional large HPC manufacturers such as IBM or Hewlett-Packard, or numerous small, specialized integrators of such systems.

In 2002 a system with a different architecture, the Earth Simulator, entered the spotlight as the new number one system on the Top500, and it managed to take the US HPC community by surprise, even though it had been announced four years earlier. The Earth Simulator, built by NEC, is based on NEC vector technology and showed unusually high efficiency on many scientific applications. This fact invigorated discussions about future architectures for high-end scientific computing systems. The first system built with a different design focus, but still with mostly conventional off-

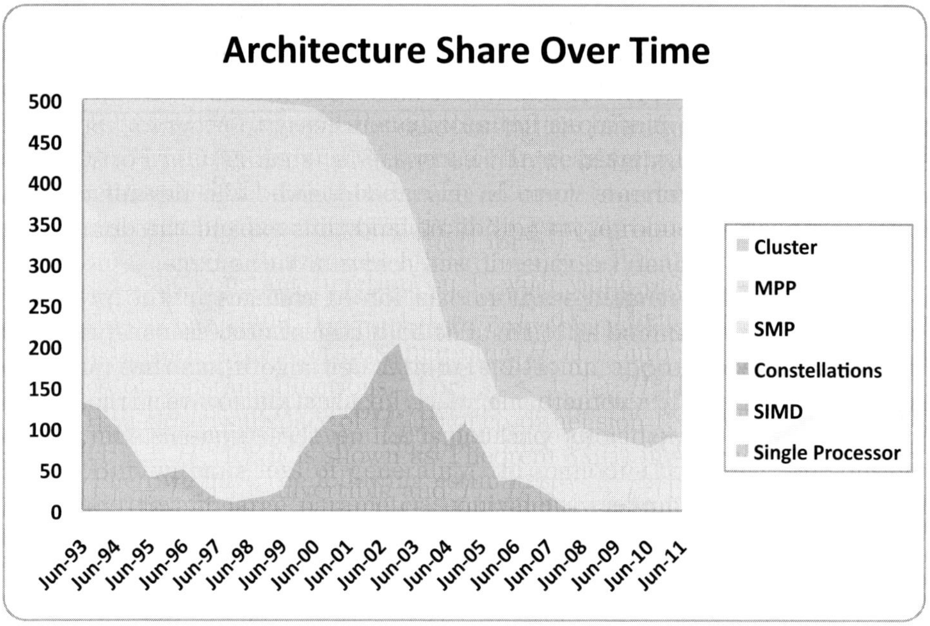

Figure 12.2. Main architectural categories seen in the TOP500. (The term 'constellations' refers to clusters of SMPs.)

the-shelf components, is the IBM BlueGene/L system. Its design focuses on a system with an unprecedented number of processors using a power-efficient design while sacrificing main memory size.

Explosion of cluster-based systems

By the end of the 1990s clusters were common in academia, but mostly as research objects, and not so much as computing platforms for applications. Most of these clusters were of comparable small scale and as a result the November 1999 edition of the TOP500 listed only seven cluster systems. This changed dramatically, as industrial and commercial customers started to deploy clusters as soon as their applications permitted them to take advantage of the better price/performance ratio of commodity-based clusters. At the same time, all major vendors in the HPC market started selling this type of cluster, fully integrated into their customer base. In November 2004 clusters became the dominant architecture in the TOP500, with 294 systems at all levels of performance (see Figure 12.2). Companies such as IBM and Hewlett-Packard sold the majority of these clusters, and a large number of them were installed at commercial and industrial sites. To some extent, the reasons for the dominance of commodity-processor systems are economic. Contemporary distributed-memory supercomputer systems based on commodity processors (such as Linux clusters) appear to be substantially more cost-effective – by roughly an order of magnitude – in delivering computing power to applications that do not have stringent communication requirements. On the other hand, there has been little progress, and perhaps regress, in making scalable systems easy to program. Software directions that were started in the early 1980s (such as CM-Fortran and High-Performance Fortran) were largely abandoned. The pay-off to finding better ways to program such systems, and thus expand the domains in which these systems can be applied, would appear to be large.

The move to distributed memory has forced changes in the programming paradigm of supercomputing. The high cost of processor-to-processor synchronization and communication requires new algorithms that minimize those operations. The structuring of an application for vectorization is seldom the best structure for parallelization on these systems. Moreover, despite some research successes in this area, without some guidance from the programmer, compilers are generally able neither to detect enough of the necessary parallelism, nor to reduce sufficiently the inter-processor overheads. The use of distributed-memory systems has led to the introduction of new programming models, particularly the message-passing paradigm, as realized in MPI, and the use of parallel loops in shared-memory subsystems, as supported by OpenMP. It has also forced significant reprogramming of libraries and applications to port onto the new architectures. Debuggers and performance tools for scalable systems have developed slowly, however, and even today most users consider the programming tools on parallel supercomputers to be inadequate.

Fortunately, there are a number of choices of communication networks available. There is generally great variation in the use of clusters and their more integrated counterparts: clusters are mostly used for capacity computing while the integrated machines are primarily used for capability computing. Traditionally, vendors of large supercomputer systems have learned to provide for capacity computing, as the precious resources of their systems were required to be used as effectively as possible. In contrast, Beowulf clusters are mostly operated through the Linux operating system, although a small minority use Microsoft Windows. These operating systems either lack the tools to make use of clusters for capacity computing, or the tools are immature. However, as clusters become on average both larger and more stable, they are being used as computational capacity servers.

Intel-ization of the processor landscape

The HPC community had already started to use commodity parts in large numbers in the 1990s. MPPs and constellations (the term 'constellations' refers to a cluster of SMPs), typically using standard workstation microprocessors, still might use custom interconnect systems. There was, however, one big exception: virtually nobody used Intel microprocessors. Lack of

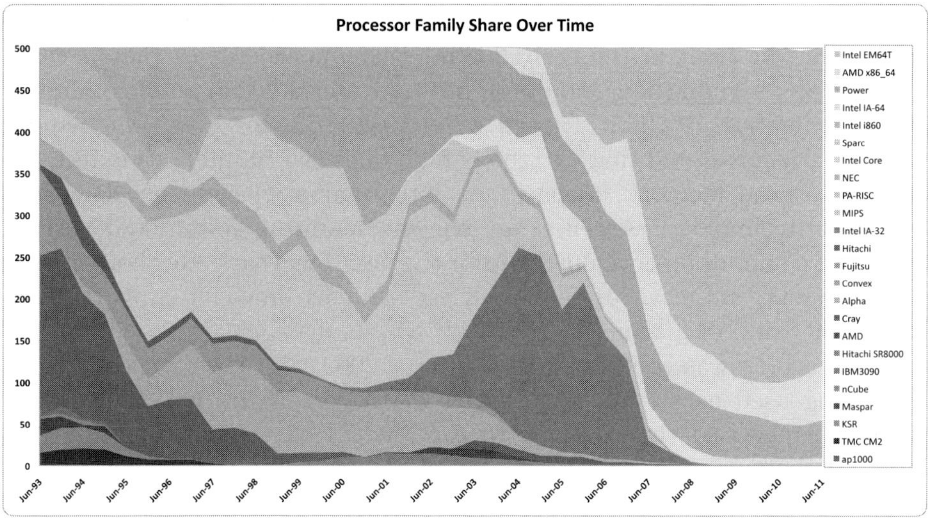

Figure 12.3. Main processor families seen in the TOP500.

performance and the limitations of a 32-bit processor design were the main reasons for this. This changed with the introduction of the Pentium 3 and especially in 2001 with the Pentium 4, which featured greatly improved memory performance due to its front-side bus and full 64-bit floating-point support. The number of systems in the TOP500 with Intel processors exploded from only six in November 2000 to 375 in June 2008 (Figure 12.3)

The Earth Simulator shock

The Earth Simulator (ES) was conceived, developed, and implemented by Dr Hajime Miyoshi, who is regarded as the Seymour Cray of Japan. Unlike his peers, he seldom attended conferences or gave public speeches. However, he was well known within the HPC community in Japan for his involvement in the development of the first Fujitsu supercomputer, and later on the Numerical Wind Tunnel (NWT) at the National Aerospace Laboratory of Japan. In 1997 he took up his post as director of the Earth Simulator Research and Development Center and led the development of the 40 Tflop/s Earth Simulator, which would serve as a powerful computational engine for global environmental simulation.

Prior to the ES, global circulation simulations were made using a 100 km grid width, although ocean-atmospheric interactive analyses were not performed. Obtaining quantitatively good predictions for the evaluation of environmental effects may require a grid width of at most 10 km or 10 times finer meshes in the x, y and z directions, and interactive simulations. Thus a supercomputer 1000 times faster and larger than a 1995 conventional supercomputer might be required. Miyoshi investigated whether such a machine could be built in the early 2000s. His conclusion was that it could be realized if several thousand of the most advanced vector supercomputers of approximately 10 Gflop/s speed were clustered using a very high-speed network. He forecasted that extremely high-density LSI integration technology, high-speed memory, all packaged into small-size, high-speed network (crossbar) technology, as well as an efficient operating system and Fortran compiler, could all be developed within the next few years. He thought that only a strong initiative project with government financial support could realize this kind of machine.

The machine was completed in February 2002, and the entire system is still being used as an end-user service. Miyoshi, as leader of the NWT project, supervised the development of NWT Fortran and organized the HPF (High Performance Fortran) Japan Extension Forum, which is used on the ES. He knew that a high-level vector/parallel language is critical for such a supercomputer.

The launch of the Earth Simulator created a substantial amount of concern in the USA that it had lost the leadership in high-performance computing. While there was certainly a loss of national pride for the USA in not being first on a list of the world's fastest supercomputers, it is important to understand the set of issues surrounding that loss of leadership. The development of the ES represents a large investment (approximately \$500m, including a special facility to house the system) and a large commitment over a long period of time. The USA has made an even larger investment in HPC in the Department of Energy Advanced Strategic Computing (ASC) programme, but the funding has not been spent on a single platform. Other important differences are as follows.

- The ES was developed for basic research and is shared internationally, whereas the ASC programme is driven by national defence and the systems have restricted domestic use.
- A large part of the ES investment supported NEC's development of their SX-6 technology. The ASC program has made only modest investments in industrial R&D.
- The ES uses custom vector processors. The ASC systems use commodity processors.
- The ES software technology largely originates from abroad, although it is often modified and enhanced in Japan. For example, significant ES codes were developed using a Japanese enhanced version of HPF. Virtually all software used in the ASC program has been developed by the USA.

Surprisingly, the Earth Simulator's number one ranking on the TOP500 list was not a matter of national pride in Japan. In fact, there is considerable resentment of the Earth Simulator in some sectors of the research community in Japan. Some Japanese researchers feel that the ES is too expensive and drains critical resources from other science and technology projects. Due to the continued economic crisis in Japan and large budget deficits, it is getting more difficult to justify government projects of this kind.

New architectures on the horizon

Interest in novel computer architectures has always been large in the HPC community, which comes as no surprise, as this field was engendered by and continues to thrive on technological innovations. Some of the concerns of recent years have been the ever-increasing space and power requirements of modern commodity-based supercomputers. In the BlueGene/L development, IBM addressed these issues by designing a very power- and space-efficient system. BlueGene/L does not use the latest commodity processors available but computationally less powerful and much more power-efficient processor versions developed, not for the PC and workstation market, but mainly for embedded applications. Together with a drastic reduction of available main memory, this provided a very dense system. To achieve the targeted extreme performance level, an unprecedented number of these processors (up to 212 992) are combined using several specialized interconnects.

There was and is considerable doubt whether such a system would be able to deliver the promised performance and would be usable as a general-purpose system. The first results of the beta-System were very encouraging, and the one-quarter size beta-System commissioned by Lawrence Livermore National Laboratory was able to claim the number one spot on the November 2004 TOP500 list.

12.4. 2005 and beyond

Three decades after the introduction of the Cray 1, the HPC market had changed considerably. It used to be a market for systems clearly different from any other computer systems. Today the HPC market is no longer an isolated niche market for specialized systems. Vertically integrated companies produced systems of any size. Components used for these systems are the same as those from an individual desktop PC up to the most powerful supercomputers. Similar software environments are available on all of these systems. This was the basis of a broad acceptance by industrial and commercial customers.

The increasing market share of industrial and commercial installations had several very critical implications for the HPC market. The manufacturers of supercomputers for numerical applications, in the market for small- to medium-size HPC systems, face strong competition from manufacturers selling their systems in the very lucrative commercial market. These systems

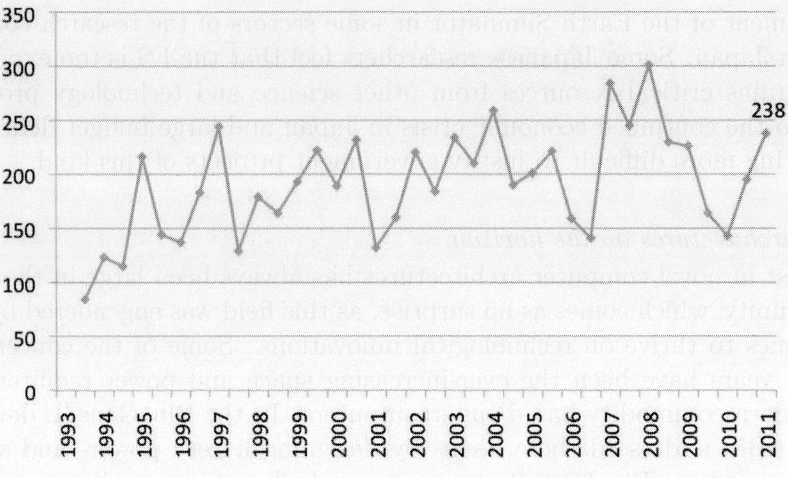

Figure 12.4. The replacement rate in the TOP500 defined as the number of systems omitted because of their performance being too small.

tend to have better price/performance ratios due to the larger production numbers of systems sold to commercial customers and the reduced design costs of medium-size systems. The market for very high-end systems itself is relatively small, and hardly grows, if at all. It cannot easily support specialized niche market manufacturers. This forces the remaining manufacturers to change the design for the very high end away from homogeneous large-scale systems towards cluster concepts based on 'off-the-shelf' components.

'Clusters' are the dominating architecture in the TOP500. In November 1999 we had only seven clusters in the TOP500, whereas in June 2011 the list included 411 cluster systems. At the same time the debate as to whether we need new architectures for very high-end supercomputers has once again increased in intensity.

Novel hybrid architectures have appeared in the TOP500 list. The number one machine in June 2008, the IBM Roadrunner, was just such a system. The Roadrunner is a hybrid design built from commodity parts. The system is composed of two processor chip architectures, the IBM PowerXCell and the AMD Opteron, which use InfiniBand interconnect. The system can be characterized as an Opteron-based cluster with Cell accelerators. Each Opteron core has a Cell chip (composed of nine cores). The Cell chip has eight vector cores and a conventional PowerPC core. The vector cores provide the bulk of the computational performance. The other hybrid design that has found some favour is one based on a linking between a commodity CPU and a graphical processing unit (GPU) accelerator. The model for GPU computing is to use a CPU and GPU together in a heterogeneous co-processing computing model. The sequential part of the application runs on the CPU and the computationally intensive part is accelerated by the GPU.

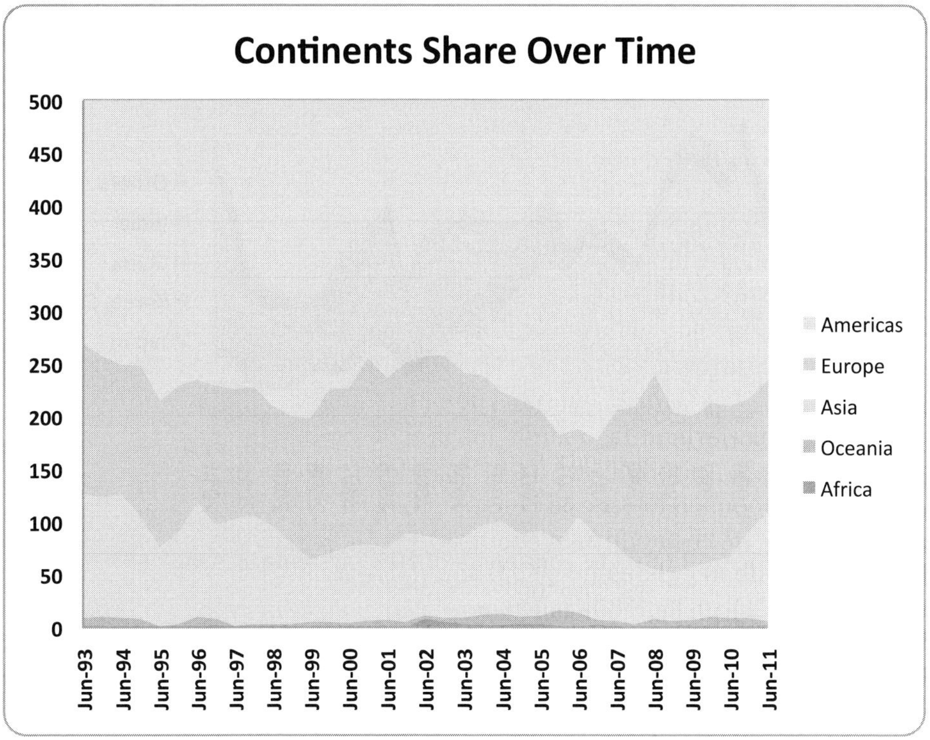

Figure 12.5. The consumers of HPC systems in different geographical regions, as seen in the TOP500.

Dynamic of the market
The HPC market is by its very nature dynamic. This is not only reflected by the coming and going of new manufacturers but especially by the need to update and replace systems quite often to keep pace with the general performance increase. This general dynamic of the HPC market is well reflected in the TOP500. In Figure 12.4 we show the number of systems that fall off the end of the list within six months due to the increase in entry-level performance. We see an average replacement rate of about 180 systems every six months, or more than half the list every year. This means that a system which is at position 100 at a given time will fall off the TOP500 within two to three years. The June 2011 list shows almost one-half replacement, with 238 systems being displaced from the previous list.

Consumer and producer
The dynamic of the HPC market is well reflected in the rapidly changing market shares of the chip or system technologies, of manufacturers, customer types or application areas. However, if we are interested in where these HPC systems are installed or produced, we see a different picture.

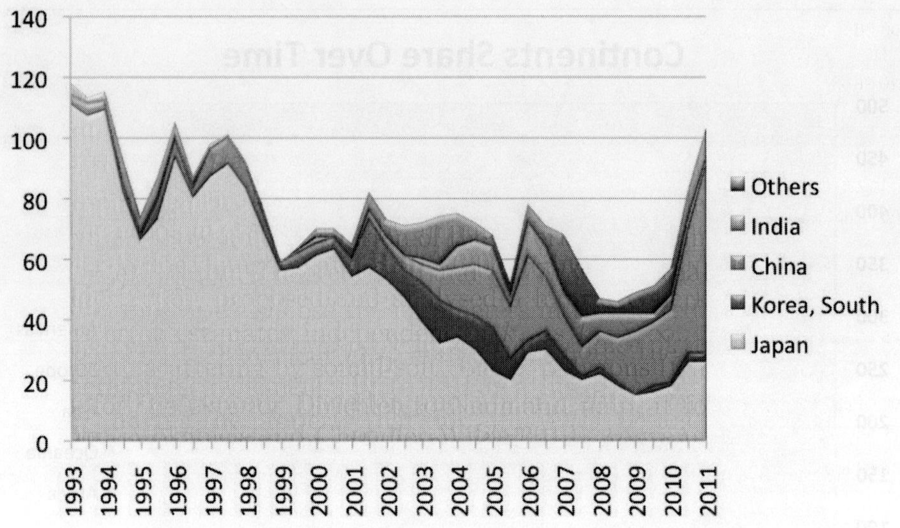

Figure 12.6. The consumers of HPC systems in Asia, as seen in the TOP500.

Plotting the number of systems installed in different geographical areas in Figure 12.5, we see a more-or-less steady distribution. The number of systems installed in the USA is about half of the list, while the number of systems in Asia is slowly increasing. Europe has steadily acquired HPC systems, as shown in Figure 12.5. While this can be interpreted as a reflection of increasing economical strength in these countries, it also highlights the fact that it is becoming easier for such countries to buy or even build cluster-based systems themselves. Figure 12.6 shows the number of HPC systems in Japan, the initial use of such systems in India, and the rapid growth of systems in China.

Performance growth

While many aspects of the HPC market change quite dynamically over time, the evolution of performance seems to follow Moore's Law quite well, as mentioned earlier. The TOP500 provides ideal data to verify an observation like this. Looking at the computing power of the individual machines presented in the TOP500 and the evolution of the total installed performance, we plot the performance of the systems at positions 1 and 500 in the list as well as the total accumulated performance. In Figure 12.7 the curve of position 500 shows on average an increase of a factor of 1.9 within one year. All other curves show a growth rate of 1.8 ± 0.05 per year.

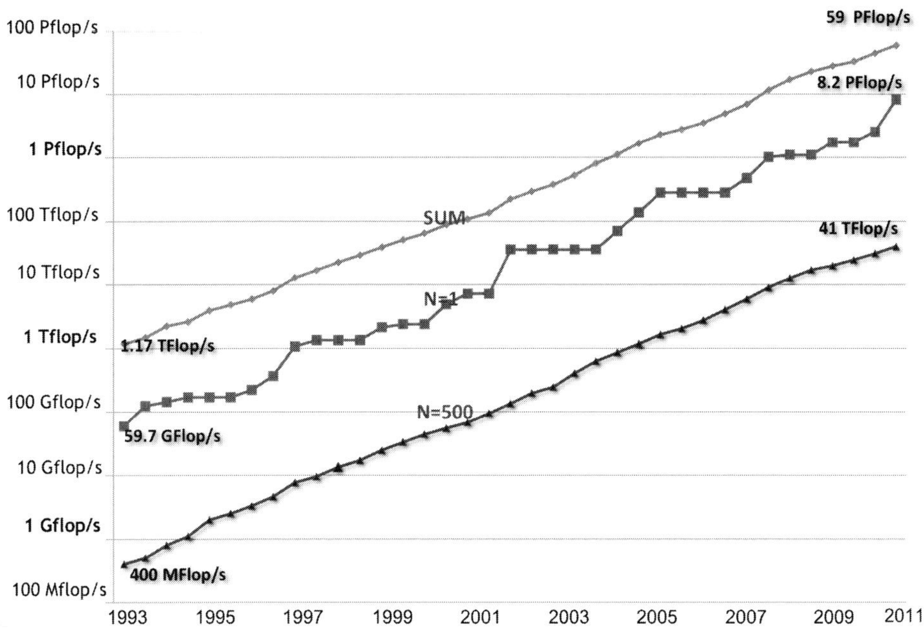

Figure 12.7. Overall growth of accumulated and individual performance, as seen in the TOP500.

Projections

Based on the current TOP500 data, which cover the last fourteen years, and the assumption that the current performance development will continue for some time to come, we can now extrapolate the observed performance and compare these values with the goals of the mentioned government programs. In Figure 12.8, we extrapolate the observed performance values using linear regression on the logarithmic scale. This means that we fit exponential growth to all levels of performance in the TOP500. This simple fitting of the data shows surprisingly consistent results. In 1999, based on a similar extrapolation (Meuer *et al.* 2011), we expected to have the first 100 TFlop/s system by 2005. We also predicted that by 2005 no system smaller than 1 TFlop/s should still be able to make the TOP500. Both of these predictions are basically certain to be fulfilled next year. Looking forward another five years to 2010, we expected to see the first petaflop system at about 2009 (Meuer *et al.* 2011). We hit the petaflop/s mark in 2008 and 10 petaflop/s in 2011.

Looking even further into the future, we could speculate that, based on the current doubling of performance every year, the first system exceeding 100 petaflop/s should be available around 2015, and we should expect an exaflop/s system in 2019, as can be seen in Figure 12.8. Indeed we see an eleven-year cycle of achieving a three-orders-of-magnitude increase in perfor-

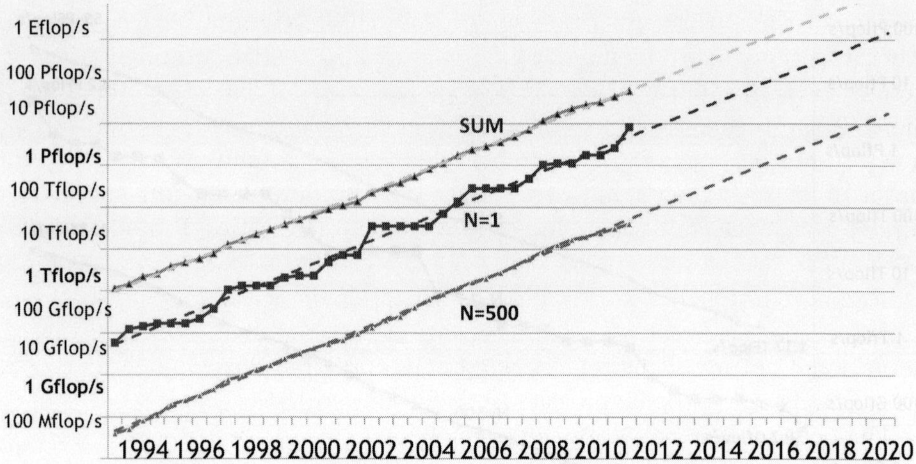

Figure 12.8. Extrapolation of recent growth rates of performance seen in the TOP500.

mance. This has been true since 1986 with the first gigaflop system, in 1997 with the first teraflop system, and in 2008 with the first petaflop system. Due to the rapid changes in the technologies used in HPC systems, there is, however, again no reasonable projection possible for the architecture of such a system in ten years. Even though the HPC market has changed quite substantially since the introduction of the Cray 1 four decades ago, there is no end in sight for these rapid cycles of re-definition: the only constant is change.

13. HPC challenges

Supercomputing capability benefits a broad range of industries, including energy, pharmaceuticals, aircraft, automobiles and entertainment. More powerful computing capability will allow these diverse industries to more quickly engineer superior new products that could improve a nation's competitiveness. In addition, there are considerable flow-down benefits that will result from meeting both the hardware and software high-performance computing challenges. These would include enhancements to smaller computer systems and many types of consumer electronics, from smartphones to cameras.

With respect to software, it seems clear that the scope of the effort to develop software for exascale must be truly international. In terms of its rationale, scientists in nearly every field now depend upon the software infrastructure of high-end computing to open up new areas of enquiry (*e.g.*, the very small, very large, very hazardous, very complex), to dramatically increase their research productivity, and to amplify the social and economic impact of their work. It serves global scientific communities who need to work together on problems of global significance and leverage distributed resources in transnational configurations. In terms of feasibility, the dimensions of the task – totally redesigning and recreating, in the period of just a few years, the massive software foundation of computational science in order to meet the new realities of extreme-scale computing – are simply too large for any one country, or small consortium of countries, to undertake all on its own.

Standardization is also a minimum requirement for broad international collaboration on development of software components. In addition, the international nature of the science will demand further development of global data management tools and standards for shared data.

The development of an exascale computing capability, with machines capable of executing $O(10^{18})$ operations per second in the 2018 time frame, will be characterized by significant and dramatic changes in computing hardware architecture from current (2011) petascale high-performance computers. From the perspective of computational science, this will be at least as disruptive as the transition from vector supercomputing to parallel supercomputing that occurred in the 1990s. Similar to that transition, the achievement of scientific application performance commensurate with the expected improvement in computing capability will require identifying and/or developing mathematical models and numerical algorithms that map efficiently onto exascale architectures, significant re-engineering of scientific application codes supported by the corresponding development of new programming models and system software appropriate for these new architectures. Achieving these increases in capability by 2018 will require a significant acceleration in the development of both hardware and software. This could be accomplished through an intensive 'co-design' effort, where system architects, application software designers, applied mathematicians, and computer scientists work interactively to characterize and produce an environment for computational science discovery that fully leverages these significant advances in computational capability.

The algorithmic challenges

Advancing science in key areas requires development of next-generation physical models to satisfy the accuracy and fidelity needs for targeted simulations. The impact of these simulation fidelity needs on requirements for computational science is twofold. First, more complex physical models must be developed to account for more aspects of the physical phenomena being modelled. Second, for the physical models being used, increases in resolution for key system variables, such as numbers of spatial zones, time steps or chemical species, are needed to improve simulation accuracy, which in turn places higher demands on computational hardware and software.

Application models represent the functional requirements that drive the need for certain numerical algorithms and software implementations. The choice of model is in part motivated by the science objectives, but it is also constrained by the computer hardware characteristics attainable in the relevant time frame. The choice and specification of system attributes (*e.g.*, peak speed or node memory capacity) tend to constrain the functional attributes able to be employed in a given physical model on that system.

Science priorities lead to science models, and models are implemented in the form of algorithms. Algorithm selection is based on various criteria, such as appropriateness, accuracy, verification, convergence, performance, parallelism and scalability.

Models and associated algorithms are not selected in isolation but must be evaluated in the context of the existing computer hardware environment. Algorithms that perform well on one type of computer hardware may become obsolete on newer hardware, so selections must be made carefully and may change over time.

Moving forward to exascale will put heavier demands on algorithms in at least two areas: the need for increasing amounts of data locality in order to perform computations efficiently, and the need to obtain much higher factors of fine-grained parallelism as high-end systems support increasing numbers of compute threads. As a consequence, parallel algorithms must adapt to this environment, and new algorithms and implementations must be developed to extract the computational capabilities of the new hardware.

As with science models, the performance of algorithms can change in two ways as application codes undergo development and new computer hardware is used. First, algorithms themselves can change, motivated by new models or performance optimizations. Second, algorithms can be executed under different specifications, *e.g.*, larger problem sizes or changing accuracy criteria. Both of these factors must be taken into account.

Significant new model development, algorithm redesign, and science application code reimplementation, supported by (an) exascale-appropriate programming model(s), will be required to effectively support the power of exascale architectures. The transition from current sub-petascale and petascale computing to exascale computing will be at least as disruptive as the transition from vector to parallel computing in the 1990s.

Uncertainty quantification will permeate the exascale science workload. The demand for predictive science results will drive the development of improved approaches for establishing levels of confidence in computational predictions. Both statistical techniques involving large ensemble calculations and other statistical analysis tools will have significantly different dynamic resource allocation requirements than in the past, and the significant code redesign required for the exascale will present an opportunity to embed uncertainty quantification techniques in exascale science applications.

New multicore-friendly and multicore-aware algorithms
Scalable multicore systems bring new computation/communication ratios. Within a node, data transfers between cores are relatively inexpensive, but temporal affinity is still important for effective cache use. Across nodes, the relative cost of data transfer is growing very large. The development of new algorithms that take these issues into account can often perform very well, as do communication-avoiding algorithms that increase the computation/communication ratio or algorithms that support simultaneous computation/communication, or algorithms that vectorize well and have a large volume of functional parallelism.

Adaptive response to load imbalance
Adaptive multiscale algorithms are an important part of the US Department of Energy portfolio since they apply computational power precisely where it is needed. However, they introduce challenging computational requirements because they introduce dynamically changing computation that results in load imbalances from a static distribution of tasks. As we move towards systems with billions of processors, even naturally load-balanced algorithms on homogeneous hardware will present many of the same daunting problems with adaptive load balancing that are observed in today's adaptive codes. For example, software-based recovery mechanisms for fault-tolerance or energy-management features will create substantial load imbalances as tasks are delayed, by rollback, to a previous state or correction of detected errors. Scheduling based on a directed acyclic graph also requires new approaches to optimizing for resource utilization without compromising spatial locality. These challenges require development and deployment of sophisticated software approaches to rebalance computation dynamically in response to changing workloads and conditions of the operating environment.

Multiple precision algorithms/software
Algorithms and applications are becoming increasingly adaptive and we have seen that various adaptivity requirements have become an essential, key component of their roadmap to exascale computing. Another aspect of this quest for adaptivity is related to the development of libraries that recognize and exploit the presence of mixed-precision mathematics. A motivation comes from the fact that, on modern architectures, the performance of 32-bit operations is often at least twice as fast as the performance of 64-bit operations. Moreover, by using a combination of 32-bit and 64-bit floating-point arithmetic, the performance of many linear algebra algorithms can be significantly enhanced while maintaining the 64-bit accuracy of the resulting solution. This can be applied not only to conventional processors but also to other technologies such as GPUs, and thus can spur the creation

of mixed-precision algorithms that more effectively utilize heterogeneous hardware.

Mixed-precision algorithms can easily provide substantial speed-up for very little code effort by mainly taking into account existing hardware properties. Earlier work has shown how to derive mixed-precision versions for various architectures and for a variety of algorithms for solving general sparse or dense linear systems of equations. Typically, a direct method is first applied in single precision in order to achieve a significant speed-up compared to double precision. Then an iterative refinement procedure aims at retrieving the lost digits. Iterative refinement can also be applied for eigenvalue and singular-value computations.

Current topics of interest include the extension and incorporation of this approach in applications that do not necessarily originate from linear algebra, and studying the robustness of mixed-precision algorithms on large-scale platforms. Indeed, the convergence of the mixed-precision iterative refinement solvers strongly depends on the condition number of the matrix at hand. The conditioning can be determined at runtime and proper precision can be selected. Ideally, the user could specify the required precision for the result and the algorithm would choose the best combination of precision on the local hardware in order to achieve it. The actual mechanics would be hidden from the user.

Fast implicit solvers

Carefully analysing complex problems, and adapting preconditioners to the underlying problem physics, is how most of the progress in this area is being made. However, it is typically the case that advanced preconditioners are composed of standard algebraic components such as advanced multigrid/multilevel methods, incomplete factorizations and basic smoothers. Furthermore, we need to renew our focus on basic iterative methods in an attempt to address bottlenecks due to collective operations (*e.g.*, dot-products) and poor kernel performance. Emphasis on block methods, recycling methods, s-step-like methods and mixed-precision formulations will be necessary to address the next generation of problems.

Communication avoiding and asynchronous algorithms

Algorithmic complexity is usually expressed in terms of the number of operations performed rather than the quantity of data movement to memory. This is antithetical to the true costs of computation, where memory movement is very expensive and operations are nearly free. To address the critical issue of communication costs, there is a need to investigate algorithms that reduce communication to a minimum. One needs to derive bandwidth and latency lower bounds for various dense and sparse linear algebra algorithms on parallel and sequential machines, *e.g.*, by extending the well-known lower bounds for the usual $O(n^3)$ matrix multiplication algorithm. Then one needs to discover new algorithms that attain these lower bounds in many cases. Second, for Krylov subspace methods such as GMRES, CG and Lanczos, one should focus on taking k steps of these methods for the same communication costs as a single step.

In a seminal paper, Chazan and Miranker (1969) studied chaotic relaxation, now usually called asynchronous relaxation, for the solution of linear systems. In chaotic relaxation, the order in which components of the solution are updated is arbitrary and the past values of components that are used in the updates are also selected arbitrarily. This is a model for parallel computation in which different processors work independently and have access to data values in local memory.

When this and subsequent research was undertaken in the late 1960s and 1970s, it was largely theoretical: the existing computers did not have the capability for massively parallel processing. Today we are at the extreme, with the next generation of machines having $O(10^9)$ program threads. We are being challenged to devise algorithms and software that can effectively exploit the parallel hardware systems that are being developed. When solving very large problems on parallel architectures the most significant concern becomes the cost per iteration of the method – typically on account of communication and synchronization overheads. This is especially the case for Krylov methods, which are the most popular class of iterative methods for large sparse systems. This means that, for the first time, totally asynchronous iterative algorithms will become competitive for a wide range of application problems. Coping with fault tolerance, load balancing, and communication overheads in a heterogeneous computation environment is a challenging undertaking for software development. In traditional synchronous algorithms each iteration can only be performed as quickly as the slowest processor permits. If a processor fails, or is less capable, or has an unduly heavy load, then this markedly impacts iteration times. The use of asynchronous methods allows one to overcome many of the communication, load-balancing and fault tolerance issues we now face and which limit our ability to scale to the extreme.

Auto-tuning

Libraries need to have the ability to adapt to the possibly heterogeneous environment in which they have to operate. The adaptation has to deal with the complexity of discovering and implementing the best algorithm for diverse and rapidly evolving architectures. This calls out for automating the process, both for the sake of productivity and for correctness. Here, productivity refers both to the development time and the user's time to solution. The objective is to provide a consistent library interface that remains the same for users independent of scale and processor heterogeneity, but which

Sensitivity analysis

Many areas of modelling and simulation are still pushing to reach high-fidelity solutions to a given set of input conditions. However, as performance and fidelity improves, it becomes possible and imperative to study the sensitivity of a model to parameter variability and uncertainty, and to seek an optimal solution over a range of parameter values. The most basic form, the forward method for either local or global sensitivity analysis, simultaneously runs many instances of the model or its linearization, leading to an embarrassingly parallel execution model. The adjoint sensitivity method, with its powerful capabilities for efficiently computing the sensitivity of an output functional with respect to perturbations in a great many parameters, is a workhorse algorithm in weather prediction and in engineering design such as shape optimization. It requires the simulation of the forward and the adjoint problem; hence its parallelization will depend on the capability for highly efficient simulation.

Multiscale/multiphysics modelling

Engineering is increasingly operating at the micro- and nanoscales to achieve objectives at the macroscale. Models of these processes are intrinsically multiscale and multiphysics. For example, electrochemically reactive surfaces play a central role in the fabrication as well as the functional capabilities of an enormous variety of technological systems. Precise control of surface processes during fabrication is required in applications including on-chip interconnections between transistors, decorative and industrial coatings, batteries for electric vehicles, thin-film photovoltaic solar devices, magnetic materials, and patterned deposits for sensors. Surface processes are occurring at the nanoscale and must be modelled by kinetic Monte Carlo methods, whereas reactions and diffusion in the electrolyte can be modelled by deterministic methods (*i.e.*, PDEs). The two computations must be dynamically linked. Such a computation is very demanding and is currently consuming huge numbers of cycles on NCSA's supercomputers, with only modest resolution of the problem domain. Simulation is only the tip of the iceberg of this type of problem, where parameter estimation and optimal design are the ultimate goals and require orders of magnitude more computation time.

Cell biology is another area where processes operating at the microscale yield change at the macroscale (phenotypical change). In microscopic systems formed by living cells, the small numbers of some reactant molecules can result in dynamical behaviour that is discrete and stochastic rather than continuous and deterministic. An analysis tool that respects these dynamical characteristics is the stochastic simulation algorithm (SSA), which applies to well-stirred (spatially homogeneous) chemically reacting systems. Usually, a large ensemble of SSA simulations is used to estimate the probability density functions of important variables in the system. This leads to an embarrassingly parallel implementation. At the same time, cells are not spatially homogeneous. Spatio-temporal gradients and patterns play an important role in many cellular processes. The modelling of stochastic diffusive transfers between subvolumes is an important challenge for parallelization.

Summary
The move to extreme-scale computing will require tools for understanding complex behaviour and for performance optimization to be based on a knowledge-oriented process. Performance models and expectations will be used to drive knowledge-based investigation and reasoning. It will raise the level at which tools inter-operate and can be integrated with the application development and execution environment. The challenges for performance analysis and tuning will grow, as performance interactions and factor analysis must involve a whole system perspective.

The co-design methodology is iterative, requiring frequent interactions among hardware architects, systems software experts, designers of programming models, and implementers of the science applications that provide the rationale for building extreme-scale systems. As new ideas and approaches are identified and pursued, some will fail. As with past experience, there may be breakthroughs in hardware technologies that result in different micro- and macro-architectures becoming feasible and desirable, but they will require rethinking of certain algorithmic and system software implementations.

13.1. Technology trends and their impact on exascale

The design of the extreme-scale platforms that are expected to become available in 2018 will represent a convergence of technological trends and the boundary conditions imposed by over half a century of algorithm and application software development. Although the precise details of these new designs are not yet known, it is clear that they will embody radical changes along a number of different dimensions as compared to the architectures of today's systems, and that these changes will render obsolete the current software infrastructure for large-scale scientific applications. The first step in developing a plan to ensure that appropriate system software and applications are ready and available when these systems come on line, so that leading-edge research projects can actually use them, is to carefully review the underlying technological trends that are expected to have such a transformative impact on computer architecture in the next decade. These factors and trends, which we summarize in this section, provide essential context for thinking about the looming challenges of tomorrow's scientific software infrastructure; by describing them, therefore, we lay the foundations for our narrative structure.

Technology trends
In developing a roadmap for the X-stack software infrastructure, the IESP has been able to draw on several thoughtful and extensive studies of impacts of the current revolution in computer architecture (Kogge *et al.* 2008, Sarkar *et al.* 2009). As these studies make clear, technology trends over the next decade – broadly speaking, increases of 1000× in capability over today's most massive computing systems, in *multiple* dimensions, as well as increases of similar scale in data volumes – will force a disruptive change in the form, function, and inter-operability of future software infrastructure components and the system architectures incorporating them. The momentous nature of these changes can be illustrated for several critical system-level parameters.

Concurrency. Moore's Law scaling in the number of transistors is expected to continue to the end of the next decade, at which point the minimal VLSI geometries will be as small as five nanometres. Unfortunately, the end of Dennard scaling means that clock rates are no longer keeping pace, and may in fact be reduced in the next few years to reduce power consumption. As a result, the exascale systems on which the X-stack will run will likely be composed of hundreds of millions of arithmetic logic units (ALUs). Assuming there are multiple threads per ALU to cover main-memory and networking latencies, applications may contain ten billion threads.

Reliability. System architecture will be complicated by the increasingly probabilistic nature of transistor behaviour due to reduced operating voltages, gate oxides, and channel widths/lengths resulting in very small noise margins. Given that state-of-the-art chips contain billions of transistors and the multiplicative nature of reliability laws, building resilient computing systems out of such unreliable components will become an increasing challenge. This cannot be cost-effectively addressed with pairing or TMR; rather, it must be addressed by X-stack software and perhaps even scientific applications.

Power consumption. Twenty years ago, HPC systems consumed less than a megawatt. The Earth Simulator was the first such system to exceed 10 MW. Exascale systems could consume over 100 MW, and few of today's computing centres have either adequate infrastructure to deliver such power or the budgets to pay for it. The HPC community may find itself measuring results in terms of power consumed, rather than operations performed. The X-stack and the applications it hosts must be conscious of this situation and act to minimize it.

Similarly dramatic examples could be produced for other key variables, such as *storage capacity*, *efficiency*, and *programmability*.

More importantly, a close examination shows that changes in these parameters are interrelated and not orthogonal. For example, scalability will be limited by efficiency, as are power and programmability. Other cross-correlations can be perceived through analysis. The US Defense Advanced Research Projects Agency (DARPA) Exascale Technology Study (Kogge *et al.* 2008) exposes power as the pace-setting parameter. Although an exact power consumption constraint value is not yet well defined, with upper limits of today's systems of the order of 5 megawatts, increases of an order of magnitude in less than 10 years will extend beyond the practical energy demands of all but a few strategic computing environments. A politico-economic pain threshold of 25 megawatts has been suggested (by DARPA) as a working boundary. With dramatic changes to core architecture design, system integration, and programming control over data movement, best estimates for CMOS-based systems at the 11-nanometre feature size is a factor of three to five times this amount. One consequence is that clock rates are unlikely to increase substantially. Among the controversial questions is how much instruction-level parallelism (ILP) and speculative operation is likely to be incorporated on a per processor core basis and the role of multi-threading in subsuming more of the fine-grained control space. Data movement across the system, through the memory hierarchy, and even for register-to-register operations will likely be the single principal contributor to power consumption, with control adding to this appreciably. Since future systems can ill afford the energy wasted by data movement that does not advance the target computation, alternative ways of hiding latency will be required in order to guarantee, as much as possible, the utility of every data transfer. Even taking into account the wastefulness of today's conventional server-level systems and the energy gains that careful engineering has delivered for systems such as BlueGene/P, an improvement of two orders of magnitude, will still be required.

As a result of these and other observations, exascale system architecture characteristics are beginning to emerge, though the details will become clear only as the systems themselves actually develop. Among the critical aspects of future systems, available by the end of the next decade, which we can predict with some confidence are the following:

- feature size of 22 to 11 nanometres, CMOS in 2018,
- total average of 25 picojoules per floating-point operation,
- approximately 10 billion-way concurrency for simultaneous operation and latency hiding,
- 100 million to 1 billion cores,
- clock rates of 1 to 2 GHz,
- multi-threaded, fine-grained concurrency of 10- to 100-way concurrency per core,

- hundreds of cores per die (varies dramatically depending on core type and other factors),
- global address space without cache coherence, and extensions to PGAS (*e.g.*, AGAS),
- 128-petabyte capacity mix of DRAM and non-volatile memory (most expensive subsystem),
- explicitly managed high-speed buffer caches; part of deep memory hierarchy,
- optical communications for distances >10 centimetres, possibly inter-socket,
- optical bandwidth of 1 terabit per second,
- systemwide latencies of the order of tens of thousands of cycles,
- active power management to eliminate wasted energy by currently unused cores,
- fault tolerance by means of graceful degradation and dynamically reconfigurable structures,
- hardware-supported rapid thread context switching,
- hardware-supported efficient message-to-thread conversion for message-driven computation,
- hardware-supported, lightweight synchronization mechanisms,
- 3D packaging of dies for stacks of four to ten dies, each including DRAM, cores, and networking.

Because of the nature of the development of the underlying technology, most of the predictions above have an error margin of +/–50 % or a factor of two, independent of specific roadblocks that may prevent us reaching the predicted value.

The list quoted above demonstrates the large variety of items that have to change significantly in order to reach the goal of a 1000× acceleration of HPC systems in the 2018–2020 timeframe. It is clear from this list that we cannot expect the current hardware technology to satisfy all these requirements. Two components that were traditionally considered less crucial than the processors are the memory system and the interconnect network. The perspective has changed drastically in this respect: data movement at low energy level has become a prime target and both the memory and the network are involved in this. We look in more detail at two of the most important directions that may help exaflop/s systems to come about: non-volatile memory and optical network components.

Non-volatile memory

The use of non-volatile memory, *i.e.*, memory that retains its contents when no current is applied, is important for two reasons. The first, obvious one

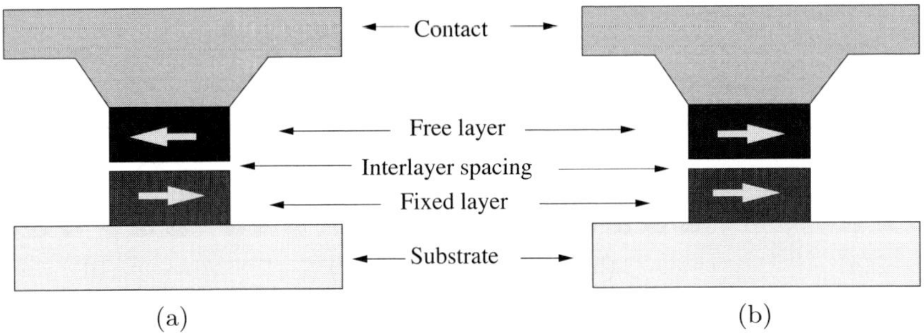

Figure 13.1. An MRAM memory cell:
(a) represents 0, (b) represents 1.

is that it is much more energy-efficient than the currently used DRAM technology, where the contents must be refreshed continuously. A second reason is connected with the present way DRAM is implemented: the feature size is currently in the range of 40 nm and it can still be shrunk somewhat by using 3D techniques instead of the planar technology that is used today (Intel and Samsung will ship 3D memory chips shortly). However, already the leak current occurring within the chips is quite significant, and it will only increase when the memory cells are more densely packed. This increases both energy costs and unreliability. Therefore new memory technologies are urgently needed.

Various interesting alternatives are actively being researched. Among them are magnetic RAM (MRAM), ferro-magnetic RAM (FRAM) and memristors (a special form of FRAM). From these three technologies spin torque transfer MRAM and FRAM are already in production, though still with a density that is not suitable for use in (HPC) memory. Both are, however, used in embedded processors and sensors where low energy consumption is of prime importance. FRAM and, consequently, memristors are based on the magnetic hysteresis effect, while MRAM is based on the giant magneto-resistive effect, as is also employed in present-day spinning disks. The MRAM implementation is, however, static. A memory cell is depicted in Figure 13.1.

When the magnetic field orientations in the fixed and free layers are opposite (Figure 13.1(a)), the total magnetic moment is much lower than when they are aligned (Figure 13.1(b)). This difference can be sensed and interpreted as a 0, resp. 1 value. The magnetic field of the free layer can be changed by a spin-polarized current, thus writing a 0 or a 1 value.

Hewlett-Packard, the first company that was able to demonstrate memristor memory, has teamed up with Hynix, a memory production company, to make commercial memristor products which are scheduled for late 2012 or early 2013.

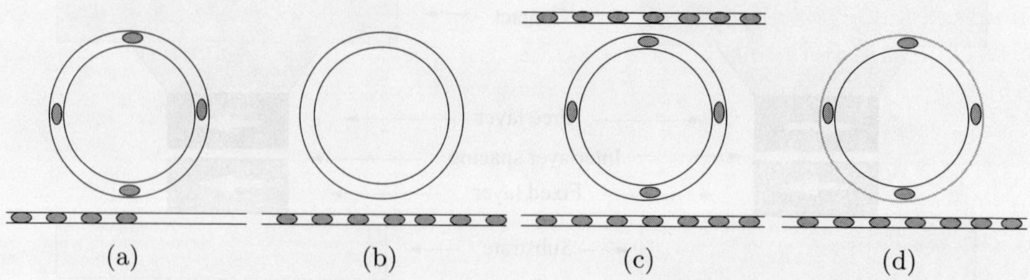

Figure 13.2. Three functions of a ring resonator. (a) It resonates in phase with the light waves and diverts the light signal. (b) It does not resonate, and lets the light signal pass. (c) Through resonation it transfers the signal to another wave guide, thus acting as a switch. (d) A doped resonator picks up a signal of a pre-defined wavelength and so acts as a signal detector.

Optical networks

Just as for memory, interconnect networks already consume a significant amount of energy, irrespective of whether they are used, because the network switches must be ready to pass on incoming messages at any time. In addition, the signalling speed of the wires that are employed start to pose problems with respect to signal integrity when the clock frequency of the communication engines increase. The increase of this frequency therefore has a negative impact both on the energy consumption (as in any electronic device) and on the reliability of the message transfer. This is why there is active research to implement the networks with optical components with as few electronic-optical transitions as possible. A big step forward in this respect is the development of the so-called ring resonator. A ring resonator is a minute glass ring with a size of $\approx 5\ \mu$m, and it is possible to make it resonate in phase with the light waves that pass along it in an optical wave guide. Figure 13.2 shows the three functions that the resonator can fulfil. According to its position and properties it can act as a modulator blocking or passing signals as desired (Figure 13.2(a, b)). When it is coupled with another wave guide and made to resonate with the right frequency, it will pass the signal on to this wave guide (Figure 13.2(c)), thus implementing a switching function. Lastly, it can act as a signal detector when the ring is doped to tune it to a signal of a desired wavelength (Figure 13.2(d)). The advantages of this optical switching are many. The bandwidth of the interconnection can be quite high because multiple wavelengths can be sent through the same wave guide. Furthermore, optical signals do not interfere with each other as electrical signals do when they are near each other. In addition, the power consumption is much lower than that of its electronic equivalent. Unfortunately there are also drawbacks. The rings are very sensitive to temperature changes, so for proper operation they must either be in an extremely well-controlled environment with respect to temperature, or provisions must be made on the rings themselves to keep their temperature constant. This is technologically possible but greatly complicates the design.

At present, optical switches are still in the laboratory phase or moving to a preproduction stage, so it will take another few years for them to emerge in commercial HPC systems. Yet there is little doubt that this path will be taken, as there are virtually no alternatives.

Science trends

A basic driver of the IESP is the fact that the complexity of advanced challenges in science and engineering continues to outpace our ability to adequately address them through available computational power. Many phenomena can be studied only through computational approaches: well-known examples include simulating complex processes in climate and astrophysics. Increasingly, experiments and observational systems are finding that not only are the data they generate exceeding petabytes and rapidly heading toward exabytes, but the computational power needed to process the data is also expected to be in exaflop/s range.

A number of reports and workshops have identified key science challenges and applications of societal interest that require computing at exaflops levels and beyond (Stevens, Zacharia and Simon 2008, US Department of Energy 2008, 2009*a*–*i*, 2010, US National Research Council Committee 2008). Here we summarize some of the significant findings on the scientific necessity of exascale computing; we focus primarily on the need for the software environments needed to support the science activities. The US Department of Energy held eight workshops in the past year that identified science advances and important applications that will be enabled through the use of exascale computing resources. The workshops covered the following topics: climate, high-energy physics, nuclear physics, fusion energy sciences, nuclear energy, biology, materials science and chemistry, and national nuclear security. The US National Academy of Sciences published the results of a study in the report 'The potential impact of high-end capability computing on four illustrative fields of science and engineering' (US National Research Council Committee 2008). The four fields were astrophysics, atmospheric sciences, evolutionary biology, and chemical separations.

Likewise, the US National Science Foundation has embarked on a petascale computing programme that has funded dozens of application teams via its Peta-Apps and PRAC programmes, across all areas of science and engineering, to develop petascale applications, and is deploying petaflops systems. It has commissioned a series of task forces to help plan for the transition from petaflops to exaflops computing facilities, to support the software development necessary, and to understand the specific science and engineering needs beyond petascale.

Similar activities are seen in Europe and Asia, all reaching similar conclusions: significant scientific and engineering challenges in both simulation and data analysis already exceed petaflops and are rapidly approaching exaflop-class computing needs. In Europe, the Partnership for Advanced Computing in Europe (PRACE) involves twenty partner countries, supports access to world-class computers, and has activities aimed at supporting multi-petaflops and eventually exaflop-scale systems for science. The European Union is also planning to launch projects aimed at petascale and exascale computing and simulation. Japan has a project to build a 10-petaflop/s system and has historically supported the development of software for key applications such as climate. As a result, scientific and computing communities, and the agencies that support them in many countries, have been meeting to plan joint activities that will be needed to support these emerging science trends.

To give a specific and timely example, a recent report, 'Science prospects and benefits with exascale computing' (Kothe 2007, p. 9) states that the characterization of abrupt climate change will require sustained exascale computing in addition to new paradigms for climate change modelling. The types of questions that could be tackled with exascale computing (and cannot be tackled adequately without it) include the following:

- How do the carbon, methane, and nitrogen cycles interact with climate change?
- How will local and regional water, ice, and clouds change with global warming?
- How will the distribution of weather events, particularly extreme events, determine regional climate change with global warming?
- What are the future sea-level and ocean circulation changes?

Among the findings of the astrophysics workshop and other studies are that exascale computing will enable cosmology and astrophysics simulations aimed at the following:

- measuring the masses and interactions of dark matter,
- understanding and calibrating supernovae as probes of dark energy,
- determining the equation of state of dark energy,
- understanding the nature of gamma-ray bursts.

Energy security

The search for a path forward in ensuring sufficient energy supplies in the face of a climate-constrained world faces a number of technical challenges, ranging from issues related to novel energy technologies, to issues related to making existing energy technologies more (economically) effective and safer, to issues related to the verification of international agreements regarding the emission (and possible sequestration) of CO_2 and other greenhouse gases. Among the science challenges are the following:

- verification of 'carbon treaty' compliance,
- improvement in the safety, security, and economics of nuclear fission,
- improvement in the efficiency of carbon-based electricity production and transportation,
- improvement in the reliability and security in the (electric) grid,
- nuclear fusion as a practical energy source.

Computational research will also play an essential role in the development of new approaches to meeting future energy requirements (*e.g.*, wind, solar, biomass, hydrogen, and geothermal), which in many cases will require exascale power.

Industrial applications, such as simulation-enhanced design and production of complex manufactured systems and rapid virtual prototyping, will also be enabled by exascale computing. To characterize materials deformation and failure in extreme conditions will require atomistic simulations on engineering time scales that are out of reach with petascale systems.

A common theme in all of these studies of the important science and engineering applications that are enabled by exaflops computing power is that they have complex structures and present programming challenges beyond just scaling to many millions of processors. For example, many of these applications involve multiple physical phenomena spanning many decades of spatial and temporal scale. As the ratio of computing power to memory grows, the 'weak scaling' that has been exploited for most of the last decade will increasingly give way to 'strong scaling', which will make scientific applications increasingly sensitive to overhead and noise generated by the X-stack. These applications are increasingly constructed from components developed by computational scientists worldwide, and the X-stack must support the integration and performance portability of such software.

Key requirements imposed by trends on the X-stack

The cited trends in technology and applications will impose severe constraints on the design of the X-stack. Below are cross-cutting issues that will affect all aspects of system software and applications at exascale.

Concurrency. A 1000× increase in concurrency for a single job will be necessary to achieve exascale throughput. New programming models will be needed to enable application groups to address concurrency in a more natural way. This capability will likely have to include 'strong scaling' because growth in the volume of main memory will not match that of the processors. This in turn will require minimizing any X-stack overheads that might otherwise become a critical Amdahl fraction.

Energy. Since much of the power in an exascale system will be expended moving data, both locally between processors and memory as well as globally, the X-stack must provide mechanisms and APIs for expressing and managing data locality. These will also help minimize the latency of data accesses. APIs also should be developed to allow applications to suggest other energy saving techniques, such as turning cores on and off dynamically, even though these techniques could result in other problems, such as more faults/errors.

Resiliency. The VLSI devices from which exascale systems will be constructed will not be as reliable as those used today. All software, and therefore all applications, will have to address resiliency in a thorough way if they are to be expected to run at scale. Hence, the X-stack will have to recognize and adapt to errors continuously, as well as provide the support necessary for applications to do the same.

Heterogeneity. Heterogeneous systems offer the opportunity to exploit the extremely high performance of niche market devices such as GPUs and game chips (*e.g.*, STI Cell) while still providing a general-purpose platform. An example of such a system today is Tokyo Tech's Tsubame, which incorporates AMD Opteron CPUs along with ClearSpeed and NVIDIA accelerators. Simultaneously, large-scale scientific applications are also becoming more heterogeneous, addressing multiscale problems spanning multiple disciplines.

I/O and memory. Insufficient I/O capability is a bottleneck today. Ongoing developments in instrument construction and simulation design make it clear that data rates can be expected to increase by several orders of magnitude over the next decade. The memory hierarchy will change based on both new packaging capabilities and new technology. Local RAM and NVRAM will be available either on or very close to the nodes. The change in memory hierarchy will affect programming models and optimization.

Relevant politico-economic trends

The HPC market is growing at approximately 11 % per year. The largest-scale systems, those that will support the first exascale computations at the end of the next decade, will be deployed by government computing laboratories to support the quest for scientific discovery. These capability computations often consume an entire HPC system and pose difficult challenges for concurrent programming, debugging and performance optimization. Thus, publicly funded computational scientists will be the first users of the X-stack and have a tremendous stake in seeing that suitable software exists, which is the *raison d'être* for IESP.

In the late 1980s, the commercial engineering market place, spanning diverse fields such as computer-aided engineering and oil reservoir modelling, used the same computing platforms and often the same software as the scientific community. This is far less the case today. The commercial workload tends to be more capacity-oriented, involving large ensembles of smaller computations. The extreme levels of concurrency necessary for exascale computing suggests that this trend may not change, so the demand for those features of the X-stack is not unique to exascale computing for scientific computing. On the other hand, the HPC vendor community is eager to work with, and leverage the research and development effort of, the IESP software community. To that end, plans for cooperation and coordination between the IESP software and the HPC vendor community are being developed.

REFERENCES[7]

C. Amza, A. L. Cox, S. Dwarkadas, P. Keleher, H. Lu, R. Rajamony, W. Yu and W. Zwaenepoel (1995), 'TreadMarks: Shared memory computing on networks of workstations', *IEEE Computer* **29**, 18–28. www.cs.rice.edu/~willy/TreadMarks/papers.htm

G. Bell (1999), The next ten years of supercomputing. In *Proc. 14th Supercomputer Conference* (H. W. Meuer, ed.), MaxionMedia (CD-ROM).

N. J. Boden, D. Cohen, R. E. Felderman, A. E. Kulawik, C. L. Seitz, J. N. Seizovic and Wen-King Su (1995), 'Myrinet: A gigabit-per-second local area network', *IEEE Micro* **15**, 29–36.

R. Chandra, L. Dagum, D. Kohr, D. Maydan, J. McDonald and R. Menon (2001), *Parallel Programming in OpenMP*, Morgan Kaufmann.

B. Chapman, G. Jost and R. van der Pas (2007), *Using OpenMP*, MIT Press.

D. Chazan and W. Miranker (1969), 'Chaotic relaxation', *Linear Algebra Appl.* **2**, 199–222.

D. E. Culler, J. P. Singh and A. Gupta (1998), *Parallel Computer Architecture: A Hardware/Software Approach*, Morgan Kaufmann.

D. W. Doerfler (2005), An analysis of the PathScale Inc. InfiniBand host channel adapter, InfiniPath. Sandia report SAND2005-5199, Sandia National Laboratories.

J. Dongarra (2011), Performance of various computers using linear equations software. www.netlib.org/benchmark/performance.ps

J. Dongarra, P. Beckman, *et al.* (2011), 'The International Exascale Software roadmap', *Internat. J. High Performance Computer Applications*, **25**, 3–60.

M. J. Flynn (1972), 'Some computer organizations and their effectiveness', *IEEE Trans. Comput.* **C-21**, 948–960.

[7] The URLs cited in this work were correct at the time of going to press, but the publisher and the authors make no undertaking that the citations remain live or are accurate or appropriate.

A. Geist, A. Beguelin, J. Dongarra, R. Manchek, W. Jaing and V. Sunderam (1994), *PVM: A Users' Guide and Tutorial for Networked Parallel Computing*, MIT Press.

W. Gropp, S. Huss-Ledermann, A. Lumsdaine, E. Lusk, B. Nitzberg, W. Saphir and M. Snir (1998), *MPI: The Complete Reference*, Vol. 2, *The MPI Extensions*, MIT Press.

D. B. Gustavson and Q. Li (1995), Local-area multiprocessor: The Scalable Coherent Interface. SCIzzL Report, Department of Computer Engineering, Santa Clara University, available via www.scizzl.com.

High Performance Fortran Forum (1993), 'High Performance Fortran language specification', *Scientific Programming*, **2**, 1–170.

R. W. Hockney and C. R. Jesshope (1988), *Parallel Computers II*, Adam Hilger.

T. Horie, H. Ishihata, T. Shimizu, S. Kato, S. Inano and M. Ikesaka (1991), AP1000 architecture and performance of LU decomposition. In *Proc. Internat. Symposium on Supercomputing*, pp. 46–55.

D. V. James, A. T. Laundrie, S. Gjessing and G. S. Sohi (1990), 'Scalable coherent interface', *IEEE Computer* **23**, 74–77. See also 'Scalable coherent interface': sunrise.scu.edu/

P. M. Kogge *et al.* (2008), ExaScale computing study: Technology challenges in achieving exascale systems. US Defense Advanced Research Projects Agency report, DARPA Information Processing Techniques Office. users.ece.gatech.edu/~mrichard/ExascaleComputingStudyReports/↩ exascale_final_report_100208.pdf

D. B. Kothe (2007), Science prospects and benefits with exascale computing. Report ORNL/TM-2007/232, National Center for Computational Sciences, Oak Ridge National Laboratory. www.nccs.gov/wp-content/media/nccs_reports/Science%20Case%20_↩ 012808%20v_final.pdf

J. Langou, J. Langou, P. Luszczek, J. Kurzuk, A. Buttari and J. J. Dongarra (2006), Exploiting the performance of 32-bit floating point arithmetic in obtaining 64-bit accuracy. In *Proc. 2006 ACM/IEEE Conference on Supercomputing*.

H. W. Meuer (1994), The Mannheim supercomputer statistics 1986–1992. TOP500 report 1993, University of Mannheim.

H. W. Meuer, E. Strohmaier, J. J. Dongarra and H. D. Simon (2011), TOP500. Available at www.top500.org.

V. Sarkar *et al.* (2009), ExaScale software study: Software challenges in extreme scale systems. US Defense Advanced Research Projects Agency report, DARPA Information Processing Techniques Office. users.ece.gatech.edu/~mrichard/ExascaleComputingStudyReports/ECSS%20↩ report%20101909.pdf

C. Schow, F. Doany and J. Kash (2010), 'Get on the optical bus'. *IEEE Spectrum*, September 2010, 31–35.

T. Shanley (2002), *InfiniBand Network Architecture*, Addison-Wesley.

H. D. Simon (1994), High performance computing in the US. TOP500 report 1993, University of Mannheim, 116–147.

M. Snir, S. Otto, S. Huss-Lederman, D. Walker and J. Dongarra (1998), *MPI: The Complete Reference*, Vol. 1, *The MPI Core*, MIT Press.

D. H. M. Spector (2000), *Building Unix Clusters*, O'Reilly.

A. J. van der Steen (1990), Exploring VLIW: Benchmark tests on a multiflow TRACE 14/300. Technical Report TR-31, Academic Computing Centre Utrecht.

A. J. van der Steen, ed. (1995), *Aspects of Computational Science*, NCF.

A. J. van der Steen (2000), An evaluation of some Beowulf clusters. Technical Report WFI-00-07, Department of Computational Physics, Utrecht University. Also available through www.euroben.nl, directory reports/.

A. J. van der Steen (2010), Overview of recent supercomputers. www.euroben.nl, directory reports/.)

T. L. Sterling, J. Salmon, D. J. Becker and D. F. Savaresse (1999), *How to Build a Beowulf*, MIT Press.

R. Stevens, T. Zacharia and H. Simon (2008), Modeling and simulation at the exascale for energy and the environment. Town Hall Meetings Report, US Department of Energy Office of Advance Scientific Computing Research. www.sc.doe.gov/ascr/ProgramDocuments/Docs/TownHall.pdf

E. Strohmaier, J. J. Dongarra, H. W. Meuer and H. D. Simon (1999), 'The marketplace of high-performance computing', *Parallel Computing* **25**, 1517.

US Department of Energy (2008), Challenges in climate change science and the role of computing at the extreme scale. Department of Energy report, *Scientific Grand Challenges Workshop Series.* www.er.doe.gov/ascr/ProgramDocuments/Docs/ClimateReport.pdf.

US Department of Energy (2009*a*), Forefront questions in nuclear science and the role of high performance computing: Summary report. Department of Energy Workshop Report, Washington DC. extremecomputing.labworks.org/nuclearphysics/PNNL_18739_↩ onlineversion_opt.pdf

US Department of Energy (2009*b*), Cross-cutting technologies for computing at the exascale. Department of Energy report, *Scientific Grand Challenges Workshop Series.* extremecomputing.labworks.org/crosscut/CrosscutWSFinalReptDraft02.pdf

US Department of Energy (2009*c*), Fusion energy science and the role of computing at the extreme scale. Department of Energy report, *Scientific Grand Challenges Workshop Series.* extremecomputing.labworks.org/fusion/PNNL_Fusion_final19404.pdf

US Department of Energy (2009*d*), Science based nuclear energy systems enabled by advanced modeling and simulation at the extreme scale. Department of Energy Workshop Report, Washington, DC. www.er.doe.gov/ascr/ProgramDocuments/Docs/SC-NEWorkshopReport.pdf

US Department of Energy (2009*e*), Discovery in basic energy sciences: The role of computing at the extreme scale. Department of Energy report, *Scientific Grand Challenges Workshop Series.*

US Department of Energy (2009*f*), Opportunities in biology at the extreme scale of computing. Department of Energy report, *Scientific Grand Challenges Workshop Series.* www.er.doe.gov/ascr/ProgramDocuments/Docs/BiologyReport.pdf

US Department of Energy (2009*g*), Scientific grand challenges in national security: The role of computing at the extreme scale. Department of Energy report, *Scientific Grand Challenges Workshop Series.* www.er.doe.gov/ascr/ProgramDocuments/Docs/NNSAGrandChallenges↩ Report.pdf

US Department of Energy (2009*h*), Architectures and technology for extreme scale computing. Department of Energy report, *Scientific Grand Challenges Workshop Series.*

US Department of Energy (2009*i*), Scientific challenges for understanding the quantum universe and the role of computing at extreme scale: Summary report. Department of Energy report, *Scientific Grand Challenges Workshop Series.* extremecomputing.labworks.org/highenergyphysics/reports/↩ HEPreport101609_final.pdf

US Department of Energy (2010), Exascale workshop panel meeting report. Department of Energy report, *Scientific Grand Challenges Workshop Series.* www.er.doe.gov/ascr/ProgramDocuments/Docs/Trivelpiece↩ ExascaleWorkshop.pdf

US National Research Council Committee (2008), The potential impact of high-end capability computing on four illustrative fields of science and engineering. US National Academy of Sciences.

G. V. Wilson (1994), Chronology of major developments in parallel computing and supercomputing. ftp://ftp.cs.toronto.edu/csri-technical-reports/312/csri312.ps.gz

P. R. Woodward (1996), 'Perspectives on Supercomputing', *Computer* **10**, 99–111.

Online resources

Co-Array Fortran: www.co-array.org/
DSM systems: www.cs.umd.edu/~keheler/dsmbiblio/dsmbiblio.html
OpenMP Application Interface, version 2.5: www.openmp.org/
Task Force on Cluster Computing: www.clustercomputing.org
Unified Parallel C: upc.gwu.edu/

Acta Numerica (2012), pp. 475–576
doi:10.1017/S0962492912000062

Secrets of image denoising cuisine*

M. Lebrun
CMLA, Ecole Normale Supérieure de Cachan,
61 Avenue du Président Wilson,
94235 Cachan CEDEX, France
E-mail: marc.lebrun.ik@gmail.com

M. Colom
Universitat de les Illes Balears,
Crta de Valldemossa, km 7.5,
07122 Palma de Mallorca, Spain
E-mail: Miguel.Colom@cmla.ens-cachan.fr

A. Buades
Universitat de les Illes Balears,
Crta de Valldemossa, km 7.5,
07122 Palma de Mallorca, Spain
and
MAP5, Université Paris Descartes,
45 Rue des Saints Pères,
75270 Paris CEDEX 06, France
E-mail: toni.buades@uib.es

J. M. Morel
CMLA, Ecole Normale Supérieure de Cachan,
61 Avenue du Président Wilson,
94235 Cachan CEDEX, France
E-mail: morel@cmla.ens-cachan.fr

Digital images are matrices of equally spaced pixels, each containing a photon count. This photon count is a stochastic process due to the quantum nature of light. It follows that all images are noisy. Ever since digital images have existed, numerical methods have been proposed to improve the signal-to-noise ratio. Such 'denoising' methods require a noise model and an image model. It is relatively easy to obtain a noise model. As will be explained in the present paper, it is even possible to estimate it from a single noisy image.

* Colour online for monochrome figures available at journals.cambridge.org/anu.

Obtaining a convincing statistical image model is quite another story. Images reflect the world and are just as complex. Thus, any progress in image denoising implies progress in our understanding of image statistics. The present paper contains an analysis of nine recent state-of-the-art methods. This analysis shows that we are probably close to understanding digital images at a 'patch' scale. Recent denoising methods use thorough non-parametric estimation processes for 8×8 patches, and obtain surprisingly good denoising results.

The mathematical and experimental evidence of two recent articles suggests that we might even be close to optimal performance in image denoising. This suspicion is supported by a remarkable convergence of all analysed methods. They certainly converge in performance. We intend to demonstrate that, under different formalisms, their image models are almost equivalent. Working in the 64-dimensional 'patch space', all recent methods estimate local 'sparse models' and restore a noisy patch by finding its likeliest interpretation, given the noiseless patches.

The story will be told in an ordered fashion. Denoising methods are complex and have several indispensable ingredients. Noise model and noise estimation methods will be explained first. The four main image models used for denoising are the Markovian–Bayesian paradigm, linear transform thresholding, so-called image sparsity, and an image self-similarity hypothesis, which will also be presented. The performance of all methods depends on three generic tools: colour transform, aggregation, and an 'oracle' step. Their recipes will also be given. These preparations will permit us to present, in a unified terminology, the complete recipes of nine different state-of-the-art patch-based denoising methods. Three quality assessment recipes for denoising methods will also be proposed and applied to compare all methods. The paper presents an ephemeral state of the art in a rapidly changing subject, but many of the presented recipes will remain useful. Most denoising recipes can be tested directly on any digital image in the web journal *Image Processing On Line*.

CONTENTS

1	Introduction	478
2	Noise	487
3	Four denoising principles	500
4	Noise reduction: generic tools	511
5	Detailed analysis of nine methods	516
6	Comparison of denoising algorithms	547
7	Synthesis	561
	References	567

Notation

- $\mathbf{i}, \mathbf{j}, \mathbf{r}, \mathbf{s}$ image pixels
- $u(\mathbf{i})$ image value at $\mathbf{i}$, written as $U(\mathbf{i})$ when the image is handled as a vector
- $\tilde{u}(\mathbf{i})$ noisy image value at $\mathbf{i}$, or $\tilde{U}(\mathbf{i})$ when the image is handled as a vector
- $\hat{u}(\mathbf{i})$ restored image value, or $\hat{U}(\mathbf{i})$ when the image is handled as a vector
- $n(\mathbf{i})$ noise at $\mathbf{i}$
- N patch of noise in vector form
- m number of pixels $\mathbf{j}$ involved in denoising a pixel $\mathbf{i}$
- P reference patch; Q, a second patch compared to P
- $\tilde{P}$, $\tilde{Q}$ noisy patches
- $\hat{P}$ restored patch
- $w(\tilde{P},\tilde{Q}) = \mathrm{e}^{-\frac{d^2(\tilde{P},\tilde{Q})}{C\sigma^2}}$ the interaction weight between P and Q
- $d(\tilde{P},\tilde{Q})$ Euclidean distance between patches (considered as vectors of their values)
- σ standard deviation of white noise at each pixel
- $\kappa \times \kappa$ dimension of patches
- $\lambda \times \lambda$ dimension of search zone for similar patches
- $\mathcal{N}(\mu, \mathbf{C})$ vectorial Gaussian distribution with mean vector μ and covariance matrix $\mathbf{C}$
- $\mathbb{P}(G)$ probability of an event G (in the image and noise stochastic models)
- $\mathbb{E}Q$ expectation (of a random patch Q)
- $\overline{P}$ empirical expectation of the patches similar to P
- Δ image Laplace operator (sum of the second derivatives in two orthogonal directions)
- $\mathrm{DCT}(n_1, n_2)$ 2D discrete cosine transform at frequencies n_1, n_2
- p percentile value of a histogram (between 0% and 100%)
- $w \times w$ block size for estimating the noise
- $\hat{\sigma}$ estimated value of the noise
- b number of bins
- h result of a high-pass filter on $\tilde{u}$
- ∇ gradient (of an image)
- M total number of pixels or patches in the image; total number of patches in the patch space
- $\mathbf{C}_P$ covariance matrix (of patches similar to P or $\tilde{P}$)
- P_1 restored patch at the first application of a denoising algorithm
- P_2 restored patch at the second application of a denoising algorithm
- $\mathcal{B} = \{G_i\}_{i=1}^{M}$ orthonormal basis of $\mathbb{R}^M$
- $\mathbf{D}$ diagonal linear operator; dictionary of patches considered as a matrix
- n_{dict} size of the dictionary (number of patches in it)
- $\mathbf{A}$ linear operator, applied to the image u encoded as a vector U
- $p(P)$ density function of patches
- K number of patch clusters; number of wavelet coefficients
- k index in K
- Ω_k patch cluster
- i index of patch P_i (in a patch cluster, in the image)

1. Introduction

Most digital images and movies are currently obtained by a CCD[1] device. The value $\tilde{u}(\mathbf{i})$ observed by a sensor at each pixel $\mathbf{i}$ is a Poisson random variable whose mean $u(\mathbf{i})$ would be the ideal image. The difference between the observed image and the ideal image, $\tilde{u}(\mathbf{i}) - u(\mathbf{i}) = n(\mathbf{i})$, is called 'shot noise'. The standard deviation of the Poisson variable $\tilde{u}(\mathbf{i})$ is equal to the square root of the number of incoming photons $\tilde{u}(\mathbf{i})$ in the pixel captor $\mathbf{i}$ during the exposure time. The Poisson noise n is the sum of thermal noise and electronic noise, which are approximately additive and white. For a motionless scene with constant lighting, $u(\mathbf{i})$ can be approached by simply accumulating photons for a long exposure time, and by taking the temporal average of this photon count, as illustrated in Figure 1.1.

Accumulating photon impacts on a surface is therefore the essence of photography. The first photograph, by Nicéphore Niépce (Chevalier, Roman and Niépce 1854), was obtained from an eight-hour exposure. The problem of long exposure is the variation of the scene due to changes in light, camera motion, and incidental motion of parts of the scene. The more these variations can be compensated for, the longer the exposure can be, and the more the noise can be reduced. If a camera is set to a long exposure time, the photograph risks motion blur. If it is taken with a short exposure, the image is dark, and enhancing it will reveal noise.

A recently available solution is to take a burst of images, each with a short exposure time, and to average them. This technique, illustrated in Figure 1.1, was evaluated recently in a paper that proposes fusing bursts of images taken by cameras (Buades, Lou, Morel and Tang 2009*b*). This paper shows that the noise reduction by this method is almost perfect: fusing m images reduces the noise by a factor of $\sqrt{m}$.

It is not always possible to accumulate photons. There are obstacles to this accumulation in astronomy, biological imaging and medical imaging. In everyday images the scene moves, and this limits the exposure time. The main limitations to any imaging system are therefore noise and blur. In this review, experiments will be conducted on photographs of scenes taken by normal cameras. Nevertheless, the image denoising problem is a common denominator of all imaging systems.

A naive view of the denoising problem would be: How do we estimate the ideal image, namely the mean $u(\mathbf{i})$, given only one sample $\tilde{u}(\mathbf{i})$ of the Poisson variable? The best estimate of this mean is of course this unique sample $\tilde{u}(\mathbf{i})$. Getting back a better estimate of $u(\mathbf{i})$ by observing only $\tilde{u}(\mathbf{i})$ is impossible. Getting a better estimate by also using the rest of the image is obviously an ill-posed problem. Indeed, each pixel receives photons coming from different sources.

[1] Charge coupled device, *i.e.*, the 'digital film' at the heart of a digital camera.

(a) (b) (c)

Figure 1.1. (a) One long-exposure image (time=0.4 s, ISO=100), one of 16 short-exposure images (time=1/40 s, ISO=1600) and their average after registration. The long-exposure image is blurry due to camera motion. (b) The middle short-exposure image is noisy. (c) The third image is about *four times* less noisy, being the result of averaging 16 short-exposure images. From Buades *et al.* (2009*b*).

Nevertheless, one hint for the solution comes from image formation theory. A well-sampled image u is band-limited (Shannon 2001). Thus, it seems possible to restore the band-limited image u from its degraded samples $\tilde{u}$, as proposed by Harris (1966). This classic Wiener–Fourier method consists in multiplying the Fourier transform by optimal coefficients to attenuate the noise. It results in a convolution of the image with a low-pass kernel.

From a stochastic viewpoint, the band-limitedness of u also implies that values $\tilde{u}(\mathbf{j})$ at neighbouring pixels $\mathbf{j}$ of a pixel $\mathbf{i}$ are positively correlated with $\tilde{u}(\mathbf{i})$. Thus, these values can be taken into account to obtain a better estimate of $u(\mathbf{i})$. Since these values are non-deterministic, Bayesian approaches are relevant and were proposed as early as Richardson (1972).

In short, there are two complementary early approaches to denoising: the Fourier method and Bayesian estimation.

The Fourier method has been extended in the past 30 years to other linear space-frequency transforms, such as the windowed discrete cosine transform (Yaroslavsky and Eden 2003) or the many wavelet transforms (Meyer 1993).

Bayesian methods were initially parametric and limited to rather restrictive Markov random field models (Geman and Geman 1984), but are now becoming non-parametric. The idea behind the recent non-parametric Markovian estimation methods is a now famous algorithm to synthesize textures from examples (Efros and Leung 1999). The underlying Markovian assumption is that, in a textured image, the stochastic model for a given pixel $\mathbf{i}$ can be predicted from a local image neighbourhood P of $\mathbf{i}$, which we shall call a 'patch'.

Algorithm 1 Non-local means (NL-means) algorithm

Input. Noisy image $\tilde{u}$, noise standard deviation σ.
Output. Denoised image $\hat{u}$.
Set parameter $\kappa \times \kappa$: dimension of patches.
Set parameter $\lambda \times \lambda$: dimension of search zone for similar patches.
Set parameter C.
for each pixel **i do**
Select a square reference sub-image (or 'patch') $\tilde{P}$ around **i**, of size $\kappa \times \kappa$.
Call $\hat{P}$ the denoised version of $\tilde{P}$ obtained as a weighted average of the patches $\tilde{Q}$ in a square neighbourhood of **i** of size $\lambda \times \lambda$. The weights in the average are proportional to

$$w(\tilde{P}, \tilde{Q}) = \mathrm{e}^{-\frac{d^2(\tilde{P}, \tilde{Q})}{C\sigma^2}},$$

where $d(\tilde{P}, \tilde{Q})$ is the Euclidean distance between patches $\tilde{P}$ and $\tilde{Q}$.
end for
Aggregation. Recover a final denoised value $\hat{u}(\mathbf{i})$ at each pixel **i** by averaging all values at **i** of all denoised patches $\hat{Q}$ containing **i**.

The assumption when recreating new textures from samples is that there are enough pixels **j** similar to **i** in a texture image $\tilde{u}$ to recreate a new but similar texture u. The construction of u is done by non-parametric sampling, amounting to an iterative copy–paste process. Let us assume that we already know the values of u on a patch P partially surrounding an unknown pixel **i**. The algorithm of Efros and Leung (1999) looks for the patches $\tilde{P}$ in $\tilde{u}$ with the same shape as P and resembling P. Then a value $u(\mathbf{i})$ is sorted among the values predicted by $\tilde{u}$ at the pixels resembling **j**. Indeed, these values form a histogram approximating the law of $u(\mathbf{i})$. This algorithm goes back to Shannon's theory of communication (Shannon 2001), in which it was used for the first time to synthesize a probabilistically correct text from a sample.

As proposed by Buades, Coll and Morel (2005*b*), an adaptation of the above synthesis principle yields an image denoising algorithm. The observed image is the noisy image $\tilde{u}$. The reconstructed image is the denoised image $\hat{u}$. The patch is a square centred at **i**, and the sorting yielding $u(\mathbf{i})$ is replaced by a weighted average of values at all pixels $\tilde{u}(\mathbf{j})$ similar to **i**. This simple change leads to the 'non-local means' (NL-means) algorithm, which can therefore be sketched in a few lines: see Algorithm 1.

Buades *et al.* (2005*b*) also proved that the algorithm gave the best possible mean square estimate if the image was modelled as an infinite stationary ergodic spatial process (see Section 5.1 for an exact statement). The algorithm was called 'non-local' because it used patches $\tilde{Q}$ that are far away from $\tilde{P}$, and *even patches taken from other images.* NL-means was not the state-of-the-art denoising method when it was proposed. As we shall see in Section 6, the algorithm of Portilla, Strela, Wainwright and Simoncelli (2003), described in Section 5.6, has better PSNR (peak signal-to-noise ratio) performance. But quality criteria show that NL-means creates fewer artifacts than wavelet-based methods. This may explain why patch-based denoising methods have flourished ever since. To date, 1500 papers have been published on non-local image processing. Patch-based methods seem to achieve the best results in denoising. Furthermore, the quality of denoised images has become excellent for moderate noise levels. Patch-based image restoration methods are used in many commercial software packages.

An exciting recent paper in this exploration of non-local methods raises the following claim (Levin and Nadler 2011): *For natural images, the recent patch-based denoising methods might well be close to optimality.* Levin and Nadler (2011) use a set of 20 000 images containing about 10^{10} patches. This paper provides a second answer to the question of absolute limits raised by Chatterjee and Milanfar (2010): 'Is denoising dead?' The Cramer–Rao-type lower bounds on the attainable RMSE (root mean square error) performance given by Chatterjee and Milanfar (2010) are actually more optimistic: they allow for the possibility of a significant increase in denoising performance. The two types of performance bounds considered by Levin and Nadler (2011) and Chatterjee and Milanfar (2010) address roughly the same class of patch-based algorithms. It is interesting to see that these same authors propose denoising methods that actually approach these bounds, as we shall see in Section 5.

The denoising method proposed by Levin and Nadler (2011) is in fact based on NL-means (Algorithm 1), with the adequate parameter C to account for a Bayesian linear minimum mean square estimation (LMMSE) of the noisy patch given a database of known patches. The only important difference is that the similar patches Q are chosen from a database of 10^{10} patches, instead of from the image itself. Furthermore, by a simple mathematical argument and intensive simulations on patch space, Levin and Nadler are able to approach *the best average estimation error* that can *ever be attained by any patch-based denoising algorithm* (see Section 5.4).

These optimal bounds are nonetheless obtained on a somewhat restrictive definition of patch-based methods. A patch-based algorithm is understood to be an algorithm that denoises each pixel by using knowledge of (a) the patch surrounding it, and (b) the probability density of all existing patches. It turns out that state-of-the-art patch-based denoising algorithms use more information taken in the image than just the patch. For example, most algorithms use the obvious but powerful trick of denoising all patches and then *aggregating* the estimate of all patches containing a given pixel to denoise it better. Conversely, these algorithms generally use much less information than a universal empirical law for patches. Nevertheless, the observation that at least one algorithm, BM3D (Dabov, Foi, Katkovnik and Egiazarian 2007), might arguably be very close to the best predicted estimation error is enlightening. Furthermore, doubling the size of the patch used in the paper by Levin and Nadler (2011) would be enough to cover the aggregation step. The difficulty is getting a faithful empirical law for 16×16 patches.

The 'convergence' of all algorithms to optimality will be corroborated here by the thorough comparison of nine recent algorithms (Section 6). These state-of-the-art algorithms seem to attain very similar qualitative and quantitative performance. Although they initially seem to rely on different principles, our final discussion will argue that these methods are equivalent.

Image restoration theory cannot be reduced to an axiomatic system, as the statistics of images are still largely unexplored territory. Therefore a complete theory or a single definitive algorithm are not possible. The problem is not fully formalized because there is no rigorous image model. Notwithstanding this limitation, rational recipes shared by all methods can be given, and the methods can be shown to rely on very few principles. More precisely, this paper will present the following recipes, and compare them whenever possible.

- Several families of noise estimation techniques (Section 2).
- The four denoising principles in competition (Section 3).
- Three techniques that improve every denoising method (Section 4).
- Nine complete and recent denoising algorithms. For these algorithms complete recipes will be given (Section 5).
- Three complementary and simple recipes to evaluate and compare denoising algorithms (Section 6).

Using the three comparison recipes, six emblematic or state-of-the-art algorithms, based on reliable and public implementations, will be compared in Section 6. This comparison is followed by a synthesis (Section 7), hopefully demonstrating that, under very different names, the state-of-the-art algorithms share the same principles.

Nevertheless, this convergence of results and techniques leaves several crucial issues unsolved. (This is fortunate, since no researcher likes finished problems.) With one exception (the BLS-GSM algorithm, Section 5.6), state-of-the-art denoising algorithms are not multiscale. High noise and low noise also remain unexplored.

In a broader perspective, the success of image denoising marks the discovery and exploration of one of the first ever densely sampled high-dimensional probability laws (numerically) accessible to mankind: the 'patch space'. For

8×8 patches, by applying a local principal component analysis (PCA) to the patches surrounding a given patch, one can deduce that this space has a dozen significant dimensions (other dimensions corresponding to small eigenvalues). Exploring its structure, as was initiated in Lee, Pedersen and Mumford (2003), seems to be the first step towards the statistical exploration of images. But, as we shall see, this local analysis of the *patch space* already enables state-of-the-art image denoising.

Most denoising and noise estimation algorithms discussed here will be available in the journal *Image Processing On Line*, http://www.ipol.im/. In this web journal, each algorithm is given a complete description along with the corresponding source code, and can be run online on arbitrary images. By the time this paper is published, most results and techniques presented here will be effortlessly verifiable and reproducible online.

This Introduction ends with a quick review of many recent contributions of interest on patch-based methods, which nevertheless fall beyond our limited scope.

1.1. *Miscellaneous 'patch-based' considerations and applications*

Statistical validity

This paper will compare patch-based algorithms according to their structure and their practical performance, which is acceptable in the absence of a satisfactory mathematical or statistical model for digital images. Nonetheless, statistical arguments have also been developed to explore the validity of denoising algorithms. The statistical validity of NL-means is discussed by Thacker, Manjon and Bromiley (2008), Kervrann, Boulanger and Coupé (2007) and Ebrahimi and Vrscay (2007), who propose a Bayesian interpretation, and by Xu, Xu and Wu (2008), who correct the bias of NL-means. Singer, Shkolnisky and Nadler (2009) give a 'probabilistic interpretation and analysis of the method viewed as a random walk on the patch space'. The most complete recent study is made in the realm of *minimax approximation theory*. The horizon class of images, which are piecewise constant with a sharp edge discontinuity (Maleki, Narayan and Baraniuk 2011*b*), is suitable for asymptotic analysis. The images are discontinuous across the edge and the edge itself is smooth, being in an $H^\alpha(C)$ class. A real function is in this class for $\alpha \geq 0$ if

$$|h^{([\alpha])}(t) - h^{([\alpha])}(s)| \leq C|t-s|^{\alpha-[\alpha]},$$

where $[\alpha]$ is the integer part of α.

The principle is to measure the expected approximation rate of a denoising algorithm applied to m noisy samples of an image u in the horizon class. This image u is given by m samples, and these samples are perturbed by a white noise with variance σ^2. A denoising algorithm delivers a corrected function $\hat{u}$. The risk function of this algorithm is defined as the expectation $R_m(u, \hat{u})$ of the mean square distance of u and $\hat{u}$. Given a class of functions $\mathcal{F}$, the *minimax risk* is defined by

$$R_m(\mathcal{F}) = \inf_{\hat{u}} \sup_{u \in \mathcal{F}} R_m(u, \hat{u}),$$

where the inf is taken over all measurable estimators. It can be proved (Mammen and Tsybakov 1995) that, for $\alpha \geq 1$,

$$R_m(H^\alpha(C)) \simeq m^{-\frac{2\alpha}{\alpha+1}}. \tag{1.1}$$

For example, for $\alpha = 2$, which corresponds to edges with bounded curvature, the optimal rate is $n^{-\frac{4}{3}}$. This result gives a kind of yardstick to measure, if not the performance, at least the theoretical limits of every denoising algorithm. This analysis has been conducted for several basic denoising methods, including NL-means. Maleki *et al.* (2011*b*) show that the decay rate is about m^{-1}, close to that obtained with wavelet threshold denoising, better than rates of elementary filters such as linear convolution, the median filter and the bilateral filter, which have rates $m^{-\frac{2}{3}}$. The decay rate of NL-means is nonetheless some distance from the optimal minimax rate of $m^{-4/3}$, which is only attained for $\alpha = 2$ by the wedgelet transform. Maleki, Narayan and Baraniuk (2011*a*) show that an anisotropic non-local means (ANLM) algorithm is near minimax optimal for edge-dominated images from the horizon class. The idea is to orient optimally rectangular thin blocks for performing the comparison. The algorithms improve on NL-means by approximately one decibel.

Other noise models

The present article focuses on algorithms removing white additive noise from digital optical images. There are other types of noise in other imaging systems. Therefore, this study cannot account for the burgeoning variety of patch-based algorithms. Improvements or adaptations of NL-means have been proposed in cryo-electron microscopy (Darbon *et al.* 2008), fluorescence microscopy (Boulanger, Sibarita, Kervrann and Bouthemy 2008), magnetic resonance imaging (MRI) (Manjón *et al.* 2008, Coupé *et al.* 2008, Wiest-Daesslé *et al.* 2008, Naegel *et al.* 2009), multispectral MRI (Manjón, Robles and Thacker 2007, Buades, Chien, Morel and Osher 2008*a*), and diffusion tensor MRI (DT-MRI) (Wiest-Daesslé *et al.* 2007).

More invariance

Likewise, several papers have explored the degree of invariance that could be applied to image patches. Zimmer, Didas and Weickert (2008) explore a rotationally invariant block matching strategy improving NL-means, and

Ebrahimi and Vrscay (2008*a*) use cross-scale (*i.e.*, down-sampled) neighbourhoods in the NL-means filter. See also the paper by Maleki *et al.* (2011*a*), mentioned above as reaching better minimax limits, for which it uses oriented anisotropic patches. Self-similarity has also been explored in the Fourier domain for MRI, by Mayer *et al.* (2008).

Fast patch methods

Several papers have proposed fast and extremely fast (linear) NL-means implementations, using block pre-selection (Mahmoudi and Sapiro 2005, Bilcu and Vehvilainen 2008), Gaussian KD-trees, to classify image patches (Adams, Gelfand, Dolson and Levoy 2009), singular value decomposition (Orchard, Ebrahimi and Wong 2008), the fast Fourier transform, to compute correlation between patches (Wang *et al.* 2006), statistical arguments (Coupé, Yger and Barillot 2006), and approximate search (Barnes, Shechtman, Finkelstein and Goldman 2009), also used for optical flow.

Other image processing tasks

The non-local denoising principle has also been expanded to most image processing tasks: *demosaicing*, the operation which transforms the 'R or G or B' raw image in each camera into an 'R and G and B' image (Buades, Coll, Morel and Sbert 2009*a*, Mairal, Elad and Sapiro 2008); *movie colourization* (Elmoataz, Lézoray, Bougleux and Ta 2008*b*, Lézoray, Ta and Elmoataz 2008); *image inpainting*, by proposing a non-local image inpainting variational framework with a unified treatment of geometry and texture (Arias, Caselles and Sapiro 2009; see also Wong and Orchard 2008); *zooming*, by a fractal-like technique where examples are taken from the image itself at different scales (Ebrahimi and Vrscay 2007); *movie flicker stabilization*, compensating for spurious oscillations in the colours of successive frames (Delon and Desolneux 2009); and *super-resolution*, an image zooming method fusing several frames from a video, or several low-resolution photographs, into a larger image (Protter, Elad, Takeda and Milanfar 2009). The main point of this super-resolution technique is that it does not give an explicit estimate of the motion, in fact allowing for multiple motions, since a block can look like several other patches in the same frame. The very same observation is made by Ebrahimi and Vrscay (2008*b*) on devising a super-resolution algorithm, and also by Elad and Datsenko (2007) and Danielyan, Foi, Katkovnik and Egiazarian (2008). Other classic image non-local applications include image *contrast enhancement*, by applying a reverse non-local heat equation (Buades, Coll and Morel 2006*a*), and *stereo vision*, by performing simultaneous non-local depth reconstruction and restoration of noisy stereo images (Heo, Lee and Lee 2007).

The link to PDEs, variational variants

The relationship of neighbourhood filters to classic local PDEs was discussed by Buades, Coll and Morel (2006*b*, 2006*c*), leading to an adaptation of NL-means which avoids the staircase effect. Non-local image-adapted differential operators and non-local variational methods were introduced by Kindermann, Osher and Jones (2006), who proposed denoising and deblurring by non-local functionals. The general goal of this development is in fact to give a variational form to all neighbourhood filters, and to give a non-local form to the total variation as well (Rudin, Osher and Fatemi 1992). Several articles on deblurring have followed this variational approach: Jung and Vese (2009), Mignotte (2008) and Gilboa and Osher (2008) for image segmentation; Boulanger *et al.* (2008) for fluorescence microscopy; Zhang, Burger, Bresson and Osher (2009) again for non-local deconvolution; and Lou, Zhang, Osher and Bertozzi (2008) for deconvolution and tomographic reconstruction. Elad and Datsenko (2007), in a paper dedicated to another notoriously ill-posed problem, that of super-resolution, view the non-local variational principle as an 'emerging powerful family of regularization techniques', and propose use of the example-based approach as a 'new regularizing principle in ill-posed image processing problems such as image super-resolution from several low resolution photographs'. A particular notion of non-local PDEs has emerged, whose coefficients are in fact image-dependent. For instance, Elmoataz *et al.* (2008*b*) view the image colourization as the minimization of a discrete partial differential functional on the weighted block graph. Thus, it can be seen either as a non-local heat equation on the image, or as a local heat equation on the space of image patches.

The geometric interpretation in a graph of patches

In an almost equivalent framework, Szlam, Maggioni and Coifman (2006) view the set of patches as a weighted graph, and the weights of the edge between two patches centred at $\mathbf{i}$ and $\mathbf{j}$, respectively, are decreasing functions of the block distances. Then a graph Laplacian can be calculated on this graph, seen as the sampling of a manifold, and NL-means can be interpreted as the heat equation on the set of blocks endowed with these weights. In the same way, the neighbourhood filter can be associated with a heat equation on the image graph (Peyré 2009). This approach has been further extended to a variational formulation on patch graphs by Elmoataz, Lézoray and Bougleux (2008*a*). In this same framework Buades *et al.* (2006*a*) proposed image contrast enhancement via a non-local reverse heat equation. Finally, still in this non-local partial differential framework, Bresson and Chan (2008) extend the Mumford–Shah image segmentation energy to contain a non-local self-similarity term replacing the usual Dirichlet term. The square of the gradient is replaced by the square of the non-local gradient.

2. Noise

2.1. Noise models

Most digital images and movies are obtained by a CCD device and the main source of noise is the so-called *shot noise*. Shot noise is inherent to photon counting. The value $\tilde{u}(\mathbf{i})$ observed by a sensor at each pixel $\mathbf{i}$ is a Poisson random variable whose mean would be the ideal image. The standard deviation of this Poisson distribution is equal to the square root of the number of incoming photons $\tilde{u}(\mathbf{i})$ in the pixel captor $\mathbf{i}$ during the exposure time. This noise is the sum of thermal noise and electronic noise, which are approximately additive and white.

For sufficiently large values of $\tilde{u}(\mathbf{i})$ (*e.g.*, $\tilde{u}(\mathbf{i}) > 1000$), the normal distribution $\mathcal{N}(\tilde{u}(\mathbf{i}), \sqrt{\tilde{u}(\mathbf{i})})$ is an excellent approximation to the Poisson distribution. If $\tilde{u}(\mathbf{i})$ is larger than 10, then the normal distribution is still a good approximation if an appropriate continuity correction is performed, namely

$$\mathbb{P}(\tilde{u}(\mathbf{i}) \leq a) \simeq \mathbb{P}(\tilde{u}(\mathbf{i}) \leq a + 0.5),$$

where a is any non-negative integer.

Nevertheless, the pixel value is *signal-dependent*, since its mean and variance depend on $\tilde{u}(\mathbf{i})$. To get back to the classic 'white additive Gaussian noise' used in most research on image denoising, a *variance-stabilizing transformation* can be applied. When a variable is Poisson-distributed with parameter $\tilde{u}(\mathbf{i})$, its square root is approximately normally distributed with expected value of about $\sqrt{\tilde{u}(\mathbf{i})}$ and variance of about $1/4$. Under this transformation, the convergence to normality is faster than for the untransformed variable. The most classic variance stabilizing transformation (VST) is the Anscombe transform (Anscombe 1948), which has the form $f(u_0) = b\sqrt{u_0 + c}$.

The denoising procedure with the standard VST procedure follows three steps:

(1) apply VST to approximate homoscedasticity,

(2) denoise the transformed data,

(3) apply an inverse VST.

Note that the inverse VST is not just an algebraic inverse of the VST, and must be optimized to avoid bias (Makitalo and Foi 2011).

Consider any additive signal-dependent noisy image, obtained, for example, by the Gaussian approximation of a Poisson variable explained above. Under this approximation, the noisy image satisfies $\tilde{u} \simeq \tilde{u} + g(\tilde{u})n$, where $n \simeq \mathcal{N}(0, 1)$. We can search for a function f such that $f(\tilde{u})$ has uniform standard deviation,

$$f(\tilde{u}) \simeq f(\tilde{u}) + f'(\tilde{u})g(\tilde{u})n.$$

Forcing the noise term to be constant, $f'(\tilde{u})g(\tilde{u}) = c$, we get

$$f'(\tilde{u}) = \frac{c}{g(\tilde{u})},$$

and integrating gives

$$f(\tilde{u}) = \int_0^{\tilde{u}} \frac{c\,\mathrm{d}t}{g(t)}.$$

When a linear variance noise model is chosen, this transformation gives an Anscombe transform. Most classical denoising algorithms can also be adapted to signal-dependent noise. This requires varying the denoising parameters at each pixel, depending on the observed value $\tilde{u}(\mathbf{i})$. Several denoising methods indeed deal directly with the Poisson noise. Wavelet-based denoising methods (Nowak and Baraniuk 1997, Kolaczyk 1999) propose adaptation of the transform threshold to the local noise level of the Poisson process. Lefkimmiatis, Maragos and Papandreou (2009) have explored a Bayesian approach without applying a VST. Deledalle, Denis and Tupin (2011*a*) argue that for high noise level it is better to adapt NL-means than to apply a VST. They propose replacing the Euclidean distance between patches with a likelihood estimate, taking into account the noise model. This distance can be adapted to each noise model, such as Poisson, Laplace or gamma noise (Deledalle, Tupin and Denis 2010*a*), and to more complex (speckle) noise occurring in radar (SAR) imagery (Deledalle, Tupin and Denis 2010*b*).

Nonetheless, dealing with white uniform Gaussian noise makes the discussion of denoising algorithms far easier. Recent papers on the Anscombe transform by Makitalo and Foi (2011) (for low-count Poisson noise) and Foi (2011) (for Rician noise) argue that, when combined with suitable forward and inverse VST transformations, algorithms designed for homoscedastic Gaussian noise work just as well as *ad hoc* algorithms based on signal-dependent noise models. This explains why, in the rest of this paper, the noise is assumed to be uniform, white and Gaussian, having previously applied a VST to the noisy image if necessary. This also implies that we deal with *raw* images, namely images as close as possible to the direct camera output before processing. Most reflex cameras, and many compact cameras nowadays give access to this raw image.

But there is definitely a need to denoise current image formats, which have undergone unknown alterations. For example, the JPEG-encoded images provided by many cameras contain noise that has been altered by a complex chain of algorithms, ending with lossy compression. Noise in such images cannot be removed by the current state-of-the-art denoising algorithms without specific adaptation. The key is to have a decent noise model. For this reason, the fundamentals of estimating noise from a single image will be given in Section 2.2.

2.2. Can noise be estimated from (just) one image?

Compared to the denoising literature, research on noise estimation is a poor relation. Few papers are dedicated to this topic. Among recent papers we can mention that of Zoran and Weiss (2009), which argues that images are scale-invariant and therefore noise can be estimated by deviation from this assumption. Unfortunately this method is not easily extendable to the estimation of scale-dependent or signal-dependent noise, such as that observed in most digital images in compressed format. As a rule of thumb, the noise model is relatively easy to estimate when the raw image comes directly from the imaging system, in which case the noise model is known and only a few parameters must be estimated. Efficient methods are described by Foi, Trimeche, Katkovnik and Egiazarian (2008) and Foi, Alenius, Katkovnik and Egiazarian (2007) for Poisson and Gaussian noise.

In this short review we will focus on methods that allow for local, signal-dependent and scale-dependent noise. Indeed, one cannot denoise an image without knowing its noise model. It might be argued that the noise model comes from knowledge of the imaging device. Nevertheless, the majority of images dealt with by the public or by scientists have lost this information. This loss is caused by format changes of all kinds, which may include resampling, denoising, contrast changes and compression. All these operations change the noise model and make it *signal- and scale-dependent.*

The question that arises is: Why are so many researchers working so hard on denoising models if their corpus of noisy images is so ill-informed?

It is common practice among image processing researchers to add the noise themselves to noise-free images to demonstrate the performance of a method. This procedure permits reliable evaluation of denoising performance, based on a controlled underlying 'true' image. Nevertheless denoising performance may, after all, critically depend on how well we are able to estimate the noise. Most images are actually encoded with lossy JPEG formats. Thus, noise is partly removed by the compression itself. Furthermore, this removal is scale-dependent. For example, the JPEG 1985 format divides the image into a disjoint set of 8×8 pixel blocks, computes their discrete cosine transform (DCT), quantizes the coefficients, and small coefficients are replaced by zero. Thus JPEG performs frequency-dependent thresholding, equivalent to a basic Wiener filter. The same is true for JPEG 2000 (based on the wavelet transform).

In addition, the Poisson noise of a raw image is signal-dependent. The typical image processing operations, demosaicing, white balance and tone curve (contrast change) alter this signal-dependency in a way that depends on the image itself.

(a) (b)

Figure 2.1. Two examples of the ten noise-free images used in the tests: (a) *computer* and (b) *traffic*.

In short:

- the noise model is different for each image,
- the noise is signal-dependent,
- the noise is scale-dependent,
- the knowledge of each dependence is crucial to proper denoising of any given image which is not raw, and for which the camera model is available.

Thus, estimating JPEG noise is a complex and risky procedure, as is well explained in Liu *et al.* (2008) and Liu, Freeman, Szeliski and Kang (2006). Danielyan and Foi (2009) argue that noise can be estimated by using a denoising algorithm. Again, this procedure is probably too risky for noise and scale-dependent signals.

This section, following Buades, Colom and Morel (2012*a*), gives a concise review and a comparison of existing noise estimation methods. The classic methods estimate white homoscedastic noise only, but they can be adapted easily to estimation of signal- and scale-dependent noise. To test the methods, a set of ten noise-free images was used. These noiseless images were obtained by taking snapshots with a reflex camera of scenes in good lighting conditions and with a low ISO. This means that the number of photons reaching each captor was very high, and the noise level therefore small. To reduce the noise level further, the average of each block of 5×5 pixels was computed, reducing the noise by a factor of 5. Since the images are RGB, taking the mean of the three channels reduces the noise by a further factor of $\sqrt{3}$. The (small) initial noise was therefore reduced by a factor of $5\sqrt{3} \simeq 8.66$, and the images can be considered noise-free. Two images from this noiseless set can be seen in Figure 2.1. The size of each image is 704×469 pixels. For the uniform-noise tests, seven noise levels were applied to these noise-free images: $\sigma \in \{1, 2, 5, 10, 20, 50, 80\}$.

Figure 2.2. Result of adding white homoscedastic Gaussian noise with $\sigma \in \{2, 5, 10, 20, 50, 80\}$ to the noise-free image *traffic*. It may need a zoom in to perceive the noise for $\sigma = 2, 5$.

Figure 2.2 shows the result of adding white homoscedastic Gaussian noise with $\sigma \in \{1, 2, 5, 10, 20, 50, 80\}$ to the noise-free image *traffic*.

This study of noise estimation proceeds as follows. In Section 2.3 we review in detail the method proposed by Buades *et al.* (2012*a*). This method has all the features of the preceding methods, so we shall be able to make a quick review (Section 2.4), followed by an overall comparison of all methods, at all noise levels. It follows that the percentile method is the most accurate. Nevertheless, the estimation of very low noise remains somewhat inaccurate, with some 20% error for noises with standard deviation below 2.

2.3. The percentile method

The percentile method, introduced in Ponomarenko *et al.* (2007), is based on the fact that the histogram of the variances of all blocks in an image is affected by the edges and textures, but this alteration appears mainly in its rightmost part. The idea of the *percentile method* is to avoid the side effect of edges and textures by taking the variance of a very low percentile of the block variance histogram, and then to infer from it the real average variance of blocks containing only noise. This correction multiplies this variance by a factor that only depends on the choice of the percentile and the block size. As usual in all noise estimation methods, to reduce the presence of deterministic tendencies in the blocks, due to the signal, the image is first high-passed. The commonly used high-pass filters are differential operators or waveforms. The typical differential operators are directional derivatives, the Δ (Laplace) operator, its iterations $\Delta\Delta$, $\Delta\Delta\Delta$, $\ldots$, the wave forms are wavelet or DCT coefficients. All of them are implemented as discrete stencils. Filtering the image with a local high-pass filter operator removes smooth variations inside blocks, which increases the number of blocks where noise dominates and on which the variance estimate will be reliable. According to the performance tests, for observed $\hat{\sigma} < 75$ the best operator is the wave associated to the highest-frequency coefficient of the transformed 2D DCT-II block with support 7×7 pixels.

The coefficient $\tilde{X}(6,6)$ of the 2D DCT-II of a 7×7 block P of the image is

$$\mathrm{DCT}(6,6) = \sum_{n_1=0}^{6} \sum_{n_2=0}^{6} F_7(n_1) F_7(n_2) P(n_1,n_2) \cos\left[\frac{\pi}{7}\left(n_1 + \frac{1}{2}\right)6\right] \cos\left[\frac{\pi}{7}\left(n_2 + \frac{1}{2}\right)6\right],$$

where

$$F_7(n) = \begin{cases} \frac{1}{\sqrt{7}} & \text{if } n = 0, \\ \sqrt{\frac{2}{7}} & \text{if } n \in \{1, \ldots, 6\}. \end{cases}$$

Therefore, the values of the associated discrete filter are

$$F_7(n_1) F_7(n_2) \cos\left[\frac{\pi}{7}\left(n_1 + \frac{1}{2}\right)6\right] \cos\left[\frac{\pi}{7}\left(n_2 + \frac{1}{2}\right)6\right], \quad n_1, n_2 \in \{0, 1, \ldots, 6\}.$$

These values must of course be normalized in order to preserve the standard deviation of the data, by dividing each value by the root of the sum of the squared filter values.

The percentile method computes the variances of overlapping $w \times w$ blocks in the high-pass filtered image. The means of the same blocks are computed from the original image (before the high-pass). These means are classified into a disjoint union of variable intervals, in such a way that each interval

Table 2.1. Percentile method results for eleven noiseless images with white homoscedastic Gaussian noise added. The last image is simply flat. The real noise variance is σ; the estimated value is $\hat{\sigma}$. The noise estimation error is remarkably low for medium and high noise. It is nevertheless larger for very low noise ($\sigma = 2$ noise is not visible with the naked eye). Indeed, most photographed objects have some micro-texture everywhere (except perhaps in blue sky, which can be fully homogeneous). Such micro-textures are widespread and hardly distinguishable from noise. The parameters of the method are a 0.5% percentile, a 21×21 pixel block size, and the DCT has support 7×7. These parameters are valid if $\hat{\sigma} < 75$. If $\hat{\sigma} \geq 75$, the best parameters are a 50% percentile, a 21×21 pixel block size and a DCT with support 3×3. Estimating the best parameters therefore requires a first estimate followed by a second one with the correct parameters.

Image / $\hat{\sigma}$	$\sigma = 1$	$\sigma = 2$	$\sigma = 5$	$\sigma = 10$	$\sigma = 20$	$\sigma = 50$	$\sigma = 80$
bag	1.34	2.33	5.26	10.36	20.30	49.87	79.96
building1	1.12	2.17	5.24	10.14	20.48	50.19	80.45
computer	1.22	2.20	5.06	10.36	20.03	50.28	80.34
dice	1.11	2.00	5.01	10.03	20.02	49.95	79.79
flowers2	1.08	2.07	5.10	9.84	20.07	49.87	79.80
hose	1.15	2.13	5.10	10.15	20.06	49.99	79.99
leaves	1.51	2.43	5.38	10.29	19.82	50.07	80.04
lawn	1.57	2.50	5.57	10.48	20.42	50.05	79.92
stairs	1.42	2.27	5.19	10.15	19.96	49.92	79.93
traffic	1.25	2.35	5.33	10.61	20.64	50.10	80.29
flat image	0.99	2.00	5.09	9.77	19.91	50.12	79.73

contains (at least) 42 000 elements. These measurements permit, for each interval of means, construction of a histogram of block variance with at least 42 000 samples, having their means in the interval. In each such variance histogram the percentile value is computed. It was observed that, for observed $\hat{\sigma} < 75$ and large images, the best results are given by the percentile $p = 0.5\%$, a block size $w = 21$ and a 7×7 support for the DCT transform. If $\hat{\sigma} \geq 75$, the percentile that should be used is the median, the block is still 21×21, but the support of the DCT should be 3×3.

This percentile value is of course lower than the real average block variance, and must be corrected by a multiplicative factor. This correction only depends on the percentile, block size and on the chosen high-pass filter. Nevertheless, the constant is not easy to calculate explicitly, but can be learned from simulations. For the 0.5% percentile, 21×21 pixel blocks and the DCT pre-filter operator with support 7×7, the empirical factor derived from pure noise images was found to be 1.249441884. In summary, to each interval of means, a standard deviation is associated. The association mean$\rightarrow$standard deviation yields a 'noise curve' associated with the image.

Algorithm 2 Percentile method algorithm

PERCENTILE. Returns a list that relates the value of the image signal with its noise level.

Input. Noisy image $\tilde{u}$; number of bins b; block dimensions $w \times w$; percentile p; filter iterations filt.

Output. List made of pairs (mean, noise standard deviation), (M, S), for each bin of grey-scale value.

$h =$ FILTER($\tilde{u}$). Apply high-pass filter to the image.
$a, v =$MEAN_FILTERED_VARIANCE($\tilde{u}, h, w$). Obtain the list of the block averages (in the original image $\tilde{u}$) and of the variances (of the filtered image h) for all $w \times w$ blocks.
Divide the block mean value list a into intervals (bins), having all the same number of elements. For each interval keep the corresponding values in v.

$S = \emptyset$; $M = \emptyset$.
for each bin **do**
  $v =$ Per(bin, p). Get the p-percentile v of the block variances whose means belong to this bin.
  $m =$ Mean[Per(bin, p)]. Get the mean of the block associated to that percentile.
  $S \leftarrow \sqrt{v}$. Store the standard deviation $\hat{\sigma}$.
  $M \leftarrow m$. Store mean.
end for
$S_c = \emptyset$. Corrected values.
for $s \in S$ **do**
  Apply correction C according to p, w and filter operator used.
  $s = Cs$. Correct direct estimate.
  $S_c \leftarrow s$.
end for
for $k = 1 \ldots$ filt **do**
  $S_c[k] =$ FILTER($S_c[k]$, filt). Filter the noise curve filt times.
end for

For each observed grey-scale value in the image, this noise curve predicts the most likely underlying standard deviation due to noise. Optionally, the noise curve obtained from real images can be filtered. Indeed, it may present some peaks when variances measured for a given grey-scale interval belong to a highly textured region. To filter the curve, the points that are above the segment that joins the points on the left and on the right are back-projected onto that segment. In general, no more than two filtering iterations are needed. For the comparative tests presented here, the curves were not filtered at all.

The pseudo-code for the percentile method is given in Algorithm 2, and the results for the white homoscedastic Gaussian noise in Table 2.1. When the image is tested for white homoscedastic Gaussian noise, only one interval for all grey-scale means is used, whereas in the signal-dependent noise case, the grey-scale interval is divided into seven bins.

The percentile method with learning

The percentile method with learning is essentially the same algorithm explained in Section 2.3, with the difference that it tries to compensate the bias caused by edges and micro-texture in the image by learning a relationship between observed values $\hat{\sigma}$ and noise real values σ. The difference value $f(\sigma) = \hat{\sigma} - \sigma$ is called the *correction*, that is, the value that must be subtracted from the direct estimate $\hat{\sigma}$ without correction to get the final estimate (which we shall still call $\hat{\sigma} \approx \sigma$). These corrections depend on the structure of real images. A mosaic of several noise-free images is shown in Figure 2.3. Simulated noise of standard deviations $\sigma = 0, \dots, 100$ was added to these noiseless images. These images were selected randomly from a large database, to be statistically representative of the natural world, with textures, edges, flat regions, dark and bright regions. The correction learned with these images is intended to be an average correction that works for a broad range of natural images. It should of course be adapted to any particular set of images. Furthermore, the correction depends on the size of the image, and must be learned for each size.

When the observed noise level is high enough ($\hat{\sigma} > 10$ for pixel intensities $u \in \{0, 1, \dots, 255\}$), the image becomes dominated by noise, that is, most of the variance measured is due to the noise, and not due to micro-textures and edges. It is therefore convenient to avoid applying the learned corrections to direct estimates $\hat{\sigma}$ when $\hat{\sigma} > 10$. Thus, for $\hat{\sigma} > 10$, only the percentile correction is applied. Table 2.2 shows the $\hat{\sigma}$ values estimated with the percentile with learning method. The correction learned with the mosaic is only applied for $\sigma \in \{1, 2, 5, 10\}$.

2.4. A crash course on all other noise estimation methods

It is easier to explain the other methods after having explained one method in detail, as above, namely the percentile method. Most noise estimation methods share the following features.

- They start by applying a high-pass filter, which concentrates the image energy on boundaries, while the noise remains spatially homogeneous.
- They compute the energy for many blocks extracted from this high-passed image.
- They estimate the noise standard deviation from the values of the standard deviations of the blocks.

Table 2.2. Percentile method with learning results, with white homoscedastic Gaussian noise added. The correction learned with the mosaic is only applied for $\sigma \in \{1, 2, 5, 10\}$. This method, being local on blocks, extends immediately to estimation of signal-dependent noise, and the performance is similar (Buades *et al.* 2012*a*).

Image / $\hat{\sigma}$	$\sigma = 1$	$\sigma = 2$	$\sigma = 5$	$\sigma = 10$	$\sigma = 20$	$\sigma = 50$	$\sigma = 80$
bag	1.15	2.11	5.05	10.26	20.06	49.68	80.05
building1	0.95	1.97	5.00	10.42	20.32	49.99	80.27
computer	1.04	2.00	4.88	10.39	20.13	50.29	80.16
dice	0.91	1.84	4.81	10.01	19.90	49.76	79.60
flowers2	0.92	1.88	4.87	9.47	20.00	49.48	79.67
hose	0.99	1.93	4.89	10.08	19.97	49.73	79.71
leaves	1.36	2.26	5.17	10.28	20.03	49.80	79.92
lawn	1.35	2.29	5.36	10.37	20.26	50.07	79.88
stairs	1.20	2.10	4.95	10.11	20.10	49.92	79.86
traffic	1.04	2.06	5.06	10.75	20.64	49.91	80.05
flat image	0.84	1.82	4.84	10.02	20.13	50.13	79.44

Figure 2.3. Mosaic used to learn the correction values in the percentile method.

- To avoid blocks contaminated by the underlying image, a statistics robust to (many) outliers must be applied. The methods therefore use the flattest blocks, which belong to a (low) percentile of the histogram of standard deviations of all blocks.

Table 2.3 shows a classification of the methods according to the preceding criteria. The 'High-pass' column shows the choice of high-pass filter, which can be a discrete differential operator of order two $\left(\frac{\partial^2}{\partial x \partial y}\right)$ in the *estimation of image noise variance* (EINV) method (Rank, Lendl and Unbehauen 1999). It is obtained as a composition of two forward discrete differences. Then we have a discrete Laplacian Δ (Olsen 1993), obtained as the difference between the current pixel value and the average of a discrete neighbourhood, an order-three operator (a difference $\Delta_1 - \Delta_2$ of two different discretizations of the Laplacian (Immerkaer 1996)), a wave associated to a DCT coefficient (Buades *et al.* 2012*a*), and sometimes a non-linear discrete differential operator as in the *median* method (Olsen 1993), which uses the difference between the image and its median value on a 3×3 block, thus equivalent to the curvature operator curv. The high-pass filter was previously applied to all pixels of the image. In the case of the DCT (Ponomarenko *et al.* 2003), the DCT is applied to a block centred on the reference pixel, and the highest-frequency coefficients are kept, for example DCT$(6,7)$, DCT$(7,6)$, DCT$(7,7)$. The most primitive methods, the *block* method (Lee 1981, Mastin 1985), the *pyramid* method (Meer, Jolion and Rosenfeld 1990) and the *scatter* method (Lee and Hoppel 1989), do not apply any high-pass filter. Nevertheless, since they compute block variances, they implicitly remove the mean from each block, which amounts to applying a high-pass filter of Laplacian type.

The 'Block' column gives the size of the block on which the standard deviation of the high-passed image is computed, which varies from 1 to 21. The *pyramid* method (Meer *et al.* 1990) uses standard deviations of blocks of all sizes and is unclassifiable. Two methods, *FNVE* (Immerkaer 1996) and the *gradient* method (Bracho and Sanderson 1985, Voorhees and Poggio 1987), do not compute any block standard deviation of the high-passed image before the final estimation.

The 'Percentile' column gives the value of the (low) percentile on which the block standard deviation are computed. When the slot contains 'all', this means that the estimator takes into account all the values.

The 'Estimator' column characterizes the estimator, for which there are several variants. The three compared percentile methods (Buades *et al.* 2012*a*) use a very low percentile 0.5% of the block standard deviations. The *average*, *median* (Olsen 1993) and *block* methods (Lee 1981, Mastin 1985) use a 1% percentile of the gradient to select the blocks for which variance is kept, while the high-pass image is a higher-order differential operator. The

Table 2.3. Table summarizing all methods. 'Block deviation' means standard deviation of block; 'at percentile 1%' means that the chosen value is the one at which the 1% percentile is attained; '3-DCT' means the three highest-frequency coefficients, namely DCT$(6,7)$, DCT$(7,6)$, DCT$(7,7)$; 'DCT 7×7' means the DCT wave associated to the highest-frequency coefficient of the 7×7 pixel support of the DCT-II transform of the block; MAD stands for *median of absolute deviation* (it is applied to the three DCT coefficients for all blocks). The methods belong to three classes. The first main class (rows 1–5) does high-pass + standard deviation of blocks + low percentile. The second class (rows 6–7) replaces the percentile by a mode of the high-pass filter histogram. Rows 8–11 are more primitive and do a simple mean of the block variances of the high-pass filtered image. The last method is unclassifiable, and performs poorly.

Method	Source	High-pass	Block	Estimator	Percentile
percentile with learning	Buades *et al.* (2012*a*), Ponomarenko *et al.* (2007)	DCT 7×7	$\underline{21}$	block deviation at percentile	0.5%
percentile	Buades *et al.* (2012*a*), Ponomarenko *et al.* 2007)	DCT 7×7	$\underline{21}$	block deviation at percentile	0.5%
block	Lee (1981), Mastin (1985)	none	7	mean of block deviation	1%
average	Olsen (1993)	Δ	3	mean of block deviation	1% of gradient histogram
median	Olsen (1993)	curv	3	mean of block deviation	1% of gradient histogram
scatter	Lee & Hoppel (1989)	none	8	block deviation at	block deviation mode
gradient	Bracho & Sanderson (1985), Voorhees & Poggio (1987)	∇	1	$\lvert\nabla\rvert$ mode	all
EINV	Rank, Lendl & Unbehauen (1999)	$\frac{\partial^2}{\partial x \partial y}$	3	deconvolution of block dev.	all
FNVE	Immerkaer (1996)	$\Delta_1 - \Delta_2$	1	RMS	all
DCT-MAD	Donoho & Johnstone (1995)	3-DCT	8	MAD of 3 DCT coefficients	all
DCT-mean	Ponomarenko *et al.* (2003)	3-DCT	8	mean of variances	all
pyramid	Meer, Jolion & Rosenfeld (1990)	none	2^L	block deviation	complex

Table 2.4. White homoscedastic Gaussian noise RMSE results for all methods and for varying σ. The pyramid tests were omitted, being incomplete. As they are obtained as an average over many noiseless images, the differences have been checked to be statistically significant. It is also clear that the ranking of the compared methods may vary with the amount of noise. Nevertheless, the ranks of methods for noises larger than 20 are irrelevant, because all of them work at an acceptable level of precision. Thus, this ranking is mainly relevant to low noise levels, $\sigma = 1, 2, 5, 10$.

Method	$\sigma = 1$	$\sigma = 2$	$\sigma = 5$	$\sigma = 10$	$\sigma = 20$	$\sigma = 50$	$\sigma = 80$
percentile	0.309	0.276	0.265	0.315	0.293	0.130	0.229
percentile learning	0.182	0.152	0.157	0.364	0.240	0.248	0.270
block	1.093	0.961	0.949	1.056	0.984	0.922	0.840
average	2.669	2.556	2.375	2.165	1.771	1.227	0.874
median	2.841	2.762	2.640	2.460	2.110	1.684	1.502
scatter	4.533	4.013	3.141	2.290	1.436	1.488	1.862
gradient	1.887	1.851	1.474	1.393	1.354	1.234	2.949
EINV	1.406	1.159	0.924	0.842	0.656	0.450	0.557
FNVE	2.738	2.231	1.357	0.767	0.397	0.196	0.225
DCT-MAD	0.858	0.721	0.533	0.356	0.239	0.296	0.583
DCT-mean	1.895	1.469	0.837	0.462	0.316	0.355	0.726

pyramid method (Meer *et al.* 1990) is quite complex, but uses all standard deviations of all possible blocks in the image. We do not provide the detailed algorithm. The *FNVE* method (Immerkaer 1996) in fact has no outlier elimination, taking simply the root mean square of all samples of the high-passed image.

Rather than using a percentile of the block variance histogram followed by a compensation factor, several methods extract a mode, considering that the mode (peak of the histogram variance) must correspond to the noise. The *gradient* method (Bracho and Sanderson 1985, Voorhees and Poggio 1987) sets $\hat{\sigma}$ to be the peak of the modulus of the gradient histogram. The *scatter* method (Lee and Hoppel 1989) also computes a mode when estimating white homoscedastic noise, namely the value at which the peak of the block standard deviations histogram is attained. The *EINV* method (Rank *et al.* 1999) does a kind of iterative deconvolution of the histogram of block variances and also extracts its mode.

All the values obtained by these methods are proportional to the noise standard deviation when the image is white noise. Thus the final step, not mentioned in the table, is to apply a correction factor to get the final estimated noise standard deviation, as explained in the percentile method (Section 2.3).

The comparison of the methods which use the highest DCT coefficients, DCT-mean (Ponomarenko *et al.* 2003) and DCT-MAD (Donoho and Johnstone 1995), where MAD stands for *median of absolute deviation*, clearly shows the advantage of a robust estimator: the estimate is obtained by averaging the three medians of absolute deviation of the three highest-frequency DCT coefficients for all blocks.

The ultimate choice of method is of course steered by the RMSE, that is, the root mean square error between the estimated value of σ and σ itself, taken over a representative set of images. As Table 2.4 shows, the ordering of methods by their RMSE is consistent and points to the percentile method as the best one. This method is further improved by learning. A good point justifying all methods is that they perform satisfactorily for all large noise values, down to $\sigma = 20$. But, with the exception of the percentile method with learning, no method performs acceptably for $\sigma < 5$.

3. Four denoising principles

In this section we will review the main algorithmic principles proposed for noise removal. All of them of course use a model for the noise, which in our study will always be Gaussian white noise. More interestingly, each principle implies a model for the ideal noiseless image. The Bayesian principle is coupled with a Gaussian (or a mixture of Gaussians) model for noiseless patches. Transform thresholding assumes that most image coefficients are high and sparse in a given well-chosen orthogonal basis, while noise remains white (and therefore with homoscedastic coefficients in any orthogonal basis). Sparse coding assumes the existence of a dictionary of patches on which most image patches can be decomposed with a sparse set of coefficients. Finally the averaging principle relies on an image self-similarity assumption. Thus four considered denoising principles are:

- Bayesian patch-based methods (Gaussian patch model),
- transform thresholding (sparsity of patches in a fixed basis),
- sparse coding (sparsity on a learned dictionary),
- pixel averaging and block averaging (image self-similarity).

As we will see in this review, the current state-of-the-art denoising recipes are in fact a smart combination of *all* these ingredients.

3.1. Bayesian patch-based methods

Let u be the noiseless ideal image, and let $\tilde{u}$ be the noisy image corrupted with Gaussian noise of standard deviation σ, so that

$$\tilde{u} = u + n. \tag{3.1}$$

Then the conditional distribution $\mathbb{P}(\tilde{u} \mid u)$ is given by

$$\mathbb{P}(\tilde{u} \mid u) = \frac{1}{(2\pi\sigma^2)^{\frac{M}{2}}} \mathrm{e}^{-\frac{\|u-\tilde{u}\|^2}{2\sigma^2}}, \tag{3.2}$$

where M is the total number of pixels in the image.

In order to compute the probability of the original image given the degraded one, $\mathbb{P}(u \mid \tilde{u})$, we need to introduce a prior on u. In the first models (Geman and Geman 1984), this prior was a parametric image model describing the stochastic behaviour of a patch around each pixel by a Markov random field, specified by its Gibbs distribution. A Gibbs distribution for an image u takes the form

$$\mathbb{P}(u) = \frac{1}{Z}\mathrm{e}^{-E(u)/T},$$

where Z and T are constants and E is called the energy function, and is given by

$$E(u) = \sum_{C \in \mathcal{C}} V_C(u),$$

where $\mathcal{C}$ denotes the set of cliques associated with the image and V_C is a potential function. The maximization of the *a posteriori* distribution is given by Bayes' formula

$$\arg\max_u \mathbb{P}(u \mid \tilde{u}) = \arg\max_u \mathbb{P}(\tilde{u} \mid u)\mathbb{P}(u),$$

which is equivalent to the minimization of $-\log \mathbb{P}(u \mid \tilde{u})$,

$$\arg\min_u \|u - \tilde{u}\|^2 + \frac{2\sigma^2}{T}E(u).$$

This energy is thus a sum of local derivatives of pixels in the image, and is therefore equivalent to a classical Tikhonov regularization (Geman and Geman 1984, Brémaud 1999).

Recent Bayesian methods have abandoned as too simplistic the global patch models formulated by an *a priori* Gibbs energy. Instead, the methods build local non-parametric patch models learned from the image itself, usually as a local Gaussian model around each given patch, or as a Gaussian mixture. The term 'patch model' is now preferred to the terms 'neighbourhood' or 'clique', previously used for the Markov field methods. In the non-parametric models the patches are larger, usually 8×8, while the cliques are often confined to 3×3 neighbourhoods. Given a noiseless patch P of u with dimension $\kappa \times \kappa$, and an observed noisy version $\tilde{P}$ of P, the same model gives

$$\mathbb{P}(\tilde{P}|P) = c \cdot \mathrm{e}^{-\frac{\|\tilde{P}-P\|^2}{2\sigma^2}}, \tag{3.3}$$

by the independence of noise pixel values, where P and $\tilde{P}$ are regarded as vectors with κ^2 components and $\|P\|$ denotes the Euclidean norm of P, *i.e.*, the Frobenius norm on $\kappa \times \kappa$ matrices. Knowing $\tilde{P}$, our goal is to deduce P by maximizing $\mathbb{P}(P|\tilde{P})$. Using Bayes' rule, we can compute this last conditional probability as

$$\mathbb{P}(P|\tilde{P}) = \frac{\mathbb{P}(\tilde{P}|P)\mathbb{P}(P)}{\mathbb{P}(\tilde{P})}. \tag{3.4}$$

Given $\tilde{P}$, this formula can in principle be used to deduce the patch P maximizing the right term, viewed as a function of P. This is only possible if we have a probability model for P, and these models will generally be learned from the image itself, or from a set of images. For example, Chatterjee and Milanfar (2012) apply a clustering method to the set of patches of a given image, and Zoran and Weiss (2011) apply it to a huge set of patches extracted from many images. Each cluster of patches is thereafter treated as a set of Gaussian samples. This permits us to associate the likeliest cluster with each observed patch, and then to denoise it with a Bayesian estimate in this cluster. A more direct way to build a model for a given patch $\tilde{P}$ is to group the patches similar to $\tilde{P}$ in the image. Assuming that these similar patches are samples of a Gaussian vector yields a standard Bayesian restoration (Lebrun, Buades and Morel 2011). We shall now discuss this particular case, where all observed patches are noisy.

Why Gaussian? As usual when we have several observations but no particular knowledge of the form of the probability density, a Gaussian model is adopted. In the case of the patches Q, similar to a given patch P, the Gaussian model has some pertinence, as it is assumed that many contingent random factors explain the difference between Q and P: other details, *e.g.*, texture, slight lighting changes, shadows. The Gaussian model, together with a combination of many such random and independent factors, is heuristically justified by the central limit theorem. Thus, for good or ill, assume that the patches Q similar to P follow a Gaussian model with (observable, empirical) covariance matrix $\mathbf{C}_P$ and (observable, empirical) mean $\overline{P}$. This means that

$$\mathbb{P}(Q) = c \cdot \mathrm{e}^{-\frac{(Q-\overline{P})^T \mathbf{C}_P^{-1}(Q-\overline{P})}{2}}. \tag{3.5}$$

From (3.2) and (3.4), for each observed $\tilde{P}$ we obtain the following equivalent problems:

$$\begin{aligned}
\max_P \mathbb{P}(P|\tilde{P}) &\Leftrightarrow \max_P \mathbb{P}(\tilde{P}|P)\mathbb{P}(P) \\
&\Leftrightarrow \max_P \mathrm{e}^{-\frac{\|P-\tilde{P}\|^2}{2\sigma^2}} \mathrm{e}^{-\frac{(P-\overline{P})^T \mathbf{C}_P^{-1}(P-\overline{P})}{2}} \\
&\Leftrightarrow \min_P \frac{\|P-\tilde{P}\|^2}{\sigma^2} + (P-\overline{P})^T \mathbf{C}_P^{-1}(P-\overline{P}).
\end{aligned}$$

This expression does not yield an algorithm. Indeed, the noiseless patch P and the patches similar to P are not observable. Nevertheless, we can observe the noisy version $\tilde{P}$ and compute the patches $\tilde{Q}$ similar to $\tilde{P}$. An empirical covariance matrix can therefore be obtained for the patches $\tilde{Q}$ similar to $\tilde{P}$. Furthermore, using (3.1) and the fact that P and the noise n are independent,

$$\mathbf{C}_{\tilde{P}} = \mathbf{C}_P + \sigma^2 \mathbf{I}, \quad \mathbb{E}\tilde{Q} = \overline{P}. \tag{3.6}$$

Note that these relations assume a search for patches similar to $\tilde{P}$, at a sufficiently large distance to include all patches similar to P but not so large that it contains outliers. Thus the safe strategy is to search for similar patches at a distance slightly larger than the expected distance caused by noise. If the above estimates are correct, our MAP (maximum *a posteriori* estimation) problem finally reduces via (3.6) to the following feasible minimization problem:

$$\max_P \mathbb{P}(P|\tilde{P}) \Leftrightarrow \min_P \frac{\|P - \tilde{P}\|^2}{\sigma^2} + (P - \overline{\tilde{P}})^T (\mathbf{C}_{\tilde{P}} - \sigma^2 \mathbf{I})^{-1} (P - \overline{\tilde{P}}).$$

Differentiating this quadratic function with respect to P and equating to zero yields

$$P - \tilde{P} + \sigma^2 (\mathbf{C}_{\tilde{P}} - \sigma^2 \mathbf{I})^{-1} (P - \overline{\tilde{P}}) = 0.$$

Taking into account that $\mathbf{I} + \sigma^2 (\mathbf{C}_{\tilde{P}} - \sigma^2 \mathbf{I})^{-1} = (\mathbf{C}_{\tilde{P}} - \sigma^2 \mathbf{I})^{-1} \mathbf{C}_{\tilde{P}}$, this yields

$$(\mathbf{C}_{\tilde{P}} - \sigma^2 \mathbf{I})^{-1} \mathbf{C}_{\tilde{P}} P = \tilde{P} + \sigma^2 (C_{\tilde{P}} - \sigma^2 \mathbf{I})^{-1} \overline{\tilde{P}},$$

and therefore

$$\begin{aligned} P &= \mathbf{C}_{\tilde{P}}^{-1} (\mathbf{C}_{\tilde{P}} - \sigma^2 \mathbf{I}) \tilde{P} + \sigma^2 \mathbf{C}_{\tilde{P}}^{-1} \overline{\tilde{P}} \\ &= \tilde{P} + \sigma^2 \mathbf{C}_{\tilde{P}}^{-1} (\overline{\tilde{P}} - \tilde{P}) \\ &= \overline{\tilde{P}} + \left[\mathbf{I} - \sigma^2 \mathbf{C}_{\tilde{P}}^{-1}\right] (\tilde{P} - \overline{\tilde{P}}) \\ &= \overline{\tilde{P}} + \left[\mathbf{C}_{\tilde{P}} - \sigma^2 \mathbf{I}\right] \mathbf{C}_{\tilde{P}}^{-1} (\tilde{P} - \overline{\tilde{P}}). \end{aligned}$$

Thus we have proved that a restored patch $\hat{P}_1$ can be obtained from the observed patch $\tilde{P}$ by the one step estimate

$$\hat{P}_1 = \overline{\tilde{P}} + \left[\mathbf{C}_{\tilde{P}} - \sigma^2 \mathbf{I}\right] \mathbf{C}_{\tilde{P}}^{-1} (\tilde{P} - \overline{\tilde{P}}), \tag{3.7}$$

which resembles a local Wiener filter.

Remark. It is easily deduced that the expected estimation error is

$$\mathbb{E}\|P - \hat{P}_1\|^2 = \mathrm{Tr}\left[\left(\mathbf{C}_P^{-1} + \frac{\mathbf{I}}{\sigma^2}\right)^{-1}\right].$$

Sections 5.2–5.6 and 5.9 will examine no less than *six Bayesian algorithms* deriving patch-based denoising algorithms from variants of (3.7). The first question, when looking at this formula, is obviously how the matrix $\mathbf{C}_{\tilde{P}}$ can be learned from the image itself. Each method proposes a different approach to learning the patch model.

Of course, other non-Gaussian Bayesian models are possible, depending on patch density assumptions. For example, Raphan and Simoncelli (2010) assume a local exponential density model for the noisy data, and give a convergence proof for the optimal (Bayes) least-squares estimator as the amount of data increases.

3.2. Transform thresholding

Classical transform coefficient thresholding algorithms, such as the DCT or wavelet denoising, use the observation that images are faithfully described by keeping only their large coefficients in a well-chosen basis. By keeping these large coefficients and setting to zero the small ones, noise should be removed and image geometry kept. By any orthogonal transform, the coefficients of a homoscedastic de-correlated noise remain de-correlated and homoscedastic. For example, the wavelet or DCT coefficients of Gaussian white noise with variance σ^2 remain Gaussian white noise with variance σ^2. Thus, a threshold on the coefficients at, say, 3σ removes most of the coefficients that are only due to noise. (The expectation of these coefficients is assumed to be zero.) The *sparsity* of image coefficients in certain bases is only an empirical observation. It is nevertheless invoked in most denoising and compression algorithms, which rely essentially on coefficient thresholds. The established image compression algorithms are based on the DCT (in the JPEG 1992 format) or, like the JPEG 2000 format (Antonini, Barlaud, Mathieu and Daubechies 1992), on biorthogonal wavelet transforms (Cohen, Daubechies and Feauveau 1992).

Let $\mathcal{B} = \{G_i\}_{i=1}^M$ be an orthonormal basis of $\mathbb{R}^M$, where M is the number of pixels of the noisy image $\tilde{U}$ (regarded here as a 'long' vector). Then we have

$$\langle \tilde{U}, G_i \rangle = \langle U, G_i \rangle + \langle N, G_i \rangle, \tag{3.8}$$

where $\tilde{U}$, U and N denote, respectively, the noisy, original and noise images. We always assume that the noise values $N(\mathbf{i})$ are uncorrelated and homoscedastic with zero mean and variance σ^2. The following calculation shows that the noise coefficients in the new basis remain uncorrelated, with zero mean and variance σ^2:

$$\mathbb{E}[\langle N, G_i \rangle \langle N, G_j \rangle] = \sum_{\mathbf{r},\mathbf{s}=1}^{M} G_i(\mathbf{r}) G_j(\mathbf{s}) \mathbb{E}[\mathbf{w}(\mathbf{r})\mathbf{w}(\mathbf{s})]$$
$$= \langle G_i, G_j \rangle \sigma^2 = \sigma^2 \delta[j-i].$$

Each noisy coefficient $\langle \tilde{U}, G_i \rangle$ is modified independently, and then the solution is estimated by the inverse transform of the new coefficients. Noisy coefficients are modified by multiplying by an attenuation factor $a(i)$, and the inverse transform yields the estimate

$$\mathbf{D}\tilde{U} = \sum_{i=1}^{M} a(i) \langle \tilde{U}, G_i \rangle G_i. \tag{3.9}$$

$\mathbf{D}$ is also called a *diagonal operator*. Noise reduction is achieved by attenuating or setting to zero small coefficients of order σ, assumed to be due to noise, while the original signal is preserved by keeping the large coefficients. This intuition is corroborated by the following result.

Theorem 3.1. The operator $\mathbf{D}_{\text{inf}}$ minimizing the mean square error (MSE),

$$\mathbf{D}_{\text{inf}} = \arg\min_{\mathbf{D}} \mathbb{E}\{\|U - \mathbf{D}\tilde{U}\|^2\},$$

is given by the family $\{a(i)\}_i$, where

$$a(i) = \frac{|\langle U, G_i \rangle|^2}{|\langle U, G_i \rangle|^2 + \sigma^2}, \tag{3.10}$$

and the corresponding expected MSE is

$$\mathbb{E}\{\|U - \mathbf{D}_{\text{inf}}\tilde{U}\|^2\} = \sum_{i=1}^{M} \frac{|\langle U, G_i \rangle|^2 \sigma^2}{|\langle U, G_i \rangle|^2 + \sigma^2}. \tag{3.11}$$

The previous optimal operator attenuates all noisy coefficients. If one restricts $a(i)$ to be 0 or 1, one gets a projection operator. In that case, a subset of coefficients is kept, and the rest are set to zero. The projection operator that minimizes the MSE under that constraint is obtained using

$$a(i) = \begin{cases} 1 & |\langle U, G_i \rangle|^2 \geq \sigma^2, \\ 0 & \text{otherwise}, \end{cases}$$

and the corresponding MSE is

$$\mathbb{E}\{\|U - \mathbf{D}_{\text{inf}}\tilde{U}\|^2\} = \sum_{i} \min(|\langle U, G_i \rangle|^2, \sigma^2). \tag{3.12}$$

A *transform thresholding* algorithm therefore keeps the coefficients with a magnitude larger than the noise, while setting the zero the rest. Note that both above-mentioned filters are 'ideal', or 'oracular' operators. Indeed, they use the coefficients $\langle U, G_i \rangle$ of the original image, which are not known. These algorithms are therefore usually called *oracle filters.* We shall discuss their implementation in the following subsections. For the moment, we shall introduce the classical thresholding filters, which approximate the oracle coefficients by using the noisy ones.

As is classical, the optimal operator (3.10) is called the *Fourier–Wiener filter* when $\mathcal{B}$ is a Fourier basis. By the use of the Fourier basis, global image characteristics may prevail over local ones and create spurious periodic patterns. To avoid this effect, the bases are usually more local, of the wavelet or block DCT type.

Sliding-window DCT

The local adaptive filters were introduced by Yaroslavsky and Eden (2003) and Yaroslavsky (1996). The noisy image is analysed in a moving window, and at each position of the window its DCT spectrum is computed and modified by using the optimal operator (3.10). Finally, an inverse transform is used to estimate only the signal value in the central pixel of the window.

This method is called the *empirical Wiener filter*, because it approximates the unknown original coefficients $\langle u, G_i \rangle$ by using the identity

$$\mathbb{E}|\langle \tilde{U}, G_i \rangle|^2 = |\langle U, G_i \rangle|^2 + \sigma^2,$$

and thus replacing the optimal attenuation coefficients $a(i)$ with the family $\{\alpha(i)\}_i$,

$$\alpha(i) = \max\left\{0, \frac{|\langle \tilde{U}, G_i \rangle|^2 - c\sigma^2}{|\langle \tilde{U}, G_i \rangle|^2}\right\}.$$

where c is a parameter, usually larger than one.

Wavelet thresholding

Let $\mathcal{B} = \{G_i\}_i$ be a wavelet orthonormal basis (Mallat 1999). The so-called *hard wavelet thresholding method* (Donoho and Johnstone 1994) is a (nonlinear) projection operator setting to zero all wavelet coefficients smaller than a certain threshold. According to the expression of the MSE of a projection operator (3.12), the performance of the method depends on the ability of the basis to approximate the image U by a small set of large coefficients. There has been a strenuous search for wavelet bases adapted to images (Pennec and Mallat 2003).

Unfortunately, image features, just like noise, can also cause many small wavelet coefficients, which are nevertheless lower than the threshold. The dramatic cancellation of wavelet (or DCT) coefficients near the image edges creates small oscillations, a Gibbs phenomenon often called *ringing.* Spurious wavelets can also be seen in flat parts of the restored image, caused by the undue cancellation of some of the small coefficients. These artifacts are sometimes called *wavelet outliers* (Durand and Nikolova 2003). These undesirable effects can be partially avoided with the use of soft thresholding (Donoho 1995), for example,

$$\alpha(i) = \begin{cases} \frac{\langle \tilde{U}, G_i\rangle - \mathrm{sgn}(\langle \tilde{U}, G_i\rangle)\mu}{\langle \tilde{U}, G_i\rangle} & |\langle \tilde{U}, G_i\rangle| \geq \mu, \\ 0 & \text{otherwise.} \end{cases}$$

The continuity of this soft thresholding operator reduces the Gibbs oscillation near image discontinuities.

Several orthogonal bases adapt better to image local geometry and discontinuities than wavelets, particularly the 'bandlets' (Pennec and Mallat 2003) and 'curvelets' (Starck, Candès and Donoho 2002). This tendency to adapt the transform locally to the image is accentuated by the methods adapting a different basis to each pixel, or selecting a few elements or 'atoms' from a huge patch dictionary to linearly decompose the local patch on these atoms. This perspective is sketched in the next subsection, on sparse coding.

3.3. Sparse coding

Sparse coding algorithms learn a redundant set $\mathbf{D}$ of vectors called a *dictionary*, and choose the right atoms to describe the current patch.

For a fixed patch size, the dictionary is encoded as a matrix of size $\kappa^2 \times n_{\text{dict}}$, where κ^2 is the number of pixels in the patch and $n_{\text{dict}} \geq \kappa^2$. The dictionary patches, which are columns of the matrix, are normalized (in the Euclidean norm). This dictionary may contain the usual orthogonal bases (*e.g.*, discrete cosine transform, wavelets, curvelets), but also patches extracted (or learned) from clean images, or even from the noisy image itself.

The dictionary permits computation of a sparse representation α of each patch P, where α is a coefficient vector of size n_{dict}^2 satisfying $P \approx \mathbf{D}\alpha$. This sparse representation α can be obtained with an ORMP (orthogonal recursive matching pursuit: Cotter, Adler, Rao and Kreutz-Delgado (1999)). The ORMP gives an approximate solution to the (NP-complete) problem

$$\arg\min_{\alpha} \|\alpha\|_0 \quad \text{such that} \quad \|P - \mathbf{D}\alpha\|_2^2 \leq \kappa^2 (C\sigma)^2, \tag{3.13}$$

where $\|\alpha\|_0$ refers to the l^0-norm of α, *i.e.*, the number of non-zero coefficients of α. This last constraint brings in a new parameter C. This coefficient multiplying the standard deviation σ guarantees that, with high probability, white Gaussian noise of standard deviation σ on κ^2 pixels has an l^2-norm lower than $\kappa C\sigma$. The ORMP algorithm is introduced in Cotter *et al.* (1999). Details of how this minimization can be achieved are given in Section 5.7 on the K-SVD algorithm. (It has been argued that the l^0-norm of the set of coefficients can be replaced by the much easier l^1 convex norm.

This remark is the starting point of the compressive sampling method; see Candès and Wakin (2008).)

In K-SVD and other current sparse coding algorithms, the previous denoising strategy is used as the first step of a two-step algorithm. The selection step is iteratively combined with an update of the dictionary, taking into account the image and the sparse codifications already computed. More details will be found in Section 5.7.

Several of our referees have objected to considering *sparse coding* and *transform thresholding* as two different denoising principles. As models, both indeed assume the *sparsity* of patches in some well-chosen basis. Nevertheless, some credit must be given to historical development. The notion of sparsity is associated with a recent and sophisticated variational principle, where the dictionary and the sparse decompositions are computed simultaneously. Transform thresholding methods existed before the term sparsity was even used. They simply pick a local wavelet or DCT basis and threshold the coefficients. In both algorithms, the sparsity is implicitly or explicitly assumed. But transform threshold methods use orthogonal bases, while the dictionaries are redundant. Furthermore, the algorithms are very different.

3.4. Image self-similarity leading to pixel averaging

The principle of many denoising methods is quite simple: they replace the colour of a pixel with an average of the colours of nearby pixels. It is a powerful and basic principle, when applied directly to noisy pixels with independent noise. If m pixels with the same colour (up to the fluctuations due to noise) are averaged, the noise is reduced by a factor of $\sqrt{m}$.

The MSE between the true (unknown) value $u(\mathbf{i})$ of a pixel $\mathbf{i}$ and the value estimated by a weighted average of pixels $\mathbf{j}$ is

$$\begin{aligned}\mathbb{E}\left\|u(\mathbf{i}) - \sum_{\mathbf{j}} w(\mathbf{j})\tilde{u}(\mathbf{j})\right\|^2 &= \mathbb{E}\left\|\sum_{\mathbf{j}} w(\mathbf{j})(u(\mathbf{i}) - u(\mathbf{j})) - \sum_{\mathbf{j}} w(\mathbf{j})n(\mathbf{j})\right\|^2 \\ &= \sum_{\mathbf{j}} w(\mathbf{j})^2(u(\mathbf{i}) - u(\mathbf{j}))^2 + \sigma^2 \sum_{\mathbf{j}} w(\mathbf{j})^2, \quad (3.14)\end{aligned}$$

where we assume that the noise, the image and the weights are independent and that the weights $\{w(\mathbf{j})\}_{\mathbf{j}}$ satisfy $\sum_{\mathbf{j}} w(\mathbf{j}) = 1$.

The above expression implies that the performance of the averaging depends on the ability to find many pixels $\mathbf{j}$ with an original value $u(\mathbf{j})$ close to $u(\mathbf{i})$. Indeed, the variance term $\sum_{\mathbf{j}} w(\mathbf{j})^2$ is minimized by a flat distribution probability $w(\mathbf{j}) = 1/m$, where m is the number of averaged pixels. The first term measures the bias caused by the fact that pixels do not have exactly the same deterministic value. Each method must find a trade-off between the bias and variance terms of equation (3.14).

Averaging spatially close pixels

An initial rather trivial idea is to average the closest pixels to a given pixel. This amounts to convolving the image with a fixed radial positive kernel. The archetype of such kernels is the Gaussian kernel.

The convolution of the image with a Gaussian kernel ensures a fixed noise standard deviation reduction factor that equals the kernel standard deviation. But nearby pixels do not necessarily share their colours. Thus, the first error term in (3.14) can quickly increase. This approach is only valid for pixels for which nearby pixels have the same colour, that is, it only works inside the homogeneous image regions but not for their boundaries.

Averaging pixels with similar colours

A simple solution to the above-mentioned dilemma is given by the sigma filter (Lee 1983) or neighbourhood filter (Yaroslavsky 1985). These filters average only nearby pixels of $\mathbf{i}$ having also a similar colour value. We shall denote these filters by YNF (Yaroslavsky neighbourhood filter). Their formula is simply

$$\mathrm{YNF}_{h,\rho}\,\tilde{u}(\mathbf{i}) = \frac{1}{C(\mathbf{i})} \sum_{\mathbf{j}\in B_\rho(\mathbf{i})} \tilde{u}(\mathbf{j})\,\mathrm{e}^{-\frac{|\tilde{u}(\mathbf{i})-\tilde{u}(\mathbf{j})|^2}{h^2}}, \tag{3.15}$$

where $B_\rho(\mathbf{i})$ is a ball of centre $\mathbf{i}$ and radius $\rho > 0$, $h > 0$ is the filtering parameter, and

$$C(\mathbf{i}) = \sum_{\mathbf{j}\in B_\rho(\mathbf{i})} \mathrm{e}^{-\frac{|\tilde{u}(\mathbf{j})-\tilde{u}(\mathbf{i})|^2}{h^2}}$$

is the normalization factor. The parameter h controls the degree of colour similarity that needs to be taken into account in the average. According to the Bayesian interpretation of the filter we should have $h = \sigma$. The filter (3.15), due to Yaroslavsky and Lee, has been reinvented several times, and has received the alternative names of *SUSAN filter* (Smith and Brady 1997) and *bilateral filter* (Tomasi and Manduchi 1998). The relatively minor difference in these algorithms is that instead of considering a fixed spatial neighbourhood $B_\rho(\mathbf{i})$, they weigh the spatial distance to the reference pixel $\mathbf{i}$ by a Gaussian.

Neighbourhood filters choose the 'neighbouring' pixels by comparing their noisy colour. The weight distribution is therefore computed by using noisy values and is not independent of the noise. Therefore the error formula (3.14) is not applicable. For a flat zone and for a given pixel with colour value a, the nearby pixels with an intensity difference lower than h will be independent and identically distributed, with a probability distribution which is the restriction of the Gaussian to the interval $(a-h, a+h)$. If the search zone (or spatial neighbourhood) is broad enough, then the average value will tend to the expectation of this random variable. Thus, the increase of the search zone, and therefore of the number of pixels being averaged beyond a reasonable value, will not increase the noise reduction capability of the filter. More precisely, the asymptotic noise reduction factor is given in the next theorem, taken from Buades (2006).

Theorem 3.2. Assume that $n(\mathbf{i})$ are independent identically distributed, with zero mean and variance σ^2. Then a noise n filtered by the neighbourhood filter YNF_h satisfies

$$\operatorname{Var} \mathrm{YNF}_{h,\rho}\, n = f\left(\frac{h}{\sigma}\right)\sigma^2,$$

where

$$f(x) = \frac{1}{(2\pi)^{3/2}} \int_{\mathbb{R}} \frac{1}{\beta^2(a,x)} (\mathrm{e}^{2xa} - 1)^2 \mathrm{e}^{(a+x)^2} \mathrm{e}^{\frac{-a^2}{2}}\, da$$

and

$$\beta(a,x) = \frac{1}{\sqrt{2\pi}} \int_{a-x}^{a+x} \mathrm{e}^{-t^2/2}\, dt.$$

The function $f(x)$ is decreasing with $f(0) = 1$ and

$$\lim_{x\to\infty} f(x) = 0$$

(see Figure 3.1). The noise reduction increases with the ratio h/σ. We see that $f(x)$ is close to zero for values of x over 2.5 or 3, that is, values of h over $2.5\,\sigma$ or $3\,\sigma$. This corresponds to the values proposed in the original papers by Lee and Yaroslavsky. However, for a Gaussian variable, the probability of observing values at a distance to the average above 2.5 or 3 times the standard deviation is very small. Thus, taking these large values excessively increases the probability of mismatching pixels that in fact

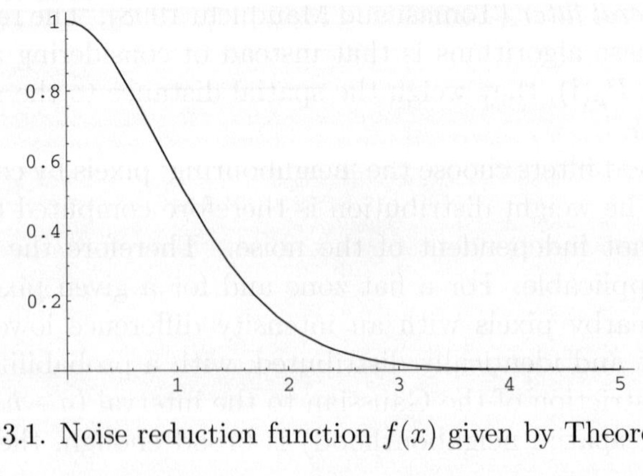

Figure 3.1. Noise reduction function $f(x)$ given by Theorem 3.2.

belong to other objects. This explains the observed decaying performance of the neighbourhood filter when the noise standard deviation or the search zone $B(\mathbf{i}, \rho)$ increase too much.

The image model underlying neighbourhood filters is that of *image self-similarity*, namely the presence in the image of pixels $\mathbf{j}$ which have the same law as $\mathbf{i}$. In Section 5.1 we will introduce the NL-means algorithm (Buades *et al.* 2005*b*), which can be seen as an extension of the neighbourhood filters attenuating their main drawbacks. In NL-means, the 'neighbourhood of a pixel $\mathbf{i}$' is defined as any set of pixels $\mathbf{j}$ in the image such that a patch around $\mathbf{j}$ looks like a patch around $\mathbf{i}$. In other words NL-means estimates the value of $\mathbf{i}$ as an average of the values of all the pixels $\mathbf{j}$ whose neighbourhood looks like the neighbourhood of $\mathbf{i}$.

4. Noise reduction: generic tools

This section describes four generic tools that permit an increase in the performance of *any* denoising principle. We shall illustrate them for DCT denoising. Starting from the application of a simple DCT transform threshold, the four generic tools will be applied successively. We shall observe a dramatic improvement of the denoising performance. This observation is valid for all denoising principles.

4.1. Aggregation of estimates

Aggregation techniques combine for any pixel a set of m possible estimates. If these estimates were independent and had equal variance, then a uniform average would reduce this estimator variance by a factor of m. Such an aggregation strategy was the main proposition of the *translation-invariant wavelet thresholding algorithm* (Coifman and Donoho 1995). This method denoises several translations of the image by a wavelet thresholding algorithm and averages these different estimates once the inverse translation has been applied to the denoised images.

An interesting case is when one is able to estimate the variance of the m estimators. Statistical arguments lead us to attribute to each estimator a weight inversely proportional to its variance (Nemirovski 2000). For most denoising methods the variance of the estimators is high near image edges. When applied without aggregation, the denoising methods leave visible 'halos' of residual noise near edges. For example, in the sliding-window DCT method, patches containing edges have many large DCT coefficients which are kept by thresholding. In flat zones, however, most DCT coefficients are cancelled and the noise is completely removed. The proposition of Guleryuz (2007) is to use the aggregation for DCT denoising, approximating the variance of each estimated patch by the number of non-zero coefficients after thresholding. In the online paper by Yu and Sapiro (2011) one can test an implementation of DCT denoising. They use an aggregation with uniform weights: 'translation invariant DCT denoising is implemented by decomposing the image to sliding overlapping patches, calculating the DCT denoising in each patch, and then aggregating the denoised patches to the image averaging the overlapped pixels. The translation invariant DCT denoising significantly improves the denoising performance, typically from about 2 to 5 dB, and removes the block artifact' (Yu and Sapiro 2011).

The same risk of 'halos' occurs with non-aggregated NL-means (Section 5.1), since patches containing edges have far fewer similar instances in the image than flat patches. Thus the non-local averaging is made over fewer samples, and the final result keeps more noise near image edges. The same phenomenon occurs with BM3D (Section 5.8) if the aggregation step is not applied (Dabov *et al.* 2007). As a consequence, an aggregation step is applied in all patch-based denoising algorithms. This weighted aggregation favours, at each pixel near an edge, the estimates given by patches which contain the pixel but do not meet the edge.

Aggregation techniques aim at superior noise reduction by increasing the number of values being averaged to obtain the final estimate, or selecting those estimates with lower variance. Kervrann and Boulanger (2008) considered the whole bias + variance decomposition in order to also adapt the search zone of neighbourhood filters or NL-means. Since the bias term depends on the original image, it cannot be computed in practice, and Kervrann and Boulanger proposed minimization of both bias and variance by choosing the smallest spatial neighbourhood attaining a stable noise reduction.

Another type of aggregation technique considers the risk estimate rather than the variance, to locally attribute more weight to the estimators with small risks. Van De Ville and Kocher (2009) give a closed-form expression of Stein's unbiased estimator of the risk (SURE) for NL-means. (See also the generalizations of the SURE estimator to the non-Gaussian case in Raphan and Simoncelli (2007).) The aim is to select globally the best bandwidth for a given image. Duval, Aujol and Gousseau (2011) also use the SURE technique to minimize the risk by selecting the bandwidth locally. Deledalle, Duval and Salmon (2012) apply the same technique to combine the results of NL-means with different window sizes and shapes. A similar treatment can be found in Raphan and Simoncelli (2010), but with the assumption of a local exponential density for the noisy patches.

4.2. Iteration and 'oracle' filters

Iterative strategies to remove residual noise would drift from the initial image. Instead, a first step denoised image can be used to improve the reapplication of the denoising method to the initial noisy image. In a second step application of a denoising principle, the denoised DCT coefficients, or the patch distances, can be computed in the first step denoised image. They are an approximation to the true measurements that would be obtained from the noise-free image. Thus, the first step denoised image is used as an 'oracle' for the second step.

For averaging filters such as neighbourhood filters or NL-means, the image u can be denoised in a first step by the method under consideration. This first step denoised image, denoted by $\hat{u}_1$, is used to compute more accurate colour distances between pixels. Thus the second step neighbourhood filter is given by

$$\mathrm{YNF}_{h,\rho}\, \tilde{u}(\mathbf{i}) = \frac{1}{C(\mathbf{i})} \sum_{\mathbf{j}\in B_\rho(\mathbf{j})} \tilde{u}(\mathbf{j})\, \mathrm{e}^{-\frac{|\hat{u}_1(\mathbf{j})-\hat{u}_1(\mathbf{i})|^2}{h^2}},$$

where $\tilde{u}$ is the observed noisy image and $\hat{u}_1$ the image previously denoised by (3.15).

Similarly, for linear transform Wiener-type methods, the image is first denoised by its classical definition, which amounts to approximating the oracle coefficients of Theorem 3.1 using the noisy ones. In a second iteration, the coefficients of the denoised image approximate the true coefficients of the noise-free image. Thus the second step filter following the first step (3.9) is given by

$$\mathbf{D}\tilde{U} = \sum_i a(i)\ \langle \tilde{U}, G_i\rangle\ G_i, \quad \text{with } a(i) = \frac{|\langle \hat{U}_1, G_i\rangle|^2}{|\langle \hat{U}_1, G_i\rangle|^2 + \sigma^2},$$

where $\hat{U}_1$ is the denoised image formed by initially applying the thresholding algorithm to the observed image $\tilde{U}$.

Alternatives and extensions: 'twicing' and Bregman iterations

In the recent review paper by Milanfar (2011), many denoising operators are formalized in a general linear framework, noting that they can be associated with a doubly stochastic diffusion matrix W with non-negative coefficients. In NL-means, for example, this matrix is obtained by the symmetrization of the matrix of the NL-means weights $w_{\tilde{P},\tilde{Q}}$ defined in Algorithm 1. Unless it is optimal, as is the case with an ideal Wiener filter, the matrix W associated with the denoising filter can be iterated. A study of MSE evolution with these iterations is proposed by Milanfar (2011) for several denoising operators, considering several different patch types (texture, edge, flat). Iteration is, however, different from the oracle iteration described above. In the oracle iteration, the matrix W is changed at each step, using the better estimate given by the previously denoised image. One does not generally observe much improvement by iterating the oracle method more than once. Milanfar (2011) points out another generic tool, used at least for total variation denoising: so-called 'twicing', a term due to Tukey (1977). Instead of repeated applications of a filter, the idea is to process the residual obtained as the difference between the estimated image and the initial image. If the residuals contain some of the underlying signal, filtering them should recover part of it. Milanfar (2011) shows that the Bregman iterations (Osher *et al.* 2004) used for improving total-variation-based denoising are a twicing, and so is the matching pursuit method used in the K-SVD filter described in Section 5.7.

4.3. Dealing with colour images

The straightforward strategy to extend denoising algorithms to colour or multivalued images is to apply the algorithm independently to each channel. The use of this simple strategy often introduces colour artifacts, easily detected by the eye. Two different strategies are observable in state-of-the-art denoising algorithms.

Depending on the algorithm formulation, a vector-valued version dealing with all colour channels simultaneously can be proposed. This solution is adopted by averaging filters such as neighbourhood filters or NL-means. These algorithms compute colour differences directly in the vector-valued image, thus yielding a unified weight configuration which is applied to each channel.

The alternative option is to convert the usual RGB image to a different colour space, where the independent denoising of each channel does not create noticeable colour artifacts. Most algorithms use the YUV system, which separates the geometric and chromatic parts of the image. This change of variables is expressed as a linear transform by multiplication of the RGB vector by the matrices

$$YUV = \begin{pmatrix} 0.30 & 0.59 & 0.11 \\ -0.15 & -0.29 & 0.44 \\ 0.61 & -0.51 & -0.10 \end{pmatrix}, \quad Y_oU_oV_o = \begin{pmatrix} \frac{1}{3} & \frac{1}{3} & \frac{1}{3} \\ \frac{1}{2} & 0 & -\frac{1}{2} \\ \frac{1}{4} & -\frac{1}{2} & \frac{1}{4} \end{pmatrix}.$$

The second colour transform to the space $Y_oU_oV_o$ is an orthogonal transform. It has the advantage of maximizing the noise reduction of the geometric component, since this component is an average of the three colours. The geometric component is perceptually more important than the chromatic ones, and the noise reduction permits better performance of the algorithm in this component. It also permits higher noise reduction on the chromatic components U_o and V_o, due to their observable regularity.

This latter strategy is adopted by transform thresholding filters, for which the design of an orthonormal basis coupling the different colour channels is not trivial.

4.4. Trying all generic tools on an example

This subsection incrementally applies the above generic denoising tools to the DCT sliding window, to illustrate how these additional tools permit drastic improvement of algorithm performance. We start with the basic DCT 'neighbourhood filter', as proposed by Yaroslavsky and Eden (2003). Its principle is to denoise a patch around each pixel, and to keep only the central denoised pixel.

Figure 4.1 displays the denoised images obtained by incrementally applying each of the following ingredients.

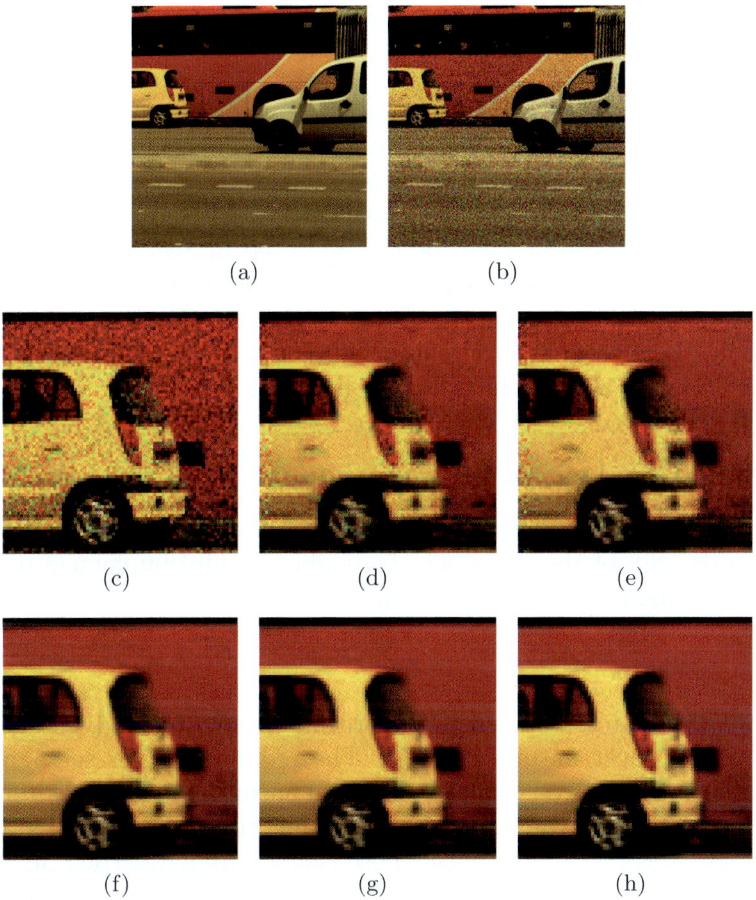

Figure 4.1. (a, b) Original and noisy images with additive Gaussian white noise of standard deviation 25. (c–h) Close-ups of denoised images by sliding DCT thresholding filters and incrementally adding use of a $Y_oU_oV_o$ colour system, uniform aggregation, variance-based aggregation and iteration with the 'oracle' given by the first step. The corresponding PSNRs are 26.85, 27.33, 30.65, 30.73, 31.25, respectively.

- A basic DCT thresholding algorithm using the neighbourhood filter technique (keeping only the central pixel of the window). Each colour channel is treated independently.
- Use of an orthogonal geometric and chromatic decomposition colour system $Y_oU_oV_o$: grey parts are better reconstructed and colour artifacts are reduced.
- Uniform aggregation: the noise reduction is superior and isolated noise points are removed.
- Adaptive aggregation using the estimator variance: the noise reduction near edges is increased and 'halo' effects are removed.
- Additional iteration using 'oracle' estimation: residual noise is totally removed and the sharpness of details is increased.

The PSNRs obtained by incrementally applying the previous strategies respectively are 26.85, 27.33, 30.65, 30.73, 31.25. This experiment illustrates how the use of these additional tools is crucial to achieving competitive results. This last version of the DCT denoising algorithm, incorporating all the proposed generic tools, will be the one used in Section 6. A complete description of the algorithm can be found in Algorithm 3. The colour version of the algorithm applies the denoising independently to each $Y_oU_oV_o$ component. This version is therefore slightly better than the version online in Yu and Sapiro (2011), which does not use the oracle step.

5. Detailed analysis of nine methods

In this section we give a detailed description and analysis of nine denoising methods. Six of them, for which reliable faithful implementations are available, will be compared in Section 6.

5.1. Non-local means

The non-local means (NL-means) algorithm tries to take advantage of the redundancy of most natural images. The redundancy (or self-similarity) hypothesis is that for every small patch in a natural image one can find several similar patches in the same image, as illustrated in Figures 5.1 and 5.2. This similarity is true for patches whose centres are within one pixel of the centre of the reference patch. In that case the self-similarity boils down to a local image regularity assumption. Such regularity is guaranteed by Shannon and Nyquist's sampling conditions, which require the image to be blurry. In a much more general sense inspired by neighbourhood filters, one can define the 'neighbourhood of a pixel $\mathbf{i}$' to be any set of pixels $\mathbf{j}$ in the image such that a patch around $\mathbf{j}$ looks like a patch around $\mathbf{i}$. All pixels in that neighbourhood can be used for predicting the value at $\mathbf{i}$, as was first

Algorithm 3 DCT denoising algorithm. DCT coefficients lower than 3σ are cancelled in the first step and a Wiener filter is applied in the 'oracle' second step. The colour DCT denoising algorithm applies the current strategy independently to each $Y_oU_oV_o$ component.

Input. Noisy image $\tilde{u}$, noise standard deviation σ.
Optional. Prefiltered image $\hat{u}_1$ for 'oracle' estimation.
Output. Denoised image.

Set parameter $\kappa = 8$: size of patches.
Set parameter $h = 3\,\sigma$: threshold parameter.

for each pixel **i** **do**

Select a square reference patch $\tilde{P}$ around **i** of size $\kappa \times \kappa$.
if $\hat{u}_1$ **then**
Select a square reference patch P_1 around **i** in $\hat{u}_1$.
end if
Compute the DCT transform of $\tilde{P}$.
if $\hat{u}_1$ **then**
Compute the DCT transform of P_1.
end if
if $\hat{u}_1$ **then**
Modify DCT coefficients of $\tilde{P}$ as

$$\tilde{P}(i) = \tilde{P}(i)\frac{P_1(i)^2}{P_1(i)^2 + \sigma^2}$$

else
Cancel coefficients of $\tilde{P}$ with magnitude lower than h.
end if
Compute the inverse DCT transform obtaining $\hat{P}$.
Compute the aggregation weight

$$w_{\tilde{P}} = 1/\{\text{number of non-zero DCT coefficients}\}.$$

end for

for each pixel **i** **do**
Aggregation. Recover the denoised value at each pixel **i** by averaging all values at **i** of all denoised patches $\hat{Q}$ containing **i**, weighted by $w_{\tilde{Q}}$.
end for

Figure 5.1. Here $q1$ and $q2$ have a large weight because their similarity windows are similar to that of p. On the other hand the weight $w(p, q3)$ is much smaller because the intensity grey-scale values in the similarity windows are very different.

shown in Efros and Leung (1999) for the synthesis of texture images. This self-similarity hypothesis is a generalized periodicity assumption. The use of self-similarities has in fact been well known in information theory from its foundation. In his paper 'A mathematical theory of communication', Shannon (2001) analysed the local self-similarity (or redundancy) of natural written language, and gave what was probably the first stochastic text synthesis algorithm. The Efros–Leung texture synthesis method adapted this algorithm to images, and NL-means (Buades, Coll and Morel 2004) seems to be the first adaptation of the same idea to denoising.[2]

NL-means denoises a square reference patch $\tilde{P}$ around $\mathbf{i}$ of dimension $\kappa \times \kappa$ by replacing it with an average of all similar patches $\tilde{Q}$ in a square neighbourhood of $\mathbf{i}$ of size $\lambda \times \lambda$. To do this, a normalized Euclidean distance between $\tilde{P}$ and $\tilde{Q}$, $d(\tilde{P}, \tilde{Q}) = \kappa^{-2} \|\tilde{P} - \tilde{Q}\|^2$ is computed for all patches $\tilde{Q}$ in the search neighbourhood. Then the weighted average is

$$\hat{P} = \frac{\sum_{\tilde{Q}} \tilde{Q}\, \mathrm{e}^{-\frac{d(\tilde{P},\tilde{Q})^2}{h^2}}}{\sum_{\tilde{Q}} \mathrm{e}^{-\frac{d(\tilde{P},\tilde{Q})^2}{h^2}}}.$$

[2] Nevertheless, some researchers have pointed out to us the report by De Bonet (1997) as giving an early discovery that intuition could use signal redundancy. This very short paper describes an experiment in a few sentences. It suggests that region redundancy on both sides of an edge can be detected, and used for image denoising. Nevertheless, no algorithm is specified in the paper.

The concentration of the noise law gives NL-means the edge over neighbourhood filters, as the number of pixels increases. Because the distances are computed on many patch samples instead of only one pixel, the fluctuations of the quadratic distance due to the noise are reduced.

Related attempts

Weissman *et al.* (2005) proposed a 'universal denoiser' for digital images, and proved that this denoiser is universal in the sense of 'asymptotically achieving, without access to any information on the statistics of the clean signal, the same performance as the best denoiser that does have access to this information'. Ordentlich *et al.* (2003) presented an implementation valid for binary images with an impulse noise, with excellent results. Awate and Whitaker (2006) also proposed a method whose principles stand close to NL-means, since the method involves comparison between patches to estimate a restored value. The objective of the algorithm is to denoise the image by decreasing the randomness of the image.

A consistency theorem for NL-means

NL-means is intuitively consistent under stationarity conditions, that is, if one can find many samples of every image detail. It can be proved (Buades *et al.* 2005*b*) that if the image is a fairly general stationary and mixing random process, for every pixel $\mathbf{i}$, NL-means converges to the conditional expectation of $\mathbf{i}$ knowing its neighbourhood, which is the best Bayesian estimate.

NL-means as an extension of previous methods

A Gaussian convolution preserves only flat zones, while contours and fine structure are removed or blurred. Anisotropic filters can instead preserve straight edges, but flat zones present many artifacts. One might consider combining these methods to improve both results. A Gaussian convolution could be applied in flat zones, while an anisotropic filter could be applied on straight edges. However, other types of filters should be designed to specifically restore corners, or curved edges, or periodic texture. Figure 5.2 illustrates how NL-means chooses the right weight configuration for each type of image self-similarity.

NL-means is easily extended to the denoising of image sequences and video, indiscriminately involving pixels belonging to a space–time neighbourhood. The algorithm favours pixels with a similar local configuration. When the similar configuration moves, so do the weights. Thus, as shown by Buades, Coll and Morel (2008*b*), the algorithm is able to follow moving similar configurations without any explicit motion computation (see Figure 5.3).

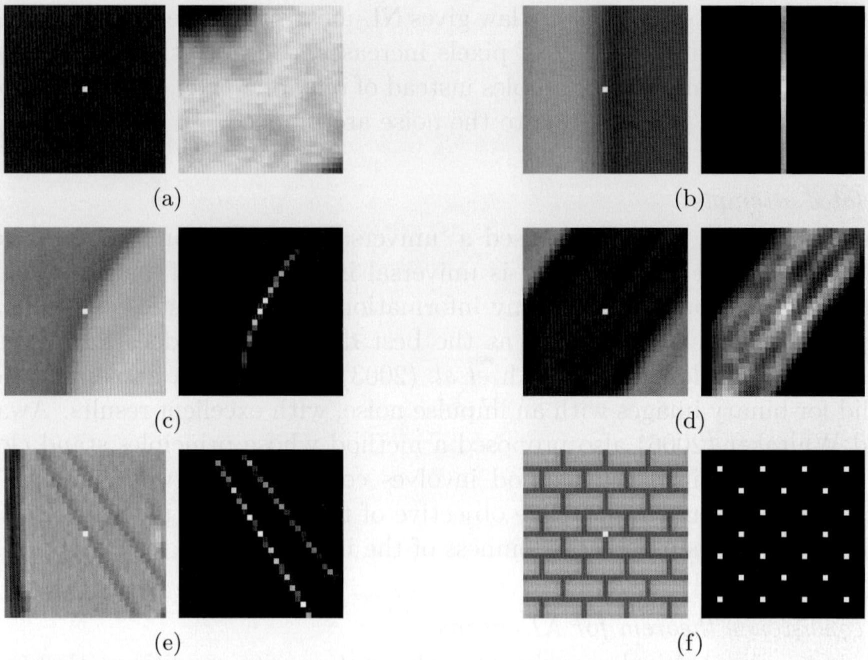

Figure 5.2. The right-hand image of each pair shows the weight distribution used to estimate a centred patch of the left-hand image by NL-means. (a) In flat zones, the weights are uniformly distributed, and NL-means acts like a low-pass isotropic filter. (b) On straight edges, the weights are distributed in the direction of the edge (as for anisotropic filters). (c) On curved edges, the weights favour pixels belonging to the same contour. (d) In a flat neighbourhood, the weights are distributed in a grey-scale neighbourhood (just as for neighbourhood filters). For (e) and (f), the weights are distributed across the more similar configurations, even though they are far away from the observed pixel. This behaviour justifies the term 'non-local'.

Indeed, this fact contrasts with previous classical movie denoising algorithms, which were motion-compensated. The underlying idea of motion compensation is the existence of a underlying 'true' image for the physical motion. Legitimate information about the colour of a given pixel should exist only along its physical trajectory. But one of the major difficulties in motion estimation is the ambiguity of trajectories, the so-called *aperture problem*. The aperture problem, viewed as a general phenomenon of movies, can be positively interpreted in the following way. There are many pixels in the succeeding or preceding frames which resemble the current pixel. Thus, it seems sound to use not just one trajectory, but rather *all similar pixels* to the current pixel across time and space, as NL-means does (see Buades *et al.* (2008*b*) for more details of this discussion).

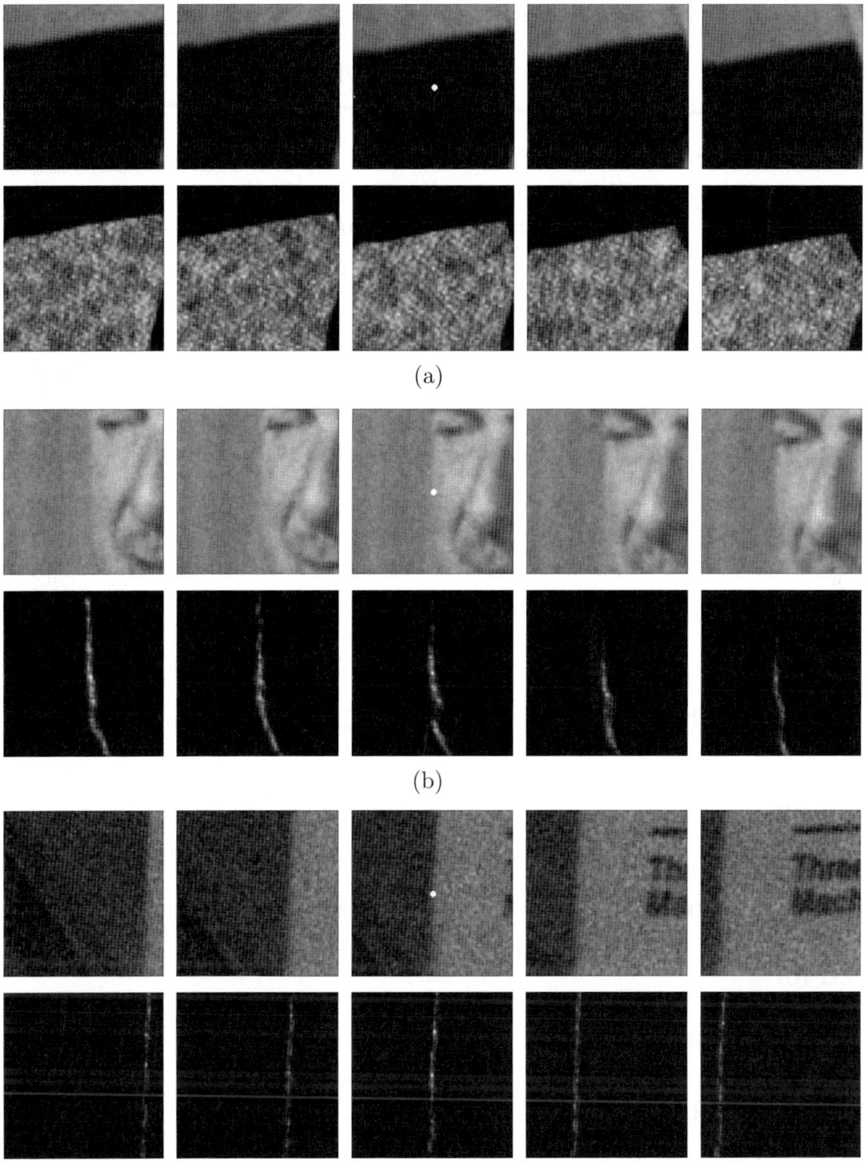

Figure 5.3. Weight distribution of NL-means applied to a movie. The upper row of each group (a), (b) and (c) shows a five-frame image sequence. The lower row shows the weight distribution used to estimate the central pixel (in white) of the middle frame is shown. The weights are equally distributed over the successive frames, including the current one. They actually involve all the candidates for the motion estimation instead of picking just one per frame. The aperture problem can be exploited for better denoising performance by using more pixels to compute the average.

Algorithm 4 NL-means algorithm (parameter values for κ, λ are indicative)

Input. Noisy image $\tilde{u}$, noise standard deviation σ.
Output. Denoised image.

Set parameter $\kappa = 3$: size of patches.
Set parameter $\lambda = 31$: size of search zone for similar patches.
Set parameter $h = 0.6\,\sigma$: bandwidth filtering parameter.

for each pixel **i do**

Select a square reference patch $\tilde{P}$ around **i** of dimension $\kappa \times \kappa$.

Set $\hat{P} = 0$ and $\hat{C} = 0$.

for each patch $\tilde{Q}$ in a square neighbourhood of **i** of size $\lambda \times \lambda$ **do**

Compute the normalized Euclidean distance between $\tilde{P}$ and $\tilde{Q}$, $d(\tilde{P}, \tilde{Q}) = \frac{1}{\kappa^2}\|\tilde{P} - \tilde{Q}\|^2$.

Accumulate $\tilde{Q}\mathrm{e}^{-\frac{d(\tilde{P},\tilde{Q})^2}{h^2}}$ to $\hat{P}$ and $\mathrm{e}^{-\frac{d(\tilde{P},\tilde{Q})^2}{h^2}}$ to $\hat{C}$.

end for

Normalize the average patch $\hat{P}$ by dividing it by the sum of weights $\hat{C}$.

end for

for each pixel x **do**
Aggregation. Recover the denoised value at each pixel **i** by averaging all values at **i** of all denoised patches $\hat{Q}$ containing **i**.
end for

5.2. *Non-local Bayesian denoising*

It is clear that (3.7) in Section 3.1,

$$\hat{P}_1 = \overline{\tilde{P}} + \left[\mathbf{C}_{\tilde{P}} - \sigma^2\mathbf{I}\right]\mathbf{C}_{\tilde{P}}^{-1}(\tilde{P} - \overline{\tilde{P}}),$$

is itself a denoising algorithm, provided we can compute the patch expectations and patch covariance matrices. We shall now explain how the non-local Bayes algorithm proposed by Lebrun *et al.* (2011) achieves this. Let $\mathcal{P}(\tilde{P})$ be the set of patches $\tilde{Q}$ similar to the patch $\tilde{P}$, which have been obtained with a suitably chosen tolerance threshold, so that we can assume that they represent noisy versions of the patches similar to P. Then, by the law of large numbers, we have

$$\begin{aligned}\mathbf{C}_{\tilde{P}} &\simeq \frac{1}{\#\mathcal{P}(\tilde{P})-1}\sum_{\tilde{Q}\in\mathcal{P}(\tilde{P})}(\tilde{Q}-\overline{\tilde{P}})(\tilde{Q}-\overline{\tilde{P}})^T,\\ \overline{\tilde{P}} &\simeq \frac{1}{\#\mathcal{P}(\tilde{P})}\sum_{\tilde{Q}\in\mathcal{P}(\tilde{P})}\tilde{Q}.\end{aligned} \tag{5.1}$$

Nevertheless, the selection of similar patches at the first step is not optimal, and can be improved in a second estimation step where the first step estimate is used as the oracle. Thus, in a second step, where all patches have been denoised at the first step, all the denoised patches can be used again to obtain an estimate $\mathbf{C}_{\hat{P}_1}$ for $\mathbf{C}_P$, the covariance of the cluster containing P, and a new estimate of $\overline{\tilde{P}}$, the average of patches similar to $\tilde{P}$. Indeed, the patch similarity is better estimated with the denoised patches. Then it follows from (3.6) and (3.7) that we can obtain a second improved denoised patch, namely

$$\hat{P}_2 = \overline{\tilde{P}}^1 + \mathbf{C}_{\hat{P}_1}\big[\mathbf{C}_{\hat{P}_1} + \sigma^2\mathbf{I}\big]^{-1}(\tilde{P} - \overline{\tilde{P}}^1), \tag{5.2}$$

where

$$\begin{aligned}\mathbf{C}_{\hat{P}_1} &\simeq \frac{1}{\#\mathcal{P}(\hat{P}_1)-1}\sum_{\hat{Q}_1\in\mathcal{P}(\hat{P}_1)}(\hat{Q}_1-\overline{\tilde{P}}^1)(\hat{Q}_1-\overline{\tilde{P}}^1)^T,\\ \overline{\tilde{P}}^1 &\simeq \frac{1}{\#\mathcal{P}(\hat{P}_1)}\sum_{\hat{Q}_1\in\mathcal{P}(\hat{P}_1)}\tilde{Q}.\end{aligned} \tag{5.3}$$

We denote the denoised patches by $\overline{\tilde{P}}$ in (5.1) and $\overline{\tilde{P}}^1$ in (5.2). Indeed, in (5.2), the denoised version of $\tilde{P}$ is computed as the average of noisy patches $\tilde{Q}$ whose denoised patch is similar to $\hat{P}_1$.

In short, the estimates (3.7) and (5.2) appear to be equivalent, but in practice they are not. $\mathbf{C}_{\hat{P}_1}$, obtained after a first denoising step, is a better estimate than $\mathbf{C}_{\tilde{P}}$. Furthermore, $\overline{\tilde{P}}^1$ is a more accurate mean than $\overline{\tilde{P}}$: it uses a better evaluation of patch similarities. Since all the above quantities are computable from the noisy image, we obtain the two-step Algorithm 5.

As pointed out by Buades, Lebrun and Morel (2012*b*), the non-local Bayes algorithm is only an interpretation (with some generic improvements such as aggregation) of the PCA-based algorithm proposed by Zhang, Dong, Zhang and Shi (2010). This paper has a self-explanatory title: 'Two-stage image denoising by principal component analysis with local pixel grouping'. It is equivalent to applying PCA on the patches similar to $\tilde{P}$, followed by a Wiener filter on the coefficients of $\tilde{P}$ for this PCA, or applying formula (3.7) with the covariance matrix of the similar patches. Indeed, the PCA simply computes the eigenvalues of the empirical covariance matrix. Thus,

Algorithm 5 Non-local Bayes image denoising

Input. Noisy image.
Output. Denoised image.

for all patches $\tilde{P}$ of the noisy image **do**

Find a set $\mathcal{P}(\tilde{P})$ of patches $\tilde{Q}$ similar to $\tilde{P}$.

Compute the expectation $\overline{\tilde{P}}$ and covariance matrix $\mathbf{C}_{\tilde{P}}$ of these patches:

$$\mathbf{C}_{\tilde{P}} \simeq \frac{1}{\#\mathcal{P}(\tilde{P})-1} \sum_{\tilde{Q}\in\mathcal{P}(\tilde{P})} (\tilde{Q}-\overline{\tilde{P}})(\tilde{Q}-\overline{\tilde{P}})^T,$$

$$\overline{\tilde{P}} \simeq \frac{1}{\#\mathcal{P}(\tilde{P})} \sum_{\tilde{Q}\in\mathcal{P}(\tilde{P})} \tilde{Q}.$$

Obtain the first-step estimate:

$$\hat{P}_1 = \overline{\tilde{P}} + \left[\mathbf{C}_{\tilde{P}} - \sigma^2\mathbf{I}\right]\mathbf{C}_{\tilde{P}}^{-1}(\tilde{P}-\overline{\tilde{P}}).$$

end for

Obtain the pixel value of the basic estimate image $\hat{u}_1$ as an average of all values of all denoised patches $\hat{Q}_1$ which contain $\mathbf{i}$.

for all patches $\tilde{P}$ of the noisy image **do**

Find a new set $\mathcal{P}_1(\tilde{P})$ of noisy patches $\tilde{Q}$ similar to $\tilde{P}$ by comparing their denoised 'oracular' versions Q_1 to P_1.

Compute the new expectation $\overline{\tilde{P}}^1$ and covariance matrix $\mathbf{C}_{\hat{P}_1}$ of these patches:

$$\mathbf{C}_{\hat{P}_1} \simeq \frac{1}{\#\mathcal{P}(\hat{P}_1)-1} \sum_{\hat{Q}_1\in\mathcal{P}(\hat{P}_1)} (\hat{Q}_1-\overline{\tilde{P}}^1)(\hat{Q}_1-\overline{\tilde{P}}^1)^T,$$

$$\overline{\tilde{P}}^1 \simeq \frac{1}{\#\mathcal{P}(\hat{P}_1)} \sum_{\hat{Q}_1\in\mathcal{P}(\hat{P}_1)} \tilde{Q}.$$

Obtain the second-step patch estimate:

$$\hat{P}_2 = \overline{\tilde{P}}^1 + \mathbf{C}_{\hat{P}_1}\left[\mathbf{C}_{\hat{P}_1} + \sigma^2\mathbf{I}\right]^{-1}(\tilde{P}-\overline{\tilde{P}}^1).$$

end for

Obtain the pixel value of the denoised image $\hat{u}(\mathbf{i})$ as an average of all values of all denoised patches $\hat{Q}_2$ which contain $\mathbf{i}$.

the method in Zhang *et al.* (2010) finds its Bayesian interpretation. A study of the compared performance of local PCA versus global PCA for two-stage image denoising (TSID) is in fact proposed by Deledalle, Salmon and Dalalyan (2011*b*).

5.3. *Patch-based locally optimal Wiener (PLOW)*

In the non-local Bayes method of Section 5.2, a local model is estimated in a neighbourhood of each patch. In the PLOW method of Chatterjee and Milanfar (2012), however, the idea is to learn from the image a sufficient number of patch clusters, in fact 15, and to apply the LMMSE estimate to each patch *after* having assigned it to one of the clusters obtained by clustering. Thus, this empirical Bayesian algorithm starts by clustering the patches by the classic K-means clustering algorithm. To take into account the fact that similar patches can have varying contrast, the inter-patch distance is photometrically neutral, and Chatterjee and Milanfar call it a 'geometric distance'. The clustering phase is accelerated by dimension reduction obtained by applying PCA to the patches. The clustering is therefore a segmentation of the set of patches, and the denoising of each patch is then performed within its cluster. Since each cluster contains patches that are geometrically similar but not necessarily photometrically similar, the method identifies the photometrically similar patches for each patch in the cluster as being those whose quadratic distance to the reference patch are within the bounds allowed by noise. Then an LMMSE estimate (Kay 1993) is obtained for the reference patch by a variant of (3.7). The algorithm uses a first phase, which performs a first denoising before constituting the clusters. Thus the main phase is actually using the first phase as an oracle to obtain the covariance matrices of the sets of patches.

5.4. *Inherent bounds in image denoising*

By 'shotgun' patch denoising methods, we mean methods that intend to denoise patches by a fully non-local algorithm, in which the patch is compared to a patch model obtained from a large or *very large* patch set. The 'sparse-land' methods intend to learn a sparse patch dictionary on which to decompose any given patch, from a single image or from a small set of images. The shotgun methods instead learn from a very large patch set extracted from tens of thousands of images (up to 10^{10} patches). Then the patch is denoised by deducing its likeliest estimate from the set of all patches. In the case of Zoran and Weiss (2011), this patch space is organized as a Gaussian mixture with about 200 components. Shotgun methods have begun to be used in several image restoration methods: for example in Hays and Efros (2007), for image inpainting, with a fairly explicit title, 'Scene completion using millions of photographs'.

Algorithm 6 PLOW denoising

Input. Image in vector form $\tilde{U}$.
Output. Denoised image in vector form $\hat{U}$.

Set parameters: patch size $\kappa \times \kappa = 11 \times 11$, number of clusters $K = 15$;
Estimate noise standard deviation $\hat{\sigma}$ by $\hat{\sigma} = 1.4826 \times \text{median}(|\nabla\tilde{U} - \text{median}(\nabla\tilde{U})|)$;
Set parameter: $h^2 = 1.75\hat{\sigma}^2\kappa^2$;
Pre-filter image $\tilde{U}$ to obtain a pilot estimate $\hat{U}_1$;
Extract overlapping patches of size $\kappa \times \kappa$, $\tilde{Q}$ from $\tilde{U}$ and $\hat{Q}_1$ from U_1;
Geometric clustering with K-means of the patches in $\hat{U}_1$ (using a variant of PCA for the patches). The distance is a geometric distance, photometrically neutral.
for each patch cluster Ω_k **do**
  Estimate from the patches $\hat{Q}_1 \in \Omega_k$ the mean patch $\overline{P}_k \simeq \sum_{\hat{Q}_1 \in \Omega_k} \hat{Q}_1$ and the cluster covariance $\mathbf{C}_P^k$.
  for each patch $\hat{Q}_{1,i} \in \Omega_k$ **do**
    Consider its associated noisy patch $\tilde{Q}_i$. Identify photometrically similar patches $\tilde{Q}_j$ in the cluster as those with a quadratic distance to $\tilde{Q}_i$ within the bounds allowed by noise, namely $\gamma^2 + 2\kappa^2\hat{\sigma}^2$, with $\gamma = \gamma(\kappa)$ a 'small' threshold.
    Compute similarity weights $w_{ij} = \mathrm{e}^{-\frac{\|\tilde{Q}_i - \tilde{Q}_j\|^2}{h^2}}$.
    Compute the LMMSE estimator for the noisy patch $\tilde{Q}_i$ (slightly more complex than usual, because the cluster contains patches that are geometrically similar but not necessarily photometrically similar):

$$\hat{Q}_i = \overline{P} + \left[\mathbf{I} - \left(\sum_j w_{ij}\mathbf{C}_P^k + \mathbf{I}\right)^{-1}\right] \sum_j \frac{w_{ij}}{\sum_j w_{ij}}(\tilde{Q}_j - \overline{P}).$$

  end for
end for
At each pixel aggregate multiple estimates from all $\hat{P}$ containing it, with weights given as inverses of the variance of each estimator.

The approach of Levin and Nadler (2011) is to define the simplest universal 'shotgun' method, where a huge set of patches is used to estimate the upper limits a patch-based denoising method would ever reach. The results support the 'near-optimality of state-of-the-art denoising results', the results obtained by the BM3D algorithm being only 0.1 decibel away from optimality for methods using small patches (typically 8×8).

To evaluate the MMSE this experiment uses a set of 20 000 images from the LabelMe dataset (Russell, Torralba, Murphy and Freeman 2008). The

Algorithm 7 Shotgun NL-means

Input. Noisy image $\tilde{u}$ in vectorial form; very large set of M patches P_i extracted from a large set of noiseless natural images.
Output. Denoised image $\hat{u}$.

for all patches $\tilde{P}$ extracted from $\tilde{u}$ **do**
Compute the MMSE denoised estimate of $\tilde{P}$:

$$\hat{P} \simeq \frac{\sum_{i=1}^{M} \mathbb{P}(\tilde{P} \mid P_i) P_i}{\sum_{i=1}^{M} \mathbb{P}(\tilde{P} \mid P_i)},$$

where $\mathbb{P}(\tilde{P} \mid P_i)$ is known from (5.4).
end for
At each pixel $\mathbf{i}$ get $\hat{u}(\mathbf{i})$ as $\hat{P}(\mathbf{i})$, where the patch P is centred at $\mathbf{i}$.
(Optional Aggregation.) For each pixel $\mathbf{j}$ of u, compute the denoised version $\hat{u}_{\mathbf{j}}$ as the average of all values $\hat{P}(\mathbf{j})$ for all patches containing $\mathbf{j}$. (This step is not considered in Levin and Nadler (2011).)

method, even though certainly impractical, is of exquisite simplicity. Given a clean patch P, the noisy patch $\tilde{P}$ with Gaussian noise, of standard deviation σ, has probability distribution

$$\mathbb{P}(\tilde{P} \mid P) = \frac{1}{(2\pi\sigma^2)^{\frac{\kappa^2}{2}}} \mathrm{e}^{-\frac{\|P-\tilde{P}\|^2}{2\sigma^2}}, \tag{5.4}$$

where κ^2 is the number of pixels in the patch. Then, given a noisy patch $\tilde{P}$, its optimal estimator for the Bayesian minimum squared error (MMSE) is, by Bayes' formula,

$$\hat{P} = \mathbb{E}[P \mid \tilde{P}] = \int \mathbb{P}(P \mid \tilde{P}) P \,\mathrm{d}P = \int \frac{\mathbb{P}(\tilde{P} \mid P)}{\mathbb{P}(\tilde{P})} \mathbb{P}(P) P \,\mathrm{d}P. \tag{5.5}$$

Using a huge set of M natural patches (with a distribution supposedly approximating the real natural patch density), we can approximate the terms in (5.5) by $\mathbb{P}(P)\,\mathrm{d}P \simeq \frac{1}{M}$ and $\mathbb{P}(\tilde{P}) \simeq \frac{1}{M} \sum_i \mathbb{P}(\tilde{P} \mid P_i)$, which in view of (5.4) yields

$$\hat{P} \simeq \frac{\frac{1}{M} \sum_i \mathbb{P}(\tilde{P} \mid P_i) P_i}{\frac{1}{M} \sum_i \mathbb{P}(\tilde{P} \mid P_i)}.$$

Thus the final MMSE estimator is simply the exact application of NL-means, denoising each patch by matching it to the huge patch database. Clearly this is not just a theoretical algorithm. Web-based application could provide a way to denoise online any image by organizing a huge patch database. The final algorithm is summarized in Algorithm 7.

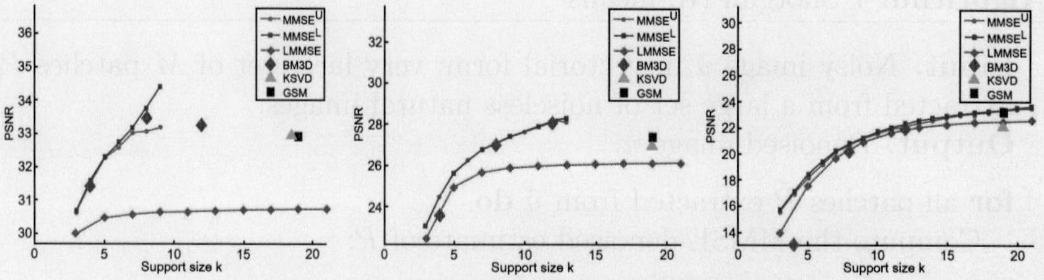

(a) $\sigma = 18$ (b) $\sigma = 55$ (c) $\sigma = 170$

Figure 5.4. From Levin and Nadler (2011). The PSNR ($-10\log_{10}$ (MMSE)) of several denoising algorithms – K-SVD (Mairal *et al.* 2009*a*), BM3D (Dabov *et al.* 2007), Gaussian scale mixture (Portilla *et al.* 2003) – compared with the PNSR predicted by MMSE^L and MMSE^U. The performance of all algorithms is bounded by the MMSE^U estimate, but BM3D approaches this upper bound to within fractional dB values. Nevertheless, the performance bounds consider more restrictive patch-based algorithms than the class to which BM3D belongs. Thus the actual distance to optimality may be greater.

However, as mentioned earlier, the main focus of Levin and Nadler (2011) is elsewhere: they use shotgun denoising to estimate universal upper and lower bounds of the attainable PSNR by any patch-based denoising algorithm. More precisely, the algorithm gives upper and lower bounds to the following problem: *Given a noisy patch* $\tilde{P}$, *and given the law* $p(P)$ *of all possible patches, find the best possible estimate* (*in the sense of MMSE*). The shotgun algorithm gives a best possible estimate for *any patch-based denoising algorithm* of this kind.

The upper bound obtained by Levin and Nadler (2011) turns out to be very close to results obtained with BM3D (see Section 5.8), and they conclude that for small window sizes, or moderate to high noise levels, the quest for the best denoising algorithm might be close to an end. More precisely, only fractions of decibels separate the current best algorithms from these demonstrated upper bounds. The EPLL method (Zoran and Weiss 2011) can be viewed as a first (slightly) more practical realization of this quasi-optimality by a shotgun algorithm, and there is no doubt that other more practical ones will follow. We now describe how the lower and upper bounds of Levin and Nadler (2011) can be estimated from a sufficient set of natural images.

The MSE for a given denoising algorithm can be obtained by randomly sampling patches P, then adding noise to generate noisy patches $\tilde{P}$, and measuring the reconstruction error $\|P - \hat{P}\|^2$. Then the mean reconstruction error is

$$\mathrm{MSE} = \int \mathbb{P}(P) \int \mathbb{P}(\tilde{P} \mid P) \|P - \hat{P}\|^2 \,\mathrm{d}\tilde{P} \,\mathrm{d}P. \tag{5.6}$$

Conversely, one can start from a noisy patch $\tilde{P}$ and measure the variance of $\mathbb{P}(P \mid \tilde{P})$ around it. According to Levin and Nadler (2011), this amounts to computing the sum of weighted distances between the restored $\hat{P}$ and all possible P explanations:

$$\mathrm{MSE} = \int \mathbb{P}(\tilde{P}) \int \mathbb{P}(P \mid \tilde{P}) \|P - \hat{P}\|^2 \,\mathrm{d}\tilde{P} \,\mathrm{d}P. \tag{5.7}$$

This last equation follows from (5.6) by Bayes' rule. For each noisy $\tilde{P}$ one can define its MMSE:

$$\mathrm{MMSE}(\tilde{P}) = \mathbb{E}[\|\hat{P} - \tilde{P}\|^2 \mid \tilde{P}] = \int \mathbb{P}(P \mid \tilde{P})(P - \hat{P})^2 \,\mathrm{d}P. \tag{5.8}$$

The main interest of this formulation is that it allows us to prove that the MMSE, out of all the denoising algorithms, is the one that minimizes the overall MSE. Indeed, differentiating (5.7) with respect to $\hat{P}$ returns the MMSE estimator (5.5). The best overall MMSE achievable by any given denoising algorithm is therefore

$$\mathrm{MMSE} = \int \mathbb{P}(\tilde{P})\mathbb{E}[\|\hat{P} - \tilde{P}\|^2 \mid \tilde{P}] \,\mathrm{d}\tilde{P} = \int \mathbb{P}(\tilde{P})\mathbb{P}(P \mid \tilde{P})(P - \hat{P})^2 \,\mathrm{d}P \,\mathrm{d}\tilde{P}. \tag{5.9}$$

The goal of Levin and Nadler (2011) is to bound the MMSE from below, ignoring of course the probability distribution $\mathbb{P}(P)$, but having enough samples of it. The main idea is to derive an upper and lower bound on the MMSE from the two MSE formulations (5.6)–(5.7). Given a set of M clean and noisy pairs $\{(P_j, \tilde{P}_j)\}$, $j = 1, \ldots, M$, and another independent set of N clean patches $\{P_i\}$, $i = 1, \ldots, N$, both randomly sampled from natural images, the proposed estimates are

$$\mathrm{MMSE}^U = \frac{1}{M} \sum_j \|\hat{P}_j - P_j\|^2 \tag{5.10}$$

and

$$\mathrm{MMSE}^L = \frac{1}{M} \sum_j \frac{\sum_i \mathbb{P}(\tilde{P}_j \mid P_i)\|\hat{P}_j - P_i\|^2}{\sum_i \mathbb{P}(\tilde{P}_j \mid P_i)}. \tag{5.11}$$

A striking feature of both estimates is that MMSE^U uses explicit knowledge of the original noise-free patch P_j, while MMSE^L does not. Since MMSE^U simply measures the error for a given denoising algorithm, it obviously provides an upper bound for the MMSE of any other denoising algorithm. As Levin and Nadler (2011) observe, MMSE^U and MMSE^L are random variables that depend on the choice of the samples. When the sample size approaches infinity, both converge to the exact MMSE. Nevertheless, Levin and Nadler (2011) give a simple proof that, for a finite sample, in expectation, MMSE^U and MMSE^L provide upper and lower bounds on the best possible MMSE. When both MMSE^U and MMSE^L coincide, they provide an accurate estimate of the optimal denoising possible with a given patch size.

For very high noise levels, Levin and Nadler (2011) also tried to apply the linear minimum mean square error (LMMSE) estimator (or Wiener filter) using only the second-order statistics of the data, by fitting a single k^2-dimensional Gaussian to the set of M image $k \times k$ patches. They conclude that even this simple approach is close to optimal for large noise.

5.5. *The expected patch log likelihood (EPLL) method*

The patch Gaussian mixture model

This other shotgun method (Zoran and Weiss 2011) is an almost literal application of the *piecewise linear estimator* (PLE) method (Yu, Sapiro and Mallat 2012) (see Section 5.9). But it is indeed shotgun, that is, it is applied to a huge set of patches instead of the image itself. A Gaussian mixture model is learned from a set of 2×10^6 patches, sampled from the Berkeley database, with their mean removed. The 200 mixture components with zero means and full covariance matrices are obtained using the EM (expectation maximization) algorithm. This training took about 30 hours with public-domain MATLAB code.[3] Thus were learned: 200 means (in fact they are all zero), 200 full covariance matrices and 200 mixing weights, which constitute the Gaussian mixture model of this set of patches. Figure 5.5 shows some six bases extracted from the Gaussian mixture. Each one shows the patches that are eigenvectors of some of the covariance matrices, sorted by eigenvalue.

Once the Gaussian mixture is learned, the denoising method maximizes the expected patch log likelihood (EPLL) while being close to the corrupted image in a way which is dependent on the (linear) corruption model. Given an image U (in vector form), the EPLL of U under prior $\mathbb{P}$ is defined by

$$\text{EPLL}_{\mathbb{P}}(U) = \sum_i \log \mathbb{P}(\mathbf{P}_i U),$$

[3] http://www.mathworks.com/matlabcentral/fileexchange/26184-em-algorithm-for-gaussian-mixture-model

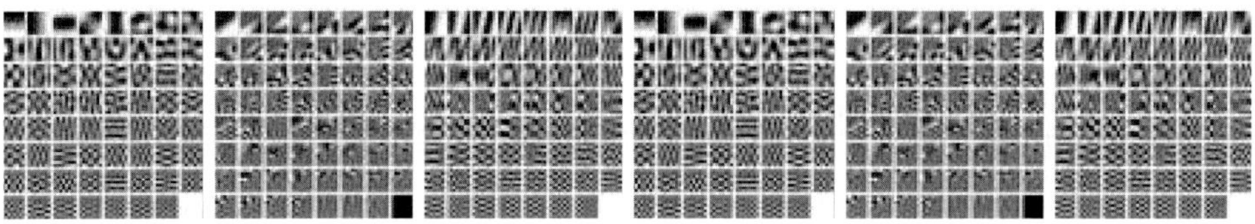

Figure 5.5. Eigenvectors of six randomly selected covariance matrices from the learned Gaussian mixture model, sorted by eigenvalue from largest to smallest (from Zoran and Weiss (2011)). The authors notice the similarity of these basis elements to DCT, but also that many seem to model texture boundaries and edges at various orientations.

where $\mathbf{P}_i$ is a matrix which extracts the ith patch P_i from the image U out of all overlapping patches, while $\log \mathbb{P}(\mathbf{P}_i X)$ is the likelihood of the ith patch under the prior $\mathbb{P}$. Assuming a patch location in the image is chosen uniformly at random, EPLL can be interpreted as the expected log likelihood of a patch in the image (up to a multiplication by $1/M$). Given a corrupted image $\tilde{U}$ in vector form and a model of image corruption of the form $\|\mathbf{A}U - \tilde{U}\|^2$, the restoration is made by minimizing

$$f_{\mathbb{P}}(U|\tilde{U}) = \frac{\lambda}{2}\|\mathbf{A}U - \tilde{U}\|^2 - \mathrm{EPLL}_{\mathbb{P}}(U).$$

According to Zoran and Weiss (2011), 'This equation has the familiar form of a likelihood term and a prior term, but note that $\mathrm{EPLL}_{\mathbb{P}}(U)$ is not the log probability of a full image. Since it sums over the log probabilities of all overlapping patches, it "double counts" the log probability. Rather, it is the expected log likelihood of a randomly chosen patch in the image.'

The optimization is made by 'half-quadratic splitting', which amounts to introducing auxiliary patch variables Z^i, $i = 1, \ldots, M$, one for each patch P_i, and minimizing the auxiliary functional

$$C_{\mathbb{P},\beta}(U, \{Z^i\}|\tilde{U}) := \frac{\lambda}{2}\|\mathbf{A}U - \tilde{U}\|^2 + \frac{\beta}{2}\sum_i \|\mathbf{P}_i U - Z^i\|^2 - \log \mathbb{P}(Z^i).$$

Solving for U given $\{Z^i\}$ amounts to the inversion

$$U = \left(\lambda \mathbf{A}^T\mathbf{A} + \beta \sum_i \mathbf{P}_i^T \mathbf{P}_i\right)^{-1} \left(\lambda \mathbf{A}^T \tilde{U} + \beta \sum_i \mathbf{P}_i^T Z^i\right).$$

In the case of denoising, $\mathbf{A}$ is simply the identity, and the above formula boils down to computing for each pixel $\mathbf{j}$ a denoised value $U(\mathbf{j})$ as a weighted average over all patches P_i containing this given pixel $\mathbf{j}$ of the noisy pixel

Algorithm 8 Patch restoration once the patch Gaussian mixture is known

for each noisy patch $\tilde{Q}$ **do**
  Compute the conditional mixture weights $\pi'_k = \mathbb{P}(k \mid \tilde{Q})$ (given by EM);
  Pick the component k with highest conditional mixing weight: $k_{\max} = \max_k \pi'_k$;
  The MAP estimate $\hat{Q}$ is a Wiener solution for the $k_{\max}$th component:
$$\hat{Q} = \left(\mathbf{C}_{k_{\max}} + \sigma^2 \mathbf{I}\right)^{-1} \left(\mathbf{C}_{k_{\max}} \tilde{Q} + \sigma^2 \mu_{k_{\max}}\right).$$
end for

value $\tilde{U}(\mathbf{j})$ and of the patch denoised values $Z_i(\mathbf{j})$:

$$U(\mathbf{j}) = \frac{\lambda \tilde{U}(\mathbf{j}) + \sum_{P_i \ni \mathbf{j}} Z_i(\mathbf{j})}{\lambda + \beta k^2}, \tag{5.12}$$

where k^2 is the patch size.

Then, solving for $\{Z_i\}$ given U amounts to solving a MAP (maximum *a posteriori*) problem of estimating the most likely patch under the prior $\mathbb{P}$, given $\mathbf{P}_i U$ and parameter β.

Once the Gaussian mixture model is known, calculating the log likelihood of a given patch is trivial:

$$\log \mathbb{P}(Q) = \log\left(\sum_{k=1}^{K} \pi_k \mathcal{N}(Q|\mu_k, \mathbf{C}_k)\right),$$

where π_k are the mixing weights for each of the mixture component, and μ_k and $\mathbf{C}_k$ are the corresponding mean and covariance matrix.

Given a noisy patch $\tilde{Q}$, the MAP estimate is computed with the procedure shown in Algorithm 8.

Zoran and Weiss (2011) comment that this is one iteration of the 'hard version' of the EM algorithm for finding the modes of a Gaussian mixture (Carreira-Perpinan 2000). The method can be used for denoising, and several experiments seem to indicate that it equals the performance of BM3D and LLSC (Mairal *et al.* 2009*a*).

5.6. Portilla et al. wavelet neighbourhood denoising (BLS-GSM)

The basic idea of this algorithm is modelling a noiseless 'wavelet coefficient neighbourhood', P, by a *Gaussian scale mixture* (GSM), which is defined by

$$P = \sqrt{z} U,$$

where U is a zero-mean Gaussian random vector and z is an independent positive scalar random variable. The wavelet coefficient neighbourhood turns out to be a patch of an oriented channel of the image at a given scale, complemented with a coefficient of the channel at the same orientation and the next lower scale. Thus, we again adopt the patch notation P. (Arguably, this method is the first patch-based method.) Using a GSM model for P estimated from the image itself, the method makes a *Bayes least-squares* (BLS) estimator. For this reason, the method will be called here BLS-GSM (Bayes least-squares estimate of Gaussian scale mixture; Portilla *et al.* called it simply BLS). Without loss of generality it is assumed that $\mathbb{E}z = 1$, and therefore the random variables U and P have similar covariances. To use the GSM model for wavelet patch denoising, the noisy input image is first decomposed into a wavelet pyramid, and each image of the pyramid will be separately denoised. The resulting denoised image is obtained by the reconstruction algorithm from the wavelet coefficients. To avoid ringing artifacts in the reconstruction, a redundant version of the wavelet transform, the so-called steerable pyramid, is used. For an $n_1 \times n_2$ image, the pyramid $\mathcal{P}$ is generated in $\log_2(\min(n_1, n_2) - 4)$ scales and eight orientations using the following procedure. First the input image is decomposed into one low-pass and eight oriented high-pass component images using two polar filters in quadrature in the Fourier domain (the sum of their squares is equal to 1). Using polar coordinates (r, θ) in the Fourier domain, the low-pass and high-pass isotropic filters are

$$l(r) = \begin{cases} 1 & 0 \leq r < 0.5, \\ \cos(\frac{\pi}{2}(-\log_2 r - 1)) & 0.5 \leq r < 1, \\ 0 & 1 \leq r \leq \sqrt{2}, \end{cases} \tag{5.13}$$

and

$$h(r) = \begin{cases} 0 & 0 \leq r < 0.5, \\ \cos(\frac{\pi}{2}(\log_2 r)) & 0.5 \leq r < 1, \\ 1 & 1 \leq r \leq \sqrt{2}. \end{cases} \tag{5.14}$$

The high-pass filter h is decomposed again into eight oriented components,

$$a_k(r, \theta) = h(r) g_k(\theta), \quad k \in [0, K-1], \tag{5.15}$$

where $K = 8$, and

$$g_k(\theta) = \frac{(K-1)}{\sqrt{K[2(K-1)]}} \left[2\cos\left(\theta - \frac{\pi k}{K}\right) \right]^{K-1}. \tag{5.16}$$

Then the steerable pyramid is generated by iteratively applying the a_k filters to the result of the low-pass filter to obtain bandpass images, and calculating the residual using the l filter followed by sub-sampling. For example, in the case of a 512×512 image we have a five-scale pyramid consisting of 49 sub-bands: eight high-pass oriented sub-bands, from $\mathcal{P}^1$ to $\mathcal{P}^8$, eight bandpass oriented sub-bands for each scale, from $\mathcal{P}^9$ to $\mathcal{P}^{48}$, in addition to one low-pass non-oriented residual sub-band, $\mathcal{P}^{49}$. (Without loss of generality we shall keep this number 49 as a landmark, but the number of course depends on the image size.) Assume now that the image has been corrupted by independent additive Gaussian noise. Therefore, a typical neighbourhood of wavelet coefficients can be represented by

$$\tilde{P} = P + N = \sqrt{z}U + N, \tag{5.17}$$

where noise, N, and P are considered to be independent. Define $p_s(i, j)$ to be the sample at position (i, j) of the sub-band $\mathcal{P}^s$, the sub-bands being enumerated as $s = 1, \ldots, 49$, for example. The neighbourhood of the wavelet coefficient $p_s(i, j)$ is composed of its spatial neighbours for the same sub-band s. It could also have contained coefficients from other sub-bands at the same scale as $p_s(i, j)$ but with different orientations, and could finally also contain sub-band coefficients from the adjacent scales, up and down. Surprisingly, the final neighbourhood is quite limited. Portilla *et al.* claim that the best efficiency is reached with a 3×3 spatial block around $p_s(i, j)$, supplemented with one coefficient at the same location and, at the next-coarser scale (considering its up-sampled parent by interpolation), with the same orientation. Hence, the neighbourhood size is 10 and contains only $\{p_s(i-1, j-1), \ldots, p_s(i+1, j+1), p_{s+8}(i, j)\}$. There are two exceptions to this. First, the neighbourhood of coarsest-scale coefficients (without any coarser scale) necessarily has only nine surrounding coefficients. Second, the boundary coefficients are processed using special steps described below. Using the observed noisy vector, $\tilde{P}$, an estimate of P can be obtained by

$$\mathbb{E}(P \mid \tilde{P}) = \int_0^\infty \mathbb{P}(z \mid \tilde{P})\mathbb{E}(P \mid \tilde{P}, z)\,\mathrm{d}z.$$

This estimate is the Bayesian denoised value of the reference coefficient. The integral is computed numerically on experimentally obtained sampled intervals of z. Here, only 13 equally spaced values of z in the interval $[\ln(z_{\min}), \ln(z_{\max})] = [-20.5, 3.5]$ are used. Therefore $\mathbb{E}(P \mid \tilde{P})$ is computed using

$$\mathbb{E}(P \mid \tilde{P}) = \sum_{i=1}^{13} \mathbb{P}(z_i \mid \tilde{P})\mathbb{E}(P \mid \tilde{P}, z_i). \tag{5.18}$$

The only question left is how to compute the conditional probability and the conditional expectation, $\mathbb{P}(z_i \mid \tilde{P})$ and $\mathbb{E}(P \mid \tilde{P}, z_i)$. For each sub-band $\mathcal{P}^s$, except for the low-pass residual $\mathcal{P}^{49}$, which remains unchanged, define $\mathbf{C}_N^s$ and $\mathbf{C}_{\tilde{P}}^s$, respectively the noise and the observation covariance matrices of the wavelet neighbourhood. If n^s denotes the size of neighbourhood at sub-band $\mathcal{P}^s$ (so n^s is 10 or 9, as explained above), $\mathbf{C}_N^s$ is an $n^s \times n^s$ matrix which can be experimentally generated by first decomposing a delta function

$\sigma\delta$ on the steerable pyramid. Here σ is the known noise variance and δ is an $n_1 \times n_2$ image defined by

$$\delta(i,j) = \begin{cases} 1 & (i,j) = \left(\frac{n_1}{2}, \frac{n_2}{2}\right), \\ 0 & \text{otherwise.} \end{cases}$$

(This covariance matrix is equal to the covariance of the white noise defined as a band-limited function obtained by randomizing uniformly the phase of the Fourier coefficients of the discrete Dirac mass δ.) Using the steerable pyramid decomposition of $\sigma\delta$, define $\mathbf{N}_s$ to be the matrix which has as its rows all neighbourhoods of the sub-band $\mathcal{P}_s$. This is a matrix with n_s columns and $(n_1-2)(n_2-2)$ rows, subtracting 2 to eliminate the boundary coefficients. The covariance matrix $\mathbf{C}_N^s$ of the neighbourhood samples for each sub-band is computed using

$$\mathbf{C}_N^s = \frac{\mathbf{N}_s^T \mathbf{N}_s}{(n_1-2)(n_2-2)}.$$

Since all the noise removal steps are calculated for each sub-band separately, in the following we skip the superscript s to simplify the notation. Similarly, but using the pyramid of observed noisy samples, $\mathbf{C}_{\tilde{P}}$ can be computed. Using (5.17) and the assumption $\mathbb{E}z = 1$, for each sub-band s we have

$$\mathbf{C}_U = \mathbf{C}_{\tilde{P}} - \mathbf{C}_N.$$

$\mathbf{C}_U$ can be forced to be positive semi-definite by setting its negative eigenvalues to zero. We can now calculate $\mathbb{E}(P \mid \tilde{P}, z_i)$. Using the fact that P and N are independent Gaussian random variables and also that the noise is additive, $\mathbb{E}(P \mid \tilde{P}, z_i)$ is simply a local Wiener estimate,

$$\mathbb{E}(P \mid \tilde{P}, z) = \frac{z\mathbf{C}_U}{z\mathbf{C}_U + \mathbf{C}_N}\tilde{P},$$

where the matrix fraction notation is understood to mean $\frac{\mathbf{C}}{\mathbf{W}} := \mathbf{C}\mathbf{W}^{-1}$. Clearly it would be cumbersome to compute as many matrix inversions as z_is. Fortunately, with a bit of linear algebra this computation can be rendered common to all z_i. Let $\{\mathbf{Q}, \mathbf{\Lambda}\}$ be the eigenvectors and eigenvalues of $\mathbf{S}^{-1}\mathbf{C}_U\mathbf{S}^{-T}$, where $\mathbf{S}_{n^s \times n^s}$ is the symmetric square root of $\mathbf{C}_N$, $\mathbf{C}_N = \mathbf{S}\mathbf{S}^T$. So we have $\mathbf{S}^{-1}\mathbf{C}_U\mathbf{S}^{-T} = \mathbf{Q}\mathbf{\Lambda}\mathbf{Q}^T$. Furthermore, set $\mathbf{M} = \mathbf{S}\mathbf{Q}$, $\mathbf{v} = \mathbf{M}^{-1}\tilde{P}$. Then we have

$$\begin{aligned} \mathbb{E}(P \mid \tilde{P}, z) &= \frac{z\mathbf{C}_U}{z\mathbf{C}_U + \mathbf{C}_N}\tilde{P} \\ &= \frac{z\mathbf{C}_U}{z\mathbf{C}_U + \mathbf{S}\mathbf{S}^T}\tilde{P} \\ &= \frac{z\mathbf{C}_U}{\mathbf{S}(z\mathbf{S}^{-1}\mathbf{C}_U\mathbf{S}^{-T} + \mathbf{I})\mathbf{S}^T}\tilde{P} \end{aligned}$$

$$
\begin{aligned}
&= \frac{z\mathbf{C}_U}{\mathbf{SQ}(z\boldsymbol{\Lambda}+\mathbf{I})\mathbf{Q}^T\mathbf{S}^T}\tilde{P} \\
&= z\mathbf{C}_U\mathbf{S}^{-T}\mathbf{Q}(z\boldsymbol{\Lambda}+\mathbf{I})^{-1}\mathbf{Q}^T\mathbf{S}^{-1}\tilde{P} \\
&= z\mathbf{S}\mathbf{S}^{-1}\mathbf{C}_U\mathbf{S}^{-T}\mathbf{Q}(z\boldsymbol{\Lambda}+\mathbf{I})^{-1}\mathbf{Q}^T\mathbf{S}^{-1}\tilde{P} \\
&= z\mathbf{SQ}\boldsymbol{\Lambda}(z\boldsymbol{\Lambda}+\mathbf{I})^{-1}\mathbf{Q}^T\mathbf{S}^{-1}\tilde{P} \\
&= z\mathbf{M}\boldsymbol{\Lambda}(z\boldsymbol{\Lambda}+\mathbf{I})^{-1}\mathbf{v}.
\end{aligned}
$$

The interesting point is that one can calculate $\mathbf{M}$, $\boldsymbol{\Lambda}$ and $\mathbf{v}$ once for each sub-band. The scalar final formulation of the above equation is

$$
\mathbb{E}(P \mid \tilde{P}, z_i)_c = \sum_{j=1}^{n^s} \frac{z_i m_{c,j}\lambda_{j,j}v_j}{z_i\lambda_{j,j}+1}, \tag{5.19}
$$

where $m_{c,j}$, $\lambda_{j,j}$ and v_j are the elements of $\mathbf{M}$, $\boldsymbol{\Lambda}$ and $\mathbf{v}$ respectively, and c is the index of the reference coefficient in the neighbourhood.

The second component of (5.18) is $\mathbb{P}(z_i \mid \tilde{P})$, which can be obtained using Bayes' rule. Here $p_z(z)$ denotes the density function of the random variable z,

$$
\mathbb{P}(z_i \mid \tilde{P}) = \frac{\mathbb{P}(\tilde{P} \mid z_i)p_z(z_i)}{\int_0^\infty \mathbb{P}(\tilde{P} \mid \alpha)p_z(\alpha)\,\mathrm{d}\alpha},
$$

or its discrete form,

$$
\mathbb{P}(z_i \mid \tilde{P}) = \frac{\mathbb{P}(\tilde{P} \mid z_i)p_z(z_i)}{\sum_{j=1}^{13} \mathbb{P}(\tilde{P} \mid z_j)p_z(z_j)}, \tag{5.20}
$$

where the density of the observed noisy neighbourhood vector $\tilde{P}$, conditioned on z_i, is a zero-mean Gaussian with covariance

$$
\mathbf{C}_{\tilde{P}|z_i} := z_i\mathbf{C}_U + \mathbf{C}_N,
$$

so that

$$
\mathbb{P}(\tilde{P} \mid z_i) = \frac{\mathrm{e}^{-\frac{1}{2}\tilde{P}^T(z\mathbf{C}_U+\mathbf{C}_N)^{-1}\tilde{P}}}{\sqrt{|z\mathbf{C}_U+\mathbf{C}_N|}}.
$$

Using the above definitions of $\mathbf{v}$ and $\boldsymbol{\Lambda}$ and the same simplifications as for $\mathbb{E}(P \mid \tilde{P}, z_i)$, we obtain

$$
\mathbb{P}(\tilde{P} \mid z_i) = \frac{\mathrm{e}^{-\frac{1}{2}\sum_{j=1}^{n^s}\frac{v_j^2}{z_j\lambda_{j,j}+1}}}{\sqrt{\Pi_{j=1}^{n^s}(z_i\lambda_{j,j}+1)}}, \tag{5.21}
$$

The only question left is the form of $p_z(z)$. Portilla *et al.* (2003), after a somewhat puzzling discussion, adopt a 'non-informative Jeffreys prior', $p_z(z) \simeq \frac{1}{z}$. Since this function cannot be a density, being non-integrable, the function is cut off to zero near $z = 0$. The algorithm of Portilla *et al.* (2003) is given here as Algorithm 9.

Algorithm 9 Portilla *et al.* wavelet neighbourhood denoising (BLS-GSM)

Input. Noisy image.
Output. Denoised image.

Parameters: $n_1 \times n_2$ the image size,
number of pyramid scales $\log_2(\min(n_1, n_2) - 4)$.
Parameter s, enumeration of all oriented channels at each scale (8 per scale).
Establish n^s, dimension of wavelet neighbourhood coefficient (10 or 9).

Apply the wavelet pyramid (5.13)–(5.16), respectively, to the noise image δ and to the observed image.
Regroup the obtained wavelet coefficients to obtain $\tilde{P}^s$, the wavelet coefficient neighbourhoods of rank s and N^s the noise wavelet coefficient neighbourhoods of rank s.
for each filter index s **do**
Compute $\mathbf{C}_N^s$ and $\mathbf{C}_{\tilde{P}}^s$, the noise and observation covariance matrices of N_s and $\tilde{P}_s$. (The subscript s is omitted hereafter.) Deduce $\mathbf{C}_U = \mathbf{C}_{\tilde{P}} - \mathbf{C}_N$.
Compute $\{\mathbf{Q}, \mathbf{\Lambda}\}$, the eigenvectors and eigenvalues of $\mathbf{S}^{-1}\mathbf{C}_U\mathbf{S}^{-T}$, where $\mathbf{S}$ is the symmetric square root of $\mathbf{C}_N$, $\mathbf{C}_N = \mathbf{S}\mathbf{S}^T$.
end for
for each wavelet coefficient neighbourhood $\tilde{P}$ and $i \in \{1, \ldots, 13\}$ **do**
Compute $\mathbf{M} = \mathbf{S}\mathbf{Q}$, $\mathbf{v} = \mathbf{M}^{-1}\tilde{P}$.
Using (5.19) obtain $\mathbb{E}(P \mid \tilde{P}, z_i)_c = \sum_{j=1}^{n^s} \frac{z_i m_{c,j} \lambda_{j,j} v_j}{z_i \lambda_{j,j} + 1}$, where $m_{c,j}$, $\lambda_{j,j}$ and v_j are the elements of $\mathbf{M}$, $\mathbf{\Lambda}$ and $\mathbf{v}$ respectively, and c is the index of the reference coefficient in the neighbourhood.
Apply (5.20) to get

$$\mathbb{P}(z_i \mid \tilde{P}) = \frac{\mathbb{P}(\tilde{P} \mid z_i) p_z(z_i)}{\sum_{j=1}^{13} \mathbb{P}(\tilde{P} \mid z_j) p_z(z_j)},$$

using the value obtaincd by (5.21) for

$$\mathbb{P}(\tilde{P} \mid z_i) = \frac{\mathrm{e}^{-\frac{1}{2}\sum_{j=1}^{n^s} \frac{v_j^2}{z_j \lambda_{j,j} + 1}}}{\sqrt{\Pi_{j=1}^{n^s}(z_i \lambda_{j,j} + 1)}}.$$

By (5.18) finally obtain $\mathbb{E}(P \mid \tilde{P}) = \sum_{i=1}^{13} \mathbb{P}(z_i \mid \tilde{P}) \mathbb{E}(P \mid \tilde{P}, z_i)$, where $p_z(z) \simeq \frac{1}{z}$ and z_i are quantized uniformly on the interval $[\ln(z_{\min}); \ln(z_{\max})] = [-20.5, 3.5]$.
end for
Reconstruct the restored image from its restored neighbourhood coefficients $\mathbb{E}(P \mid \tilde{P})$ by the inverse steerable pyramid.

As we shall see in Section 7, in spite of its formalism, this method is in fact extremely similar to other patch-based Bayesian methods. It has received a more recent extension, reaching state-of-the-art performance, in Lyu and Simoncelli (2009): this paper proposes an extension of the above method modelling the wavelet coefficients as a global random field of Gaussian scale mixtures.

5.7. K-SVD

The K-SVD method was introduced by Aharon, Elad and Bruckstein (2006), whose whole objective was to optimize the quality of sparse approximations of vectors in a learnt dictionary. Although they noted the relevance of the technique to image processing tasks, it was Elad and Aharon (2006) who made a detailed study of the denoising of grey-scale images. Since then, the adjustment to colour images has been treated by Mairal *et al.* (2008). Let us note that Mairal *et al.* (2008) proved that the K-SVD method can also be useful in other image processing tasks, such as non-uniform denoising, demoisaicing and inpainting. For a detailed description of K-SVD the reader is referred to Mairal (2010) and Mairal *et al.* (2009*b*).

The algorithm is divided into three steps. In the two first steps an optimal dictionary and a sparse representation are built for each patch in the image, using among other tools a singular value decomposition (SVD). In the last step, the restored image is built by aggregating the computed sparse representations of all image patches. The algorithm requires an initialization of the dictionary, which is updated during the process. The dictionary initialization may contain the usual orthogonal basis (*e.g.*, the discrete cosine transform, or wavelets), or patches from clean images, or even from the noisy image itself.

The first step looks for sparse representations of all patches of size κ^2 in the noisy image in vector form $\tilde{U}$ using a fixed dictionary $\mathbf{D}$. A dictionary is represented as a matrix of size $\kappa^2 \times n_{\text{dict}}$, with $n_{\text{dict}} \geq \kappa^2$, whose columns (the 'atoms of the dictionary') are normalized in the Euclidean norm. For each noisy patch $\mathbf{R}_{\mathbf{i}}\tilde{U}$ (where the index $\mathbf{i}$ indicates that the top left corner of the patch is the pixel $\mathbf{i}$, and $\mathbf{R}_{\mathbf{i}}$ is the matrix extracting the patch vector from $\tilde{U}$), a 'sparse' column vector $\alpha_{\mathbf{i}}$ (of size n_{dict}) is calculated by optimization. This vector of coefficients should have only a few non-zero coefficients, the distance between $\mathbf{R}_{\mathbf{i}}\tilde{U}$ and its sparse approximation $\mathbf{D}\alpha_{\mathbf{i}}$ remaining as small as possible. The dictionary allows one to compute a sparse representation $\alpha_{\mathbf{i}}$ of each patch $\mathbf{R}_{\mathbf{i}}\tilde{U}$. These sparse vectors are assembled in a matrix α with κ^2 rows and N_p columns, where N_p is the number of patches of dimension κ^2 of the image.

More precisely, an ORMP (orthogonal recursive matching pursuit) gives an approximate solution of the (NP-complete) problem

$$\arg\min_{\alpha_{\mathbf{i}}} \|\alpha_{\mathbf{i}}\|_0 \quad \text{such that} \quad \|\mathbf{R}_{\mathbf{i}}\tilde{U} - \mathbf{D}\alpha_{\mathbf{i}}\|_2^2 \leq \kappa^2(C\sigma)^2, \tag{5.22}$$

Algorithm 10 K-SVD algorithm for grey-scale images

Input. Noisy image $\tilde{u}$, $\tilde{U}$ in vector form, noise standard deviation σ; dimension of patches κ^2 (number of pixels); dictionary size n_{dict}; iteration number K of the dictionary optimization; initial patch dictionary $\mathbf{D}_{\text{init}}$ as matrix with n_{dict} columns and κ^2 rows.
Output. Image in vector form $\hat{U}$.

Collect all noisy patches of dimension κ^2 in column vectors $\mathbf{R_i}\tilde{U}$.
Set $\hat{\mathbf{D}} = \mathbf{D}_{\text{init}}$.

for k=1 to K **do**

An ORMP is applied to the vectors $\mathbf{R_i}\tilde{U}$ in such a way that a vector of sparse coefficients $\hat{\alpha}_\mathbf{i}$ is obtained verifying $\mathbf{R_i}\tilde{U} \approx \mathbf{D}\hat{\alpha}_\mathbf{i}$.

Introduce $\omega_l = \{\, \mathbf{i} \,|\, \hat{\alpha}_\mathbf{i}(l) \neq 0 \,\}$;
For $\mathbf{i} \in \omega_l$, obtain the residual

$$e^l_\mathbf{i} = \mathbf{R_i}\tilde{U} - \hat{\mathbf{D}}\hat{\alpha}_\mathbf{i} + \hat{d}_l\hat{\alpha}_\mathbf{i}(l);$$

Put these column vectors together in a matrix $\mathbf{E}_l$. Values $\hat{\alpha}_\mathbf{i}(l)$ are also assembled in a row vector denoted by $\hat{\alpha}^l$ for $\mathbf{i} \in \omega_l$;
Update $\hat{d}_l$ and $\hat{\alpha}^l$ as solutions of the minimization problem:

$$(\hat{d}_l, \hat{\alpha}^l) = \arg\min_{d_l, \alpha^l} \|\mathbf{E}_l - d_l\alpha^l\|^2_F.$$

A truncated SVD is applied to the matrix $\mathbf{E}_l$. It partially provides $\mathbf{U}$, $\mathbf{V}$ (orthogonal matrices) and $\mathbf{\Delta}$ (filled in with zeros except on its first diagonal), such that $\mathbf{E}_l = \mathbf{U\Delta V}^T$. Then $\hat{d}_l$ is defined again as the first column of $\mathbf{U}$ and $\hat{\alpha}^l$ as the first column of $\mathbf{V}$ multiplied by $\mathbf{\Delta}(1,1)$.
end for

Aggregation. For each pixel the final result $\hat{U}$ in vector form is obtained thanks to the weighted aggregation

$$\hat{U} = \Big(\lambda\mathbf{I} + \sum_\mathbf{i} \mathbf{R}_\mathbf{i}^T\mathbf{R_i}\Big)^{-1}\Big(\lambda\tilde{U} + \sum_\mathbf{i} \mathbf{R}_\mathbf{i}^T\hat{\mathbf{D}}\hat{\alpha}_\mathbf{i}\Big).$$

where $\|\alpha_\mathbf{i}\|_0$ refers to the l^0-norm of $\alpha_\mathbf{i}$, *i.e.*, the number of non-zero coefficients of $\alpha_\mathbf{i}$. The additional constraint guarantees that the residual has an l^2-norm lower than $\kappa C\sigma$. C is a user parameter.

The second step tries to update one by one the columns of the dictionary $\mathbf{D}$ and the representations α, to improve the overall fidelity of the patch approximation. The goal is to decrease the quantity

$$\sum_{\mathbf{i}} \|\mathbf{D}\alpha_{\mathbf{i}} - \mathbf{R}_{\mathbf{i}}\tilde{U}\|_2^2 \tag{5.23}$$

while keeping the sparsity of the vectors $\alpha_{\mathbf{i}}$. We will denote by $\hat{d}_l$ ($1 \leq l \leq n_{\text{dict}}$) the columns of the dictionary $\hat{\mathbf{D}}$. First, the quantity (5.23) is minimized without taking care of the sparsity. The atom $\hat{d}_l$ and the coefficients $\hat{\alpha}_{\mathbf{i}}(l)$ are modified to make the approximations of all the patches more efficient. For each $\mathbf{i}$, introduce the residual

$$e^l_{\mathbf{i}} = \mathbf{R}_{\mathbf{i}}\tilde{U} - \hat{\mathbf{D}}\hat{\alpha}_{\mathbf{i}} + \hat{d}_l\hat{\alpha}_{\mathbf{i}}(l), \tag{5.24}$$

which is the error committed by deciding to omit $\hat{d}_l$ from the representation of the patch $R_{\mathbf{i}}\tilde{U}$. Thus $e^l_{\mathbf{i}}$ is a vector of size κ^2.

These residuals are grouped together in a matrix $\mathbf{E}_l$ (whose columns are indexed by $\mathbf{i}$). The values of the coefficients $\hat{\alpha}_{\mathbf{i}}(l)$ are also grouped in a row vector denoted by $\hat{\alpha}^l$. Therefore, $\mathbf{E}_l$ is a matrix of size $\kappa^2 \times N_p$ (recall that N_p is the total number of patches in the image) and $\hat{\alpha}^l$ is a row vector of size N_p. We must try to find a new $\hat{d}_l$ and a new row vector $\hat{\alpha}^l$ minimizing

$$\sum_{\mathbf{i}} \|\hat{\mathbf{D}}\hat{\alpha}_{\mathbf{i}} - \hat{d}_l\hat{\alpha}_{\mathbf{i}}(l) + d_l\alpha^l - \mathbf{R}_{\mathbf{i}}\tilde{U}\|_2^2 = \|\mathbf{E}_l - d_l\alpha^l\|_F^2, \tag{5.25}$$

where the squared Frobenius norm $\|\mathbf{M}\|_F^2$ refers to the sum of the squared elements of $\mathbf{M}$. This Frobenius norm is also equal to the sum of the squared (Euclidean) norms of the columns, and it is plausible that minimizing (5.25) amounts to reducing the approximation error caused by $\hat{d}_l$. It is well known that the minimization of such a Frobenius norm consists of a rank-one approximation, which always admits a solution, obtained in practice using the SVD. Using the SVD of $\mathbf{E}_l$,

$$\mathbf{E}_l = \mathbf{U}\boldsymbol{\Delta}\mathbf{V}^T,$$

where $\mathbf{U}$ and $\mathbf{V}$ are orthogonal matrices and $\boldsymbol{\Delta}$ is a diagonal matrix whose diagonal elements are non-negative and decreasing, the updated values of $\hat{d}_l$ and $\hat{\alpha}^l$ are, respectively, the first column of $\mathbf{U}$ and the first column of $\mathbf{V}$ multiplied by $\boldsymbol{\Delta}(1,1)$.

After K iterations of these two steps, a denoised patch $\hat{\mathbf{D}}\hat{\alpha}_{\mathbf{i}}$ is available for each patch position $\mathbf{i}$, where $\hat{\mathbf{D}}$ is the final updated dictionary. The third and last (aggregation) step consists in merging the denoised versions of all patches of the image in order to obtain a global estimate. This is achieved by solving the minimization problem

$$\hat{U} = \arg\min_{U_0 \in \mathbb{R}^M} \lambda\|U_0 - \tilde{U}\|_2^2 + \sum_{\mathbf{i}} \|\hat{\mathbf{D}}\hat{\alpha}_{\mathbf{i}} - \mathbf{R}_{\mathbf{i}}U_0\|_2^2$$

using the closed formula

$$\hat{U} = \left(\lambda \mathbf{I} + \sum_{\mathbf{i}} \mathbf{R}_{\mathbf{i}}^T \mathbf{R}_{\mathbf{i}}\right)^{-1} \left(\lambda \tilde{U} + \sum_{\mathbf{i}} \mathbf{R}_{\mathbf{i}}^T \hat{\mathbf{D}} \hat{\alpha}_{\mathbf{i}}\right). \tag{5.26}$$

For each pixel this amounts to averaging its initial noisy value with the average of all estimates obtained with all patches containing it. The parameter λ controls the trade-off between these two values and thus measures the fidelity to the initial noisy image.

Mairal *et al.* (2008) proposed direct extension of the algorithm to vector-valued images, instead of converting the colour image to another colour system decorrelating geometry and chromaticity. The previous algorithm is applied to column vectors which are a concatenation of the R,G,B values. In this way, the algorithm, when updating the dictionary, takes into account the inter-channel correlation. We shall describe the algorithm for grey-scale images: the colour version simply requires an adaptation of the Euclidean norm to the colour space.

5.8. BM3D

BM3D is a sliding-window denoising method extending DCT denoising and NL-means. Instead of adapting locally a basis or choosing from a large dictionary, it uses a fixed basis. The main difference from DCT denoising is that a set of similar patches are used to form a three-dimensional block, which is filtered by using a three-dimensional transform, hence the name *collaborative filtering*. The method has four steps: (a) finding the image patches similar to a given image patch and grouping them in a three-dimensional block, (b) three-dimensional linear transform of the three-dimensional block, (c) shrinkage of the transform spectrum coefficients, and (d) inverse three-dimensional transformation. This three-dimensional filter therefore filters out simultaneously all two-dimensional image patches in the three-dimensional block. By attenuating the noise, collaborative filtering reveals even the finest details shared by the grouped patches. The filtered patches are then returned to their original positions and an adaptive aggregation procedure is applied by taking into account the number of kept coefficients per patch during the thresholding process (see Section 4 for more details of aggregation).

The first collaborative filtering step is much improved in a second step using an oracle Wiener filtering. This second step mimics the first step, with two differences. The first difference is that it compares the filtered patches instead of the original patches, as described in Section 4. The second difference is that the new three-dimensional group (built with the unprocessed image samples, but using the patch distances of the filtered image) is processed by an oracle Wiener filter, using coefficients from the

Algorithm 11 BM3D first iteration algorithm for grey-scale images

Input. Noisy image $\tilde{u}$, noise standard deviation σ.
Output. Basic estimate $\hat{u}_1$ of the denoised image.

Set parameter $\kappa \times \kappa = 8 \times 8$: dimension of patches.
Set parameter $\lambda \times \lambda = 39 \times 39$: size of search zone for similar patches.
Set parameter $N_{\max} = 16$: maximum number of similar patches retained during the grouping part.
Set parameter $s = 3$: step in both rows and columns between two reference patches.
Set parameter $\lambda_{3\mathrm{D}} = 2.7$: coefficient used for the hard thresholding.
Set parameter $\tau = 2500$ (if $\sigma > 40, \tau = 5000$): threshold used to determine similarity between patches.

for each pixel **i**, with a step s in rows and columns **do**
  Select a square reference patch $\tilde{P}$ around **i** of size $\kappa \times \kappa$.
  Look for square patches $\tilde{Q}$ in a square neighbourhood of **i** of size $\lambda \times \lambda$ having a distance to $\tilde{P}$ lower than τ.
  if there are more than $N_{\max}$ similar patches **then**
    keep only the $N_{\max}$ closest similar patches to $\tilde{P}$ according to their Euclidean distance
  else
    keep 2^p patches, where p is the largest integer such that 2^p is smaller than the number of similar patches
  end if
  A 3D group $\mathcal{P}(\tilde{P})$ is built with those similar patches.
  A biorthogonal spline wavelet (Bior 1.5) is applied on every patch in $\mathcal{P}(\tilde{P})$.
  A Walsh–Hadamard transform is then applied along the third dimension of the 3D group $\mathcal{P}(\tilde{P})$.
  A hard thresholding with threshold $\lambda_{3\mathrm{D}}\sigma$ is applied to $\mathcal{P}(\tilde{P})$. An associated weight $w_{\tilde{P}}$ is computed:

$$w_{\tilde{P}} = \begin{cases} (N_{\tilde{P}})^{-1} & N_{\tilde{P}} \geq 1, \\ 1 & N_{\tilde{P}} = 0, \end{cases}$$

  where $N_{\tilde{P}}$ is the number of retained (non-zero) coefficients.
  The estimate $\hat{u}_1^{\tilde{Q},\tilde{P}}$ for each pixel **i** in similar patches $\tilde{Q}$ of the 3D group $\mathcal{P}(\tilde{P})$ is then obtained by applying the inverse of the Walsh–Hadamard transform along the third dimension, followed by the inverse of the biorthogonal spline wavelet on every patches of the 3D group.
end for
for each pixel **i do**
  Aggregation. Recover the denoised value at **i** by averaging all estimates of all patches $\tilde{Q}$ in all 3D groups $\mathcal{P}(\tilde{P})$ containing **i**, using the weights $w_{\tilde{P}}$.
end for

Algorithm 12 BM3D second iteration algorithm for grey-scale images

Input. Noisy image $\tilde{u}$; noise standard deviation σ; basic estimate $\hat{u}_1$ obtained at the first step.
Output. Final denoised image $\hat{u}$.

Set parameter $\kappa \times \kappa = 8 \times 8$ (up to 12 for high noise level): dimension of patches.
Set parameter $\lambda \times \lambda = 39 \times 39$: size of search zone for similar patches.
Set parameter $N_{\max} = 32$: maximum number of similar patches retained during the grouping part.
Set parameter $s = 3$: step in both rows and columns between two reference patches.
Set parameter $\tau = 400$ (if $\sigma > 40, \tau = 3500$): threshold used to determinate similarity between patches.

for each pixel **i**, with a step s in rows and columns **do**

Take the square reference patches $\tilde{P}$ and $\hat{P}_1$ centred at **i**, of size $\kappa \times \kappa$ in the initial and basic estimation images.

Look for square patches $\hat{Q}_1$ in a square neighbourhood of **i** of size $\lambda \times \lambda$ having a distance lower than τ in the basic estimate image $\hat{u}_1$.

Select the number of similar patches as done in the first step.

Two 3D groups $\mathcal{P}(\tilde{P})$ and $\mathcal{P}(\hat{P}_1)$ are built with those similar patches, one from the noisy image $\tilde{u}$ and one from the basic estimate image $\hat{u}_1$.

A 3D transform (denoted by $\tau_{3\mathrm{D}}$) is applied on both 3D groups:

- First a 2D DCT is applied on every patch contained in $\mathcal{P}(\tilde{P})$ and $\mathcal{P}(\hat{P}_1)$;
- Then a Walsh–Hadamard transform is applied along the third dimension of $\mathcal{P}(\tilde{P})$ and $\mathcal{P}(\hat{P}_1)$.

Compute the Wiener coefficient $\omega_{\tilde{P}} = \frac{|\tau_{3\mathrm{D}}(\mathcal{P}(\hat{P}_1)|^2}{|\tau_{3\mathrm{D}}(\mathcal{P}(\hat{P}_1)|^2+\sigma^2}$.

The Wiener collaborative filtering of $\mathcal{P}(\tilde{P})$ is realized as the element-by-element multiplication of the 3D transform of the noisy image $\tau_{3\mathrm{D}}(\mathcal{P}(\tilde{P}))$ with the Wiener coefficients $\omega_{\tilde{P}}$.

An associated weight $w_{\tilde{P}}$ is computed:

$$w_{\tilde{P}} = \begin{cases} (\|\omega_{\tilde{P}}\|_2)^{-2} & \|\omega_{\tilde{P}}\|_2 > 0, \\ 1 & \|\omega_{\tilde{P}}\|_2 = 0. \end{cases}$$

The estimate $\hat{u}_2^{\tilde{Q},\tilde{P}}$ for each pixel **i** in similar patches $\tilde{Q}$ of the 3D group $\mathcal{P}(\tilde{P})$ is then obtained by applying the inverse of the 1D Walsh–Hadamard transform along the third dimension, followed by the inverse of the 2D DCT on every patch of the 3D group.

end for

for each pixel **i do**

Aggregation. Recover the denoised value $\hat{u}(\mathbf{i})$ at **i** by averaging all estimates of patches $\tilde{Q}$ in all 3D groups $\mathcal{P}(\tilde{P})$ containing **i**, using the weights $\omega_{\tilde{P}}$.

end for

denoised image obtained at the first step to approximate the true coefficients given by Theorem 3.1. The final aggregation step is identical to that of the first step.

The algorithm is extended to colour images via the $Y_oU_oV_o$ colour system. The previous strategy is applied independently to each channel, with the exception that similar patches are always selected by computing distances in the channel Y_o.

Here we have described the basic implementation given in the original paper, and which will also be used in Section 6. However, BM3D has several more recent variants that improve its performance. As for NL-means, there is a variant with shape-adaptive patches (Dabov, Foi, Katkovnik and Egiazarian 2009). In this algorithm, called BM3D-SAPCA, the sparsity of image representation is improved in two respects. First, it employs image patches (neighbourhoods) which can have a data-adaptive shape. Second, the PCA bases are obtained by eigenvalue decomposition of empirical second-moment matrices that are estimated from groups of similar adaptive-shape neighbourhoods. This method improves BM3D, especially in preserving image details and introducing very few artifacts. The anisotropic shape-adaptive patches are obtained using the 8-directional LPA-ICI techniques (Katkovnik, Egiazarian and Astola 2006).

The very recent development of BM3D is presented in Katkovnik, Danielyan and Egiazarian (2011) and Danielyan, Katkovnik and Egiazarian (2012), where it is generalized to become a generic image restoration tool, including deblurring.

5.9. *The piecewise linear estimation (PLE) method*

The ambitious Bayesian restoration model proposed by Yu, Sapiro and Mallat (2010, 2012) is a general framework for restoration, including denoising, deblurring, and inpainting. An image is decomposed into overlapping patches $\tilde{P}_i = \mathbf{A}_i P_i + N_i$, where $\mathbf{A}_i$ is the degradation operator restricted to the patch i, P_i is the original patch, $\tilde{P}_i$ the degraded one, and N_i the noise restricted to the patch. Since we are only studying the denoising problem, we shall take $\mathbf{A}_i$ to be the identity. The (straightforward) extension including a linear perturbation operator is beyond our scope.

The patch density law is modelled as a mixture of Gaussian distributions $\{\mathcal{N}(\mu_k, \mathbf{C}_k)\}_{1\leq k\leq K}$ parametrized by their means μ_k and covariance matrices $\mathbf{C}_k$. Thus each patch $\tilde{P}_i$ is assumed to be independently drawn from one of these Gaussians with an unknown index k and the density function

$$p(P_i) = \frac{1}{(2\pi)^{\frac{\kappa^2}{2}} |\mathbf{C}_{k_i}|^{\frac{1}{2}}} \mathrm{e}^{-\frac{1}{2}(P_i-\mu_k)^T \mathbf{C}_{k_i}^{-1}(P_i-\mu_k)}.$$

Algorithm 13 Piecewise linear estimation (PLE)

Input. Noisy image $\tilde{u}$ given by the family of its noisy patches $(\tilde{P}_i)_i$; initial set of 19 Gaussian models $\mathcal{N}(\mu_k, \mathbf{C}_k)$ obtained as: (a) the 18 PCAs of the patches of 18 synthetic edge images, each with a different orientation, (b) a Gaussian model with a diagonal covariance matrix on the DCT basis.
Output. Denoised image $\hat{u}$.

E-STEP
for all patches $\tilde{P}_i$ of the noisy image **do**

for each k **do**
Estimate the MAP of P_i knowing k: $P_i^k = \mu_k + (\mathbf{I} + \sigma^2 \mathbf{C}_k^{-1})^{-1} \tilde{P}_i$.
end for
Select the best Gaussian model k_i for P_i as
$k_i = \arg\min_k \big(\|P_i^k - \tilde{P}_i\|^2 + \sigma^2 (P_i^k - \mu_k)^T \mathbf{C}_k^{-1} (P_i^k - \mu_i) + \log |\mathbf{C}_k| \big)$.
Obtain the best estimate of P_i knowing the Gaussian models $(\mu_k, \mathbf{C}_k)$, $\hat{P}_i = P_i^{k_i}$.
end for

M-STEP
for all k **do**
Compute the expectation μ_k and covariance matrix $\mathbf{C}_k$ of each Gaussian by

$$\mu_k = \frac{1}{\#\mathcal{C}_k} \sum_{i \in \mathcal{C}_k} \hat{P}_i, \quad \mathbf{C}_k = \frac{1}{\#\mathcal{C}_k - 1} \sum_{i \in \mathcal{C}_k} (\hat{P}_i - \mu_k)(\hat{P}_i - \mu_k)^T.$$

end for
Iterate **E-STEP** and **M-STEP**
Aggregation. Obtain the pixel value of the denoised image $u(\mathbf{i})$ as a weighted average of all values of all denoised patches P_i which contain $\mathbf{i}$.

Estimating all patches P_i from their noisy observations $\tilde{P}_i$ amounts to solving the following problems:

- estimating the Gaussian parameters $(\mu_k, \mathbf{C}_k)_{1 \leq k \leq K}$ from the degraded data $\tilde{P}_i$,
- identifying the index k_i of the Gaussian distribution generating the patch P_i,
- estimating P_i from its corresponding Gaussian distribution $(\mu_{k_i}, \mathbf{C}_{k_i})$ and from its noisy version $\tilde{P}_i$.

In consequence PLE (Yu *et al.* 2012) has two distinct steps in the estimation procedure. In an E-step (E for estimate), the Gaussian parameters $(\mu_k, \mathbf{C}_k)_k$ are known, and for each patch the maximum *a posteriori* (MAP) estimate

$\hat{P}_i^k$ is computed with each Gaussian model. Then the best Gaussian model k_i is selected to obtain the estimate $\hat{P}_i = \hat{P}_i^{k_i}$.

In the M-step (M for model), the Gaussian model selection k_i and the signal estimates $\hat{f}_i$ are assumed to be known for all patches i, and again permit estimation of the Gaussian models $(\mu_k, \mathbf{C}_k)_{1\leq k\leq K}$. According to the terminology of Section 4.2, this subsection gives the *oracle* allowing us to estimate the patches by a Wiener-type filter in the E-step.

For each image patch with index i, the patch estimate and its model selection are obtained by maximizing the log *a posteriori* probability $\mathbb{P}(P_i \,|\, \tilde{P}_i, k)$,

$$\begin{aligned}(\hat{P}_i, k_i) &= \arg\max_{P,k} \log \mathbb{P}(P_i \mid \tilde{P}_i, \mathbf{C}_k) \qquad (5.27)\\ &= \arg\max_{P,k} \big(\log \mathbb{P}(\tilde{P}_i \mid P_i, \mathbf{C}_k) + \log \mathbb{P}(P_i \mid \mathbf{C}_k)\big)\\ &= \arg\min_{P_i,k} \big(\|P_i - \tilde{P}_i\|^2 + \sigma^2 (P_i - \mu_k)^T \mathbf{C}_k^{-1} (P_i - \mu_k) + \sigma^2 \log |\mathbf{C}_k|\big)\end{aligned}$$

where the second equation follows from Bayes' rule and the third one assumes white Gaussian noise with diagonal covariance matrix $\sigma^2\mathbf{I}$ (of the dimension of the patch) and $P_i \simeq \mathcal{N}(\mu_k, \mathbf{C}_k)$. This minimization can be made first over P_i, which amounts to a linear filter, and then over k, which is a simple comparison of a small set of real values. The index k being fixed, the optimal P_i^k satisfies

$$P_i^k = \arg\min_{P_i} \big(\|P_i - \tilde{P}_i\|^2 + \sigma^2 (P_i - \mu_k)^T \mathbf{C}_k^{-1} (P_i - \mu_i) + \log |\mathbf{C}_k|\big),$$

and therefore

$$P_i^k = \mu_k + (\mathbf{I} + \sigma^2 \mathbf{C}_{k_i}^{-1})^{-1} (\tilde{P}_i - \mu_k),$$

which is the formula (5.2) already seen in Section 5.2. Then the best Gaussian model k_i is selected as

$$k_i = \arg\min_{k} \big(\|P_i^k - \tilde{P}_i\|^2 + \sigma^2 (P_i^k - \mu_k)^T \sigma_k^{-1} (P_i^k - \mu_i) + \log |\mathbf{C}_k|\big).$$

Assuming now that, for each patch P_i, the model k_i and estimate $\hat{P}_i$ are known, the next question is to give the maximum likelihood estimate for $(\mu_k, \mathbf{C}_k)$ for each k, given all the patches assigned to the kth cluster $\mathcal{C}_k$, namely,

$$(\mu_k, \mathbf{C}_k) = \arg\max_{\mu_k, \mathbf{C}_k} \log \mathbb{P}(\{\hat{P}_i\}_{i\in\mathcal{C}_k} \mid \mu_k, \mathbf{C}_k).$$

This yields the empirical estimate

$$\mu_k = \frac{1}{\#\mathcal{C}_k} \sum_{i\in\mathcal{C}_k} \hat{P}_i, \quad \mathbf{C}_k = \frac{1}{\#\mathcal{C}_k - 1} \sum_{i\in\mathcal{C}_k} (\hat{P}_i - \mu_k)(\hat{P}_i - \mu_k)^T,$$

which are the estimates (5.3) also used in Section 5.2.

Finally, the above MAP-EM algorithm is iterated and Yu *et al.* observe

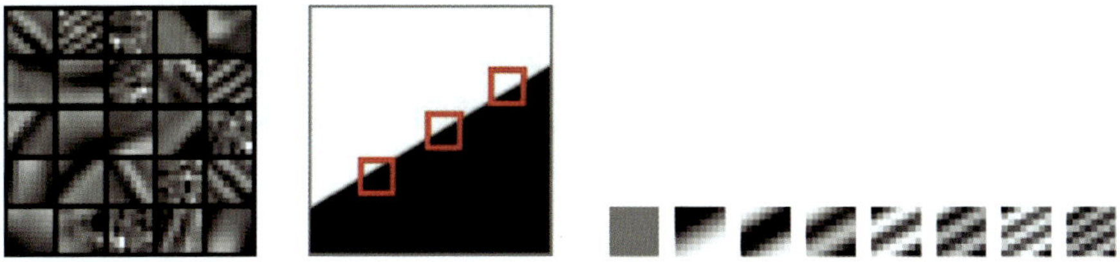

(a) (b) (c)

Figure 5.6. From Yu *et al.* (2012). (a) Typical dictionary atoms learned from the classic image Lena with K-SVD. (b, c) The numerical procedure to create one of the oriented PCAs. (b) A synthetic edge image. Patches 8×8 touching the edge are used to calculate an initial PCA basis. (c) The first 8 patches of the PCA basis (ordered by the larger eigenvalue).

that the MAP probability of the observed signals $\mathbb{P}(\{\hat{P}_i\}_i \mid \{\tilde{P}_i\}_i, \{\mu_k, \mathbf{C}_k\}_k)$ always increases. The clusters and the patch estimates converge. Nevertheless, this algorithm requires a good initialization. Noting that having the adequate Gaussians describing the patch space amounts to having a good set of PCA bases for intuitive patch clusters, Yu *et al.* create 19 orthogonal bases in the following way. One of them, say $k = 0$, is the classic DCT basis and corresponds to the 'texture cluster'. The others are obtained by fixing 18 uniformly sampled directions in the plane. For each direction, PCA is applied to a set of patches extracted from a synthetic image containing an edge in that direction. The PCA yields an oriented orthonormal basis. In short, the initial clusters segment the patch set into 18 classes of patches containing an edge or an oriented texture and one class containing the more isotropic patches.

The study by Yu *et al.* (2012) gives an interpretation of the patch dictionary methods such as K-SVD and fuses them with Bayesian methods and the Wiener method. In particular, Yu *et al.* show how the K-SVD method actually learns patches that are quite similar to oriented patches obtained by the above procedure, as illustrated in Figure 5.6. This analysis provides the framework for the synthesis proposed in Section 7.

6. Comparison of denoising algorithms

In this section we shall compare the following 'state-of-the-art' denoising algorithms:

- the sliding DCT filter as specified in Algorithm 3,
- the wavelet neighbourhood Gaussian scale mixture (BLS-GSM) algorithm as specified in Algorithm 9,
- the classical vector-valued NL-means as specified in Algorithm 4,

- the BM3D algorithm as specified in Algorithms 11 and 12,
- the K-SVD denoising method as described in Algorithm 10, and
- the non-local Bayes algorithm as specified in Algorithm 5.

These algorithms have been chosen for two reasons. First they have a public and completely transparent code available, which is in agreement with their present description. Second, they all represent distinct denoising principles and therefore illustrate the methodological progress and the diversity of denoising principles.

The comparison, using public-domain IPOL algorithms when possible, will be based on four quantitative and qualitative criteria: visualization of the *method noise*, namely the part of the image that the algorithm has taken out as noise; the visual verification of the *noise-to-noise* principle; the *mean square error* or *PSNR* tables; and last but but not least, the *visual quality* of the restored images, which must of course be the ultimate criterion. It is easily seen that a single criterion is not enough to judge a restoration method. A good denoising solution must have high performance under all mentioned criteria.

6.1. 'Method noise'

The difference between the original image and its filtered version shows the 'noise' removed by the algorithm. The procedure was introduced by Buades, Coll and Morel (2005*a*), who called this difference *method noise.* They pointed out that the method noise should look like a noise, at least in the case of additive white noise. A visual inspection of this difference tells us which geometrical features or details have been unduly removed from the original. Only human perception is able to detect these unduly removed structures in the method noise. Furthermore, for several classical algorithms, such as Gaussian convolution, anisotropic filters, neighbourhood filters or wavelet thresholding algorithms, a closed formula permits mathematical analysis of the method noise, and thus gives an explanation of the observed structure of image differences when applying the denoising method (Buades *et al.* 2005*b*). Such an analysis is unfortunately not available, and not easy for the state-of-the-art algorithms compared in this section. The degree of complexity of each method does not allow for a mathematical study of the method noise. Therefore the evaluation of this criterion will be based only on visual inspection.

When the standard deviation of the added noise is higher than contrast in the original image, a visual exploration of the method noise is nevertheless not reliable. Image features in the method noise may be hidden in the removed noise. For this reason, the evaluation of the method noise should not rely on experiments where white noise with standard deviation larger than 5 or 10 has been added to the original.

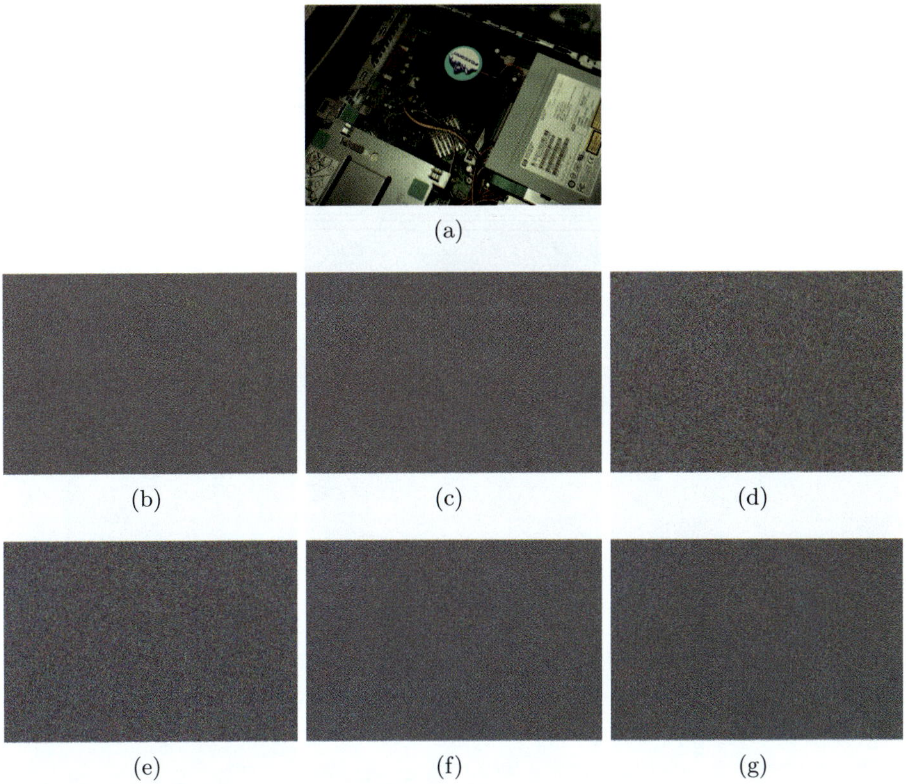

Figure 6.1. Method noise. The noisy image was obtained by adding Gaussian white noise of standard deviation 5. (a) Slightly noisy image, (b) DCT sliding window (StD = 4.69), (c) BLS-GSM (StD = 4.28), (d) NL-means (StD = 5.78), (e) K-SVD (StD = 5.67), (f) BM3D (StD = 4.25) and (g) non-local Bayes (StD = 4.28). All methods have a difference similar to white noise even if the magnitudes of the NL-means and K-SVD differences are larger. This is corroborated by the standard deviation of each residual noise. Due to the thresholding nature of DCT, BLS-GSM, BM3D and NL-Bayes, which make little change to the coefficients larger than those predicted by noise, noise is not removed in textured and edge zones. This can be easily seen in Figure 6.2, where a piece of the residual noise has been enlarged.

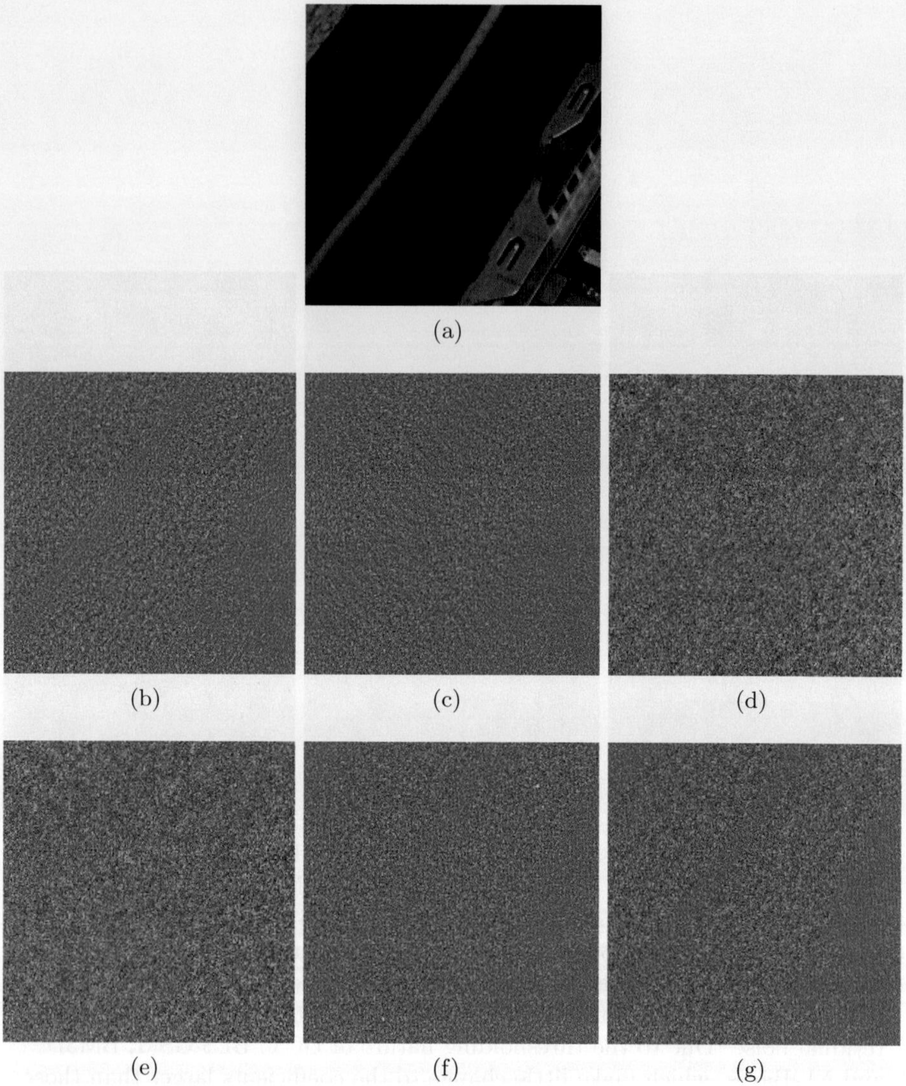

Figure 6.2. Enlargement of the method noise difference of Figure 6.1. (a) Slightly noisy image, and the method noise for (b) DCT sliding window, (c) BLS-GSM, (d) NL-means, (e) K-SVD, (f) BM3D and (g) non-local Bayes. The amplitude of the noise removed by NL-means and K-SVD is uniform all over the image, while it is region-dependent for the rest of the algorithms. Threshold-based algorithms prefer to keep noisy values nearly untouched on highly textured or edge zones.

Figure 6.1 displays the method noise for the state-of-the-art algorithms compared in this section, when Gaussian white noise of standard deviation $\sigma = 5$ has been added. The image differences have been rescaled from $[-4\sigma, 4\sigma]$ to $[0, 255]$ for visualization purposes, and values outside this range have been saturated. In a first visual inspection, we see that all methods have a difference similar to white noise. This is an outstanding property of these algorithms, which is not shared by classical denoising techniques such as anisotropic filtering, total variation minimization or wavelet thresholding (see Buades *et al.* (2005*b*) for a more detailed study). We also see immediately that the magnitude of the method noise of NL-means and K-SVD is larger than for the rest of the methods. This is corroborated by the standard deviation of each residual noise (see Figure 6.1), which is $\simeq 5.7$ for NL-means and K-SVD, $\simeq 4.7$ for DCT denoising, and $\simeq 4.25$ for the other algorithms. DCT-denoising, BLS-GSM, BM3D and non-local Bayes keep the transform coefficients that are larger than those predicted by noise. This explains why they remove little noise in textured or edge regions. This fact can be easily seen in Figure 6.2, where a piece of the residual noise of Figure 6.1 has been enlarged. The amplitude of the noise removed by NL-means and K-SVD is uniform all over the image, while it depends on the underlying image for the rest of the algorithms.

6.2. The 'noise-to-noise' principle

The *noise-to-noise* principle, introduced by Buades *et al.* (2008*b*), requires a denoising algorithm to transform white noise into white noise. This paradoxical requirement seems to be the best way to characterize artifact-free algorithms. The transformation of white noise into any correlated signal creates structure and artifacts. Only white noise is perceptually devoid of structure, as was pointed out by Attneave (1954).

The noise-to-noise of classical denoising algorithms was studied by Buades *et al.* (2008*b*), who showed that neighbourhood filters and, asymptotically, NL-means transform white noise into white noise. The convolution with a Gauss kernel keeps the low frequencies and cancels the high ones. Thus, the filtered noise actually shows big grains due to its prominent low frequencies. Noise filtered by a wavelet or DCT thresholding is no longer white noise. The few coefficients with a magnitude larger than the threshold are spread all over the image. The pixels which do not belong to the support of one of these coefficients are set to zero. The visual result is a constant image with superimposed wavelets or cosines if the DCT is used. The mathematical analysis of the rest of the algorithms is not feasible due to its degree of complexity. Thus, only a visual inspection of this filtered noise is possible.

The methodology adopted to process the *noise-to-noise* and to exhibit it is as follows.

Table 6.1.

Method	PSNR	RMSE
NL-Bayes	45.45	1.36
BM3D	45.03	1.43
NL-means	41.45	2.16
TV denoising	41.06	2.26
DCT denoising	40.91	2.30
K-SVD	38.44	3.05

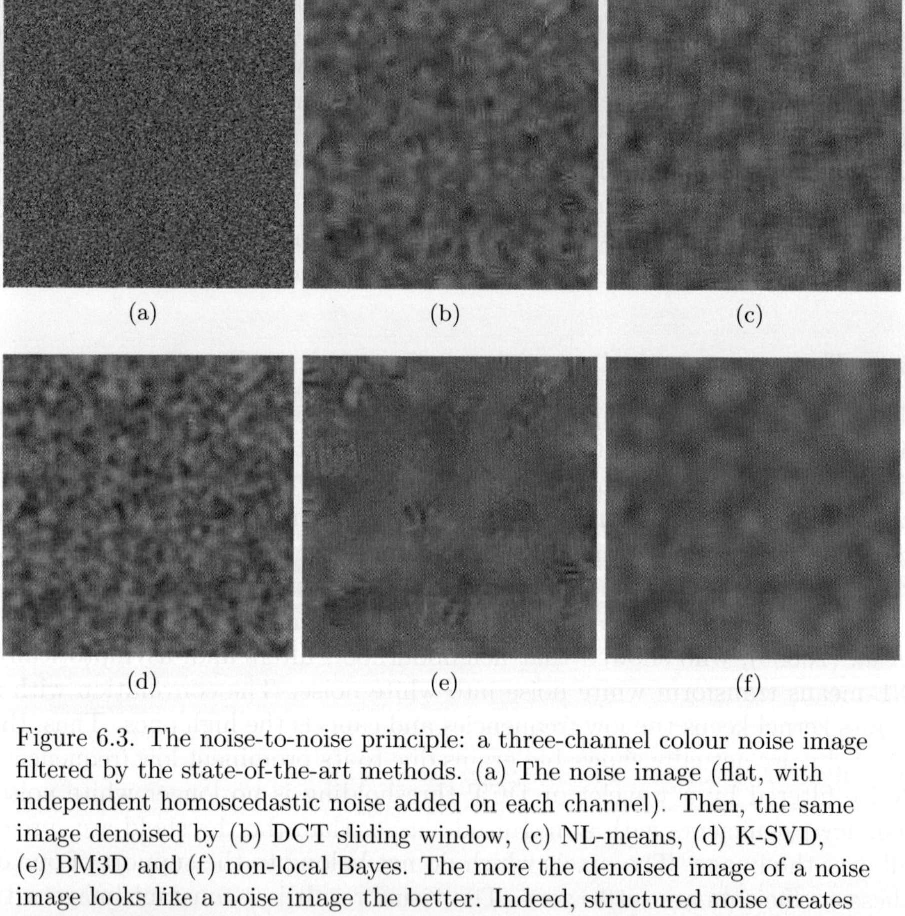

(a) (b) (c)

(d) (e) (f)

Figure 6.3. The noise-to-noise principle: a three-channel colour noise image filtered by the state-of-the-art methods. (a) The noise image (flat, with independent homoscedastic noise added on each channel). Then, the same image denoised by (b) DCT sliding window, (c) NL-means, (d) K-SVD, (e) BM3D and (f) non-local Bayes. The more the denoised image of a noise image looks like a noise image the better. Indeed, structured noise creates artifacts. BLS-GSM was not compared because we lack a colour version for this algorithm. None of the methods gives a satisfactory result: they all create a lower-frequency oscillation or local artifacts for DCT and BM3D. Only a multiscale version could cope with the low-frequency remaining noise.

- Since most recent methods process colour images (except BLS-GSM), the *noise-to-noise* is applied on a colour flat image, *i.e.*, an image with three channels of slightly different values:[4] $RGB = (127, 128, 129)$.
- To reduce the variations due to the random nature of the noise, the tests are performed on relatively large noise images. The chosen size is 1024×1024. The PSNR and RMSE results then become fairly independent of the simulated noise.
- Noise is added to each channel independently. It is therefore a colour noise, and its standard deviation is equal to 30 on each channel of the flat original image.
- All compared algorithms are processed on this noisy image.
- The denoised image is displayed. The mean on every channel is set to 128, and the difference to this mean is enhanced by a factor of 5. A small part with size 256×256 of the denoised image is shown in close-up in Figure 6.3.

The results in PSNR and RMSE are summarized in Table 6.1. The 'order' of performance of the methods is almost respected, except for TV denoising, which shows a really good result compared to K-SVD. Figure 6.3 displays the filtered noise images by several state-of-the-art algorithms.

As expected, threshold-based methods present noticeable artifacts, in particular DCT denoising and BM3D. The NL-means result reflects the size of the search zone, and therefore leaves behind a low-frequency oscillation. Despite its good results, TV denoising presents a lot of artifacts which do not look like noise, and are uglier than the K-SVD artifacts. Only non-local Bayes has no artifacts. Indeed, it detects flat patches and replaces them with their mean. This trick could in fact be applied to all algorithms. Last but not least, each method leaves a sizeable low-frequency noise, which could be removed with a multiscale approach.

6.3. Comparing visual quality

The visual quality of the restored image is obviously a necessary, if not sufficient, criterion to judge the performance of a denoising algorithm. It permits control of the absence of artifacts and the correct reconstruction of edges, texture and fine structure. Figures 6.5–6.7 display the noisy and denoised images for the algorithms under comparison for noise standard deviations of 20, 30 and 40.

Figure 6.5 presents an image with straight edges and flat and fine structures with a noise of standard deviation 20. The main artifacts are noticeable in the DCT, BLS-GSM and K-SVD denoised images. These are the

[4] Values are different on each channel in order to force the algorithm to consider this image as a colour image, and not a grey-scale image with a single channel.

(a) alley (b) computer (c) dice

(d) flowers (e) girl (f) traffic

(g) trees (h) Valldemossa

Figure 6.4. A set of noiseless images used for the comparison tests.

most local algorithms and therefore have more trouble removing the low frequencies of the noise. As a consequence, the denoised images present many low-frequency colour artifacts in flat and dark zones. These artifacts are noticeable for all these algorithms even though each uses a different strategy to deal with colour images. DCT uses the $Y_oU_oV_o$, K-SVD a vector-valued algorithm and BLS-GSM is applied independently to each RGB component. NL-means does not suffer from these noise low-frequency problems, but it leaves some isolated noise points on non-repetitive structures, mainly on corners. These isolated noise points could be attenuated by using the $Y_oU_oV_o$ colour space instead of the vector-valued algorithm. In this experiment, BM3D and non-local Bayes give similar performance, superior to the rest of algorithms.

Figures 6.6 and 6.7 again illustrate the low-frequency colour artifacts of DCT, BLS-GSM and K-SVD. In these figures, DCT and BLS-GSM also suffer from a strong Gibbs effect near all image boundaries. This Gibbs effect is nearly unnoticeable in the denoised image by K-SVD, since the use of the whole dictionary permits better reconstruction of edges when the right atoms are present in the dictionary. The image denoised by NL-means

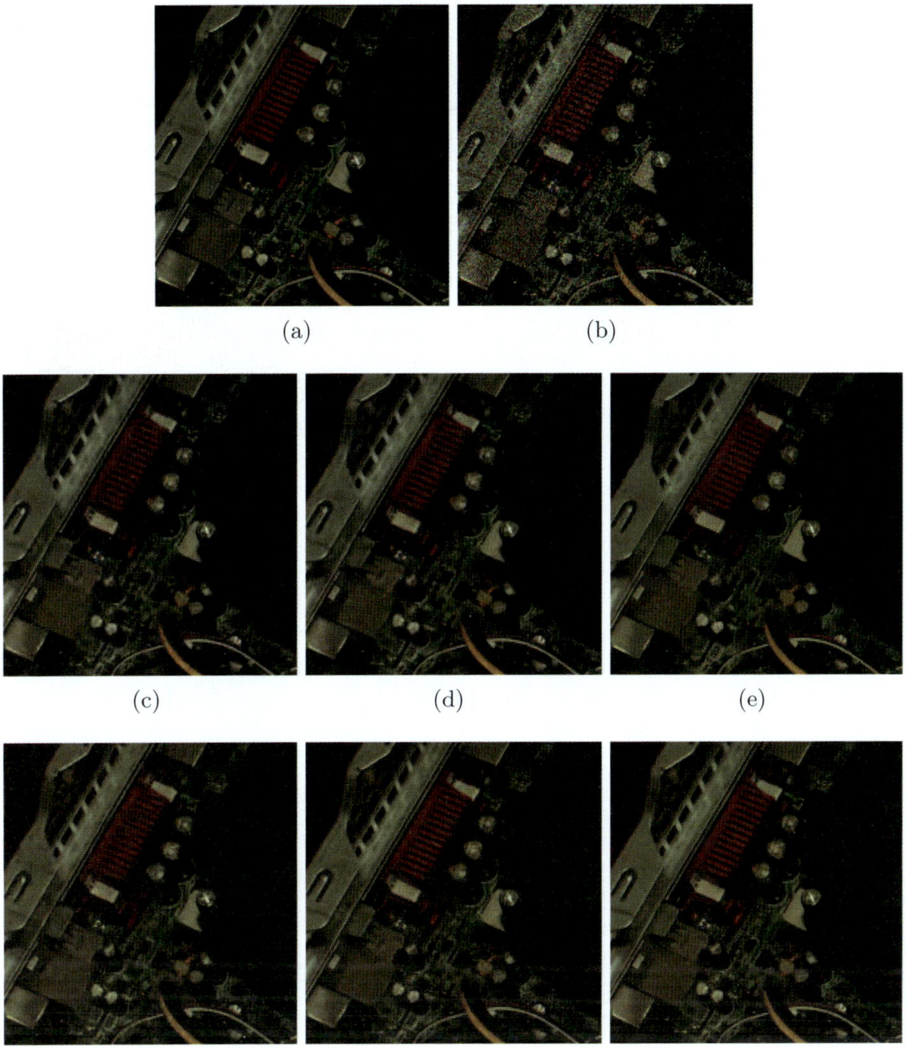

(a) (b)

(c) (d) (e)

(f) (g) (h)

Figure 6.5. Comparison of visual quality. The noisy image was obtained by adding Gaussian white noise of standard deviation 20. (a) Original, (b) noisy, (c) DCT sliding window, (d) BLS-GSM, (e) NL-means, (f) K-SVD, (g) BM3D and (h) non-local Bayes.

(a) (b)

(c) (d) (e)

(f) (g) (h)

Figure 6.6. Comparison of visual quality. The noisy image was obtained by adding Gaussian white noise of standard deviation 30. (a) Original, (b) noisy, (c) DCT sliding window, (d) BLS-GSM, (e) NL-means, (f) K-SVD, (g) BM3D and (h) non-local Bayes.

(a) (b) (c) (d) (e) (f) (g) (h)

Figure 6.7. Comparison of visual quality. The noisy image was obtained by adding Gaussian white noise of standard deviation 40. (a) Original, (b) noisy, (c) DCT sliding window, (d) BLS-GSM, (e) NL-means, (f) K-SVD, (g) BM3D and (h) non-local Bayes.

has no visual artifacts but is more blurred than those given by BM3D and non-local Bayes, which clearly have superior performance to the rest of the algorithms. The image denoised by BM3D has some Gibbs effect near edges, which sometimes degrades the visual quality of the solution. The non-local Bayes image shows no artifacts. It often preserves better textures than BM3D, in which trees and vegetation can be slightly blurred by the use of the linear transform threshold.

In short, the visual quality of DCT, BLS-GSM and K-SVD is inferior to that of NL-means, BM3D and non-local Bayes, because of strong colour noise low frequencies in flat zones, and of a Gibbs effect. NL-means does not show noticeable artifacts but the denoised image is more blurred than those of BM3D and non-local Bayes. BM3D still has some Gibbs effect due to the use of a single basis for all pixels and a slightly inferior noise reduction, compared to non-local Bayes.

6.4. Comparing by PSNR

The mean square error is the square of the Euclidean distance between the original image and its estimate. In the denoising literature an equivalent measurement, up to a decreasing scale change, is the PSNR:

$$\mathrm{PSNR} = 10\log_{10}\left(\frac{255^2}{\mathrm{MSE}}\right).$$

These numerical quality measurements are the most objective, since they do not rely on any visual interpretation. Tables 6.2 and 6.3 display the PSNR of state-of-the-art denoising methods using the images in Figure 6.4 and several values of σ from 2 to 40.

Before jumping to conclusions, we would like to point out that such a PSNR comparison is merely informative, and cannot lead to an objective ranking of algorithms. Indeed, what is really needed is a comparison of denoising principles. To compare them, these denoising principles must be implemented in denoising recipes containing several ingredients. Since the PSNR difference between recipes is very close, the way generic tools are implemented, and the degree of sophistication with which each principle is implemented, do matter. For example, two of our readers, Alessandro Foi and Vladimir Katkovnik,[5] have pointed out to us that an experimental analysis carried out exclusively on colour images does not permit a comparison between the different strategies devised to take advantage of spatial redundancy. They suggest complementing the denoising results on colour images with experiments on grey-scale images. Then it would be possible to: (1) compare the degree of success of these different denoising principles in

[5] Personal communication.

Table 6.2. PSNR table for $\sigma = 2$, 5 and 10. Only the three first digits are actually significant; the last one may vary with different white noise realizations.

	NL-Bayes	BM3D	BLS-GSM	K-SVD	NL-means	DCT denoising
				$\sigma = 2$		
alley	45.42	44.95	–	41.51	42.75	44.58
computer	45.96	45.22	44.69	44.52	44.03	44.54
dice	49.00	48.86	48.59	47.79	48.51	48.39
flowers	47.77	47.31	47.12	47.09	46.36	47.05
girl	47.56	47.40	47.14	47.28	46.96	46.76
traffic	45.33	44.56	44.15	43.80	43.55	44.26
trees	43.51	43.07	–	42.05	42.22	42.95
Valldemossa	45.17	44.68	44.41	40.08	43.33	44.50
mean	46.22	45.76	–	44.27	44.71	45.37
				$\sigma = 5$		
alley	39.24	38.95	–	38.45	37.18	38.37
computer	40.69	39.98	39.30	39.58	38.86	39.03
dice	46.09	45.80	45.21	45.27	45.12	45.22
flowers	43.44	42.99	42.76	43.09	42.05	42.78
girl	44.26	44.03	43.70	43.59	43.44	43.36
traffic	39.70	38.67	38.10	38.75	37.50	38.21
trees	36.70	36.10	–	35.61	34.69	35.76
Valldemossa	38.73	38.33	38.02	37.87	35.94	37.94
mean	41.11	40.61	–	40.28	39.35	40.08
				$\sigma = 10$		
alley	35.05	34.82	–	34.29	33.53	34.22
computer	36.58	36.28	35.47	35.79	35.44	35.34
dice	43.30	43.02	42.21	41.71	42.06	42.22
flowers	39.52	39.49	39.10	39.31	38.49	39.03
girl	41.69	41.45	41.14	40.29	40.42	40.55
traffic	34.93	34.54	33.92	34.69	33.89	34.11
trees	36.70	36.10	–	35.61	29.42	30.92
Valldemossa	38.73	38.33	38.02	37.87	32.02	33.45
mean	37.06	36.83	–	36.31	35.66	36.23

Table 6.3. PSNR table for $\sigma = 20$, 30 and 40.

	NL-Bayes	BM3D	BLS-GSM	K-SVD	NL-means	DCT denoising
			$\sigma = 20$			
alley	<u>31.36</u>	31.23	–	30.55	29.94	30.21
computer	<u>33.08</u>	32.71	31.89	31.96	31.59	31.45
dice	<u>40.19</u>	39.93	39.00	37.23	38.17	38.67
flowers	<u>35.87</u>	35.85	35.34	35.24	34.56	34.89
girl	<u>38.92</u>	38.71	38.49	36.36	36.81	37.27
traffic	<u>31.14</u>	30.83	30.14	30.70	30.12	29.98
trees	<u>27.22</u>	26.92	–	26.88	26.28	26.27
Valldemossa	<u>29.81</u>	29.57	26.97	29.08	28.37	28.91
mean	<u>33.45</u>	33.22	–	32.25	31.98	32.20
			$\sigma = 30$			
alley	<u>29.42</u>	29.33	–	28.60	27.58	28.25
computer	<u>31.00</u>	30.67	29.90	29.84	28.98	29.20
dice	<u>38.20</u>	37.88	37.05	36.52	37.18	35.89
flowers	33.67	<u>33.73</u>	33.19	33.54	32.66	32.46
girl	<u>37.12</u>	36.97	36.91	35.38	35.54	34.67
traffic	<u>29.08</u>	28.87	28.20	28.60	27.40	27.87
trees	<u>24.95</u>	24.64	–	24.52	23.29	23.83
Valldemossa	<u>27.51</u>	27.30	26.97	26.80	25.55	26.48
mean	<u>31.37</u>	31.17	–	30.48	29.77	29.83
			$\sigma = 40$			
alley	<u>28.16</u>	28.08	–	27.29	26.30	27.14
computer	<u>29.55</u>	29.15	28.52	28.25	27.31	27.44
dice	<u>36.91</u>	36.28	35.50	34.49	35.31	33.06
flowers	31.94	<u>32.10</u>	31.68	31.90	30.99	30.80
girl	<u>36.09</u>	35.62	35.61	33.73	34.03	32.01
traffic	<u>27.67</u>	27.50	26.93	27.19	26.01	26.49
trees	<u>23.35</u>	23.17	–	23.06	21.91	22.46
Valldemossa	<u>27.51</u>	25.78	25.50	25.28	24.10	25.08
mean	<u>30.15</u>	29.71	–	28.90	28.25	28.05

exploiting spatial redundancy, and (2) evaluate the effectiveness of the various ways in which these grey-scale algorithms are extended to colour data.

In short, Foi and Katkovnik do not share our analysis or our conclusions drawn from the experimental results, because these results are very much influenced by the way colour data are treated, while many of the conclusions are applied concerning the relative effectiveness in exploiting spatial redundancy.

For the same reasons, they also disagree with the taxonomy summarized in Table 7.1 (page 564), where it seems that the extension to colour is to be considered as a feature of a particular algorithm. Some methods are applied to colour data in a very simple non-adaptive way and thus cannot be expected to fully decorrelate the colour channels. For instance, this is the case for BM3D, which uses a YUV/Opp colour transformation. Data-adaptive colour transformations for multispectral data are considered in Danielyan, Foi, Katkovnik and Egiazarian (2010). This adaptive method provides substantially better results than a standard colour transformation.

Another reason for being cautious is that all methods in existence do in fact have variants, and we are using the basic algorithms as stated in their original papers. For example, it is shown in Hou, Zhao, Yang and Cheng (2011) that BM3D can be slightly improved for heavy noise > 40 by changing the method parameters.

In other words, the following PSNR comparison of colour images must be taken 'as is': it gives some hints, and these hints depend on the particular implementation of the denoising principles. We observe in the results that DCT denoising, BLS-GSM, K-SVD and NL-means have a similar PSNR. The relative performance of the methods depends on the kind of image and on noise level σ. On average, K-SVD and BLS-GSM are slightly superior to the other two, even if this is not the case visually, where K-SVD and BLS-GSM have poor visual quality compared to NL-means. In all cases, BM3D and non-local Bayes have better PSNR performance than the others. Because of superior noise reduction in flat zones and the presence of fewer artifacts of non-local Bayes, the PSNR of BM3D is slightly inferior to non-local Bayes. BM3D seems to retain the best conservation of detail. Some ringing artifacts near boundaries can probably be eliminated by the same trick as for non-local Bayes, namely detecting and giving a special treatment to flat three-dimensional groups.

7. Synthesis

We have showed that all methods either already use or should adopt the same three *generic denoising tools* described in Section 4. Since all methods denoise not just the pixel but a whole neighbourhood, they give several evaluations for each pixel. Thus, they all use an aggregation step. There is only one method for which the aggregation is not explicitly stated as such, the wavelet neighbourhood (BLS-GSM) algorithm. Nevertheless, closer examination shows that it denoises not just one but some 49 wavelet channels for a 512×512 image. The applied wavelet transform is redundant. Thus, an aggregation is implicit in its final reconstruction step from all channels. BLS-GSM is also patch-based. Indeed, each 'wavelet neighbourhood' contains a 3×3 patch of a wavelet channel, complemented by one more sample from the down-scale channel sharing the same orientation. Thus, like the others, this algorithm builds Bayesian estimates of patches. The difference is that the patches belong to the wavelet channels. Each one of these channels is denoised separately, before the reconstruction of the image from its wavelet channels.

In short, even if the BLS-GSM formalization initially looks different from the other algorithms, it relies on similar principles: it estimates patch models to denoise them, and aggregates the results. But it is also the only multiscale algorithm among those considered here. Indeed, it denoises the image at all scales. Furthermore, it introduces a scale interaction. These features are neglected in the other algorithms and might make a significant difference in future algorithms.

Why is its performance slightly inferior to that of the current state-of-the-art algorithms? First of all, this algorithm, like many wavelet-based algorithms, has not proposed a good solution to deal with colour. Applying the colour space tool of Section 4.3 can probably bring about a PSNR improvement. Portilla *et al.* (2003) do not specify whether there is an aggregation step, but a first aggregation step is possible (the second aggregation being implicit in the reconstruction step from all redundant channels). Indeed, each wavelet channel patch contains ten coefficients, and these coefficients are therefore estimated ten times. These estimates might be aggregated.

7.1. The synoptic table

Table 7.1 shows a synopsis of the ten methods that have been thoroughly discussed. The classification criteria are as follows.

The method's denoising principle
Our task here is to show that, in spite of the different language used for each method, the underlying principles are, in fact, very similar. The dominant principle is to compute a linear minimum least-squares estimator (LMMSE) after building a Bayesian patch model. As a matter of fact, even if this is not always explicit, *all* methods follow the same LMMSE estimator principle very closely. For example, the DCT threshold is simply a Wiener thresholding version of the Bayesian LMMSE. This threshold is used because the DCT of the underlying noiseless image is unknown. The same argument

applies for NL-means, which was interpreted as an LMMSE in Section 5.1. A close examination of K-SVD can convince the user that this algorithm is very close to PLE, PLOW or EPLL, and conversely. Indeed, the patch clustering performed in these three algorithms interprets the patch space as a redundant dictionary. Each cluster is treated by a Bayesian estimator as a Gaussian vector, for which an orthogonal eigenvector basis is computed. This basis is computed from the cluster patches by PCA. Thus, PLE, PLOW and EPLL actually deliver a dictionary, which is the union of several orthogonal bases of patches. For each noisy patch, PLE, PLOW and EPLL select one of the bases, for which the patch will be sparse. In short, like K-SVD, they compute a sparse representation for each patch in an overcomplete dictionary. In this argument, we follow the simple and intelligent interpretation proposed for the PLE method by Yu *et al.* (2010, 2012), who summarize their method as follows:

> An image representation framework based on structured sparse model selection is introduced in this work. The corresponding modeling dictionary is comprised of a family of learned orthogonal bases. For an image patch, a model is first selected from this dictionary through linear approximation in a best basis, and the signal estimation is then calculated with the selected model. The model selection leads to a guaranteed near optimal denoising estimator. The degree of freedom in the model selection is equal to the number of the bases, typically about 10 for natural images, and is significantly lower than with traditional overcomplete dictionary approaches, stabilizing the representation.

From the algorithmic viewpoint, EPLL is a variant of PLE, but used in a different setting. The comparison of these two almost identical Gaussian mixture models is of particular interest. EPLL is applied to a huge set of patches (of the order of 10^{10}) united in some 200 clusters. PLE is applied with 19 clusters, each learned from some 64 patches. Thus, the open question is: How many clusters and how many learning patches are truly necessary to obtain the best PSNR? The disparity between these figures is certainly too large to be realistic.

Finally, we must wonder if transform thresholding methods fit into the united view of all algorithms. The Bayesian–Gaussian estimate used by most algorithms mentioned can be interpreted as a Wiener filter on the eigenvector basis of the Gaussian. It sometimes includes a threshold (to avoid negative eigenvalues for the covariance matrix of the Gaussian vector). Thus, the only difference between Bayesian–Gaussian methods and the classic transform thresholding is that in the Bayesian methods the orthogonal basis is adapted to each patch. Therefore, they appear to be a direct extension of transform thresholding methods, and have logically replaced them. BM3D combines several linear transform thresholds (2D bior 1.5, 2D DCT, 1D Walsh–Hadamard), applied to the three-dimensional block obtained by grouping similar patches. Clearly, it has found, by a

Table 7.1. Synoptic table of all methods considered.

Method	Denoising principle	Patches	Size	Aggr.	Oracle	Col.
DCT	transform threshold	one	8	yes	yes	yes
NL-means	average	neighbourhood	3	yes	yes	no
NL-Bayes	Bayes	neighbourhood	3–7	yes	yes	yes
PLOW	Bayes, 15 clusters	image	11	yes	yes	yes
shotgun-NL	Bayes	10^{10} patches	3–20	yes	no	no
EPLL	Bayes, 200 clusters	2×10^{10} patches	8	yes	yes	yes
BLS-GSM	Bayes in GSM	image	3	yes	no	no
K-SVD	sparse dictionary	image	8	yes	yes	yes
BM3D	transform threshold	neighbourhood	8–12	yes	yes	yes
PLE	Bayes, 19 clusters	image	8	yes	yes	yes

rather systematic exploration, the right two-dimensional orthogonal bases, and therefore does not need to estimate them for each patch group.

We shall now reunite two groups of methods that are only superficially different. NL-means, non-local Bayes, shotgun-NL, and BM3D denoise a patch after comparing it with a group of similar patches. The other five patch-based Bayesian methods *do not perform a search for similar patches.*

These other patch methods, PLE, PLOW, EPLL, BLS-GSM and K-SVD, globally process the 'patch space' and construct patch models. Nevertheless, this difference is easily reduced. Indeed, PLE, PLOW and EPLL segment the patch space into a sufficient number of clusters, each one endowed with a rich structure (an orthonormal basis). Thus, the patches contributing to the denoising of a given patch estimation are not compared with each other, but they are compared with the clusters. Similarly, dictionary-based methods such as K-SVD propose over-complete dictionaries learned from the image or from a set of images. Finding the best elements of the dictionary to decompose a given patch, as K-SVD does, amounts to classifying this patch. This is suggested by Yu *et al.* (2012), for PLE: the dictionary is a list of orthogonal bases which are initiated by sets of oriented edges. Each basis is therefore associated with an orientation (plus one associated with the DCT). Thus PLE is very similar to BLS-GSM, which directly applies a set of oriented filters. Another link between the Bayesian method and sparse modelling is elaborated in Zhou *et al.* (2011).

Patches

The 'Patches' column in Table 7.1 indicates the number of patches used for the denoising method, and where they are found. Trivial DCT uses only the current patch to denoise it. NL-means, non-local Bayes and BM3D compare the reference patch with a few hundred patches in its spatial neighbourhood.

PLE, PLOW, BLS-GSM and K-SVD compare each noisy patch to a learned model of all image patches. Finally, shotgun-NL and EPLL involve a virtually infinite number of patches in the estimation. Surprisingly enough, the performance of all methods is relatively similar. Thus, the huge numbers used to denoise in shotgun-NL and EPLL clearly depend on the fact that the patches were not learned from the image itself, and their number can arguably be considerably reduced.

Size (of patches)
The 'Size' column compares the patch sizes. All methods without exception try to deduce the correct value of a given pixel $\mathbf{i}$ by using a neighbourhood of $\mathbf{i}$ called a patch. This patch size goes from 3×3 to 8×8, with a strong dominance of 8×8 patches. Nevertheless, the size of the patches obviously depends on the amount of noise and should be adapted to the noise standard deviation. For very large noises, a size of 8×8 can be insufficient, while for small noises small patches might be better. As a matter of fact, all articles focus on noise standard deviations around 30 (most algorithms are tested for σ between 20 and 80). Little work has been done on small noise (below 10). For large noise, above 50, most algorithms do not deliver a satisfactory result and most papers show denoising results for $20 \leq \sigma \leq 40$. This may also explain the homogeneity of the patch size.

Aggregation, oracle, colour
A good sign of maturity of the methods is that the three generic improvement tools described in Section 4 are used by most methods. When a 'no' is present in the table in these three columns, this indicates that the method can probably be substantially improved with little effort by using the corresponding tool. Shotgun-NL and BLS-GSM can probably gain some decibels via aggregation and via the oracle strategy.

Algorithms compared by complexity and information
Current research focuses on obtaining the best possible results, and perhaps the optimal results. We have followed this path and have completely ignored complexity issues in this comparison. For example, the 'shotgun' patch methods are not reproducible in an acceptable time. But 'all is fair in love and war'. The question of how to obtain the best acceptable results must be solved first, by every possible means, before fast algorithms are devised. On the other hand, complexity does not seem to be a serious obstacle. Indeed, several of the mentioned algorithms are already realizable, and five of them are even functioning online at *Image Processing On Line.* At least two of them give state-of-the-art results. Thus, we hold the view that complexity is not a central issue in the current debate. Another question that emerged in this study is the *amount of information needed to*

achieve optimal denoising. Here, we have observed that the methods divide into two groups. The simplest one (DCT denoising) uses only one image patch and gets results only 1 dB from optimal results. The classic non-local methods only use a larger neighbourhood of a given pixel, in spite of their 'non-local' epithet. Then, an intermediate class of methods uses all image patches simultaneously. The shotgun methods use virtually all possible existing image patches. The fact that the performance gap between them is so small seems to indicate that all obtain a decent estimate of the 'patch space' around each given image patch. This also means that, arguably, there is enough information for that in just one image.

7.2. Conclusion

Most patch-based image denoising methods follow one paradigm: they unite the transform thresholding method and a Markovian–Bayesian estimation. This unification is complete when the patch space is assumed to be a Gaussian mixture. Each Gaussian is associated with its orthonormal basis of patch eigenvectors. Thus, transform thresholding (or a Wiener filter) is applied to these local orthogonal bases.

This method seems to be almost optimal. It yields satisfactory results for an interval of standard deviations ranging from 5 to 40. (These figures are valid for current image formats, with a range of $[0, 255]$.) Small noise (below 5) and large noise (above 50) are largely unexplored. They may require new tools or a different theory. Are they important? The answer is yes, because for several applications, such as photogrammetry and stereovision, the precision varies inversely with the signal-to-noise ratio. Thus, even with good-quality stereo pairs, it is relevant to remove even more noise. For high noise, near-optimal denoising can be obtained by applying a global Wiener filter using the second-order statistics of the image. Nevertheless, all existing filters produce too many artifacts for large noise.

The multiscale aspect of denoising has been explored only on three dyadic scales (since most patch methods use 8×8 patches). It may be insufficient. The success of denoising methods suggests that the statistical exploration of images has been advancing slowly. The exploration of the huge 'patch space' is only starting. Its structure remains largely unknown, and we know little of its geometry. Its representation by a sum of Gaussians, or by a Gaussian scale mixture, is only a first rough approximation.

Are image denoising algorithms close to achieving their optimal bounds? For the ranges of noise that we have tested, the image visual improvement obtained by state-of-the-art denoising methods is undeniable and sometimes spectacular. For movies, which are much more redundant, this effect is even more impressive. Can we deduce from the arguments developed by Levin and Nadler (2011) and Chatterjee and Milanfar (2010) that the current methods are close to optimality? In these papers optimal bounds for *all* patch-based methods are proposed. Nevertheless, a closer examination shows that existing methods are probably further away from optimality than is understood. Indeed, all state-of-the-art patch-based methods use the aggregation step, which doubles the size of the neighbourhood effectively used. Thus, in a fair comparison, the shotgun Bayesian estimate (Levin and Nadler 2011) should use 16×16 patches. We might be facing the curse of dimensionality.

Acknowledgements

The authors wish to thank Richard Baraniuk, Alessandro Foi, Vladimir Katkovnik, Peyman Milanfar, Boaz Nadler, Boshra Rajaei, Guillermo Sapiro, Eero Simoncelli, Yair Weiss and Guoshen Yu for valuable comments, which have been systematically integrated into the text. Research was partly financed by the MISS project of Centre National d'Etudes Spatiales, the US Office of Naval Research under grant N00014-97-1-0839, and by the European Research Council's 'Twelve Labours' advanced grant.

REFERENCES[6]

A. Adams, N. Gelfand, J. Dolson and M. Levoy (2009), Gaussian KD-trees for fast high-dimensional filtering. In *ACM Transactions on Graphics: TOG*, Vol. 28, ACM, p. 21.

M. Aharon, M. Elad and A. M. Bruckstein (2006), 'K-SVD: An algorithm for designing overcomplete dictionaries for sparse representation', *IEEE Trans. Signal Processing* **54**, 4311–4322.

F. Anscombe (1948), 'The transformation of Poisson, binomial and negative-binomial data', *Biometrika* **35**, 246–254.

M. Antonini, M. Barlaud, P. Mathieu and I. Daubechies (1992), 'Image coding using wavelet transform', *IEEE Trans. Image Processing* **1**, 205–220.

P. Arias, V. Caselles and G. Sapiro (2009), A variational framework for non-local image inpainting. In *Proc. 7th International Conference on Energy Minimization Methods in Computer Vision and Pattern Recognition*, Springer, pp. 345–358.

F. Attneave (1954), 'Some informational aspects of visual perception', *Psych. Rev.* **61**, 183–193.

S. Awate and R. Whitaker (2006), 'Unsupervised, information-theoretic, adaptive image filtering for image restoration', *IEEE Trans. Patt. Anal. Machine Intell.* **28**, 364–376.

[6] The URLs cited in this work were correct at the time of going to press, but the publisher and the authors make no undertaking that the citations remain live or are accurate or appropriate.

C. Barnes, E. Shechtman, A. Finkelstein and D. Goldman (2009), PatchMatch: A randomized correspondence algorithm for structural image editing. In *ACM Transactions on Graphics: TOG*, Vol. 28, ACM, p. 24.

R. Bilcu and M. Vehvilainen (2008), Combined non-local averaging and intersection of confidence intervals for image denoising. In *15th IEEE International Conference on Image Processing*, pp. 1736–1739.

J. Boulanger, J. Sibarita, C. Kervrann and P. Bouthemy (2008), Non-parametric regression for patch-based fluorescence microscopy image sequence denoising. In *5th IEEE International Symposium on Biomedical Imaging: From Nano to Macro, 2008*, pp. 748–751.

R. Bracho and A. Sanderson (1985), Segmentation of images based on intensity gradient information. In *Proc. IEEE Computer Society Conference on Computer Vision and Pattern Recognition*, pp. 19–23.

P. Brémaud (1999), *Markov Chains: Gibbs Fields, Monte Carlo Simulation, and Queues*, Vol. 31 of *Texts in Applied Mathematics*, Springer.

X. Bresson and T. Chan (2008), Non-local unsupervised variational image segmentation models. UCLA CAM Report 08-67.

A. Buades (2006), Image and film denoising by non-local means. PhD thesis, Universitat Illes Balears.

A. Buades, A. Chien, J. Morel and S. Osher (2008*a*), 'Topology preserving linear filtering applied to medical imaging', *SIAM J. Imaging Sci.* **1**, 26–50.

A. Buades, B. Coll and J. Morel (2004), Image data process by image noise reduction and camera integrating the means for implementing this process. French Patent 0404837.

A. Buades, B. Coll and J. Morel (2005*a*), 'A non-local algorithm for image denoising', *IEEE Computer Vision and Pattern Recognition* **2**, 60–65.

A. Buades, B. Coll and J. Morel (2005*b*), 'A review of image denoising algorithms, with a new one', *Multiscale Model. Simul.* **4**, 490–530.

A. Buades, B. Coll and J. Morel (2006*a*), Image enhancement by non-local reverse heat equation. Preprint CMLA 2006-22.

A. Buades, B. Coll and J. Morel (2006*b*), 'Neighborhood filters and PDE's', *Numer. Math.* **105**, 1–34.

A. Buades, B. Coll and J. Morel (2006*c*), 'The staircasing effect in neighborhood filters and its solution', *IEEE Trans. Image Processing* **15**, 1499–1505.

A. Buades, B. Coll and J. Morel (2008*b*), 'Nonlocal image and movie denoising', *Internat. J. Computer Vision* **76**, 123–139.

A. Buades, B. Coll, J. Morel and C. Sbert (2009*a*), 'Self-similarity driven color demosaicking.', *IEEE Trans. Image Processing* **18**, 1192–1202.

A. Buades, M. Colom and J. Morel (2012*a*), 'Multiscale signal dependent noise estimation', *Image Processing On Line* (www.ipol.im).

A. Buades, M. Lebrun and J. Morel (2012*b*), 'Implementation of the 'non-local Bayes' image denoising algorithm', *Image Processing On Line* (www.ipol.im).

A. Buades, Y. Lou, J. Morel and Z. Tang (2009*b*), A note on multi-image denoising. In *Local and Non-Local Approximation in Image Processing, 2009: LNLA 2009*, IEEE, pp. 1–15.

E. Candès and M. Wakin (2008), 'An introduction to compressive sampling', *IEEE Signal Process. Magazine* **25**, 21–30.

M. Carreira-Perpinan (2000), 'Mode-finding for mixtures of Gaussian distributions', *IEEE Trans. Patt. Anal. Machine Intell.* **22**, 1318–1323.

P. Chatterjee and P. Milanfar (2010), 'Is denoising dead?', *IEEE Trans. Image Processing* **19**, 895–911.

P. Chatterjee and P. Milanfar (2012), 'Patch-based near-optimal image denoising', *IEEE Trans. Image Processing* **21**, 1635–1649.

C. Chevalier, G. Roman and J. Niépce (1854), *Guide du Photographe*, C. Chevalier.

A. Cohen, I. Daubechies and J. Feauveau (1992), 'Biorthogonal bases of compactly supported wavelets', *Comm. Pure Appl. Math.* **45**, 485–560.

R. Coifman and D. Donoho (1995), Translation-invariant de-noising. In *Wavelets and Statistics*, Lecture Notes in Statistics, Springer, pp. 125–125.

S. Cotter, R. Adler, R. Rao and K. Kreutz-Delgado (1999), 'Forward sequential algorithms for best basis selection', *IEE Proc. Vision, Image and Signal Processing* **146**, 235–244.

P. Coupé, P. Yger and C. Barillot (2006), Fast non local means denoising for 3D MRI images. In *Medical Image Computing and Computer-Assisted Intervention: MICCAI 2*, pp. 33–40.

P. Coupé, P. Yger, S. Prima, P. Hellier, C. Kervrann and C. Barillot (2008), 'An optimized blockwise nonlocal means denoising filter for 3-D magnetic resonance images', *IEEE Trans. Medical Imaging* **27**, 425–441.

K. Dabov, A. Foi, V. Katkovnik and K. Egiazarian (2007), 'Image denoising by sparse 3D transform-domain collaborative filtering', *IEEE Trans. Image Processing* **16**, 2080–2095.

K. Dabov, A. Foi, V. Katkovnik and K. Egiazarian (2009), BM3D image denoising with shape-adaptive principal component analysis. In *Proc. Workshop on Signal Processing With Adaptive Sparse Structured Representations: SPARS 09*, Vol. 49.

A. Danielyan and A. Foi (2009), Noise variance estimation in nonlocal transform domain. In *Proc. International Workshop on Local and Non-Local Approximation in Image Processing: LNLA 2009*, pp. 41–45.

A. Danielyan, A. Foi, V. Katkovnik and K. Egiazarian (2008), Image and video super-resolution via spatially adaptive block-matching filtering. In *Proc. International Workshop on Local and Non-Local Approximation in Image Processing*.

A. Danielyan, A. Foi, V. Katkovnik and K. Egiazarian (2010), Denoising of multispectral images via nonlocal groupwise spectrum-PCA. In *Proc. Fifth European Conference on Colour in Graphics, Imaging, and Vision and of the 12th International Symposium on Multispectral Colour Science*, pp. 261–266.

A. Danielyan, V. Katkovnik and K. Egiazarian (2012), 'BM3D frames and variational image deblurring', *IEEE Trans. Image Processing* **21**, 1715–1728.

J. Darbon, A. Cunha, T. Chan, S. Osher and G. Jensen (2008), Fast nonlocal filtering applied to electron cryomicroscopy. In *5th IEEE International Symposium on Biomedical Imaging: From Nano to Macro*, pp. 1331–1334.

J. De Bonet (1997), 'Noise reduction through detection of signal redundancy', *Rethinking Artificial Intelligence*.

C.-A. Deledalle, L. Denis and F. Tupin (2011*a*), 'NL-InSAR: Nonlocal interferogram estimation', *IEEE Trans. Geoscience and Remote Sensing* **49**, 1441–1452.

C.-A. Deledalle, V. Duval and J. Salmon (2012), 'Non-local methods with shape-adaptive patches (NLM-SAP)', *J. Math. Imaging Vision*, to appear.

C.-A. Deledalle, J. Salmon and A. Dalalyan (2011*b*), Image denoising with patch based PCA: Local versus global. In *Proc. British Machine Vision Conference*, BMVA Press, pp. 25.1–25.10.

C.-A. Deledalle, F. Tupin and L. Denis (2010*a*), Poisson NL means: Unsupervised non local means for Poisson noise. In *17th IEEE International Conference on Image Processing: ICIP*, pp. 801–804.

C.-A. Deledalle, F. Tupin and L. Denis (2010*b*), Polarimetric SAR estimation based on non-local means. In *IEEE International Geoscience and Remote Sensing Symposium: IGARSS*, pp. 2515–2518.

J. Delon and A. Desolneux (2009), Flicker stabilization in image sequences. hal.archives-ouvertes.fr.

D. Donoho (1995), 'De-noising by soft-thresholding', *IEEE Trans. Inform. Theory* **41**, 613–627.

D. Donoho and I. Johnstone (1995), 'Adapting to unknown smoothness via wavelet shrinkage', *J. Amer. Statist. Assoc.* **90**, 1200–1224.

D. Donoho and I. Johnstone (1994), 'Ideal spatial adaptation by wavelet shrinkage', *Biometrika* **81**, 425–455.

S. Durand and M. Nikolova (2003), Restoration of wavelet coefficients by minimizing a specially designed objective function. In *Proc. IEEE Workshop on Variational, Geometric and Level Set Methods in Computer Vision*, pp. 145–152.

V. Duval, J. Aujol and Y. Gousseau (2011), 'A bias-variance approach for the nonlocal means', *SIAM J. Imaging Sci.* **4**, 760.

M. Ebrahimi and E. Vrscay (2007), Solving the inverse problem of image zooming using 'self-examples'. In *Image Analysis and Recognition: ICIAR 2007*, Vol. 4633 of *Lecture Notes in Computer Science*, Springer, pp. 117–130.

M. Ebrahimi and E. Vrscay (2008*a*), Examining the role of scale in the context of the non-local-means filter. In *Proc. International Conference on Image Analysis and Recognition: ICIAR 2008*, Vol. 5112 of *Lecture Notes in Computer Science*, Springer, pp. 170–181.

M. Ebrahimi and E. Vrscay (2008*b*), Multi-frame super-resolution with no explicit motion estimation. In *Proc. 2008 International Conference on Image Processing, Computer Vision, and Pattern Recognition: IPCV 2008*, pp. 455–459.

A. Efros and T. Leung (1999), Texture synthesis by non parametric sampling. In *Proc. International Conference on Computer Vision*, Vol. 2, pp. 1033–1038.

M. Elad and M. Aharon (2006), 'Image denoising via sparse and redundant representations over learned dictionaries', *IEEE Trans. Image Processing* **15**, 3736–3745.

M. Elad and D. Datsenko (2007), 'Example-based regularization deployed to super-resolution reconstruction of a single image', *Comput. J.* **52**, 15–30.

A. Elmoataz, O. Lézoray and S. Bougleux (2008*a*), 'Nonlocal discrete regularization on weighted graphs: A framework for image and manifold processing', *IEEE Trans. Image Processing* **17**, 1047–1060.

A. Elmoataz, O. Lézoray, S. Bougleux and V. Ta (2008*b*), Unifying local and nonlocal processing with partial difference operators on weighted graphs. In *International Workshop on Local and Non-Local Approximation in Image Processing*, pp. 11–26.

A. Foi (2011), Noise estimation and removal in MR imaging: The variance-stabilization approach. In *IEEE International Symposium on Biomedical Imaging: From Nano to Macro*, pp. 1809–1814.

A. Foi, S. Alenius, V. Katkovnik and K. Egiazarian (2007), 'Noise measurement for raw-data of digital imaging sensors by automatic segmentation of non-uniform targets', *IEEE Sensors J.* **7**, 1456–1461.

A. Foi, M. Trimeche, V. Katkovnik and K. Egiazarian (2008), 'Practical Poissonian–Gaussian noise modeling and fitting for single-image raw-data', *IEEE Trans. Image Processing* **17**, 1737–1754.

S. Geman and D. Geman (1984), 'Stochastic relaxation, Gibbs distributions and the Bayesian restoration of images', *IEEE Pat. Anal. Mach. Intell.* **6**, 721–741.

G. Gilboa and S. Osher (2008), 'Nonlocal linear image regularization and supervised segmentation', *Multiscale Model. Simul.* **6**, 595–630.

O. Guleryuz (2007), 'Weighted averaging for denoising with overcomplete dictionaries', *IEEE Trans. Image Processing* **16**, 3020–3034.

S. Harris (1966), 'Image evaluation and restoration', *J. Optical Soc. Amer.* **56**, 569–570.

J. Hays and A. Efros (2007), Scene completion using millions of photographs. In *ACM Transactions on Graphics: TOG*, Vol. 26, ACM, p. 4.

Y. Heo, K. Lee and S. Lee (2007), Simultaneous depth reconstruction and restoration of noisy stereo images using non-local pixel distribution. In *IEEE Conference on Computer Vision and Pattern Recognition*, pp. 1–8.

Y. Hou, C. Zhao, D. Yang and Y. Cheng (2011), 'Comments on "Image denoising by sparse 3D transform-domain collaborative filtering" ', *IEEE Trans. Image Processing* **20**, 268–270.

J. Immerkaer (1996), 'Fast noise variance estimation', *Computer Vision and Image Understanding* **64**, 300–302.

M. Jung and L. Vese (2009), Nonlocal variational image deblurring models in the presence of Gaussian or impulse noise. In *Proc. SSVM 2009*, Vol. 5567 of *Lecture Notes in Computer Science*, Springer, pp. 401–412.

V. Katkovnik, K. Egiazarian and J. Astola (2006), *Local Approximation Techniques in Signal and Image Processing*, SPIE.

V. Katkovnik, A. Danielyan and K. Egiazarian (2011), Decoupled inverse and denoising for image deblurring: Variational BM3D-frame technique. In *Proc. IEEE International Conference on Image Processing: ICIP 2011*, pp. 3453-3456.

S. Kay (1993), *Fundamentals of Statistical Signal Processing: Estimation Theory*, Prentice-Hall.

C. Kervrann and J. Boulanger (2008), 'Local adaptivity to variable smoothness for exemplar-based image regularization and representation', *Internat. J. Computer Vision* **79**, 45–69.

C. Kervrann, J. Boulanger and P. Coupé (2007), Bayesian non-local means filter, image redundancy and adaptive dictionaries for noise removal. In *Proc. SSVM '07*, Vol. 4485 of *Lecture Notes in Computer Science*, Springer, pp. 520–532.

S. Kindermann, S. Osher and P. Jones (2006), 'Deblurring and denoising of images by nonlocal functionals', *Multiscale Model. Simul.* **4**, 1091–1115.

E. Kolaczyk (1999), 'Wavelet shrinkage estimation of certain Poisson intensity signals using corrected thresholds', *Statist. Sin.* **9**, 119–135.

M. Lebrun, A. Buades and J. Morel (2011), 'Study and analysis of NL-PCA', *Image Processing On Line* (www.ipol.im).

A. Lee, K. Pedersen and D. Mumford (2003), 'The nonlinear statistics of high-contrast patches in natural images', *Internat. J. Computer Vision* **54**, 83–103.

J. Lee (1981), 'Refined filtering of image noise using local statistics', *Computer Graphics and Image Processing* **15**, 380–389.

J. Lee (1983), 'Digital image smoothing and the sigma filter', *Computer Vision, Graphics, and Image Processing* **24**, 255–269.

J. Lee and K. Hoppel (1989), Noise modelling and estimation of remotely-sensed images. In *Proc. International Geoscience and Remote Sensing Symposium*, IEEE, pp. 1005–1008.

S. Lefkimmiatis, P. Maragos and G. Papandreou (2009), 'Bayesian inference on multiscale models for Poisson intensity estimation: Application to photo-limited image denoising', *IEEE Trans. Image Processing* **18**, 1724–1741.

A. Levin and B. Nadler (2011), Natural image denoising: Optimality and inherent bounds. In *IEEE Conference on Computer Vision and Pattern Recognition: CVPR 2011*, pp. 2833–2840.

O. Lézoray, V. Ta and A. Elmoataz (2008), 'Nonlocal graph regularization for image colorization', *19th International Conference on Pattern Recognition: ICPR 2008*, pp. 1–4.

C. Liu, W. Freeman, R. Szeliski and S. Kang (2006), Noise estimation from a single image. In *IEEE Computer Society Conference on Computer Vision and Pattern Recognition 2006*, Vol. 1, pp. 901–908.

C. Liu, R. Szeliski, S. Kang, C. Zitnick and W. Freeman (2008), 'Automatic estimation and removal of noise from a single image', *IEEE Trans. Patt. Anal. Machine Intell.* **30**, 299–314.

Y. Lou, X. Zhang, S. Osher and A. Bertozzi (2008), Image recovery via nonlocal operators. UCLA CAM Report 08-35.

S. Lyu and E. Simoncelli (2009), 'Modeling multiscale subbands of photographic images with fields of Gaussian scale mixtures', *IEEE Trans. Patt. Anal. Machine Intell.* **31**, 693–706.

M. Mahmoudi and G. Sapiro (2005), 'Fast image and video denoising via nonlocal means of similar neighborhoods', *IEEE Signal Processing Letters* **12**, 839–842.

J. Mairal (2010), Représentations parcimonieuses en apprentissage statistique, traitement d'image et vision par ordinateur. PhD thesis, Ecole Normale Supérieure de Cachan.

J. Mairal, F. Bach, J. Ponce, G. Sapiro and A. Zisserman (2009*a*), Non-local sparse models for image restoration. In *IEEE 12th International Conference on Computer Vision*, pp. 2272–2279.

J. Mairal, F. Bach, J. Ponce, G. Sapiro and A. Zisserman (2009*b*), Non-local sparse models for image restoration.. In *ICCV'09*, pp. 2272–2279.

J. Mairal, M. Elad and G. Sapiro (2008), 'Sparse representation for color image restoration', *IEEE Trans. Image Processing* **17**, 53–69.

M. Makitalo and A. Foi (2011), 'Optimal inversion of the Anscombe transformation in low-count Poisson image denoising', *IEEE Trans. Image Processing* **20**, 99–109.

A. Maleki, M. Narayan and R. Baraniuk (2011*a*), Anisotropic nonlocal means denoising. Arxiv preprint arXiv:1112.0311.

A. Maleki, M. Narayan and R. Baraniuk (2011*b*), Suboptimality of nonlocal means on images with sharp edges. In *49th Annual Allerton Conference on Communication, Control, and Computing*, pp. 299-305.

S. Mallat (1999), *A Wavelet Tour of Signal Processing*, Academic press.

E. Mammen and A. Tsybakov (1995), 'Asymptotical minimax recovery of sets with smooth boundaries', *Ann. Statist.* **23**, 502–524.

J. Manjón, J. Carbonell-Caballero, J. Lull, G. García-Martí, L. Martí-Bonmatí and M. Robles (2008), 'MRI denoising using non-local means', *Medical Image Analysis* **12**, 514–523.

J. Manjón, M. Robles and N. Thacker (2007), Multispectral MRI denoising using non-local means. In *Proc. MIUA*, Vol. 7, pp. 41–45.

G. Mastin (1985), 'Adaptive filters for digital image noise smoothing: An evaluation', *Computer Vision, Graphics, and Image Processing* **31**, 103–121.

G. Mayer, E. Vrscay, M. Lauzon, B. Goodyear and J. Mitchell (2008), 'Self-similarity of images in the Fourier domain, with applications to MRI', In *Proc. International Conference on Image Analysis and Recognition: ICIAR 2008*, Vol. 5112 of *Lecture Notes in Computer Science*, Springer, pp. 43–52.

P. Meer, J. Jolion and A. Rosenfeld (1990), 'A fast parallel algorithm for blind estimation of noise variance', *IEEE Trans. Patt. Anal. Machine Intell.* **12**, 216–223.

Y. Meyer (1993), *Wavelets: Algorithms and Applications*, SIAM.

M. Mignotte (2008), 'A non-local regularization strategy for image deconvolution', *Pattern Recognition Letters* **29**, 2206–2212.

P. Milanfar (2011), 'A tour of modern image filtering', *IEEE Signal Processing Magazine*. Preprint at http://users.soe.ucsc. edu/~milanfar/publications/.

B. Naegel, A. Cernicanu, J. Hyacinthe, M. Tognolini and J. Vallée (2009), 'SNR enhancement of highly-accelerated real-time cardiac MRI acquisitions based on non-local means algorithm', *Med. Image Anal.* **13**, 598–608.

A. Nemirovski (2000), Topics in non-parametric statistics. In *Lectures on Probability Theory and Statistics: Saint-Flour 1998*, Vol. 1738 of *Lecture Notes in Mathematics*, Springer, pp. 85–277.

R. Nowak and R. Baraniuk (1997), 'Wavelet-domain filtering for photon imaging systems', *IEEE Trans. Image Processing* **8**, 666–678.

S. Olsen (1993), 'Estimation of noise in images: An evaluation', *CVGIP: Graphical Models and Image Processing* **55**, 319–323.

J. Orchard, M. Ebrahimi and A. Wong (2008), Efficient non-local-means denoising using the SVD. In *Proc. IEEE International Conference on Image Processing*, pp. 1732–1735.

E. Ordentlich, G. Seroussi, S. Verdu, M. Weinberger and T. Weissman (2003), A discrete universal denoiser and its application to binary images. In *International Conference on Image Processing*, Vol. 1, pp. 117–120.

S. Osher, M. Burger, D. Goldfarb, J. Xu and W. Yin (2004), Using geometry and iterated refinement for inverse problems 1: Total variation based image restoration. Department of Mathematics, UCLA, 90095, 04–13.

E. Pennec and S. Mallat (2003), Geometrical image compression with bandelets. In *Proc. SPIE 2003*, Vol. 5150, pp. 1273–1286.

G. Peyré (2009), 'Manifold models for signals and images', *Computer Vision and Image Understanding* **113**, 249–260.

N. Ponomarenko, V. Lukin, S. Abramov, K. Egiazarian and J. Astola (2003), Blind evaluation of additive noise variance in textured images by nonlinear processing of block DCT coefficients. In *Proc. SPIE*, Vol. 5014, p. 178.

N. Ponomarenko, V. Lukin, M. Zriakhov, A. Kaarna and J. Astola (2007), An automatic approach to lossy compression of AVIRIS images. In *IEEE International Geoscience and Remote Sensing Symposium*, pp. 472–475.

J. Portilla, V. Strela, M. Wainwright and E. Simoncelli (2003), 'Image denoising using scale mixtures of Gaussians in the wavelet domain', *IEEE Trans. Image Processing* **12**, 1338–1351.

M. Protter, M. Elad, H. Takeda and P. Milanfar (2009), 'Generalizing the non-local-means to super-resolution reconstruction', *IEEE Trans. Image Processing* **18**, 36–51.

K. Rank, M. Lendl and R. Unbehauen (1999), 'Estimation of image noise variance', *IEE Proc. Vision, Image and Signal Processing* **146**, 80–84.

M. Raphan and E. Simoncelli (2007), Learning to be Bayesian without supervision. In *Advances in Neural Information Processing Systems*, Vol. 19, pp. 1145–1152.

M. Raphan and E. Simoncelli (2010), An empirical Bayesian interpretation and generalization of NL-means. Technical Report TR2010-934, Courant Institute of Mathematical Sciences.

W. Richardson (1972), 'Bayesian-based iterative method of image restoration', *J. Optical Soc. Amer.* **62**, 55–59.

L. Rudin, S. Osher and E. Fatemi (1992), 'Nonlinear total variation based noise removal algorithms', *Physica D* **60**, 259–268.

B. Russell, A. Torralba, K. Murphy and W. Freeman (2008), 'LabelMe: A database and web-based tool for image annotation', *Internat. J. Computer Vision* **77**, 157–173.

C. Shannon (2001), A mathematical theory of communication. In *ACM SIGMOBILE Mobile Computing and Communications Review*, Vol. 5, pp. 3–55.

A. Singer, Y. Shkolnisky and B. Nadler (2009), 'Diffusion interpretation of nonlocal neighborhood filters for signal denoising', *SIAM J. Imaging Sci.* **2**, 118–139.

S. Smith and J. Brady (1997), 'SUSAN: A new approach to low level image processing', *Internat. J. Computer Vision* **23**, 45–78.

J. Starck, E. Candès and D. Donoho (2002), 'The curvelet transform for image denoising', *IEEE Trans. Image Processing* **11**, 670–684.

A. Szlam, M. Maggioni and R. Coifman (2006), A general framework for adaptive regularization based on diffusion processes on graphs. Yale technical report.

N. Thacker, J. Manjon and P. Bromiley (2008), A statistical interpretation of non-local means. In *5th International Conference on Visual Information Engineering*, IEEE, pp. 250–255.

C. Tomasi and R. Manduchi (1998), Bilateral filtering for gray and color images. In *Sixth International Conference on Computer Vision, 1998*, IEEE, pp. 839–846.

J. Tukey (1977), *Exploratory Data Analysis*, Addison-Wesley.

D. Van De Ville and M. Kocher (2009), 'Sure-based non-local means', *IEEE Signal Processing Letters* **16**, 973–976.

H. Voorhees and T. Poggio (1987), Detecting textons and texture boundaries in natural image. In *Proc. First IEEE International Conference on Computer Vision*, pp. 250–258.

J. Wang, Y. Guo, Y. Ying, Y. Liu and Q. Peng (2006), Fast non-local algorithm for image denoising. In *2006 IEEE International Conference on Image Processing*, pp. 1429–1432.

T. Weissman, E. Ordentlich, G. Seroussi, S. Verdu and M. Weinberger (2005), 'Universal discrete denoising: Known channel', *IEEE Trans. Inform. Theory* **51**, 5–28.

N. Wiest-Daesslé, S. Prima, P. Coupé, S. Morrissey and C. Barillot (2007), Non-local means variants for denoising of diffusion-weighted and diffusion tensor MRI. In *Proc. 10th International Conference on Medical image Computing and Computer-Assisted Intervention: MICCAI 2007*, Vol. 4792 of *Lecture Notes in Computer Science*, Springer, pp. 344–351.

N. Wiest-Daesslé, S. Prima, P. Coupé, S. Morrissey and C. Barillot (2008), Rician noise removal by non-local means filtering for low signal-to-noise ratio MRI: Applications to DT-MRI. In *Medical Image Computing and Computer-Assisted Intervention: MICCAI 2008*, Vol. 5242 of *Lecture Notes in Computer Science*, Springer, pp. 171–179.

A. Wong and J. Orchard (2008), A nonlocal-means approach to exemplar-based inpainting. In *15th IEEE International Conference on Image Processing*, pp. 2600–2603.

H. Xu, J. Xu and F. Wu (2008), On the biased estimation of nonlocal means filter. In *2008 IEEE International Conference on Multimedia and Expo*, pp. 1149–1152.

L. Yaroslavsky (1985), *Digital Picture Processing*, Springer.

L. Yaroslavsky (1996), Local adaptive image restoration and enhancement with the use of DFT and DCT in a running window. In *Proc. SPIE*, Vol. 2825, p. 2.

L. Yaroslavsky and M. Eden (2003), *Fundamentals of Digital Optics*, Birkhäuser.

G. Yu and G. Sapiro (2011), 'DCT image denoising: A simple and effective image denoising algorithm', *Image Processing On Line* (www.ipol.im).

G. Yu, G. Sapiro and S. Mallat (2010), Image modeling and enhancement via structured sparse model selection. In *17th IEEE International Conference on Image Processing: ICIP 2010*, pp. 1641–1644.

G. Yu, G. Sapiro and S. Mallat (2012), Solving inverse problems with piecewise linear estimators: From Gaussian mixture models to structured sparsity. *IEEE Trans. Image Processing*, to appear.

L. Zhang, W. Dong, D. Zhang and G. Shi (2010), 'Two-stage image denoising by principal component analysis with local pixel grouping', *Pattern Recognition* **43**, 1531–1549.

X. Zhang, M. Burger, X. Bresson and S. Osher (2009), Bregmanized nonlocal regularization for deconvolution and sparse reconstruction. UCLA CAM Report 09-03.

M. Zhou, H. Yang, G. Sapiro, D. B. Dunson and L. Carin (2011), Dependent hierarchical beta process for image interpolation and denoising. In *Proc. Journal of Machine Learning Research*, pp. 883–891.

S. Zimmer, S. Didas and J. Weickert (2008), A rotationally invariant block matching strategy improving image denoising with non-local means. In *Proc. 2008 International Workshop on Local and Non-Local Approximation in Image Processing*, IEEE, pp. 135–142.

D. Zoran and Y. Weiss (2009), Scale invariance and noise in natural images. In *IEEE 12th International Conference on Computer Vision*, pp. 2209–2216.

D. Zoran and Y. Weiss (2011), From learning models of natural image patches to whole image restoration. In *2011 IEEE International Conference on Computer Vision* (ICCV), pp. 479–486.

Acta Numerica (2012), pp. 577–671
doi:10.1017/S0962492912000074

Pricing futures by deterministic methods*

Olivier Pironneau
Université Pierre et Marie Curie,
UMR CNRS 7598 Laboratoire Jacques Louis Lions,
F-75252 Paris, France
E-mail: olivier.pironneau@upmc.fr

In this article we will focus on only a small part of financial mathematics, namely the use of partial differential equations for pricing futures. Even within this narrow range it is hard to be systematic and complete, or even to do better than existing books such as Wilmott, Howison and Dewynne (1995), Achdou and Pironneau (2005), or software manuals such as Lapeyre, Martini and Sulem (2010). So this article may be valuable only to the extent that it reflects ten years of teaching, conferences and interaction with the protagonists of financial mathematics.

Also, because the theory of partial differential equations is not always well known, we have chosen a pragmatic approach and left out the details of the theory or the proofs of some results, and refer the reader to other books. The numerical algorithms, on the other hand, are given in detail.

CONTENTS

Introduction 579

PART 1: Modelling financial assets and derivatives

1.1 Financial assets 580
1.2 Models with local volatilities 582
1.3 Stochastic volatility models 583
1.4 Models with jump-diffusion processes 584
1.5 Financial derivatives 586
1.6 Conclusion 595

* Colour online for monochrome figures available at journals.cambridge.org/anu.

PART 2: SDE-based numerical methods

2.1 Principles of numerical simulation by Monte Carlo 596
2.2 The Longstaff–Schwartz algorithm for Bermudas 598
2.3 Methods using the process PDF for basket options 599
2.4 A binomial tree for American options 600

PART 3: Some partial differential equations in finance

3.1 From SDEs to PDEs and PIDEs 601
3.2 Fokker–Planck and Kolmogorov equations 601
3.3 Some existence and regularity results 610

PART 4: Numerical methods

4.1 A finite difference method for the Black–Scholes PDE 616
4.2 Spectral methods 620
4.3 Finite element methods 621
4.4 Variational methods for American options 625
4.5 A stochastic volatility model solved by FEM 629
4.6 Asian puts 631
4.7 Automatically adapted meshes 633
4.8 Greeks 635
4.9 Reduced-order modelling 637

PART 5: Calibration

5.1 Introduction 645
5.2 Dupire's equation 646
5.3 Calibration of the local volatility 651
5.4 Calibration of implied volatility 654
5.5 Calibration on a reduced basis 656

PART 6: Open-source tools

6.1 Premia 656
6.2 QuantLib 658
6.3 FreeFem++ 659
6.4 CUDA and other tools 664

Conclusion 667

Acknowledgements 667

References 668

Introduction

Except for the pioneering work of Louis Bachelier (1900), financial mathematics is relatively new. Perhaps the right landmarks are the paper by Black and Scholes (1973) and the work of Merton (1976), which led to their Nobel Prize. Forty years later, because of the explosion of applications, more and more mathematicians, physicists, economists and computer scientists have joined the fun. Despite several global crises the subject is exploding, and the literature around it.

Banks, insurance companies, *etc.*, have also been very innovative in inventing new contingent claims, and since no one sees the global picture, every day there seems to be a new challenge to solve. The field is huge: the derivatives market is worth many times the value of the underlying assets. The domain is very lively, with career opportunities unheard of in other fields of applied mathematics, and it attracts students too, because there is plenty to do and still many jobs, despite systemic instabilities.

It is true that the field of financial mathematics is in its infancy, the models are crude and focused on a small set of futures; global models are just being initiated with game theory and utility functions (Lions and Lasry 2007). The lack of global perspective among scientists and their inability to predict systemic risks have given 'quants' a bad name. A French politician has even accused financial mathematicians of 'crimes against humanity'. Let's be reasonable: if there is a culprit it is the internet. It has given 'banksters' a Kalashnikov – electronic trading – and mathematicians are only providing the system with tiny bullet-proof shields.

Futures are modelled by using advanced stochastic tools and then priced using simple Monte Carlo algorithms or analytical formulae of limited application. But the days of closed-form solutions are numbered because models are increasingly complex. Numerical algorithms will have to be more efficient because speed is becoming a dominant factor with high-frequency trading: 40 % of transactions are done automatically by 'robots', computer programs taking instant decisions based on overnight calculations and in-house know-how. Some banks currently double their computing facilities every six months and try to shorten the wires between their IT centres and stock markets to improve transaction speed (this is not a joke). At the current rate they will soon be the biggest users of high-performance computing (HPC).

Monte Carlo is often slow, but it is akin to the stochastic models and it is highly parallel and easy to implement in hybrid computers (GPUs). Alternatives are binomial or trinomial trees, fast-Fourier-transform-based methods, when the characteristic functions of the stochastic processes are known, and the partial differential equations derived via Itô's lemma.

Most option pricing problems can be formulated as partial differential equations (PDEs), partial integro-differential equations (PIDEs) or partial differential inequalities (PDIs). They result in clean and fast algorithms but are cursed by the dimensionality of the problem and hard parallelization procedures.

As a coauthor of Achdou and Pironneau (2005, 2009) I do not intend to present the same material here or to present it in the same way. Here mathematical details are not given, only the essence of the methods, the more recent developments and, in more detail, the numerical algorithms.

The material is divided into six parts.

- Part 1 starts with an introduction to financial assets, then a presentation of some of the models, followed by a list of the most popular derivatives.
- Part 2 begins with a short section on Monte Carlo methods and then deals with deterministic methods based on the knowledge of the probability density of the processes, or their characteristic functions.
- In Part 3 we shall derive the partial differential equations of finance and present some mathematical results for their existence and regularity.
- Then, in Part 4, the main algorithms based on PDE formulations are given. As in Achdou and Pironneau (2005), more emphasis is placed on variational methods with mesh adaptivity in mind. Some acceleration techniques, such as proper orthogonal decomposition and sparse grids, are presented.
- Calibration is dealt with briefly in Part 5, with a more consistent section on Dupire's equations.
- Part 6 is a review of some numerical tools and toolboxes for financial derivatives.

PART ONE
Modelling financial assets and derivatives

Pricing with mathematical and computational tools is practised by institutions or individuals who observe a financial asset, *i.e.*, a share, a bond, a commodity, *etc.*, and wish to draw some conclusion about their future value and the risk involved in dealing with them.

1.1. Financial assets

No arbitrage assumes that the asset's market price is public, fluid, and that it is not possible to make a profit in a loop of buys and sells without taking the risk of losing something. The current pricing theory assumes also that there is a riskless asset, a *numéraire* (money in one currency for example), which can be borrowed or lent at a fixed *interest rate* r, possibly a function of time t. Consequently any riskless asset of price S_t at time t obeys the following:

$$\mathrm{d}S_t = S_t r(t)\,\mathrm{d}t, \quad S_0 \text{ known, } i.e., \ S_t = S_0 \mathrm{e}^{\int_0^t r(s)\,\mathrm{d}s}. \tag{1.1.1}$$

By convention $t = 0$ is today, and the price of the asset today, being public, is known.

Obviously most financial assets are not riskless, and for lack of information (or understanding) their variations are assumed to be random. The simplest model with randomness is

$$\mathrm{d}S_t = S_t(\mu_t\,\mathrm{d}t + \sigma_t\,\mathrm{d}W_t), \quad S_0 \text{ known}, \tag{1.1.2}$$

where $\mu_t = \mathbb{E}[\,\mathrm{d}S_t/S_t|S_0]$ is the tendency (or *drift*) of S_t, and σ_t, the *volatility* at time t, measures the amplitude of the random variations, and W_t is assumed to be a Brownian motion for simplicity. This last assumption will be relaxed later; $\mathbb{E}[\cdot|\cdot]$ is the conditional expected value in the same probability space where W_t is defined.

Remark 1.1.1. Recall that a Brownian motion W_t (also called a Wiener process) has zero mean and variance such that $\mathbb{E}[(W_t - W_s)^2] = t - s$, $t > s$, and $W_t - W_s$ is the Gaussian random variable $W_t - W_s = \mathbf{N}(0,1)\sqrt{t-s}$: its probability density function (PDF) is $\theta \to \exp\big(-\frac{\theta^2}{2(t-s)}\big)/\sqrt{2\pi(t-s)}$ and so its cumulative distribution function (CDF) is

$$\begin{aligned} \mathcal{P}(W_t - W_s < x) &= \frac{1}{\sqrt{2\pi(t-s)}}\int_{-\infty}^{x} \exp\left(-\frac{\theta^2}{2(t-s)}\right)\mathrm{d}\theta \\ &= \frac{1}{2}\left(1 + \operatorname{erf}\left(\frac{x}{\sqrt{2(t-s)}}\right)\right). \end{aligned} \tag{1.1.3}$$

Notice that the model is set in a probability space, about which not much is said. No arbitrage implies in this case the existence of a *risk-neutral probability measure* under which W_t is still a Brownian motion but[1] $\mathbb{E}[\,\mathrm{d}S_t/S_t|S_0] = r(t)$. Numerically, $\mathbb{E}$ is often replaced by a statistical average (Central Limit Theorem, Theorem 2.1.1), so the probability measure does not appear explicitly in the numerical computations. Hence, for practical purposes, the model for S_t is

$$\mathrm{d}S_t = S_t(r(t)\,\mathrm{d}t + \sigma_t\,\mathrm{d}W_t), \quad S_0 \text{ known}. \tag{1.1.4}$$

[1] If μ is a random variable in a probability space of measure $\mathcal{P}$ and $\mathcal{P}^*$ is another probability measure in the same space, then $\mathbb{E}^*[\mu] = \mathbb{E}\big[\frac{\mathrm{d}\mathcal{P}^*}{\mathrm{d}\mathcal{P}}\mu\big]$, where $\frac{\mathrm{d}\mathcal{P}^*}{\mathrm{d}\mathcal{P}}$ is the Radon–Nikodym derivative of $\mathcal{P}^*$ with respect to $\mathcal{P}$. Obviously, if ρ and ρ^* are the PDF of $\mathcal{P}$ and $\mathcal{P}^*$, $\frac{\mathrm{d}\mathcal{P}^*}{\mathrm{d}\mathcal{P}} = \frac{\rho^*}{\rho}$.

The precise meaning of an SDE (stochastic differential equation) such as (1.1.4) requires the theory of stochastic integrals: the reader is referred to Lamberton and Lapeyre (1997) for more details. From a practical numerical point of view it suffices to view (1.1.4) as the limit of (2.1.3) below, when $\delta t \to 0$.

With Itô's lemma (3.2.7), it is not hard to show that (1.1.4) has an explicit solution when r and σ are constant.[2]

Proposition 1.1.2. The random process $X_t = \log S_t$ satisfies

$$\mathrm{d}X_t = \left(r - \frac{1}{2}\sigma_t^2\right)\mathrm{d}t + \sigma_t\,\mathrm{d}W_t, \tag{1.1.5}$$

and when σ_t and r are constant then

$$S_t = S_0 \mathrm{e}^{(r-\frac{\sigma^2}{2})t+\sigma(W_t-W_0)}. \tag{1.1.6}$$

Proof. By Itô's lemma,

$$\begin{aligned}\mathrm{d}\ln S_t &= (\ln S_t)'\,\mathrm{d}S_t + \frac{1}{2}(\ln S_t)''(\sigma S_t)^2\,\mathrm{d}t \\ &= \frac{\mathrm{d}S_t}{S_t} - \frac{\sigma^2 S_t^2\,\mathrm{d}t}{2S_t^2} = r\,\mathrm{d}t + \sigma\,\mathrm{d}W_t - \frac{1}{2}\sigma^2\,\mathrm{d}t.\end{aligned}$$

Hence $X_t = \ln S_t$ satisfies (1.1.5), and so

$$X_t - X_0 = \left(r - \frac{1}{2}\sigma^2\right)t + \sigma(W_t - W_0). \qquad \square$$

Remark 1.1.3. This shows that $Y_t = S_t \mathrm{e}^{-rt}$ is a martingale, *i.e.*,

$$\mathbb{E}(Y_t|Y_\tau) = Y_\tau \quad \forall \tau < t.$$

The concept of martingales (see Lamberton and Lapeyre (1997) for details) is important for arbitrage-free markets.

1.2. Models with local volatilities

In practice the Black–Scholes model with constant volatility is too simple. So it is common practice to *calibrate* the model by finding a so-called *local volatility surface* $\sigma(S,t)$ for (1.1.4) in order to reproduce the observations.[3]

One such local volatility model, known as CEV (constant elasticity of variance) has been proposed by Cox (1975):

$$\mathrm{d}S_t = S_t\big(r\,\mathrm{d}t + \delta S_t^{\frac{\beta}{2}-1}\,\mathrm{d}W_{1t}\big), \quad S_0 \text{ known}, \quad 0 < \beta < 2. \tag{1.2.1}$$

[2] If r is not constant, the same formula holds with $\int_0^t r(\tau)\,\mathrm{d}\tau$ instead of rt.

[3] Hence it is important to remember that a mathematical result or a numerical method for (1.1.4) confined to constant volatilities is of limited interest.

It is of numerical interest because it will lead to a semi-analytical solution, as the probability density function of S_T, knowing that $S_t = S$, is (see Schroder (1989))

$$\rho(S_T|S_t) = (2-\beta)k^{\frac{1}{1-2\beta}}x^{\frac{1}{4-2\beta}}w^{\frac{1-2\beta}{4-2\beta}}\mathrm{e}^{-x-w}I_{\frac{1}{2-\beta}}(2\sqrt{xw}), \qquad (1.2.2)$$

with $\tau = T - t$, I_a the modified Bessel function of the first kind of order a, and

$$k = \frac{2r}{\delta^2(2-\beta)(\mathrm{e}^{r(2-\beta)\tau}-1)}, \quad x = kS^{2-\beta}\mathrm{e}^{r(2-\beta)\tau}, \quad w = kS_T^{2-\beta}. \qquad (1.2.3)$$

1.3. Stochastic volatility models

By assuming a stochastic volatility defined by a scalar or vector SDE, one obtains a large class of models with fewer parameters than with a general local volatility surface.

As before,

$$\mathrm{d}S_t = S_t(\mu_t\,\mathrm{d}t + \sigma_t\,\mathrm{d}W_{1t}), \qquad (1.3.1)$$

where μ_t is the tendency of S_t, taken to be r_t, the interest rate, if a risk-neutral probability exists, and W_{1t} is a Brownian motion. Now we assume that σ_t is a stochastic process satisfying an SDE driven by a second Brownian motion W_{2t}, correlated to W_{1t}: $\mathbb{E}[\mathrm{d}W_{1t}\,\mathrm{d}W_{2t}] = \rho\,\mathrm{d}t$,

$$\mathrm{d}v_t = \kappa(m - v_t)\,\mathrm{d}t + \beta\,\mathrm{d}W_{2t}, \quad \sigma_t = f(v_t), \qquad (1.3.2)$$

where κ, β, m are given, and are possibly functions of v_t and t, f is a given positive function and (v_t) is the driving process. Popular choices include the following.

- The log-normal process: $\kappa < 0, \beta$ constant, $m = 0$.
- The mean reverting Ornstein–Uhlenbeck (OU) process: $\kappa > 0, \beta, m > 0$ constant.
- The Cox–Ingersoll–Ross process (CIR): $\beta = \lambda\sqrt{v_t}$, $\kappa > 0, m > 0, \lambda$ constant.

The volatility β of v_t is commonly referred to as *volvol* (volatility of the volatility). In the case of OU and CIR , m is the limit of v_t when $t \to \infty$ (the process is said to be mean reverting), κ controls the rate at which the limit is reached, and β the randomness about this limit.

When the Brownian motion W_{2t} is correlated to W_{1t}, it can be written as a linear combination of W_{1t} and an independent Brownian motion Z_t:

$$\hat{Z}_t = \rho W_{1t} + \sqrt{1-\rho^2}Z_t, \qquad (1.3.3)$$

where the *correlation factor* ρ lies in $[-1, 1]$.

Table 1.3.1. Frequently used models of stochastic volatilities.

Authors	ρ	$f(y)$	v_t process
Hull and White (1987)	$\rho = 0$	$f(y) = \sqrt{y}$	log-normal
Stein and Stein (1991)	$\rho = 0$	$f(y) = \lvert y\rvert$	mean reverting OU
Heston (1993)	$\rho \neq 0$	$f(y) = \sqrt{y}$	CIR

Table 1.3.1 summarizes some popular choices for $f(v)$ and ρ.

Following Bervoets (2006), with $\tau = t - t_0$, the characteristic function of $X_t = \ln S_t$, knowing S_{t_0} and v_{t_0}, when S_t obeys Heston's model, is

$$\Phi(z) = \exp\left(\mathrm{i}zr\tau + \frac{\rho}{\lambda}(v_t - v_{t_0} - \kappa m\tau)\right)\phi\left(z\left(\frac{\kappa\rho}{\lambda} - \frac{1}{2}\right) + \frac{1}{2}\mathrm{i}z^2(1-\rho^2)\right), \tag{1.3.4}$$

where $\Phi(z) := \mathbb{E}[\mathrm{e}^{\mathrm{i}zX_t}]$ and

$$\phi(z) = \gamma(z)\frac{\mathrm{e}^{-\frac{1}{2}(\gamma(z)-\kappa)\tau}(1-\mathrm{e}^{-\kappa\tau})}{\kappa(1-\mathrm{e}^{-\gamma(z)\tau})} \times \mathrm{e}^{\frac{v_{t_0}+v_t}{\kappa^2}\left[\frac{\kappa(1+\mathrm{e}^{-\kappa\tau})}{1-\mathrm{e}^{-\kappa\tau}}\right]} \frac{I_{\frac{d}{2}-1}\left(\sqrt{v_{t_0}v_T}\frac{4\gamma(z)\mathrm{e}^{-\frac{1}{2}\gamma(z)\tau}}{\lambda^2(1-\mathrm{e}^{-\kappa\tau})}\right)}{I_{\frac{d}{2}-1}\left(\sqrt{v_{t_0}v_T}\frac{4\kappa\mathrm{e}^{-\frac{1}{2}\kappa\tau}}{\lambda^2(1-\mathrm{e}^{-\kappa\tau})}\right)}, \tag{1.3.5}$$

with $\gamma(z) = \sqrt{\kappa^2 - 2\lambda^2\mathrm{i}z}$, $d = 4m/\lambda^2$ and $I_a(x)$ the modified Bessel function of the first kind.

1.4. Models with jump-diffusion processes

To account for sudden discontinuous variations of S_t, one adds random jumps to W_t. Jump processes were introduced by P. Lévy (see Bertoin (1996)). For details the reader is referred to Cont and Tankov (2003). Below we present, without mathematical rigour, the minimum needed to generate a Lévy process numerically.

1.4.1. Poisson process

A positive random variable τ with exponential law (ELVR) is defined by the probability law

$$\mathcal{P}(\tau > y) = \mathrm{e}^{-\lambda y}. \tag{1.4.1}$$

Let $\{\tau_i\}$ be a set of independent ELVRs with the same λ. Let

$$T_n = \sum_1^n \tau_i, \quad N_t = \sum_{n\geq 1} \mathbf{1}_{t\geq T_n}. \tag{1.4.2}$$

By definition, N_t is a *Poisson process.*

It is very useful in queuing theory: for instance, it can be used to model the number of buses which arrive at one bus station.

Proposition 1.4.1. $N_t - N_s$ and N_{t-s} have the same law:

$$P(N_t = n) = \mathrm{e}^{-\lambda t}\frac{(\lambda t)^n}{n!}.$$

Furthermore, if it is known that $N_T = n$, then $T_1, \ldots, T_n$ are uniformly distributed on $(0, T)$.

Definition 1.4.2. A *compound Poisson* process on f is $Z_t = \sum_1^{N_t} v_i$, where $\{v_i\}$ are independent random variables with law f.

Consequently a numerical method to generate a compound Poisson process Z_t is as follows.

(1) Generate N_T by (1.4.1), (1.4.2).
(2) Generate $\{U_i\}_1^{N_T}$ uniformly distributed on $(0, T)$.
(3) Generate N_T random variables $\{v_i\}_1^{N_T}$ with law f.
(4) Set $Z_t = \sum_1^{N_T} v_i \mathbf{1}_{U_i \le t}$.

1.4.2. Jump-diffusion process

Let W_t be a Brownian process and let Z_t be a compound Poisson process. Let $\sigma, \gamma \in \mathbb{R}$. Then

$$X_t = \gamma t + \sigma W_t + Z_t \tag{1.4.3}$$

is, by definition, a *jump-diffusion process.*

Theorem 1.4.3. Let $\nu(x) = \lambda f(x)$. The *characteristic function* of X_t is

$$\mathbb{E}(\mathrm{e}^{\mathrm{i}uX_t}) = \exp\left(t\left[\mathrm{i}\gamma u - \frac{\sigma^2 u^2}{2} + \int_{\mathbb{R}} (\mathrm{e}^{\mathrm{i}ux} - 1 - \mathrm{i}u\mathbf{1}_{|x|<1})\nu(\mathrm{d}x)\right]\right). \tag{1.4.4}$$

Conversely, if $\nu \in L^1(R)$ and $\int_{-1}^{1} x^2\nu(\mathrm{d}x) < \infty$, (1.4.4) defines the characteristic function of a *generalized* jump-diffusion process. Formula (1.4.4) is referred to as the *Lévy–Khinchin formula.*

Theorem 1.4.4. The infinitesimal generator of the semi-group associated to X_t is

$$\begin{aligned} L^X\phi &:= \lim_{t\to 0}\frac{1}{t}\mathbb{E}(\phi(x + X_t) - \phi(x)) \qquad (1.4.5)\\ &= \frac{\sigma^2}{2}\partial_{xx}\phi + \gamma\partial_x\phi + \int_{\mathbb{R}} (\phi(x+y) - \phi(x) - y\mathbf{1}_{|y|<1}\partial_x\phi(x))\nu(\mathrm{d}y). \end{aligned}$$

The following property will be useful later.

Proposition 1.4.5. The process e^{X_t} is a martingale if and only if

$$\gamma+\frac{\sigma^2}{2}+\int_{\mathbb{R}}(e^x-1-x\mathbf{1}_{|x|<1})\nu(dx)=0, \tag{1.4.6}$$

and with this value for γ the infinitesimal generator of X_t is

$$L^X\phi=\frac{\sigma^2}{2}(\partial_{xx}\phi-\partial_x\phi)+\int_{\mathbb{R}}(\phi(x+y)-\phi(x)-(e^y-1)\partial_x\phi(x))\nu(dy). \tag{1.4.7}$$

1.5. Financial derivatives

1.5.1. European options

Given a financial asset S_t, an option on S is a contract for buying or selling S at a future time T. Such a product is also called a *financial derivative* or a *future* on S.

The European contract is the simplest: it gives its owner the right to buy (call) or sell (put) S at time T (maturity) at price K (the strike).

Such contracts have a market value: they too can be bought or sold. No arbitrage implies that the fair value C_t (resp. P_t) of a European call (resp. put) at time $t<T$ is the expected value of the owner's profit at time T discounted by the interest rate r down to time t:

$$C_t=e^{-r(T-t)}\mathbb{E}[(S_T-K)^+|S_t],\quad P_t=e^{-r(T-t)}\mathbb{E}[(K-S_T)^+|S_t]. \tag{1.5.1}$$

Recall the notation $a^+=\max(a,0)$.

Note that C_t-P_t pays $(S_T-K)^+-(K-S_T)^+$ at maturity, which is equal to (S_T-K). Because it is riskless, no arbitrage implies that they must be equal at time t discounted by r, namely

$$C_t-P_t=e^{-r(T-t)}(S_T-K). \tag{1.5.2}$$

This relation is known as the *put–call parity*.

Calling $P(S,t)$ the price of an option with maturity T and pay-off function $P_0(S)$, and assuming that r and $\sigma>0$ are constant, the Black–Scholes formula is

$$P(S,t)=e^{-r(T-t)}\mathbb{E}^*\big(P_0\big(Se^{r(T-t)}e^{\sigma(W_T-W_t)-\frac{\sigma^2}{2}(T-t)}\big)\big), \tag{1.5.3}$$

and since, under P^*, W_T-W_t is a centred Gaussian distribution with variance $T-t$,

$$P(S,t)=\frac{1}{\sqrt{2\pi}}e^{-r(T-t)}\int_{\mathbb{R}}P_0\big(Se^{(r-\frac{\sigma^2}{2})(T-t)+\sigma x\sqrt{T-t}}\big)e^{-\frac{x^2}{2}}\,dx. \tag{1.5.4}$$

When the option is a European option, a more explicit formula can be deduced from (1.5.4). Take, for example, a call:

$$C(S,t) = \frac{1}{\sqrt{2\pi}} \int_{-d_2}^{+\infty} \big(S\mathrm{e}^{-\frac{\sigma^2}{2}(T-t)+\sigma x\sqrt{T-t}} - K\mathrm{e}^{-r(T-t)}\big)\mathrm{e}^{-\frac{x^2}{2}}\, \mathrm{d}x$$
$$= \frac{1}{\sqrt{2\pi}} \int_{-\infty}^{d_2} \big(S\mathrm{e}^{-\frac{\sigma^2}{2}(T-t)-\sigma x\sqrt{T-t}} - K\mathrm{e}^{-r(T-t)}\big)\mathrm{e}^{-\frac{x^2}{2}}\, \mathrm{d}x, \quad (1.5.5)$$

where

$$d_1 = \frac{\log\big(\frac{S}{K}\big) + \big(r + \frac{\sigma^2}{2}\big)(T-t)}{\sigma\sqrt{T-t}} \quad \text{and} \quad d_2 = d_1 - \sigma\sqrt{T-t}. \quad (1.5.6)$$

Finally, introducing the cumulative distribution function for $\mathbf{N}(0,1)$,

$$N(d) = \frac{1}{\sqrt{2\pi}} \int_{-\infty}^{d} \mathrm{e}^{-\frac{x^2}{2}}\, \mathrm{d}x, \quad (1.5.7)$$

and using (1.5.5), (1.5.6), we obtain the *Black–Scholes formula.*

Proposition 1.5.1. When σ and r are constant, the price of the call is given by

$$C(S,t) = SN(d_1) - K\mathrm{e}^{-r(T-t)}N(d_2), \quad (1.5.8)$$

and the price of the put is given by

$$P(S,t) = -SN(-d_1) + K\mathrm{e}^{-r(T-t)}N(-d_2), \quad (1.5.9)$$

where d_1 and d_2 are given by (1.5.6) and N is given by (1.5.7).

Proposition 1.5.2. With CEV volatilities $\sigma_t = \delta S_t^{\frac{\beta}{2}-1}$,

$$C_t = S_t Q\bigg(2kK^{2-\beta}; 2 + \frac{2}{2-\beta}, 2x\bigg)$$
$$- K\mathrm{e}^{-r(T-t)}\bigg(1 - Q\bigg(2x; \frac{2}{2-\beta}, 2kK^{2-\beta}\bigg)\bigg), \quad (1.5.10)$$

where Q is the non-central Chi-square distribution (see Schroder (1989)) with x, k defined by (1.2.3). When $\beta = 4/3$, the Q needed are

$$Q(z;1,\kappa) = N(\sqrt{\kappa} - \sqrt{z}) + N(-\sqrt{\kappa} - \sqrt{z})$$
$$Q(z;3,\kappa) = Q(z;1,\kappa) + \frac{1}{\sqrt{\kappa}}[N'(\sqrt{\kappa} - \sqrt{z}) - N'(\sqrt{\kappa} + \sqrt{z})]$$
$$Q(z;5,\kappa) = Q(z;1,\kappa) + \frac{1}{\sqrt{\kappa}^3}[(\kappa - 1 + \sqrt{\kappa z})N'(\sqrt{\kappa} - \sqrt{z})$$
$$- (\kappa - 1 - \sqrt{\kappa z})N'(\sqrt{\kappa} + \sqrt{z})], \quad (1.5.11)$$

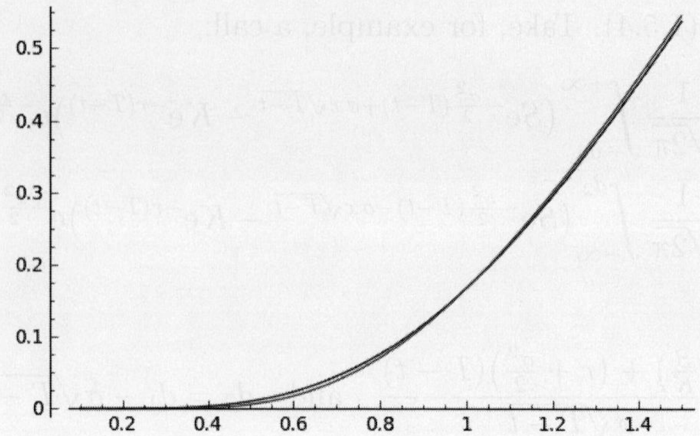

Figure 1.5.1. European call with CEV volatility, $\beta = \frac{4}{3}$, $\delta = 0.3$, $T = 2$, $r = 0$, $K = 1$. The integrated software Sage was used (www.sage.org) on (1.5.10). A comparison with the Black–Scholes solution is given with $\sigma = 0.3$.

where $N'(x) = e^{\frac{x^2}{2}}/\sqrt{2\pi}$ is the standard normal density function and

$$N(x) = \frac{1 + \operatorname{erf}(\frac{x}{\sqrt{2}})}{2}$$

is the normal probability distribution.

Proof. See Schroder (1989). □

A numerical simulation shows $S_t \to C_t$ in Figure 1.5.1 when $t = 0$, $T = 2$, $\delta = 0.3$, $\beta = 4/3$, $K = 1$ and $r = 0$.

```
// Sage script
s=0.6;S=var('S'); z =var('z');
d=0.3; tau=2; y= 9/d^2/tau; x=y*S^(2/3);
p=sqrt(y)+sqrt(x); m = sqrt(x)-sqrt(y);
N(z)=(1+erf(z/sqrt(2)))/2;
Np(z) =  exp(-z^2/2)/sqrt(2*pi);
Q1= N(m)+N(-p)+x^(-3/2)*( (x-1+sqrt(x*y))*Np(m)
                        -(x-1-sqrt(x*y))*Np(p));
Q2= N(-m)+N(-p)+y^(-1/2)*(Np(-m)-Np(p));
C=S*Q1-1+Q2;
v=(S*(1-erf(-ln(S)/s - s/4))-1-erf(ln(S)/s - s/4))/2;
for i in range (1,15) : print  numerical_approx(C(0.1*i));
A=plot(C,(0.1,1.5),color='blue'); B=plot(v,(0.1,1.5),color='red'); A+B;
```

Proposition 1.5.3. If we use Heston's stochastic volatility model (1.3.1), (1.3.2), (1.3.3), knowing that $S = S_0, v = v_0$ at $t = 0$, then the European call is

$$
\begin{aligned}
C &= \frac{1}{2}(S_0 - K\mathrm{e}^{-rT}) + \frac{1}{\pi}\int_0^\infty (\mathrm{e}^{rT} f_1 - K f_2)\,\mathrm{d}u,\\
f_j &= \mathrm{Re}\frac{\phi(u+(j-2)\mathrm{i})}{\mathrm{i}uK^{\mathrm{i}u}},\\
\phi(u) &= \mathrm{e}^{rT} S_0^{\mathrm{i}u}\left(\frac{1-g\mathrm{e}^{-dT}}{1-g}\right)^{-2\frac{m\kappa}{\lambda^2}} \exp\left(\frac{m\kappa T}{\lambda^2}(\kappa - \rho\lambda \mathrm{i}u - d)\right)\\
&\quad\times \exp\left(\frac{v_0}{\lambda^2}(\kappa - \rho\lambda\mathrm{i}u + d)\frac{1-\mathrm{e}^{dT}}{1-g\mathrm{e}^{dT}}\right),
\end{aligned}
$$

$$
\text{where } d = \sqrt{(\rho\lambda u\mathrm{i} - \kappa)^2 + \lambda^2(\mathrm{i}u + u^2)}, \quad g = \frac{\kappa - \rho\lambda\mathrm{i}u - d}{\kappa - \rho\lambda\mathrm{i}u + d}. \tag{1.5.12}
$$

The proof is due to Cox, Ingersoll and Ross (1985), and revised by Albrecher, Mayer, Schoutens and Tistaert (2007). A summary can be found in Mikhailov and Nögel (2003), and hints on how to compute the complex integral in Albrecher *et al.* (2007).

```
# Sage script for Heston model by a complex integral when K=1
umax=1000; N=10000; du=umax/N;
S0=1; T=2;Y0=0.0175; r=0.025;
kappa=1.5768;m=0.0398; lambd=0.5751;rho=-0.5711;
aa= m*kappa*T/lambd^2; bb= -2*m*kappa/lambd^2;
P=0;
for i in range (1,N) :
     u2=i*du
     u1=u2-I
     a1=rho*lambd*u1*I
     a2=rho*lambd*u2*I
     d1=sqrt((a1-kappa)^2+lambd^2*(u1*I+u1^2))
     d2=sqrt((a2-kappa)^2+lambd^2*(u2*I+u2^2))
     g1=(kappa-a1-d1)/(kappa-a1+d1)
     g2=(kappa-a2-d2)/(kappa-a2+d2)
     b1=exp(u1*I*(log(S0)+r*T)) *( (1-g1*exp(-d1*T))/(1-g1) )^bb
     b2=exp(u2*I*(log(S0)+r*T)) *( (1-g2*exp(-d2*T))/(1-g2) )^bb
     phi1=b1*exp(aa*(kappa-a1-d1)+Y0*(kappa-a1-d1)*(1-exp(-d1*T))/
       (1-g1*exp(-d1*T))/lambd^2)
     phi2=b2*exp(aa*(kappa-a2-d2)+Y0*(kappa-a2-d2)*(1-exp(-d2*T))/
       (1-g2*exp(-d2*T))/lambd^2)
     P+= real((phi1-phi2)/(u2*I))*du
print numerical_approx((S0-exp(-r*T))/2+P/pi)
```

1.5.2. *Barrier option*

Barriers are introduced in case the market goes wild. The contracts include a clause that cancels the deal if S_t crosses over a specified price S_M or under S_m.

1.5.3. *American and Bermuda options*

With American options the owner of the contract can exercise his right and force the deal any time before maturity T. To decide to exercise the option at time τ one must compare the profit gain now, namely $S_\tau - K$ for a call, with the expected profit if one waits, namely $\mathrm{e}^{-r(T-\tau)}\mathbb{E}(S_T - K)^+$. Thus the value of the American put option is

$$P_t = \sup_{\tau \in \mathcal{T}(t,T)} \mathbb{E}[\mathrm{e}^{-r(\tau-t)}(K - S_\tau)^+ | S_t], \tag{1.5.13}$$

where $\mathcal{T}$ is the set of stopping times in (t, T) adapted to the probability space.

Bermuda option
The mechanism of Bermudas is similar to Americans except that the decisions to exercise are made only at fixed, given time intervals – for instance every month. The Bermuda islands are between America and Europe, hence the name.

1.5.4. *Path-dependent options*

A path-dependent or Asian (or Asiatic) option tries to smooth the oscillations of the underlying asset at maturity. Some average of S_t over a time period is used in place of S_T in the pay-off $(A_T - K)^+$ for a call, with arithmetic averaging, for instance

$$A_t = \frac{1}{t}\int_0^t S_\tau \,\mathrm{d}\tau,$$

or geometric averaging,

$$A_t = \exp\left(\frac{1}{t}\int_0^t \log(S_\tau)\,\mathrm{d}\tau\right).$$

Another example is the call with floating strike with pay-off $(S_T - A_T)^+$. No arbitrage gives

$$C_t = \mathrm{e}^{-r(T-t)}\mathbb{E}[(S_T - A_T)^+ | S_t, A_t]. \tag{1.5.14}$$

Another class of Asian options, the so-called *lookback options*, involve the extremal values of the asset price for $t \le T$. The floating strike lookback call has a pay-off of $(S_T - \min_{0\le\tau\le T} S_\tau)^+$, whereas the lookback put has a pay-off of $(\max_{0\le\tau\le T} S_\tau - S_T)^+$. Similarly, the fixed strike lookback call (resp. put) has a pay-off of $(\max_{0\le\tau\le T} S_\tau - K)^+$ (resp. $(K - \min_{0\le\tau\le T} S_\tau)_+$). One can also define lookback options on averages.

1.5.5. Basket options

Consider d risky assets S_{it}, $i = 1, \ldots, d$, each modelled by a separate SDE:

$$dS_{it} = S_{it}(r\,dt + \sigma_i\,dW_{it}). \tag{1.5.15}$$

However, the standard Brownian motions (W_{it}), $1 \le i \le d$, are possibly correlated: $\rho_{i,j} := \frac{1}{t}\mathbb{E}[W_{it}W_{jt}]$. As usual r is the interest rate. In what follows, the notation $\mathbf{S}$ is used for the vector $(S_1, \ldots, S_d)^T$.

Remark 1.5.4. As in the one-dimensional case, if σ_i is constant for all i:

$$S_{it} = S_{i0}\exp\biggl(\biggl(r - \frac{1}{2}\sigma_i^2\biggr)t + \sigma_i(W_{it} - W_{0t})\biggr). \tag{1.5.16}$$

A European option with pay-off $\pi(\mathbf{S}_T)$ at maturity T will be priced by

$$P_t = e^{-r(T-t)}\mathbb{E}[\pi(\mathbf{S}_T)|\mathbf{S}_t]. \tag{1.5.17}$$

The pay-off function could be a function of a weighted sum of the assets, *i.e.*, in the case of a put:

$$\pi(\mathbf{S}) = \biggl(K - \sum_{i=1}^{d}\alpha_i S_i\biggr)^+.$$

As in the scalar case, the put–call parity holds:

$$C_t - P_t = \sum_{i=1}^{d}\alpha_i S_{it} - Ke^{-r(T-t)}.$$

The pay-off could also be a function of $\max_{i=1,\ldots,d} S_i$: these options are called *best-of* options. The best-of put option yields

$$\pi(\mathbf{S}) = \Bigl(K - \max_{i=1,\ldots,d} S_i\Bigr)^+.$$

In contrast with the previous case, there is no put–call parity for best-of options.

For a detailed mathematical treatment of the problem in the presence of jumps, see Farkas, Reich and Schwab (2006).

Remark 1.5.5. Another presentation can be made with independent Brownian motions $\tilde{W}_{it}$ and

$$dS_{it} = S_{it}\biggl(r\,dt + \sum_{j=1}^{i}\sigma_{ij}\,d\tilde{W}_{jt}\biggr), \tag{1.5.18}$$

with $\sum_{k\le i,j}\sigma_{ik}\sigma_{jk} = \sigma_i\sigma_j\rho_{ij}$.

Then, when r, σ_{ij} are constant,

$$S_{iT}(\mathbf{X}) = S_{i0} \exp\left(\left(r - \frac{1}{2}\sum_j \sigma_{ij}^2\right)T + \sum_j \sigma_{ij} X_j \sqrt{T}\right), \tag{1.5.19}$$

where $\mathbf{X} = \{X_j\}_1^d$ are d normal random variables. The computation of $\pi(\mathbf{S}_0) := \mathrm{e}^{-rT}\mathbb{E}[(K - H(\mathbf{S}_T))^+|\mathbf{S}_0]$ reduces to the multiple integral

$$\pi(\mathbf{S}) = \mathrm{e}^{-rT} \int_{\mathbb{R}^d} (K - H(\mathbf{S}_T(\mathbf{X})))^+ \prod_j \frac{\mathrm{e}^{\frac{-x_j^2}{2}}}{\sqrt{2\pi}} \,\mathrm{d}X_1, \ldots, \mathrm{d}X_d \tag{1.5.20}$$

$$= \frac{\mathrm{e}^{-rT}}{(2\pi)^{\frac{d}{2}}} \int_{\mathbb{R}^d} (K - H(S_{i0}\,\mathrm{e}^{(r - \frac{1}{2}\sum_j \sigma_{ij}^2)T + \sum_j \sigma_{ij} X_j \sqrt{T}}))^+ \,\mathrm{e}^{-\frac{|\mathbf{X}|^2}{2}} \,\mathrm{d}\mathbf{X}.$$

1.5.6. Convertible bonds

Historically, bonds issued by a company, whose share is denoted here by S_t, were refunded in cash. Convertible bonds can be refunded in shares of the company. When this happens the debts of the company are reduced. Conversely, the bond's owner can participate in the growth of the company.

The market for convertible bonds is also very large (>\$500 billion in 2002). The challenge here is to price such a convertible bond and possibly tell when it is best to convert bonds into shares. The difficulty comes from the spread between the put price B_p and the call price B_c.

Following Ayache, Forsyth and Vetzal (2003) we assume that S_t is driven by a standard SDE such as (1.1.1); the bond is exchangeable for κS_t.

Like any contingent claim on S_t, no arbitrage implies that the value of the convertible bond is

$$V_0 = \mathrm{e}^{-r(T-t)}\mathbb{E}[\max\{B_T, \kappa S_T\}|S_0],$$

where B_T is the agreed refund value of the bond at maturity. However, there are constraints, due to the convertibility of V_t into κS_t or B_c and of B_p into V_t: if $B_c < \kappa S_t$ then $V_t = \kappa S_t$, otherwise

$$\max\{B_p, \kappa S_t\} \le V_t \le B_c. \tag{1.5.21}$$

So the situation is similar to American options (though more complex), and even more so if the bond is itself a risky asset.

1.5.7. Interest rates, swaps and swaptions

The fluctuation of interest rate r cannot be ignored for assets with long maturity, such as zero coupon bonds sometimes used in retirement plans. The receiver of a given sum today, B_0, signs an obligation to pay a certain amount B_T at time T, and no yearly interest (zero coupon) is paid in between. A fair price for the bond would be $B_0 = \mathbb{E}[\exp(\int_0^T r_s \,\mathrm{d}s) B_T | r_0]$.

The simplest model to describe the fluctuations of the interest rate is the one-factor model of Vasicek:

$$dr_t = \lambda(m - r_t)\,dt + \eta\,dW_{1t}, \quad r_0 \text{ given.} \tag{1.5.22}$$

Similar problems arise in conjunction with derivatives on S_t when T is large. SABR (stochastic alpha, beta, rho: see Brigo and Mercurio (2007)) can be used for S_t,

$$\frac{dS_t}{S_t} = r_t\,dt + \sigma_t S_t^{\beta-1}\,dW_{2t}, \quad S_0 \text{ given,} \tag{1.5.23}$$

combined with Vasicek and Hull–White as in Chen, Grzelak and Oosterlee (2011),

$$\sigma_t = g(t)\kappa_t, \quad \frac{d\kappa_t}{\kappa_t} = h(t)\,dW_{3t}, \quad \kappa_0 \text{ given,} \tag{1.5.24}$$

with given functions $g(t), h(t)$, constant parameters $0 < \beta < 1$, and $\lambda, \eta, m > 0$, and given correlations $\rho_{ij}\,dt = \mathbb{E}[\,dW_{it}\,dW_{jt}]$, $i, j = 1, 2, 3$.

Another way is to use a Heston–Hull–White model in conjunction with (1.5.22):

$$\begin{aligned} \frac{dS_t}{S_t} &= r_t\,dt + \sqrt{\mu_t}\,dW_{2t}, \quad S_0 \text{ given,} \\ d\mu_t &= \kappa(\overline{\mu} - \mu_t)\,dt + \gamma\sqrt{\mu_t}\,dW_{1t}, \quad \mu_0 \text{ given.} \end{aligned} \tag{1.5.25}$$

Interest rate swap

An interest rate swap is an agreement between the *payer* A and the *receiver* B to exchange interest rate payments on a notational (imaginary) amount, under some conditions on time and amount.

Typically A must pay the LIBOR floating interest rate (London interbank offered rate) while B must pay a fixed interest rate on the same notational amount.

If A wants the security of B's rate, he has to pay B a fixed coupon (called the *fixed leg*), which is the interest rate that B has to pay plus a premium, say semi-annually.

In exchange B will give to A the LIBOR rate plus possibly a fixed coupon, the two making the *floating leg* paid at some time interval δ (quarterly, for instance). Naturally the value of a swap is the difference between the two legs; at time 0 it is usually 0.

The modelling begins with an assumption on the short interest rate r_t, for instance the SDE (1.5.22). Next the dematerialized sum on which the swap is based could be modelled as a zero coupon bond B_t driven by

$$\frac{dB_t}{B_t} = r_t\,dt + \sigma_t^B\,dW_t^B, \quad B_0 \text{ given.} \tag{1.5.26}$$

If L_t is the value of the LIBOR, then $B_{t+\delta} = B_t/(1+\delta L_t)$. The LIBOR is also modelled by an SDE,

$$\frac{\mathrm{d}L_t}{L_t} = \mu_t \,\mathrm{d}t + \sigma_t^L \,\mathrm{d}W_t^L, \quad L_0 \text{ given.} \tag{1.5.27}$$

Model coherence requires us to relate σ_t^B to L_t and σ_t^L, thus making σ^B stochastic even if σ^L is a fixed number.

If the times of swaps are $\{T_i\}_1^I$, and $B_t(T)$ denotes the value at t of the zero coupon fixed at T, then the value of the swap is

$$V_t(T, T_I) = \frac{B_t(T) - B_t(T_I)}{\sum_1^I \delta B_t(T_i)}.$$

Interest rate derivatives, swaptions
An interest rate derivative is a future where the underlying asset is the right to pay or receive a notional amount at a given interest rate. The interest rate derivative market is huge, more than \$500 trillion in 2010. Interest rate derivatives are used to hedge interest rate risk. Swaptions are one such tool. A swaption is a future on the swap: it is priced as

$$SW = B_t(T)\mathbb{E}\left[\sum_1^I \mathrm{e}^{-\int_T^{T_i} r_s \,\mathrm{d}s} \delta (V_T(T, T_I) - K)^+\right],$$

where T is the maturity of the contract and K the strike. More details can be found in El Karoui and E. Gobet (2011) and Xuan Luo (2007).

1.5.8. CDSs: credit default swaps

A CDS is a type of insurance contract against the event that a firm or a sovereign defaults on its debt. The seller of the swap will refund some of the loss to the buyer in case of default, in exchange for annuities (spread) up to the maturity of the contract.

In 2008, CDSs (there were \$58 trillion of these, twice the US stock market) and CDOs triggered the domino effect in the subprime crisis that led to the bankruptcy of AIG and Lehman Brothers.

In an interesting article, Ang and Longstaff (2011) model the contract assuming two causes trigger default: one due to a jump in the model for the underlying asset, a bond, and characterized by the intensity of the jump, ξ_t, and the other due to a systemic jump, also characterized by the intensity of the jump λ_t. Both are modelled by a mean reverting process:

$$\begin{aligned} \mathrm{d}\xi_t &= a(\xi_m - \xi_t)\,\mathrm{d}t + c\sqrt{\xi_t}\,\mathrm{d}W_{1t}, \\ \mathrm{d}\lambda_t &= \alpha(\lambda_m - \lambda)\,\mathrm{d}t + \sigma\sqrt{\lambda}\,\mathrm{d}W_{2t}. \end{aligned} \tag{1.5.28}$$

For the second cause the probability of default given a systemic jump, γ, is specific to the bond.

Ang and Longstaff (2011) show that the instantaneous probability of default is $\gamma\lambda + \xi$. Assuming that, in case of default, the bond holder recovers $1 - w$ of the par value of the bond, the protection leg of the CDS and the spread leg are

$$w\mathbb{E}\left[\int_0^T \mathrm{e}^{-rt}(\gamma\lambda + \xi)\mathrm{e}^{-\int_0^t p_\tau \,\mathrm{d}\tau}\,\mathrm{d}t\right] \quad \text{and} \quad s\mathbb{E}\left[\int_0^T \mathrm{e}^{-rt}\mathrm{e}^{-\int_0^t p_\tau \,\mathrm{d}\tau}\,\mathrm{d}t\right]. \tag{1.5.29}$$

A formula for s, the spread paid by the buyer of the default protection, is found by writing that both legs are equal (no arbitrage).

1.5.9. CDOs: collateralized debt obligations

Companies who own a hundred or more underlying assets structure them into a single financial product of the same value. Yet the product is usually divided into several tranches, each with its own default risk.

Following Joshi (2011), let us consider a very simple model for CDOs with a single tranche. Each asset has its own SDE,

$$\mathrm{d}S_{it} = S_{it}(r\,\mathrm{d}t + \sigma_i\,\mathrm{d}W_{it}), \quad \text{with } W_{it} = \rho Z_t + \sqrt{1-\rho^2}Z_{it},$$

where Z_t and Z_{it} are independent Wiener processes. We assume also that each asset defaults according to a Poisson process with deterministic intensity λ_i (*cf.* (1.4.1)). Then the cumulative default probability $\mathcal{P}_i$ and hence the correlated defaults τ_i are

$$\mathcal{P}_i(\tau) = 1 - \mathrm{e}^{-\int_0^\tau \lambda_i(s)\,\mathrm{d}s}, \quad \tau_i = \mathcal{P}_i^{-1}(\mathcal{P}(W_i)),$$

where $\mathcal{P}(x) = \left(1 + \mathrm{erf}\left(\frac{x}{\sqrt{2}}\right)\right)/2$ is the normal CDF. If λ_j is independent of time,

$$\tau_i = \frac{1}{\lambda_i}\ln\left(\frac{1}{2}(1 - \mathrm{erf}(W_i))\right).$$

Once the default times are known, the pay-off can be calculated. For example, the contract can say that if 10 assets default before maturity, full refund of these will be given.

The difficulty, however, is that very few paths of a Monte Carlo method will lead to a non-zero pay-off: importance sampling must be used.

1.6. Conclusion

In the beginning the models were simple, but now some contain several correlated processes, no longer Gaussian, leading to integro-differential systems in high dimensions with inequalities and unusual boundary conditions.

Analytical solutions to the models are no longer available: numerical simulations are necessary.

So far each *future* has been simulated separately but, the threat of systemic risk being real, it is expected that the next generation of models will involve multiple financial assets and derivatives. If so, massive computing power will be necessary.

PART TWO
SDE-based numerical methods

In this section we intend to give a pragmatic account of a few algorithms based on the stochastic formulation of the problems.

2.1. Principles of numerical simulation by Monte Carlo

Notice that it is easy to simulate (1.1.4) or (1.1.6) numerically. One way is to use the C-function `rand()` which, each time it is called, returns a different equiprobable integer between `0` and `RAND_MAX` (a machine-dependent large integer) and the Box–Muller formula to generate a normal Gaussian random variable,

$$\mathbf{N}_{0,1} = \sqrt{2\ln\frac{1}{x}}\cos 2\pi y, \tag{2.1.1}$$

where x and y are generated by `rand()/RAND_MAX`, then set

$$W_t - W_0 = \mathbf{N}_{0,1}\sqrt{t}, \quad \delta W_t = \mathbf{N}_{0,1}\sqrt{\delta t}, \tag{2.1.2}$$

and then use (1.1.6) directly or, if σ is not constant, use a finite difference approximation of (1.1.4):

$$S_{t+\delta t} = S_t(1 + r\delta t + \sigma_t \delta W_t). \tag{2.1.3}$$

Any mean quantity function of S_T such as $C = \mathrm{e}^{-rT}\mathbb{E}[(S_T - K)^+]$ can be computed by using the law of large numbers:

$$C \approx \frac{\mathrm{e}^{-rT}}{N}\sum_{n=1}^{N}\max\{S^n_{J\delta t} - K, 0\}, \tag{2.1.4}$$

where $\{S^n_t\}$ are N independent computations with (2.1.3). There are two sources of numerical error on C: one due to the Euler scheme which is first-order accurate only, and another due to the replacement of the mean with respect to the risk-neutral probability by a sample average. The following addresses the second source of errors.

Theorem 2.1.1. (Central Limit Theorem) Let x be a random variable with probability density p, expectation $\mathbb{E}[x]$ and variance

$$\mathrm{var}(x) = \mathbb{E}[(x - \mathbb{E}([x])^2].$$

Let $\{x_i\}_1^N$ be N samples of x. The approximation

$$\mathbb{E}[x] := \int_{-\infty}^{+\infty} p(x)x\,\mathrm{d}x \approx \mathbb{E}_N[x] := \frac{1}{N}\sum_1^N x_i \tag{2.1.5}$$

satisfies, for all $c_1 < 0 < c_2$,

$$\lim_{N\to\infty} \mathcal{P}\left(\mathbb{E}[x]-\mathbb{E}_N[x] \in \left(c_1\sqrt{\frac{\mathrm{var}(x)}{N}}, c_2\sqrt{\frac{\mathrm{var}(x)}{N}}\right)\right) = \frac{1}{\sqrt{2\pi}}\int_{c_1}^{c_2} \mathrm{e}^{-\frac{x^2}{2}}\,\mathrm{d}x, \tag{2.1.6}$$

where $\mathcal{P}(y \in Y)$ stands for the probability that y belongs to Y.

The error due to the Euler scheme is as follows.

Theorem 2.1.2. Assume that r, σ are bounded and uniformly Lipschitz-continuous in time and space (implying at most linear growth at infinity). Then the $L^\infty(L^p)$-error, $p < 1$, is bounded by $c\sqrt{\delta t(1-\ln\delta t)}$, for some constant c.

A proof is available in Pagès (2008), for instance.

Theorem 2.1.3. (Talay and Tubaro 1996) Let $f \in C^\infty(\mathbb{R})$. If r and σ are uniformly Lipschitz in time and Lipschitz in space, the error on $\mathbb{E}(f(S_T))$ when S_T is computed by (2.1.3) is of order $O(\delta t)$.

Remark 2.1.4. Note that $f : S_T \to (K - S_T)^+$ is not in $C^\infty(\mathbb{R})$. Consequently the precision of (2.1.4) is $O(\sqrt{\delta t}) + O(\frac{1}{\sqrt{N}})$. However, if f were regular it would be $O(\delta t) + O(\frac{1}{\sqrt{N}})$.

Remark 2.1.5. There exist schemes more precise than (2.1.3) (see Remark 2.1.6), but in the author's experience the improvement seems more theoretical than actual. On the other hand there are many extremely effective *variance reduction* techniques which reduce the constants multiplying the term $O(\frac{1}{\sqrt{N}})$.

Most options can be simulated by the Monte Carlo method. For instance, a Monte Carlo simulation to price an Asian option (1.5.14) with arithmetic averaging is as follows.

Choose $N, \delta t$, set $C_0 = 0$.
For$(n = 0\,; n \leq N;\ n = n+1)$

  S_0 given, set $A = 0$.
  For$(t = 0\,; t \leq T;\ t = t + \delta t)$

    Call the random generator to simulate $W_{t+\delta t} - W_t$.
    Compute $S_{t+\delta t} = S_t(1 + r\delta t + \sigma(W_{t+\delta t} - W_t))$.
    Do $A = A + \frac{\delta t}{2t}(S_t + S_{t+\delta t})$.

  Compute $\frac{\mathrm{e}^{-rT}}{N}(S_T - A_T)^+$ and add it to C_0.

Multi-level Monte Carlo

The most common method to speed up Monte Carlo is *variance reduction*, whereby one searches for an equivalent random process x with larger var(x), thus giving a better estimate in (2.1.6). Quasi-Monte Carlo is also popular but multi-level Monte Carlo achieves the same efficiency with less programming cost.

Following Giles (2008), a multi-level technique similar to multigrid is used. While standard Monte Carlo costs $O(\epsilon^{-3})$ operations (using the optimistic Theorem 2.1.2 with $\delta t = O(\epsilon)$ and $N = O(\epsilon^{-2})$) to reach a mean-square precision $O(\epsilon^2)$, multi-level Monte Carlo does the same in $O(\epsilon^{-2}\ln^2\epsilon)$.

So consider a multiple set of simulation with $\delta t = \frac{T}{2m}$, $m = 0,\dots,M$ to compute the pay-off P^m, and write

$$\mathbb{E}[P^M] = \mathbb{E}[P^0] + \sum_1^M \mathbb{E}[P^m - P^{m-1}].$$

The key point is to approximate $\mathbb{E}[P^m - P^{m-1}]$ by the law of large numbers with N_m simulations using the same Brownian path:

$$\mathbb{E}[P^m - P^{m-1}] \approx \frac{1}{N_m}\sum_{i=1}^{N_m}(P^{m,i} - P^{m-1,i}).$$

Remark 2.1.6. The theory suggests using the Milstein scheme, which is $O(\delta t)$ in the L^∞-norm, and gives a more robust error estimator and $O(\epsilon^2)$ operations:

$$S_{t+\delta t} = S_t\big(1 + r\delta t + \sigma(W_{t+\delta t} - W_t)\big) + \frac{S_t^2\sigma\partial_t\sigma}{2}\big((W_{t+\delta t} - W_t)^2 - \delta t\big).$$

2.2. The Longstaff–Schwartz algorithm for Bermudas

We follow the presentation of Clément, Lamberton and Pottier (2002). For clarity assume $r = 0$. After discretization on the discrete early exercise set times, $\{t_j\}_1^J$, (1.5.13) becomes

$$P_j = \max_{k\geq j}\mathbb{E}[(K - S_k)^+|S_j], \quad j = 1,\dots,J.$$

Following the usual dynamic programming paradigm, the same can be written as

$$P_J = (K - S_J)^+, \quad P_j = \max\{(K - S_j)^+, \mathbb{E}[P_{j+1}|S_j]\}.$$

It can also be rewritten so as to compute the exercise times τ_j:

$$\begin{aligned}
&\tau_J = T,\ P_J = (K - S_J)^+;\\
&\text{for } (\ j = J \text{ down to } j = 1)\\
&\qquad \text{if } (K - S_j)^+ \geq \mathbb{E}[P_{j+1}|S_j] \text{ do } \tau_j = t_j,\ P_j = (K - S_j)^+\\
&\qquad \text{else do } \tau_j = \tau_{j+1},\ P_j = \mathbb{E}[P_{j+1}|S_j]
\end{aligned} \tag{2.2.1}$$

Longstaff and Schwartz (1992) suggest using a least-squares projection on a polynomial basis to compute $\mathbb{E}[P_{j+1}|S_j]$.

For this, S_t is computed on a set of Monte Carlo paths indexed by i, giving S_j^i. Since $\mathbb{E}[P_{j+1}|S]$ is known from the previous iteration at the sample points $\{S = S_{j+1}^i\}_i$, then $\mathbb{E}[P_{j+1}|S = S_j]$ is computed by interpolation.

To project a function $y \to f(y)$ on $\{1, y, y^2, \ldots, y^n\}$ from a known set of values $f(y_i)$, $i = 1, \ldots, I$, one solves

$$\min_{a_1,\ldots,a_n} \sum_{i=1}^{I} \left\| \sum_{m=1}^{n} a_m y_i^m - f(y_i) \right\|^2 .$$

An implementation in C++ written by Tobias Lipp is available upon request by email to tobias.lipp@upmc.fr or olivier.pironneau@upmc.fr.

2.3. Methods using the process PDF for basket options

2.3.1. Basket options with constant volatilities

When the volatilities of the elements of a basket are constant, the computation of an option on the basket reduces to a multiple integral (1.5.20). For a basket with many elements ($d \gg 1$), we will show below how to use sparse grids to compute this integral efficiently.

2.3.2. Numerical simulation by Fourier transform

Take the case of a vanilla European call. $X_T = \ln(S_T)$ is a martingale so the PDF is $\rho(X_T - X_0)$. As above, the option is

$$C_0(X_0) = \mathrm{e}^{-rT} \int_{\mathbb{R}} (\mathrm{e}^{X_T} - K)^+ \rho(X_T - X_0)\, \mathrm{d}X_T.$$

As the right-hand side is a convolution, a Fourier transform changes the integral into a product. However the pay-off $(\mathrm{e}^x - \mathrm{e}^k)^+$ is not a k-integrable function on $\mathbb{R}$ so its Fourier transform is undefined. With an appropriate damping factor e^{ak}, Carr and Madan (1999) arrived at

$$\mathcal{F}[\mathrm{e}^{ak} C_0] = \frac{\mathrm{e}^{-rT}\Phi(z - (a+1)\mathrm{i})}{(a\mathrm{i} - z)(z - \mathrm{i}(1+a))}. \tag{2.3.1}$$

When X_t is driven by (1.1.5), with constant volatility, the characteristic function is

$$\Phi(z) = \mathrm{e}^{\mathrm{i}z(r - \frac{\sigma^2}{2})T - \frac{\sigma^2}{2}Tz^2}.$$

However, this so-called 'CONV' method applies to more complex situations, whenever Φ is known for a monotone function of S_t, as is the case with Lévy processes (see Lord, Fang, Bervoets and Oosterlee (2007)). The computational cost is a single Fourier transform.

Basket options

The method also applies to basket options. With (4.3.6) the joint characteristic function is

$$\Phi(\vec{z}) := \mathbb{E}[e^{i\vec{z}\cdot\vec{S}}|\vec{S}_0] = S_0 e^{iT\sum_i(r-\frac{\sigma_i^2}{2}z_i)-\frac{T}{2}\sum_{i,j}z_iz_j\sigma_i\sigma_j\rho_{ij}}. \tag{2.3.2}$$

Damping factors do not seem to work, so the domain is truncated, $\mathbb{R}^d \approx D := (-L, L)^d$, and the following is computed approximately by using two FFTs:

$$C_0 = e^{-rT}\mathcal{F}^{-1}\left[\mathcal{F}\left[\left(\sum_i \alpha_i e^{y_i} - K\right)^+ \mathbf{1}_D(\vec{y})\right]\Phi(-\vec{z})\right]. \tag{2.3.3}$$

Bermuda options

The same method is applied backwards in time on each interval $[t_i, t_{i+1}]$, where the contract cannot be exercised early. Then the pay-off is compared with the constraint ($(S-K)^+$ for a call) and the process is continued unless early exercise is better.

2.4. A binomial tree for American options

Trees are perhaps among the most popular tools for numerical finance. They are easy to understand and close to the probabilistic model, but their convergence is not so easy to establish. There are parallels between binomial trees and explicit finite difference methods. Because it is so well documented, only one example is given here, for the computation of an American option.

The time interval to maturity $(0, T)$ is divided into smaller intervals of size δt, days, for example. It is also assumed that the underlying asset is S_0 initially and can only take a finite set of values $u^j S_0$, indexed by $j = 1, 2, \ldots, J$. After n days, the underlying asset takes one of its possible values, S_j^n. Then it could either go up with probability p the next day to $S_{j+1}^{n+1} = uS_j^n$ or down to $S_{j-1}^{n+1} = \frac{1}{u}S_j^n$ with probability $1-p$. At time $T = N\delta t$, the probability p_j^N that $S_T = u^j S_0$ is known.

The American call is computed backwards in time from its value at T, knowing that S_T is $u^j S_0$, $C_j^N = (u^j S_0 - K)^+$:

$$C_j^n = \max\{e^{-r\delta t}(pC_{j+1}^{n+1} + (1-p)C_{j-1}^{n+1}), (S_0 u^j - K)^+\}. \tag{2.4.1}$$

Equivalence with a finite difference method

It will be shown later that C is also such that $C(T, x) = (e^x - K)^+$ and

$$\min\left\{-\partial_t C - \frac{\sigma^2}{2}\partial_{xx}C - \left(r - \frac{\sigma^2}{2}\right)\partial_x C + rC, C - (e^x - K)^+\right\} = 0,$$

where $x = \ln S$. Applying an explicit Euler scheme in time and central differences in x, the problem becomes

$$\min\left\{-\frac{1}{\delta t}(C_j^{n+1} - C_j^n) - \frac{\sigma^2}{2\delta x^2}(C_{j+1}^{n+1} - 2C_j^{n+1} + C_{j-1}^{n+1}) \right. \tag{2.4.2}$$
$$\left. - \left(r - \frac{\sigma^2}{2}\right)\frac{C_{j+1}^{n+1} - C_{j-1}^{n+1}}{2\delta x} + rC_j^{n+1}, C_j^{n+1} - (\mathrm{e}^{j\delta x} - K)^+\right\} = 0.$$

When $\delta x = \sigma\sqrt{\delta t}$ and $p = \frac{1}{2} + \frac{\sqrt{\delta t}}{2\sigma}(r - \frac{\sigma^2}{2})$, and u such that $S_0 u^j = j\delta x$, then both schemes are identical up to higher-order terms in δt. See Jiang and Dai (1999) for more details.

The website finance.bi.no/~bernt/gcc_prog/recipes/recipes/ of Bernt Arne Ødegaard has more applications of binomial trees with the corresponding C++ codes.

PART THREE
Some partial differential equations in finance

3.1. From SDEs to PDEs and PIDEs

Numerical methods based on stochastic differential equations are usually easy to understand and very close to the probabilistic model. On the other hand they can be slow. The duality between the stochastic and deterministic formulations has been known at least since the discovery of molecular agitation as the underlying principle of temperature. Indeed, the heat equation is the macroscopic result of Brownian molecular agitation. The general rule is that if a deterministic quantity is a function of a Markov process, then there is a deterministic equation for it. Kolmogorov and Itô have laid the foundations and given the mathematical tools to understand and derive the (possibly integral) partial differential equations and inequalities in finance. Dupire has also contributed to the field, as we shall see.

3.2. Fokker–Planck and Kolmogorov equations

Recall that, almost by definition, the conditional probability density function $x \to \rho(x)$ of a random variable X, knowing that an event A has occurred, is related to the conditional expectation of $f(X_t)$, knowing A, by

$$\mathbb{E}(f(X)|A) = \int_{\mathbb{R}} f(x)\rho(x|A)\,\mathrm{d}x. \tag{3.2.1}$$

Consider the SDE

$$\mathrm{d}X_t = \mu_t\,\mathrm{d}t + \sigma_t\,\mathrm{d}W_t, \quad X_0 \text{ given.} \tag{3.2.2}$$

Kolmogorov (see Kolmogoroff 1931) showed that, when μ_t and σ_t are explicit functions of t and X_t, the PDF $X, T \to \rho(X_T|x,t)$ of X_τ, knowing that $X_t = x$, satisfies

$$\partial_T \rho - \partial_{XX}\left(\frac{\sigma(X,T)^2}{2}\rho\right) + \partial_X(\mu\rho(X,T)) = 0, \quad \rho|_t \text{ given}, \quad T > t. \quad (3.2.3)$$

This equation is known as the Kolmogorov forward or *Fokker–Planck equation* of X_T. It can be used to compute a European call, for instance,

$$C_0 = \mathrm{e}^{-(r-\frac{\sigma^2}{2})T}\int_{\mathbb{R}^+} \rho(X,T)(\mathrm{e}^X - K)^+ \,\mathrm{d}X,$$

where ρ is the solution of (3.2.3) with initial condition $\rho(x,0) = \delta(\mathrm{e}^x - S_0)$, the Dirac mass at 0 (see (1.1.5)).

Conversely, if the statistics of X_T are known instead of those of X_t, the *Kolmogorov backward equation* may be used to compute $x, t \to \rho$:

$$\partial_t \rho + \frac{\sigma(x,t)^2}{2}\partial_{xx}\rho + \mu(x,t)\partial_x\rho = 0, \quad \rho|_T \text{ known}, \quad t < T. \quad (3.2.4)$$

From this the *Feynman–Kac formula* can be derived. Let u be the solution of

$$\partial_t u + \frac{\sigma(x,t)^2}{2}\partial_{xx}u + \mu(x,t)\partial_x u - ru = 0, \quad u|_T \text{ known}, \quad t < T. \quad (3.2.5)$$

Let X_t be a solution of the SDE (3.2.2). Then

$$u(x,t) = \mathbb{E}[\mathrm{e}^{-r(T-t)}u(X_T,T)|X_t = x]. \quad (3.2.6)$$

Corollary 3.2.1. The European call (1.5.1) can be computed by

$$C_t = u(\log(S_t), t),$$

where u is the solution of (3.2.5) with $u(x,T) = (\mathrm{e}^x - K)^+$ and $\mu = r - \frac{1}{2}\sigma^2$.

Proof. The Feynman–Kac formula can be applied to (1.1.5) with $u(x,t) = C_t|X_t = x$. It shows that if u satisfies (3.2.5) with $u(x,T) = (\mathrm{e}^x - K)^+$, then (3.2.6) holds and it is the definition of C_t. □

3.2.1. Itô's lemma

Let X_t be a solution of the SDE (3.2.2) and let $f(x,t)$ be twice differentiable in x and once in t. Then Itô's lemma says

$$\mathrm{d}f(X_t,t) = \frac{\partial f}{\partial x}(X_t,t)\,\mathrm{d}X_t + \frac{\partial f}{\partial t}(X_t,t)\,\mathrm{d}t + \frac{\sigma_t^2}{2}\frac{\partial^2 f}{\partial x^2}(X_t,t)\,\mathrm{d}t. \quad (3.2.7)$$

Note that when f is non-random, *i.e.*, $f = \bar{f} := \mathbb{E}[f(X_\tau, \tau|X_t = x]$, then by taking the expected value of (3.2.7) and (3.2.2),

$$\mathrm{d}\bar{f}(X_t,t) = \frac{\partial \bar{f}}{\partial x}(X_t,t)\mu\,\mathrm{d}t + \frac{\partial \bar{f}}{\partial t}(X_t,t)\,\mathrm{d}t + \frac{\mathbb{E}[\sigma_t^2]}{2}\frac{\partial^2 \bar{f}}{\partial x^2}(X_t,t)\,\mathrm{d}t. \quad (3.2.8)$$

For more details, see for example Øksendal (1998).

3.2.2. Application to European options

Let us apply (3.2.7) to

$$C(S,t) := \mathbb{E}[\mathrm{e}^{-r(T-t)}(S_T - K)^+ | S_t = S], \quad \text{with } \mathrm{d}S_t = S_t(r\,\mathrm{d}t + \sigma_t\,\mathrm{d}W_t).$$

Then

$$\begin{aligned}\mathrm{d}C &= \frac{\partial C}{\partial S}(S_t,t)\,\mathrm{d}S_t + \frac{\partial C}{\partial t}(S_t,t)\,\mathrm{d}t + \frac{S_t^2\sigma_t^2}{2}\frac{\partial^2 C}{\partial S^2}(S_t,t)\,\mathrm{d}t \\ &= \left(rS_t\frac{\partial C}{\partial S} + \frac{\partial C}{\partial t} + \frac{S_t^2\sigma_t^2}{2}\frac{\partial^2 C}{\partial S^2}\right)\mathrm{d}t + \frac{\partial C}{\partial S}S_t\sigma_t\,\mathrm{d}W_t + o(\mathrm{d}t). \qquad (3.2.9)\end{aligned}$$

Note that while C_t is not random, $\mathrm{d}C$ is.

3.2.3. Application to hedging

Is it possible to build a riskless portfolio π_t made of κ parts of S_t plus one C_t? If so, no arbitrage requires that it evolves like a riskless asset, hence $\mathrm{d}\pi_t = r\pi_t\,\mathrm{d}t$. On the other hand

$$\begin{aligned}\mathrm{d}\pi_t = \kappa\,\mathrm{d}S_t + \mathrm{d}C_t &= \kappa S_t(r\,\mathrm{d}t + \sigma_t\,\mathrm{d}W_t) + \frac{\partial C}{\partial S}S_t\sigma_t\,\mathrm{d}W_t \\ &+ \left(rS_t\frac{\partial C}{\partial S} + \frac{\partial C}{\partial t} + \frac{S_t^2\sigma_t^2}{2}\frac{\partial^2 C}{\partial S^2}\right)\mathrm{d}t. \qquad (3.2.10)\end{aligned}$$

Accordingly π_t is non-random only if the factor of $\mathrm{d}W_t$ vanishes, namely $\kappa = -S\partial_S C$. Furthermore, the above simplifies to

$$r\pi_t\,\mathrm{d}t = r(\kappa S_t + C_t)\,\mathrm{d}t = \kappa S_t r\,\mathrm{d}t + \left(rS_t\frac{\partial C}{\partial S} + \frac{\partial C}{\partial t} + \frac{S_t^2\sigma_t^2}{2}\frac{\partial^2 C}{\partial S^2}\right)\mathrm{d}t.$$

Consequently we have the following result.

Theorem 3.2.2. A European call C_t on S, knowing S_t, can be priced by solving for $C(S,t)$ the backwards-in-time parabolic PDE

$$\begin{aligned}\partial_t C + \frac{S^2\sigma^2}{2}\partial_{SS}C + rS\partial_S C - rC = 0, &\quad \forall S,t \in \mathbb{R}^+ \times (0,T), \\ C(S,T) = (S-K)^+, &\quad \forall S \in \mathbb{R}^+, \qquad (3.2.11)\end{aligned}$$

and then set $C_t = C(S_t,t)$. Furthermore the portfolio $\kappa S_t + C_t$ is riskless when $\kappa = -S\partial_S C(S_t,t)$.

Remark 3.2.3. (1) Equation (3.2.11) is known as the Black–Scholes PDE.

(2) By a change of coordinate $x = \log(S)$, it is seen that (3.2.11) and (3.2.5) are the same.

(3) Hedging is not necessary to derive (3.2.11); it is not an additional hypothesis to the Black–Scholes model. Since it leads to an extra result on riskless portfolios, it is generally preferred over the Feynman–Kac formula to derive the PDEs of finance.

(4) A similar equation can be derived for the European put, either with a riskless portfolio made of a put and the asset, or by using the put–call parity on (3.2.11).

(5) More generally, if the premium is $\phi(S_T)$ instead of $(S_T - K)^+$ and/or if r, σ depend on time and/or S_t, the same equation (3.2.11) holds but with $C(S,T) = \phi(S)$.

3.2.4. Application to basket options

Recall (1.5.15) written under the risk-neutral probability law, namely that the pay-off ϕ of the option is a function of

$$\mathrm{d}S_{it} = S_{it}(r\,\mathrm{d}t + \sigma_i\,\mathrm{d}W_{it}), \quad \text{with } \rho_{i,j}t := \mathbb{E}[W_{it}W_{jt}]. \tag{3.2.12}$$

No arbitrage and Itô calculus say that

$$\begin{aligned} \mathrm{e}^{r\,\mathrm{d}t}\phi &= \mathrm{d}\phi \\ &= \mathbb{E}\left[\partial_t\phi\,\mathrm{d}t + \sum_i \partial_{S_i}\phi\,\mathrm{d}S_{it} + \frac{1}{2}\sum_{ij}\partial_{S_iS_j}\phi S_iS_j\,\mathrm{d}W_{it}\,\mathrm{d}W_{jt}|S_{i0}\right] \\ &= \left[\partial_t\phi + \sum_i rS_i\partial_{S_i}\phi + \frac{1}{2}\sum_{ij} S_iS_j\sigma_{it}\sigma_{jt}\rho_{ij}\partial_{S_iS_j}\phi\right]\mathrm{d}t. \end{aligned} \tag{3.2.13}$$

With $\tau = T - t$, this equation can be written in compact form as

$$\partial_\tau\phi - r\vec{S}\cdot\nabla\phi - \Xi : \nabla\nabla\phi + r\phi = 0, \quad \phi(\vec{S},0) \text{ given}, \tag{3.2.14}$$

with $\Xi_{ij} = S_iS_j\sigma_{it}\sigma_{jt}\rho_{ij}/2$ and the notation $A : B = \sum_{ij} A_{ij}B_{ij}$ as the trace of the matrix AB^T.

3.2.5. Application to Asian options

With $A_t = \frac{1}{t}\int_0^t S_\tau\,\mathrm{d}\tau$ and $\mathrm{d}S_t = S_t(r\,\mathrm{d}t + \sigma_t\,\mathrm{d}W_t)$, $S(0) = S_0$, an Asian call $C_t = \mathrm{e}^{-r(T-t)}\mathbb{E}[(A_T - K)^+|S_t]$ satisfies

$$\begin{aligned} \mathrm{e}^{r\,\mathrm{d}t}C_t = \mathrm{d}C_t &= \partial_t C\,\mathrm{d}t + \partial_A C\,\mathrm{d}A_t + \frac{1}{2}\partial_{AA}C(\mathrm{d}A)^2 + \partial_S C\,\mathrm{d}S_t \\ &\quad + \frac{1}{2}\partial_{SS}C(\mathrm{d}S)^2 + \partial_{AS}C\,\mathrm{d}S\,\mathrm{d}A \\ &= \partial_t C\,\mathrm{d}t + \partial_A C(S - A)\frac{\mathrm{d}t}{t} + \partial_S C(Sr\,\mathrm{d}t + S\sigma\,\mathrm{d}W_t) \\ &\quad + \frac{1}{2}\partial_{SS}C(\sigma S)^2\,\mathrm{d}t + o(\mathrm{d}t), \end{aligned} \tag{3.2.15}$$

because $\mathrm{d}(tA_t) = t\,\mathrm{d}A_t + A_t\,\mathrm{d}t = S_t\,\mathrm{d}t$.

As for vanilla European calls, a portfolio $\pi_t = \kappa S_t + C_t$ with variation $\kappa\,\mathrm{d}S_t + \mathrm{d}C_t$ will be riskless only if $\kappa = -S\partial_S C(S_t, t)$, and so by saying that $\mathrm{d}\pi = r\pi\,\mathrm{d}t$, the following is found for all $S \in \mathbb{R}^+, A \in \mathbb{R}^+, t \in (0,T)$:

$$\partial_t C + \frac{S-A}{t}\partial_A C + rS\partial_S C + \frac{\sigma^2 S^2}{2}\partial_{SS}C - rC = 0,$$
$$C(S,A,T) = (A-K)^+. \tag{3.2.16}$$

Let us set $C(S,A,t) = u(\mathrm{e}^{r(T-t)}S, tA, T-t)$. Then, in terms of $x = \mathrm{e}^{r(T-t)}S, y = tA, \tau = T-t$, the PDE for u is

$$\partial_\tau u - x\mathrm{e}^{-r\tau}\partial_y u - \frac{\sigma^2 x^2}{2}\partial_{xx}u + ru = 0,$$
$$u(x,y,0) = \left(\frac{y}{T} - K\right)^+, \quad \forall x,y,\tau \in \mathbb{R}^+ \times \mathbb{R}^+ \times (0,T). \tag{3.2.17}$$

Additional boundary conditions are needed, such as

$$u(x,0,\tau) = 0, \quad u(0,y,\tau) = \mathrm{e}^{-r\tau}\left(\frac{y}{T} - K\right)^+, \quad \forall x,y,\tau \in \mathbb{R}^+ \times \mathbb{R}^+ \times (0,T). \tag{3.2.18}$$

An asymptotic condition can be found if $yT > K$ and $x \to \infty$. Then

$$u \sim \mathrm{e}^{-r\tau}\left[y + \frac{1-\mathrm{e}^{-r\tau}}{r}x - KT\right]/T.$$

3.2.6. *From jump processes to partial integro-differential equations*

The *generalized Black–Scholes model* uses a jump-diffusion process X_t, given by (1.4.3) instead of the Wiener process W_t. Hence consider the following model for the underlying asset S_t:

$$\mathrm{d}S_t = S_t^-(r\,\mathrm{d}t + \mathrm{d}X_t), \quad S_0 \text{ given, } i.e., S_t = S_0\mathrm{e}^{rt+X_t}. \tag{3.2.19}$$

A corollary to (1.4.6) is that $\mathrm{e}^{rt}S_t$ is a martingale if and only if

$$\gamma + \frac{\sigma^2}{2} + \int_{\mathbb{R}}(\mathrm{e}^x - 1 - x\mathbf{1}_{|x|<1})\nu(\mathrm{d}x) = 0. \tag{3.2.20}$$

Now, because (1.4.5) holds, it can be proved (Tankov and Voltchkova 2009) that a call option on S_t with pay-off $(S_t - K)^+$, namely

$$C_t = \mathrm{e}^{-r(T-t)}\mathbb{E}[(S_T - K)^+ | S_t = S),$$

is also given by $C(S,t)$, where C is the solution of

$$\partial_t C + \frac{\sigma^2 S^2}{2}\partial_{SS}C + r\partial_S C - rC \\ + \int_{\mathbb{R}} [C(S\mathrm{e}^y, t) - C(S,t) - S(\mathrm{e}^y - 1)\partial_S C(S,t)]\nu(\mathrm{d}y) = 0, \\ C(S,T) = (S-K)^+. \tag{3.2.21}$$

Remark 3.2.4. With $x = \ln S$, $\tau = T - t$ and $u(x,\tau) = C(\mathrm{e}^x, T-\tau)$ then (3.2.22) becomes

$$\partial_\tau u - \frac{\sigma^2}{2}\partial_{xx}u - \left(r - \frac{\sigma^2}{2}\right)\partial_x u + ru \\ - \int_{\mathbb{R}} [u(x+y,\tau) - u(x,\tau) - (\mathrm{e}^y - 1)\partial_x u(x,\tau)]\nu(\mathrm{d}y) = 0, \\ u(x,0) = (\mathrm{e}^x - K)^+. \tag{3.2.22}$$

Theorem 3.2.5. Assume $\sigma > 0$ and, for some $\beta \in (0,2)$,

$$\int_{|y|>1} \mathrm{e}^{2y}\nu(\mathrm{d}y) < \infty, \quad \liminf_{\epsilon\to 0} \epsilon^{-\beta}\int_{-\epsilon}^{\epsilon} |y|^2\nu(\mathrm{d}y) > 0. \tag{3.2.23}$$

Then $C(S,t) = \mathrm{e}^{-r(T-t)}\mathbb{E}[(S_T - K)^+ | S_t = S]$ is continuous on $\mathbb{R}^+ \times [0,T]$, C^1 in t and C^2 in S on $\mathbb{R}^+ \times (0,T)$ and satisfies (3.2.22).

Proof. The leading idea is to apply Itô's formula to the martingale $\hat{C}_t = \mathrm{e}^{r\tau}C(S_t,t)$ with $\tau = T - t$.

The second condition in (3.2.23) implies that X_t has a smooth C^2 density and so $C(S,t)$ has the regularity mentioned in the theorem.

The fact that $\int_{|y|>1}\mathrm{e}^{2y}\nu(\mathrm{d}y) < \infty$ implies that there exists J_X such that

$$\mathrm{d}S_t = S_t\left(\sigma\,\mathrm{d}W_t + \int_{\mathbb{R}} (\mathrm{e}^x - 1)J_X(\mathrm{d}x\,\mathrm{d}t)\right).$$

Skipping the details (see Cont and Voltchkova (2004), Cont and Tankov (2003)),

$$\mathrm{d}\hat{C}_t = a\,\mathrm{d}t + \mathrm{d}M_t, \tag{3.2.24}$$

with

$$\mathrm{e}^{-r\tau}a = -rC + \partial_t C + \frac{\sigma^2 S^2}{2}\partial_{SS}C + rS\partial_S C \\ + \int_{\mathbb{R}} (C(S\mathrm{e}^x, t) - C(S,t) - S(\mathrm{e}^x - 1)\partial_S C(S,t))\nu(\mathrm{d}x), \\ \mathrm{e}^{-r\tau}\,\mathrm{d}M_t = \partial_S C\sigma S\,\mathrm{d}W_t + \int_{\mathbb{R}} (C(S\mathrm{e}^x, t) - C(S,t)J_X(\mathrm{d}t\,\mathrm{d}x). \tag{3.2.25}$$

Then one shows that M_t is a square-integrable martingale and so is $\hat{C}_t - M_t$. Thus, by taking the expected value of (3.2.24), we obtain that $a = 0$. $\square$

Popular choices for ν are $\nu(\mathrm{d}y) = k(y)\,\mathrm{d}y$ with the *variance gamma model*,

$$k(y) = \frac{\omega}{|y|}\mathrm{e}^{-y^+\eta^+ - y^-\eta^-}, \tag{3.2.26}$$

with $\eta^\pm > 2$, the *CGMY model* (Carr, Geman, Madan and Yor 2002),

$$k(y) = \frac{\omega}{|y|^{1+Y}}\mathrm{e}^{-y^+\eta^+ - y^-\eta^-}, \tag{3.2.27}$$

with $0 < Y < 2,\ 0 < \eta^-,\ 2 < \eta^+$, and *Merton's model*,

$$k(x) = \frac{\lambda}{\sqrt{2\pi}\delta}\exp\left(-\frac{(y-\mu)^2}{2\delta^2}\right). \tag{3.2.28}$$

Theorem 3.2.6. In the case of Merton's model C, the call on S at $t = 0$, can be computed by

$$C = \mathrm{e}^{-rT}\sum_0^\infty \frac{\mathrm{e}^{-\lambda T}(\lambda T)^n}{n!}\mathrm{e}^{r_n T} C_{BS}(S_0 \mathrm{e}^{n\delta^2/T}, K, T, \sigma_n, r_n),$$

where C_{BS} refers to the Black–Scholes formula and

$$r_n = r - \lambda(\mathrm{e}^{\mu+\delta^2/2} - 1) + \mu n T, \quad \sigma_n = \sqrt{\sigma^2 + n\frac{\delta^2}{2}}.$$

Remark 3.2.7. Note that the variance gamma model and the CGMY model do not satisfy the second hypothesis in (3.2.23). Another proof must be made, using viscosity solutions, for instance.

3.2.7. Lévy copula

An extension to basket options can be found in Farkas *et al.* (2006). The following PIDE is found, in variational form (see below):

$$\int_{\mathbb{R}^d} \partial_\tau u(x,\tau)v(x)\,\mathrm{d}x + \int_{\mathbb{R}^d} \nabla v^T \Xi \nabla u - \int_{\mathbb{R}^d} v\mathbf{b}\cdot\nabla u = J(u,v), \quad \forall v \in V, \tag{3.2.29}$$

where

$$J(u,v) = \frac{1}{2}\int_{x\in\mathbb{R}^d}\int_{y\in\mathbb{R}^d}(u(x+y)-u(x))(v(x+y)-v(x))\tilde{\nu}(\mathrm{d}y)\,\mathrm{d}x.$$

Here V is a Sobolev space with anisotropic weight, Ξ is the variance-correlation matrix (see (3.2.14)) and $\mathbf{b}$ is the drift in log price variables, made up of the usual drift proportional to r, plus the term proportional to $\nabla\Xi$, plus the part of the integro-differential term corresponding to $(e^y - 1)\partial_x u$ in the scalar case. The symmetry of J corresponds to a special case in the general theory.

3.2.8. From constrained SDEs to PDIs

American option

Let P_t be an American put on S_t with strike K and maturity T. We have seen above that unless $P_t \geq (K - S_t)^+$ the contract ceases to exist. As for a European call, the put $u(S_t, t) = e^{-(T-t)}\mathbb{E}[(K - S_T)^+|S_t]$ can be estimated by computing $\mathbb{E}[\,du]$ with Itô's lemma. Intuitively, no arbitrage implies that that $\mathbb{E}\,du \leq ru\,dt$, so in addition to $P(S,T) = (K - S_T)^+$, we have

$$\partial_t P + \frac{\sigma^2 S^2}{2}\partial_{SS}P + rS\partial_S P - rP \leq 0, \quad \text{or} \quad P(S,t) \geq (K - S)^+, \tag{3.2.30}$$

and at every S, t, one of the two, at least, is an equality. This can be summarized by the requirement that, for all $S > 0, t \in (0,T)$,

$$\min\left\{-\partial_t P - \frac{\sigma^2 S^2}{2}\partial_{SS}P - rS\partial_S P + rP, P - (K - S)^+\right\} = 0. \tag{3.2.31}$$

Convertible bond

Suppose the bond has a probability of default p with a fraction of return R. Let V denote the convertible, B the bond component, q the dividend and S the share. Then Hull (2003) suggests the following:

$$\begin{aligned}
&\partial_t V + \frac{\sigma^2 S^2}{2}\partial_{SS}V + (r - q)S\partial_S V - rV = sB, \quad V_T = \max\{B_T, \kappa S_T\},\\
&\partial_t B + \frac{\sigma^2 S^2}{2}\partial_{SS}B + rS\partial_S B - (r + s)B = 0, \quad B_T \text{ given},
\end{aligned} \tag{3.2.32}$$

where

$$\text{if } B_c < \kappa S, \text{ then } V = \kappa S, \text{ else } \max\{B_p, \kappa S_t\} \leq V_t \leq B_c, \tag{3.2.33}$$

where the spread s is related to p and R by $s = p(1 - R)$. Assuming B_c constant in time, for simplicity, the ambiguity between the constraints and the PDE is resolved by:

$$\begin{aligned}
&\text{if } B_c < \kappa S \text{ then } V = \kappa S \text{ else}\\
&\max\left\{\max\{B_p, \kappa S\} - V, \partial_t V + \frac{\sigma^2 S^2}{2}\partial_{SS}V + (r - q)S\partial_S V - rV + sB\right\}\\
&\quad = 0.
\end{aligned} \tag{3.2.34}$$

3.2.9. PDEs for interest rates and swaps

Options on interest rates

Letting $\bar{r}(t)$ denote the mean interest rate at T and coming back to (1.5.23) and (1.5.22), an option on S_t with pay-off

$$C(S,r,\sigma,t)=\mathbb{E}\left[\mathrm{e}^{-\frac{1}{T}\int_t^T r(s)\,\mathrm{d}s}\phi(S_T)|S_t=S,r_t=r,\sigma_t=\sigma\right]$$

will satisfy, by Itô's lemma and the no arbitrage rule,

$$\begin{aligned}C(S,r,\sigma,t)r\,\mathrm{d}t=\ &\mathrm{d}C=\partial_t C\,\mathrm{d}t+\partial_S C\mathbb{E}[\mathrm{d}S]+\partial_r C\mathbb{E}[\mathrm{d}r]+\partial_\sigma C\mathbb{E}[\mathrm{d}\sigma]\\ &+\frac{1}{2}\partial_{SS}C\mathbb{E}[\mathrm{d}S^2]+\frac{1}{2}\partial_{rr}C\mathbb{E}[\mathrm{d}r^2]+\frac{1}{2}\partial_{\sigma\sigma}C\mathbb{E}[\mathrm{d}\sigma^2]\\ &+\partial_{Sr}C\mathbb{E}[\mathrm{d}S\,\mathrm{d}r]+\partial_{S\sigma}C\mathbb{E}[\mathrm{d}S\,\mathrm{d}\sigma]+\partial_{r\sigma}C\mathbb{E}[\mathrm{d}S\,\mathrm{d}\sigma]\\ =\ &\partial_t C\,\mathrm{d}t+Sr\,\mathrm{d}t\partial_S C+\lambda(m-r)\,\mathrm{d}t\partial_r C\\ &+\frac{1}{2}\sigma^2S^{2\beta}\rho_{22}\,\mathrm{d}t\partial_{SS}C+\frac{1}{2}\eta^2\rho_{11}\,\mathrm{d}t\partial_{rr}C+\frac{1}{2}\gamma^2\rho_{33}\,\mathrm{d}t\partial_{\sigma\sigma}C\\ &+\sigma S^\beta\eta\rho_{21}\,\mathrm{d}t\partial_{Sr}C+\gamma\sigma S^\beta\rho_{23}\,\mathrm{d}t\partial_{S\sigma}C+\gamma\eta\rho_{31}\,\mathrm{d}t\partial_{r\sigma}C.\end{aligned}\tag{3.2.35}$$

So the PDE of the problem is in $\mathbb{R}^{+3}$: find C with $C(S,r,\sigma,T)=\phi(S)$ and

$$\begin{aligned}&\partial_t C-rC+Sr\partial_S C+\lambda(m-r)\partial_r C\\ &\quad+\frac{1}{2}\sigma^2S^{2\beta}\rho_{22}\partial_{SS}C+\frac{1}{2}\eta^2\rho_{11}\partial_{rr}C+\frac{1}{2}\gamma^2\rho_{33}\partial_{\sigma\sigma}C\\ &\quad+\sigma S^\beta\eta\rho_{21}\partial_{Sr}C+\gamma\sigma S^\beta\rho_{23}\partial_{S\sigma}C+\gamma\eta\rho_{31}\partial_{r\sigma}C=0.\end{aligned}\tag{3.2.36}$$

Credit default swaps

We have seen above that the spread s paid by the buyer is a deterministic function of the probability of default, which is itself a linear function of the two SDEs (1.5.28). Itô calculus gives

$$\begin{aligned}\mathrm{d}s&=\partial_t s\,\mathrm{d}t+\partial_\xi s\mathbb{E}[\mathrm{d}\xi]\\ &\quad+\partial_\lambda s\mathbb{E}[\mathrm{d}\lambda]+\frac{1}{2}\partial_{\xi\xi}s\mathbb{E}[\mathrm{d}\xi^2]+\partial_{\xi\lambda}s\mathbb{E}[\mathrm{d}\xi\,\mathrm{d}\lambda]+\frac{1}{2}\partial_{\lambda\lambda}s\mathbb{E}[\mathrm{d}\lambda^2]\\ &=\mathrm{d}t\bigg(\partial_t s+a(\xi_m-\xi)\partial_\xi s+\alpha(\lambda_m-\lambda)\partial_\lambda s\\ &\qquad+\frac{c^2\xi}{2}\partial_{\xi\xi}s+c\sigma\sqrt{\xi\lambda}\partial_{\xi\lambda}s+\frac{\sigma^2\lambda}{2}\partial_{\lambda\lambda}s\bigg).\end{aligned}\tag{3.2.37}$$

If the buyer's spread evolves with the current interest rate then the following PDE holds for s:

$$\partial_t s + a(\xi_m - \xi)\partial_\xi s + \alpha(\lambda_m - \lambda)\partial_\lambda s$$
$$+ \frac{c^2\xi}{2}\partial_{\xi\xi} s + c\sigma\sqrt{\xi\lambda}\partial_{\xi\lambda} s + \frac{\sigma^2\lambda}{2}\partial_{\lambda\lambda} s = rs,$$
$$s(\xi, \lambda, T) = \phi(\xi, \lambda) \quad \text{given by (1.5.29).} \tag{3.2.38}$$

3.3. Some existence and regularity results

3.3.1. Existence results for the Black–Scholes PDE

By Theorem 3.2.2 a plain vanilla put P can be computed by solving

$$\partial_t P + \frac{\sigma^2 S^2}{2}\partial_{SS} P + rS\partial_S P - rP = 0, \text{ or } P(S,t) \geq (K - S)^+, \tag{3.3.1}$$

for all $S, t \in \mathbb{R}^+ \times (0, T)$. For clarity we assume r constant (but not σ).

Change of variables
Let

$\tau := T - t$, the time to maturity,

$y := e^{r(T-t)}\dfrac{S}{K}$, known as the *forward moneyness*,

$x := \ln y$ (*i.e.*, $S = Ke^{x-(r-d)\tau}$), the *log forward moneyness variable.*

Proposition 3.3.1. Let $P(S, t)$ be a solution of (3.3.1), and let

$$v(y, \tau) := \frac{e^{r\tau}}{K}P(yKe^{-r\tau}, T - \tau), \quad u(x, \tau) := \frac{e^{r\tau}}{K}P(Ke^{x-r\tau}, T - \tau). \tag{3.3.2}$$

Then

$$\partial_\tau v - \frac{\sigma^2 y^2}{2}\partial_{yy} v = 0 \quad \text{in } \mathbb{R}^+ \times (0, T), \quad v(y, 0) = (1 - y)^+, \tag{3.3.3}$$

$$\partial_\tau u + \frac{\sigma^2}{2}\partial_x u - \frac{\sigma^2}{2}\partial_{xx} u = 0 \quad \text{in } \mathbb{R} \times (0, T), \quad u(x, 0) = (1 - e^x)^+. \tag{3.3.4}$$

Proof. Note that $(S - K)^+ = K(\frac{S}{K} - 1)^+$ and $S^2\partial_{SS} = (\frac{S}{K})^2\partial_{\frac{S}{K}\frac{S}{K}}$. Finally note that

$$K\partial_\tau v(y, \tau) = \partial_\tau\big(e^{r\tau}P(yKe^{-r\tau}, T - \tau)\big)$$
$$= re^{r\tau}P(S, t) - e^{r\tau}\partial_t P(S, t) - e^{r\tau}rKye^{-r\tau}\partial_S P(S, t)$$
$$= e^{r\tau}(rP - \partial_t P - S\partial_S P). \tag{3.3.5}$$

To obtain the equation for u, one makes the change of variable $x = \ln y$ in (3.3.3):

$$\partial_x = \partial_x(y)\partial_y = \partial_x(\mathrm{e}^x)\partial_y = y\partial_y,$$
$$\partial_y(\partial_y) = \mathrm{e}^{-x}\partial_x(\mathrm{e}^{-x}\partial_x) = (\mathrm{e}^{-x})^2(\partial_{xx} - \partial_x). \qquad \square$$

A direct proof of existence of solution for (3.3.3) or (3.3.4) is possible but hard to generalize to the multi-dimensional case, so we prefer to use the variational framework, at the cost of imposing some regularity on σ.

Variational formulation
Consider

$$H = \{w \in L^2(\mathbb{R}^+) : y\partial_y w \in L^2(\mathbb{R}^+)\}. \tag{3.3.6}$$

It can be shown (Achdou and Pironneau 2005) that H is a Hilbert space with the scalar product and norm

$$\langle a, b\rangle = \int_{\mathbb{R}^+} [ab + y^2\partial_y a\partial_y b]\,\mathrm{d}y, \quad \|a\| = \left(\int_{\mathbb{R}^+} [a^2 + y^2(\partial_y a)^2]\,\mathrm{d}y\right)^{\frac{1}{2}},$$

for $a, b \in H$. The variational formulation of (3.3.3) is obtained by multiplying it by $w \in H$, integrating it on $\mathbb{R}^+$ and integrating the second derivative in y by parts: find $v \in L^2(0, T; H)$ with $v(y, 0) = (1 - y)^+$, such that, for all $w \in H$,

$$\int_{\mathbb{R}^+} \left[w(y)\partial_\tau v(y,\tau) + \partial_y\left(w(y)\frac{\sigma^2 y^2}{2}\right)\partial_y v(y,\tau)\right] \mathrm{d}y = 0, \text{ a.e. in } (0,T). \tag{3.3.7}$$

We recall the result established in Achdou and Pironneau (2005).

Theorem 3.3.2. If there exist three positive constants $\sigma_m, \sigma_M, C_\sigma$ such that

$$\sigma_m \le \sigma(S,t) \le \sigma_M, \quad |S\partial_S\sigma(S,t)| \le C_\sigma, \tag{3.3.8}$$

then (3.3.7) has a solution in $L^2(0, T; H)$ and the solution is unique. Furthermore $S \to P(S, t)$ is positive bounded by $K\mathrm{e}^{-r(T-t)}$ and convex almost everywhere.

Proof. The proof relies on an extension of the usual theory of parabolic equations, where the coercivity of the bilinear form is established with Gårding's inequality (see Achdou and Pironneau (2005)) for the bilinear form of the PDE:

$$a(u, u) \ge \alpha\|u\|^2 - \beta\|u\|^2_{L^2(\mathbb{R}^+)}.$$

Indeed, by replacing the time derivative by a time difference of size $\delta\tau$ and renaming $v^m(y) := v(y, m\delta\tau)$, (3.3.7) becomes

$$\int_{\mathbb{R}+}\left[w\frac{v^m - v^{m-1}}{\delta t} + \partial_y\left(w\frac{\sigma^2 y^2}{2}\right)\partial_y v^m\right]\mathrm{d}y = 0.$$

It is a problem of the following type: find $v^m \in H$ such that

$$a(v^m, w) = (f, w) \quad \forall w \in H,$$

where

$$a(v,w) = \int_{\mathbb{R}+}\left[w\frac{v}{\delta t} + \partial_y\left(w\frac{\sigma^2 y^2}{2}\right)\partial_y v\right]\mathrm{d}y \quad \text{and} \quad f = \frac{v^{m-1}}{\delta\tau}.$$

The fundamental result of the variational framework is that if a is bilinear continuous and satisfies Gårding's inequality for some strictly positive α, β and if $f \in H$, then the solution v exists and is unique. Here

$$\begin{aligned}
a(v,v) &\geq \int_{\mathbb{R}+}\left[\frac{v^2}{\delta\tau} + \frac{\sigma_m^2}{2}y^2(\partial_y v)^2\right] + \int_{\mathbb{R}+}\left(\partial_y\frac{y\sigma^2}{2}\right)v(y\partial_y v)\\
&\geq \min\left\{\frac{1}{\delta\tau}, \frac{\sigma_m^2}{2}\right\}\|v\|^2 - \left(\max_y \partial_y\frac{y\sigma^2}{2}\right)\|v\|\|v\|_0\\
&\geq \min\left\{\frac{1}{\delta\tau}, \frac{\sigma_m^2}{2}\right\}\|v\|^2 - \left(\frac{\sigma_m^2}{4}\|v\|^2 + \lambda\|v\|_0^2\right) \qquad (3.3.9)
\end{aligned}$$

for some bounded λ as long as $\partial_y(y\sigma^2)$ is bounded too, pointwise. □

The following measures the effect of *localization*, *i.e.*, the error due to the replacement of $\mathbb{R}^+$ in (3.3.1) by $(0, S_M)$. Naturally a boundary condition is needed: $P(S_M, t) = 0$ for all $t \in (0, T)$.

Proposition 3.3.3. With the hypotheses of Theorem 3.3.2, the $L^\infty(\mathbb{R}^+ \times (0,T))$-error between the solution of (3.3.7) and the solution of (3.3.7) in which $\mathbb{R}^+$ and H are replaced by $(0, S_M)$ and $\{w \in H : w(S_M) = 0\}$, decays faster than any exponential $\mathrm{e}^{-\eta S_M}$ as $S_M \to \infty$.

Proof. See Achdou and Pironneau (2005). □

A further change of variables

Assume σ constant. Note that

$$\frac{\mathrm{d}}{\mathrm{d}\tau}u\left(\tau, x + \frac{\sigma^2}{2}\tau\right) = \partial_\tau u + \frac{\sigma^2}{2}\partial_x u.$$

Let $w(x,\tau) = u\big(\tau, x + \frac{\sigma^2}{2}\tau\big)$, where u is the solution of (3.3.4). Then

$$\partial_\tau w - \frac{\sigma^2}{2}\partial_{xx} w = \partial_\tau u + \frac{\sigma^2}{2}\partial_x u - \frac{\sigma^2}{2}\partial_{xx} u = 0,$$

and $w(x,0) = (1-\mathrm{e}^x)^+$. Finally note that changing $\tau \to \frac{\tau\sqrt{2}}{\sqrt{\sigma}}$ transforms the equation into the canonical heat equation. The Green's function of the one-dimensional heat equation is known,

$$\partial_\tau G - \partial_{xx} G = 0, \quad G(x,0) = \delta(x-x_0) \;\Rightarrow\; G(x,t) = \frac{1}{\sqrt{4\pi t}}\mathrm{e}^{-\frac{|x-x_0|^2}{4t}},$$

so $w = \int_{\mathbb{R}^+} G(x-x')(1-\mathrm{e}^{x'})^+\,\mathrm{d}x'$. Hence we have a closed-form solution for (3.3.4). This leads to the Black–Scholes formula for v. It can also be written in terms of a more C-friendly form with erf functions, as follows.

Black–Scholes formula for a vanilla put:

$$\begin{aligned} v_\sigma(y,\tau) &= \frac{y}{2}\left(1 + \mathrm{erf}\left(\frac{\ln y}{\sigma\sqrt{2\tau}} + \sigma\sqrt{\frac{\tau}{8}}\right)\right) \\ &\quad - \frac{1}{2}\left(1 + \mathrm{erf}\left(\frac{\ln y}{\sigma\sqrt{2\tau}} - \sigma\sqrt{\frac{\tau}{8}}\right)\right) + 1 - y \\ &\text{and} \quad \mathrm{erf}(y) = \frac{2}{\sqrt{\pi}}\int_0^y \mathrm{e}^{-x^2}\,\mathrm{d}x. \end{aligned} \tag{3.3.10}$$

3.3.2. Existence results for basket options

The above results extend without difficulty to basket option. Using the moneyness prices $y_i = \mathrm{e}^{r\tau} S_i/K$, (3.2.14) becomes

$$\begin{aligned} &\partial_\tau u - \Xi' : \nabla\nabla u = 0, \quad u(y,0) = \phi'(y), \\ &\text{with} \quad {\Xi'}_{ij} = \frac{y_i y_j}{2}\sigma_i\sigma_j\rho_{ij}, \quad i,j = 1,\ldots,d. \end{aligned} \tag{3.3.11}$$

The variational formulation is as follows: find

$$u \in V := \{v \in L^2(\mathbb{R}^{+d}) : y_i\partial_i u \in L^2(\mathbb{R}^{+d})\}$$

such that

$$\int_{\mathbb{R}^{+d}} \left[w\partial_\tau u + \sum_{ij}\partial_i(w{\Xi'}_{ij})\partial_j u\right] = 0 \quad \forall w \in V. \tag{3.3.12}$$

Theorem 3.3.4. If $\{\sigma_i\sigma_j\rho_{ij}\}$ is a positive definite matrix and if there exists $\sigma_m, \sigma_M, C_\sigma$ with $0 < \sigma_m \le \sigma_i \le \sigma_M$, $|y_i\sigma_i| < C_\sigma$, for all $i, y_i > 0$, then (3.3.12) has exactly one solution.

The proof is established exactly as in the one-dimensional case.

3.3.3. Existence result for an Asian option

Consider (3.2.16) and the following change of variable:

$$C(S,A,t) = Su(z,t)|_{z=\frac{K}{S}-\frac{t}{T}\frac{A}{S}}.$$

We have

$$\partial_A C = -\frac{t}{T}\partial_z u, \quad \partial_S C = u - \frac{1}{S}\left(K - \frac{At}{T}\right)\partial_z u, \quad \partial_t C = S\partial_t u - \frac{A}{T}\partial_z u,$$

$$\partial_{SS} C = \frac{1}{S^3}\left(K - A\frac{t}{T}\right)^2 \partial_{zz} u.$$

This leads to

$$\partial_t u + \frac{z^2\sigma^2}{2}\partial_{zz}u - \left(\frac{1}{T} + rz\right)\partial_z u = 0, \quad u(z,T) = z^+, \quad \forall z,t \in \mathbb{R}\times(0,T). \tag{3.3.13}$$

Appropriate behaviour for $z \gg 1$ is required: $u(z) \sim e^{rt}z$.

Note that the drift term gives a contribution $b(u,u)$ to the bilinear form of the variational formulation, which is compatible with Gårding's inequality:

$$b(u,u) = \int_{\mathbb{R}}\left(\frac{1}{T} + rz\right)\partial_z u^2 = -\int_{\mathbb{R}} ru^2 \geq -r\|u\|_0^2.$$

Existence and uniqueness hold for this problem too. However, there is a technical difficulty due to the fact that the second-order term vanishes at $z = 0$, which is in the middle of the domain of z. Existence can be established by using the concept of viscosity solutions of the PDE (see Barles (1997)).

3.3.4. Analysis of the PDE for stochastic volatilities

It is shown in Achdou and Pironneau (2005, 2009) that the PDE for a European put P_t on S driven by the stochastic volatility model (1.3.1), (1.3.2) is

$$\begin{aligned}
&\partial_\tau u - \frac{f(y)^2S^2}{2}\partial_{SS}u - \rho\beta S f(y)\partial_{Sy}u - \frac{\beta^2}{2}\partial_{yy}u - rS\partial_S u \\
&\qquad + \left(\beta[\rho\frac{\mu - r}{f(y)} + \gamma\sqrt{1-\rho^2}] - \alpha(m-y)\right)\partial_y u + ru = 0, \\
&u(S,y,0) = (K-S)^+.
\end{aligned} \tag{3.3.14}$$

As usual, $\tau = T - t$ and $u(S,y,\tau) = P|_{T-\tau,S_0=S,Y_0=y}$. The return on volatility at risk γ is a function of S, y, τ, which is part of the model and which comes from the fact that the PDE derived by Itô calculus and hedging from (1.3.1), (1.3.2) is not unique.

With $x = \ln y,\ v(x,y,\tau) = u(\mathrm{e}^x, y, \tau)$, the PDE becomes

$$\partial_\tau v - \frac{f(y)^2}{2}\partial_{xx}v - \rho\beta f(y)\partial_{xy}v - \frac{\beta^2}{2}\partial_{yy}v - (r - \frac{f(y)^2}{2})\partial_x v$$
$$+ \left(\beta[\rho\frac{\mu - r}{f(y)} + \gamma\sqrt{1-\rho^2}] - \alpha(m-y)\right)\partial_y v + rv = 0,$$
$$v(x,y,0) = (K - \mathrm{e}^x)^+. \tag{3.3.15}$$

The characteristic polynomial $\pi(X,Y)$ of the second-order terms satisfies

$$2\pi(X,Y) = f(y)^2X^2 + 2\rho\beta XY + \beta^2Y^2.$$

The discriminant $D = \beta^2(\rho^2 - f^2)$ must be negative to secure a strongly parabolic PDE (3.3.15):

$$|\rho| < |f|. \tag{3.3.16}$$

This condition is violated by the Cox–Ingersoll–Ross model because $f(y) = \sqrt{y}$; numerical difficulties are expected (change of type from elliptic to hyperbolic could cause discontinuous solutions and non-uniqueness). On the other hand, when the condition is satisfied, the problem can be studied in the variational space

$$V = \{w : w\sqrt{1+y^2},\ \partial_y w,\ S|y|\partial_S w \in L^2(\mathbb{R}^+ \times \mathbb{R})\}.$$

Let $\Omega = \mathbb{R}^+ \times \mathbb{R}$. Find $u \in L^2(0,T;V)$ such that, for all $w \in V$,

$$\int_\Omega (w\partial_\tau u + ruw) + \int_\Omega \left[\frac{f(y)^2}{2}\partial_S(wS^2)\partial_S u + \partial_y\left(w\frac{\beta^2}{2}\right)\partial_y u\right] \tag{3.3.17}$$
$$+ \frac{1}{2}\int_\Omega [\beta f(y)\partial_S(w\rho S)\partial_y u + S\partial_y(w\rho\beta f(y))\partial_S u]$$
$$+ \int_\Omega \left[-rS\partial_S u + \left(\beta[\rho\frac{\mu-r}{f(y)} + \gamma\sqrt{1-\rho^2}] - \alpha(m-y)\right)\partial_y u\right]w = 0.$$

3.3.5. Partial integro-differential variational equations

Consider a basket $S_{1t}, S_{2t}, \ldots, S_{dt}$ described by d SDEs with Lévy processes. Let $X_{it} = \ln S_{it}$ and let $g(X_t)$ be the pay-off of the option. By an argument close to that made for Theorem 3.2.5, we show that

$$\partial_\tau C + \Xi_{ij}\partial_{x_ix_j}C + r_i\partial_{x_i}C - rC$$
$$+ \int_\mathbb{R} [C(x+y,t) - C(x,t) - (\mathrm{e}^{y_i} - 1)\partial_{x_i}C(x,t)]\nu(\mathrm{d}y) = 0,$$
$$C(x,T) = g(x), \tag{3.3.18}$$

where r_i are the tendencies of the S_i (there may not be a risk-neutral d-dimensional measure that brings all r_i to the value r) and Ξ is computed from the volatilities σ_i and the correlation ρ_{ij} of the Brownian parts. For instance, if σ_1, σ_2 are the volatilities of S_1 and S_2 and ρ is the correlation of the Brownian motions, then

$$\Xi = \begin{pmatrix} \sigma_1^2 & \frac{2\rho\sigma_1\sigma_2}{1+\rho^2} \\ \frac{2\rho\sigma_1\sigma_2}{1+\rho^2} & \sigma_2^2 \end{pmatrix}. \tag{3.3.19}$$

In variational form, for $u(x,\tau) = C(T-\tau, \mathrm{e}^x)$, the PDE reads as follows: for all $v \in V$, the set of square-integrable functions with square-integrable derivatives in $\mathbb{R}^d$, we have

$$\int_{\mathbb{R}^d} (v\partial_\tau u + \partial_{x_i}(\Xi_{ij}v)\partial_{x_j}u - r_i v\partial_{x_i}u + rvu) = \tag{3.3.20}$$
$$\int_{x,y\in\mathbb{R}^d} \big[(u(x+y) - u(x))v(x) - (\mathrm{e}^{y_i} - 1)\partial_{x_i}u\ v(x)\big]\nu(\mathrm{d}y)\,\mathrm{d}x.$$

As mentioned before, this vectorial PIDE can be written as in (3.2.29), and when the integral is symmetric with respect to u and v, existence and uniqueness have been obtained by Farkas *et al.* (2006) (see also Tankov and Voltchkova (2009)).

PART FOUR
Numerical methods

4.1. A finite difference method for the Black–Scholes PDE

One of the oldest numerical methods to integrate a PDE is to replace the partial derivatives by finite differences on a grid: namely, given $x, t \to f(x,t)$ and a step size δx and a time step δt, f_i^m is the computed approximation of $f(j\delta x, m\delta t)$. For instance (3.3.3) for a vanilla put, written with the moneyness variable y and time to maturity τ, is approximated on the lattice $y_j = j\delta y$ by

$$\begin{aligned} \frac{P_j^{m+1} - P_j^m}{\delta\tau} - \frac{y_j^2 {\sigma_j^{m+\frac12}}^2}{2\delta y^2}\big(P_{j+1}^{m+\frac12} - 2P_j^{m+\frac12} + P_{j-1}^{m+\frac12}\big) &= 0, \quad j = 1, \ldots, J-1, \\ P_j^0 = (1-y_j)^+, \quad j &= 0, \ldots, J, \\ P_0^m = 1, \quad P_J^m = 0, \quad m &> 0, \end{aligned} \tag{4.1.1}$$

with the convention that $f^{m+\frac12} := \frac12(f^{m+1} + f^m)$. This is certainly the best among the easily implemented schemes because at each stage m it involves a tridiagonal system, which is very fast to solve by Gauss factorization.

When σ is not time-dependent, stability of this Crank–Nicolson scheme is easy to establish.

Proposition 4.1.1. When σ does not depend on t, scheme (4.1.1) is unconditionally stable and second-order in space and time.

Proof. The proof requires us to replace the non-homogeneous boundary conditions by zero at the cost of adding a non-zero right-hand side. So let $w(y) = a(y-b)(J\delta y - y)^2$ when $y > \frac{1}{2}$ and $w(y) = 1 - y$ otherwise, with a, b such that $w(\frac{1}{2}) = \frac{1}{2}$ and $w'(\frac{1}{2}) = -1$. The scheme for $u = P - w$ is

$$\frac{u_j^{m+1} - u_j^m}{\delta\tau} - \frac{y_j^2 {\sigma_j}^2}{2\delta y^2}(u_{j+1}^{m+\frac{1}{2}} - 2u_j^{m+\frac{1}{2}} + u_{j-1}^{m+\frac{1}{2}}) = y_j^2 \sigma_j f_j, \quad j = 1, \ldots, J-1,$$
$$u_j^0 = (1 - y_j)^+ - w(y_j), \quad j = 0, \ldots, J,$$
$$u_0^m = 0, \quad u_J^m = 0, \quad m > 0, \tag{4.1.2}$$

where $f_j = \frac{1}{2}(w_{j+1} - 2w_j + w_{j-1})/\delta y^2$ is the centred finite difference approximation of $\frac{1}{2}w''$. For what follows it is important to notice that $u_j^0 = 0$ when $j\delta y < \frac{1}{2}$.

Let (4.1.2) be multiplied by $u_j^{m+\frac{1}{2}}\delta\tau/(y_j^2\sigma_j^2)$ and summed in j:

$$\sum_{j=1}^{J-1} \frac{{u_j^{m+1}}^2 - {u_j^{m}}^2}{y_j^2\sigma_j^2}$$
$$= \frac{\delta\tau}{2\delta y^2}\sum_{j=1}^{J-1} u_j^{m+\frac{1}{2}}(u_{j+1}^{m+\frac{1}{2}} - 2u_j^{m+\frac{1}{2}} + u_{j-1}^{m+\frac{1}{2}}) + \sum_{j=1}^{J-1}\delta\tau f_j u_{j+\frac{1}{2}}^m$$
$$= -\frac{\delta\tau}{2\delta y^2}\left(\sum_{j=0}^{J-2}(u_{j+1}^{m+\frac{1}{2}} - u_j^{m+\frac{1}{2}})^2 + {u_{J-1}^{m+\frac{1}{2}}}^2 + (u_1^{m+\frac{1}{2}} - u_0^{m+\frac{1}{2}})u_0^{m+\frac{1}{2}}\right)$$
$$+ \delta\tau\sum_{j=1}^{J-1} f_j u_j^{m+\frac{1}{2}}$$
$$\le \delta\tau\sum_{j=1}^{J-1} f_j u_j^{m+\frac{1}{2}} \le \delta\tau\left(\sum_{j=1}^{J-1} y_j^2\sigma_j^2 {f_j}^2\right)^{\frac{1}{2}}\left(\sum_{j=1}^{J-1}\frac{{u_j^{m+\frac{1}{2}}}^2}{y_j^2\sigma_j^2}\right)^{\frac{1}{2}}. \tag{4.1.3}$$

This is an inequality of the type $A^{m+1} - A^m \le \delta\tau F\sqrt{A^{m+\frac{1}{2}}}$, which leads to the boundedness of A^M because

$$A^{m+1} - A^m = (\sqrt{A^{m+1}} + \sqrt{A^m})(\sqrt{A^{m+1}} - \sqrt{A^m})$$
$$\text{and} \quad \sqrt{A^{m+1}} + \sqrt{A^m} \ge \sqrt{2}\sqrt{A^{m+\frac{1}{2}}}.$$

So finally, after a summation in m,

$$\sum_{j=1}^{J-1} \frac{u_j^{M2}}{y_j^2\sigma_j^2} \leq \sum_{j>\frac{1}{2\delta y}}^{J-1} \frac{u_j^{02}}{y_j^2\sigma_j^2} + 2\left[\sum_{m=1}^{M-1} \delta\tau \left(\sum_{j=1}^{J-1} y_j^2\sigma_j^2 f_j^2\right)^{\frac{1}{2}}\right]^2.$$

The scheme is shown to be second-order accurate by a Taylor expansion in time and space of $u(j\delta y, m\delta\tau)$ in the usual way (Richtmyer and Morton 1967). □

Remark 4.1.2. In a different context, Rannacher (1984) showed that for problems with non-smooth initial conditions (which is the case here since the second derivative of u^0 is a Dirac mass), the Crank–Nicolson scheme is not second-order and may have oscillations near $\tau = 0$. It can be cured by starting with four smaller time steps of implicit Euler. More precisely, to integrate $\dot{u} + Au = 0$, one should use

$$\left(I + \frac{\delta\tau}{2}A\right)u^{m+\frac{1}{2}} = u^m, \qquad m = 0, \frac{1}{2}, 1, \frac{3}{2} \quad \text{(Euler)}, \tag{4.1.4}$$

$$\left(I + \frac{\delta\tau}{2}A\right)u^{m+1} = \left(I - \frac{\delta\tau}{2}A\right)u^m, \quad m = 2, 3, \ldots \quad \text{(Crank–Nicolson)}.$$

4.1.1. Alternative directions for basket options

Consider a basket option, with pay-off $\phi = (K - S_{1T} - S_{2T})^+$, on S_1, S_2. By a change to the moneyness variables $y = y_1, y_2$ and time to maturity, the PDE (3.2.14) is written in $\mathbb{R}^+ \times \mathbb{R}^+ \times (0, T)$ as

$$\partial_\tau u - \Xi : \nabla\nabla u = 0, \quad u(y, 0) = \phi(y). \tag{4.1.5}$$

Assuming $\Xi_{12} = \Xi_{21}$, Crank–Nicolson centred differences would lead to

$$\begin{aligned}
&\frac{u_{ij}^{m+1} - u_{ij}^m}{\delta\tau} - \frac{\Xi_{11ij}^{m+\frac{1}{2}}}{\delta y_1^2}\left(u_{i+1\,j}^{m+\frac{1}{2}} - 2u_{ij}^{m+\frac{1}{2}} + u_{i-1\,j}^{m+\frac{1}{2}}\right) \\
&\quad - \frac{\Xi_{12ij}^{m+\frac{1}{2}}}{2\delta y_1 \delta y_2}\left(u_{i+1\,j+1}^{m+\frac{1}{2}} - u_{i+1\,j-1}^{m+\frac{1}{2}} - u_{i-1\,j+1}^{m+\frac{1}{2}} + u_{i-1\,j-1}^{m+\frac{1}{2}}\right) \\
&\quad - \frac{\Xi_{22ij}^{m+\frac{1}{2}}}{\delta y_2^2}\left(u_{i\,j+1}^{m+\frac{1}{2}} - 2u_{ij}^{m+\frac{1}{2}} + u_{i\,j+1}^{m+\frac{1}{2}}\right) = 0.
\end{aligned} \tag{4.1.6}$$

The linear system generated by this scheme is pentadiagonal, and so an efficient numerical solution requires the use of a linear algebra library and/or writing a complex computer program. An alternative is to use a fractional step method.

Following Ikonen and Toivanen (2009), if the cross-derivative is approximated by

$$(\partial_{y_1y_2}u)_{ij} \approx \frac{1}{2\delta y_1\delta y_2}\left[2u_{ij} - u_{i-1\ j+1} - u_{i+1\ j-1} - \sum_{k=1,2}\delta y_k^2(\partial_{y_ky_k}u)_{ij}\right] \tag{4.1.7}$$

and centred finite differences for $\partial_{y_ky_k}u$, then, when $\Xi_{12} \leq 0$, the matrices of the linear systems are M-matrices, and the scheme is second-order and unconditionally stable. The scheme is not superior to (4.1.6) but it is appropriate to AD (alternate directions).

With an implicit Euler time scheme, (4.1.7) leads to

$$\frac{u^{m+1} - u^m}{\delta\tau} + A_1u^{m+1} + A_cu^{m+1} + A_2u^{m+1} = 0, \tag{4.1.8}$$

where A_i contains the approximation of $\partial_{y_iy_i}u$ and A_c the rest of the cross-derivative $\partial_{y_1y_2}u$.

AD is a fractional time step method (where $m + \frac{1}{3}$ is no longer the value at $(m + \frac{1}{3})\delta t$):

$$\begin{aligned}
\frac{u^{m+\frac{1}{3}} - u^m}{\delta\tau} + A_1u^{m+\frac{1}{3}} + A_cu^m + A_2u^m &= 0,\\
\frac{u^{m+\frac{2}{3}} - u^m}{\delta\tau} + A_1u^{m+\frac{1}{3}} + A_cu^{m+\frac{2}{3}} + A_2u^{m+\frac{1}{3}} &= 0,\\
\frac{u^{m+1} - u^m}{\delta\tau} + A_1u^{m+\frac{2}{3}} + A_cu^{m+\frac{2}{3}} + A_2u^{m+1} &= 0.
\end{aligned} \tag{4.1.9}$$

Each equation is, as it were, one-dimensional, so yields a tridiagonal linear system. The first one involves a horizontal stencil, the second one a diagonal stencil and the third a vertical stencil.

The same idea can used with Crank–Nicolson time discretization. Strang symmetrization can be added to obtain a fully second-order scheme.

4.1.2. Numerical method for PIDEs

The finite difference method applies equally well to PIDEs such as (3.2.21), but to avoid solving a full linear system it is wise to use a semi-implicit time scheme whereby the integro-differential term is discretized explicitly.

For instance, in the case of a finite difference scheme for (3.2.21), at each time step the right-hand side of the linear system is $P^{m+1}\delta t + v^{m+1}\delta t$, and

$$\begin{aligned}
v &= \int_{\mathbb{R}}(P(Se^y) - P(S) - S(e^y - 1)\partial_S P(S))\nu(dy)\\
&\approx \sum_{j=0}^{N}\nu\left(\log\frac{S_j}{S_i}\right)[P_j - P_i - (P_{i+1} - P_i)]\frac{S_j - S_i}{(j + \frac{1}{2})\delta x}.
\end{aligned} \tag{4.1.10}$$

Furthermore, the sums can be restricted to $|i-j|<M$, M large, when $\nu(y)$ decays exponentially with y.

Theorem 4.1.3. The scheme is stable if $\delta t \le C\frac{\delta x}{\lambda+\delta x}$.

For a proof the reader is sent to Cont and Voltchkova (2005).

4.2. Spectral methods

As shown by Jackson, Jaimungal and Surkov (2008), Fourier transform of the PIDE is very powerful for pricing options on assets modelled with a jump process when σ is constant. Consider (3.2.22) written with $x=\ln S$, $\tau = T-t$ and $v(x,\tau)=\mathrm{e}^{r\tau}C(\mathrm{e}^x, T-\tau)$:

$$\begin{aligned}&\partial_\tau v - r\partial_x v - \frac{\sigma^2}{2}\partial_{xx}v\\ &\quad - \int_{\mathbb{R}} \big[v(x+y,\tau)-v(x,\tau)-y\mathbf{1}_{|y|<1}\partial_x v(x,\tau)\big]\nu(\mathrm{d}y)=0,\\ &v(x,0)=v^0 := (\mathrm{e}^x-K)^+.\end{aligned} \tag{4.2.1}$$

Let

$$\mathcal{F}(u) = \int_{\mathbb{R}} \mathrm{e}^{-\mathrm{i}\omega x}u(x)\,\mathrm{d}x \quad \text{and} \quad \mathcal{F}^{-1}(\hat{u}) = \frac{1}{2\pi}\int_{\mathbb{R}} \mathrm{e}^{\mathrm{i}\omega x}\hat{u}(\omega)\,\mathrm{d}\omega.$$

Applying the operator $\mathcal{F}$ to (4.2.1) gives

$$\partial_\tau \hat{v} - \Psi\hat{v} = 0 \text{ in } \mathbb{R}, \quad \hat{v}(\omega,0)=\mathcal{F}(\mathrm{e}^x-K)^+,$$

where Ψ is

$$\Psi(\omega) = \mathrm{i}r\omega - \frac{\sigma^2}{2}\omega^2 + \int_{\mathbb{R}} (\mathrm{e}^{\mathrm{i}\omega y} - 1 - \mathrm{i}y\omega\mathbf{1}_{|y|<1})\nu(\mathrm{d}y).$$

So the solution is

$$v^T(x) = \mathcal{F}^{-1}[\{\mathcal{F}v^0\}(\omega)\mathrm{e}^{\Psi(\omega)T}].$$

It is numerically approximated by using discrete fast Fourier transforms, so the solution is computed with only two of those.

The problem must be localized into (x_m, x_M) and a grid is chosen with $\delta x = \frac{1}{N}(x_M - x_m)$, $x_k = x_m + k\delta x$. Similarly a mesh in the frequency domain is chosen, but to avoid aliasing errors it is best to use the Nyquist critical frequency so that $\omega_M(x_M - x_m) = \frac{N}{2}$, $\delta\omega = \frac{2}{N}\omega_M$, $\omega_j = j\delta\omega$. Then $\mathcal{F}v$ is approximated by

$$\hat{v}_j = \mathcal{F}v(\omega_j) \approx \sum_{k=0}^{N-1} \mathrm{e}^{-\mathrm{i}\omega_j x_k} v_k \delta x = \mathrm{e}^{\mathrm{i}\omega_j x_m}\mathtt{FFT}(v), \quad v_k = v(x_k), \tag{4.2.2}$$

where $\mathtt{FFT}(v)$ is a numerical fast Fourier transform algorithm. Finally

$$v^T = \mathtt{FFT}^{-1}(\mathrm{e}^{\Psi T}\mathtt{FFT}(v^0)).$$

4.3. Finite element methods

Finite difference methods have one limitation: they are not easy to adapt to arbitrary meshes. Yet for a given precision, the gain in computing time obtained by using a mesh adapted to the problem is tremendous. While it is not impossible to use arbitrary meshes with finite difference methods, the formulae are rapidly inextricable, so it is better to use a finite element method.

Let us consider a system of equations such as (3.3.14) for stochastic volatility models, which we write symbolically as

$$\begin{aligned}
&\partial_\tau u - \Xi : \nabla\nabla u + b\cdot\nabla u + ru = 0,\\
&u(x,\tau) = u_\Gamma \quad \forall x\in\Gamma_1, \quad n^T\Xi\nabla u(x,t) = 0 \quad \forall x\in\partial\Omega\backslash\Gamma_1,\\
&u(x,0) = \phi(x) \quad \forall x\in\Omega, \quad \forall\tau\in(0,T).
\end{aligned} \tag{4.3.1}$$

For (3.3.14),

$$\Xi_{11} = f(x_2)x_1^2, \quad \Xi_{12} = \Xi_{21} = \frac{1}{2}\rho\beta x_1 f(x_2), \quad \Xi_{22} = \beta^2/2,$$

$$b_1 = rx_1, \quad b_2 = \beta\left[\rho\frac{\mu - r}{f(x_2)} + \gamma\sqrt{1-\rho^2}\right] - \alpha(m - x_2).$$

The domain Ω is a d-dimensional Cartesian product of $\mathbb{R}$ or $\mathbb{R}^+$. In the case of (3.3.14), $d = 2$ and $\Omega = \mathbb{R}^+\times\mathbb{R}$.

Note that this formalism contains a wide variety of PDEs in finance. For example, a plain vanilla European put with local volatility corresponds to

$$d = 1, \quad \Omega = \mathbb{R}^+, \quad \Xi = \sigma^2(x,\tau)x^2/2, \quad b = -rx, \quad \phi(x) = (K - x)^+.$$

The variational formulation of (4.3.1) consists in seeking $u - \tilde{u}_\Gamma \in V_0$ such that

$$\int_\Omega \left[w\partial_\tau u + \nabla\cdot(w\Xi\nabla u) + wb\cdot\nabla u + rwu\right]\mathrm{d}x = 0 \quad \forall w\in V_0,$$
$$\text{where } u|_{\tau=0} = \phi. \tag{4.3.2}$$

As before, V is a space of square-integrable functions with weighted square-integrable derivatives. In the case of (3.3.14),

$$V = \left\{v : v\sqrt{1 + f(y)^2}, \frac{\partial v}{\partial y}, x|f(y)|\frac{\partial v}{\partial x} \in L^2(\Omega\times(0,T)))\right\}, \tag{4.3.3}$$

where V_0 is the subspace of V of functions which are zero on Γ_1. Finally $\tilde{u}_\Gamma$ is an extension of u_Γ in V. We write (4.3.2) formally as follows: find

$u - \tilde{u}_\Gamma \in V_0$ such that $u|_{\tau=0} = \phi$ and

$$\frac{\mathrm{d}}{\mathrm{d}\tau}(w,u) + a_\tau(w,u) = 0 \quad \forall w \in V_0. \tag{4.3.4}$$

It goes without saying that

$$(w,u) = \int_\Omega wu\,\mathrm{d}x, \quad a_\tau(w,u) = \int_\Omega [\nabla\cdot(w\Xi\nabla u) + wb\cdot\nabla u + rwu]\,\mathrm{d}x.$$

4.3.1. Galerkin method

The Galerkin method obtains a discrete version of (4.3.4) by replacing V with a finite-dimensional space $V_h \subset V$. The following is needed for V_h:

- given $\epsilon > 0$ and $v \in V$, there exists $v_h \in V_h$ such that $\|v_h - v\| \le \epsilon$,
- accumulation points of sequences $v_h \in V_h,\ h \to 0$ belong to V.

So consider the problem of finding $u_h \in V_h$ with $u_h - u_{\Gamma h} \in V_{0h}$ and

$$\frac{\mathrm{d}}{\mathrm{d}\tau}(w_h,u_h) + a_\tau(w_h,u_h) = 0. \tag{4.3.5}$$

If $\{w^i\}_1^N$ is a basis of V_h, then (4.3.5) is equivalent to

$$u_h(x,\tau) = u_{\Gamma h}(x) + \sum_1^N u_j(\tau)w^j(x),$$

$$\frac{\mathrm{d}}{\mathrm{d}\tau}\left(w^i, \sum_1^N u_j w^j\right) + a_\tau\left(w^i, \sum_1^N u_j w^j\right) = 0, \quad i = 1,\dots,N, \tag{4.3.6}$$

which is a system of differential equations

$$M\dot{U} + A(\tau)U = 0, \tag{4.3.7}$$

with $M_{ij} = (w^j, w^i)$, $A_{ij}(\tau) = a_\tau(w^j, w^i)$. A discrete time-stepping scheme still has to be applied to (4.3.7), for instance the Crank–Nicolson scheme

$$M\frac{U^{m+1} - U^m}{\delta t_m} + A^{m+\frac{1}{2}}U^{m+\frac{1}{2}} = 0, \tag{4.3.8}$$

where $A^m = A(t_m)$ and $A^{m+\frac{1}{2}} = A(\frac{1}{2}(\tau_m + \tau_{m+1}))$.

4.3.2. Finite elements in 1D

Consider a plain vanilla put with local volatility and 'localize' the problem by approximating $\mathbb{R}^+$ by $\Omega := (0,L)$, L large. Then take

$$V_0 = \{v \in L^2(0,1) : x\partial_x v \in L^2(0,1)\}, \quad V_0 = \{v \in V : v(L) = 0\}.$$

The easiest way is to choose a mesh $\cup_1^I[x_{i-1},x_i]=[0,L]$ and take V_h to be the space of piecewise linear functions,

$$V_{0h}=\{v_h \text{ continuous, linear on each subinterval } [x_{i-1},x_i],\ v_h(L)=0\}.$$

The canonical basis for V_{0h} is the set of $\{w^i\}_0^{I-1}$, uniquely defined by

$$w^i(x_j)=\delta_{ij},\quad w^i\in V_{0h},\quad i,j=0,\dots,I-1.$$

With this basis, called the nodal basis (or the hat functions), the integrals M_{ij} and A_{ij} can be computed exactly, or up to quadrature error due to σ if not piecewise polynomial on the intervals $[x_{i-1},x_i]$ (see Achdou and Pironneau (2005)).

It is easy to show that the matrices A and M are tridiagonal, so a resolution of the linear system (4.3.8) at each time step is best done with an LU factorization. In the end the computational complexity is the same as for a finite difference method. In fact both methods are identical when the mesh is uniform.

Convergence in the V-norm can be proved to be of the order $h^2+\delta t^2$ in the L^2-norm and of order $h+\delta t^2$ in the V-norm

Most interesting is the following *a posteriori* estimate.

Proposition 4.3.1. (Achdou and Pironneau 2005) We have the following *a posteriori* estimate:

$$[[u-u_{h,\delta t}]](t_n)\le c(u_0)\delta t+c\frac{\mu}{\sigma_m^2}$$

$$\times\left(\sum_{m=1}^{n}{\eta_m}^2+(1+\rho)^2\max(2,1+\rho)\sum_{m=1}^{n}\frac{\delta t_m}{\sigma_m^2}\prod_{i=1}^{m-1}(1-2\lambda\delta t_i)\sum_{\omega\in\mathcal{T}_{mh}}{\eta_{m,\omega}}^2\right)^{\frac12}$$

where μ is the continuity constant of a_t, $\rho=\max_{2\le n\le N}\frac{\delta t_n}{\delta t_{n-1}}$, $\mathcal{T}_{mh}$ is the partition of $(0,L)$ in small intervals ω at time t_m, and

$$\eta_m^2=\delta t_m e^{-2\lambda t_{m-1}}\frac{\sigma_m^2}{2}|u_h^m-u_h^{m-1}|_V^2,$$

$$\eta_{m,\omega}=\frac{h_\omega}{x_{\max}(\omega)}\left\|\frac{u_h^m-u_h^{m-1}}{\delta t_m}-rx\frac{\partial u_h^m}{\partial x}+ru_h^m\right\|_{L^2(\omega)},$$

and h_ω is the size of ω.

Note that everything is known after the computation of u_h. The mesh can then be adapted by using the error indicator η_m to adjust δt and $\eta_{m,\omega}$ to adjust h. Figure 4.3.1 shows the computed *a posteriori* error and the actual error for a vanilla put with constant volatility and comparison with the Black–Scholes analytic formula.

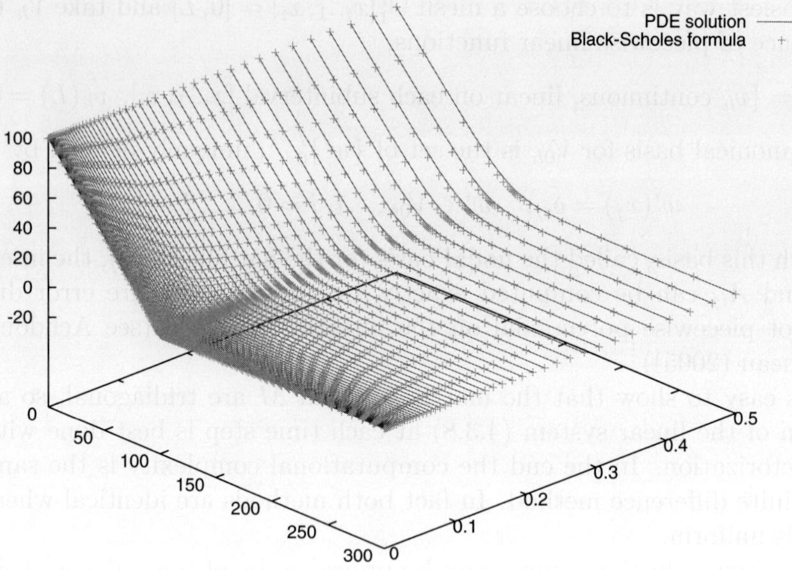

(a)

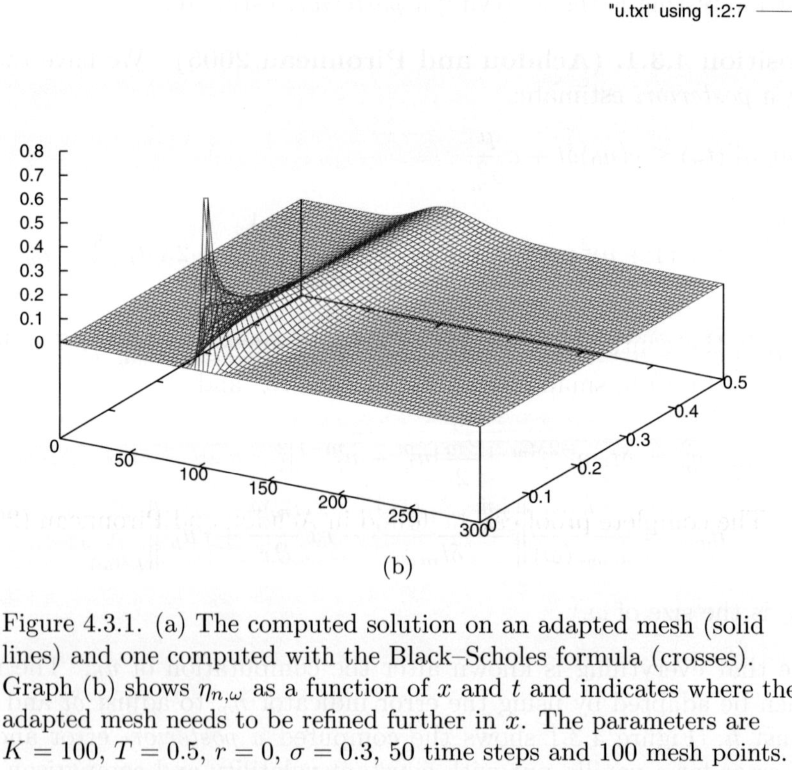

(b)

Figure 4.3.1. (a) The computed solution on an adapted mesh (solid lines) and one computed with the Black–Scholes formula (crosses). Graph (b) shows $\eta_{n,\omega}$ as a function of x and t and indicates where the adapted mesh needs to be refined further in x. The parameters are $K = 100$, $T = 0.5$, $r = 0$, $\sigma = 0.3$, 50 time steps and 100 mesh points.

4.4. Variational methods for American options

Referring to (3.2.31), an American put on S can be valued by solving

$$\partial_t P + \frac{\sigma^2 S^2}{2}\partial_{SS}P + rS\partial_S P - rP \le 0,$$
$$P - (K-S)^+ \ge 0 \quad \forall S \in \mathbb{R}^+, \quad t \in (0,T), \tag{4.4.1}$$

with, at all (S,t), at least one of the two inequalities satisfied as an equation. Let $(u,v) := \int_0^\infty uv\,\mathrm{d}S$ and $a,b : V \times V \to R$ be the two bilinear forms

$$a(u,v) := \int_{\mathbb{R}^+}\left(\frac{\sigma^2 S^2}{2}\partial_S u \partial_S v + rvu\right),$$
$$b(u,v) := \int_{\mathbb{R}^+}\left(\partial_S\left(\frac{\sigma^2 S^2}{2}\right) - rS\right)v\partial_S u, \tag{4.4.2}$$

where V is the Sobolev space

$$V = \{v \in L^2(\mathbb{R}^+) : S\partial_S v \in L^2(\mathbb{R}^+)\},$$

which has already been used for European options. The problem becomes: find $u \in L^2(0,T;V)$ with $u(S,T) = \phi(S) := (K-S)^+$ and a.e. in $(0,T)$, $u \ge \phi$ and

$$(\partial_t u, v-\phi) - a(u, v-\phi) - b(u, v-\phi) \le 0,$$
$$\forall v \in V^\phi = \{v \in V : v \ge \phi\}. \tag{4.4.3}$$

Let us call

$$V^+ := \{v \in V : v(S) \ge 0, \text{ a.e. in } \mathbb{R}^+\}.$$

Let $w = v - \phi$, $\tau = T - t$, to prove the following.

Proposition 4.4.1. The variational inequality is as follows: find $u \in L^2(0,T;V)$ with $u(S,T) = \phi(S) := (K-S)^+$, and a.e. in $(0,T)$

$$u - \phi \in V^+, \quad \partial_\tau(u,w) + a(u,w) + b(u,w) \ge 0 \quad \forall w \in V^+ \tag{4.4.4}$$

has exactly one solution. Further, for regular σ, u is a solution of (4.4.1).

Proof. The complete proof can be found in Achdou and Pironneau (2005). Let us give only a simple argument. Let (4.4.3) be discretized in time by

$$\left(\frac{u^{m+1}-u^m}{\delta t}, v-\phi\right) - a(u^m, v-\phi) - b(u^{m+1}, v-\phi) \le 0 \quad \forall v \in V^\phi \tag{4.4.5}$$

with $M\delta t = T$. As u^M is known, (4.4.5) is solved backwards in time. With

$$A(u,v) = \left(\frac{u}{\delta t}, v\right) + a(u,v) \quad \text{and} \quad L(v) := \left(\frac{u^{m+1}}{\delta t}, v\right) - b(u^{m+1}, v),$$

the problem becomes: find $u^m \in V^\phi$ such that

$$A(u^m, v-\phi) \geq L(v-\phi) \quad \forall v \in V^\phi. \tag{4.4.6}$$

Let us show that it is equivalent to u^m being the solution of

$$\min_{u \in V^\phi} \frac{1}{2} A(u,u) - L(u). \tag{4.4.7}$$

The minimum exists and is unique because the functional is strictly convex and V^ϕ is a closed cone in V. By definition, if u^m is the solution,

$$\frac{1}{2}A(u^m+w, u^m+w) - L(u^m+w) \geq \frac{1}{2}A(u^m, u^m) - L(u^m) \; \forall w : \; u^m + w \in V^\phi.$$

Take $w = \lambda v$, $\lambda > 0$ and let $\lambda \to 0$. At the limit

$$A(u^m, v) \geq L(v) \quad \forall v \in V, \quad \text{such that } \lim_{\lambda \to 0} u^m + \lambda v \geq \phi.$$

If u^m is smooth, by integration by parts, this is also

$$\int_\Omega \left(\frac{u^m - u^{m+1}}{\delta t} + r u^m - \partial_S \left(\frac{\sigma^2 S^2}{2} \partial_S u^m \right) - \left(rS - \partial_S \frac{\sigma^2 S^2}{2} \right) \partial_S u^{m+1} \right) v \geq 0,$$

which implies that the integrand is positive everywhere. Consider an S where $u^m(S) > \phi(S)$. Then, for any $v \in V$, there exists an interval (S_1, S_2) containing S where $\lim_{\lambda \to 0} u^m + \lambda v \geq \phi$ holds if the support of v is in (S_1, S_2). This in turn implies that at this S, t either $u^m(S) = \phi(S)$ or

$$\frac{u^m - u^{m+1}}{\delta t} + r u^m - \partial_S \left(\frac{\sigma^2 S^2}{2} \partial_S u^m \right) - \left(rS - \partial_S \frac{\sigma^2 S^2}{2} \right) \partial_S u^{m+1} = 0.$$

To complete the proof one must show that the finite difference scheme converges when $\delta t \to 0$. Note that there will be a stability condition because the first-order terms have been discretized explicitly in time (*i.e.*, at t^{m+1} instead of t^m or $t^{m+\frac{1}{2}}$). □

4.4.1. Discretization by FEM

Variational methods lead naturally to Galerkin and finite element methods. First the problem is localized: $\mathbb{R}^+ \approx (0, L)$ with L large, typically 3 times K. Then V is approximated by the piecewise linear continuous finite element space

$$V_{0h} = \{ v_h \in C^0(0, L) : v_h|_{T_k} \in P^1, \quad v_h(L) = 0 \}, \tag{4.4.8}$$

and V^ϕ by

$$V_{0h}^\phi = \{ v_h \in V_{0h} : v_h \geq \phi_h \}, \tag{4.4.9}$$

where ϕ_h is the continuous piecewise linear interpolate of ϕ on the triangulation, uniquely defined by

$$\phi_h \in V_{0h}, \quad \phi_h(q^j) = \phi(q^j), \quad \text{for all vertices } q^j \text{ of the triangulation.}$$

The discretized problem is as follows: find $\{u_i^m\}_{i=1}^I$ such that

$$\left(\frac{u_h^{m+1}-u_h^m}{\delta t}, v_h-\phi_h\right) - a(u_h^m, v_h-\phi_h) - b(u_h^m, v_h-\phi_h) \le 0 \quad \forall v_h \in V_{0h}^{\phi},$$

$$\text{with } u_h^m(x) = \sum_{i=1}^I u_i^m w^i(x). \tag{4.4.10}$$

Note that $w^i(x) \ge 0\ \forall x$, so

$$v_h \ge \phi_h \Leftrightarrow v_j \ge \phi_j \quad \text{for all vertices } q^j.$$

Thus, and because $w^j(x) \ge 0$ for all j, (4.4.10) is also

$$\left(\frac{u_h^{m+1}-u_h^m}{\delta t} w^j\right) - a(u_h^m, w^j) - b(u_h^m, w^j) \le 0, \quad j = 1, \ldots, I,$$

which we write as

$$\tilde{a}(u_h, w^j) \ge (f, w^j), \quad j = 1, \ldots, I, \tag{4.4.11}$$

$$u_h(x) = \sum_1^I u_i w^i(x) \ge \phi_h(x),$$

$$\text{where } \tilde{a}(u,v) = \frac{1}{\delta t}(u,v) + a(,v) + b(u,v), \quad f = \frac{u_h^{m+1}}{\delta t}. \tag{4.4.12}$$

Ito and Kunisch (2006) suggested reformulating (4.4.11) as follows.

Proposition 4.4.2. For any given constant $c > 0$, problem (4.4.11) is equivalent to

$$\begin{aligned} &\tilde{a}(u_h, w_h) + (\lambda, w_h) = (f, w_h), \quad \forall w_h \in V_{0h}, \\ &\lambda(x) - \min\{0, \lambda(x) + c(u_h(x) - \phi_h(x))\} = 0, \quad \forall x. \end{aligned} \tag{4.4.13}$$

Proof. The last equality is equivalent to

$$\begin{aligned} &\lambda \le 0, \quad \lambda \le \lambda + c(u_h - \phi_h), \\ i.e., \quad &u_h \ge \phi_h, \quad \lambda \le 0, \quad \lambda(x)(u_h(x) - \phi(x)) = 0, \quad \forall x. \end{aligned}$$

The last statement means that for each x one of the two inequalities at least one is an equality. The rest of the proof is a classical argument using Lagrange multipliers:

$$\begin{aligned} &\tilde{a}(u_h, w_h) + (\lambda, w_h) = (f, w_h), \quad \lambda(x) \le 0, \quad w_h(x) \ge 0 \\ &\text{implies } \tilde{a}(u_h, w_h) \ge (f, w_h), \quad \forall w_h \ge 0. \end{aligned} \tag{4.4.14}$$

Recall that an equation such as $F(x) = 0$ (with $F : \mathbb{R}^n \to \mathbb{R}^n$) can be solved by Newton's method:

$$x_{k+1} = x_k - G(x_k)^{-1} F(x_k),$$

with $G = F'$ the Jacobian of F. Hintermüller, Kovtunenko and Kunisch (2004) observed that Newton's algorithm converges even if F is not differentiable, provided that there exists G such that

$$\text{for all } x, \quad \lim_{\|h\|\to 0} \|F(x+h) - F(x) - G(x+h)h\| = 0,$$

a property which is satisfied by $F(x) = \max\{0, x\}$, for instance with $G(x) = \max\{0, x\}/x$.

4.4.2. Semi-smooth Newton algorithm

Let $A_{ij} := \tilde{a}(w^i, w^j)$. Newton's algorithm applied to (4.4.13) gives the following.

(1) Choose $c > 0$, u_0, $\lambda_0 \le 0$, set $k = 0$.

(2) Determine

$$A_k := \{x : \lambda_k(x) + c(u_k(x) - \phi(x)) < 0\}.$$

(3) Set

$$u_{k+1} = \arg \min_{u\in H^1(\mathbb{R}^+)} \left\{\frac{1}{2}a(u,u) - (f,u) : u = \phi \text{ on } A_k\right\}.$$

(4) Set

$$\lambda_{k+1} = f - Au_{k+1}.$$

Set $k = k + 1$ and go to (2).

The beauty of the method is that its implementation is almost painless from an implicit finite element or finite difference solver with the following. To compute the solution v_h of

$$\bar{a}(v_h, w^j) = (\bar{f}, w^j) \quad \forall j \notin K, \quad v_h(q^j) = \phi(q^j), \quad j \in K, \quad j = 1, \dots, I,$$

- Compute $A_{ij} = \tilde{a}(w^i i, w^j)$ and $f_j = (f, w^j)$, for all i, j.
- Reset $A_{ii} = 10^{30}$, $f_i = 10^{30}\phi(q^j)$, for all $i \in K$.
- Solve $Av = f$.

Although it looks as if the modification of A would ruin its condition number, in reality it does not, because it is essentially equivalent to the elimination of the rows and columns of indices in K. It would be equivalent to row and column elimination if, instead of 10^{30}, the largest integer number that the computer can store is used so that the addition of $\sum_{j\neq i} A_{ij}v_j$ to $A_{ii}v_i$ does not affect the significant digits.

Theorem 4.4.3. Algorithm 4.4.2 converges locally super-linearly provided A is an M-matrix.

The proof can be found in Hintermüller *et al.* (2004). When $b = 0$, A is an M-matrix if mass-lumping (a reduced integration formula with Gauss points at the vertices) is applied to the part of A which comes from $(w^i, w^j)/\delta t$, and provided the triangulation has no angle greater than $\pi/2$. If this is not the case, the method still converges provided u_0, λ_0 is not too far from the solution. Hintermüller *et al.* (2004) also showed that the semi-smooth Newton method converges if applied to the continuous problem, before discretization, but the precision is proportional to c^{-1}, so c must be large.

4.5. A stochastic volatility model solved by FEM

We now return to (4.3.1) for the stochastic volatility model (3.3.14). When $f(0) = 0$, the problem is complicated by the fact that the PDE degenerates at $y = 0$. In the Stein–Stein case $f(y) = |y|$, at least, it shown in Achdou and Tchou (2002) and Achdou and Pironneau (2009) that the solution exists and is unique if the data are smooth, and if

$$r(t) \geq r_0 > 0, \quad \alpha^2 > 2\beta^2, \quad 2\alpha^2\eta(1-\eta) > \beta^2.$$

4.5.1. Finite elements for a 2D stochastic volatility model

As above we localize the problem in $\Omega := (0, L_x) \times (-L_y, L_y)$. No boundary condition is needed on the axis $x = 0$ since the PDE is degenerate there. Notice also that u is expected to tend to zero as x tends to infinity. For large y no arguments seem to give a boundary condition. However, if L_y is chosen such that

$$\alpha(L_y \pm m) \gg \beta\left[\rho\frac{\mu - r}{f(L_y)} + \gamma\sqrt{1-\rho^2}\right],$$

then for $y \sim \pm L_y$, the coefficient of the advection term in the y direction, and as the advective effects will dominate the diffusive ones, even if a wrong boundary condition is imposed, it will not propagate inside the domain. Hence we may apply a Neumann condition.

Now Ω is triangulated into a set of non-intersecting triangles $\{T_k\}_1^K$ with vertices q^j satisfying

$$\begin{gathered}\Omega_h := \cup_{k=1}^K T_k \approx \overline{\Omega}, \quad T_k \cap T_l = \emptyset \quad \text{or a vertex, or an entire edge;} \\ \text{vertices on } \Gamma_h = \partial\Omega_h \text{ are on } \Gamma\text{; corners of } \Gamma \text{ are vertices of } \Gamma_h. \end{gathered} \tag{4.5.1}$$

V_0 is approximated by V_{0h}, the space of continuous piecewise linear functions on the triangulation which vanish at $x = L_x$:

$$V_{0h} := \{v_h \in C^0(\Omega_h) : v_h|_{T_k} \in P^1(T_k)\ k = 1, \ldots, K,\ v_h(L_x, y) = 0\}.$$

The basis $\{w^i\}_1^I$ is uniquely defined by

$$w^i \in V_{0h}, \quad w^i(q^j) = \delta_{ij} \quad \forall j$$

Table 4.5.1. Comparison of CPU time for the LU algorithm and SuperLU for a product put option on a uniform mesh and 200 time steps (computed by N. Lantos).

Mesh size	Gauss-LU (s)	Relative error	SuperLU (s)	Relative error
101×101	10.094	3.532 %	2.39	3.076 %
126×126	14.547	1.338 %	4.016	1.797 %
151×151	22.313	0.751 %	6.203	0.489 %
176×176	31.985	1.131 %	8.735	0.790 %
201×201	43.938	0.432 %	12.109	0.670 %

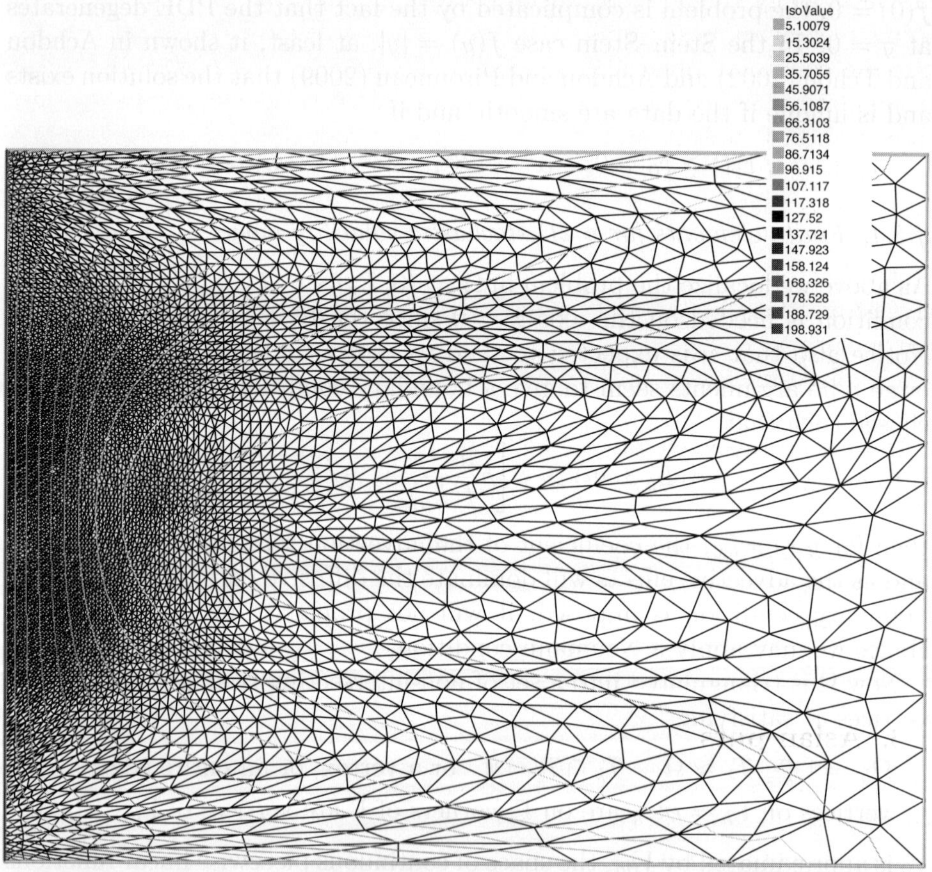

Figure 4.5.1. Solution of (4.3.1) for a put with maturity $T = 1$ and strike $K = 100$, with $r = 0.02$, $\beta = \alpha = 1$ and $m = 0.2$. We have used the automatic mesh adaptivity of FreeFem++; the contours and the triangulation are shown.

for all index i such that the corresponding vertex q^i is not on the boundary $x = L_x$, *i.e.*, $q_1^i \neq L_x$.

As in the one-dimensional case, the variational formulation (4.3.4) is discretized by the Galerkin procedure (4.3.5), leading to (4.3.7) when

$$u_h^m(x,y) = \sum_1^I u_i^m w^i(x,y).$$

4.5.2. Numerical implementation

Integrals of polynomial expressions of x, y can be evaluated exactly on triangles (formula (4.19) in Achdou and Pironneau (2005)). The difficulty is in the strategy to solve the linear systems arising from (4.3.7). Multigrid seems to give the best results in term of speed (Ikonen and Toivanen 2008), but the computer program may not be so easy to write for an arbitrary triangulation. Since the number of vertices are not likely to exceed a few thousand, fast multifrontal libraries such as SuperLU (Xiaoye Li, Demmel and Gilbert 1999) can be used with minimal programming effort. Table 4.5.1 shows the gain in time over more classical algorithms.

Alternatively one could also use a preconditioned conjugate gradient iterative method. Another way is to use FreeFem++ (Hecht, Pironneau, Le Hyaric and Otsuka 2008), which is a general two-dimensional PDE solver particularly well suited to these option pricing models, where the coefficients are involved functions and one may want to try to change them often. Section 6.3, at the end of this article, is devoted to FreeFem++.

Results for the Stein–Stein model obtained with FreeFem++ are displayed in Figure 4.5.1.

Finite element solution of 3D problems

A basket option on three correlated assets has been computed by the same finite element method in 3D. The method and results are documented in Section 6.3.2.

4.6. Asian puts

Consider an Asian put with fixed strike whose pay-off is $P_0(S,y) = (K-y)^+$, where y is the average value of the asset in time, $y = \frac{1}{\tau}\int_0^\tau S(\tau)\,\mathrm{d}\tau$. The price of the option is found by solving, for all $\{x, y, \tau\} \in \mathbb{R}^+ \times \mathbb{R}^+ \times (0, T]$,

$$\begin{aligned}
&\partial_\tau u - \partial_x\left(\frac{\sigma^2 x^2}{2}\partial_x u\right) - (\sigma^2 - r)x\partial_x u + \frac{y-x}{T-\tau}\partial_y u + ru = 0,\\
&u(x,y,0) = (K-y)^+,\ \partial_x u|_{+\infty,y,\tau} \approx \begin{cases} \frac{\tau}{T}\mathrm{e}^{-r\tau} & \text{if } y < \frac{KT}{T-\tau},\\ \frac{1-r\mathrm{e}^{-r\tau}}{Tr} & \text{otherwise.}\end{cases}
\end{aligned} \tag{4.6.1}$$

The problem is hard because the second-order part of the differential operator in y is missing, and because the convective velocity $v = \left((\sigma^2 - r)x, \frac{y-x}{T-t}\right)^T$ tends to infinity as $t \to T$.

Convective terms are not easy to discretize with finite element methods. Ideally, an equation such as

$$\partial_\tau u + v_1 \partial_x u + v_2 \partial_y u = 0, \quad u(x,y,0) = u^0 \tag{4.6.2}$$

on an unstructured mesh ought to be discretized by a finite volume method rather than finite elements. There are ways to combine finite volume and finite element methods (Dervieux 1996), and more recently discontinuous Galerkin methods have also been shown to be well adapted, and to any order of accuracy, but numerically expensive. With triangular elements and polynomials of degree one, we can use streamline Galerkin least-squares or characteristic Galerkin for unwinding. Our preference is for the second scheme because there is no *ad hoc* parameter to tune.

The characteristic Galerkin finite element method is as follows:

$$(\partial_\tau u + v \cdot \nabla u)|_{x,y,\tau^{m+1}} \approx \frac{1}{\delta\tau}(u^{m+1}(q) - u^m(Q)),$$

$$\text{with} \quad Q = q - \delta\tau v\left(q - v^m(q)\frac{\delta\tau}{2}, \tau^{m+\frac{1}{2}}\right) \quad \text{and} \quad q = (x,y)^T. \tag{4.6.3}$$

Boundary conditions for (4.6.2) ought to be u, given at time zero and on the part of the boundary where $v \cdot n < 0$, n being the outer normal to the boundary.

In the case of (4.6.1) the convective velocity is

$$v_1 = -(\sigma^2 - r)x, \quad v_2 = \frac{y - x}{T - \tau}.$$

After localization the domain of the PDE is $(0, L_x) \times (0, L_y)$. Hence $v \cdot n$ is

$$\begin{aligned} v \cdot n|_{x=0} &= -\frac{y}{T-\tau} < 0, & v \cdot n|_{x=L_x} &= \frac{y - L_x}{T - \tau} < 0, & &\text{if } L_y < L_x, \\ v \cdot n|_{y=0} &= (\sigma^2 - r)x > 0, & v \cdot n|_{y=L_y} &= (\sigma^2 - r)x < 0, & &\text{if } \sigma^2 > r. \end{aligned} \tag{4.6.4}$$

This shows that boundary conditions are needed everywhere except at $y = 0$. One possibility is to use a discounted value of the initial condition at $x = 0$ and a condition compatible with the asymptotic behaviour expected for x or y large:

$$u(x,y,\tau) = \mathrm{e}^{-r\tau}K \text{ at } x = 0, \quad u(L_x, y, \tau) = \mathrm{e}^{-r\tau}(K - y)^+, \quad u(x, L_y, \tau) = 0.$$

The discontinuous P^2 reconstruction with characteristic convection of the gradient and of the value at the centre of gravity is applied. The domain is localized to $(0, L) \times (0, L)$, with $L = 250$, *i.e.*, greater than twice the

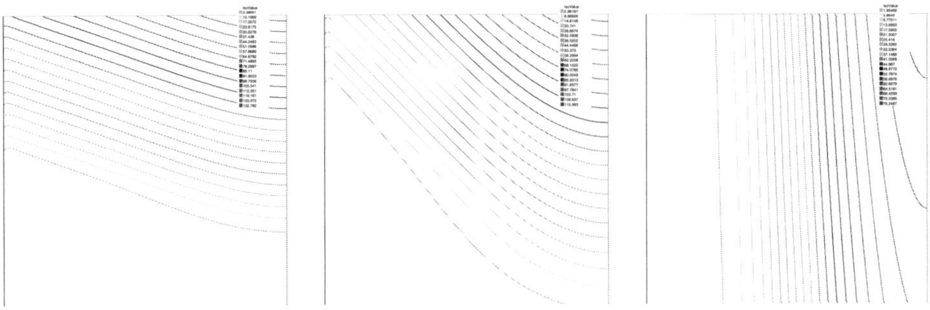

Figure 4.6.1. Solution of (4.6.1) for an Asian option of maturity $T = 4$ and strike $K = 100$ when the interest rate is 0.03 and the volatility 0.3. The three figures show the contours at times 0.96, 1.96 and 3.96.

contract price $K = 100$, in this example; the interest rate is $r = 3\,\%$ and the volatility $\sigma = 0.3$. The square domain is triangulated by a uniform 50×50 mesh and we have taken 50 time steps over the 4-year period of this example ($T = 4$). The results are shown in Figure 4.6.1 at times $t = 0.96$, 1.96 and 3.96.

4.7. Automatically adapted meshes

When an *a posteriori* error estimate is available, mesh adaptation based on this example is a sure way to reduce the error below a given threshold; when it is not available, local mesh adaptation by metric control is a powerful tool for building meshes well suited to the problem to be solved (George and Borouchaki 1998).

Delaunay criterion. A *Delaunay mesh* is one where all inner edges are such that the circle defined by one triangle having the edge as a side is such that the third vertex of the other triangle having the same edge as a side is outside the circle.

A quick way to make a Delaunay mesh is as follows.

4.7.1. Delaunay mesh generator

(1) Unmark all edges.

(2) Loop on the unmarked inner edges until all are marked.

- For each edge identify the two triangles which have this edge as a side. If the edge fails the Delaunay test above, then swap the edge with the other diagonal of the quadrilateral made by the two triangles, and update the triangulation by replacing the two old triangles by the two new ones.
- Mark the new edge and unmark the four edges of the quadrilateral.

It can be shown that the loop stops in finite time because the minimum angle in the triangulation is strictly increased at each loop, and that the resulting mesh is Delaunay if the domain is convex.

If the domain is not convex, a complicated recovery of the boundary edges must be performed, but still the mesh has similar good qualities in most cases (see George and Borouchaki (1998)). In fact it can be shown that the Delaunay mesh is the nearest to a quasi-equilateral mesh, in the sense that the smallest angle in the triangulation is maximized by the edge swaps so as to satisfy the Delaunay criterion.

4.7.2. Anisotropic Delaunay mesh generator

Note that if the local metric is Euclidean, the mesh elements are isotropic in the sense that there is no preferred direction in the algorithm.

To obtain an anisotropic mesh one must replace circles by ellipses in the Delaunay criterion.

Given a positive definite matrix $M(x)$, we define a variable metric by $\|x-y\|^2 = (x-y)^T M(x)(x-y)$. M-circles, *i.e.*, circles with respect to the variable metric, are ellipses in the Euclidean space.

A given triangulation is said to satisfy the M-Delaunay criterion if, for all inner edges, the quadrangle made by the two adjacent triangles is such that the fourth vertex is outside the M-circle passing through the other three vertices.

The adaptive Delaunay mesh generator of George and Borouchaki (1998) is based on six steps.

(1) Discretize the boundary of Ω, the domain for which we seek a triangulation.
(2) Build an M-Delaunay triangulation of the convex hull of all the boundary nodes (no internal points).
(3) Add an internal point to all edges which are longer (in the variable metric) than the prescribed length.
(4) Rebuild the triangulation with internal nodes, using the Delaunay criterion and the variable metric.
(5) Remove the triangles which are outside Ω.
(6) Go to (3) until the elements have the required quality.

So anisotropy is controlled by M, the local metric.

4.7.3. Metric adapted to the solution of the PDE

As stated earlier, if we want the mesh to be adapted to the solution, we need to define the metric at each point of the domain and use it in the Delaunay algorithm above.

The definition of the metric is based on the Hessian of the state variables of the problem. Indeed, for a P^1 Lagrange discretization of a variable u, the interpolation error is bounded by

$$\mathcal{E} := |u - \Pi_h u|_0 \leq ch^2 |D^2 u|_0, \tag{4.7.1}$$

where h is the element size, $\Pi_h u$ the P^1 interpolation of u and $D^2 u$ its Hessian matrix. This matrix is symmetric and therefore diagonalizable:

$$D^2 u = \begin{pmatrix} \partial^2 u/\partial x^2 & \partial^2 u/\partial x \partial y \\ \partial^2 u/\partial x \partial y & \partial^2 u/\partial y^2 \end{pmatrix} = \mathcal{R} \begin{pmatrix} \lambda_1 & 0 \\ 0 & \lambda_2 \end{pmatrix} \mathcal{R}^{-1}. \tag{4.7.2}$$

Using this information, we introduce the following metric tensor $\mathcal{M}$:

$$\mathcal{M} = \mathcal{R} \begin{pmatrix} \tilde{\lambda}_1 & 0 \\ 0 & \tilde{\lambda}_2 \end{pmatrix} \mathcal{R}^{-1}, \tag{4.7.3}$$

where

$$\tilde{\lambda}_i = \min\left(\max\left(|\lambda_i|, \frac{1}{h_{\max}^2} \right), \frac{1}{h_{\min}^2} \right), \tag{4.7.4}$$

with $h_{\min}$ and $h_{\max}$ being the minimal and maximal edge lengths allowed in the mesh.

Now, if we generate, by a Delaunay procedure, an equilateral mesh with edges of length 1 in the metric $\mathcal{M}/(c\mathcal{E})$, the interpolation error $\mathcal{E}$ is equidistributed over the edges of length a_i if

$$\frac{1}{c\mathcal{E}} a_i^T \mathcal{M} a_i = 1. \tag{4.7.5}$$

4.8. Greeks

Probably due to the fact that mathematicians have used Greek letters to name the partial derivatives of option prices with respect to the relevant parameters, these are called the *Greeks*. Let P be the price of a vanilla European put.

- The *delta*, δ, is its derivative with respect to the stock price S: $\partial_S P$.
- The Θ, or *time-decay*, is its derivative with respect to time: $\partial_t P$.
- The *vega* κ is its derivative with respect to the volatility $\partial_\sigma P$.
- The *rho*, ρ, is its derivative with respect to the interest rate, $\partial_r P$.
- The *eta*, η, is its derivative with respect to the strike $\partial_K P$.
- The *gamma*, γ, is the rate of change of its delta: $\gamma = \partial_{SS} P = \partial_S \delta$.

Computing Greeks from the SDE of the asset can be challenging, but the same is very easily done with the PDE of the option. For an illustration, take the vanilla put on S_t written in moneyness y and time to maturity τ:

$$\partial_\tau u - \frac{y^2\sigma^2}{2} \partial_{yy} u = 0, \quad u(y, 0) = (1 - y)^+.$$

A finite element or finite difference method numerical solution will give the δ with order-one precision. If more precision is needed, by differentiation of the PDE with respect to y, $v = \partial_y u$ satisfies

$$\partial_\tau v - \frac{y^2\sigma^2}{2}\partial_{yy}v = (y\sigma^2 + y^2\sigma\partial_y\sigma)\partial_{yy}u, \quad v(y,0) = H(y-1) - 1,$$

with $H(x)$ the Heaviside function, which is zero for $x < 0$ and one otherwise. Thus the delta is $\delta = Kv(\mathrm{e}^{-r(T-t)S/K}, T-t)$. Further,

$$\partial_\tau v - \frac{y^2\sigma^2}{2}\partial_{yy}v - (y\sigma^2 + y^2\sigma\partial_y\sigma)\partial_y v = 0, \quad v(y,0) = H(y-1) - 1.$$

The Θ can be computed by $\Theta = y^2\sigma^2\partial_y v/2$ or by differentiating the equation for v.

The vega satisfies

$$\partial_\tau\kappa - \frac{y^2\sigma^2}{2}\partial_{yy}\kappa = y^2\sigma\partial_{yy}u, \quad \kappa(y,0) = 0.$$

Alternatively, as explained in Achdou and Pironneau (2005, 2009), if the numerical solver for u is written in C++, all Greeks can be computed at no programming cost by automatic differentiation using operator overloading and template classes.

Greeks for American options

American options have been recast as variational inequalities such as (4.4.4):

$$u - \phi \in V^+ : \partial_\tau(u,w) + a(u,w) \geq (f,w) \quad \forall w \in V^+ = \{w \in V : w \geq 0\},$$

where $(\cdot,\cdot)$ is the scalar product of L^2, $a(\cdot,\cdot)$ is a continuous coercive bilinear form on the Hilbert space V and $f \in V$. Another form of the same is, for some operator A,

$$u - \phi \in V^+ : \partial_\tau u + Au \geq f \quad \forall w \in V^+ = \{w \in V : w \geq 0\}.$$

To compute the sensitivity of u with respect to a parameter α, it is best to return to the definition of the derivative and work out the equation/inequality for the increments. For instance,

$$u(\alpha) \geq \phi(\alpha), \quad u(\alpha+\delta\alpha) \geq \phi(\alpha+\delta\alpha) \Rightarrow u'_\alpha \geq \phi'_\alpha \quad \text{at } S,\tau,$$

where $u(\alpha) = \phi(\alpha)$. Similarly

$$\partial_\tau u'_\alpha + Au'_\alpha + \lambda' = f'_\alpha - A'_\alpha u,$$

with $\lambda' \leq 0$ at all S,τ where $\partial_\tau u + Au = f$ and $\lambda' = 0$ at all other points. But by (4.4.4) we know that, at the point where $\partial_\tau u + Au \neq f$, we have $u = \phi$, and consequently u'_α is a solution of

$$u'_\alpha \geq \phi'_\alpha, \quad \partial_\tau u'_\alpha + Au'_\alpha \geq f'_\alpha - A'_\alpha u,$$

or equivalently

$$u'_\alpha - \phi'_\alpha \in V^+ : \partial_\tau(u'_\alpha, w) + a(u'_\alpha, w) \geq (f'_\alpha, w) - a'_\alpha(u, w) \quad \forall w \in V^+.$$

The semi-smooth Newton algorithm can be used for u'_α exactly as for u.

4.9. Reduced-order modelling

Derivatives such as European calls and puts on one or many assets are computed a very large number of times every day. When a closed-form solution is not available, an alternative to Monte Carlo simulation is to solve the Black–Scholes partial differential equation.

A complement to the finite element or finite difference method for a numerical solution of the problem is to construct an appropriate basis, smaller in size but with larger support. Proper orthogonal decomposition (POD) is the usual tool for this.

Cont, Lantos and Pironneau (2011) showed that a set of rescaled calls with constant volatilities forms a reduced basis with a similar performance to POD.

If τ is the time to maturity, K is the strike, r the interest rate, and $y = \mathrm{e}^{r\tau}\frac{S}{K}$ the moneyness, a put P can be priced by solving

$$\partial_\tau v_\sigma - \frac{\sigma^2 y^2}{2}\partial_{yy} v_\sigma = 0, \quad v_\sigma(0) = (1-y)^+, \tag{4.9.1}$$

where $v_\sigma = P\mathrm{e}^{r\tau}$.

When σ is constant, the Black–Scholes formula provides the analytical solution

$$v_\sigma = \frac{y}{2}\left(1 + \operatorname{erf}\left(\frac{\ln y}{\sigma\sqrt{2\tau}} + \sigma\sqrt{\frac{\tau}{8}}\right)\right) - \frac{1}{2}\left(1 + \operatorname{erf}\left(\frac{\ln y}{\sigma\sqrt{2\tau}} - \sigma\sqrt{\frac{\tau}{8}}\right)\right) + 1 - y \tag{4.9.2}$$

where $\operatorname{erf}(y) = \frac{2}{\sqrt{\pi}}\int_0^y \mathrm{e}^{-x^2}\,\mathrm{d}x$. Note that only $\sigma\sqrt{\tau}$ appears, so *changing the volatility is equivalent to changing* τ. Hence a basis of snapshots in time for (4.9.1) can also be

$$v_\sigma(y,\tau) = v_\Sigma(y,\tau) + \sum_{i=1}^{I} a_i(\tau)\bar{w}_i(y) \quad \text{with } \bar{w}_i := v_{\sigma_i}(y,T) - v_\Sigma(y,T), \tag{4.9.3}$$

where $\{\sigma_i\}_{i\in I}$ is an appropriate set of constant volatilities and Σ is a chosen reference volatility. The PDE satisfied by $u := v_\sigma - v_\Sigma$ is

$$\partial_\tau u - \frac{\sigma^2 y^2}{2}\partial_{yy} u = -\left(\partial_\tau v_\Sigma - \frac{\sigma^2 y^2}{2}\partial_{yy} v_\Sigma\right), \quad u(0) = 0. \tag{4.9.4}$$

Applying Galerkin's method to (4.9.4) with (4.9.3) will give the $a_i(\tau)$.

However, it is numerically much easier to work with $\hat{w}_i := \partial_{yy} v_{\sigma_i}(y,T)$ instead of $\bar{w}_i = v_{\sigma_i}(y,T) - v_\Sigma(y,T)$, and since ∂_{yy} is a bijection with appropriate boundary conditions it should not make much of a difference!

As we shall see, a good choice is $\sigma_i = (2cT\,i)^{-1/2}$, $i = 1, 2, \ldots$, where c is a chosen constant. A quick computation shows that $\hat{w}_i$ is proportional to

$$w_i := \sqrt{y}\mathrm{e}^{-\alpha_i \ln^2 y}, \quad \text{with } \alpha_i = c\,i.$$

This will be the basis chosen here, and $c = 1$ in the numerical tests.

The POD algorithm. In this framework POD entails the following.

(1) Choose $I \ll 100$.

(2) Apply the Gram–Schmidt algorithm to orthogonalize the $\{w_i\}_{i\in I}$ with respect to the $L^2(\mathbb{R}^+)$ or $H^1(\mathbb{R}^+)$ scalar product $(\cdot,\cdot)$. Let $\{p_i\}_{i\in I}$ be the result.

(3) Solve (4.9.4) on this basis by the Galerkin method.

(4) If the result is not precise enough, add a new element in I and go back to step (2).

The 'not precise enough' of step (4) needs in principle an *a posteriori* estimate to become mathematically sound. The selection of a 'good new basis vector' is also difficult in step (4), and one way is to order the vectors p_i so as to have a decreasing sequence of eigenvalues of the correlation matrix (w_i, w_j) (see Prud'homme *et al.* (2002) for details).

4.9.1. The Galerkin approximation

Mathematical result

Decomposition (4.9.3) gives a v_σ which satisfies the initial and boundary conditions of the Black–Scholes equations (4.9.1); it also has the right exponential behaviour when $y \to \infty$. Consider $y \to u(y)/\sqrt{y}$ and, as before, $u = v_\sigma - v_\Sigma$.

Note that $\mathrm{e}^{-ic\ln^2 y} = f(y)^i$ with $f(y) = \mathrm{e}^{-c\ln^2 y}$. By the Stone–Weierstrass theorem, $\{f(y)^n\}_{n=0,1,\ldots}$ is a basis if $y \to f(y)$ separates points. However, $\mathrm{e}^{-c\ln^2 y}$ is a separating function only on the interval $(0,1)$ or on $(1,\infty)$ but not on $(0,\infty)$. Thus we need a symmetry assumption about 1, $u(\frac{1}{y}) = \frac{1}{y}u(y)$, for (4.9.3) to work. Fortunately, if $\sigma(\frac{1}{y},\tau) = \sigma(y,\tau)$ then it is easy to show that u has the required symmetry. If symmetry does not hold, then another set of vectors must be added to the basis, such as $\{\sqrt{y}w^i(y)\}_{i=1,2,\ldots}$. More details can be found in Cont *et al.* (2011).

Implementation

Note that

$$\partial_{yy} w^i = \frac{\mathrm{e}^{-\alpha_i \ln^2 y}}{y\sqrt{y}}\left(4\alpha_i^2 \ln^2 y - \frac{1}{4} - 2\alpha_i\right), \tag{4.9.5}$$

$$\partial_\tau v_\Sigma - \frac{\sigma^2 y^2}{2}\partial_{yy} v_\Sigma = \frac{(\Sigma^2-\sigma^2)y^2}{2}\partial_{yy} v_\Sigma = \frac{\mathrm{e}^{-\frac{\Sigma^2\tau}{8}}}{2\Sigma\sqrt{2\pi\tau}}(\Sigma^2-\sigma^2)\sqrt{y}\mathrm{e}^{-\frac{\ln^2 y}{2\Sigma^2\tau}}.$$

So (4.9.3) is

$$\sum_1^I \dot{a}_i \sqrt{y}\mathrm{e}^{-\alpha_i \ln^2 y} - \sum_1^I a_i \mathrm{e}^{-\alpha_i \ln^2 y}\sigma^2\sqrt{y}\left(2\alpha_i^2 \ln^2 y - \frac{1}{8} - \alpha_i\right)$$
$$= -\frac{\mathrm{e}^{-\frac{\Sigma^2\tau}{8}}}{2\Sigma\sqrt{2\pi\tau}}(\Sigma^2-\sigma^2)\sqrt{y}\mathrm{e}^{-\frac{\ln^2 y}{2\Sigma^2\tau}}. \tag{4.9.6}$$

For the Galerkin method we multiply this equation by $\gamma(y)\sqrt{y}\mathrm{e}^{-\alpha_j \ln^2 y}$ (the extra factor $\gamma(y)$ is a weight) and integrate over $\mathbb{R}^+$. We obtain an ODE system for $a = (a_1(\tau), \ldots, a_I(\tau))^T$:

$$M\dot{a} + Ba = F, \tag{4.9.7}$$

with

$$M_{ij} := \int_0^\infty \mathrm{e}^{-(\alpha_i+\alpha_j)\ln^2 y}\gamma(y) y\,\mathrm{d}y,$$
$$B_{ij} := -\int_0^\infty \mathrm{e}^{-(\alpha_i+\alpha_j)\ln^2 y}\sigma^2\left(2\alpha_i^2 \ln^2 y - \frac{1}{8} - \alpha_i\right)\gamma(y) y\,\mathrm{d}y,$$
$$F_j := -\frac{\mathrm{e}^{-\frac{\Sigma^2\tau}{8}}}{2\Sigma\sqrt{2\pi\tau}}\int_0^\infty \mathrm{e}^{-(\alpha_j+\frac{1}{2\Sigma^2\tau})\ln^2 y}(\Sigma^2-\sigma^2)\gamma(y) y\,\mathrm{d}y. \tag{4.9.8}$$

The case σ constant

When σ is constant and $\gamma(y) = y^{-2}$,

$$M_{ij} = \sqrt{\frac{\pi}{\alpha_i+\alpha_j}},$$
$$B_{ij} = \frac{\sqrt{\pi}\sigma^2}{(\alpha_i+\alpha_j)^{\frac{1}{2}}}\left(\frac{\alpha_i\alpha_j}{\alpha_i+\alpha_j} + \frac{1}{8}\right),$$
$$F_j = -\frac{\mathrm{e}^{-\frac{\Sigma^2\tau}{8}}(\Sigma^2-\sigma^2)}{2\sqrt{2\alpha_j\Sigma^2\tau+1}}. \tag{4.9.9}$$

The non-constant case

If σ is a function of y and τ, with $\sigma(\frac{1}{y},\tau)=\sigma(y,\tau)$, it is best to express it using an exponential basis, such as

$$\sigma(x,t)=\sigma_0+\sum_1^J \sigma_j(t)\mathrm{e}^{-\alpha_j \ln^2 y}, \tag{4.9.10}$$

because then all integrals can be computed analytically.

Another choice for γ

Later on we will change variables to $z=\mathrm{e}^{-\ln^2 y}$. Then, in view of the fact that $\mathrm{d}z=\frac{2}{y}\mathrm{e}^{-\ln^2 y}\ln\frac{1}{y}\,\mathrm{d}y$, we will need to choose

$$\gamma(y)=\frac{2}{y^2}|\ln y|,$$

so as to have

$$\mathrm{e}^{-\ln^2 y}y\gamma(y)\,\mathrm{d}y=\mathrm{e}^{-\ln^2 y}\frac{2}{y}|\ln y|\,\mathrm{d}y=\,\mathrm{d}z.$$

Then, with σ constant,

$$\begin{aligned}
M_{ij}&=\frac{1}{\alpha_i+\alpha_j},\\
B_{ij}&=\frac{\sigma^2}{\alpha_i+\alpha_j}\left[\frac{1}{8}+\alpha_i-\frac{2\alpha_i^2}{\alpha_i+\alpha_j}\right],\\
F_j&=-(\Sigma^2-\sigma^2)\frac{\mathrm{e}^{-\frac{\Sigma^2\tau}{8}}\sqrt{2\Sigma^2\tau}}{2\sqrt{\pi}(1+2\Sigma^2\tau\alpha_j)}.
\end{aligned} \tag{4.9.11}$$

The computation of the integrals relies on the following:

$$\begin{aligned}
\int_0^\infty \mathrm{e}^{-k\ln^2 y}\gamma(y)y\,\mathrm{d}y&=2\int_0^1 z^k\,\mathrm{d}z=\frac{2}{k+1},\\
\int_0^\infty \mathrm{e}^{-k\ln^2 y}\ln^2\frac{1}{y}\gamma(y)y\,\mathrm{d}y&=2\int_0^1 z^k\ln\frac{1}{z}\,\mathrm{d}z=\frac{2}{(k+1)^2}.
\end{aligned} \tag{4.9.12}$$

4.9.2. Numerical tests

The method was thoroughly tested by Cont *et al.* (2011), but since we are concerned with a minor variation of the original, we retested the method for the computation of a vanilla call of volatility $\sigma=\sqrt{0.9}$, using $\Sigma=\sqrt{0.5}$ and I ranging from 5 to 30 (using multi-precision arithmetic when $I>20$). The maturity is $T=2$ and 10 time steps are used to integrate (4.9.7); with 20 time steps the precision is not significantly improved.

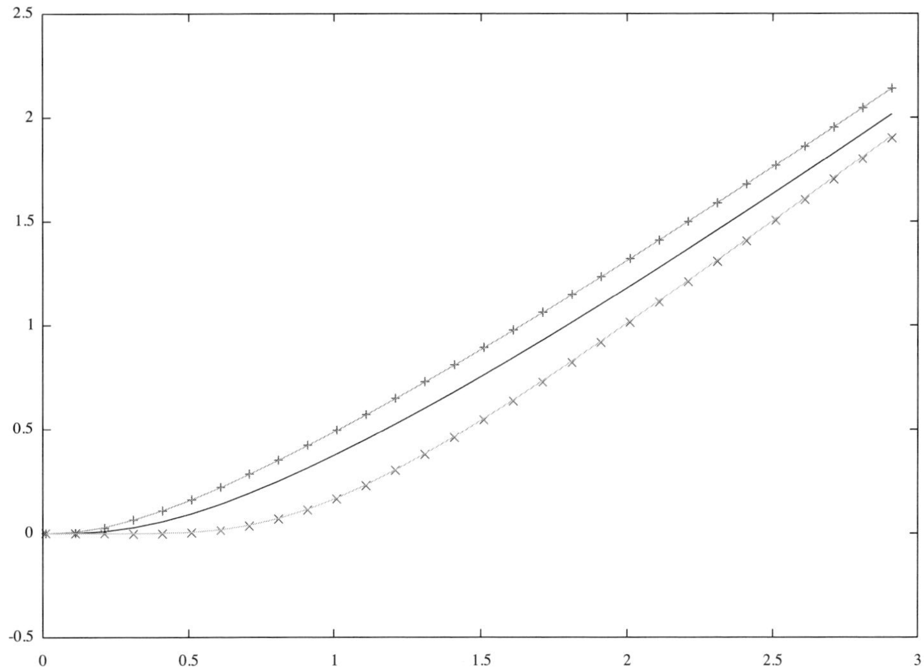

Figure 4.9.1. Here $\Sigma = \sqrt{0.5}$, and the call for this volatility is the middle curve. The call computed by the reduced basis method overlaps perfectly, even with $I = 5$ here. The call is computed analytically for both $\sigma = \sqrt{0.9}$ (the highest curve and the plus marks) and $\sigma = \sqrt{0.3}$ (the lowest curve and the cross marks).

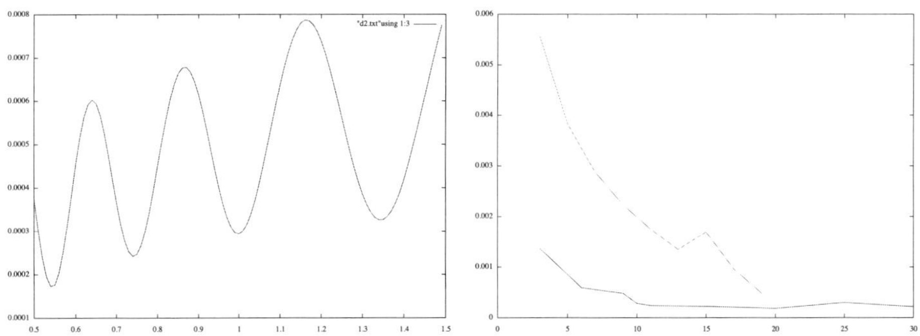

(a) (b)

Figure 4.9.2. (a) Errors with $I = 10$ (difference between the analytical Black–Scholes formula and this method) in the interval $y \in (0.5, 1.5)$ are shown to be below the 0.1 % accepted standard. (b) The upper curve is the L^1-error versus ten times the time to maturity τ. The lower curve is the L^1-error at $\tau = T$ versus the number of basis functions I.

Figures 4.9.1 and 4.9.2 show that the precision of the method is well within the three digits required by banks. Calls are computed from puts by the put–call parity relation. The linear system is solved with the SVD module of Press, Teukolsky, Vettering and Flannery (1992). However, ill-conditioning is such that even SVD fails when I is larger than 20 or so. An analytical orthogonalization of the basis is necessary, as used in spectral methods when one replaces the natural polynomial basis z^n by the Legendre polynomials.

4.9.3. Sparse grids

To compute one price with a d-dimensional SDE, Monte Carlo algorithms are not cursed by the dimension d in the sense that their computational complexity is proportional to d^2. With PDEs, even storing the solution costs $O(N^d)$ operations on a uniform grid of size N in each direction. Sparse grids can potentially deal with this dimensional curse.

The idea that underlies sparse grids is due to Smolyak (1963). It is possible to compute $\int_\Omega f$, $\Omega \subset \mathbb{R}^d$, by a quadrature formula, where the quadrature points are on a manifold of dimension $\mathbb{R}^{d-1}$. Smolyak's method achieves a precision proportional to $N^{-r}|\ln N|^{(d-1)(r+1)}$ with a grid of size N when $f \in C^r(\Omega)$.

This is a recursive argument. Indeed, if the quadrature points were on the boundary, for instance, then the sum which approximates the integral could be seen as an approximation of another integral on the $\partial\Omega$, which itself could be approximated by quadrature with points on a manifold of dimension $d-2$, *etc.*

Let $Q_i^{(1)}$ be I quadrature rules for $g : \mathbb{R} \to \mathbb{R}$ with quadrature points x_k^i and weights w_k^i:

$$Q_i^{(1)} g = \sum_{k=0}^{n_i-1} w_k^i g(x_k^i), \quad i = 1, \ldots, I, \quad Q_0^{(1)} g = 0.$$

For example, for a quadrature on $(-L, +L)$,

$$n_i = 2^i, \quad w_k^i = \frac{2L}{n_i}, \quad x_k^i = -L + \frac{2(k+\frac{1}{2})L}{n_i}, \quad k = 0, \ldots, n_i - 1.$$

Smolyak computes $Q_I^{(d)} f$ by

$$Q_i^{(p)} f = \sum_{j=1}^{i} (Q_j^{(1)} - Q_{j-1}^{(1)}) \otimes Q_{i-j+1}^{(p-1)} f, \quad i = 1, \ldots, I, \quad p = 2, \ldots, d,$$
$$Q_0^{(p)} f = 0. \tag{4.9.13}$$

Note that if $x = (x^1, x^2, \ldots, x^d)^T$, then $Q_i^{(p)}$ is an approximation of the p-multiple integral of f with respect to $x^1, \ldots, x^p$, so it is a function of $x^{p+1} \cdots x^d$, and $Q_j^{(1)} \otimes$ is an approximate integration with respect to the first free variable, namely x^{p+1}. More details can be found in Gerstner and Griebel (1998).

Remark 4.9.1. Equivalently

$$Q_i^{(p)} f = \sum D_{j_1} \otimes \cdots \otimes D_{j_p} f \quad \text{with } D_k = Q_k^{(1)} - Q_{k-1}^{(1)}. \tag{4.9.14}$$

The method is easy to apply to the computation of a basket option by (1.5.20). Even though the regularity is $r = 0$, the method gives reasonably good results. The following program implements (4.9.14). Results are shown in Figure 4.9.3. Improvement can be obtained by using recursively embedded grids and computing directly $Q_k - Q_{k-1}$ instead of Q_k and Q_{k-1} separately.

```
#include <iostream>
#include <cmath>
#include <ctime>
using namespace std;
#define d 5           // dimension of the integration space
const double L=2*atan(1.);    // size of the cube (-L,L)^d;
const int m=4;        // max number of points of quadrature in each dim
int *y, ik[d];        // container of a distribution of integers with sum<m+d
double *z;            // will be all the quadrature points in R^d, each in turn

void explore(int p, int s); // finds y[] s.t. sum of y<=s,
double f(double *y);        // the function to integrate
void QId(int* y);           // the 1d integration procedure

int *ni,*kd; // working array
double somme=0, L2=2*L, pi=4*atan(1.0), xpi=sqrt(pow(2*pi,d));

double f(double *x){ double aux=1;
for (int i=0;i<d;i++) aux*=cos(x[i]/2); return aux;}
/*  double f(double *z){ double aux=0, aux1=0, pid2=1;
        for (int i=0;i<d;i++){ aux+= z[i]; aux1+=z[i]*z[i];pid2*=2*pi;}
        return max(1-exp(aux),0.)*exp(-aux1/2)/sqrt(pid2);
} */ // change to this f for finance
void explore(int p, int s){ // p ends recursion
    if(p>0)
        for(y[p]=0;y[p]<=s;y[p]++)
              explore(p-1,s-y[p]);
    else {
            y[p]=s;// here p==0
        QId(y);
    }
}
```

```
void QId(int* y){
    double dd;
    for( ik[0]=0;ik[0]<=1;ik[0]++) // to add dimensions add loops below
      for( kd[0]=0;kd[0]<ni[y[0]-ik[0]];kd[0]++)
#if(d>1)
    for( ik[1]=0;ik[1]<=1;ik[1]++) for( kd[1]=0;kd[1]<ni[y[1]-ik[1]];kd[1]++)
#endif
#if(d>2)
    for( ik[2]=0;ik[2]<=1;ik[2]++) for( kd[2]=0;kd[2]<ni[y[2]-ik[2]];kd[2]++)
#endif
#if(d>3)
    for( ik[3]=0;ik[3]<=1;ik[3]++) for( kd[3]=0;kd[3]<ni[y[3]-ik[3]];kd[3]++)
#endif
#if(d>4)
    for( ik[4]=0;ik[4]<=1;ik[4]++) for( kd[4]=0;kd[4]<ni[y[4]-ik[4]];kd[4]++)
#endif
#if(d>5)
    for( ik[5]=0;ik[5]<=1;ik[5]++) for( kd[5]=0;kd[5]<ni[y[5]-ik[5]];kd[5]++)
#endif
#if(d>6)
    for( ik[6]=0;ik[6]<=1;ik[6]++) for( kd[6]=0;kd[6]<ni[y[6]-ik[6]];kd[6]++)
#endif
      {
        dd=1;
        for(int k=0;k<d;k++){
            int iy=y[k]-ik[k], nk = ni[iy];
            if(iy>=1){
              z[k] = -L + (L2*kd[k]+L)/nk;
              dd *= ik[k]? -L2/nk : L2/nk;
            } else dd=0;
            if(dd==0)break;
          }
         if(dd!=0)   somme += f(z)*dd;
      }
 }
int main(){
    y = new int[d];
    ni = new int[m+1];
    kd = new int[d];
    z = new double[d];
    ni[0]=1; for(int k=1;k<=m;k++) ni[k] =2*ni[k-1];  // ni[k]=2^k;
    time_t time0=clock();
    for(int n=1;n<m+d;n++) explore(d-1,n); // y[] with sum_0^{d-1} y[j]=n
    cout <<"{\sc cpu}=\t"<<double(clock()-time0)/CLOCKS_PER_SEC
        <<"\t relative error = \t"<<fabs(1-pow(4*sin(L/2),d)/somme)<<"
        "<<somme<<endl;
    return 0;
}
```

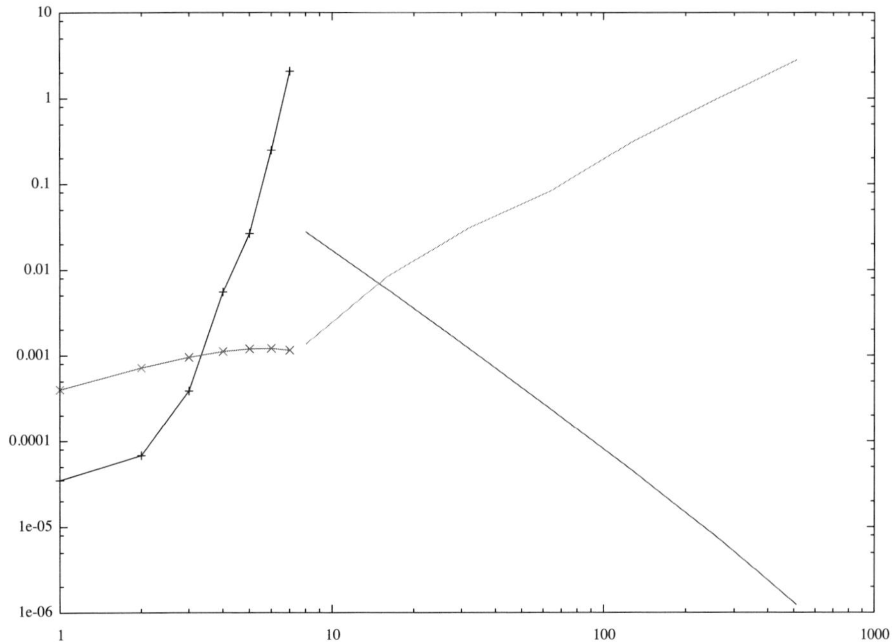

Figure 4.9.3. Left steeply ascending curve: log-log CPU time (in seconds) versus d, the dimension, at $I = 4$ obtained by the sparse grid implementation described above and applied to $f(x) = \Pi_1^d \cos(\frac{x_i}{2})$, $x \in [-\frac{\pi}{2}, \frac{\pi}{2}]^d$, $L = \frac{\pi}{2}$. Left slightly ascending curve (log-log): precision versus d at $I = 4$, showing only a mild degradation of the precision with dimension for a fixed N in each direction. Right descending curve (log-log): precision versus N at $d = 5$. Right ascending curve (log-log): CPU time versus N at $d = 5$.

PART FIVE
Calibration

5.1. Introduction

Whatever the quality of the modelling, the simulations usually do not reproduce the market data. Besides, some parameters, like volatilities, are not easy to estimate. On the other hand many assets and options are observable, and available, albeit possibly at a cost (for example, Reuters provide such data, as do Bloomberg).

Consider the problem of building an accurate model for a vanilla European call on S_t. Today, at time $t = 0$, we can observe

- the underlying asset S_0,
- the values $(c_i)_{i\in I}$ of calls $(C_{K_i,T_i}(S_0, 0))_{i\in I}$ of maturity T_i and strike K_i.

We wish to use the Black–Scholes PDE with a local (*i.e.*, non-constant) volatility, so we must find $x, t \to \sigma(x,t)$ such that the $\{C_i\}_{i\in I}$ computed by

$$\begin{aligned}&\partial_t C_i + \frac{\sigma^2 S^2}{2}\partial^2_{SS} C_i + rS\partial_S C_i - rC_i = 0,\\ &C_i(S,T_i) = (S-K_i)^+, \quad \text{for all } t \in [0,T_i[,\ x \in \mathbb{R}^+, \end{aligned} \tag{5.1.1}$$

satisfy $C_i(S_0, 0) = c_i$

The most natural way to proceed is to solve this *inverse problem* by least-squares:

$$\min_{\sigma\in\mathcal{S}} \sum |C_i(S_0,0) - c_i|^2 \quad \text{subject to (5.1.1).}$$

This problem has three difficulties.

(1) It is ill-posed unless $\mathcal{S}$ is severely constrained or a *Tikhonov regularization* is added to the cost function (see Crépey (2003)).
(2) It has infinitely many solutions, as we shall see.
(3) It is numerically expensive to solve.

Curse number (3) is problematic. One reason is that it is an optimal control problem with card(I) PDEs. Working with the moneyness variables

$$\tau = T - t, \quad y = \mathrm{e}^{r\tau}\frac{S}{K}, \quad Ku(y,\tau) = \mathrm{e}^{r\tau} C(K\mathrm{e}^{-r\tau}y, T-\tau)$$

reduces the number of PDEs to one, and the problem becomes

$$\begin{aligned}&\min_{\sigma\in\mathcal{S}} \sum \left|\mathrm{e}^{rT_i}K_i u\left(\mathrm{e}^{rT_i}\frac{S_0}{K_i}, T_i\right) - c_i\right|^2\\ &\text{subject to} \quad \partial_\tau u - \frac{\sigma^2 y^2}{2}\partial_{yy} u = 0, \quad u(y,0) = (y-1)^+, \end{aligned} \tag{5.1.2}$$

where $\mathcal{S} = \{\sigma(y,\tau) : \sigma_m \le \sigma(y,\tau) \le \sigma_M\}$. If a Tikhonov regularization of the form

$$\|\sigma\|_0^2 + \|\partial_y\sigma\|_0^2 + \|\partial_\tau\sigma\|_0^2 + \|\partial_{y\tau}\sigma\|_0^2$$

is added to the cost function, then the problem is well posed and solvable numerically via a conjugate gradient method, for example (see Achdou and Pironneau (2005)).

However, the explicit dependence of C on K is specific to this problem and the method is not general. Dupire had a better idea!

5.2. Dupire's equation

Consider a call priced by (3.2.11). To stress explicitly the dependence on the parameters K, T we write the solution as $C_{K,T}(S,t)$.

Dupire (1997) observed that an equation for $C_{S,t}(K,T)$ can be derived, where S,t are now parameters:

$$\partial_\tau C - \frac{1}{2}\sigma^2 K^2 \partial_{KK} C + rK\partial_K C = 0,$$
$$C(K,t_0) = (S-K)^+ \quad \forall K > 0, \tau \geq t_0. \tag{5.2.1}$$

Dupire's proof (Dupire 1997) uses the Green's function of the differential operator in (3.2.11). We shall give a proof which can be extended to any financial object governed by a *linear* system. The proof relies on a duality argument, such as the forward and backward Kolmogorov equation. The proof also shows that a double primitive of the dual of the Black–Scholes PDE is involved. Thus it is misleading to use K,T as variables in (5.2.1).

Proposition 5.2.1. (Dupire) Let v be a solution in $\mathbb{R}^+ \times (t_0,T)$ of

$$\partial_t v - \frac{1}{2}\sigma^2 S^2 \partial_{SS} v + rS\partial_S v = 0, \quad v(S,t_0) = (S_0 - S)^+. \tag{5.2.2}$$

Then $C_{K,T}(S_0,t_0) = v(K,T)$ and $p = \partial_{SS} v$ is a solution of the adjoint of (3.2.11):

$$\partial_t p - \partial_{SS}\left(\frac{\sigma^2 S^2}{2} p\right) + \partial_S(rSp) + rp = 0, \ p(t_0) = \delta(S-S_0). \tag{5.2.3}$$

Proof. An integration by parts in time and Green's formula in space applied to (3.2.11), multiplied by p and integrated over $\mathbb{R}^+ \times (t_0,T)$, yields

$$C(S_0,t_0) = \int_0^\infty \mathrm{d}S\left[C_T p(T) + \int_{t_0}^T \left(p\frac{\sigma^2 S^2}{2}\partial_S C - C\partial_S\left(\frac{\sigma^2 S^2}{2}p\right)\right)\mathrm{d}t\right], \tag{5.2.4}$$

where $C_T(S)$ is the option's premium. By the properties of C and p at zero and infinity the boundary term vanishes. Let v be the double primitive of p, *i.e.*, $\partial_{SS} v = p$. Then (5.2.3) integrated twice becomes (5.2.2) for an appropriated choice of the integration constants. □

5.2.1. Example of generalization: Dupire for binary call options

The premium is one monetary unit if $S_T > K$ and zero otherwise. It can be treated by the same method. The adjoint equation is integrated once only, and the result is

$$u(S_0,t_0) = \int_0^\infty u_T \partial_S w \,\mathrm{d}S = -\int_0^\infty w\partial_S u_T \,\mathrm{d}S + [u_T w]_0^\infty = -w(K,T),$$
$$\text{with} \quad \partial_t w - \partial_S\left(\frac{\sigma^2 S^2}{2}\partial_S w\right) + rS\partial_S w + rw = 0, \ w(t_0) = \mathbf{1}_{S>S_0} - 1. \tag{5.2.5}$$

5.2.2. Dupire's equation for barrier options

Consider a European barrier option which ends if $S_t \notin (S_m e^{rt}, S_M e^{rt})$. The change of variable $c(s,t) = e^{-rt} C(se^{rt}, t)$ reduces (3.2.11) with barriers to

$$\partial_t c + \frac{1}{2}\sigma^2 s^2 \partial_{ss} c = 0, \quad c(s,T) = (s - Ke^{-rT})^+,$$
$$c(S_m, t) = c(S_M, t) = 0, \quad \forall s, t \in (S_m, S_M) \times (0,T). \tag{5.2.6}$$

Proposition 5.2.2. Assume that $S_m < Ke^{-rT} < S_M$. If c satisfies (5.2.6) then

$$c(s_0, t_0) = v(Ke^{-rT}, T) + (S_M - Ke^{-rT})^+ \partial_S v|_{S_M,T}, \tag{5.2.7}$$

where v is the solution of

$$\partial_t v - \frac{1}{2}\sigma^2 s^2 \partial_{ss} v = 0, \quad v(s, t_0) = (s_0 - S)^+,$$
$$v(S_m, t) = (s_0 - S_m)^+, \quad v(S_M, t) = 0 \quad \forall t \in (t_0, T). \tag{5.2.8}$$

Proof. Now the boundary conditions in (5.2.3) with $r = 0$ are

$$p(S_m, t) = p(S_M, t) = 0$$

for all t. For v, the double primitive of p, it translates into $\partial_{ss} v = 0$, which in turn implies $\partial_t v = 0$, which means that v is constant at the barriers. Naturally $C(S_0, t_0)$ can be recovered from c by choosing $s_0 = S_0 e^{rt_0}$. □

5.2.3. Dupire's equation for options on Lévy-driven assets

If a Poisson–Lévy process is used in the Black–Scholes model, as in Cont and Tankov (2003), a term appears on the right-hand side of (3.2.11):

$$\int_{\mathbb{R}} \left(C(Se^y, t) - C(S,t) - S(e^y - 1)\frac{\partial C}{\partial S} \right) k(y)\, dy, \tag{5.2.9}$$

where k, the kernel of the process, is usually singular at the origin and decaying fast at infinity.

When multiplied by p and integrated in S, with $z = Se^{-y}$, this term can be transformed into an integral of $C(S,t)\chi(S,t)$ for an appropriate χ leading to the integro-differential equation for p. The resulting equation is again integrated twice and leads to the following result.

Proposition 5.2.3. A Poisson–Lévy-driven option $C_{K,T}(S,t)$, *i.e.*, a solution of (3.2.11) with (5.2.9) on the right-hand side, satisfies Dupire's identity

$$C_{K,T}(S_0, t_0) = v(K, T),$$

where v is the solution of

$$\begin{aligned}
&\partial_t v - \frac{\sigma^2 S^2}{2}\partial_{SS}v + rS\partial_S v \qquad (5.2.10)\\
&\quad - \int_{\mathbb{R}} \big(\mathrm{e}^y(v(S\mathrm{e}^{-y},t) - v(S,t)) + (\mathrm{e}^y - 1)S\partial_S v(S,t)\big)k(y)\,\mathrm{d}y = 0,\\
&v(S,t_0) = (S_0 - S)^+, \quad v(0,t) = S_0, \ \lim_{S\to\infty} v(S,t) = 0 \quad \forall t \in (t_0,T).
\end{aligned}$$

Theorem 5.2.4. Consider a call on S modelled by

$$\begin{aligned}
&\partial_t C + \frac{\sigma^2 S^2}{2}\partial_{SS}v + rS\partial_S C - rC \qquad (5.2.11)\\
&\quad + \int_{\mathbb{R}} \big(C(S\mathrm{e}^y,t) - C(S,t) - S(\mathrm{e}^y - 1)\partial_S C(S,t)\big)\nu(\mathrm{d}y) = 0,\\
&C(S,T) = (S-K)^+.
\end{aligned}$$

Dupire's identity holds, *i.e.*, $C_{K,T}(S_0,0) = v(K,T)$, when v is the solution of

$$\begin{aligned}
&\partial_t v - \frac{\sigma^2 S^2}{2}\partial_{SS}v + rS\partial_S v \qquad (5.2.12)\\
&\quad - \int_{\mathbb{R}} \big(\mathrm{e}^y(v(S\mathrm{e}^{-y},t) - v(S,t)) + (\mathrm{e}^y - 1)S\partial_S v(S,t)\big)\nu(\mathrm{d}y) = 0,\\
&v(S,0) = (S_0 - S)^+, \quad v(0,t) = S_0, \quad \lim_{S\to\infty} v(S,t) = 0 \quad \forall t \in (0,T).
\end{aligned}$$

The proof has (5.2.11) multiplied by p and integrated in S by parts. At one stage a change of integration variable is necessary, $z = S\mathrm{e}^{-y}$. Then, as for the Black–Scholes equation, one introduces v such that $\partial_{SS}u = p$.

Now assuming that $C_i = C_{K_i,T_i}(S_0,0)$ are market observations, then to calibrate the model one may solve

$$\min \sum_{i\in I} |v(K_i,T_i) - C_i|^2,$$

where the optimization variables are the parameters of the model, *i.e.*, $\lambda, \gamma, \sigma, \mu, \delta$ in the case of Merton's model, for example.

5.2.4. Dupire's identity in the discrete case

Extension to a more complex option is possible as long as the PDE is linear. For instance, Dupire's argument does not work with American options. Furthermore, it is not always possible to find a double primitive of the adjoint equation like (5.2.3) which, as it stands, is numerically very hard because of the singularity of the initial condition. On the other hand, at the discrete level the same difficulty is not present for the variational approach.

For simplicity assume that the coefficients in (5.2.3) do not depend on time, and consider a time discretization with an Euler implicit scheme with time step δt, and a space discretization with piecewise linear finite elements in the variable S. The vertices of the grid are called q^i, $0 \le i \le I$, and we set $L = q^I$. The scheme can be written in matrix form,

$$(\mathbf{M} + \mathbf{A})\mathbf{u}^n - \mathbf{M}\mathbf{u}^{n+1} = 0, \tag{5.2.13}$$

where $\mathbf{u}^n$ is the vector of the nodal values of the piecewise linear function u_h^n which approximates $u(\cdot, n\delta t)$. We have

$$\mathbf{M}_{ij} = \frac{1}{\delta t}\int_0^L w^i w^j,$$

$$\mathbf{A}_{ij} = \int_0^L \left(\partial_S\left(\frac{y^2\sigma^2}{2}w^j\right)\partial_S w^i + \frac{y^2\sigma^2}{2}w^j\partial_S w^i + r w^i w^j\right), \tag{5.2.14}$$

where the hat functions w^i are by definition piecewise linear continuous and take values 0 at q^j, $j \ne i$ and 1 at q^i. Given a vector $\mathbf{p}^0$, introduce the sequence of vectors $\mathbf{p}^n$ obtained by iterating

$$(\mathbf{A} + \mathbf{M})^T\mathbf{p}^{n+1} - \mathbf{M}^T\mathbf{p}^n = 0, \quad \mathbf{p}^0 \text{ given}. \tag{5.2.15}$$

Now notice that (5.2.13) multiplied by $(\mathbf{p}^{n+1})^T$ gives

$$\begin{aligned} 0 &= (\mathbf{p}^{n+1})^T(\mathbf{A} + \mathbf{M})\mathbf{u}^n - (\mathbf{p}^{n+1})^T\mathbf{M}\mathbf{u}^{n+1} \\ &= (\mathbf{p}^n)^T\mathbf{M}\mathbf{u}^n - (\mathbf{p}^{n+1})^T\mathbf{M}\mathbf{u}^{n+1}, \end{aligned} \tag{5.2.16}$$

where the last equality used (5.2.8). Summing up over all n gives

$$(\mathbf{p}^0)^T\mathbf{M}\mathbf{u}^0 = (\mathbf{p}^N)^T\mathbf{M}\mathbf{u}^N. \tag{5.2.17}$$

Let

$$\mathbf{p}_i^0 \text{ the solution of } \mathbf{M}^T\mathbf{p}_i^0 = \delta_{ij} \quad j = 0, \ldots, I, \tag{5.2.18}$$

and let $\mathbf{p}_i^N$ be given by (5.2.15). Then (5.2.17) gives the discrete Dupire equation

$$u_i^0 = (\mathbf{p}_i^N)^T\mathbf{M}\mathbf{u}^N. \tag{5.2.19}$$

Remark 5.2.5. It is not necessary to solve for $\mathbf{p}^0$; the knowledge of $\mathbf{M}^T\mathbf{p}^0$ is sufficient to initialize (5.2.15).

5.3. Calibration of the local volatility

We return to the calibration of the local volatility, but now with Dupire's equation.

Knowing the spot price S_0 and some values $\{c_i\}_1^d$ of calls on S_t, we find the solution $\sigma(S,t)$ of

$$\min_{\sigma\in\mathcal{S}}\sum|u(K_i,T_i)-c_i|^2,$$
$$\text{where}\quad \partial_t u-\frac{\sigma^2x^2}{2}\partial^2_{xx}u+rx\partial_x u=0,\quad u(x,0)=(S_0-x)^+. \tag{5.3.1}$$

The problem has an obvious candidate solution. Let $\tilde{u}$ be a C^2 interpolation of $\{K_i,T_i,c_i\}_I$. Then

$$\sigma^2=\frac{\partial_t\tilde{u}+rx\partial_x\tilde{u}}{\frac{1}{2}\partial^2_{xx}\tilde{u}} \tag{5.3.2}$$

is a solution.

Obviously these (infinitely many) solutions are not robust with respect to the data, in the sense that a small change in one c_i can cause a large change of σ. This can be fixed, however. For instance, Bézier splines are less sensitive to parameter change. Yet the main reason for not using this simple construction is that it is quite specific to the one-dimensional case. It would not work with a basket option, for instance, and it is usually not a good idea to invest numerically in a solution limited in scope.

Stabilization of (5.3.1) is achieved either by adding a Tikhonov regularization or by restricting the space $\mathcal{S}$. When the second solution is used and the optimization problem is solved by conjugate gradients, using automatic differentiation for the sensitivities (see Achdou and Pironneau (2005)), it is not that expensive if the dimension of $\mathcal{S}$ is small. For example, suppose $\sigma(x,t)$ is C^1, bilinear in an (x,t) parallelogram, grows large outside and is parametrized by eight values $\{z_k\}_{k=0}^7$, giving the following values:

$$S_{11}=z_0^2,\quad S_{21}=z_1^2,\quad S_{12}=S_{11}+z_2^2,\quad S_{22}=S_{21}+z_3^2,$$
$$\sigma_{1j}=\frac{z_{3+j}^2}{1+z_{3+j}^2},\quad \sigma_{2j}=\frac{z_{5+j}^2}{1+z_{5+j}^2},\quad j=1,2. \tag{5.3.3}$$

Let S_i, σ_i, $i=1,2$ be the following linear functions of t:

$$S_i=S_{i1}\biggl(1-\frac{t}{T}\biggr)+S_{i2}\frac{t}{T},\quad \sigma_i=\sigma_{i1}\biggl(1-\frac{t}{T}\biggr)+\sigma_{i2}\frac{t}{T}. \tag{5.3.4}$$

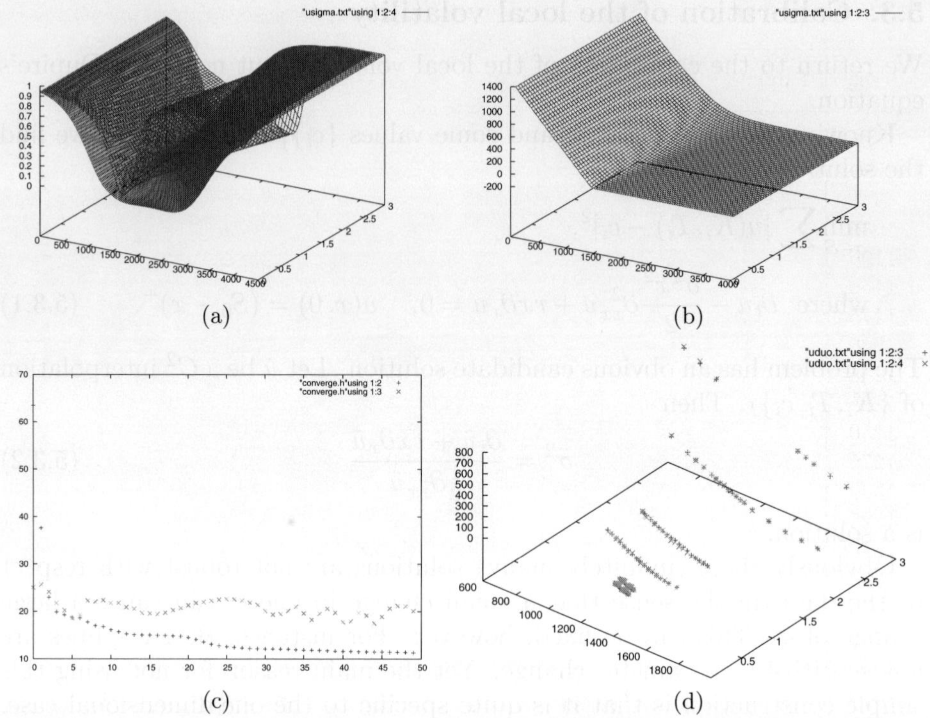

Figure 5.3.1. Calibration by least-squares and a spline parametrization of the local volatility on the SPX index of 21 Dec. 2006 at $S_0 = 1418.3, r = 3\,\%$. (a) The local volatility surface, and (b) the corresponding option price versus strike and maturity. (c) Convergence of the conjugate gradient algorithm. The functional (+) decays rapidly in less than five iterations and very slowly afterwards, while the norm of the functional gradient (×) decays erratically. (d) The data and the model's values at the same K_i, T_i points.

The volatility surface ($a = \sigma(0,t)$) is given by

$$\sigma(S,t) = \begin{cases} a + \left(2\frac{\sigma_1 - a}{S_1} - \frac{\sigma_2-\sigma_1}{S_2-S_1}\right)S + \left(\frac{\sigma_2-\sigma_1}{S_2-S_1} - \frac{\sigma_1-a}{S_1}\right)\frac{S^2}{S_1} & \text{if } (S < S_1), \\ \sigma_2\frac{S-S_1}{S_2-S_1} + \sigma_1\frac{S_2-S}{S_2-S_1} & \text{if } S_1 \le S \le S_2, \\ \sigma_2 + (S-S_2)\frac{\sigma_2-\sigma_1}{S_2-S_1} + \left((S-S_2)\frac{\sigma_2-\sigma_1}{S_2-S_1}\right)^2 & \text{if } S > S_2. \end{cases} \tag{5.3.5}$$

The parametrization is such that, for all values of $\{z_i\}_0^7$, σ_{ij} is positive and bounded from below and above and smooth, so no Tikhonov regularization is needed: the optimization is now done with respect to $\{z_i\}_0^7$, unconstrained.

Results on the SPX index of 21 December 2006 at $S_0 = 1418.3, r = 3\,\%$ for some fifty observations are shown in Figure 5.3.1. The table of calls used is shown in Table 5.3.1.

Table 5.3.1. The prices of a family of calls on the same asset (Eurostoxx50).

Strike	1 month	2 months	6 months	12 months	24 months	36 months
700						733
800						650.6
900						569.8
1000					467.8	
1100					385.3	
1150					345.4	
1175			265.2			
1200			242	266.1	306.6	
1215				253.4		
1225			219	245		
1250			196.6	224.2	269.2	
1275			174.5	203.9	251	
1300			152.9	184.1	233.2	
1325			131.9	164.9	215.8	
1350			111.7	146.3	198.9	
1365			100			
1375	50.6	60	92.5	139	182.6	
1380	46.1	55.8				
1385	41.8	51.8				
1390	37.5	47.9				
1395	33.4	44				
1400	29.4	40.3	74.5	128.4	166.7	215.9
1405	25.6	36.7				
1410	21.9	33.2				
1415	18.7	29.8				
1420	15.4	26.6				
1425	12.7	23.8	58	111.4	151.5	
1430	10	20.7				
1435	8	18.2				
1440	6.3	15.7				
1445	4.4	13.4				
1450	3.1	11.3	43.3	95.2	136.9	187.5
1455	2.05	9.6				
1460	1.45	7.9				
1475			30.6	80.2		
1500			20.3	54	109.6	160.8
1525			12.6	42.7		
1550			7.5	33		
1575				24.7		
1600			1.95	18.2	64.5	113.9
1700					32.7	75.7
1800					15.5	
1900					5.2	

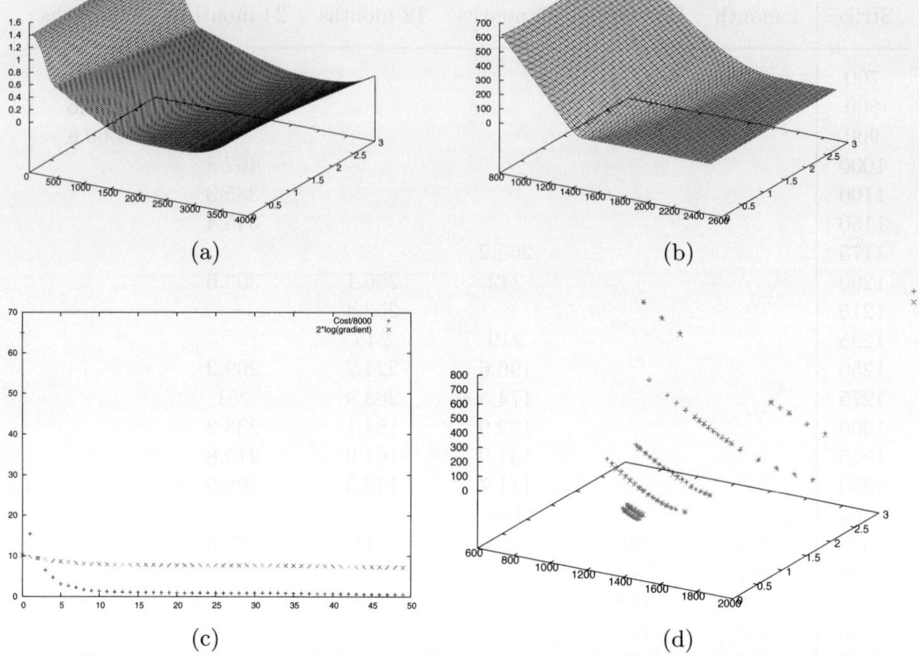

Figure 5.4.1. The same as Figure 5.3.1, except that the implied volatility surface is the parameter of the calibration problem.

5.4. Calibration of implied volatility

Given a spot price S_0, and an option price u at $t_0 = 0$, the Black–Scholes formula can be solved for σ_I:

$$C_{BS}(S_0, r, K, T, \sigma_I) = u(K, T), \tag{5.4.1}$$

where K, T are the strike and maturity of u and r is the interest rate.

The solution is called the *implied volatility.* In practice a Newton algorithm inverts the formula in less than five iterations.

Now, to fit a set of discrete observations we solve

$$\min_{\sigma_I \in \mathcal{B}} \sum |C_{BS}(S_0, r, K_i, T_i, \sigma_I) - u_i|^2. \tag{5.4.2}$$

The local volatility σ_L is recovered from (5.3.2) in terms of the computed $u(K,T) := C_{BS}(S_0, r, K, T, \sigma_I)$.

The method is very fast because no PDE is solved, but it seems hard to extend to multi-dimensional cases. Results with this method are shown in Figure 5.4.1.

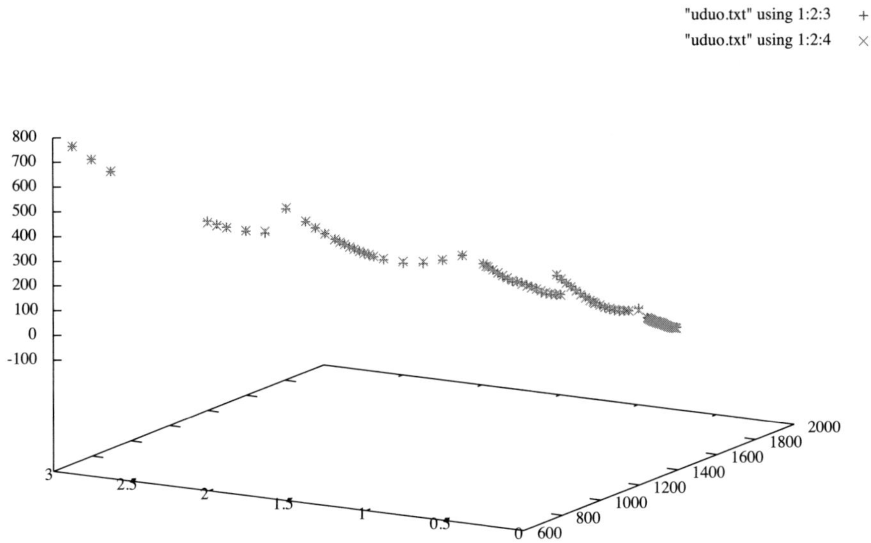

(a)

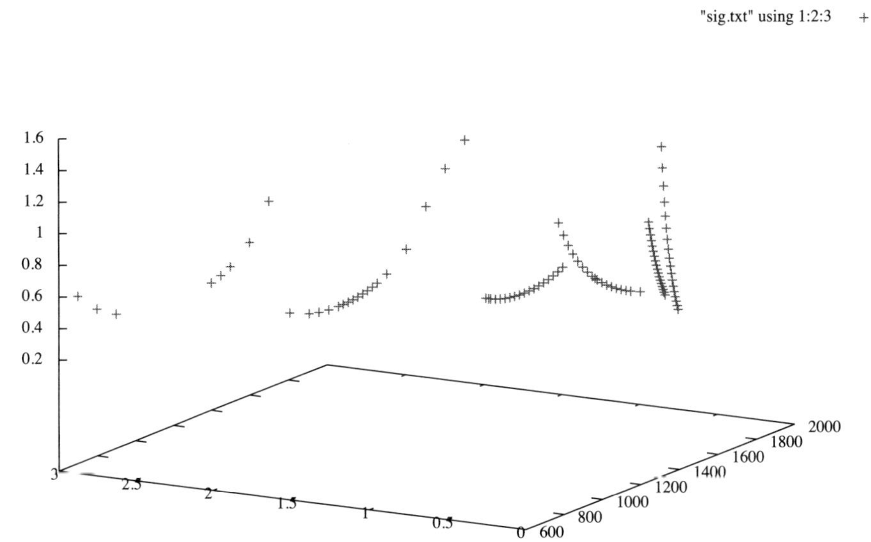

(b)

Figure 5.4.2. The same as in Figure 5.3.1 but calibrated on a reduced basis. (a) Observed prices and model predictions for the calls at the observation points; errors are hardly visible (between \$1 and \$9) and on average \$4 at each point. (b) The local volatility recovered from the Dupire equation at the observed points. For both graphs the horizontal axis is time, the vertical axis is the price of calls and the other axis is the strike.

5.5. Calibration on a reduced basis

As before, one observes at $t = 0$ some calls $\{u_j\}_{j=1}^{J}$, all based on the same asset S; these have strikes K_j and maturity T_j.

Consider Dupire's equation,

$$\partial_T u_\sigma - \frac{\sigma^2}{2}\partial_{KK} u_\sigma + r\partial_K u_\sigma = 0, \quad u_\sigma(K,0) = (S-K)^+. \tag{5.5.1}$$

One wishes to choose the function $\sigma(.,.)$ to reproduce these calls. Formulated in terms of a least-squares optimization problem, we have

$$\sigma = \arg\min_\sigma \sum_{j=1}^{J} |u_\sigma(K_j,T_j) - u_j|^2. \tag{5.5.2}$$

When a decomposition of the general solution of (5.5.1) is used, but with K, T as variables, then (5.5.2) is a sum of independent problems at each time T_j. For each T' one solves

$$\alpha(T') = \arg\min_\alpha \sum_{j:T_j=T'} |u(K_j,T';\alpha) - u_j|^2, \quad \text{subject to}$$

$$u(K_j,T';\alpha) = u_\Sigma(K_j,T') + \sum_{i=1}^{I} \alpha_i\big[u_{\sigma_i}(K_j,T_M) - u_\Sigma(K_j,T_M)\big], \tag{5.5.3}$$

where $T_M = \max T_j$ is the reference time chosen to build the basis. The volatility surface is recovered from Dupire's equation and

$$u_\sigma(K,T) = u(K,T;\alpha).$$

The derivatives with respect to K are computed analytically.

The method is tested on the data of Table 5.3.1. Results are shown in Figure 5.4.2 for $r = 0.03$ and the spot price is $S_0 = 1418.3$.

PART SIX
Open-source tools

6.1. Premia

Premia (see www-rocq.inria.fr/mathfi/Premia/) is a platform and a consortium for pricing financial derivatives. Recently it has been integrated with NSP, a Matlab-like software by a member of INRIA. The C++ functions are written and maintained by the members of the consortium (including several banks and hedge funds) under the leadership of INRIA and the ENPC laboratory in France.

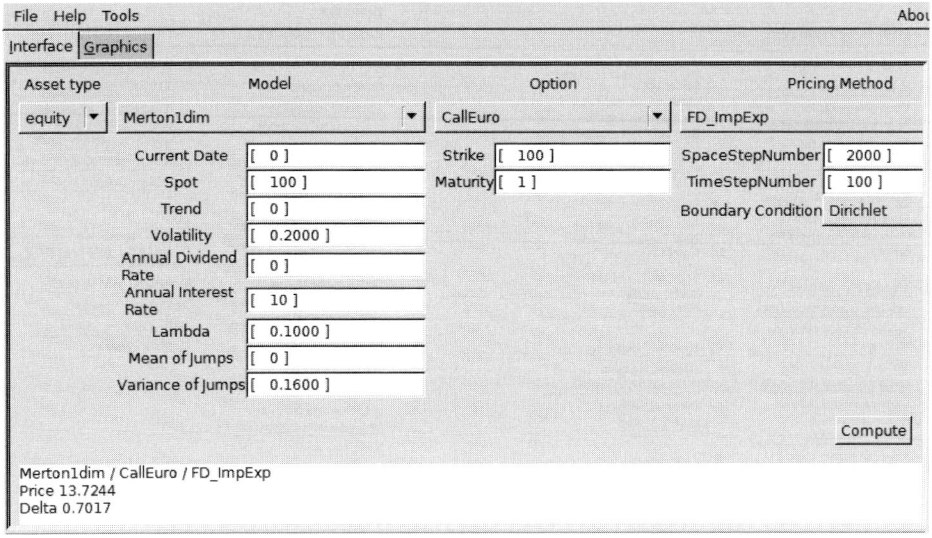

Figure 6.1.1. Under NSP, Premia is mouse/menu-driven, making the learning curve very short. This implementation runs under Windows, Linux and Mac OS X.

Premia solves the following.

Equity derivatives

- The Black–Scholes model (up to dimension 10), stochastic volatility models (Hull–White, Heston, Fouque–Papanicolaou–Sircar), models with jumps (Merton, Kou, tempered stable processes, variance gamma, normal inverse Gaussian), Bates model.
- For high-dimensional American options, Premia provides the most recent Monte Carlo algorithms: Longstaff–Schwartz, Andersen–Broadie, Barraquand–Martineau, Tsitsklis–Van Roy, Broadie–Glassermann, quantization methods, Malliavin calculus-based methods.
- Dynamic hedging for Black–Scholes and jump models is available.
- Calibration algorithms for some models with jumps, local volatility and stochastic volatility.

Interest rate derivatives

- Heath–Jarrow–Morton and LIBOR market models (LMM): affine models, Hull–White, CIR++, Black–Karasinski, squared Gaussian, Bhar–Chiarella, Li–Ritchken–Sankara–Subramanian, jump diffusion LMM, Markov functional LMM, LMM with stochastic volatility.
- Premia provides a calibration toolbox for LIBOR market models using implied volatilities for a database of swaptions and caps.

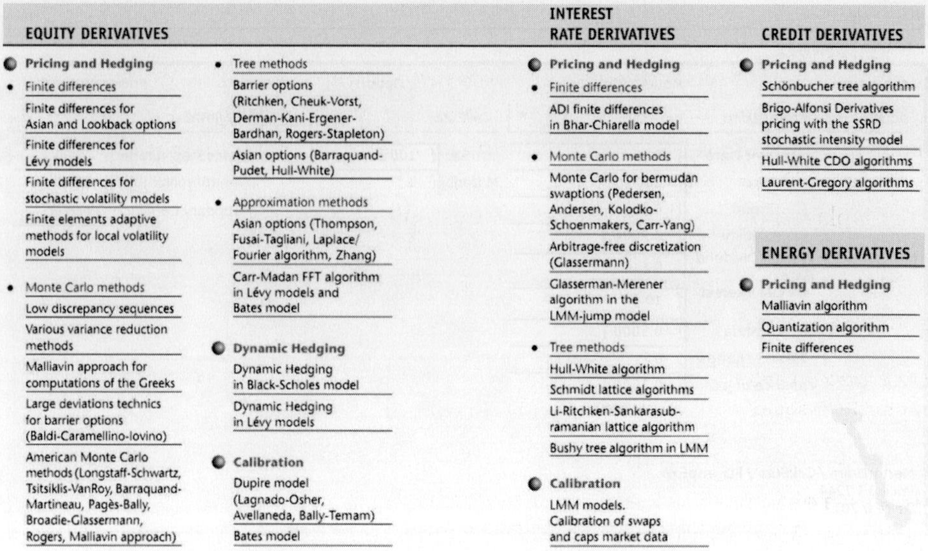

Figure 6.1.2. The main functions and algorithms of Premia (from www-rocq.inria.fr/mathfi/Premia).

Credit derivatives: CDSs, CDOs

- Reduced-form models and copula models are considered.
- Premia provides a toolbox for pricing CDOs using the most recent algorithms (Hull–White, Laurent–Gregory, El Karoui–Jiao, Yang–Zhang).

Hybrid products

- PDE solver for pricing derivatives on hybrid products, *e.g.*, options on inflation and interest or change rates.

Energy derivatives: swing options

- Mean reverting and jump models are considered.
- Premia provides a toolbox for pricing swing options using finite differences, the Monte Carlo Malliavin-based approach and quantization algorithms.

The main solution methods are listed in Figure 6.1.2.

6.2. QuantLib

QuantLib (see quantlib.org) is an open-source C++ library of functions and a class hierarchy to build complex finance solvers. It contains many functions to generate random numbers and variables of all types. There are also several functions for interest rate models, Black–Scholes pricers, Ornstein–Uhlenbeck, Heston, *etc.* It can be used as a base to build more complex solvers.

6.3. FreeFem++

FreeFem++ (see www.freefem.org) is a general-purpose finite element package in the public domain to solve 2D and 3D partial differential equations. It has its own language/script, which in some sense is an extension of the C++ language. During execution the FreeFem script is interpreted and graphics are displayed on the fly; input/output can also be made using C++ keywords such as `cout<<`.

Most of the time in finance the domain of the PDE is a rectangle. It is generated by the script

```
real Lx=2, Ly=3;
mesh th = square(10,20,[Lx*x,Ly*y]);
```

It generates a uniform 10×20 triangulation (mesh) of the rectangle $(0, Lx)\times(0, Ly)$; `real,mesh,square` are keywords of the language and `Lx,Ly,th` are variables. Mesh adaptivity is built in and called by

```
mesh Th = adaptmesh(th, max(1.-x*y,0.), err=0.005);
plot(Th);
```

Here the new mesh `Th` is adapted to the function $\max(1 - x * y, 0)$ in the sense that there are more triangles in places where the Hessian of $(1-xy)^+$ is large, namely where $x * y < 1$. The optional parameter `err` controls the precision and the number of triangles generated. The result is displayed in the last line.

Piecewise polynomial finite element functions of degree one on `Th` are defined by

```
fespace Vh(Th,P1);
Vh u,v,uold=max(1.-x*y,0.);
```

Here `u`, `v` are declared only while `uold` is declared and assigned to the piecewise linear continuous interpolation of $(1-xy)^+$ on the vertices of `Th`; `u` will be used for u^{m+1} and `uold` for u^m, the values of the basket option at times to maturity $(m+1)\delta t$ and $m\delta t$.

The Black–Scholes PDE (4.3.1) for a basket with two assets x, y, discretized in time by a semi-implicit Euler scheme and P^1 continuous elements in space, is entered in variational form as

```
problem basket(u,v) = int2d(Th) ( u*v/dt
     + sigmax^2*x^2*dx(u)*dx(v)/2 + sigmay^2*y^2*dy(u)*dy(v)/2
     + sigmax*sigmay*x*y*rho*dx(u)*dy(v) )
     - int2d(Th)(uold*v/dt + sigmax*x*dx(uold)*v+sigmay*y*dy(uold)*v);
```

Here `problem,int2d,x,y,dx,dy` are keywords and `sigmax,sigmay,rho,dt` are parameters or functions which must be pre-defined.

Figure 6.3.1. Result of a computation of a basket option using FreeFem++. The mesh is adapted to the pay-off $(1 - xy)^+$. The isovalues range from 1 to 0 with decrements 0.05. The parameters of the option can be read in the program below.

Finally, the time loop to compute u at maturity T is

```
for(real t=0;t<T;t+=dt){
   basket;  uold = u;
}
```

6.3.1. Two-dimensional Black–Scholes equation

The complete script and results are given below for a basket put option with multiplicative pay-off:

```
real Lx=2, Ly=3;
mesh th = square(10,20,[Lx*x,Ly*y]);
mesh Th = adaptmesh(th, max(1.-x*y,0.),err=0.005);
fespace Vh(Th,P1);
Vh v,u, uold=max(1.-x*y,0.);
plot(Th,wait=1,ps="Th.ps");
real dt=0.01, sigmax=0.5, sigmay=0.3, rho=-0.5, T=1;
```

```
problem basket(u,v) = int2d(Th) ( u*v/dt
     + sigmax^2*x^2*dx(u)*dx(v)/2 + sigmay^2*y^2*dy(u)*dy(v)/2
     + sigmax*sigmay*x*y*rho*dx(u)*dy(v) )
     - int2d(Th)(uold*v/dt - sigmax*x*dx(uold)*v - sigmay*y*dy(uold)*v);

for(real t=0;t<T;t+=dt){
   basket;  uold = u;
}
plot(u);
```

Remark 6.3.1. A large number of finite elements are built into FreeFem++, including discontinuous Galerkin methods.

FreeFem++ was not designed to solve one-dimensional problems, yet it can be done by making them 2D and invariant with respect to the added coordinate. The following is a script to compute a plain vanilla put and compare it with the Black–Scholes formula:

```
real Lx=4, Ly=0.1;
mesh Th = square(150,2,[Lx*x,Ly*y]);
fespace Vh(Th,P1);
Vh v,u, uold=max(1.-x,0.);

plot(Th,wait=1);

real dt=0.01, sigmax=0.5,  T=1;
problem basket(u,v) = int2d(Th) ( u*v/dt
     + sigmax^2*x^2*dx(u)*dx(v)/2  )
     - int2d(Th)(uold*v/dt - sigmax^2*x*dx(uold)*v) + on(2,u=0);

for(real t=0;t<T;t+=dt){
   basket;  uold = u;
}
plot(u);
ofstream fich("u.txt");
real x0=0,x1=Lx, y0=0;
for(real t=0;t<=1;t+=0.05){
real c=x0+t*x1;
real BS=0.5*c*(1+ erf(log(c)/(sigmax*sqrt(2*T)) + sigmax * sqrt(T/8.)))
      -0.5*(1+erf(log(c)/(sigmax*sqrt(2*T)) -sigmax * sqrt(T/8.)))+1-c;
    real error = u(c,y0)-BS;
fich<<c<<"    "<< u(c,y0)<<"    "<<BS<<"    "<<error<<endl;
}
// display "u.txt" by gnuplot: plot "u.txt"using 1:2 w l
// display the error by plot "error.txt"using 1:4 w l
```

Using FreeFem++ to compute a one-dimensional option requires about four times the CPU of a hand-written C++ program.

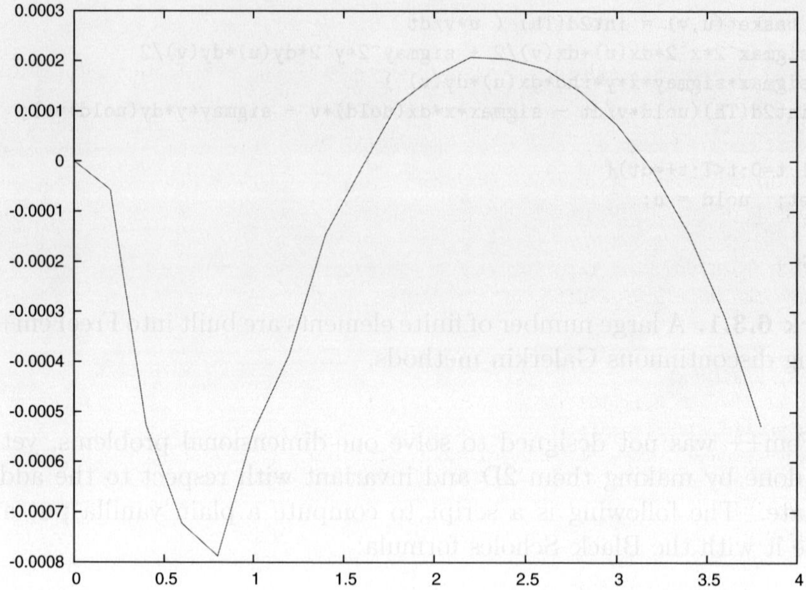

Figure 6.3.2. Result of a computation of a plain vanilla put option using FreeFem++. The plot displays the difference between the computation and the Black–Scholes analytical solution. The error is well below 0.1 %.

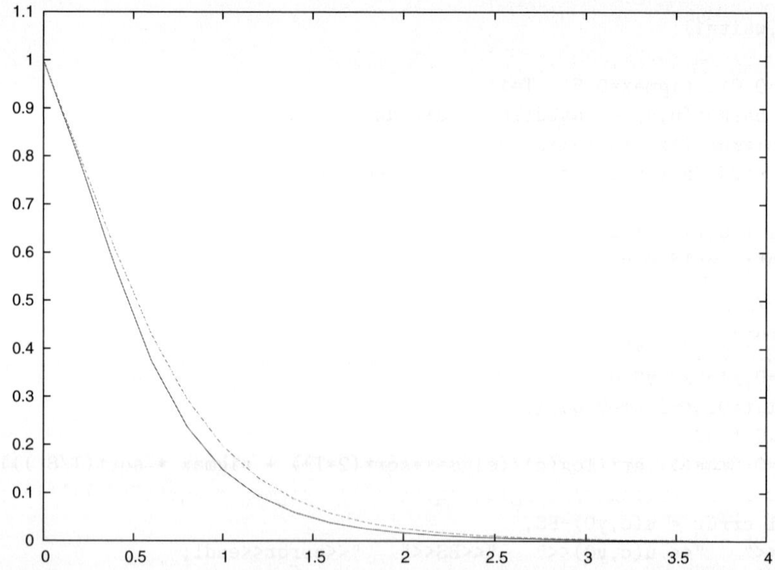

Figure 6.3.3. Computation of (one-dimensional) put with FreeFem++: comparison between a local volatility CEV with $\sigma y^{\frac{1}{4}}$, $\sigma = 0.5$ and a plain vanilla put with σ at maturity 1. The results are displayed in moneyness y. There are 150 y-points and 150 time steps.

6.3.2. *Three-dimensional basket*

The same problem with three assets in the basket can be solved with FreeFem++. The script is given below for a put on a basket with linear payoff $(1 - x/2 - y/3 - z/4)^+$. A fully implicit Euler scheme is used combined with a characteristic Galerkin method (keyword `convect`). The mesh is built by extrusion of the mesh of a square. Results are shown in Figure 6.3.4.

```
load "msh3" load "tetgen" load "mshmet" load "medit"
// improve the search of a point in the mesh
//build initial mesh
int nn  = 6;
real Lx=2,Ly=3,Lz=4;
int[int] l1111=[1,2,2,1],l01=[0,1],l02=[0,2],l1122=[1,1,2,2];//   label numbering
mesh3 Th3=buildlayers(square(nn,nn,region=0,label=l1111,[x*Lx,y*Ly]),
      nn,  zbound=[0,Lz],  labelmid=l1122,   labelup = l02,  labeldown = l01);

fespace Vh(Th3,P1);
Vh u,v,uold,usol,h;

real dt = 0.02, sigmax=0.3,sigmay=0.2,sigmaz=0.1;
real sigmax2=sigmax^2/2,sigmay2=sigmay^2/2,sigmaz2=sigmaz^2/2;
real U=x*sigmax2, V=y*sigmay2, W=z*sigmaz2;

problem Poisson(u,v,solver=CG) = int3d(Th3)( u*v/dt
+ x^2*sigmax2*dx(u)*dx(v) + y^2*sigmay2*dy(u)*dy(v) + z^2*sigmaz2*dz(u)*dz(v) )
    - int3d(Th3)( convect([U,V,W],-dt,uold)*v/dt) + on(2,u=0);

real errm=1e-3;// level of error for mesh adaptivity
int iimax=2;
for(int ii=0; ii<=iimax; ii++){ // iterations of mesh adaption
   uold = max(0., 1.-x/2.-y/3.-z/4.);
for(int k=0;k<50;k++){
Poisson;
uold=u;
}
  cout <<" u min, max = " <<  u[].min << " "<< u[].max << endl;
  h=0. ;// for resizing h
  h[]=mshmet(Th3,u,normalization=1,aniso=0,nbregul=1,hmin=1e-3,hmax=0.3,err=errm);
  cout <<" nb of vertices " << Th3.nv << endl;
  plot(u,wait=0,nbiso=15);
  errm *= 0.5;// change the level of error
  cout << " Th3" << Th3.nv < " " << Th3.nt << endl;
  if(ii != iimax) Th3=tetgreconstruction(Th3,switch="raAQ",sizeofvolume=h*h*h/6.);
}
medit("U-adap-iso",Th3,u,wait=1); // for interactive graphics
ofstream fich("u.txt");          // for gnuplot display by plot"u.txt"using 1:2 w l
real x0=0, y0=0, z0=0, x1=Lx, y1=Ly,z1=Lz; // the cube diagonal
for(real t=0;t<=1;t+=0.05)
fich<<t<<"    "<< u(x0+t*x1,y0+t*y1,z0+t*z1)<<endl;

// CPU= 144" end with NUMBER OF GIVEN VERTICES      12156
```

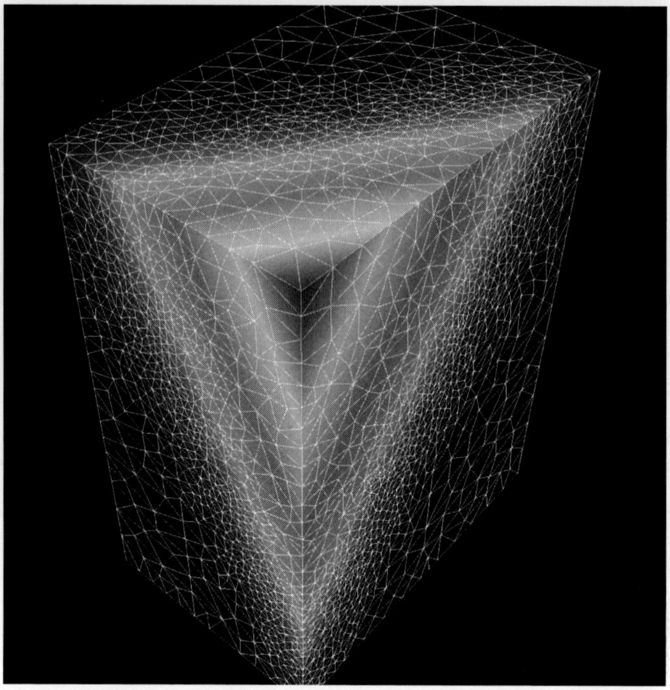

Figure 6.3.4. Put option on a basket of three assets computed via FreeFem++.

6.4. CUDA and other tools

Several programming environments provide tools such as Matlab and Mathematica for option pricing. The open-source project Sage will have several too (see www.sagemath.org/doc/reference/finance.html). An example with the Black–Scholes formula is displayed in Figure 6.4.1.

CUDA is an add-on language to C++, and is the property of NVIDIA corporation, but free: using CUDA, the programmer can make sure that part of the program runs on NVIDIA's GPU (graphic processor unit), when available (see also OpenCL). GPUs have many computing units, though less powerful than Intel's i7, for instance. In finance a speed up of 100 with GPUs is not rare. We give here a very simple example for illustration.

Given a random number generator to compute realizations (n) of a normal random variable X, the aim is to compute

$$C = \frac{\mathrm{e}^{-rT}}{N}\sum_n \max(S_T^{(n)} - K)^+, \quad \text{with } S_T^{(n)} = S_0\mathrm{e}^{(r-\frac{\sigma^2}{2}T+\sigma\sqrt{T}X^{(n)}},$$

when N is large, and accelerate the computation with GPU.

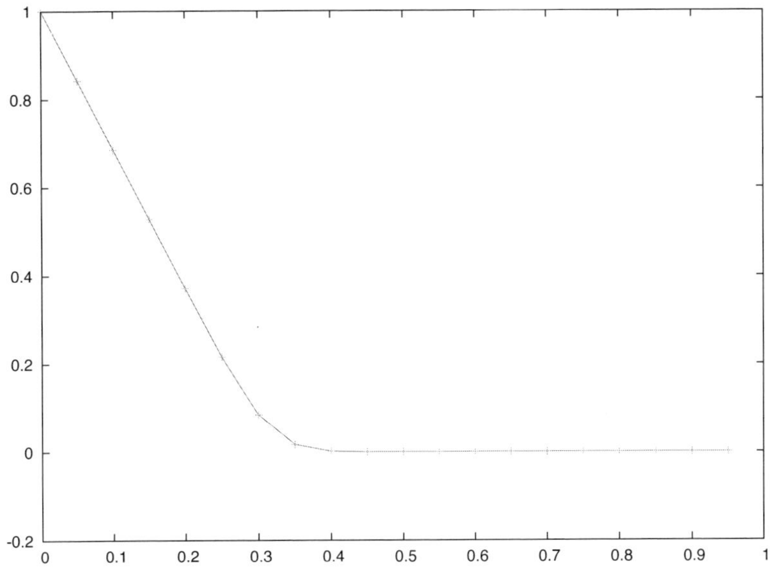

Figure 6.4.1. Put option computed by using Sage.

First we need a random number generator that can execute on the GPU (called `__device__` in CUDA) and some initialization constants:

```
#define PI 3.14159265358979323846264338327950288f
#define IA 16807
#define IM 2147483647
#define AM (float)(1.0/IM)
#define IQ 127773
#define IR 2836
#define MASK 123459876

const int NbBlocs = 50000;
const int NbThreads = 500;
const long N  = NbBlocs*NbThreads;
const float  K = 220, S0=100, r=0.02f, sig=0.3f, T=1.0f;
const float R = (r-sig*sig/2)*T;
const float q = 0;

__device__
void ran0(long*idum, float *res)
{
   long k;
   float ans;
   *idum^=MASK;
   k=(*idum)/IQ;
   *idum=IA*(*idum-k*IQ)-IR*k;
   if(*idum<0)*idum+=IM;
   ans=AM*(*idum);
```

```
    *idum^=MASK;
    *res = ans;
}
```

Then we shall use the Box–Muller formula. The following function will be run on the GPU and the CPU for comparison:

```
__host__ __device__
void BS(float *x, float *y){
   float xx, yy;   // Box-Muller formula for normal r.v.
   float u1 = *x, u2 = *y;
   xx = sqrtf(-2.0f * logf(u1)) * cosf(2.0f * PI * (u2));
   *y = S0*expf(R+sig*sqrtf(T)*xx);   // S_T
   yy = sqrtf(-2.0f * logf(u1)) * sinf(2.0f * PI * (u2));
   *x = S0*expf(R+sig*sqrtf(T)*(yy + q * xx));  // S_T
}
```

On input it takes two independent random variables x and y, uniformly distributed on $[0, 1]$; on output it returns two values of $S_T^{(n)}$. In standard C, given two arrays `a1,a2`, the task would be achieved by a call to `BScpu` and calls to the C-function `rand()`:

```
void BScpu(float *a1, float *a2, int I) {
    srandom(time(NULL));
   for ( int n = 0 ; n < N ; ++n ){             // fills a1,a2
     a1[n] = (rand() + 0.5f)/(RAND_MAX + 1.0f);// with random
     a2[n] = (rand() + 0.5f)/(RAND_MAX + 1.0f);// nb in (0,1)
   for ( int i = 0 ; i < I ; ++i )
     BS(a1+i, a2+i);   // same as BS(&a1[i],&a2[i])
}
```

The same can be done on the GPU by:

```
__global__
void BSgpu(float *a1, float *a2, long *idum, long N) {
   int i = blockDim.x*blockIdx.x + threadIdx.x;
   if ( i < N ) {
     idum[i] = -1-i;     // used as seeds for the random gen
     ran0(idum+i,a1+i);  // fills array a1 with r.v.
     ran0(idum+i,a2+i);  // fills array a2 with r.v.
     BS(a1+i, a2+i);
   }
}
```

However, arrays `a1,a2` must be allocated on the GPU (by `CUDAMALLOC`), and after the computations the results must be sent back to the CPU (by `cudaMemcpy`). Finally, the put is computed by a reduction into a sum of all the values in `a1,a2` (here A1 and A2):

```
   cudaMalloc( (void **) &B1, taille);   // allocate B in GPU RAM
   cudaMalloc( (void **) &B2, taille);   // allocate B in GPU RAM
   cudaMalloc( (void **) &idum, taillelong);
```

```
BSgpu<<<NbBlocs, NbThreads>>>(B1,B2,idum, N);
cudaMemcpy(A1, B1, taille, cudaMemcpyDeviceToHost); // B1 -> A1
cudaMemcpy(A2, B2, taille, cudaMemcpyDeviceToHost);
float gput=0;
for ( int n = 0 ; n < N ; ++n )
  gput += fmax(K-A1[n]-A2[n],0.0f);
```

Notice the peculiar syntax to call `BSgpu()`. The modifiers `NbBlocs` and `NbThreads` influence the computing time on the GPU because they specify how the data are arranged in blocks and the number of threads.

Conclusion

We hope to have convinced the reader that computing financial assets by deterministic methods, in particular by solving PDEs, PDIs and PIDEs, can have advantages: certainly precision, maybe speed, and a sound treatment of singularities if any. Yet many pricing problems, such as complex contracts with many discrete dates, are hard to formulate in terms of partial differential equations.

In computational fluid dynamics, it took many years to convince aeroplane manufacturers to use variational methods instead of finite difference methods. The difficulty was in training engineers in this new field but the advantage was that an entire aeroplane could be discretized by an arbitrary mesh instead of the painful task of matching uniform blocks, as needed by finite difference methods. Here in finance the domain of integration is simple, usually $(0, L_1) \times \cdots \times (0, L_d)$, and so the advantage of variational methods is not so much in the mesh but rather in the *optimal* mesh: a mesh where the points are where they are needed for precision and for resolving singularities, if any.

Parallelism continues to grow on the hardware side, and certainly it is in favour of Monte Carlo methods; variational methods can be parallelized as well, but are more complex and much harder to implement.

In the end, the method which provides the best ratio of speed over precision will win, and the result will only be known after several years of numerical experimentation. Meanwhile innovation will perfect all methods: it is just the beginning!

Acknowledgements

Because of the rules of the game, this article contains some of my own research. I would like to thank my research collaborators, Yves Achdou, Rama Cont, Nicolas Dicesare, Frédéric Hecht, Grégoire Loeper and Nicolas Lantos.

REFERENCES[4]

Y. Achdou and O. Pironneau (2005), *Computational Methods for Option Pricing*, Vol. 30 of *Frontiers in Applied Mathematics*, SIAM.

Y. Achdou and O. Pironneau (2009), *Mathematical Modeling and Numerical Methods in Finance*, Vol. XV of *Handbook of Numerical Analysis* (P. G. Ciarlet *et al.*, eds), North-Holland.

Y. Achdou and N. Tchou (2002), 'Variational analysis for the Black and Scholes equation with stochastic volatility', *M2AN: Math. Model. Numer. Anal.* **36**, 373–395.

H. Albrecher, P. Mayer, W. Schoutens, and J. Tistaert (2007), 'The little Heston trap', *Wilmott Magazine*, January 2007, 83–91.

L. Andersen and J. Andreasen (2000), 'Jump diffusion models: Volatility smile fitting and numerical methods for pricing', *Review of Derivatives Research* **4**, 231–262.

A. Ang and F. Longstaff (2011), Systemic sovereign credit risk: Lessons from the U.S. and Europe. NBER Working Paper No. 16982, April 2011. Also available at www.scribd.com/doc/55649615/7/MODELING-CDS-SPREADS-WITH-SYSTEMIC-RISK.

E. Ayache, P. A. Forsyth and K. R. Vetzal (2003), 'The valuation of convertible bonds with credit risk', *J. Derivatives* **11**, 9–29.

L. Bachelier (1900), 'Théorie de la spéculation', *Annales Scientifiques de l'Ecole Normale Supérieure* **3**, #17.

G. Barles (1997), Convergence of numerical schemes for degenerate parabolic equations arising in finance. In *Numerical Methods in Finance* (L. C. G. Rogers and D. Talay, eds), Cambridge University Press, pp. 1–21.

A. Bentata and R. Cont (2012), 'Forward equations for option prices in semimartingale models', *Finance & Stochastics*, to appear.

J. Bertoin (1996), *Lévy Processes*, Cambridge University Press, Cambridge.

F. Bervoets (2006), Model risk for exotic equity options. Master's thesis, Delft University of Technology, Netherlands.

F. Black and M. Scholes (1973), 'The pricing of options and corporate liabilities', *J. Political Economy* **81**, 637–654.

D. Brigo and F. Mercurio (2002), 'Lognormal-mixture dynamics and calibration to market volatility smiles', *Internat. J. Theoret. Appl. Finance* **5** 427–446.

D. Brigo and F. Mercurio (2007), *Interest Rate Models: Theory and Practice. With Smile, Inflation and Credit*, second edition, Springer Finance.

P. Carr and D. B. Madan (1999), 'Option valuation using the fast Fourier transform', *J. Comput. Finance* **2**, 61–73.

P. Carr, H. Geman, D. Madan and M. Yor (2002), 'The fine structure of asset returns: An empirical investigation', *J. Business.* **75**, 305–332.

B. Chen, L. A. Grzelak and C. W. Oosterlee (2011), Calibration of a stochastic interest rate SAB. In *Proc. Quantitative Methods in Financial and Insurance Mathematics*, Lorentz Center, Leiden, the Netherlands. To appear.

[4] The URLs cited in this work were correct at the time of going to press, but the publisher and the authors make no undertaking that the citations remain live or are accurate or appropriate.

E. Clément, D. Lamberton and P. Protter (2002), 'An analysis of a least square regression method for American option pricing', *Finance & Stochastics* **6**, 449–471.

R. Cont and P. Tankov (2003), *Financial Modelling with Jump Processes*, Chapman & Hall.

R. Cont and E. Voltchkova (2004), Integro-differential equations for option prices in exponential Lévy models. CMAPX report R.I.547, Ecole Polytechnique.

R. Cont and E. Voltchkova (2005), 'A finite difference scheme for option pricing in jump diffusion and exponential Lévy models', *SIAM J. Numer. Anal.* **43**, 1596–1626.

R. Cont, N. Lantos and O. Pironneau (2011), 'A reduced basis for option pricing', *SIAM J. Math. Finance* **2**, 287–316.

J. Cox (1975), Notes on option pricing I: Constant elasticity of variance diffusions. Working Paper, Stanford University.

J. Cox, J. Ingersoll and S. Ross (1985), 'A theory of the term structure of interest rates', *Econometrica* **53**, 389–408.

S. Crépey (2003), 'Calibration of the local volatility in a generalized Black–Scholes model using Tikhonov regularization', *SIAM J. Math. Anal.* **34**, 1183–1206.

A. Dervieux (1996), About the basic methods. In *Handbook of Computational Fluid Dynamics* (R. Peyret, ed.), Academic Press, pp. 1–23.

B. Dupire (1997), Pricing and hedging with smiles. In *Mathematics of Derivative Securities* (M. Dempster and S. Pliska, eds), Cambridge University Press, pp. 103–111.

N. El Karoui and E. Gobet (2011), *Les Outils Stochastiques des Marchés Financiers: Une Visite Guidée de Einstein à Black–Scholes*, Ecole Polytechnique.

W. Farkas, N. Reich and C. Schwab (2006), Anisotropic stable Lévy copula processes? Analytical and numerical aspects. ETH Research Report No. 2006-08, Zurich.

P.-L. George and H. Borouchaki (1998), *Delaunay Triangulation and Meshing: Application to Finite Elements* (trans. P. J. Frey and S. A. Canann), Editions Hermès.

T. Gerstner and M. Griebel (1998), 'Numerical integration using sparse grids', *Numer. Algorithms* **18**, 209–232.

M. Giles (2008), 'Multilevel Monte Carlo path simulation', *Operations Research* **56**, 981–986.

P. Glasserman (2003), *Monte Carlo Methods in Financial Engineering*, Vol. 53 of *Stochastic Modelling and Applied Probability*, Springer.

F. Hecht, O. Pironneau, A. Le Hyaric and K. Otsuka (2008), The FreeFem++ documentation. UPMC Publications. See also www.freefem.org.

S. L. Heston (1993), 'A closed-form solution for options with stochastic volatility with application to bond and currency options', *Review of Financial Studies* **6**, 327–343.

M. Hintermüller, V. A. Kovtunenko and K. Kunisch (2004), 'Semismooth Newton methods for a class of unilaterally constrained variational problems', *Adv. Math. Sci. Appl.* **14**, 513–535.

J. C. Hull (2003), *Options, Futures and Other Derivatives*, fifth edition, Prentice-Hall.

J. C. Hull and A. White (1987), 'The pricing of options on assets with stochastic volatilities', *J. Finance* **42**, 281–300.

S. Ikonen and J. Toivanen (2008), 'Efficient numerical methods for pricing American options under stochastic volatility', *Numer. Methods Partial Diff. Equations* **24**, 104–126.

S. Ikonen and J. Toivanen (2009). 'Operator splitting methods for pricing American options under stochastic volatility', *Numer. Math.* **113**, 299–324.

K. Ito and K. Kunisch (2006), 'Parabolic variational inequalities: The Lagrange multiplier approach', *J. Math. Pures Appl.* **85**, 415–449.

K. R. Jackson, S. Jaimungal, and V. Surkov (2008), 'Fourier space time-stepping for option pricing with Lévy models', *J. Comput. Finance* **12**, 1–29.

L. Jiang and M. Dai (1999), Convergence of binomial tree method for American options. In *Partial Differential Equations and their Applications* (C. Hua and L. Rodino, eds), World Scientific, pp. 106–108.

M. S. Joshi (2011), *More Mathematical Finance*, Pilot Whale Press, Melbourne.

A. Kolmogoroff (1931), 'Über die analytischen Methoden in der Wahrscheinlichkeitsrechnung', *Math. Annalen* **104**, 415–458.

D. Lamberton and B. Lapeyre (1997), *Introduction au Calcul Stochastique Appliqué à la Finance*, Ellipses.

B. Lapeyre, C. Martini and A. Sulem (2010), The Premia consortium and software. www.premia.fr.

A. Lewis (2001), A simple option formula for general jump-diffusion and other exponential Lévy processes. www.optioncity.net.

P.-L. Lions and J.-M. Lasry (2007), 'Mean field games', *Japan. J. Math.* **2**, 229–260.

F. A. Longstaff and E. S. Schwartz (1992), 'Interest rate volatility and the term structure: A two-factor general equilibrium model', *J. Finance* **47**, 1259–1282.

R. Lord, F. Fang, F. Bervoets and C. W. Oosterlee (2007), A fast and accurate FFT-based method for pricing early-exercise options under Lévy processes. Available at SSRN: ssrn.com/abstract=966046

A.-M. Matache, T. von Petersdoff and C. Schwab (2004), 'Fast deterministic pricing of Lévy driven assets', *J. Math. Model. Numer. Anal.* **38**, 37–72.

R. C. Merton (1976), 'Option pricing when underlying stock returns are discontinuous', *J. Financial Econ.* **3**, 125–144.

S. Mikhailov and U. Nögel (2003), Heston's stochastic volatility model implementation, calibration and some extensions. *Wilmott Magazine*, July 2003.

B. Øksendal (1998), *Stochastic Differential Equations*, Springer.

G. Pagès (2008), Méthodes de Monte-Carlo. Notes de cours Master 2, Université Pierre et Marie Curie. www.proba.jussieu.fr/pageperso/pages/

O. Pironneau (2009), 'Calibration of options on a reduced basis', *J. Comput. Appl. Math.* **232**, 139–147.

W. H. Press, S. A. Teukolsky, W. T. Vettering and B. P. Flannery (1992), *Numerical Recipes in C*, second edition, Cambridge University Press.

C. Prud'homme, D. Rovas, K. Veroy, Y. Maday, A. T. Patera and G. Turinici (2002), 'Reliable real-time solution of parametrized partial differential equations: Reduced-basis output bound methods', *J. Fluid Engng* **124**, 70–80.

R. Rannacher (1984), 'Finite element solution of diffusion problems with irregular data', *Numer. Math.* **43**, 309–327.

R. D. Richtmyer and Morton (1967), *Difference Methods for Initial-Value Problems*, second edition, Wiley–Interscience.

E. W. Sachs and M. Schu (2008), Reduced order models (POD) for calibration problems in finance. In *Numerical Mathematics and Advanced Applications: Proc. ENUMATH 2007* (K. Kunisch, G. Of, and O. Steinbach, eds), Springer, pp. 735–742.

M. Schroder (1989), 'Computing the Constant Elasticity of Variance option pricing formula', *J. Finance* **44**, 211–219.

S. A. Smolyak (1963), 'Quadrature and interpolation formulas for tensor products of certain classes of functions', *Dokl. Akad. Nauk. USSR* **4**, 240–243.

E. Stein and J. Stein (1991), 'Stock price distributions with stochastic volatility: An analytic approach', *Rev. Financ. Stud.* **4**, 727–752.

D. Talay and L. Tubaro (1996), *Probabilistic Models for Nonlinear Partial Differential Equations*, Vol. 1627 of *Lecture Notes in Mathematics*, Springer.

P. Tankov and E. Voltchkova (2009), Jump-diffusion models: A practitioner's guide. *Banque et Marchés*, no. 99.

P. Wilmott, S. Howison, and J. Dewynne (1995), *The Mathematics of Financial Derivatives: A Student Introduction*, Cambridge University Press.

Xiaoye S. Li, J. W. Demmel and J. R. Gilbert (1999), The SuperLU library. crd.lbl.gov/~xiaoye/SuperLU.

Xuan Luo (2007), Bermudan swaption hedging in Black–Karasinski model. PhD thesis, New York University.

